Advances in Intelligent Systems and Computing

Volume 239

Series Editor

Janusz Kacprzyk, Warsaw, Poland

For further volumes:
http://www.springer.com/series/11156

Advances in Intelligent Systems and Computing

Volume 280

Álvaro Herrero · Bruno Baruque
Fanny Klett · Ajith Abraham
Václav Snášel · André C.P.L.F. de Carvalho
Pablo García Bringas · Ivan Zelinka
Héctor Quintián · Emilio Corchado
Editors

International Joint Conference SOCO'13-CISIS'13 -ICEUTE'13

Salamanca, Spain, September 11th–13th, 2013
Proceedings

 Springer

Editors
Álvaro Herrero
Department of Civil Engineering
University of Burgos
Burgos, Spain

Bruno Baruque
Department of Civil Engineering
University of Burgos
Burgos, Spain

Fanny Klett
German Workforce ADL Partnership
 Laboratory
Waltershausen, Germany

Ajith Abraham
Machine Intelligence Research Labs
 (MIR Labs)
Scientific Network for Innovation
 and Research Excellence
Auburn, Washington, USA

Václav Snášel
Department of Computer Science
VŠB-TU Ostrava
Faculty of Elec. Eng. and Computer Science
Ostrava, Czech Republic

André C.P.L.F. de Carvalho
Department of Computer Science
University of Sao Paulo
Sao Carlos, Brazil

Pablo García Bringas
DeustoTech Computing
University of Deusto
Bilbao, Spain

Ivan Zelinka
VŠB-TU Ostrava
Department of Computer Science
Faculty of Elec. Eng. and Computer Science
Ostrava, Czech Republic

Héctor Quintián
Universidad de Salamanca
Salamanca, Spain

Emilio Corchado
Universidad de Salamanca
Salamanca, Spain

ISSN 2194-5357   ISSN 2194-5365   (electronic)
ISBN 978-3-319-01853-9   ISBN 978-3-319-01854-6   (eBook)
DOI 10.1007/978-3-319-01854-6
Springer Cham Heidelberg New York Dordrecht London

Library of Congress Control Number: 2013946039

Printed on acid-free paper

Springer is part of Springer Science+Business Media (www.springer.com)

Preface

This volume of Advances in Intelligent and Soft Computing contains accepted papers presented at SOCO 2013, CISIS 2013 and ICEUTE 2013, all conferences held in the beautiful and historic city of Salamanca (Spain), in September 2013.

Soft computing represents a collection or set of computational techniques in machine learning, computer science and some engineering disciplines, which investigate, simulate, and analyse very complex issues and phenomena.

After a through peer-review process, the 8th SOCO 2013 International Program Committee selected 40 papers which are published in these conference proceedings, and represents an acceptance rate of 41%. In this relevant edition a special emphasis was put on the organization of special sessions. Four special sessions were organized related to relevant topics as: Systems, Man, and Cybernetics, Data Mining for Industrial and Environmental Applications, Soft Computing Methods in Bioinformatics, and Soft Computing Methods, Modelling and Simulation in Electrical Engineering.

The aim of the 6th CISIS 2013 conference is to offer a meeting opportunity for academic and industry-related researchers belonging to the various, vast communities of Computational Intelligence, Information Security, and Data Mining. The need for intelligent, flexible behaviour by large, complex systems, especially in mission-critical domains, is intended to be the catalyst and the aggregation stimulus for the overall event.

After a through peer-review process, the CISIS 2013 International Program Committee selected 23 papers which are published in these conference proceedings achieving an acceptance rate of 39%.

In the case of 4th ICEUTE 2013, the International Program Committee selected 11 papers which are published in these conference proceedings.

The selection of papers was extremely rigorous in order to maintain the high quality of the conference and we would like to thank the members of the Program Committees for their hard work in the reviewing process. This is a crucial process to the creation of a high standard conference and the SOCO, CISIS and ICEUTE conferences would not exist without their help.

SOCO'13, CISIS'13 and ICEUTE'13 enjoyed outstanding keynote speeches by distinguished guest speakers: Prof. Hojjat Adeli - Ohio State University (USA), Prof. Hujun Yin - University of Manchester (UK), and Prof. Manuel Graña – University of Pais Vasco (Spain).

SOCO'13 has teamed up with the International Journal of Neural Systems (WORLD SCIENTIFIC), Integrated Computer-Aided Engineering (IOS PRESS), Computer-Aided Civil and Infrastructure Engineering (IOS PRESS), Neurocomputing (ELSEVIER), Journal of Applied Logic (ELSEVIER), Applied Soft Computing (ELSEVIER) and DYNA (FEDERACION ASOCIACIONES INGENIEROS INDUSTRIALES ESPANA) journals for a suite of special issues and fast track including selected papers from SOCO'13.

For this CISIS'13 special edition, as a follow-up of the conference, we anticipate further publication of selected papers in a special issue of the prestigious Logic Journal of the IGPL Published by Oxford Journals.

Particular thanks go as well to the Conference main Sponsors, IEEE-SecciónEspaña, IEEE Systems, Man and Cybernetics-CapítuloEspañol, AEPIA, Ayto. of Salamanca, University of Salamanca, World Federation of Soft Computing, MIR Labs, IT4 Innovation Centre of Excellence, The International Federation for Computational Logic, Ministerio de Economía y Competitividad (TIN 2010-21272-C02-01), Junta de Castilla y León (SA405A12-2), INMOTIA, REPLENTIA, and HIDROGENA, who jointly contributed in an active and constructive manner to the success of this initiative. We want also to extend our warm gratitude to all the Special sessions chairs for their continuing support to the SOCO, CISIS and ICEUTE Series of conferences.

We would like to thank all the special session organizers, contributing authors, as well as the members of the Program Committees and the Local Organizing Committee for their hard and highly valuable work. Their work has helped to contribute to the success of the SOCO 2013, CISIS 2013 and ICEUTE 2013 events.

September, 2013

The editors
Álvaro Herrero
Bruno Baruque
Fanny Klett
Ajith Abraham
Václav Snášel
André C.P.L.F. de Carvalho
Pablo García Bringas
Ivan Zelinka
Héctor Quintián
Emilio Corchado

SOCO 2013

Organization

General Chair

Emilio Corchado — University of Salamanca, Spain

Honorary Chair

Alfonso Fernández Mañueco — Mayor of Salamanca, Spain
Costas Stasopoulos — Director-Elect. IEEE Region 8
Antonio Bahamonde — President of the Spanish Association for Artificial Intelligence, AEPIA
Pilar Molina — IEEE Spanish Section President, Spain

International Advisory Committee

Ashraf Saad — Armstrong Atlantic State University, USA
Amy Neustein — Linguistic Technology Systems, USA
Ajith Abraham — Machine Intelligence Research Labs -MIR Labs, Europe
Jon G. Hall — The Open University, UK
Paulo Novais — Universidade do Minho, Portugal
Antonio Bahamonde — President of the Spanish Association for Artificial Intelligence (AEPIA)
Michael Gabbay — Kings College London, UK
Isidro Laso-Ballesteros — European Commission Scientific Officer, Europe
Aditya Ghose — University of Wollongong, Australia
Saeid Nahavandi — Deakin University, Australia
Henri Pierreval — LIMOS UMR CNRS 6158 IFMA, France

Industrial Advisory Committee

Rajkumar Roy	The EPSRC Centre for Innovative Manufacturing in Through-life Engineering Services, UK
Amy Neustein	Linguistic Technology Systems, USA
Francisco Martínez	INMOTIA, Spain

Program Committee Chair

Emilio Corchado	University of Salamanca, Spain
Fanny Klett	Director of German Workforce Advanced Distributed Learning Partnership Laboratory, Germany
Ajith Abraham	Machine Intelligence Research Labs (Europe)
Vaclav Snasel	University of Ostrava, Czech Republic
André C.P.L.F. de Carvalho	University of São Paulo, Brazil

Program Committee

Abdelhamid Bouchachia	Bournemouth University, UK
Abraham Duarte	Universidad Rey Juan Carlos, Spain
Alberto Freitas	University of Porto, Portugal
Alexis Enrique Marcano Cedeño	Polytechnic University of Madrid, Spain
Alfredo Cuzzocrea	ICAR-CNR and University of Calabria, Italy
Álvaro Herrero	University of Burgos, Spain
Ana Figueiredo	Instituto Superior de Engenharia do Porto, Portugal
Andrea Schaerf	University of Udine, Italy
Angel Arroyo	University of Burgos, Spain
Anna Bartkowiak	University of Wroclaw, Poland
Antonio Bahamonde	University of Oviedo, Spain
Antonio Peregrin	University of Huelva, Spain
Ashish Umre	University of Sussex, UK
Ashraf Saad	Armstrong Atlantic State University, USA
Aureli Soria-Frisch	Starlab Barcelona S.L., Spain
Ayeley Tchangani	Université de Toulouse, France
Benjamin Ojeda Magaña	Universidad de Guadalajara, Mexico
Benoit Otjacques	Centre de Recherche Public - Gabriel Lippmann, Luxembourg
Borja Sanz	S3Lab - University of Deusto, Spain
Bruno Baruque	University of Burgos, Spain
Camelia Chira	Babes-Bolyai University, Romania
Carlos Laorden	DeustoTech Computing - S3Lab, University of Deusto, Spain
Carlos Pereira	ISEC, Portugal
Cesar Analide	University of Minho, Portugal
Cesar Hervas	University of Córdoba, Spain

Jorge Díez	University of Oviedo, Spain
José Riquelme Santos	University of Sevilla, Spain
Jose Gamez	University of Castilla-La Mancha, Spain
Jose Luis Calvo Rolle	University of Coruña, Spain
Jose Luis Casteleiro	University of Coruña, Spain
José M. Benítez	University of Granada, Spain
Jose M. Molina	University Carlos III of Madrid, Spain
José Ramón Villar	University of Oviedo, Spain
José Valente De Oliveira	Universidade do Algarve, Portugal
Josef Tvrdik	University of Ostrava, Czech Republic
Jose-Maria Pena	Polytechnic University of Madrid, Spain
Juan Álvaro Muñoz Naranjo	University of Almería, Spain
Juan Manuel Corchado	University of Salamanca, Spain
Juan Gomez Romero	University Carlos III of Madrid, Spain
Laura García-Hernández	University of Córdoba, Spain
LenkaLhotska	Czech Technical University in Prague, Czech Republic
León Wang	University of Kaohsiung, Taiwan
Luciano Stefanini	University of Urbino "Carlo Bo", Italy
Luis Correia	University of Lisbon, Portugal
Luís Nunes	Instituto Universitário de Lisboa, Portugal
Luis Paulo Reis	University of Minho, Portugal
M. Chadli	University of Picardie Jules Verne, France
Maciej Grzenda	Warsaw University of Technology, Poland
Manuel Graña	University of Basque Country, Spain
Marcin Paprzycki	IBS PAN and WSM, Poland
Marco Cococcioni	University of Pisa, Italy
Maria JoãoViamonte	Instituto Superior de Engenharia do Porto, Portugal
Maria Jose Del Jesus	University of Jaén, Spain
María N. Moreno García	University of Salamanca, Spain
Mario Giovanni C.A. Cimino	University of Pisa, Italy
Mario Koeppen	Kyushu Institute of Technology, Japan
Marius Balas	AurelVlaicu University of Arad, Romania
Martin Macas	Czech Technical University in Prague, Czech Republic
Martin Stepnicka	University of Ostrava, Czech Republic
Mehmet EminAydin	University of Bedfordshire, UK
Michael Vrahatis	University of Patras, Greece
Michal Wozniak	Wroclaw University of Technology, Poland
Miguel Angel Patricio	University Carlos III of Madrid, Spain
Milos Kudelka	VŠB-Technical University of Ostrava, Czech Republic
Miroslav Bursa	Czech Technical University in Prague, Czech Republic
Mitiche Lahcene	University of Djelfa, France

Mohammed Eltaweel	Arab Academy for Science, Technology & Maritime Transport, Egypt
Nineta Polemi	University of Pireaus, Greece
Noelia Sanchez-Maroño	University of Coruña, Spain
Oliviu Matei	North University of Baia Mare, Romania
Oscar Fontenla-Romero	University of Coruña, Spain
Oscar Luaces	University of Oviedo, Spain
Paulo Moura Oliveira	UTAD University, Portugal
Paulo Novais	University of Minho, Portugal
Pavel Brandstetter	VŠB-Technical University of Ostrava, Czech Republic
Pavel Kromer	VŠB-Technical University of Ostrava, Czech Republic
Pavol Fedor	Technical University of Košice, Slovak Republic
Pedro M. Caballero Lozano	CARTIF Centro Tecnológico, Spain
Petr Novak	VŠB-Technical University of Ostrava, Czech Republic
Petr Palacky	VŠB-Technical University of Ostrava, Czech Republic
Petr Pošík	Czech Technical University in Prague, Czech Republic
Petrica Claudiu Pop	North University of Baia Mare, Romania
Raquel Redondo	University of Burgos, Spain
Richard Duro	University of Coruña, Spain
Robert Burduk	Wroclaw University of Technology, Poland
Roman Senkerik	TBU in Zlin, Czech Republic
Rosa Basagoiti	Mondragon University, Spain
Rui Sousa	University of Minho, Portugal
Sara Silva	INESC-ID, Portugal
Sebastián Ventura	University of Córdoba, Spain
Stefano Pizzuti	ENEA, Italy
Sung-Bae Cho	Yonsei University, South Korea
Tzung-Pei Hong	National University of Kaohsiung, Taiwan
Urko Zurutuza	Mondragon University, Spain
Valentina Casola	University of Naples "Federico II", Italy
Valentina Emilia Balas	Aurel Vlaicu University of Arad, Romania
Verónica Tricio	University of Burgos, Spain
Vicente Martin-Ayuso	Polytechnic University of Madrid, Spain
Wei-Chiang Hong	Oriental Institute of Technology, Taiwan
Wilfried Elmenreich	University of Klagenfurt, Austria
Zdenek Peroutka	Regional Innov. Center for Elect. Eng. (RICE), Czech Republic
Zita Vale	GECAD - ISEP/IPP, Portugal

Special Sessions

Systems, MAN, and Cybernetics

Emilio Corchado	University of Salamanca, Spain (Organiser)
Héctor Quintián	University of Salamanca, Spain (Organiser)
Jiri Dvorsky	VŠB-Technical University of Ostrava, Czech Republic
Mario Giovanni C.A. Cimino	University of Pisa, Italy
Miroslav Bursa	Czech Technical University in Prague, Czech Republic
Robert Burduk	Wroclaw University of Technology, Poland
Valentina Casola	University of Naples "Federico II", Italy

Data Mining for Industrial and Environmental Applications

Alicia Troncoso	Pablo de Olavide University of Seville, Spain (Organiser)
Francisco Martínez Álvarez	Pablo de Olavide University of Seville, Spain (Organiser)
María Martínez-Ballesteros	University of Sevilla, Spain (Organiser)
Jorge García-Gutiérrez	University of Sevilla, Spain (Organiser)
Cristina Rubio-Escudero	University of Sevilla, Spain
Daniel Mateos	University of Sevilla, Spain
José C. Riquelme	University of Sevilla, Spain

Soft Computing Methods in Bioinformatics

Camelia Chira	Instituto Tecnológico de Castilla y León, Spain & Babes-Bolyai University, Romania (Organiser)
José Ramón Villar Flecha	University of Oviedo, Spain (Organiser)
Emilio Corchado	University of Salamanca, Spain (Organiser)
Javier Sedano	Instituto Tecnologico de Castilla y León, Spain (Organiser)
Antonio Mucherino	University of Rennes 1, France
Carlos Prieto	Instituto de Biotecnologia de León, Spain
Carmen Vidaurre	Technical University Berlin, Germany
Elena Bautu	University of Constanta, Romania
Horacio Pérez-Sánchez	Catholic University of Murcia,Spain
Mihaela Breaban	Al.I.Cuza University of Iasi, Romania
Nima Hatami	University of California, Italy
Sameh Kessentini	Universite de Technologie de Troyes, France
Sorin Ilie	University of Craiova, Romania

Soft Computing Methods, Modelling and Simulation in Electrical Engineering

Ajith Abraham   Machine Intelligence Research Labs, Europe
                                (Organiser)
Václav Snášel   VŠB-Technical University of Ostrava,
                                Czech Republic (Organiser)
Pavel Brandstetter   VŠB-Technical University of Ostrava,
                                Czech Republic (Organiser)
Abdelhamid Bouchachia   Bournemouth University, UK
Abraham Duarte   University Rey Juan Carlos, Spain
Ana Figueiredo   Instituto Superior de Engenharia do Porto, Portugal
Daniela Perdukova   Technical University of Košice, Slovak Republic
Giuseppe Cicotti   University of Naples, Italy
Gregg Vesonder   AT&T Labs - Research
Ivo Neborak   VŠB-Technical University of Ostrava,
                                Czech Republic
Jan Vittek   University of Žilina, Slovak Republic
Jaroslava Zilkova   Technical University of Košice, Slovak Republic
Jesus Garcia   University Carlos III of Madrid, Spain
Jiri Koziorek   VŠB-Technical University of Ostrava,
                                Czech Republic
Libor Stepanec   POLL, Ltd., Czech Republic
Martin Kuchar   POLL, Ltd., Czech Republic
Milan Zalman   Slovak University of Technology in Bratislava,
                                Slovak Republic
Oliviu Matei   North University of Baia Mare, Romania
Pavol Fedor   Technical University of Košice, Slovak Republic
Pavol Spanik   University of Žilina, Slovak Republic
Pedro M. Caballero Lozano   CARTIF Centro Tecnológico, Spain
Petr Palacky   VŠB-Technical University of Ostrava,
                                Czech Republic
Petr Chlebis   VŠB-Technical University of Ostrava,
                                Czech Republic
Rui Sousa   University of Minho, Portugal
Sara Silva   INESC-ID, Portugal
StanislavRusek   VŠB-Technical University of Ostrava,
                                Czech Republic
Tatiana Radicova   Slovak University of Technology in Bratislava,
                                Slovak Republic
Urko Zurutuza   Mondragon University, Spain
Verónica Tricio   University of Burgos, Spain
Vitezslav Styskala   VŠB-Technical University of Ostrava,
                                Czech Republic
Zdenek Peroutka   University of West Bohemia, Czech Republic
Zita Vale   GECAD - ISEP/IPP, Portugal

Organising Committee

Emilio Corchado	University of Salamanca, Spain
Álvaro Herrero	University of Burgos, Spain
Bruno Baruque	University of Burgos, Spain
Héctor Quintián	University of Salamanca, Spain
Roberto Vega	University of Salamanca, Spain
Jose Luis Calvo	University of Coruña, Spain
Ángel Arroyo	University of Burgos, Spain
Laura García-Hernández	University of Córdoba, Spain

CISIS 2013

Organization

General Chair

Emilio Corchado   University of Salamanca, Spain

Honorary Chair

Alfonso Fernández Mañueco	Mayor of Salamanca, Spain
Costas Stasopoulos	Director-Elect. IEEE Region 8
Antonio Bahamonde	President of the Spanish Association for Artificial Intelligence, AEPIA
Pilar Molina	IEEE Spanish Section President, Spain

International Advisory Committee

Antonio Bahamonde	President of the Spanish Association for Artificial Intelligence (AEPIA)
Michael Gabbay	Kings College London, UK
Ajith Abraham	Machine Intelligence Research Labs - MIR Labs Europe

Program Committee Chair

Emilio Corchado	University of Salamanca, Spain
Ajith Abraham	VSB-Technical University of Ostrava, Czech Republic
Pablo García	DeustoTech, Spain
Álvaro Herrero	University of Burgos, Spain
Václav Snasel	VSB-Technical University of Ostrava, Czech Republic
Ivan Zelinka	VSB-Technical University of Ostrava, Czech Republic

Program Committee

Alberto Peinado	Universityof Malaga, Spain
Amparo Fuster-Sabater	Institute of Applied Physics (C.S.I.C.), Spain
Ana I. González-Tablas	University Carlos III of Madrid, Spain
Andre Carvalho	University of Sao Paulo, Brazil
Ángel Arroyo	University of Burgos, Spain
Ángel Martín Del Rey	Universityof Salamanca, Spain
Antonio Zamora	University of Alicante, Spain
Araceli Queiruga Dios	University of Salamanca, Spain
Bartosz Krawczyk	Wroclaw University Of Technology, Poland
Borja Sanz	S3Lab - University of Deusto, Spain
Bruno Baruque	University of Burgos, Spain
Carlos Pereira	ISEC, Portugal
Carlos Laorden	Deusto Tech Computing - S3Lab, University of Deusto, Spain
Constantino Malagón	Nebrija University, Spain
Cristina Alcaraz	University of Malaga, Spain
Dario Forte	Dflabs, Italy
David G. Rosado	University of Castilla-La Mancha, Spain
Davide Leóncini	University of Genoa, Italy
Debasis Giri	Haldia Institute of Technology, India
Domingo Gomez	University of Cantabria, Spain
Eduardo Carozo Blumsztein	Montevideo University, Uruguay
Enrico Appiani	SELEX Elsag spa, Italy
Enrique De La Hoz De La Hoz	University of Alcala, Spain
Enrique Daltabuit	UNAM, Mexico
Eva Volna	University of Ostrava, Czech Republic
Fausto Montoya	CSIC, Spain
Félix Brezo	S3lab, University of Deusto, Spain
Fernando Tricas	Universityof Zaragoza, Spain
Francisco Valera	University Carlos III of Madrid, Spain
Francisco Herrera	University of Granada, Spain
Francisco José Navarro-Ríos	University of Granada, Spain
Gabriel López	University of Murcia, Spain
Gabriel Diaz Orueta	IEEE Spain, Spain
Gerald Schaefer	Loughborough University, UK
Gerardo Rodriguez Sanchez	University of Salamanca, Spain
Guillermo Morales-Luna	CINVESTAV-IPN, Mexico
Gustavo Isaza	University of Caldas, Colombia
Héctor Quintián	University of Salamanca, Spain
Hugo Scolnik	University of Buenos Aires, Argentina
Igor Santos	DeustoTech, University of Deusto, Spain

Paulo Moura Oliveira	UTAD University, Portugal
Pavel Kromer	VŠB-Technical University of Ostrava, Czech Republic
Pedro Pablo	University of the Basque Country, Spain
Pino Caballero-Gil	University of La Laguna, Spain
Rafael Alvarez	University of Alicante, Spain
Rafael Corchuelo	University of Seville, Spain
Rafael M. Gasca	University of Seville, Spain
Ramon Rizo	University of Alicante, Spain
Raquel Redondo	University of Burgos, Spain
Raúl Durán	University of Alcalá, Spain
Reinaldo N. Mayol Arnao	UPB, Colombia
Ricardo Contreras	University of Concepción, Chile
Robert Burduk	Wroclaw University of Technology, Poland
Roberto Uribeetxeberria	Mondragon University, Spain
Rodolfo Zunino	University of Genoa, Italy
Roman Senkerik	Tomas Bata University, Czech Republic
Rosaura Palma-Orozco	CINVESTAV - IPN, Mexico
Salvador Alcaraz	University Miguel Hernández, Spain
Simone Mutti	Università degli Studi di Bergamo, Italy
SorinStratulat	Universite Paul Verlaine - Metz, France
Teresa Gomes	University of Coimbra, Portugal
Tomas Olovsson	Chalmers University of Technology, Sweden
Tomasz Kajdanowicz	Wroclaw University of Technology, Poland
Urko Zurutuza	Mondragon University, Spain
Valentina Casola	University of Naples "Federico II", Italy
Victoria Lopez	University Complutense of Madrid, Spain
Vincenzo Mendillo	University Central of Venezuela, Venezuela
Walter Baluja García	CUJAE, Cuba
Wei Wang	University of Luxembourg, Luxembourg
WenjianLuo	University of Science and Technology of China, China
Wojciech Kmiecik	Wroclaw University of Technology, Poland
Zuzana Oplatkova	Tomas Bata University, Czech Republic

Organising Committee

Emilio Corchado	University of Salamanca, Spain
Álvaro Herrero	University of Burgos, Spain
Bruno Baruque	University of Burgos, Spain
Diego Martínez	University of Burgos, Spain
Francisco José Güemes	University of Burgos, Spain

Héctor Quintián   University of Salamanca, Spain
Roberto Vega   University of Salamanca, Spain
Jose Luis Calvo   University of Coruña, Spain
Ángel Arroyo   University of Burgos, Spain
Laura García-Hernández   University of Córdoba, Spain

ICEUTE 2013

Organization

Honorary Chairs

Alfonso Fernández Mañueco Mayor of Salamanca, Spain
Costas Stasopoulos Director-Elect. IEEE Region 8
Antonio Bahamonde President of the Spanish Association for
Artificial Intelligence, AEPIA
Pilar Molina IEEE Spanish Section President, Spain

International Advisory Committee

Jean-yves Antoine Université François Rabelais, France
Reinhard Baran Hamburg University of Applied Sciences, Germany
Fernanda Barbosa Instituto Politécnico de Coimbra,Portugal
Bruno Baruque University of Burgos, Spain
Emilio Corchado University of Salamanca, Spain
Wolfgang Gerken Hamburg University of Applied Sciences, Germany
Arnaud Giacometti Université François Rabelais, France
Helga Guincho Instituto Politécnico de Coimbra, Portugal
Álvaro Herrero University of Burgos, Spain
Patrick Marcel Université François Rabelais, France
Gabriel Michel University Paul Verlaine - Metz, France
Viorel Negru West University of Timisoara, Romania
Jose Luis Nunes Instituto Politécnico de Coimbra, Portugal
Salvatore Orlando Ca' Foscari University,Italy
Veronika Peralta Université François Rabelais, France
Carlos Pereira Instituto Politécnico de Coimbra, Portugal
Teppo Saarenpä Turku University of Applied Sciences, Finland
Sorin Stratulat University Paul Verlaine - Metz, France

Program Committee Chair

Emilio Corchado   University of Salamanca, Spain
Bruno Baruque   University of Burgos, Spain
Fanny Klett   Distributed Learning Partnership Laboratory,
                           Germany

Program Committee

Agostino Cortesi   Universita' Ca' Foscari di Venezia, Italy
Álvaro Herrero   University of Burgos, Spain
Ana Rosa Borges   INESC Coimbra, Portugal
Anabela Gomes   University of Coimbra, Portugal
Ángel Arroyo   University of Burgos, Spain
Carlos Pereira   ISEC, Argentina
DraganSimic   University of Novi Sad, Serbia
Fernanda Correia Barbosa   University of Aveiro, Portugal
Héctor Quintián   University of Salamanca, Spain
Ivan Zelinka   VŠB-Technical University of Ostrava,
                                Czech Republic
Jean-Yves Antoine   UniversitFranois Rabelais de Tours, France
Jiri Dvorsky   VŠB-Technical University of Ostrava,
                                Czech Republic
Jon Mikel Zabala Iturriagagoitia   CIRCLE, Lund University, Sweden
Jose Luis Casteleiro   University of Coruña, Spain
Jose Luis Calvo Rolle   University of Coruña, Spain
Laura García-Hernández   University of Córdoba, Spain
Leticia Curiel   University of Burgos, Spain
Mª Belén Vaquerizo   University of Burgos, Spain
Maria José Marcelino   University of Coimbra, Portugal
Milos Kudelka   VŠB-Technical University of Ostrava,
                                Czech Republic
Pavel Kromer   VŠB-Technical University of Ostrava,
                                Czech Republic
Sorin Stratulat   Universite Paul Verlaine, Metz, France
Vaclav Snasel   VŠB-Technical University of Ostrava,
                                Czech Republic
Verónika Peralta   University of Tours, France
Viorel Negru   West University of Timisoara, Romania
Wolfgang Gerken   Hamburg university of Applied Sciences, Hamburg
Zdeněk Troníček   CTU in Prague, Czech Republic

Special Sessions

Web 2.0

Thomas Connolly	University of the West of Scotland, UK(Organiser)
Gavin Baxter	University of the West of Scotland, UK (Organiser)
Carole Gould	University of the West of Scotland, UK
Gary McKenna	University of the West of Scotland, UK
Mario Soflano	University of the West of Scotland, UK
Thomas Hainey	University of the West of Scotland, UK
Wallace Gray	University of the West of Scotland, UK
Yaelle Chaudy	University of the West of Scotland, UK

Organising Committee

Emilio Corchado	University of Salamanca, Spain
Álvaro Herrero	University of Burgos, Spain
Bruno Baruque	University of Burgos, Spain
Héctor Quintián	University of Salamanca, Spain
Roberto Vega	University of Salamanca, Spain
Jose Luis Calvo	University of Coruña, Spain
Ángel Arroyo	University of Burgos, Spain
Laura García-Hernández	University of Córdoba, Spain

Contents

Intelligent Systems

Applications

Classification and Clustering Methods

Special Sessions

Systems, Man, and Cybernetics

Data Mining for Industrial and Environmental Applications

Soft Computing Methods in Bioinformatics

Soft Computing Methods, Modeling and Simulation in Electrical Engineering

CISIS 2013

General Track

Applications of Intelligent Methods for Security

Infrastructure and Network Security

ICEUTE 2013

General Track

Special Session

Web 2.0

Parsimonious Support Vector Machines Modelling for Set Points in Industrial Processes Based on Genetic Algorithm Optimization

Andrés Sanz-García, Julio Fernández-Ceniceros, Fernando Antoñanzas-Torres, and F. Javier Martínez-de-Pisón-Ascacibar

EDMANS Research Group, University of La Rioja, Logroño, Spain
{andres.sanz,julio.fernandezc,fjmartin}@unirioja.es,
{antonanzas.fernando}@gmail.com
http://www.mineriadatos.com

Abstract. An optimization based on genetic algorithms for both feature selection and model tuning is presented to improve the prediction of set points in industrial lines. The objective is the development of an automatic procedure that efficiently generates parsimonious prediction models with higher generalisation capacity. These models can achieve higher accuracy in predictions, maintaining the high quality of products while working with continual changes in the production cycle. The proposed method deals with three strict restrictions: few individuals per population, low number of holds and runs in model validation procedure and a reduced number of maximum generations. To fullfill these restrictions, we propose to include in the optimization the reranking of the individuals by their complexity when no significant difference is found between the values of their fitness functions. The method is applied to develop support vector machines for predicting three temperature set points in the annealing furnace of a continuous hot-dip galvanising line. The results demonstrate the rerank makes more efficiently and easily the process of obtaining parsimonious models without reducing performance.

Keywords: Genetic Algorithm, Optimization, Support Vector Machine, Galvanising Line, Parsimony Criterion.

1   Introduction

Continuous hot dip galvanising line (CHDGL) is a key process from an economic and strategic point of view for steel industry to produce rolled flat steel products. The difficulty of its properly control is widely known to become more problematic when an increase of production is required [1]. Nowadays, a CHDGL is fairly automated and the tuning of its parameters stands as the most remarkable problem when trying to increase both plant flexibility or yield. This is due to the high amount of inherent non-linearities and input variables that galvanising involves. Therefore, plant engineers require better prediction models that help in adjusting the CHDGL to different running operation conditions.

Á. Herrero et al. (eds.), *International Joint Conference SOCO'13-CISIS'13-ICEUTE'13*,
Advances in Intelligent Systems and Computing 239,
DOI: 10.1007/978-3-319-01854-6_1, © Springer International Publishing Switzerland 2014

In this paper, the problem addressed is the necessity of accurately predicting these set points when dealing with new products. The adjustment has traditionally been done conducting a number of in-plant process trials. However, in today´s competitive markets, this is really inefficient due to their excessive costs. Other feasible alternative is the use of mathematical models that integrate physics and mechanics of the galvanising. Nevertheless, these models often require high computation times and involve great difficulties for swiftly adjusting the parameters of each component of the CHDGL to novel product specifications. Another approach is based on large volumes of data collected from the processes to create data-driven models. Taking advantage of this "stored information", we proposed soft computing (SC) methods as a clear alternative to parametric ones to create better performing models.

SC is based on combining intelligent systems with computational techniques to solve inexact and complex problems [12]. It involves several computational methods such as support vector machines (SVMs) [13], artificial neural networks (ANNs) [5], and genetic algorithms (GAs)[9]. These are stochastic and very suited to deal with real engineering problems [4]. Several papers have reported reliable SC models for predicting galvanising set points. For instance, ANNs are applied by [6] for coating control with significant improvements. Multilayer perceptron (MLP) models are developed by [10] for estimating the velocity set point of coils inside the continuous anneling furnace (CAF) on a CHDGL. However, finding the best model´s topology and its setting parameters is still a challenging task. An approach is the use of genetic algorithms (GAs), which are inspired by the law of nature [2]. Through the principles of biological evolution, GAs have the capacity to provide an efficient optimization process that can find the optimal model [3]. Indeed, many authors have combined MLPs with GAs showing correctly tuned MLP can predict optimal set points for control systems. Selecting the less complex model is a common topic in machine learning in order to achieve higher generalization capacity in predictions. For instance, a methodology is proposed by [11] for creating parsimonious models with lower prediction errors and higher generalization capacity than those generated by other methods. However, these methods are not completely automatic and need high computation times. Robust parsimonious models are still required to estimate set points when dealing with changing conditions in industrial plants. Parsimony can be achieved by several ways, being the selection of models with lower complexity and less inputs very commonly used.

The article proposes a new GA-based optimization to create better overall parsimonous models for predicting set points in industrial plants. The support vector machine (SVM) is selected as a promising regression technique but yielding higher training times than ANNs. Indeed, their training is usually automated by using GA but this leads to even higher computational times [11]. For that reason, our optimization should work with few individuals per generation and low number of generations. We consider different strategies but the key point is the complexity of each model for evaluating them after an initial evaluation. So, the individuals are reranked based on complexity when differences in their cost functions are not statistically significant. The proposal is applied for predicting three CAF temperature set points on a CHDGL but it might be extended to other sections of CHDGLs, i.e. the chromatic process or the zinc coating bath.

2 Description of the Industrial Process

Galvanised steel is broadly used by many industries such us automotive or appliance due to its corrosion resistance in many environments. The most widely known coating processes is the CHDGL, in which preheated steel coils are submerged in a pot of molten zinc at constant temperature. The process is divided into five independent sections that perform specific treatments on the steel strip during the galvanising (see Fig. 1). The goal is to produce flat steel products with high quality coating. Note that a deeper description of the galvanising line studied is provided in [7].

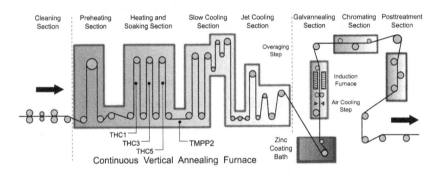

Fig. 1. Simplified scheme of a continuous hot dip galvanising line [11]

This is specifically focused on the CAF, which is a critical component for maintaining the quality of the galvanised products. A CAF contains five separate areas: a pre-heating and heating areas, a holding area, a slow cooling area, a rapid cooling area and finally, an overaging area for ensuring the carbon precipitation to minimize the effects of steel aging. The standard control of a CAF consists in controlling the temperature settings in each area while keeping the velocity of the steel strip within a pre-established range of values. This control can be properly achieved using accurate models for predicting three temperature set points (*THC1*, *THC3* and *THC5*) in the CAF. The optimization proposed was validated in a CAF of an CHDGL operating in the north of Spain.

3 Methodology

An optimization based on GA is proposed for tuning the parameters of the models and an input feature selection (FS) in order to obtain better overall parsimonious models. The main goal is to improve the efficiency of the process mainly fulfilling three restrictions: lower number of individuals per population, less computational cost in validation process (lower number of repetitions n and folds k for repeated n times k-fold CV) and a reduced number of maximum generations required to obtain a reliable model.

Our proposal is mainly based on classical GA but including a criterion to rearrange the individuals of each generation after their evaluation. The new step is next to the evaluation and ranking steps of classical GA. The new criterion measures the complexity of models to reorder these after the initial standard ranking based on their fitness function (J). Using this approach, two models that are not significantly different in terms of J can switch their position when the first is more complex than the latter. This algorithm for reranking the models is named *ReRank function* and is detailed below. This function looks over the complete list of individuals comparing consecutively comparing the first with the second, the second to the third and so on. For each comparison, it computes the p-value resulting from testing the values of both fitness functions. In case of p-vale obtained is higher than the parameter α and the complexity of the second model is lower than the first, both model will swap their positions.

Algorithm of ReRank function

```
program ReRank ()
    input    G(J, Model-Complexity): Individuals sorted by J
    const    NUMINDIV = cte.; alpha = cte.;
    var    PosFirst, PosSecond: 0..NUMINDIV;
    begin
      PosFirst := 0;
      repeat
        PosFirst := PosFirst+1;
        repeat
          PosSecond := PosFirst+1;
          p-value := test(G[PosFirst](J),G[PosSecond](J))
          if p-value > alpha AND G[PosSecond](Size)<G[PosFirst](Size)
            Swap(G[PosFirst],G[PosSecond])
          end if
        until (p-value <= alpha OR PosSecond == NUMINDIV)
      until (PosFirst == NUMINDIV-1)
end.
```

The complete optimization procedure is reported as *GeneticAlgorithm function* and is divided into two steps. First, individuals are sorted according to their (J) based on k-fold CV errors (see *Rank function*). Then the individuals without a significantly lower J than the next one are re-sorted according to their complexity (*ReRank function*). To this end, a nonparametric Wilcoxon ranked test is proposed with a level of significance α. The aim of this procedure is to rearrange individuals with similar J by their complexity.

Classical Genetic Algorithm modified with ReRank() function included

```
program GeneticAlgorithm ()
  const MAXGEN: cte., alpha: cte.;
  var Gen: 0..MAXGEN, P[Gen]: void;
  begin
    Gen := 0; Initialization (P[Gen]);
    J := Evaluation (P[Gen]); Write(P[Gen],J);
    repeat
      P[Gen] := Rank (P[Gen],J);
      P[Gen] := ReRank (P[Gen], Model-Complexity);
      Gen := Gen + 1;
      P[Gen] := Selection (P[Gen-1]);
      P[Gen] := P[Gen] + Crossover (P[Gen]);
      P[Gen] := P[Gen] + Mutation (P[Gen]);
      J := Evaluation (P[Gen]); Write(P[Gen],J);
    until (end condition is not fullfilled OR Gen == MAXGEN)
end.
```

4   Case of Study: Initial Settings and Specifications

Three common genetic operators are used for selecting the elitist individuals: *random uniform selection*, *roulette selection by rank*, and *tournament selection* with $k = 3$ individuals. Another well-known method like *Roulette selection by cost* is discarded due to the possibility that the operator cancels the effects of previously explained reranking of individuals according to their complexity (*ReRank* function). The crossover operator selected is named *heuristic blending* [8]. Other crossover operators based on swapping variables between parents randomly selected or using break points to combine a part of each parent chromosome are discarded. This is due to the fact that a small rate of elitism, such as four individuals, creates a large number of elitist individuals with similar chromosomes in few number of generations. The operator used is defined in Eq.1:

$$p_{new} = \beta (p_{mn} - p_{dn}) + p_{mn} \tag{1}$$

where p_{mn} and p_{dn} are the n_{th} variable in mother and father chromosomes respectively, p_{new} is the new single offspring variable, and β a random number chosen on the interval $[-0.1, 1.1]$. A different β is generated for each variable included. Finally, a mutation percentage of 10% is applied to all the experiments.

The input FS is controlled by using a binary-coded array of 19 bits, where 1 indicated that the attribute is considered and 0 the opposite. A real-coded chromosome is defined with 21 values that involve all the information related to the training parameters and the inputs selected for each individual. The 21 chromosome values, with its intervals are the following:

- Base-10 logarithm of the complexity parameter C of SVM within the interval: $[-3.\hat{9}, 3.\hat{9}]$. So, minimum and maximum C values are $10^{-3.\hat{9}} \simeq 0.0001$ and $10^{3.\hat{9}} \simeq 9999.97$, respectively.
- Gamma parameter of RBF kernel within the interval: $[0.000001, 0.999999]$

The optimization process is carried out using a maximum number of generations of $G = 30$ with a population size of 16 individuals. The individuals are firstly ranked according to J an then selected by applying an elitism percentage of 25%. So, the best four models are selected as progenitors for creating the next generation. The rest of the individuals are discarded.

The fitness function J is the average of the root mean squared error in valida-tion process ($RMSE_{val}$). In the experiments, the validation process was carried out using a 4-fold CV with 5 repetitions, i.e. the $RMSE_{val}$ was the average of 20 values. In this case, $RMSE$ is more suitable than mean absolute error (MAE) because of our intentions to look for models adjusted to all the cases in the dataset, reducing the quantity of residuals with high value. For each rep-etition, a 80% of the database is used for the 4-fold CV and the rest (20%) is used to obtain the testing error where the root mean squared error in testing process ($RMSE_{tst}$) is the average of 5 testing errors, one per each repetition. Next to this step, individuals without statistical significance between J were re-ranked according their model complexity using $ReRank$ $function$. After that, the comparison of the results from different models according to their cost values is carried out using Wilcoxon non-parametric paired test with $\alpha = 0.05$. Besides, model complexity is calculated as:

$$Complexity = 10^6 N_{FS} + N_{SV} \qquad (2)$$

where N_{FS} is the number of input features and N_{SV} the number of principal vectors in SVM. Two models with the same N_{FS} are reranked according to their N_{SV} if there is not statistical significance between their J. Obviously, N_{FS} are considered more important than N_{SV} in Eq.2.

To accelerate convergence, Latin Hypercube is used to define the first gener-ation, ensuring a uniform distribution of initial individuals.

5 Results and Discussion

The dataset is formed by a total number of $48,017$ instances of 721 different types of coils with 19 inputs and 3 outputs. Several attributes were previously removed for not having enough metallurgical relevance in a preprocessing phase carried out by [7]. The dataset is then homogenised by sampling the same number of cases for each type of coil to increase the accuracy predicting all set points. The inputs are WIDTHCOIL, THICKCOIL, VELMED, TMPP1, TMPP2CNG, C, Mn, Si, S, P, Al, Cu, Ni, Cr, Nb, V, Ti, B and N. The outputs involve three main temperature set points THC1, THC3 and THC5. The idea is to apply the new proposal for tuning model's parameters and selecting the best features to obtain a overall and parsimonious model capable to work well with new operation conditions. The numerical prediction SVMs use radial basis functions (RBF) kernels due to excellent results obtained in the preliminary tests. The statistical software for programming and testing all the experiments was the statistical software R 2.15.1 with e1071 package for calculating the SVM models.

Table 1 shows a summary of the principal results with the $RMSE_{tst}$ and N_{FS} for the best individual of the last generation for each out variable *THC1*, *THC3* and *THC5* and the method used for selecting the individuals (*uniform, roulette by rank, tournament*). The columns 3rd and 4th show the results obtained using a classical GA optimization, and the columns 5th and 6th present the results of the new proposal. Last column represents the difference of N_{FS} obtained between both methodologies and also the value of this relative difference in percentage between parentheses.

Table 1. Results for each configuration: with complexity and without complexity

| Setting | | Without | | With | | With vs Without |
Output	Selection	$RMSE_{tst}$	N_{FS}	$RMSE_{tst}$	$N_{features}$	$Diff.(\%)$
THC1	Uniform	0.035	7	0.035	5	2 (29%)
THC3	Uniform	0.034	6	0.034	4	2 (33%)
THC5	Uniform	0.033	8	0.034	4	4 (50%)
THC1	RankRoulette	0.036	7	0.035	6	1 (14%)
THC3	RankRoulette	0.035	6	0.033	5	1 (17%)
THC5	RankRoulette	0.034	6	0.035	6	0 (0%)
THC1	Tournament	0.035	7	0.036	5	2 (29%)
THC3	Tournament	0.035	5	0.034	4	1 (20%)
THC5	Tournament	0.034	6	0.035	4	2 (33%)

In case of *uniform selection method* is used, the mean reduction in N_{FS} is around 37%, with similar $RMSE_{tst}$ in both classical method and our proposal. In addition, a reduction of 50% in N_{FS} values is achieved for the model preicting *THC5,* only using four inputs against the model obtained with the classical methodology (8 inputs). Similar results are obtained with the *tournament selection* method achieving a mean N_{FS} reduction around 27% and identical number of features for the *THC1, THC3,* and *THC5* models (5, 4, and 4 inputs respectively). Nevertheless, with the *roulette selection method* reduction mean was only a 10% and with THC5 there was not reduction. Also, it might be observed that final models obtained with this selection method were less parsimonious than the others. The *roulette method by rank* did not work properly with only four parents because first two individuals has a high probability to be selected as parents for the next generation. In spite of this, also with the *roulette selection method* there is a slight reduction in the number of features.

In Table 2, a statistical comparison between both methods is presented where the *p*-values are obtained with the Wilcoxon paired test with level of significance $\alpha = 0.05$. In this table it might be deduced that there is a statistical significance between the number of features obtained with a classical genetic optimization and the proposed methodology (*p-value*=0.01264). Due to the fact that there is not statistical significance between $RMSE_{tst}$ (*p-value*=0.8203) we can conclude that including *ReRank function* allows to achieve models with less number of features with similar overall accuracy. Summarizing, the best overall parsimonious model is achieved.

Table 2. Statistical comparison between both methods

Description	$RMSE_{tst}$	Num.Attrib
Mean using classical opt (without complex.)	3.45%	6.4
Mean using proposal (with complex.)	3.45%	4.7
p-value (paired Wilcoxon-test)	0.821	0.013
Statistical significance ($\alpha = 0.05$)	NO	YES

In Figure 2, N_{FS} evolution of elitist individuals for THC5 models are presented with *random uniform selection* procedure. White box plots show the N_{FS} evolution for classical genetic optimization and gray boxplots the N_{FS} evolution with the new proposal. In this case, the use of *ReRank function* allows to obtain an stable and reduced number of inputs against classical methodology in which even N_{FS} is increased in the last generations.

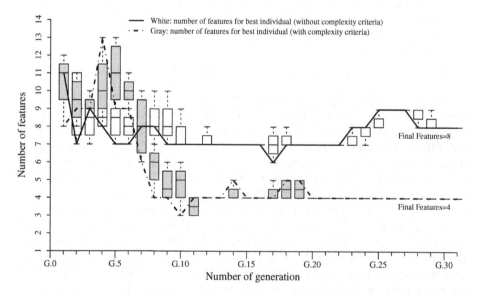

Fig. 2. Evolution of N_{FS} elitist individuals for predicting *THC5* using *random uniform selection* with (gray box plots) and without (white) *ReRank function*

Figures 3.a and 3.b show the evolution of $RMSE_{val}$ and $RMSE_{tst}$ predicting temperature *THC5*. These are obtained by using *random uniform selection* and *tournament selection* respectively. Comparing both figures, the second seems to have higher rate of convergence finding parsimonious models than the first one. Indeed, *random uniform selection* evidently allows higher variability in the complexity of models for the initial generations. Taking into consideration that only a few number of elitist individuals (4) are selected, the figures demonstrate that the first is more easygoing in model complexity than the second one. However, future experiments should consider the use of a higher elitism value and other operators like *tournament* or *roulette*.

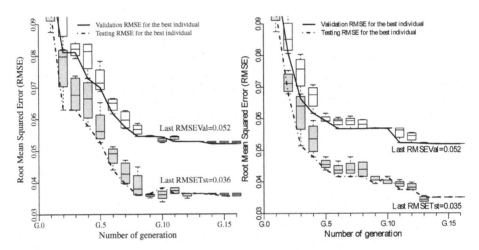

Fig. 3. Evolution of $RMSE_{val}$ and $RMSE_{tst}$ of temperature *THC5* (expressed as per unit value) for the elitist individuals using *ReRank function* and *random uniform selection* (left) or *tournament selection* (right)

6   Conclusions

A new GA optimization is presented to develop parsimonious overall SVMs for predicting three temperature set points in the CAF of a HDGL. Due to SVMs are associated with high training time, the methodology proposed is oriented to find accurate models with three restrictions: a low number of population individuals, a validation process with a small number of calculations and a reduced number of final generations to achieve the optimum model.

As a result of the application of this methodology, SVM algorithms might be very useful in the estimation of continuous line processes in the metalurgic industry. Our proposal is based on a second reranking step between pairs of preranked individuals according to their model complexity only if there is not a statistical significance between the fitness function (J) of both. The methodology proposed is also compared with classical genetic optimization and with different selection methods. The results have showed that new methodology helps to obtain models with a lower number of features but with similar overall precision, leading to the selection of SVM in this kind of problems.

Acknowledgments. We would like to convey our gratitude to the *Autonomous Government of La Rioja* for the continuous encouragement by the means of the "*Tercer Plan Riojano de Investigacion y Desarrollo de la Rioja*" on the project FOMENTA 2010/13, and to the *University of La Rioja* and *Santander Bank* for the project API11/13.

References

1. Bian, J., Zhu, Y., Liu, X.H., Wang, G.D.: Development of hot dip galvanized steel strip and its application in automobile industry. Journal of Iron and Steel Research International 13(3), 47–50 (2006)
2. Corchado, E., Abraham, A., Carvalho, A.: Hybrid intelligent algorithms and applications. Information Sciences 180(14), 2633–2634 (2010)
3. Corchado, E., Graña, M., Wozniak, M.: Editorial: New trends and applications on hybrid artificial intelligence systems. Neurocomputing 75(1), 61–63 (2012)
4. Corchado, E., Herrero, A.: Neural visualization of network traffic data for intrusion detection. Applied Soft Computing (2010)
5. Haykin, S.: Neural networks: a comprehensive foundation. Prentice Hall (1999)
6. Lu, Y.Z., Markward, S.: Development and application of an integrated neural system for an hdcl. IEEE Transactions on Neural Networks 8(6), 1328–1337 (1997)
7. Martínez-De-Pisón, F.J.: Optimización mediante técnicas de minería de datos del ciclo de recocido de una línea de galvanizado. Ph.D. thesis, Mechanical Department. University of La Rioja, Logroño, Spain (2003)
8. Michalewicz, Z., Janikow, C.Z.: Handling constraints in genetic algorithms. In: ICGA, pp. 151–157 (1991)
9. Mitchell, M.: An introduction to genetic algorithms. The MIT Press (1998)
10. Pernía-Espinoza, A., Castejón-Limas, M., González-Marcos, A., Lobato-Rubio, V.: Steel annealing furnace robust neural network model. Ironmaking and Steelmaking 32(5), 418–426 (2005)
11. Sanz-García, A., Fernández-Ceniceros, J., Fernández-Martínez, R., Martínez-de Pisón, F.J.: Methodology based on genetic optimisation to develop overall parsimony models for predicting temperature settings on an annealing furnace. Ironmaking & Steelmaking, 1–12 (November 2012)
12. Sedano, J., Curiel, L., Corchado, E., de la Cal, E., Villar, J.: A soft computing method for detecting lifetime building thermal insulation failures. Integrated Computer-Aided Engineering 17(2), 103–115 (2010)
13. Vapnik, V.N.: The nature of statistical learning theory. Springer-Verlag New York, Inc., New York (1995)

A Multi-level Filling Heuristic for
the Multi-objective Container Loading Problem

Yanira González, Gara Miranda, and Coromoto León

Dpto. Estadística, I.O. y Computación, Universidad de La Laguna,
Avda. Astrofísico Fco. Sánchez s/n, 38271 La Laguna, Santa Cruz de Tenerife, Spain
{ygonzale,gmiranda,cleon}@ull.es

Abstract. This work deals with a multi-objective formulation of the Container Loading Problem which is commonly encountered in transportation and wholesaling industries. The goal of the problem is to load the items (boxes) that would provide the highest total volume and weight to the container, without exceeding the container limits. These two objectives are conflicting because the volume of a box is usually not proportional to its weight. Most of the proposals in the literature simplify the problem by converting it into a mono-objective problem. However, in this work we propose to apply multi-objective evolutionary algorithms in order to obtain a set of non-dominated solutions, from which the final users would choose the one to be definitely carried out. To apply evolutionary approaches we have defined a representation scheme for the candidate solutions, a set of evolutionary operators and a method to generate and evaluate the candidate solutions. The obtained results improve previous results in the literature and demonstrate the importance of the evaluation heuristic to be applied.

Keywords: Container Loading Problem, Multi-objective Optimisation, Evolutionary Algorithms.

1 Introduction

The *Container Loading Problem* (CLP) belongs to an area of active research and has numerous applications in the real world, particularly in container transportation and distribution industries. When solving the CLP, normally, the goal is to distribute a set of rectangular pieces (boxes) in one large rectangular object (container) so as to maximize the total volume of packed boxes, i.e., trying to obtain a pattern of packaging that uses the container space as much as possible. However, in many real-world situations there are many other issues that may be taken into account [1]. Sometimes, to pack all the boxes in a container you must take care of their orientation, for example when their content is fragile. Other times, a good load distribution in the container should be made for a proper transportation, or a specific order for later shipment landed is needed. However, a rather common aspect in the scope of this problem is the weight limit of the containers, since they normally can't exceed a certain weight for their transportation. The rented trucks to transport the shipment are paid according to the total weight they can transport regardless of the total volume. Thus, the decision maker prefers to load and ship a shipment with high total weight rather than a shipment with low total weight.

Á. Herrero et al. (eds.), *International Joint Conference SOCO'13-CISIS'13-ICEUTE'13*,
Advances in Intelligent Systems and Computing 239,
DOI: 10.1007/978-3-319-01854-6_2, © Springer International Publishing Switzerland 2014

For that reason, in this work we consider the problem which simultaneously tries to maximize both objectives: the weight and volume utilization.

This way, the problem can be stated as a multi-objective optimization problem, trying to optimize the pieces layout inside the container so that the volume is maximized at the same time that the weight, without exceeding the weight limits. The formulation of the here addressed problem is as follows. We have a container with known width W, length L, height H, maximum weight P_{max}, and a set of N rectangular boxes. These boxes belong to one of the sets of boxes types $\mathcal{D} = \{T_1 \ldots T_m\}$, where the i-th type T_i is of dimensions $w_i \leq W$, $l_i \leq L$ and $h_i \leq H$. Associated with each T_i type exists a weight $p_i < P_{max}$, a volume v_i, a demand b_i, and a number of orientations allowed $o_i \in [1, 6]$. Ultimately, the aim is to find a packing into the container without overlapping, with x_i boxes of type T_i maximizing the total volume and weight of packaging:

$$\max \sum_{i=1}^{m} x_i v_i \text{ and } \max \sum_{i=1}^{m} x_i p_i$$

$$\text{subject to } x_i \in [0, b_i] \text{ and } \sum_{i=1}^{m} x_i p_i \leq P_{max} \text{ and } \sum_{i=1}^{m} x_i v_i \leq W \times L \times H$$

In this work, the problem will be solved using the following assumptions:

- Each box is placed in the container floor or on top of another box.
- The stability of the distribution of the boxes is not considered, since it is assumed the use of filler material to prevent potential problems.

In the literature, although there are some isolated works that use exact algorithms to deal with the CLP, most studies focus on providing solutions using heuristics and metaheuristics, because, computationally, the CLP is a NP-hard problem [2]. Although, many approaches deal with single-objective formulations of the CLP, the works dealing with a multi-objective formulation of the problem are almost non-existent [1]. In fact, the only known work in the literature that addresses the here studied multi-objective problem uses a simulated annealing algorithm hybridized with filling heuristics [3]. In [3] the authors consider the same objectives as the ones here studied: the weight maximization of the boxes loaded in the container and the maximization of the total volume occupied inside the container. The method proposed in [3] applies different weights to the two considered objectives, obtaining - at each execution - a single and different solution. To achieve a complete set of candidate solutions, the algorithm must be executed several times. However, in [4] we have applied multi-objective evolutionary algorithms to obtain a complete set of candidate solutions in a single execution. This way, with only one execution of the algorithm, the user obtain a set of possible solutions from which the most appropriate one can be selected at each moment.

In [4] we have been able to improve the results given in [3], but we haven't analysed the influence of the different design aspects of our approach. As a first attempt to analyse the design decisions adopted for this problem, in this work we propose a new evaluation heuristic which establishes a less restrictive criterion when placing the pieces within the available space. The remaining content of this paper is organised as follows. Section 2

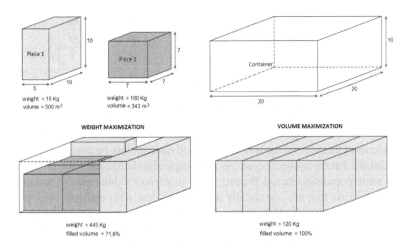

Fig. 1. Example of conflict objectives

gives a general overview of multi-objective optimization. A description of the applied multi-objective approach is given in section 3. The experimental results are presented in section 4. Finally, the conclusions and some lines of future work are given in section 5.

2   Multi-objective Optimization

Multi-objective optimization problems (MOPs) [5,6] arise in most real-world disciplines. While in single-objective optimization the optimal solution is usually clearly defined, this does not hold for MOPs. Instead of a single optimum, there is rather a set or front of alternative trade-offs, known as *Pareto-optimal* front, constituted by the non-dominated solutions. These solutions are optimal in the sense that no other solutions in the search space are superior to them when all objectives are considered. The final aim when dealing with MOPs is to obtain a non-dominated solution set which, in the best case, will coincide with the Pareto-optimal front. From the resulting final solution set, a human decision maker will be able to select a suitable compromise solution. In most optimization problems in the real world, the objectives are often involved in conflict between them, i.e., there is no single solution to optimize all objectives simultaneously. In the case of the multi-objective CLP, you might think a priori that the total volume maximization implies a maximization of weight. However, many times the size of the pieces or boxes to pack in the container is not proportional to their weight. That is, a box can be large and the content thereof may be lighter than the content of a smaller box (see Figure 1). So, the problem has two goals with at least some degree of conflict.

Evolutionary algorithms (EAs) have shown great promise for calculating solutions to large and difficult optimization problems and have been successfully used across a wide variety of real-world applications [7]. In fact, when applied to MOPs, EAs seem to perform better than other blind search strategies. Although this statement must be qualified with regard to the no free lunch theorems for optimization [8], to date there are few, if any, alternatives to EA-based multi-objective optimization [9]. The use of EAs to

		G₁	G₂	G₃		Gs-2	Gs-1	Gs
piece type	t_i	t2	t4	t1	⋯	t2	t3	t1
number of pieces	n_i	4	1	2	⋯	2	2	3
pieces orientation	r_i	1	0	1	⋯	4	0	2

Fig. 2. Scheme to represent a candidate solution

solve problems of this special nature has been motivated mainly because they are able to capture multiple Pareto-optimal solutions in a single simulation run - which is possible thanks to their population-based feature - and to exploit similarities of solutions by recombination. EAs that are specifically designed to deal with multiple objective functions are known as MOEAs [10]. When designing MOEAs two major problems must be addressed [11]: how to accomplish fitness assignment and selection in order to guide the search towards the Pareto-optimal set, and how to maintain a diverse population in order to prevent premature convergence and achieve a well distributed trade-off front. Many alternatives have been proposed in an attempt to adhere to such design goals [12].

3   Multi-objective Approach

Since MOEAs have shown promise for solving other multi-objective problems in the area of cutting and packing [13], we have already tried to prove their effectiveness in this particular problem [4]. To apply MOEAs to the CLP, we have designed a scheme representation for the candidate solutions, a set of evolutionary operators, and of course, a method for the evaluation of the candidate solutions. However, we would like to analyse whether the design of the representation scheme or the placement heuristics to evaluate the solution objectives have a decisive impact on the overall approach. For this reason, we have designed and tested a new filling heuristic to evaluate the candidate solutions.

3.1   Representation of Candidate Solutions

For the representation of candidate solutions, we have defined the structure in Figure 2. Considering $s \in [1, N]$ the chromosome size - which varies according to the grouping of pieces - an individual will consist of a sequence of genes $G_1, \ldots, G_s$ where each gene consists of three elements (t_i, n_i, r_i), with $t_i \in [t_1, t_m]$, $n_i \in [1, b_i]$ and $r_i \in [0, o_i)$. So, it is a sequence formed by piece type, number of pieces of that type and rotation for those pieces. This will determine the order and orientation in which the pieces will be put inside the container. Each piece has allowed two, four or six possible orientations, as determined by the input instance. Thus, to indicate the orientation r_i of a set of pieces we use a digit from 0 to 5, indicating a rotation of the piece with respect to its original orientation. A valid chromosome must contain all the pieces of each different type. Using this representation, the chromosome size may vary, depending on how the pieces of the same type are grouped:

$$\forall j \in [1, m] \sum_{k=1}^{s} z_{jk} = b_j \begin{Bmatrix} z_{jk} = n_k & \text{if } t_k = t_j \\ z_{jk} = 0 & \text{if } t_k \neq t_j \end{Bmatrix}$$

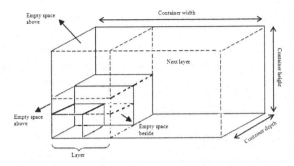

Fig. 3. Creation of holes

3.2 Generation of the Initial Population

For the initial generation of individuals, first a type of piece t_i for which $x_i < b_i$, must be randomly generated. Then, the number of pieces of that type, $n_i \leq b_i - x_i$, to be considered at the current gene and the rotation allowed for this set of pieces $r_i \in [0, o_i)$ are also randomly generated. These steps must be carried out until all the pieces have been entered in the chromosome, i.e., $\forall i \in [1, m]$ $x_i = b_i$.

3.3 Evaluation of the Objectives

Each chromosome represents a certain order - and orientation - in which pieces are to be placed in the container. However, based on this information, it's still necessary to apply a placement or filling heuristic, in order to decide where exactly to locate each item. Once the filling heuristic is applied, and thus, we know exactly which items have been loaded into the container, we can calculate both optimization objectives: total volume and weight. The filling heuristic presented in this paper is called *Multiple-Level Filling Heuristic* (MLFH) and it is a modification of the *Single-Level Filling Heuristic* (SLFH) implemented in [4]. Both heuristics are based on the creation and management of piece levels or layers within the container. Such levels or layers identify empty spaces inside the container, and thus, they represent areas where to locate items. The difference between SLFH and MLFH is that SLFH never uses the next layer (Figure 3) if there are other empty spaces that can allocate items. However, in MLFH all layers are available to accommodate pieces, i.e., if a given piece doesn't fit into the current empty spaces, it is located into the next and unused layer, thus creating a new set of empty spaces. MLFH works as follows:

- The first piece in the chromosome is introduced at the bottom left corner of the container. This piece will determine the dimensions of the spaces to be generated: one **in front** of the box, other **above**, and other **beside** the placed item (Figure 3).
- The following piece in the chromosone is selected and placed into the container according to the next guidelines:
    - First, we try to fill the space created in front of the box. If there is no available space for our current piece in the front-level, we check for available space in

the above-level. If even then, the current piece doesn't fit, we try to locate it
into the level beside the box.

- When the piece fits into a level, the box is placed at the bottom left corner of
  the level. In such a case, it must be checked if the box fits into the level without
  leaving empty space. If so, the given level is now completed and thus, we can
  avoided it from our list of open or available levels to locate items. If not, it
  means that there is still some remaining free space infront, above and/or beside
  the already placed box, and so it's necessary to create a set of new - in front,
  above, and/or beside - levels.
- At any moment, when checking for a given type of level, e.g., above level, first
  we must consider the most recently created above-level.

– When the current analysed box doesn't fit in any of the available levels, the procedure finishes, and no more items are loaded into the container. At this moment,
it's possible to compute the value of the objectives (total volume and weight) by
adding the volume and weight of all the loaded pieces.

3.4 Operators

Taking into account an individual represented by the chromosome $C1 =
(G1_1, \ldots, G1_{s_1})$ and another individual represented by the chromosome $C2 =
(G2_1, \ldots, G2_{s_2})$, a one point crossover operator has been implemented as follows:

– First, we select a random point from each individual, $p_1 \in [1, s_1]$ and $p_2 \in [1, s_2]$.
– Then, the two parts after the selected points are exchanged, so that, $C1 =
  (G1_1, \ldots,$
  $G1_{p_1}, G2_{p_2+1}, \ldots, G2_{s_2})$ and $C2 = (G2_1, \ldots, G2_{p_2}, G1_{p_1+1}, \ldots, G1_{s_1})$.
– The first part of each individual, $G1_1, \ldots, G1_{p_1}$ and $G2_1, \ldots, G2_{p_2}$, is modified
  to respect the total number of pieces of each type present in the chromosome. That
  is, we remove any extra piece types if there are too many pieces of this type, or we
  add if missing (Figure 4).

For the mutation we have introduced three different types of movements on the chromosome, each of which applied under the probability of mutation:

– *Add one gene*: a type of piece is randomly generated $t_x \in [t_1, t_m]$. Then, all genes
  with this type of piece are searched, and we select one with more than one associated piece. A number of pieces $n_x < n_y$ is chosen from the selected gene
  $G_y = (t_y, n_y, r_y)$ with $t_y = t_x$, so that the pieces are distributed between that gene
  $(t_y, n_y - n_x, r_y)$ and the new one (t_x, n_x, r_x). The orientation is chosen from those
  allowed for that type of piece $r_x \in [0, o_x)$. Finally, we choose the position of the
  chromosome in which to insert the new gene, moving the rest to the right.
– *Remove a gene*: a position within the chromosome is randomly selected. If the
  piece type t_x of the selected gene $G_x = (t_x, n_x, r_x)$ appears more times in that
  chromosome, then a gene $G_y = (t_y, n_y, r_y)$ is randomly selected from among the
  same type, i.e. $t_y = t_x$, and the number of pieces will be increased with the number
  of pieces of the first gene to be removed ($n_y = n_y + n_x$). As it is possible that both
  genes do not have the same type of orientation of pieces ($r_x \neq r_y$), one of them is
  randomly selected. Finally, G_x is eliminated by compaction to the left.

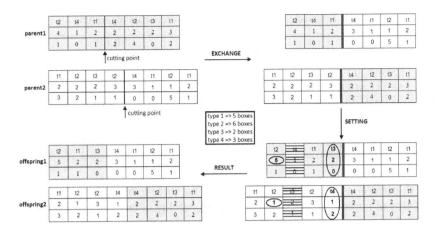

Fig. 4. Crossover operator

– *Change a gene*: a random position of the chromosome is selected and the type of orientation is randomly changed within the possible orientations for the piece type.

Note that for the generation of random numbers, the implementation makes usage of the rand() and srand() functions from the *C Standard General Utilities Library*.

4   Experimental Evaluation

In order to solve the multi-objective CLP, we have applied the evolutionary algorithms which are provided by METCO [14], a pluging-based tool which incorporates a set of multi-objective schemes to tackle MOPs. Through the use of METCO, the implementation from scratch of MOEAs is prevented, since it provides implementations of some of the most widely used MOEAs in the literature. The experimental evaluation was carried out using a dedicated cluster with 20 nodes of dual core and Debian GNU/Linux operating system. Both, METCO framework and the problem approach, have been implemented using C++ and compiled with *gcc 4.1.3* and MPICH *1.2.7*. For the computational study, we have used the real test problem instance proposed in [3]. The instance has 12 types of products with different sizes and associated weight, however, there is no information about the boxes orientations. Therefore, and because some of the boxes contain liquids, it is assumed that you can only rotate the base of the boxes. This problem instance is solved in [3] by the application of a simulated annealing algorithm with filling heuristic, and assigning different weights, from 0 to 1, to each objective. Since our knowledge this is the only work dealing with the here presented multi-objective formulation of the problem. For this reason, we will use such a work as a reference to analyse the results here obtained.

For the experimental evaluation, we have checked the behaviour of three of the most-known MOEAs: NSGA-II, SPEA2, and an adaptive version of IBEA. We checked the algorithms with different population sizes (20, 25, 50, 100), mutation rates (0.1, 0.2, 0.3) and crossover rates (0.7, 0.8, 0.9). However, after such an intitial tunning, and in

Table 1. Statistical comparison among evolutionary algorithms

MOEAS	SLFH			MLFH		
	NSGA-II	SPEA2	IBEA	NSGA-II	SPEA2	IBEA
NSGA2-SLFH	↔	↔	↑	↔	↔	↓
SPEA2-SLFH	↔	↔	↑	↓	↔	↓
IBEA-SLFH	↓	↓	↔	↓	↔	↓
NSGA2-MLFH	↔	↑	↑	↔	↔	↑
SPEA2-MLFH	↔	↔	↑	↔	↔	↔
IBEA-MLFH	↑	↔	↑	↓	↓	↔

order to perform a deeper comparison among the three evolutionary algorithms, the probability of mutation was fixed to 0.3, the probability of crossover was fixed to 0.9, the population size was fixed to 20, the fitness scaling factor was fixed to 0.001 and the stopping criterion was fixed to 30000 evaluations, i.e., 1500 generations. The three algorithms were tested with both filling heuristics: SLFH and MLFH.

Since we are dealing with stochastic algorithms, each execution was repeated 30 times. In order to provide the results with confidence, comparisons have been performed following the next statistical analysis. First, a Kolmogorov-Smirnov test is performed in order to check whether the values of the results follow a normal (gaussian) distribution or not. If so, the Levene test checks for the homogeneity of the variances. If samples have equal variance, an ANOVA test is done; otherwise a Welch test is performed. For non-gaussian distributions, the non-parametric Kruskal-Wallis test is used to compare the medians of the algorithms. A confidence level of 95% is considered, which means that the differences are unlikely to have occurred by chance with a probability of 95%. The analysis is performed using the hypervolume [15] metric. Table 1 shows the statistical comparison between the considered approaches. The symbol ↑ is used to denote that differences between the models are statistically significant and that the model in the left column obtains a higher median and mean value. In the cases in which the opposite occurs, the symbol ↓ is used. Finally, for the cases in which the differences have not been statistically significant, the symbol ↔ is used. NSGA-II and SPEA2 have obtained statistically better results than IBEA. Moreover, approaches which use MLFH have obtained statistically better results than the approaches using SLFH.

Since the differences between NSGA-II and SPEA2 are not significant, we have choosen NSGA-II to calculate the average highest values obtained for each objective. From each of the 30 executions of NSGA-II, the final solution set is analysed in order to select the highest volume. The 30 selected values are considered to calculate the

Table 2. Comparison of objectives: volume and weight

Objective	MLFH			SLFH			Dereli and Sena 2010 [3]			
	Sol. Volume	Sol. Weight	Std. Dev.	Sol. Volume	Sol. Weight	Std. Dev.	Sol. 1	Sol. 2	Sol. 3	Sol. 4
Volume(%)	91.40%	89.95%	0.20%	91.13%	90.87%	0.36%	87.51%	86.13%	86.09%	85.96%
Weight(kg)	7198.85	7199.99	2.68	7197.39	7199.52	1.49	6713.37	7157.06	7150.41	7151.37

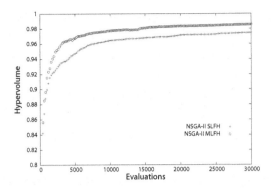

Fig. 5. Hypervolume: MLFH vs. SLFH

average highest volume obtained by our approach. A similar process is repeated for the weight objective. Table 2 shows the solution with the highest-average volume and the solution with the highest-average weight. For both cases, the value of the other objective is also shown. Equivalent results obtained with the SLFH are also shown in the Table. Finally, the four solutions obtained by the approach presented in [3] are also included in the Table. Clearly, the results obtained with MLFH improve the results given by the referenced approaches, achieving higher packaging volume and weight than the highest obtained by the related works. In order to perform a deeper comparison among the two filling heuristics, Figure 5 shows the NSGA-II hypervolume when applying SLFH and when applying MLFH. As we can see, the NSGA-II version which applies the modified filling heuristic outperforms the original approach using the single-level heuristic.

5 Conclusions and Future Work

In this work, we have presented a modification of the Single-Level Filling Heuristic (SLFH) proposed in [4]. In opposition to the SLFH, the modified heuristic MLFH manages a dynamic set of layers where to place a given item. These filling heuristics are imprescindible for the evaluation of the individuals or chromosomes which represent the candidate solutions. Such a representation is defined for the application of MOEAs to the solution of a multi-objective formulation of the Container Loading Problem. The final goal is to optimise the usage of the container so as to maximise the total packed volume, as well as maximise the total used weight, without exceeding the container weight and volume limits. The obtained solutions show the great effectiveness of MOEAs when applied to such kind of real-world problems and also the importance of the representation and evaluation strategies designed for the problem.

In order to further analyse the design issues for the usage of MOEAs, it would be interesting to compare and test more variation operators, and even try to design new representations for the candidate solutions. For doing such an extensive study, it would be necessary to use more problem instances. The problem instance here used for the computational experiments is the only real instance defined in the related works. However, the obtained Pareto fronts demonstrated that there is not a high degree of conflict among the two objectives. For this reason, it would be interesting to define new sets of problem instances for the multi-objective Container Loading Problem.

Acknowledgment. This work was funded by the Spanish Ministry of Science and Technology as part of the 'Plan Nacional de I+D+i' (TIN2011-25448). The work of Yanira González was funded by grant CAJACANARIAS by post-graduate.

References

1. Bortfeldt, A., Wäscher, G.: Container Loading Problems - A State-of-the-Art Review. Otto-von-Guericke-Universität Magdeburg, Working Paper 1 (April 2012)
2. Scheithauer, G.: Algorithms for the Container Loading Problem. In: Operations Research Proceedings 1991, pp. 445–452. Springer (1992)
3. Dereli, T., Sena Das, G.: A Hybrid Simulated Annealing Algorithm for Solving Multi-objective Container Loading Problems. Applied Artificial Intelligence: An International Journal 24(5), 463–486 (2010)
4. de Armas, J., González, Y., Miranda, G., León, C.: Parallelization of the Multi-Objective Container Loading Problem. In: IEEE World Congress on Computational Intelligence (WCCI), Brisbane, Australia, pp. 155–162 (June 2012)
5. Sawaragi, Y., Nakayama, H., Tanino, T.: Theory of multiobjective optimization. Academic Press, Orlando (1985)
6. Steuer, R.E.: Multiple Criteria Optimization: Theory, Computation and Application. John Wiley, New York (1986)
7. Eiben, A.E.: In: Bäck, T., Fogel, D., Michalewicz, M. (eds.) Handbook of Evolutionary Computation. IOP Publishing Ltd. and Oxford University Press (1998)
8. Wolpert, D.H., Macready, W.G.: No free lunch theorems for optimization. IEEE Transactions on Evolutionary Computation 1(1), 67–82 (1997)
9. Horn, J.: In: Bäck, T., Fogel, D.B., Michalewicz, Z. (eds.) Handbook of Evolutionary Computation. Institute of Physics Publishing (1997)
10. Coello, C.A., Lamont, G.B., Van Veldhuizen, D.A.: In: Goldberg, D.E., Koza, J.R. (eds.) Evolutionary Algorithms for Solving Multi-Objective Problems. Genetic and Evolutionary Computation. Springer (2007)
11. Zitzler, E., Deb, K., Thiele, L.: Comparison of multiobjective evolutionary algorithms: Empirical results. Evolutionary Computation 8(2), 173–195 (2000)
12. Coello Coello, C.A.: An Updated Survey of Evolutionary Multiobjective Optimization Techniques: State of the Art and Future Trends. In: Proceedings of the Congress on Evolutionary Computation, vol. 1, pp. 3–13. IEEE Press (1999),
    citeseer.ist.psu.edu/coellocoello99updated.html
13. de Armas, J., Miranda, G., Leon, C., Segura, C.: Optimisation of a Multi-Objective Two-Dimensional Strip Packing Problem based on Evolutionary Algorithms. International Journal of Production Research 48(7), 2011–2028 (2009)
14. León, C., Miranda, G., Segura, C.: METCO: A Parallel Plugin-Based Framework for Multi-Objective Optimization. International Journal on Artificial Intelligence Tools 18(4), 569–588 (2009)
15. Zitzler, E., Thiele, L.: Multiobjective Optimization Using Evolutionary Algorithms - A Comparative Case Study. In: Eiben, A.E., Bäck, T., Schoenauer, M., Schwefel, H.-P. (eds.) PPSN 1998. LNCS, vol. 1498, pp. 292–301. Springer, Heidelberg (1998)

Multiple Choice Strategy Based PSO Algorithm with Chaotic Decision Making – A Preliminary Study

Michal Pluhacek, Roman Senkerik, and Ivan Zelinka

Tomas Bata University in Zlin, Faculty of Applied Informatics, T.G. Masaryka 5555,
760 01 Zlin, Czech Republic
{pluhacek,senkerik,zelinka}@fai.utb.cz

Abstract. In this paper, it is proposed the utilization of chaotic pseudo random number generators based on six selected discrete chaotic maps to enhance the performance of newly proposed multiple choice strategy based PSO algorithm. This research represents a continuation of previous successful experiments with the fusion of the PSO algorithm and chaotic systems. The performance of proposed algorithm is tested on a set of four test functions. Obtained promising results are presented, discussed and compared against the basic PSO strategy with inertia weight.

Keywords: PSO, Chaos, Optimization, Swarm.

1 Introduction

Since several past decades the evolutionary computation techniques (ECTs) have been developed and used to solve complex optimization problems [1-10]. One of the most popular ECT is the Particle Swarm Optimization algorithm (PSO) [1, 3, 7-13]. All ECTs are inspired in nature and use stochastic operations. One of the newly spreading approaches for generation of pseudo-random numbers for these stochastic operations is by the utilization of chaotic sequences. [11, 14]. In this presented preliminary small-scale study, the possibilities of using the chaotic sequences to enhance the performance of newly proposed Multiple choice strategy for PSO algorithm are investigated.

2 Particle Swarm Optimization Algorithm

The PSO algorithm is inspired by the natural swarm behavior of birds and fish. It was introduced by Eberhart and Kennedy in 1995 [1] as an alternative to other metaheuristics, such as Ant Colony Optimization [2], Genetic Algorithms (GAs) [4] or Differential Evolution (DE) [5]. Each particle in the population represents a possible solution of the optimization problem which is defined by its cost function. In each iteration, a new location (combination of cost function parameters) of the particle is calculated based on its previous location and velocity vector (velocity vector contains particle velocity for each dimension of the problem).

Á. Herrero et al. (eds.), *International Joint Conference SOCO'13-CISIS'13-ICEUTE'13*,
Advances in Intelligent Systems and Computing 239,
DOI: 10.1007/978-3-319-01854-6_3, © Springer International Publishing Switzerland 2014

One of the disadvantages of the original PSO algorithm was poor local search ability. Another problem was the rapid acceleration of particles, which causes them to abandon the defined area of interest. For this reasons, several modifications of the PSO were introduced. The main principles of the PSO algorithm and its modifications are detailed in [1–3, 12, 13]. Within this research, the linear decreasing inertia weight strategy for PSO was utilized [12]. This strategy was first introduced in 1998 [12] in order to improve the local search capability of PSO. The selection of inertia weight strategy of PSO was based on numerous previous experiments [11]. Several modifications of inertia weight strategy are well described in [13]. In this study, linear decreasing inertia weight [12 13] is used. Default values of all PSO parameters were chosen according to the recommendations given in [1–3, 12, 13].

Inertia weight is designed to influence the velocity of each particle differently over time [12, 13]. In the beginning of the optimization process, the influence of inertia weight factor w is minimal. As the optimization continues, the value of w is decreasing, thus the velocity of each particle is decreasing, since w is always the number less than one and it multiplies the previous velocity of particle in the process of new velocity value calculation. Inertia weight modification PSO strategy has two control parameters w_{start} and w_{end}. A new w for each iteration is given by Eq. 1, where i stand for current iteration number and n for the total number of iterations.

$$w = w_{start} - \frac{\left((w_{start} - w_{end}) \cdot i\right)}{n} \tag{1}$$

The main PSO formula that determines a new velocity and thus the position of each particle in the next iteration (or migration cycle) is given by Eq. 2.

$$v(i+1) = w \cdot v(i) + c_1 \cdot Rand \cdot (pBest - x(i))$$
$$+ c_2 \cdot Rand \cdot (gBest - x(i)) \tag{2}$$

where:
$v(i + 1)$ - New velocity of a particle.
$v(i)$ - Current velocity of a particle.
c_1, c_2 - Priority factors.
$pBest$ - Best solution found by a particle.
$gBest$ - Best solution found in a population.
$x(i)$ - Current position of a particle.
$Rand$ - Random number, interval (0, 1).

The new position of a particle is then given by Eq. 3, where x(i + 1) is the new position:

$$x(i+1) = x(i) + v(i+1) \tag{3}$$

3   Multiple Choice Particle Swarm Optimization Algorithm (MC-PSO)

Within this newly proposed strategy the original way (Eq. 2) of calculating the particle velocity for the next generation is altered. At first, three numbers b1, b2 and b3 are defined at the start of the algorithm. These numbers represent limit values for different rules, so they should follow the pattern: b1 < b2 < b3. In this research, the following values were used: b1 = 0.1, b2 = 0.6, b3 = 0.7. Furthermore during the calculation of a new velocity of an each particle a random number r is generated from the interval < 0, 1 >. In this study, the random number r is generated by means of chaotic pseudo random number generator based on selected discrete chaotic maps. Finally, the new velocity is calculated following these four simple rules: If $r \leq b1$, the new velocity of particle is given by Eq. 4:

$$v(i+1) = 0 \qquad (4)$$

If $b_1 < r \leq b_2$, the new velocity of particle is given by Eq. 5:

$$v(i+1) = w \cdot v(i) + c \cdot Rand \cdot (x_r(i) - x(i)) \qquad (5)$$

Where $x_r(i)$, is the position of randomly chosen particle.
If $b_2 < r \leq b_3$, the new velocity of particle is given by Eq. 6:

$$v(i+1) = w \cdot v(i) + c \cdot Rand \cdot (pBest - x(i)) \qquad (6)$$

Finally, if $b_3 < r$, the new velocity of particle is given by Eq. 7:

$$v(i+1) = w \cdot v(i) + c \cdot Rand \cdot (gBest - x(i)) \qquad (7)$$

The priority factors c1 and c2 from original PSO formula (Eq. 2) are replaced within this novel approach with single parameter c. Within this new strategy parameter c defines not the priority (which is given by b1, b2 and b3 setting) but the overstep value. Within this research, c was set to 2.

The pseudo-code of the new proposed MC-PSO strategy is depicted in Fig. 1.

1. **Initialize population**
2. **Evaluate and assign** *pBest* **and** *gBest*
3. **REPEAT UNTIL stopping condition met:**
    4. **Calculate inertia weight constant** *w*
    5. **FOR each individual:**
        6. *r* = **RANDOM[0,1]**
        7. **IF** *(r ≤ b₁)* *v(t+1) = 0*
        8. **IF** *(b₁< r ≤ b₂)* *v(t+1) = w • v(t) + c •* **RANDOM[0,1]** *• (xᵣ(t) - x(t))*
        9. **IF** *(b₂ < r ≤ b₃)* *v(t+1) = w • v(t) + c •* **RANDOM[0,1]** *• (pBest- x(t))*
        10. **IF** *(b₃ < r)* *v(t+1) = w • v(t) + c •* **RANDOM[0,1]** *• (gBest - x(t))*
        11. **Evaluate and assign pBest and gBest**
12. **End**

Fig. 1. MC-PSO pseudo-code

4   Chaotic Maps

This section contains the description of six discrete chaotic maps used as the chaotic pseudo random generators for the decision making within the MC-PSO algorithm. Initial and, in comparison with this research, different concept of embedding chaotic dynamics into the evolutionary algorithms is given in [14].

4.1   Lozi Map

The Lozi map is a simple discrete two-dimensional chaotic map. The map equations are given in (8). The parameters used in this work are: a = 1.7 and b = 0.5 with respect to [15]. For these values, the system exhibits typical chaotic behavior and with this parameter setting it is used in the most research papers and other literature sources.

$$X_{n+1} = 1 - a|X_n| + bY_n$$
$$Y_{n+1} = X_n \tag{8}$$

4.2   Dissipative Standard Map

The Dissipative Standard map is a two-dimensional chaotic map. The parameters used in this work are b = 0.6 and k = 8.8 as suggested in [15]. The map equations are given in Eq. 9.

$$X_{n+1} = X_n + Y_{n+1} \pmod{2\pi}$$
$$Y_{n+1} = bY_n + k \sin X_n \pmod{2\pi} \tag{9}$$

4.3   Arnold's Cat Map

The Arnold's Cat map is a simple two dimensional discrete system that stretches and folds points (x, y) to (x+y , x+2y) mod 1 in phase space. The map equations are given in Eq. 10. This map was used with parameter k = 0.1.

$$X_{n+1} = X_n + Y_n \pmod{1}$$
$$Y_{n+1} = X_n + kY_n \pmod{1} \tag{10}$$

4.4   Sinai Map

The Sinai map is a simple two dimensional discrete system similar to the Arnold's Cat map. The map equations are given in Eq. 11. The parameter used in this work is δ = 0.1 as suggested in [15].

$$X_{n+1} = X_n + Y_n + \delta \cos 2\pi Y_n \,(\text{mod } 1)$$
$$Y_{n+1} = X_n + 2Y_n \,(\text{mod } 1) \tag{11}$$

4.5   Burgers Map

The Burgers map is a discretization of a pair of coupled differential equations The map equations are given in Eq. 12 with control parameters a = 0.75 and b = 1.75 as suggested in [15].

$$X_{n+1} = aX_n - Y_n^2$$
$$Y_{n+1} = bY_n + X_n Y_n \tag{12}$$

4.6   Tinkerbell Map

The Tinkerbell map is a two-dimensional complex discrete-time dynamical system given by Eq. 13 with following control parameters: a = 0.9, b = -0.6, c = 2 and d = 0.5 [15].

$$X_{n+1} = X_n^2 - Y_n^2 + aX_n + bY_n$$
$$Y_{n+1} = 2X_n Y_n + cX_n + dY_n \tag{13}$$

5   Test Functions

Four well-known and basic static test functions were used in this study: 1st De Jong´s function, 2nd De Jong´s function, Rastrigin´s function and Schwefel´s function.
    The 1st De Jong´s function is given by Eq. 14.

$$f(x) = \sum_{i=1}^{\dim} x_i^2 \tag{14}$$

The 2nd De Jong´s function is given by Eq. 15.

$$f(x) = \sum_{i=1}^{\dim-1} 100(x_i^2 - x_{i+1})^2 + (1 - x_i)^2 \tag{15}$$

Rastrigin´s function is given by Eq. 16.

$$f(x) = 10\dim + \sum_{i=1}^{\dim} x_i^2 - 10\cos(2\pi x_i) \tag{16}$$

Schwefel´s function is given by Eq. 17.

$$f(x) = \sum_{i=1}^{\dim} -x_i \sin(\sqrt{|x|}) \tag{17}$$

6   Results

The optimization experiments results are presented in this section. The simple statistical overview is given in Tables 1 – 4 and the graphical comparison of time evolution of gBest value history on Figures 2 – 5. The bold numbers within all tables represent the best value. Since the MC-PSO strategy is primarily aimed to improve the performance of PSO algorithm in higher-dimensional problems, all optimization experiments were performed with setting dimension = 100. Eight different versions of PSO algorithm were used: Basic PSO with inertia weight (PSO Weight) MC-PSO without any chaotic attributes, (MC-PSO) and MC-PSO with decision making based on six different chaotic pseudo random number generators (MC-PSO Lozi, MC-PSO Disi, MC-PSO Arnold, MC-PSO Sinai, MC-PSO Burger, MC-PSO Tinker). Each experiment was repeated 50 times. All algorithms were set as follows: Population size = 50, Iterations = 5000, wstart = 0.9 and wend = 0.4.

Table 1. Results for 1st De Jong´s function

Dim: 100	PSO Weight	MC-PSO	MC-PSO Lozi	MC-PSO Disi	MC-PSO Arnold	MC-PSO Sinai	MC-PSO Burger	MC-PSO Tinker
Mean Value:	4.51504	1.01309	2.9108	0.365606	1.21338	0.95798	**0.0843405**	5.2837
Std. Dev.:	1.00244	0.202977	0.500936	0.0640452	0.173721	0.147289	0.0180132	0.743263
Median:	4.36371	0.99574	2.86291	0.362654	1.20426	0.948753	0.0820493	5.30146
Worst result:	6.81426	1.47736	4.03755	0.576457	1.61133	1.34585	0.123476	7.25066
Best result:	2.47183	0.674095	2.1342	0.261599	0.797293	0.679793	**0.0476482**	3.81958

Table 2. Results for 2nd De Jong´s function

Dim: 100	PSO Weight	MC-PSO	MC-PSO Lozi	MC-PSO Disi	MC-PSO Arnold	MC-PSO Sinai	MC-PSO Burger	MC-PSO Tinker
Mean Value:	1113.23	241.137	446.451	153.013	270.962	251.098	**113.049**	698.709
Std. Dev.:	262.066	33.833	57.4678	17.2257	38.1637	38.7536	10.0665	93.2292
Median:	1088.36	238.4	447.228	151.763	267.717	248.917	110.397	680.877
Worst result:	1813.97	358.884	627.436	217.432	377.14	342.231	166.099	911.099
Best result:	644.741	191.873	319.409	126.026	206.625	194.375	**105.069**	562.514

Table 3. Results for Rastrigin´s function

Dim: 100	PSO Weight	MC-PSO	MC-PSO Lozi	MC-PSO Disi	MC-PSO Arnold	MC-PSO Sinai	MC-PSO Burger	MC-PSO Tinker
Mean Value:	427.245	344.091	492.739	233.992	374.449	346.593	**168.016**	629.297
Std. Dev.:	49.2834	49.3686	56.6294	35.9781	46.9436	52.0409	47.3005	55.9192
Median:	429.614	347.511	491.975	234.111	373.528	338.147	165.783	627.121
Worst result:	527.35	444.083	643.4	311.512	482.34	500.125	322.021	737.733
Best result:	308.793	233.903	372.491	163.093	289.479	261.12	**99.1245**	508.52

Table 4. Results for Schwefel´s function

Dim: 100	PSO Weight	MC-PSO	MC-PSO Lozi	MC-PSO Disi	MC-PSO Arnold	MC-PSO Sinai	MC-PSO Burger	MC-PSO Tinker
Mean Value:	-13013.6	-17457	-16346.5	-19048.4	-17319.6	-17953.7	**-20872.6**	-13915
Std. Dev.:	851.511	918.114	1023.23	1278.93	1428.84	1137.75	1419.25	1168.41
Median:	-12919.9	-17409.4	-16267.8	-19352.3	-17177.5	-18032.5	-21054.4	-13910.5
Worst result:	-10471.5	-15256.2	-13911.8	-16181.3	-14131.9	-15696.8	-17848.8	-11514.6
Best result:	-15064.7	-18917.9	-18792.9	-22194.5	-20527.9	-21091.2	**-24885.2**	-17154

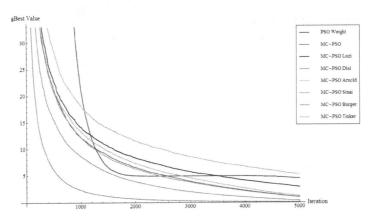

Fig. 2. gBest history for 1st De Jong´s function

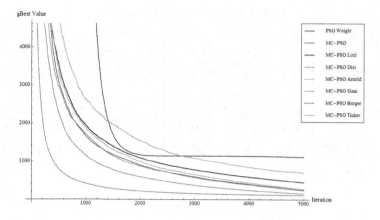

Fig. 3. gBest history for 2nd De Jong´s function

7   Brief Analysis of the Results

Based on the results presented in previous section it seems that the different chaotic number generators have an impact on both the performance in terms of solution quality and the behavior (convergence speed, premature convergence avoidance etc.) of the MC-PSO algorithm when dealing with higher dimension optimization problems. This impact varies for different chaotic maps and could be both very negative (e.g. MC-PSO Tinker) or very positive (e.g. MC-PSO Burger). The results particularly of MC-PSO Burger version are very promising and should be examined more closely in future studies. The Mersenne Twister algorithm with default automatic setting pseudo-random number generator was applied to represent traditional pseudo-random number generators in comparisons.

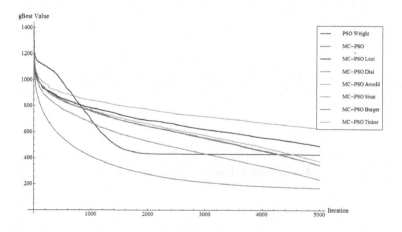

Fig. 4. gBest history for Rastrigin´s function

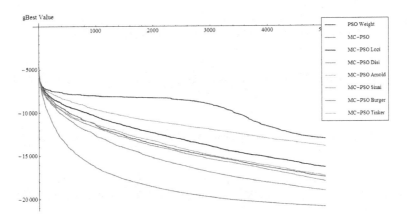

Fig. 5. gBest history for Schwefel´s function

8   Conclusion

In this paper the novel Multiple Choice strategy for PSO algorithm with chaos driven decision making was investigated. Six different chaotic maps were utilized as pseudo-random number generators and implemented into the most crucial part of the MC-PSO strategy, which determines the exact action of each single particle for the next iteration. Presented results support the presumption that the chaotic sequences could influence the performance of such algorithm. As the scale of this study is very small, more research will be needed to strengthen these claims. Nevertheless the main aim of this study is to inform about these interesting results of this novel approach and proposed MC-PSO algorithm.

Acknowledgements. This work was supported by Grant Agency of the Czech Republic-GACR P103/13/08195S, by the project Development of human resources in research and development of latest soft computing methods and their application in practice, reg. no. CZ.1.07/2.3.00/20.0072 funded by Operational Program Education for Competitiveness, co-financed by ESF and state budget of the Czech Republic; European Regional Development Fund under the project CEBIA-Tech No. CZ.1.05/2.1.00/03.0089 and by Internal Grant Agency of Tomas Bata University under the project No. IGA/FAI/2013/012.

References

1. Kennedy, J., Eberhart, R.: Particle Swarm Optimization. In: Proceedings of IEEE International Conference on Neural Networks, vol. IV, pp. 1942–1948 (1995)
2. Dorigo, M.: Ant Colony Optimization and Swarm Intelligence. Springer (2006)

3. Eberhart, R., Kennedy, J.: Swarm Intelligence. The Morgan Kaufmann Series in Artificial Intelligence. Morgan Kaufmann (2001)
4. Goldberg, D.E.: Genetic Algorithms in Search Optimization and Machine Learning, p. 41. Addison Wesley (1989) ISBN 0201157675
5. Storn, R., Price, R.: Differential evolution - a simple and efficient heuristic for global optimization over continuous spaces. Journal of Global Optimization 11, 341–359 (1997)
6. Zelinka: SOMA - self organizing migrating algorithm. In: Babu, B.V., Onwubolu, G. (eds.) New Optimization Techniques in Engineering, ch. 7, vol. 33. Springer (2004) ISBN: 3-540-20167X
7. Beghi, A., Cecchinato, L., Cosi, G., Rampazzo, M.: A PSO-based algorithm for optimal multiple chiller systems operation. Applied Thermal Engineering 32, 31–40 (2012) ISSN 1359-4311
8. Yu, Y.-Z., Ren, X.-Y., Du, F.-S., Shi, J.-J.: Application of Improved PSO Algorithm in Hydraulic Pressing System Identification. International Journal of Iron and Steel Research 19(9), 29–35 (2012) ISSN 1006-706X
9. Arani, B.O., Mirzabeygi, P., Panahi, M.S.: An improved PSO algorithm with a territorial diversity-preserving scheme and enhanced exploration–exploitation balance. Swarm and Evolutionary Computation (January 9, 2013) ISSN 2210-6502
10. Zamani, K.N.: Optimization of optical absorption coefficient in asymmetric double rectangular quantum wells by PSO algorithm. Optics Communications (January 8, 2013) ISSN 0030-4018
11. Pluhacek, M., Senkerik, R., Davendra, D., Kominkova Oplatkova, Z., Zelinka, I.: On the behavior and performance of chaos driven PSO algorithm with inertia weight. Computers and Mathematics with Applications (in press, 2013), doi:10.1016/j.camwa.2013.01.016
12. Shi, Y.H., Eberhart, R.C.: A modified particle swarm optimizer. In: IEEE International Conference on Evolutionary Computation, Anchorage Alaska, pp. 69–73 (1998)
13. Nickabadi, A., Ebadzadeh, M.M., Safabakhsh, R.: A novel particle swarm optimization algorithm with adaptive inertia weight. Applied Soft Computing 11(4), 3658–3670 (2011) ISSN 1568-4946
14. Caponetto, R., Fortuna, L., Fazzino, S., Xibilia, M.G.: Chaotic sequences to improve the performance of evolutionary algorithms. IEEE Transactions on Evolutionary Computation 7(3), 289–304 (2003)
15. Sprott, J.C.: Chaos and Time-Series Analysis. Oxford University Press (2003)

Optimization of Solar Integration in Combined Cycle Gas Turbines (ISCC)

Javier Antoñanzas-Torres, Fernando Antoñanzas-Torres,
Enrique Sodupe-Ortega, and F. Javier Martínez-de-Pisón

EDMANS Research Group, University of La Rioja, Logroño, Spain
fernando.antonanzas-torres@unirioja.es
http://www.mineriadatos.com

Abstract. The estimation of the optimum number of loops to operate an integrated solar combined cycle gas turbine (ISCC) represents a complex problem and a very time demanding operation, which must be calculated in near-real time and as a result, it is hardly possible to be solved with regular ISCC production models. This problem is addressed evaluating different soft computing techniques, concluding that the BAG-REPT metamodel fits best generating MAE^{test} of 4.19% and $RMSE^{test}$ of 8.75%. This model presents much lower time than regular ISCC production models and might be used as a decision tool for feasibility assessments and also in pre-design stages of new ISCC projects.

Keywords: ISCC, Bagging, REPT, Combined cycle, Decission Support System.

1 Introduction

The rise of natural gas price, the growing electricity demand and the higher ecological conscience of our society lead to achieve higher efficiency in electricity generation with combined cycle gas turbines (CCGT). Although these plants are in the top of the efficiency of thermal power plants, their efficiency might be improved significantly by integrating solar concentrated thermal energy in combined cycles (ISCC). The fundament of the ISCC is consequence of the efficiency drop in the gas turbine and the subsequent yield loss in both gas and steam turbines under known meteorological conditions based on high air temperatures, low atmospheric pressure and low relative humidity, which generally coincide with those moments with higher solar radiation [1], when solar thermal energy can be integrated in the steam turbine and therefore, operate it full loaded. The operation of the solar field is based on concentrating normal direct irradiation on the focus of parabolic troughs and transferring this heat to a heat transfer fluid up to 390°C. Steam is generated in the solar steam generator with this heat and then injected in the heat recovery steam generator to displace latent vaporization heat and boost the steam turbine. The ISCC requires the solar field as a regular concentrated solar power plant (CSP), since the other components of the plant are provided by the CCGT [2].

The global installed and under development ISCC solar capacity since the first installed plant in 2010 is about 245MW [3]. Nevertheless, the ISCC technology presents some issues that need to be solved to operate them more efficiently and simplify the

Á. Herrero et al. (eds.), *International Joint Conference SOCO'13-CISIS'13-ICEUTE'13*,
Advances in Intelligent Systems and Computing 239,
DOI: 10.1007/978-3-319-01854-6_4, © Springer International Publishing Switzerland 2014

electricity yield forecast of the close future (1-3 days). In this line, one of the most important subjects is the optimization of the number of loops of the solar field to generate solar thermal energy, minimising solar thermal energy waste (denoted as dumping) in near-to-real time, since production models are very time demanding. In this paper, the optimization of the number of loops is addressed through the design of a production model, serving as input generator to train a model to derive the maximum potential number of loops of the solar field. The soft computing (SC) techniques are used to provide an intelligent system to solve this inexact and complex problem and reduce the computational time of the optimization [4] [5] [6]. Eventually, the bagging-REPT is selected and tuned among different algorithms as the best technique in the forecasting of the potential maximum number of loops hour by hour and provides useful information to the ISCC operator given certain predictions of meteorological variables.

2 Production Model

The production model is assessed via the thermodynamic analysis of both the CCGT and the solar field under known meteorological conditions. The solar thermal power is modelled related to the direct normal irradiation and solar geometry on an hourly basis based on a production model [2]. Many other characteristics of the solar field with LS-3 troughs are considered as constants such as the reflectivity of facets, transmissivity and absorptivity of the tubes, interception factor and mirror cleanness amongst others, considering usual values as shown in Table 1. Eventually, throughout the analysis of a solar field with a given number of loops of solar collector assemblies (SCA), the model provides the solar thermal power injectable ($Psol_{int}$). However, under variations on direct normal irradiation or number of loops this solar thermal power changes.

Table 1. Technical specifications of the solar field

Type.of.collector	LS-3
Length of SCA	99m
SCAs per loop	4
Collector orientation	South
Mirror cleanliness	98%
Transmissivity	95.50%
Reflectivity	94%
Interception factor	99.70%
Absorptivity	95.50%
Efficiency peak	85.50%
Width	5.76m
Focal distance	1.71m
Opening area	545m2
HTF	Therminol VP1
HTF_{inlet} nominal temperature	293°C
HTF_{oulet} nominal temperature	393°C

This production model is applied in 17 existing combined cycles in Spain with different sizes of gas and steam turbines and meteorological conditions, providing inputs to derive the optimum maximum number of loops.

3   Data

Meteorological hourly data is collated from several freely available and without restrictions databases, choosing the year 2005 as the representative meteorological year amongst 1983-2005. Direct horizontal irradiation is obtained from the satellite derived Climate Monitoring Satellite Application Facility (CM SAF)[1], and then transformed into direct normal irradiation (DNI) based on the solar geometry of each CCGT position. Relative humidity and air temperatures are collated from nearby meteorological stations from the Spanish Agroclimatic Service for Agriculture (SIAR)[2] and air pressure is obtained from the Spanish Agency of Meteorology (AEMET)[3]. Additionally, other variables related to the CCGT are extracted from the production model explained in the Production model chapter and collated with meteorological data to build a database. These variables related to the CCGT are the potential solar integration in the combined cycle ($Psol_{sc1}$) and the instantaneous power of the steam and gas turbine (Pst_{sc0} and $Pgas_{sc0}$, respectively). The $Psol_{sc1}$ denotes the heat required by the steam turbine to operate full load as a consequence of the load drop derived from high air temperature, low humidity and low pressure. The output data to be estimated is the maximum number of loops of the solar field, previously calculated via iterative seeking with the production model. Table 2 shows the range of the inputs database.

Table 2. Database characteristics

Variable	Minimum value	Maximum value
DNI (W/m2)	0	1000
Temperature (°C)	-11.3	43.05
Humidity (%)	5.22	100
Pressure (mmHg)	943.7	1036.8
Psol_sc1 (MW)	0	8.83
Pst_sc0 (MW)	93.92	309.13
Pgas_sc0 (MW)	171.6	621
Loop	0	20

Each observation of the database corresponds with the mean values in an hour of production. As a result, the total number of observations of the database is 148,920 (365 days x 24 hours x 17 CCGT).

---

[1] http://www.cmsaf.eu

[2] http://www.marm.es/siar

[3] http://www.aemet.es

4   Method

The methodology below addressed leads to select the best model in the estimation of the number of loops. Initially in Subsection 4.1, main variables and possible instances of solar integration are selected by the experts. On a second stage, in Subsection 4.2 a comparative of different soft computing techniques is performed and then a design of experiments is also developed for a tune-fitting of the best model.

4.1   Pre-processing Stage

Initially, those cases in which solar integration potential is not possible are removed from the database. Observations with air temperature lower than 15°C (the efficiency of the gas turbine decreases with temperatures higher than 15°C), $Psol_{sc1}$ equal to zero, or observations without any direct normal irradiation available are removed from the database. The database is shrunk to 51,458 observations after this filter (34.55% of the initial size) and then, variables are normalized between 0 and 1. Afterwards, the database is divided in a testing database composed with the instances of 3 CCGT located in different climates, Toledo (dry continental), Algeciras (dry Mediterranean) and Tarragona (Mediterranean) accounting 8,992 observations and a training-validation database composed with the other 14 CCGT (42,466 observations).

4.2   Model Selection

Different techniques such as linear regression, multilayer perceptron (MLP), k-nearest neighbour (k-NN), M5P, RepTree (REPT), bagging bootstrap aggregating (BAG) and additive regression (AR) are evaluated, selecting eventually the one with lower repeated 10-fold cross validation mean absolute error (MAE) [7] [8][4].

- Linear regression: adapted linear regression algorithm with variable selection based on the Akaike algorithm [9].
- MLP: feed forward artificial neural network with back propagation learning algorithm and sigmoid transfer function in all nodes except the output node in which an unthresholded linear unit is selected [8]. The evaluation is performed considering a number of neurons in the hidden layer between 1-10.
- IBk: a k-nearest neighbour-derived algorithm with distance weighting integrated, in which k is the number of neighbours, defined from 1-3 [10].
- M5P: a decision tree with simple linear regression in leaves derived from the Quinlan's M5 algorithm [4].
- REPT: a regression tree pruned based on variance and a reduced-error pruning algorithm [11].
- BAG (bagging): meta-model or ensemble model that averages a number of based learners fitted with a set of different training datasets obtained by bootstrapping with replacement [12]. In this study, REPT and MLP with different hidden neurons are considered based learners (BAG-REPT and BAG-MLP).

Table 3. Parameter specifications for base learners and meta-models

Algorithm	Parameter	Values
LR	Eliminate collinear attributes	True
LR	Attribute selection	M5P method
MLP	Learning rate	0.3
MLP	Momentum	0.2
MLP	Num. hidden neurons	1, 2,..., 10
MLP	Decay learning rate	True
MLP	Validation set size	20%
MLP	Validation threshold	15, 25,..., 55
IBK	K neighbours	1, 2, 3
IBK	Distance weighting	False
M5P	Min. instances/leafs	2, 4,..., 80
M5P	Prune	True
M5P	Unsmoothed model	False
REPT	Min. variance proportion	0.001
REPT	Prune	True
BAG-REPT	Num. iterations	10, 20, 30, 50, 100
BAG-MLP	Num. iterations	10
BAG-MLP	Num. hidden neurons (MLP)	1, 2, 3, 5, 10
AR-LR	Num. iterations	10
AR-MLP	Num. iterations	10
AR-MLP	Num. hidden neurons (MLP)	1, 2, 3, 5

– AR (additive regression): ensemble model that enhances the performance of a regression model by fitting a new model on the residuals of the previous iteration [13]. Eventually, the output is obtained collating the prediction of the whole set of models. The learning rate, also denoted as shrinkage prevents over-fitting with a smoothing effect and a consequent increase of the learning time. MLP and LR are selected as based learners defining AR-MLP and AR-LR, respectively.

The predictor model is selected according to the average 10-fold cross-validation MAE with 10 runs amongst the configurations of Table 3.

5 Results

Table 4 shows the results of the mean 10-fold cross-validation MAE with 10 runs and root average square error (RMSE)-(ordered according to the mean MAE)- and their mean, maximum, minimum and standard deviation. Since the output variable is normalized between 0 and 1, RMSE and MAE can be interpreted in terms of percentages.

Some observations might be extracted from the results of Table 4:

1. Models and meta-models based on regression trees M5P and REPT obtain lower errors than those based on linear regression, MLP or IBK. This might be explained due to the high heterogeneity and complexity of the database. As a result, regression tree-derived models are able to better fit each situation of the instance space than the other techniques.

Table 4. Repeated 10-fold cross validation errors

Algorithm	MAE_{mean}	MAE_{max}	MAE_{min}	MAE_{sd}	$RMSE_{mean}$	$RMSE_{max}$	$RMSE_{min}$	$RMSE_{sd}$
BAG100-REPT	0.044	0.044	0.044	0.000045	0.092	0.093	0.092	0.000178
BAG50-REPT	0.044	0.044	0.044	0.000068	0.093	0.093	0.092	0.000214
BAG30-REPT	0.044	0.045	0.044	0.000074	0.093	0.093	0.092	0.000228
BAG20-REPT	0.045	0.045	0.044	0.000095	0.093	0.094	0.093	0.000248
BAG10-REPT	0.045	0.045	0.045	0.000103	0.094	0.095	0.093	0.000336
M5P	0.048	0.049	0.048	0.000184	0.097	0.097	0.096	0.000428
REPT	0.050	0.050	0.049	0.000202	0.103	0.103	0.101	0.000723
IBK1	0.054	0.055	0.054	0.000238	0.151	0.152	0.150	0.000698
IBK2	0.055	0.056	0.055	0.000227	0.134	0.135	0.133	0.000735
BAG10-MLP5	0.055	0.053	0.056	0.016032	0.103	0.101	0.103	0.018044
IBK3	0.057	0.057	0.057	0.000158	0.129	0.129	0.128	0.000347
BAG10-MLP10	0.058	0.056	0.058	0.034482	0.104	0.102	0.104	0.033731
BAG10-MLP3	0.059	0.056	0.060	0.000900	0.107	0.104	0.108	0.001590
MLP5	0.065	0.065	0.064	0.000321	0.112	0.113	0.111	0.000471
MLP6	0.065	0.066	0.064	0.000407	0.112	0.113	0.110	0.000615
AR-MLP5	0.065	0.066	0.065	0.000372	0.113	0.114	0.112	0.000495
MLP7	0.066	0.067	0.065	0.000789	0.112	0.113	0.111	0.000695
MLP8	0.066	0.067	0.065	0.000590	0.112	0.113	0.111	0.000610
MLP9	0.067	0.067	0.066	0.000485	0.112	0.113	0.111	0.000420
MLP4	0.067	0.068	0.066	0.000592	0.114	0.115	0.114	0.000306
MLP10	0.067	0.068	0.066	0.000610	0.113	0.114	0.112	0.000619
AR-MLP3	0.068	0.069	0.068	0.000114	0.116	0.116	0.116	0.000069
AR-MLP2	0.071	0.071	0.071	0.000138	0.118	0.118	0.118	0.000126
MLP3	0.072	0.072	0.072	0.000084	0.118	0.118	0.118	0.000056
BAG10-MLP2	0.086	0.083	0.087	0.003560	0.128	0.125	0.129	0.004371
AR-MLP1	0.091	0.098	0.088	0.003177	0.148	0.159	0.144	0.005015
MLP2	0.095	0.095	0.095	0.000042	0.137	0.137	0.137	0.000080
BAG10-MLP1	0.118	0.115	0.119	0.004197	0.199	0.196	0.200	0.004350
MLP1	0.128	0.128	0.128	0.000055	0.209	0.209	0.209	0.000024
LR	0.133	0.133	0.133	0.000007	0.214	0.214	0.214	0.000013
AR-LR	0.133	0.133	0.133	0.000007	0.214	0.214	0.214	0.000013

2. Particularly, the first 5 BAG-REPT meta-models analysed improve significantly the results obtained with simple M5P and REPT. It is also remarkable that these 5 meta-models show similar and stable MAE_{mean} and $RMSE_{mean}$ with any number of based learners (number of iterations) according to the standard deviation of both errors (MAE_{sd} and $RMSE_{sd}$).

3. Models based on classificatory techniques such as IBK show MAE_{mean} and $RMSE_{mean}$ in a range of 1% higher related to the BAG-REPT meta-models. The low errors of IBK can be explained by the high probability of similar instances both in training and testing and also in the cross validation since the problem many of the instances (meteorological and the output) are very similar.

4. Artificial neural network MLP and meta-models (AR-MLP and BAG-MLP) show significantly higher training times and errors than the previous techniques.

Eventually, it is concluded that the BAG-REPT technique is selected to build the prediction model. Finally, parameters are adjusted considering number of iterations between 1-100, minimum variance proportion (minVarianceprop) between 0.0001-0.1 and tree depth between 1-15. The number of iterations designates the number of based learners considered in the meta-model BAG. The *minVarianceprop* parameter denotes the minimum proportion of the variance that should be present in a node to divide it in two branches of the REPT. The tree depth parameter corresponds with the levels of the tree and controls the complexity of the based learner by disenabling the REPT pruning. The selection of the optimum model has been developed with a design of experiments with 100 models defined with an optimal Latin Hypercube Sampling [14]. This methodology leads to seek 100 instances homogeneously distributed in the Euclidean space constituted by these 3 parameters and their afore-mentioned ranges, whose results are shown in Figure 1.

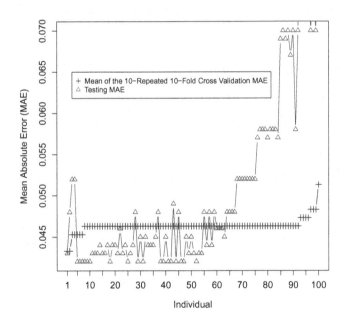

Fig. 1. Design of experiments with BAG-REPT ordered by the mean validation MAE

Table 5 presents the best 20 individuals ordered according the mean 10-fold cross-validation MAE with 10 runs (MAE_{mean}^{val}). The last column corresponds with the testing MAE with the testing database. As it can be deduced from the Table 5, a certain level of instability within the MAE^{test} is observed in the first individuals, which is later stabilized between individuals 5-10. It is also remarkable that the higher the MAE^{val}_{mean} the higher and more instable the MAE^{test}.

Table 5. Ranking of best 20 individuals on mean MAE^{val}

Name	Iterations	minVarianceProp	Tree depth	MAE^{val}_{mean}	MAE^{test}
Indiv004	8	0.0083	12	0.0433	0.0430
Indiv001	14	0.0045	8	0.0433	0.0480
Indiv003	27	0.0014	7	0.0453	0.0520
Indiv002	12	0.0060	7	0.0453	0.0520
Indiv010	38	0.0035	13	0.0453	0.0420
Indiv014	13	0.0054	14	0.0453	0.0420
Indiv013	35	0.0062	13	0.0453	0.0420
Indiv011	37	0.0091	12	0.0463	0.0420
Indiv022	30	0.0085	13	0.0463	0.0420
Indiv024	15	0.0095	14	0.0463	0.0420
Indiv016	23	0.0071	11	0.0463	0.0430
Indiv012	5	0.0084	13	0.0463	0.0430
Indiv019	26	0.0094	11	0.0463	0.0430
Indiv018	3	0.0041	14	0.0463	0.0440
Indiv021	29	0.0034	11	0.0463	0.0430
Indiv026	21	0.0095	12	0.0463	0.0430
Indiv023	24	0.0047	10	0.0463	0.0440
Indiv031	22	0.0072	12	0.0463	0.0420
Indiv017	11	0.0076	10	0.0463	0.0440
Indiv027	22	0.0098	10	0.0463	0.0440

The properties of the final model are derived from individuals 5-10 medians. Eventually, a number of 32.5 iterations, minVarianceProp of 0.0074 and tree depth of 13 are chosen to build the BAG-REPT model. Both the validation and testing errors are shown in Table 5, standing out for their low MAE^{test} of 4.19%.

Table 6. Errors of the BAG-REPT selected

MAE^{val}	MAE^{test}	$RMSE^{val}$	$RMSE^{test}$
0.0446	0.0419	0.0929	0.0875

The actual number of loops of the first 100 hours of the testing database is represented against the predicted values in Figure 2.

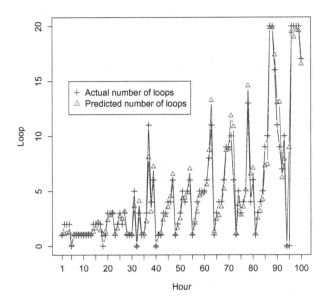

Fig. 2. Actual number of loops vs predicted number of loops

6 Conclusions

The estimation of the optimum number of loops to operate an ISCC represents a complex problem and a very time demanding operation. ISCC operators need to make predictions of the required number of loops in near-to-real time, which is hardly possible with their production models. This problem is addressed evaluating different soft computing techniques, concluding that the BAG-REPT metamodel fits best generating MAE^{test} of 4.19% and $RMSE^{test}$ of 8.75%, which is in the range of the intrinsic uncertainty of sensors, such as the pyrheliometer 5% to measure direct normal irradiation reducing operational time from up to hours to the reason of seconds. The model built might be also a useful decision support tool for feasibility assessments and pre-design stages of new ISCC projects to optimize their number of loops based on locally measured variables (meteorological) and estimated variables (with a regular ISCC production model). We leave for future work the implementation of economic variables of plant operation and investment as a decision tool within the optimization of the number of loops for the pre-design of new ISCC.

Acknowledgments. The authors thank to the University of La Rioja for the FPI-UR 2012 grant from which portions of this research have been financed.

References

1. Baghernejad, A., Yaghoubi, M.: Exergy analysis of an integrated solar combined cycle system. Renewable Energy 35, 2157–2164 (2010) 1
2. Antonanzas-Torres, F., Sodupe-Ortega, E., Sanz-Garcia, A., Fernandez-Martinez, R., Martinez-de-Pison-Ascacibar, F.J.: Technical feasibility assessment of integrated solar combined cycle power plants in Ciudad Real (Spain) and Las Vegas (USA). In: Proc. 16th International Congress on Project Engineering, Valencia, Spain (2012) 1, 2
3. NREL, National Renewable Energy Laboratory (2013), http://www.nrel.gov/csp/solarpaces/ 1
4. Sanz-Garcia, A., Antonanzas-Torres, F., Fernandez-Ceniceros, J., Martinez-de-Pison, F.J.: Overall models based on ensemble methods for predicting continuous annealing furnace temperature settings. Ironmaking and Steelmaking (2013), doi:10.1179/1743281213Y. 0000000104 1, 4.2
5. de Oliveira Penna Tavares, G., Pacheco, M.A.C.: A Genetic algorithm applied to a main sequence stellar model. In: Corchado, E., Kurzyński, M., Woźniak, M. (eds.) HAIS 2011, Part I. LNCS (LNAI), vol. 6678, pp. 32–42. Springer, Heidelberg (2011) 1
6. Pop, P.C., Matei, O., Sitar, C.P., Chira, C.: A genetic algorithm for solving the generalized vehicle routing problem. In: Corchado, E., Graña Romay, M., Manhaes Savio, A. (eds.) HAIS 2010, Part II. LNCS (LNAI), vol. 6077, pp. 119–126. Springer, Heidelberg (2010) 1
7. Martinez-de-Pison, F.J., Pernia, A.V., Blanco, J., Gonzalez, A., Lostado, R.: Control model for an elastomer extrusion process obtained via a comparative analysis of data mining and artificial intelligence techniques. Polymer-Plastics Technology and Engineering 49, 779–790 (2010) 4.2
8. Martinez-de-Pison, F.J., Pernia, A.V., Gonzalez, A., Lopez-Ochoa, L.M., Ordieres, J.B.: Optimum model for predicting temperature settings on hot dip galvanising line. Ironmaking and Steelmaking 37(3), 187–194 (2010) 4.2
9. Akaike, H.: A new look at the statistical model identification. IEEE Transactions on Automatic Control 16, 716–723 (1974) 4.2
10. Cover, T., Hart, P.: Nearest neighbor pattern classification. IEEE Transactions on Information Theory 13, 21–27 (1967) 4.2
11. Wang, Y., Witten, I.H.: Induction of model trees for predicting continuous classes. In: Proc. 9th European Conference on Machine Learning, Prague, Czech Republic, pp. 128–137 (April 1997) 4.2
12. Breiman, L.: Bagging predictors. Machine Learning 24, 123–140 (1996) 4.2
13. Stochastic gradient boosting. Stanford University, Standford (1999) 4.2
14. Mckay, M., Beckman, R., Conover, W.: A comparison of three method for selecting values of input variables in the analysis of output from a computer code. Tecnometrics 21, 239–245 (1979) 5

Performance of Chaos Driven Differential Evolution on Shifted Benchmark Functions Set

Roman Senkerik[1], Michal Pluhacek[1], Ivan Zelinka[2],
Zuzana Kominkova Oplatkova[1], Radek Vala[1], and Roman Jasek[1]

[1] Tomas Bata University in Zlin , Faculty of Applied Informatics,
Nam T.G. Masaryka 5555, 760 01 Zlin, Czech Republic
[2] Technical University of Ostrava, Faculty of Electrical Engineering and Computer Science,
17. listopadu 15,708 33 Ostrava-Poruba, Czech Republic

Abstract. This research deals with the extended investigations on the concept of a chaos-driven evolutionary algorithm Differential Evolution (DE). This paper is aimed at the embedding of set of six discrete dissipative chaotic systems in the form of chaos pseudo random number generator for DE. Repeated simulations were performed on the set of two shifted benchmark test functions in higher dimensions. Finally, the obtained results are compared with canonical DE.

Keywords: Differential Evolution, Deterministic chaos, Optimization.

1 Introduction

These days the methods based on soft computing such as neural networks, evolutionary algorithms, fuzzy logic, and genetic programming are known as powerful tool for almost any difficult and complex optimization problem. Ant Colony (ACO), Genetic Algorithms (GA), Differential Evolution (DE), Particle Swarm Optimization (PSO) and Self Organizing Migration Algorithm (SOMA) are some of the most potent heuristics available.

Recent studies have shown that Differential Evolution [1] has been used for a number of optimization tasks, [2], [3] has explored DE for combinatorial problems, [4] has hybridized DE whereas [5] - [7] has developed self-adaptive DE variants.

This paper is aimed at investigating the chaos driven DE. Although a several of papers have been recently focused on the connection of DE and chaotic dynamics either in the form of hybridizing of DE with chaotic searching algorithm [8] or in the form of chaotic mutation factor and dynamically changing weighting and crossover factor in self-adaptive chaos differential evolution (SACDE) [9], the focus of this paper is the embedding of chaotic systems in the form of chaos number generator for DE and its comparison with the canonical DE.

This research is an extension and continuation of the previous successful initial application based experiment with chaos driven DE [10] – [12] with simple test functions in low dimensions.

Á. Herrero et al. (eds.), *International Joint Conference SOCO'13-CISIS'13-ICEUTE'13*,
Advances in Intelligent Systems and Computing 239,
DOI: 10.1007/978-3-319-01854-6_5, © Springer International Publishing Switzerland 2014

The primary aim of this work is not to develop a new type of pseudo random number generator, which should pass many statistical tests, but to try to use and test the implementation of natural chaotic dynamics into evolutionary algorithm as a chaotic pseudo random number generator.

The chaotic systems of interest are the simple discrete dissipative systems. Six different chaotic maps were selected as the chaos pseudo random number generators for DE based on the successful results obtained with DE [10] or PSO algorithm [13].

Firstly, DE is explained. The next sections are focused on the used chaotic systems and test function. Results and conclusion follow afterwards.

2   Differential Evolution

DE is a population-based optimization method that works on real-number-coded individuals [14]. DE is quite robust, fast, and effective, with global optimization ability. It does not require the objective function to be differentiable, and it works well even with noisy and time-dependent objective functions. Description of the used DE-Rand1Bin strategy (both for Chaos DE and Canonical DE) is presented in (1). Please refer to [14] - [17] for the detailed description of the used DERand1Bin strategy (both for Chaos DE and Canonical DE) as well as for the complete description of all other strategies.

3   Chaotic Maps

This section contains the description of discrete dissipative chaotic maps used as the chaotic pseudo random generators for DE. In this research, direct output iterations of the chaotic maps were used for the generation of real numbers in the process of crossover based on the user defined CR value and for the generation of the integer values used for selection of individuals. The initial concept of embedding chaotic dynamics into the evolutionary algorithms is given in [18]. Following chaotic maps were used: Burgers (1), Delayed logistic (2), Dissipative standard (3), Ikeda (4), Lozi (5) and Tinkerbell (6).

The Burgers mapping is a discretization of a pair of coupled differential equations which were used by Burgers [19] to illustrate the relevance of the concept of bifurcation to the study of hydrodynamics flows. The map equations are given in (1) with control parameters $a = 0.75$ and $b = 1.75$ as suggested in [20].

$$X_{n+1} = aX_n - Y_n^2$$
$$Y_{n+1} = bY_n + X_nY_n$$
(1)

The Delayed Logistic is a simple two-dimensional discrete system similar to the one-dimensional Logistic Equation. The map equations are given in (2). The parameter used in this work is $A = 2.27$ as also suggested in [20].

$$X_{n+1} = AX_n(1-Y_n)$$
$$Y_{n+1} = X_n$$
(2)

The Dissipative Standard map (3) is a two-dimensional chaotic map. The parameters used in this work are $b = 0.1$ and $k = 8.8$ as suggested in [20].

$$X_{n+1} = X_n + Y_{n+1} \pmod{2\pi}$$
$$Y_{n+1} = bY_n + k\sin X_n \pmod{2\pi}$$
(3)

In physics and mathematics, the Ikeda map is a discrete-time dynamical system given by the complex map as a model of light going around across a nonlinear optical resonator. A 2D real example is given by (4). For the following values $\alpha = 6$, $\beta = 0.4$, $\gamma = 1$ and $\mu = 0.9$ [20] this system has a chaotic attractor.

$$X_{n+1} = \gamma + \mu(X_n \cos\phi + Y_n \sin\phi)$$
$$Y_{n+1} = \mu(X_n \sin\phi + Y_n \cos\phi)$$
$$\phi = \beta - \alpha/\left(1 + X_n^2 + Y_n^2\right)$$
(4)

The Lozi map is a discrete two-dimensional chaotic map. The map equations are given in (5). The parameters used in this work are: $a = 1.7$ and $b = 0.5$ as suggested in [20]. For these values, the system exhibits typical chaotic behavior and with this parameter setting it is used in the most research papers and other literature sources.

$$X_{n+1} = 1 - a|X_n| + bY_n$$
$$Y_{n+1} = X_n$$
(5)

The Tinkerbell map is a two-dimensional complex discrete dynamical system given by (6) with following control parameters: $a = 0.9$, $b = -0.6$, $c = 2$ and $d = 0.5$ [20].

$$X_{n+1} = X_n^2 - Y_n^2 + aX_n + bY_n$$
$$Y_{n+1} = 2X_n Y_n + cX_n + dY_n$$
(6)

4   Benchmark Functions

For the purpose of evolutionary algorithms performance comparison within this presented research, the set of two shifted test functionsrepresenting both simple unimodal and very complex multi modal example was selected: the shifted 1st De Jong's function (7) and the shifted Ackley's original function in the form (8).

$$f(x) = \sum_{i=1}^{\dim}(x_i - shift_i)^2$$
(7)

Function minimum: Position for E_n: $(x_1, x_2 \ldots x_n) = \textbf{\textit{shift}}$; Value for E_n: $y = 0$
Function interval: <-5.12, 5.12>.

$$f(x) = -20 \exp\left(-0.02\sqrt{\frac{1}{D}\sum_{i=1}^{D}(x_i - shift_i)^2}\right) - \exp\left(\frac{1}{D}\sum_{i=1}^{D}\cos 2\pi(x_i - shift_i)\right) +$$
$$+ 20 + \exp(1) \tag{8}$$

Function minimum:Position for E_n: $(x_1, x_2...x_n) = shift$; Value for E_n: $y = 0$
Function interval: <-30, 30>.

Shift$_i$is a random number from the 50% range of function interval. *Shift* vector is randomly generated before each run of the optimization process.

5   Results

The novelty of this approach represents the utilization of discrete chaotic maps as the pseudo random number generator for DE. In this paper, the canonical DE strategy DERand1Bin and the Chaos DERand1Bin strategy driven by six different chaotic maps (ChaosDE) were used. The parameter settings for both canonical DE and ChaosDE were obtained analytically based on numerous experiments and simulations (see Table 1).

Table 1. Parameter set up for canonical DE and ChaosDE

DE Parameter	Value
Popsize	75
F	0.8
Cr	0.8
Dimensions	30
Generations	100•D = 3000
Max Cost Function Evaluations (CFE)	225000

Experiments were performed in an environment of *Wolfram Mathematica*, canonical DE therefore used the built-in *Mathematica software* pseudo random number generator. The default *Mathematica Software* pseudo random number generator - extended cellular automaton generator "*Extended CA*" with default automatic setting was applied to represent traditional pseudorandom number generators in comparisons.All experiments used different initialization, i.e. different initial population was generated in each run of Canonical or Chaos driven DE.

Within this research, one type of experiment was performed. It utilizes the maximum number of generations fixed at 3000 generations. This allowed the possibility to analyze the progress of DE within a limited number of generations and cost function evaluations.

The statistical results of the experiments are shown in Tables 2 and 4, which represent the simple statistics for cost function values, e.g. average, median, maximum values, standard deviations and minimum values representing the best individual solution for all 50 repeated runs of canonical DE and ChaosDE.

Tables 3 and 5 compare the progress of ChaosDe and Canonical DE. These tables contain the average CF values for the generation No. 750, 1500, 2250 and 3000 from all 50 runs.

The bold values within the all Tables 2-5 depict the best obtained result.

The main aim of the optimization was to find the global extreme (minimum) of the set of two shifted test function in higher dimensions (for $D = 30$).

Table 2. Simple results statistics for the shifted 1st De Jong's function – 30D

DE Version	Avg. CF	Median CF	Max CF	Min CF	Std.Dev.
Canonical DE	0.115704	0.115841	0.188099	0.05772	0.029188
Burgers Map	**7.44E-17**	**9.61E-18**	**8.16E-16**	**5.4E-20**	**1.68E-16**
Delayed Logistic	2.07E-10	7.41E-11	1.88E-09	3.96E-12	3.53E-10
Dissipative	0.03174	0.030072	0.055251	0.012981	0.010001
Ikeda	1.99E-06	1.18E-06	7.69E-06	8.51E-08	1.88E-06
Lozi	9.18E-07	5.59E-07	4.85E-06	4.07E-08	1.08E-06
Tinkerbelt	6.99E-16	2.01E-16	9.53E-15	5.93E-18	1.5E-15

Table 3. Comparison of progress towards the minimum for the shifted 1st De Jong's function

DE Version	Generation No. 750	Generation No. 1500	Generation No. 2250	Generation No. 3000
Canonical DE	10.73251	2.334888	0.538142	0.115704
Burgers Map	0.011418	**2.31E-07**	**3.89E-12**	**7.44E-17**
Delayed Logistic	0.134328	0.000161	2.03E-07	2.07E-10
Dissipative	8.23385	1.310565	0.207738	0.03174
Ikeda	1.202665	0.014816	0.000181	1.99E-06
Lozi	1.033973	0.009797	8.93E-05	9.18E-07
Tinkerbelt	**0.004223**	2.74E-07	1.75E-11	6.99E-16

Table 4. Simple results statistics for the shifted Ackleys's function – 30D

DE Version	Avg. CF	Median CF	Max CF	Min CF	Std.Dev.
Canonical DE	3.375169	3.423801	3.802878	2.648132	0.25055
Burgers Map	0.36171	**1.84E-08**	2.220113	**3.32E-09**	0.620909
Delayed Logistic	**5.12E-05**	3.42E-05	**0.000268**	5.89E-06	**5.22E-05**
Dissipative	2.5766	2.506757	3.144533	1.961304	0.281449
Ikeda	0.006048	0.005318	0.01901	0.0014	0.003586
Lozi	0.004134	0.003131	0.01267	0.001331	0.002636
Tinkerbelt	1.243854	1.155149	4.294839	1.22E-08	1.199894

Obtained numerical results given in Table 2-5 and graphical comparisons in Figs. 1-4 support the claim that all Chaos DE versions driven by six selected chaotic maps have given better overall results in comparison with the canonical DE version. From the presented data it follows, that Chaos DE driven by Burgers Map has given the best overall results. Very promising results were obtained also through utilization of Delayed Logistic,Lozi and Tinkerbelt map. The last mentioned chaotic maphas unique properties with connection to DE: strong progress towards global extreme, but weak overall statistical results, like average CF value and std. dev.

Table 5. Comparison of progress towards the minimum for the shifted Ackleys's function

DE Version	Generation No. 750	Generation No. 1500	Generation No. 2250	Generation No. 3000
Canonical DE	12.01697	7.619339	4.802582	3.375169
Burgers Map	**0.806666**	0.362769	0.361715	0.36171
Delayed Logistic	2.971642	**0.056344**	**0.001645**	**5.12E-05**
Dissipative	10.96532	6.21261	3.790617	2.5766
Ikeda	6.280929	1.459155	0.083709	0.006048
Lozi	5.841436	1.055847	0.050686	0.004134
Tinkerbelt	1.662962	1.24514	1.243861	1.243854

The graphical comparisons of the time evolution of CF values for the best individual solutions (the solution with the minimal final cost function value) for Chaos DE with six chaotic maps and canonical DERand1Bin strategy are depicted in Fig. 1

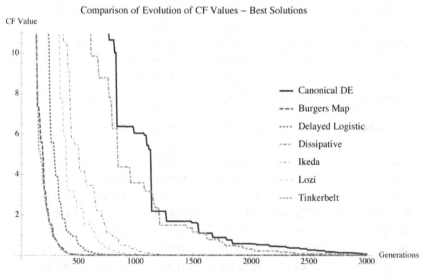

Fig. 1. Comparison of the time evolution of CF values for the best individual solutions, i.e. the solutions with the minimal final cost function value. Shifted 1st De Jong's function, $D = 30$.

(for shifted 1st De Jong's function) and Fig 3 (for shifted Ackley's function). Finally the Fig. 2 and 4 confirms the robustness of Chaos DE driven by chaotic Burgers map in finding the best solutions for all 50 runs.

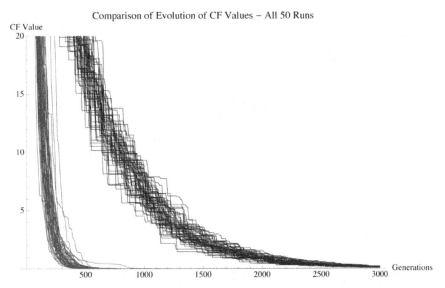

Fig. 2. Comparison of the time evolution of CF values for all 50 runs of canonical DE (blue) and ChaosDE driven by Burgers map (red). Shifted 1st De Jong's function, $D = 30$.

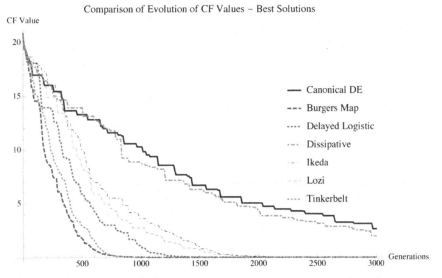

Fig. 3. Comparison of the time evolution of CF values for the best individual solutions, i.e. the solutions with the minimal final cost function value. Shifted Ackley's function, $D = 30$.

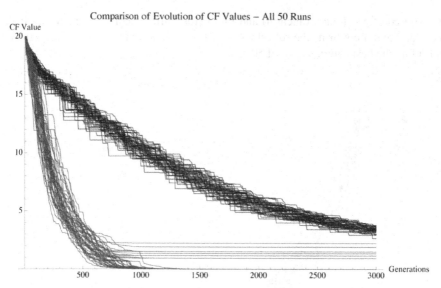

Fig. 4. Comparison of the time evolution of CF values for all 50 runs of canonical DE (blue) and ChaosDE driven by Burgers map (red). Shifted Ackley's function, $D = 30$.

6   Conclusion

In this paper, chaos driven DERand1Bin strategy was tested and compared with canonical DERand1Bin strategy on the set of two shifted benchmark functions. Based on obtained results, it may be claimed, that the developed ChaosDE gives considerably better results than other compared heuristics.

Future plans include experiments with benchmark functions in higher dimensions, testing of different chaotic systems and obtaining a large number of results to perform statistical tests.

Furthermore chaotic systems have additional parameters, which can by tuned. This issue opens up the possibility of examining the impact of these parameters to generation of random numbers, and thus influence on the results obtained using differential evolution.

Acknowledgements. This work was supported by Grant Agency of the Czech Republic- GACR P103/13/08195S, by the project Development of human resources in research and development of latest soft computing methods and their application in practice, reg. no. CZ.1.07/2.3.00/20.0072 funded by Operational Program Education for Competitiveness; by the European Regional Development Fund under the project CEBIA-Tech No. CZ.1.05/2.1.00/03.0089, by the Technology Agency of the Czech Republic under the Project TE01020197, and by Internal Grant Agency of Tomas Bata University under the project No. IGA/FAI/2013/012.

References

1. Price, K.: An Introduction to Differential Evolution. In: Corne, D., Dorigo, M., Glover, F. (eds.) New Ideas in Optimization, pp. 79–108. McGraw-Hill, London (1999) ISBN 007-709506-5
2. Tasgetiren, M.F., Suganthan, P.N., Pan, Q.K.: An Ensemble of Discrete Differential Evolution Algorithms for Solving the Generalized Traveling Salesman Problem. Applied Mathematics and Computation 215(9), 3356–3368 (2010)
3. Onwubolu, G., Davendra, D. (eds.): Differential Evolution: A handbook for Permutation-based Combinatorial Optimization. Springer, Germany (2009)
4. Das, S., Konar, A., Chakraborty, U.K., Abraham, A.: Differential evolution with a neighborhood based mutation operator: a comparative study. IEEE Transactions on Evolutionary Computations 13(3), 526–553 (2009)
5. Qin, A.K., Huang, V.L., Suganthan, P.N.: Differential evolution algorithm with strategy adaptation for global numerical optimization. IEEE Transactions on Evolutionary Computations 13(2), 398–417 (2009)
6. Zhang, J., Sanderson, A.C.: JADE: Self-adaptive differential evolution with fast and reliable convergence performance. In: Proceedings of IEEE Congress on Evolutionary Computation, Singapore, pp. 2251–2258 (2007)
7. Zhang, Sanderson, A.C.: Self-adaptive multiobjective differential evolution with direction information provided by archived inferior solutions. In: Proceedings of IEEE World Congress on Evolutionary Computation, Hong Kong, pp. 2801–2810 (2008)
8. Liang, W., Zhang, L., Wang, M.: The Chaos Differential Evolution Optimization Algorithm and its Application to Support Vector Regression Machine. Journal of Software 6(7), 1297–1304 (2011)
9. Zhenyu, G., Bo, C., Min, Y., Binggang, C.: Self-Adaptive Chaos Differential Evolution. In: Jiao, L., Wang, L., Gao, X.-b., Liu, J., Wu, F. (eds.) ICNC 2006. LNCS, vol. 4221, pp. 972–975. Springer, Heidelberg (2006)
10. Davendra, D., Zelinka, I., Senkerik, R.: Chaos driven evolutionary algorithms for the task of PID control. Computers & Mathematics with Applications 60(4), 1088–1104 (2010) ISSN 0898-1221
11. Senkerik, R., Davendra, D., Zelinka, I., Pluhacek, M., Oplatkova, Z.: An Investigation on the Chaos Driven Differential Evolution: An Initial Study. In: Proceedings of 5th International Conference on Bioinspired Optimization Methods and Their Applications, BIOMA 2012, pp. 185–194 (2012) ISBN 978-961-264-043-9
12. Senkerik, R., Davendra, D., Zelinka, I., Pluhacek, M., Oplatkova, Z.: An Investigation on the Differential Evolution Driven by Selected Discrete Chaotic Systems. In: Proceedings of the 18th International Conference on Soft Computing, MENDEL 2012, pp. 157–162 (2012) ISBN 978-80-214-4540-6
13. Pluhacek, M., Senkerik, R., Davendra, D., Kominkova Oplatkova, Z., Zelinka, I.: On the behavior and performance of chaos driven PSO algorithm with inertia weight. Computers & Mathematics with Applications (article in press, 2013), ISSN 0898-1221, doi:10.1016/j.camwa.2013.01.016
14. Storn, R., Price, K.: Differential evolution - a simple and efficient heuristic for global optimization over continuous spaces. Journal of Global Optimization 11, 341–359 (1997)
15. Price, K.: An Introduction to Differential Evolution. In: Corne, D., Dorigo, M., Glover, F. (eds.) New Ideas in Optimization, pp. 79–108. McGraw-Hill, London (1999) ISBN 007-709506-5

16. Price, K., Storn, R.: Differential evolution homepage (2001),
    http://www.icsi.berkeley.edu/~storn/code.html
17. Price, K., Storn, R., Lampinen, J.: Differential Evolution - A Practical Approach to Global Optimization. Springer (2005) ISBN: 3-540-20950-6
18. Caponetto, R., Fortuna, L., Fazzino, S., Xibilia, M.: Chaotic sequences to improve the performance of evolutionary algorithms. IEEE Trans. Evol. Comput. 7(3), 289–304 (2003)
19. ELabbasy, E.M., Agiza, H.N., EL-Metwally, H., Elsadany, A.A.: Bifurcation analysis, chaos and control in the Burgers mapping. International Journal of Nonlinear Science 4(3), 171–185 (2007)
20. Sprott, J.C.: Chaos and Time-Series Analysis. Oxford University Press (2003)

A Neural Network Model for Energy Consumption Prediction of CIESOL Bioclimatic Building

Rafael Mena Yedra[1,*], Francisco Rodríguez Díaz[1],
María del Mar Castilla Nieto[1], and Manuel R. Arahal[2]

[1] University of Almería Agrifood Campus of International Excellence, ceiA3
CIESOL, Joint Center University of Almería - CIEMAT,
rafael.mena@ual.es
[2] University of Sevilla, Dpto. Ingeniería de Sistemas y Automática,
Escuela Técnica Superior de Ingeniería

Abstract. Energy efficiency in buildings is a topic that is being widely studied. In order to achieve energy efficiency it is necessary to perform both, a proper management of the electric demand, and an optimal exploitation of renewable sources, using for that appropriate control strategies. The main objective of this paper is to develop a short term predictive model, based on neural networks, of the electricity demand for the CIESOL research center. The performed experiments, using different techniques for weather forecast, show a quick prediction with acceptable final results for real data, obtaining a maximum root mean squared error of 5 % in validation data, with a short-term prediction horizon of 60 minutes.

Keywords: Electric demand prediction, Predictive model, Neural network

1 Introduction

Nowadays, the global energetic model is unsustainable in economic, social and environmental terms, where a *sustainable model* can be defined as that which "satisfies the actual needs without compromising the future generations ability to satisfy their own needs" [1]. Consequently, due to the high impact of the production and consumption of electricity, it becomes increasingly necessary a proper electricity demand management to achieve energy efficiency. At the same time, the optimal exploitation of the renewable energy sources is another fundamental aspect with the aim to shift the load peaks whenever it is possible. In this line, in the past twenty years a special emphasis has been made in the construction of buildings with bioclimatic architecture, which can benefit from the solar energy and the natural air flow for the use in natural heat and passive cooling thus reducing the intensive electricity consumption. In this paper, the

---

* Corresponding author.

Á. Herrero et al. (eds.), *International Joint Conference SOCO'13-CISIS'13-ICEUTE'13*, 51
Advances in Intelligent Systems and Computing 239,
DOI: 10.1007/978-3-319-01854-6_6, © Springer International Publishing Switzerland 2014

CDdI-ARFRISOL-CIESOL (CIESOL) building has been chosen for the study and development of the model based on neural networks for the short-term prediction of electric power demand.

There are numerous techniques that has been applied to the task of electricity demand prediction, among them there are engineering methods, statistical methods and artificial intelligence methods. A more complete review of building's energy prediction techniques may be viewed at [2], and a more general review of energy prediction techniques in [3].

Quantifying energy demand in buildings is quite complex in general and more specifically in the present case, being CDdI-ARFRISOL-CIESOL a solar energy research center, with a wide variety of energy and construction types. For this, in this paper a characterization of load in different conditions and categorization of such conditions has been developed. In the end, a prediction model based on Artificial Neural Networks (ANN) has been obtained following a concrete methodology for the proper architecture and structure selection for a NARX model and the selection of the embedding delay and the embedding dimension to reconstruct the state space. The choice of ANNs has been made because of the complexity of the problem and by its distinctive features: learning, self-adaptative, fault tolerance, flexibility and real time response [4]. Finally, the predictive model obtained has been tested with an assessment using a battery of tests spanning working and non-working days, different temperature conditions for winter and summer, different radiation conditions with cloudy and sunny days and also nightly consumption, and, at the same time it has been done several comparison for the weather input data for the model by using a simple hold out predictor (lazy predictor), a double exponential method and some ANNs.

This paper is organized as follows. In Section 2, CDdI-ARFRISOL-CIESOL building is briefly described. Section 3 shows an analysis of the energy demand profile of the building used as case study. In Section 4, the methodology used to develop the model based on neural networks is widely described. Finally in Section 5, the results of different experiments are showed and discussed.

2   Scope of the Research: The CDdI-ARFRISOL-CIESOL Building

The CDdI-ARFRISOL-CIESOL building is a solar energy research center located in Almeria, in the South East of Spain (Fig. 1). This building, known by its acronym *CIESOL* for its name in Spanish *[Centro de Investigación de la Energía Solar]*, has been designed and built within the framework of research PSE-ARFRISOL [5]. It has been built with bioclimatic architecture criteria, for which it takes advantage of the benefits of solar energy and natural air flow for its use in natural heating and passive cooling. Furthermore, the building has an air conditioning system based on solar cooling connected to a solar collector array, a hot water storage, boiler and an absorption chiller with its cooling tower. On the other hand, the building has a photovoltaic power plant with a peak power of $9\,kW$ to supply green power to the electricity grid of the building.

Fig. 1. The CDdI-ARFRISOL-CIESOL building.

In addition, this building has all its enclosures monitored by a broad network of sensors, whose data is stored in a database through different acquisition systems. However, this data may have any irregularities just as mismeasure, noise in the signals, discontinuities through the time series and different scales and range of values, so it was mandatory to make use of some preprocessing techniques like filtering, interpolation or normalization.

3 Analysis of the Energy Demand Profile of the CDdI-ARFRISOL-CIESOL Building

In general, from an energy demand point of view, a building can be considered as a complex system composed of different types of elements, such as: HVAC (Heating, Ventilation and Air Conditioning) systems, illumination, etc. More specifically, in this building one of most expensive systems, in energy consumption terms, is the solar cooling installation. Hence, it is necessary to perform a preliminary analysis of the data in order to obtain some conclusions. This analysis is focused on the study of demand profile patterns for different seasons and types of days, working and non-working days, using for that certain statistical parameters like arithmetic mean, standard deviation, minimum and maximum of the electric power demand over different periods of time.

First of all, a comparison between the consumption of a working day and a non-working day had been performed. On the one hand, energy consumption in the working day starts to raise about at 8.00 am and it begins to decrease around 8.00 pm, so it can be established that energy consumption along working days is characterized by an office schedule. On the other hand, energy consumption in a non-working day is stationary. Due to inductive and research facilities, the energy consumption is around $20\,kW$. A comparison between the working and non-working days from the statistical parameters point of view, can be observed in Table 1.

The other analysis performed inside the building was a comparison of the energy consumption through a week, from Monday to Sunday, along different seasons in a year. All the seasons have a similar pattern, according to the energy demand profiles of working and non-working days commented previously. In addition, through the comparison of different seasons, as it is shown in Table 1, it can be observed that energy consumption for Spring and Summer is higher

Table 1. Descriptive parameters for a working and non-working day, and for weekly energy consumption for each season [in kW]

	$\bar{x}$	σ	min	max
Working day	25.42	4.67	17.91	34.73
Non-working day	20.32	0.96	17.66	23.84

Winter	26.45	4.55	18.93	39.47
Summer	27.89	8.47	16.06	60.77
Autumn	22.24	4.60	14.83	42.45
Spring	25.33	7.35	15.33	59.31

than for Winter and Autumn. This behaviour is derived from the typical semidesertical mediterranean climate of Almería, which requires a greater use of the HVAC system for refrigeration than for heating. Furthermore, in Table 1 it can be perceived that both, the peak demand and the standard deviation, are similar for, on one hand, Spring and Summer, and, on the other hand, Winter and Autumn.

Finally, the building has a dynamic nature, since it is focused to research, and therefore, it has a complex behaviour. In fact, there are days where energy consumption oscillates very quickly with a variance of approximately $40\,kW$ in short periods of time. After the performed analysis, a preliminary list of potential variables had been selected, see Table 2.

Table 2. Preliminary list of potential variables

Variable name	Unit	Measurement range
Type of day (Working day / Non-working day)	Boolean	$\{0, 1\}$
Hour of the day (Sine)	Numeric	$\sin(\frac{1}{24}2\pi h)$, $h = [0, 23]$
Hour of the day (Cosine)	Numeric	$\cos(\frac{1}{24}2\pi h)$, $h = [0, 23]$
Outdoor temperature	$[^{\circ}C]$	$[-5 .. 50]$
Outdoor humidity	$[\%]$	$[0 .. 100]$
Outdoor solar radiation	$[W/m^2]$	$[0 .. 1440]$
Outdoor wind speed	$[m/s]$	$[0 .. 22]$
Outdoor wind direction	$[^{\circ}]$	$[0 .. 360]$
Electric power demand	$[kW]$	$[0 .. 85]$
Electric power injected by the photovoltaic plant	$[kW]$	$[0 .. 9]$

4   A Neural Network Model for the Energy Demand of the CDdI-ARFRISOL-CIESOL Building

4.1   Data-Sets Construction

In order to obtain a proper neural network model, a set of historic data acquired at the building has been used. More specifically, the historic data set is composed by 700000 points from 09/01/2010 to 02/29/2012, with a sample time equal to 1 minute. In addition, the historic data has been split into three different data sets for train, validate and test the Neural Network model. The data division has been made manually due to discontinuities in time series and trying to get an equilibrated amount of data for each season and data set. The first data set is the *Training Data Set* (318340 data points) and it is used for obtaining the ANN parameters using a Levenberg-Marquardt algorithm based on gradient descent [6]. The second one is the *Validation Data Set* (89158 data points), which is necessary in the training and post-training processes, in order to prevent the *over-training* in the ANN. The final data set is the *Test Data Set* (107264 data points) whose objective is serving as an independent data set for benchmarking and testing purposes.

4.2   Architecture and Structure Selection

In the one-step-ahead prediction task, the ANN must estimate the next time series state, without being fedback to the model input's regressor. Namely, the input regressor contains only real data points from the time series. To achieve a larger prediction horizon, which is known as multistep-ahead prediction, model's output should be fedback to the input regressor to obtain a recursive prediction for a number of time steps in the future. In this case, the input regressor components, previously composed by actual time series values, are progressively replaced by predicted values forming a recurrent architecture.

In [7] it is noted that the learning of long temporal dependencies with gradient descent methods is more effective using a recurrent architecture named Nonlinear Autoregressive with eXogenous input (NARX) [8] than with the use of other models. This may be due to the fact that the NARX architecture is characterized by having as input vector two tapped delay lines, one for the input signal and the another one for the output signal. The NARX scheme can be observed in Fig. 2, and it is mathematically expressed as it is shown in Eq. 1.

$$y(n+1) = f\left(u(n), u(n-1), \ldots, u(n-d_u+1);\right.$$
$$\left.y(n), y(n-1), \ldots, y(n-d_y+1)\right) \tag{1}$$

where $u(n) \in \mathbb{R}$ and $y(n) \in \mathbb{R}$ are the input and output signals in the timestep n ($\bar{u}(n)$ and $\bar{y}(n)$ are in the form of vector), and $d_u \geq 1$, $d_y \geq 1$ and $d_u \leq d_y$, are the memory order for input and output signals, respectively. The non-linear mapping function f can be approximated, for example, by using a standard ANN like the multilayer Perceptron. The resulting connection architecture is called NARX network. According to Takens' theorem [9,10], the state space of a dynamical system can be reconstructed by:

$$\hat{x}(n+1) = [x(n), x(n-\tau), \ldots, x(n-(d_E-1)\tau)] \tag{2}$$

where $\hat{x}(n+1)$ is the state space of the system reconstructed at the timestep $(n+1)$, d_E is the embedding dimension and τ is the embedding delay. Therefore, it is necessary to known the order of the model, this is the embedding dimension, d_u and d_y for input and outputs signals, and the embedding delay τ.

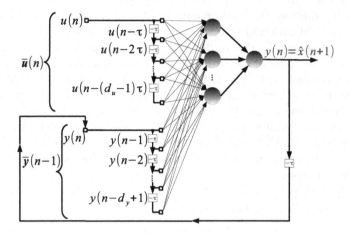

Fig. 2. Non-linear autoregressive with exogenous inputs (NARX) architecture

4.3   ANN Inputs and Size

After several data analysis techniques and model tests the initial list of potential variables, see Table 2, has been reduced to the following ones: type of day, hour of the day (sine and cosine), outdoor temperature, outdoor solar radiation and the electric power consumed by the building, that is, the electric power demand added up with the electric power supplied by the photovoltaic plant.

The next step is to select the order of the model inputs. For this purpose, and according to Equation 2, it is necessary to obtain the embedding delay τ and the embedding dimension d in order to reconstruct the state space.

More specifically, both for the input variables as the target variable used for feedback, an embedding delay $\tau = 1$ has been used due to the system variability and the lack of delay in the system's response.

The method used to find a proper embedding dimension d is the False Neighbors Method [11]. After some tests, it can be deduced that the optimal embedding dimension is around $d = 4$ for both the temperature and the solar radiation. The inputs *type of day* and *hour of the day* have an embedding dimension $d = 1$. Embedding dimension for the feedback variable has been established in $d_y = 20$ according to model tests.

Finally, as ANN with only one hidden layer with tangent hyperbolic activation function and an output neuron with lineal activation function are universal approximators [12]. The number of neurons in the hidden layer has been established in $n = 15$ according to results of model tests.

5   Results and Discussion

Furthermore, for the prediction process, in the timestep $t = k$, the ANN generates a prediction for the time interval from $t = k + 1$ to $t = k + N$, where N

is the prediction horizon. Therefore, in the timestep $t = k$, the ANN receives as inputs for the prediction interval: the type and hour of the day, and a prediction of the disturbances, outdoor temperature and solar radiation. Hence, the ANN generates a recursive prediction feeding back the output in a closed loop. Besides, in this paper, different ways for the prediction of disturbances, outdoor temperature and solar radiation, are used:

1. Real inputs, using the real known inputs for comparison purposes.
2. Lazy man weather prediction mode, the inputs are set to the last known value.
3. Auxiliary ANNs, using two small autorregressive ANNs to estimate the future values.
4. Double exponential method (DES) [13], an analytic technique to estimate the future values.

The battery of tests has been designed with the aim to check the ANN response with a representative set of conditions including working and non-working days, different temperature conditions for winter and summer, different radiation conditions with cloudy and sunny days and also a test for nightly consumption. The obtained results are organized in the following way. From Figs. 3a - 3c, the dynamic evolution of the real measured power demand (in black) and the provided for the ANN model developed in this paper using each one of the approximations for the disturbances prediction commented previously. In addition, in Table 3, a comparison among different tests based on several statistical parameters has been performed. More specifically, these statistical parameters are the coefficient of variation of the root mean square deviation in % (CV-RMSE), the mean absolute percentage error in % (MAPE), the mean absolute error in kW (MAE), the standard deviation of the error in kW (σ) and the maximum error in kW.

In Fig. 3a, results of $N = 60$ minutes demand prediction for a Winter cloudy working day are shown. As it can be seen, results are very similar for the Real, lazy and ANNs disturbances setting modes, but not for the DES setting mode, where the output is very inaccurate because of a bad prediction of the outdoor temperature and solar radiation disturbances. In general, results are good with a maximum CV-RMSE of 3 %.

In Fig. 3b, it can be seen the output of $N = 60$ minutes demand prediction for a Summer cloudy working day. In this case, prediction using real data for disturbances and generated with ANNs are more accurate and respond better to the system dynamics because of a better knowledge of the environment status (temperature and radiation). However, when lazy setting mode is used, the prediction output is more static although the average is smoother. For the DES setting mode, the output is considerably worse. The CV-RMSE is around the 4.5 %.

Finally, it has been done an experiment for the nightly energy consumption, test (C), which can be seen in Fig. 3c. In this case, the prediction outputs of electric demand for the real, lazy and ANNs setting mode are the same because the temperature and radiation are not so much decisive factors when they are

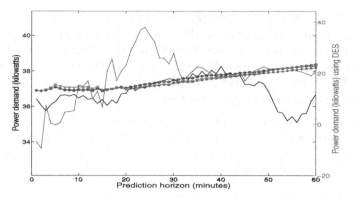

(a) Cloudy winter working day (02/15/2011 10:30 am).

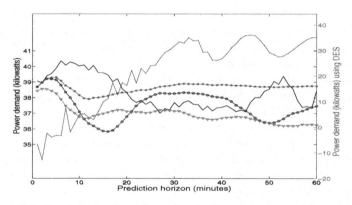

(b) Cloudy summer working day (06/30/2011 10:20 am).

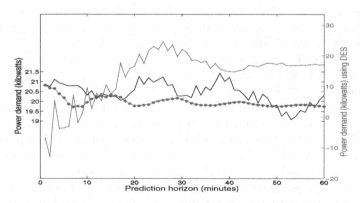

(c) Nightly electric consumption (06/29/2011 12:00 pm).

Fig. 3. Real power demand (black) and prediction outputs using real (blue with circle markers), lazy disturbances (green with star markers) and predicted with ANNs (magenta with triangle markers) and with DES (red with dot markers) for different tests

Table 3. Results

	Cloudy winter working day					
	CV-RMSE (%)	MAPE (%)	Variation (kW)	MAE (kW)	σ (kW)	Max Error (kW)
Real	3.0582	2.0433	[35.09 − 38.27]	0.73	0.85	3.13
Lazy	2.8985	2.0177	[35.09 − 38.27]	0.72	0.78	2.97
ANNs	3.0699	2.1316	[35.09 − 38.27]	0.77	0.83	3.12
DES	55.5792	49.8696	[35.09 − 38.27]	18.31	9.21	45.32
	Cloudy summer working day					
	CV-RMSE (%)	MAPE (%)	Variation (kW)	MAE (kW)	σ (kW)	Max Error (kW)
Real	4.5622	3.4514	[37.01 − 40.36]	1.34	1.13	3.86
Lazy	3.1464	2.7707	[37.01 − 40.36]	1.06	0.58	2.13
ANNs	4.5382	3.6694	[37.01 − 40.36]	1.42	1.00	3.43
DES	58.0064	43.4166	[37.01 − 40.36]	16.95	14.53	51.61
	Nightly electric consumption					
	CV-RMSE (%)	MAPE (%)	Variation (kW)	MAE (kW)	σ (kW)	Max Error (kW)
Real	3.4245	2.7279	[19.08 − 21.43]	0.56	0.41	1.57
Lazy	3.4263	2.7284	[19.08 − 21.43]	0.56	0.41	1.57
ANNs	3.4132	2.7222	[19.08 − 21.43]	0.56	0.41	1.56
DES	51.7685	34.8219	[19.08 − 21.43]	7.16	7.83	33.52

similar, while can being others like the type of day or the hour of the day. In spite of this, for the DES setting mode the prediction output is worse because the temperature and radiation predictions were probably anomalous for that day. The CV-RMSE is around the 3.4 % .

6   Conclusions

In this work, a model based on neural networks for the electric demand prediction of a solar energy research center, the CDdI-ARFRISOL-CIESOL building has been presented. The obtained ANN model allows to generate a quick prediction of the building electric demand. In order to test its properly behaviour, different tests have been performed. These tests include different types of days and energy consumption patterns. The results obtained from different tests are acceptable, since they show a maximum CV-RMSE of 5 % when real data is used for the disturbances inputs (temperature, radiation). However, for different modes of disturbance inputs (lazy, ANNs, DES) the results are slightly worse, specially when using the DES algorithm. An effort could be made to improve the disturbances prediction methods based on ANNs since they show promising results. As future works, disturbance predictions methods based on ANNs will be improved.

Acknowledgements. The authors are very grateful to Andalusia Regional Government (Spain), for financing this work through the Programme "Formación de personal docente e investigador predoctoral en las Universidades Andaluzas, en áreas de conocimiento deficitarias por necesidades docentes (FPDU 2009)". This is a programme cofinanced by the European Union through the European Regional Development Fund (ERDF). This work has been partially funded by the following projects: PSEARFRISOL PS-120000-2005-1 and DPI2010-21589-C05-04 (financed by the Spanish Ministry of Science and Innovation and EU-ERDF funds).

References

1. United Nations Department of Economic & Social Affairs, W.E.C.: World Energy Assessment - Energy and the Challenge of Sustainability. United Nations Development Programme (2000)
2. Zhao, H., Magoulès, F.: A review on the prediction of building energy consumption. Renewable and Sustainable Energy Reviews 16(6), 3586–3592 (2012)
3. Suganthi, L., Samuel, A.A.: Energy models for demand forecasting - a review. Renewable and Sustainable Energy Reviews (2011)
4. Castilla, M., Álvarez, J.D., Ortega, M.G., Arahal, M.R.: Neural network and polynomial approximated thermal comfort models for hvac systems. Building and Environment (2012)
5. Bosqued, A., Palero, S., San Juan, C., Soutullo, S., Enríquez, R., Ferrer, J.A., Martí, J., Heras, J., Guzmán, J.D., Jiménez, M.J., Bosqued, R., Heras, M.R.: ARFRISOL, bioclimatic architecture and solar cooling project. In: Proceedings of PLEA 2006 Passive and Low Energy Architecture, Geneva, Switzerland (2006)
6. More, J.: The Levenberg-Marquardt algorithm: implementation and theory. Numerical Analysis, 105–116 (1978)
7. Lin, T., Horne, B.G., Giles, C.L.: How embedded memory in recurrent neural network architectures helps learning long-term temporal dependencies. Neural Networks 11(5), 861–868 (1998)
8. Lin, T., Horne, B.G., Tino, P., Giles, C.L.: Learning long-term dependencies in NARX recurrent neural networks. IEEE Transactions on Neural Networks 7(6), 1329–1338 (1996)
9. Takens, F.: Detecting strange attractors in turbulence. In: Rand, D., Young, L.S. (eds.) Dynamical Systems and Turbulence, Warwick 1980. Lecture Notes in Mathematics, vol. 898, pp. 366–381. Springer, Heidelberg (1981)
10. Kantz, H., Schreiber, T.: Nonlinear time series analysis, vol. 7. Cambridge University Press (2004)
11. Kennel, M.B., Brown, R., Abarbanel, H.D.I.: Determining embedding dimension for phase-space reconstruction using a geometrical construction. Physical Review 45(6), 3403 (1992)
12. Hornik, K., Stinchcombe, M., White, H.: Multilayer feedforward networks are universal approximators. Neural Networks 2(5), 359–366 (1989)
13. Pawlowski, A., Guzmán, J., Rodríguez, F., Berenguel, M., Sánchez, J.: Application of time-series methods to disturbance estimation in predictive control problems. In: IEEE International Symposium on Industrial Electronics (ISIE), pp. 409–414. IEEE (2010)

ACO Using a MCDM Strategy for Route Finding in the City of Guadalajara, México

Leopoldo Gómez-Barba[1], Benjamín Ojeda-Magaña[1],
Rubén Ruelas[1], and Diego Andina[2]

[1] Departamento de Sistemas de Información CUCEA, Universidad de Guadalajara,
C.P. 45100, Zapopan, Jalisco, México
`lgomez@cucea.udg.mx`, `benojed@hotmail.com`,
`ruben.ruelas@cucei.udg.mx`
[2] E.T.S.I. de Telecomunicación, Universidad Politécnica de Madrid,
Avda. Complutense 30, Madrid 28040, Spain
`d.andina@upm.es`

Abstract. Car traffic problems become more important every day and the shortest distance route is not always the best because streets are saturated with cars, nearly at peak hours; when an event occurs or car traffic overpasses a threshold, streets reduce their fluent flow capacity, thus car drivers must seek new alternatives for their routes. In this work we propose an approach to synthesize street network characteristics and affectations on the streets (as number of available lanes, maximum allowed speed, etc.) into a significative value useful for route decision for path finding. Such variables are processed through a Multi-Criteria Decision Making (MCDM) method, which return a value for each street, that represents a level of quality to hold traffic flow, this is used with an Ant Colony Optimization (ACO) algorithm in order to find alternative routes that use those most fluid streets, those with the "better" characteristics; So, our contribution is the synthesizing of many street characteristics into a value that denote the quality of the street in the search space, helpful for path finding purposes. We use information from the city of Guadalajara, México, to perform experiments and show the advantages of this proposal.

1 Introduction

An important rising problem closely related to high growth cities, including its social and economic impact, is the car mobility, which occasionally generates over demand congesting some segments of the streets network, car drivers mostly choose to use just few alternatives to travel because these arteries tend to have many of the desired characteristics for the commuter, this approaches the principle of the 80/20 described by Jiang *et al* [1], that states twenty percent of the roads gives service to eighty percent of the population. So, many innovative efforts are arising to tackle and understand the way to improve the solutions to these kinds of problems [2].

Path searching problems on street network mostly focus on minimizing distance or time (this last is often a composite measure affected by several factors), this is done using popular path finding algorithms as Dijkstra [3] which have been widely used, with the disadvantage of need a large computational burden, unless a refinement is used;

Á. Herrero et al. (eds.), *International Joint Conference SOCO'13-CISIS'13-ICEUTE'13*,
Advances in Intelligent Systems and Computing 239,
DOI: 10.1007/978-3-319-01854-6_7, © Springer International Publishing Switzerland 2014

however, new path searching algorithms have been emerging on the last two decades on the brand of metaheuristics, such techniques include but are not limited to: Ant Colony Optimization, Tabu Search, Simulated Annealing, Genetic Algorithms, among other, described in [4] for example, although these do not guarantee to find the global optimum, they offer very good solutions with less computational effort, so they become attractive for implementation when low computational capabilities are present (as in smartphones) and fast response is required.

Before driving through the city streets, many people evaluate several factors about the streets and its environment to decide which is the best path to take, considerations as if it is a straight path, if streets support traffic conditions, if certain average speed can be achieved, how much time it would take, among others. Our proposal takes available information about streets characteristics and city streets environment and synthesize it into a numerical value for each street segment, this value represents a level of quality of the street, so the bigger this value the better the street is; thus, this value could help to guide a step by step decision process that want to take those better streets, as in the case of a route search. To derive a single value for each alternative considering the several available street characteristics we used a multi-criteria decision making method capable of integrating both qualitative and quantitative criteria, derivation of the value is a prior process that become an attribute for each street which, as a consequence, enhance the speed of search algorithms to find solutions due to a kind of branch pruning of the search space, when low value streets have poor chance to be selected, specially when using a stochastic selection rule as the one of ACO. Although many configurations can be carried out, in this work we set up the most attractive example where derived values represents the importance of the streets to hold traffic flow.

Experimentation was carried out using information about the city of Guadalajara, the second most important city in México, with about 4.2 millions of people and 1.7^1 millions of vehicles approximately, along the problem of having a less fluid traffic through streets which affects the city and citizens through noise pollution, air pollution, because of the vehicles wasting most of its travel time stopped because of traffic congestion.

The rest of this paper is structured as follows. In section 2 we describe generalities about the AHP and ACO algorithms. In section 3 we present our method to find new possibly routes and in 4 we describe some experimentation and obtained results. Finally, we draw conclusions about our proposal in section 5.

2   Theoretical Framework

2.1   Multi-criteria Decision Making

When dealing with decision making problems with multiple criteria, it is helpful to dispose of a procedure to organize the attributes according to each *criteria*, in order to rank the finite set of *alternatives* depending on the satisfaction of the whole criterion, or as a support to find the solution that best fits a desired goal [5] giving a *hierarchy* of the alternatives.

---

[1] http://cuentame.inegi.org.mx/monografias/informacion/
jal/poblacion/

The Analytic Hierarchy Process (AHP) is a multi criteria decision making method, developed by Thomas Saaty in the 1970Át's, that deals with both qualitative and quantitative criteria, what is one of its strengths. This method returns a preference value for each alternative, so that they could be ranked or only identify those better alternatives [6] [7] [8].

This multi criteria method is based on four steps[7] that include the *problem modeling* required for defining the structure and identify both criteria and alternatives which are the input for the mathematical process, *Judgement scales* used as a way to weight criteria and alternatives according to a preference intensity based on a ratio or scale, *pair-wise comparisons* to differentiate intensity of the elements among them obtained from a quotient of two quantities with the same units, and *priorities derivation* which return a numerical value according the "importance" derived from the mathematical process, for a better description refer [6] or [7].

2.2 Ant Colony Optimization

Ant colony metaheuristic paradigm was first introduced by Marco Dorigo in the 1990's. Ants are insects that work on a social environment with a structured organization that helps the colony to solve complex problems for a single ant [9]. The paradigm has been inspired by the behavior of real ants when searching for food. The resulting algorithm is actually useful for solving combinatorial problems among which we can find internet routing, vehicle routing, scheduling planning, and many more [10]. The search process is determined by an *heuristic* and *pheromone* values associated to the possible alternatives, where the pheromone is an indirect communication mechanism based on the principle called *Stigmergy* (term introduced by Grassé [11]), and serves to reinforce the most attractive alternatives, while the rest are forgotten.

Artificial ants find solutions to complex problems by moving through the edges of a graph $G=(V,E)$. They begin their search from a starting point to a final destination, keeping in *mind* the visited points for further strengthening or weakening partial solutions. Each ant of the colony move according to a stochastic selection rule which depends on the heuristic and the pheromone. The selection of the best alternative, at each moment, is based on (1) [9].

$$ P^k_{ij} = \frac{\tau^\alpha_{ij} \cdot \eta^\beta_{ij}}{\sum_{c_{il}} \tau^\alpha_{il} \cdot \eta^\beta_{il}} \quad for\ those \quad c_{il} \in N(s^p) \tag{1} $$

where P^k_{ij} represents the probability that the ant k choose the arc from node i to node j. This value is calculated with the pheromone τ associated to the path (i,j) and the heuristic function η; α, and β control the importance of the τ and η respectively. l represents the nodes that have not been visited yet. At each iteration, the function must only include the visible or feasible alternatives, that is, the neighbors of the actual $N(s^p)$. A similar expression is provided by the proportion-sum Independence theorem from Lukacs [12]. For a complete description refer[9].

3 Proposed Method

When modeling a path search process, the common objective function is to minimize a weight value whose popular dimensions are distance or time with some variations on them; however, to consider several characteristics the problem gets complicated. Our proposed method integrates many street network data from a street map by synthesizing multiple criteria into a single value for each street which represents a level of quality of the streets to provide a service, in this case to hold traffic flow, this value is then used for a path searching algorithm to deviate the search process over those streets with the better quality value (which in this case are those most fluid streets) that give car drivers better chance to move fluently over the overall route; such method is given below, as steps:

Assessment and path process

-      [**Step 1.**] *LoadMap ← ExternalSource*
-      [**Step 2.**] Generate a simplified graph from an original map
-      [**Step 3.**] Define street attributes: static (length, number of lanes, etc.), and dynamic (speed, etc.)
-      [**Step 4.**] Apply the multi-criteria decision making method (AHP)
-      [**Step 5.**] Assign the priority (range or weight) on each street of the simplified graph
-      [**Step 6.**] Apply the ACO algorithm using derived priorities as heuristic

if path with other criteria values is needed **then**
     repeat process from step 4
else[stop]
end if

In a **first step** we selected a portion of the city map of Guadalajara, México, represented as a graph $G=(V,E)$, including several characteristics. For this work we use *Open-StreetMaps*[2] (OSM), where raw data is provided as an XML file. Then, an incidence matrix was directly derived from the simplified graph with vertices $V=(v_1, v_2, ..., v_n)$, where each one represents a joint point of two or more streets (2).

$$v_i = e_j \cap e_k | e_j, e_k \in E : e_j \neq e_k \tag{2}$$

Vertices have a single coordinate vector, *(Latitude,Longitude)*, that geo-references each node and serves as a label. Edges $E=(e_1, e_2, ..., e_m)$ denote the streets and each edge e_m has a relationship between two vertices $(v_i, v_j) | v_i, v_j \in V$. Each edge has associated some attributes, as the *length of the street*, representing, in this case, the distance d_{ij} between two vertices (v_i, v_j) or intersections.

Second Step: As the OSM map contained a higher resolution than needed, it was necessary to eliminate the intermediate vertices between two consecutive join points (v_i, v_j). This simplification was carried out on the XML file, to get a simplified graph.

---

[2] http://www.openstreetmap.org/, access date: June-02-2012

Third Step: Attributes were gathered from OSM and other sources. They are: *length of the street, number of lanes, sense of circulation* (one way / two ways), as well as the corresponding *speed*. Last attribute was constrained to a constant value for all streets, and set equal to 60 km/h (from the driving city standars); these attributes are directly related to the criteria in the AHP process.

Fourth Step: This procedure is performed as the rules revised on section 2.1. AHP was selected because its capacity to combine qualitative and quantitative criteria, useful when alternatives are compared using their quantitative implicit characteristics and user subjective perspective, which is performed to assess and generate ranges that allow select alternatives based on multiple criteria.

Fifth Step: After former step, the MCDM process, has been performed, it returns priority values for each street which are associated to each edge on the simplified graph.

Sixth Step: Finally, the path searching ACO algorithm is applied using the priority values, whose values influence the exploration phase during the search process.

4 Experiments

In a first stage we gather information about the case of study, and then the AHP method was applied. On a second stage, the priority values calculated previously were used in the ACO heuristic as weighting parameters. The selected map, being part of the Guadalajara city of México, is bounded by the (20.6747199,-103.339321) and (20.7089919,-103.373707) coordinates. This map encloses 1937 junctions (nodes) with a triad for each one containing an *Id*, corresponding to a *latitude* and a *longitude*. The AHP method was implemented using the python programming language, over the 6950 streets (edges) of the considered map. Simulations were carried out in a Macbook Pro with an Intel Core 2 Duo processor, running at 2.26 Ghz, and with 4 GB at 1067 Mhz of DDR3 RAM memory. For visualization we have used *Gephi*[3] and GPS Visualizer[4]

4.1 Street Assessment for the Case of Study

Length, number of lanes, one or two ways, and *speed* were used as criteria in the AHP process, and a pairwise comparison was carried out over these criteria using a subjective preference, which is based on authors' perspective, only two values, 1 and 2, were used to represent equality and a relative preference respectively. Although there are only two values in the scale, the last one could be interpreted as strong in Saaty's scale[6]. The resulting values from a pairwise comparison are arranged in Table 1, where the last column of the expanded matrix contains the preference values of the criteria.

Once the values of Table 1 were calculated, the next step is the assessment of the value corresponding to each street. This is done through a pairwise comparison whose resulting values are collected in a matrix for each criterion. However, instead of using the SaatyÁt's scale for each street, we have used the implicit value for each feature,

---

[3] https://gephi.org/, last access: Jan-20-2013
[4] http://www.gpsvisualizer.com, last access: feb-07-2013

Table 1. Preference matrix for street assessment

	length	# lanes	# ways	vel	preference
length	1	0.5	2	1	**0.2371153**
# lanes	2	1	2	0.5	**0.28727607**
# ways	0.5	0.5	1	0.5	**0.13569132**
vel	1	2	2	1	**0.33991732**

Table 2. Resulting range values for each street

street id	length	# lanes	# ways	vel	range
1222122854	17.7146	2.0	2.0	60.00	**0.000118**
1228021984	326.5441	2.0	1.0	60.00	**0.000224**
1228429738	87.6124	2.0	1.0	60.00	**0.000131**
1332321536	173.63	6.0	2.0	60.00	**0.000245**
1333422342	70.44	8.0	2.0	60.00	**0.000237**
4104121510	45.96	4.0	1.0	60.00	**0.000147**
4104321277	111.62	4.0	1.0	60.00	**0.000173**
...	...	...	...	...	**...**

so this process can be automatically executed. Hence, if we do a comparison according to the *sense of the streets*, for streets with *one way and two ways*, the resulting value is 0.5 (1way/2way). When the *speed* criterion is considered, the calculated value is given by the relation of maximum speeds in each street. Comparison matrices are calculated for *length*, *number of lanes*, *sense* and *maximum speed*.

At the end of the MCDM process we get a set of values, as those of Table 2, and as the examples represented in Figure 1.

Former values represent levels of importance or attractiveness, and they are useful when searching for an alternative. In fact, they are incorporated as parameters into the heuristic of the ACO algorithm (see equation (1)).

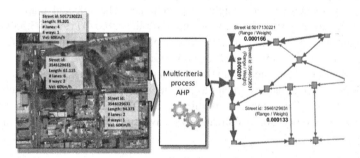

Fig. 1. Representation of the street assessment over a simplified graph

4.2 Probability of Choice for the ACO

The iterative process of search in the ACO algorithm is based on a selection rule that depends on the probability resulting from the product of a pheromone value, associated to each possible alternative, and the heuristic given by (1). In Fig. 2 two examples are shown, one under a distance (Fig. 2a), and the other mixing several attributes (Fig. 2b), which are being transformed into a probability value given by equation (1), and representing a level of attractiveness on the algorithm. However, these selected criterion does not take into account other characteristics, as the *capacity of streets*, and the particularities of each of them, as it could be the *timing of traffic lights*, *left turns*, *preference over other streets*, etc., which in our case are *length*, *# lanes*, *# ways* and *velocity*. This information, can drive the search process through more interesting solutions, as a more realistic environment is considered.

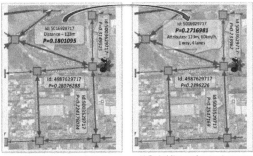

a) Probabilites under distance   b) Probabilites under street assessent

Fig. 2. Probabilities under a) distance, b) street assessment

As stated before first we gather information about the case of study, and apply AHP method, then priority values were calculated to be used in the ACO heuristic as weight parameters.

The parameters for the ACO algorithm execution were a set of 30 ants, 50 iterations, and the values $\alpha = 1$, $\beta = 6$ and $\rho = 0.6$, as suggested by Gaertner [13]. According to the results, whose behavior is shown in Fig. 3 a), only about 7 % of ants found a complete route from origin to destination. Fig. 3 b) shows the results for the same example, except that in this case the streets value assessment was included under a shortest time objective. This time about 30 % of ants found a solution. This is understandable as the street values focus the search through the nearest most important alternatives, and there should be important streets in any populated areas, no matter the journey we are interested on. Many test were carried out, two of them are described below:

First, a well established route was selected in order to compare our proposal against googleÂt's solution. Fig. 4 a) and 4 b) show the corresponding results for the path whose length is 2 627 m.

Second, while this study was conducted there was an accident that forced the closure of four of the eight lanes of an avenue, blocking time was approximately six hours long. In a non-peak hour, it took 15 minutes more than normal, to pass the three blocks affected by the accident. This occurrence was incorporated into the approach of streets value assessment such that, under these conditions, other alternatives could be more interesting. Two main alternative routes were found, the results are shown in Fig. 4 c), with a length of 3 351 m, and Fig. 4 d), with a length of 3 095 m. Both routes were shorter in time than the normal route, because the perturbation was avoided. These results are in accordance with the capacity variation of the avenue to provide the expected service. Under this condition the range of the avenue decreases and it becomes less important while the nearest avenues and streets gain in importance over the previous one. So, they emerge as the most attractive under such condition.

The behavior of the solutions quality (based on travel time to move along the route) found after 150 times of repeating the process with the ACO algorithm can be seen in Fig. 5, where several solutions were found considering the shortest time path with better characteristics; Fig. 5 a) shows the behavior under a route that is conducted over a continuity of important streets and no perturbations included (Fig. 4 b)) when a route

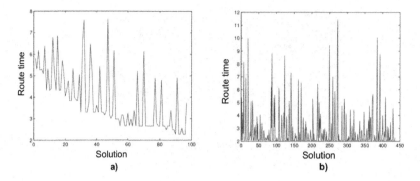

Fig. 3. Searching behavior: a) without street assesment, b) with street assessment

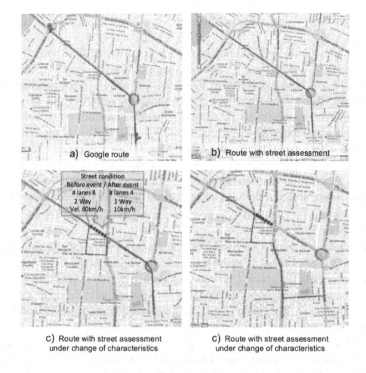

Fig. 4. a) Google route, b) Route with street assessment and normal conditions, c) Route with street assessment and a perturbation (first suggestion), d) Route with street assessment and a perturbation (second suggestion)

can be found by identifying important streets (a dominant optimal), these solutions are kind of stable more than 70% of the them laid on the "best route", with a path time of 2.62 min. to travel from source to target. On Fig. 5 b) is shown the behavior when added the perturbation described above, more variety of solutions were found due to a kind

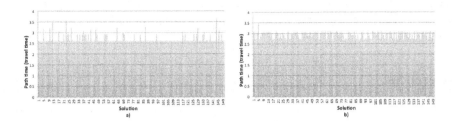

Fig. 5. Behavior of proposal - a) Without disturbance on streets, b) With disturbance on streets

of homogeneity among the solutions, most of them range from 2.74min. to 3.09min. for path time. having three main paths solutions that avoid those streets affected by the perturbation (as in Fig. 4 c) and d)), response time to draw a path was on average 5.2 seconds when no perturbation were found while time was 6 seconds when perturbation was found; also those solutions as in Fig. 4 b) were more stable than those of perturbations, this is due to more local optima.

As the results clearly show, the approach proposed in this work allows to find interesting routes according to the infrastructure and conditions of the city, to plan alternative routes depending on the traffic load and the time of the day, and the possibility to take into account the particular events that can be present at any moment, which can save time, stress, money, and reduce pollution to cite some important examples.

5   Conclusions

This work proposal is an approach to find the most fluid alternative routes in a city given a set of preference criteria for the route, which consider streets particular conditions. This is possible through an approach that summarizes streets characteristics and conditions into a numerical value that, in this case, represents the capacity for traffic flow of each particular street at a given moment, resulting values are then used on an ACO algorithm; so, good enough alternatives can be found for a journey, depending on a selected criterion as it could be distance or time for example. As a case of study we have used a portion of the city of Guadalajara, México, and we have shown what happen to some streets priorities when a random event occurs (in this case an accident). Two main different alternatives have been found that avoid the area in conflict. In both cases, the solutions are longer in distance but shorter in time during the time the affectation lasted. Moreover, logical results and similar to those of Google were found under normal conditions of car traffic. In forthcoming papers we are going to combine the streets value assessment proposal with other graph analysis techniques, including known ranking algorithms, as page rank, to improve and test the robustness of the approach proposed in this work. We are also planning a software development for smartphones as a way to gather data (provided by volunteers) about perturbations on streets related to vehicular load, state of the streets (repairs, maintenance, modifications, etc.), accidents, among others.

Acknowledgment. We would like to thank the Mexican Council of Science and Technology -CONACyT- and IBM by the fundings received through the University of Guadalajara.

References

1. Jiang, B.: Street hierarchies: A minority of streets account for a majority of traffic flow. Int. J. Geogr. Inf. Sci. 23(8), 1033–1148 (2009)
2. Naphade, M., Banavar, G., Harrison, C., Paraszczak, J., Morris, R.: Smarter cities and their innovation challenges. Computer 44(6), 32–39 (2011)
3. Dijkstra, E.W.: A note on two problems in connexion with graphs. Numer. Math. 1, 269–271 (1959)
4. Blum, C., Roli, A.: Metaheuristics in combinatorial optimization: Overview and conceptual comparison. ACM Comput. Surv. 35, 268–308 (2003)
5. Jurgen Branke, K.M., Deb, K., Slowinski, R.: Multiobjective optimization - interactive and evolutionary approaches. Springer, Heidelberg (2008)
6. Saaty, T.L.: Decision making with the analytic hierarchy process. Int. J. Services Sciences 1(1), 83–98 (2008)
7. Ishizaka, A., Labib, A.: Review of the main developments in the analytic hierarchy process. Expert Syst. Appl. 38(11), 14336–14345 (2011)
8. Anderson, D.R., Sweeney, D.J., Williams, T.A., Camm, J.D., Martin, K.: An introduction to management science: quantitative approaches to decision making, 13th edn. South-Western Cengage Learning UK (2012)
9. Dorigo, M., Stutzle, T.: Ant Colony Optimization. MIT Press, Cambridge (2004)
10. Marco Dorigo, M.B., Stutzle, T.: Ant colony optimization - artificial ants as a computational intelligence technique. Universit Libre de Bruxelles. Tech. Rep. TR/IRIDIA/2006-023 (2006)
11. Grasse, P.P.: La reconstruction du nid et les coordinations interindi- viduelles chez bellicositermes natalensis et cubitermes sp. la theorie de la stigmergie: Essai dinterpretation du comportement des termites constructeurs. Insectes Sociaux 6, 41–81 (1959)
12. Lukacs, E.: A characterization of the gamma distribution. The Annals of Mathematical Statistics 26(2), 319–324 (1955)
13. Gaertner, D., Clark, K.: On optimal parameters for ant colony optimization algorithms. In: Proceedings of the Int. Conference on Artificial Intelligence (2005)

Microcontroller Implementation of a Multi Objective Genetic Algorithm for Real-Time Intelligent Control

Martin Dendaluce[2], Juan José Valera[2], Vicente Gómez-Garay[2],
Eloy Irigoyen[1,2], and Ekaitz Larzabal[2]

[1] Computational Intelligence Group
Department of Systems Engineering and Automatic Control
University of the Basque Country (UPV/EHU), ETSI, 48013 Bilbao
[2] Intelligent Control Research Group
Department of Systems Engineering and Automatic Control
University of the Basque Country (UPV/EHU), ETSI, 48013 Bilbao
{mdendaluce001,elarzabal001}@ikasle.ehu.es
{juanjose.valera,vicente.gomez,eloy.irigoyen}@ehu.es

Abstract. This paper presents an approach to merge three elements that are usually not thought to be combined in one application: evolutionary computing running on reasonably priced microcontrollers (µC) for real-time fast control systems. A Multi Objective Genetic Algorithm (MOGA) is implemented on a 180MHz µC.A fourth element, a Neural Network (NN) for supporting the evaluation function by predicting the response of the controlled system, is also implemented. Computational performance and the influence of a variety of factors are discussed. The results open a whole new spectrum of applications with great potential to benefit from multivariable and multiobjective intelligent control methods in which the hybridization of different soft-computing techniques could be present. The main contribution of this paper is to prove that advanced soft-computing techniques are a feasible solution to be implemented on reasonably priced µC -based embedded platforms.

Keywords: Soft-Computing, Intelligent Control, Multi Objective Genetic Algorithm, NSGA-II, Artificial Neural Network, Microcontroller.

1 Introduction

Soft-computing techniques such as Genetic Algorithms (GA), as well as other intelligent techniques like NNs and Fuzzy Logic (FL), have already proven to be a serious alternative not only to complement but also to replace conventional algorithms in a wide diversity of applications[1].

Intelligent control schemes based on soft-computing techniques are showing to be an excellent solution for dealing with the difficulty of optimizing nonlinear and multi objective control systems. They also adapt conveniently to diverse real-world issues such as system parameter alterations, operation point variations, measurement imprecision, noise, unforeseen perturbations and the overall present nonlinearities and non-convexities[1, 2]. Trying to control these aspects through conventional

Á. Herrero et al. (eds.), *International Joint Conference SOCO'13-CISIS'13-ICEUTE'13,*
Advances in Intelligent Systems and Computing 239,
DOI: 10.1007/978-3-319-01854-6_8, © Springer International Publishing Switzerland 2014

hard-computing methods usually will increase the solution complexity up to unviable levels, forcing the designer to accept assumptions and simplifications on the control problem. Therefore, the control problem transformations –which are needed to solve the problemby conventional hard-computing methods in reasonable sampling times - can ensure stability on the closed-loop response under some assumptions, but can lose optimality in their response when variable and frequent changes in the system operation point are required. All these aspectsare likely to deteriorate the final result and could impede the development of an enhanced controller which would provide added value to the final control system.

NN and FL have already been implemented on embedded platforms for real-time control applications, often also in hybrid algorithms combining them witheach other (e.g. Neuro-Fuzzy) or with classical control techniques like PI controllers[3–6].

GAs are also well known for being able to offer great results when correctly tuned, but unlike the previous cases,they are usually used for offline optimization of a wide variety of designs and parameter determination[7, 8]. It is not likely to see GAs running on an embeddedtarget platforms, and when done so, they typically control slow systemsor optimize parameters of another controller according to variations [9–13].

One of the main drawbacks for the application of GAs and Evolutionary Algorithms(EA) on real-time control applicationsis their high computational cost, their complexity and their variability, which may also make it difficult to guarantee the searching convergence, i.e. the control action convergence inside the sampling time window.Additionally, ensuringstability of the closed-loop response becomes an arduous task (but however a challenge)that should be done by using Finite Markov Chain analysis[14]. These difficulties, together with the fact of lower demand for complex optimization methods running on targets, have prevented them from moving onto embedded platforms, hindering the appearance of new applications.

The core of this work is the MOGA which is desired to be proven as feasible for fast real-time intelligent control applications under the umbrella of a nonlinear model predictive strategy [15]. A NN for supporting the evaluation function by predicting the response of the control systems in a short horizon is also implemented. The aim of our research is to contribute to the applicability of intelligent & hybrid control schemes - using soft-computing techniques – but considering constricted electronic control units based on μC or digital signal processors (DSP).

2   A Real-Time MOGA Based Hybrid Controlscheme

GAs are stochastic optimization algorithms inspired in evolutionary genetics and natural selection, in which a population of solutions evolves through various generations under the influence of randomness and specimen survival, reproduction and mutation [16, 17].The resultisa method that handles very well nonlinear and complex systems, does not require excessive model simplifications and has less tendency to get stuck in local optima as happens with conventional methods, therefore being very successful in finding optimum or near-optimum solutions. Nevertheless, it is an iterative method in whicha series of operators have to be applied to a population of individuals (candidates) that also need to be evaluated at each iteration, which means the computational complexity is a problem. The application of GAs over multiobjective optimization problems leads to MOGA techniques and methods where fitness functions with multiple objectives are considered[18, 19].

For the implementation approach presented in this work, the well-knownfast and elitist MOGA algorithm NSGA-II (Nondominated Sorting Genetic Algorithm) has been selected[20], see Fig. 1. Apart from its high relevancy in the current state of the art, its reasonable and foreseeable computational cost together with its fast convergence propertymaintaining a good spread of solutions makes it very suitable for working as solver in real-time optimization problems.The algorithm complexity of the NSGA-II is given by the expression$O(MN^2)G$, where M stands for the amount of objectives, N for the population size and G for the number of generations. According to the previous expression the time T required to execute the searching procedure is given by $T = C(MN^2)G$ where the factor C represents all the ignored aspects after using the asymptotic approach,O, which in this case is dependant of the platform performance, the fitness functions used and the issues related to the programming and implementation of the algorithm. As the impact of the population number N is squared, this variable should be tried to be kept low, thusimproving the searching results and convergence by adequately tuning the MOGA operators and the number of generations required, as far as the characteristics of the problem to be solved allow it.

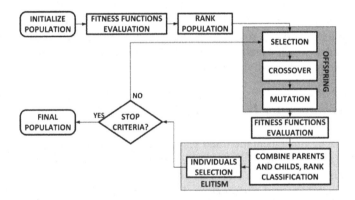

Fig. 1. Diagram representing the NSGA-II Multi Objective Genetic Algorithm

The interesting concept of using a MOGA optimizer as main controller and supporting it on NN based predictions as fitness functions is a feasible idea for handling the multiobjective nonlinear model predictive control [15]. Further enhancement of the controller can be obtained by implementing a final solution selection mechanism (among the Pareto's set) based on FL or Expert Systems as proposed in[15], see Fig. 2. Both mechanisms (NN and FL) can be implemented in an efficient and uncomplicated manner on μC platforms.

The implementation of the NN is a very critical point that must be implemented with high efficiency, as it needs to be executed hundreds or thousands of times per second, therefore consuming a considerable part of the total computation time.

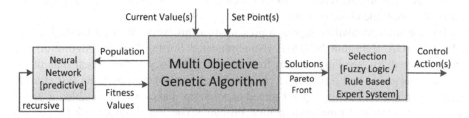

Fig. 2. The proposed Real-Time MOGA based Intelligent Hybrid scheme

Supporting the previous concept by transferring the task of executing the NN to a FPGA platform is a very interesting approach, due to their parallel processing capabilities with potential to rapidly run predictions of various solutions simultaneously [21]. The need for this hybrid hardware architecture solution will depend on the complexity of the NN and the time and cost boundary conditions.

Another interesting approach is the usage of downsized GAs. Micro-GAs offer a lightweight alternative for cases where higher computational costs are not acceptable[22]. They represent an optimization solution able to offer a compromise between results and time constraints which might be adequate in certain applications. Also in this direction, techniques based on small population combining Particle Swarm Optimization with evolutionary mechanisms have been proposed[23]. When adequately set up and tuned, which usually is a delicate task in GAs, smaller and simpler algorithms can also provide satisfying results.

3   Microcontroller Implementation

For the NSGA-II implementation and testing, a high performance floating-pointμC was selected(ARM$^{®}$ Cortex^{TM}-R4F - 180 MHz - 3 MB of flash memory - 256 KB RAM). This is one of the μCs that were considered as interesting – in addition to its processing capabilities and performance -because it is available as a redundant dual-core controller conceived,among others, for safety critical control functions, as it occurs in many automotive applications. This makes it well suited for promising application fields in the transportation sector, such as the relevant trend of Advanced Driver Assistance Systems(ADAS). However, other μCsand DSPs offering performances up to GHz range and parallel processing can be candidates forapplications in which higher data processing speeds are required[24].

An important consideration for the software development was to avoid bounding the solution to a specific target platform and/or software development tool which could create undesirable dependencies and workflow limitations. Therefore a simple programmingscheme has been selectedin order to provide a high code portability and flexibility to work with other platforms and algorithms. A single primary task executes the algorithm on a timer-based determinist and time-constrained loop.

The programming abstraction has been kept low: it is written in C and no RTOS has been used.Algorithms coming from a code generation tool canalso be easily integrated.

During the adaptation of the original code into embedded platform friendly C, diverse functions were suppressed or substituted, for example those regarding to interface, visualization and storage operations. Programming optimizations such as reducing data type casting operations and avoiding unnecessary intermediate results have also been applied. Regarding to the compiler, optimizations affecting registers, local code and global code were selected, together with function inlining and the most speed oriented speed-VS-code-size balance.

Regarding to the evaluation function, different approaches were taken. First two non-representative multiobjective optimization problems were defined: simple (1) and heavy (2) cost functions. Each one is composed by two fitness (objective) functions (f_0, f_1) with different mathematical complexity.

$$f_0 = x_0^2 + 5x_0x_1^2 + 4x_2 + {x_3x_4}/x_5 \;\;;\;\; f_1 = x_5^4 + 8x_2x_3^2 + 9x_1 + {x_0x_1}/x_2 \quad (1)$$

$$f_0 = e^{\sqrt{x_0^2 + 5x_0x_1^2}/x_5} + \sqrt{x_2^2 + \sqrt{x_3x_4}} \cdot (\frac{x_5}{123456.789})^5 \quad (2)$$

$$f_1 = e^{543212345/x_1\sqrt{x_5^2 + 5x_4^2/x_3}} + \sqrt{x_0^2 + \sqrt{x_4x_5}} \cdot x_4^{1/x_5}$$

A function which represents a practical problem was also tested: the model of a helicopter-like Twin Rotor MIMO System (Feedback Instruments Ltd.). This system has two voltage inputs to control the power of each of the two rotors which are rotated 90° respect to each other. As the structure is fixed to the ground through an articulation, it offers 2 degrees of freedom: the outputs are the pitch and the yaw angles.

This system is represented not by simple mathematical expressions, but bymore complex functions for which two different approaches have been implemented. In both of them$H$recursive function calls are executed for predicting the response over the prediction horizon.There are two objectives and their functions are chosen to be the calculation of the quadratic error over H regarding to the pitch and yaw angle setpoints. This can be easily extended by, for example, adding the control action energy consumption. The MOGA chromosome contains 10 real variables, as the MOGA has to obtain 2 control actions over a selected horizon of $H = 5$ steps-ahead.

The first approachis a relatively complex nonlinear Simulink™ model which represents the differential equations and was translated into a C function using code-generation. It contains 3 transfer functions, 4 integrators and a series of nonlinear mathematical expressions andtrigonometric functions.

The second and especiallyrelevant approach, which leads to the interesting hybrid concept with a NN, isimplemented instead of the previous model in order to be trained with the real system inputs and outputs so that it better reflects the non linearities. It is based on a NARX topology containing a layer with 8 neurons and a hidden layer with 12 neurons, plus the corresponding input and output layers. It is also integrated into the recursive execution based predictive context.

Inevitably the µC implementation must be adapted to the specific application considering the control specifications and requirements. This not only does mean optimizing the size and complexity of the algorithm itself, but also the required range and resolution of the variables being used. Once theplatform is established, platform specific low level coding must be considered in order to optimize the execution performance.Especially on routines that are executed many times per iteration- such as the fitness function or many of the evolutionary operations in EAs or GAs - should be taken care to be efficient, as they represent a significant fraction of the overall computation time. On high performance devices, the correct usage of resources such as the cache and the pipeline, and when possible executing more than one instruction per cycle(or even fullparallel processing)should beimplemented. Combining the previous actions with compiler/linker level optimizations and other actions, such as function inlining, significant performance improvements are obtained.

Aiming to test and analyze different soft-computing algorithms and intelligent control strategies, several systems would be needed. As setting up physical models would suppose an excessive effort, simulating the corresponding plant models on a parallel hardware platform was chosen as an agile, cost effective and safe alternative using the Hardware in the Loop (HiL) testing approach, see Fig. 3. For the first tests, a cost effective 32 bit floating point DSC running at 150 MHz was selected.It offers the possibility of directly embedding the automatically generated code from Simulink™ models, and additionally monitor and interact with them through a PC based GUI. It will enable to run the plant model in real-time with differential integration steps between one and three orders of magnitude faster than the controller sample time.

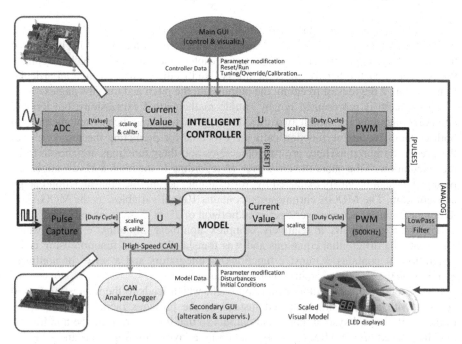

Fig. 3. HiL based Testing Setup. ARM® Cortex™-180 MHz as Intelligent Controller; DSC-150 MHz as Emulator of the system to be controlled

4   Results

As it was expected, the results obtained from this implementation have shown that MOGAsare computationally costly compared to other algorithms such as FL, NN or PID controllers, of which some simple versions were also implemented in a separate experiment for testing and comparative purposes, with resulting times in the units to hundreds of μs range. More specifically, the NN implemented for the hybrid-predictive-fitness scheme averaged at 34.7μswith its NARX topology and 8 neurons in the main layer and 12 neurons in the hidden layer.

The numbers corresponding to the most representative cases of the MOGA are presented in Table 1.A *big* and a *small*MOGA were adequately dimensioned thinking of complex and simpler problems, whereas the MOGA for the two rotor system was adjusted to allow a fast sample rate of 100mswith a prediction horizon of 5 steps.

Table 1. Summary of the most significant results of the MOGA implementation

Fitness function	Big MOGA test		Small MOGA test		Twin Rotor test	
	Simple(1)	Heavy(2)	Simple(1)	heavy(2)	Nonlinear Model	NN
Objectives	2	2	2	2	2	2
Population	60	60	15	15	15	15
Generations	60	60	20	20	24	24
Chromosome size	6	6	6	6	10	10
Prediction horizon	-	-	-	-	5	5
Constraints	-	-	-	-	-	-
Code size	79.0 KB	79.7KB	78.4 KB	79.2 KB	101.8 KB	104.3 KB
RAM occupation	85.5 KB	85.5 KB	11.6 KB	11.6 KB	14.9 KB	13.5KB
T_{cycle}Avg./Worst [ms]	**922**/970	**998**/1085	**21.5**/24.2	**28.1**/32.3	**74.6**/82.1	**88.1**/96.1
C factor (see section 2)	0.002134	0.002310	0.002389	0.003122	0.006907	0.008157
t_{eval}Average	0.335 μs	21.8 μs	0.335 μs	21.8 μs	136.2 μs	173.5 μs
C' factor(see eq. 3)	0.002130	0.002130	0.002378	0.002378	0.002371	0.002371

It can be observedthat more RAM than Flash memory is needed, in spite of the speed oriented and flash consuming optimization settings. More generations barely increase Flash nor RAM usage, but a bigger population once again dramatically increases the RAM consumption: without including the Twin Rotor nonlinear model (or the NN), the maximum possible population is 113.

Regarding to the cycle times - the really critical point - the nonlinear model and NN based predictive evaluation functions are considerably more costly that the *Simple* and *Heavy*mathematical expressions (eqs. 1 and 2). Nevertheless, as the most significant part of the computation time is still taken by the MOGA itself, the relative growth of the total computation time is still acceptable.It can be also noted that, although the model is slightly faster than the NN in this implementation, it has some simplifications so that the NN might be able to reflect nonlinearities better.

A major advantage of having a generic expression of the computational cost is the fact that instead of relying on large tables constructed through extensive testing, the execution time of a MOGA with some specific parameters can be extrapolated with sufficient precision basing on a reduced set of test results and the expression shown in section 2: $T = C(MN^2)G$. The resulting C values obtained are shown in Table 1. The difference between C valuesis due to the fact that the N^2 factor applies to the evolutionary algorithm, but the fitness function evaluation is only called $M \cdot N \cdot G$ times. Therefore this should be considered if the execution cost of the algorithm as a whole, including the fitness functions, is to be considered.

Consequently the following new expression is proposed:

$$T = C'(MN^2)G + C'(NF)G \;\; \rightarrow \;\; \boldsymbol{T = C'[N[MN + F]]G} \qquad (3)$$

Being $F = t_{eval}/C$ and t_{eval} the total computation time of one fitness function call (including all the objectives and therefore not$M$ dependant in this representation). Here the C' value corresponds to the MOGA running without any fitness evaluation.

In this way, knowing the computation time of the fitness expression (which can be relatively easily obtained in diverse manners) and the C' value which reflects the platform/implementation performance, the computation time of a MOGA of any desired population, generation and objective number can be easily and reliably obtained.

A small difference between the new C' values can still be seen, which is explained by the fact that there are a series of operations in the loops that introduce offsets and different dependencies to the M, N, G magnitudes, hereby slightly distorting the relation. Their values can be obtained by a combination of code analysis and measurements, but it turns into anunnecessarilycomplex analysis.

The results obtained in this µC implementation may be compared with those obtained in a NSGA-II implementation on industrial systems by Larzabal et al. [25]. This was done on industrial platforms with much higher performance, size and cost. The resulting cycle time for a 1 GHz Intel® Celeron® PLC was under 0.15s for a population of 70, 100 generations and 1 objective. This gives $C = 0.00031$. Fasterplatforms obviously provide a lower C number due to their higher operating frequency. Therefore representing these values in a frequency independent unit could provide an interesting benchmark. Calculating $B = C \times f_{processor}(MHz)$ the result is $B = 0.38\sim0.43$ for the 180 MHz ARM® based µC and $B = 0.31$ for the 1 GHz Industrial PLC platform. The overall performance of the PLC is better, in spite of the additional cost of running a RTOS, as it benefits from higher processing capacity and other advantages such as cache and faster memory.

5   Conclusions and Future Work

The experimental results obtained in this work prove that the idea of embedding a NSGA-II MOGA on a reasonably priced µC - which hasconsiderable restrictions in terms of computational power and memory - is possible and reasonable. Therefore, this contribution is meant to support the future use of MOGAs not only for off-line optimization, but also as main controller in fast real-time control systems.

After a first phase where rather simple mathematical equations with reduced capacity to represent real world systems were used as fitness function, more complex nonlinear models have been introduced in order toenable to control dynamic systems,based on a recursive predictive strategy. Anonlinear model with differential equation systemswas inserted into this scheme. Finally, following the hybrid soft-computing concept,a NN which can representcomplex nonlinearities with considerable accuracy was also implemented into the predictive structure.

The resulting cycle times are reasonably low and encouragingto follow this line and exploit its potential, as they enable the application of complex intelligent control methods on relatively fast dynamic systems.

Sufficient degrees of freedom are still available to improve the application of complex soft-computing algorithms and techniques, thus enabling the development of new systems and control concepts. Fine algorithm tuning and the code optimization for specific µCs or DSPs could still push the overall performance.

A new and more detailed expression to predict - in a reliable and simple manner - the total computation time for the MOGA has also been presented. In this formula the influence of the fitness functions time cost has been treated independently and included as a new term in the expression.

Future work will develop a more refined implementation and a more exhaustive study of alternative solutions regarding both to the algorithmia and the µC or DSP based platforms. Following the intelligent control research line, next works will also focus on hybrid intelligent controllers' implementation and their validation through the HiL setup. Transferring time consuming parallel operations to a FPGA to optimize the controller implementation will also be investigated.

References

1. Rudas, I.J., Fodor, J.: Intelligent Systems. International Journal of Computers, Communications and Control 3(Spl. Iss.), 132–138 (2008)
2. Zadeh, L.A.: Fuzzy Logic, Neural Networks, and Soft Computing. Communications of the ACM 37(3), 77–84 (1994)
3. Del Campo, et al.: Efficient Hardware/Software Implementation of an Adaptive Neuro-Fuzzy System. IEEE Transactions on Fuzzy Systems 16(3), 761–778 (2008)
4. Velagic, et al.: Microcontroller Based Fuzzy-PI Approach Employing Control Surface Discretization. In: 2012 20th Mediterranean Conference on Control Automation, MED, Piscataway, NJ, USA, pp. 638–645 (2012)
5. Jung, S., Su Kim, S.: Hardware Implementation of a Real-Time Neural Network Controller With a DSP and an FPGA for Nonlinear Systems. IEEE Transactions on Industrial Electronics 54(1), 265–271 (2007)
6. Ravi, S., Sudha, M., Balakrishnan, P.A.: Design and Development of a Microcontroller Based Neuro Fuzzy Temperature Controller. In: 2012 International Conference on Informatics, Electronics &Vision (ICIEV), Piscataway, NJ, USA, pp. 103–107 (2012)
7. Yousefpoor, et al.: THD Minimization Applied Directly on the Line-to-Line Voltage of Multilevel Inverters. IEEE Transactionson Industrial Electronics 59, 373–380 (2012)
8. Fleming, P., Purshouse, R.: Evolutionary Algorithms in Control Systems Engineering: a Survey. Control Engineering Practice 10(11), 1223–1241 (2002)

9. Mamdoohi, et al.: Realization of Microcontroller-Based Polarization Control System with Genetic Algorithm. In: 2009 IEEE 9th Malaysia International Conference on Communications (MICC), Piscataway, NJ, USA, pp. 774–779 (2009)
10. Krishnan, et al.: Parallel Distributed Genetic Algorithm Development Based on Microcontrollers Framework. In: First International Conference on Distributed Framework and Applications, DFmA 2008, Piscataway, NJ, USA, pp. 35–40 (2008)
11. Mininno, E., et al.: Real-Valued Compact Genetic Algorithms for Embedded Microcontroller Optimization. IEEE Transaction on Evolutionary Computing 12, 203–219 (2008)
12. Cao, et al.: DSP Implementation of the Particle Swarm and Genetic Algorithms for Real-Time Design of Thinned Array Antennas. IEEE Antennas and Wireless Propagation Letters 11, 1170–1173 (2012)
13. Kwak, M., Shin, T.S.: Real-Time Automatic Tuning of Vibration Controllers for Smart Structures by Genetic Algorithm. In: Proc. of SPIE, USA, vol. 3667, pp. 679–690 (1999)
14. Goldberg, D., Segrest, P.: Finite Markov Chain Analysis of Genetic Algorithms. In: Proceedings of the Second International Conference on Genetic Algorithms and their Application, pp. 1–8. L. Erlbaum Associates Inc., Hillsdale (1987)
15. Valera Garcia, J.J., et al.: Intelligent Multi-Objective Nonlinear Model Predictive Control (iMO-NMPC): Towards the 'On-line' Optimization of Highly Complex Control Problems. Expert Systems with Applications 39(7), 6527–6540 (2012)
16. Holland, J.H.: Adaptation in Natural and Artificial Systems:. Univ. Michigan Press (1975)
17. Goldberg, D., Holland, J.: Genetic Algorithms and Machine Learning. Machine Learning 3(2-3), 95–99 (1988)
18. Coello Coello, C.A.: Evolutionary Multi-Objective Optimization: a Historical View of the Field. IEEE Computational Intelligence Magazine 1(1), 28–36 (2006)
19. Konak, et al.: ulti-Objective Optimization using Genetic Algorithms: A tutorial. Reliability Engineering & System Safety 91(9), 992–1007 (2006)
20. Deb, et al.: A Fast and Elitist Multiobjective Genetic Algorithm: NSGA-II. IEEE Transactions on Evolutionary Computation 6(2), 182–197 (2002)
21. Martin, P.: A hardware implementation of a genetic programming system using FPGAs and Handel-C. Genetic Programming and Evolvable Machines 2(4), 317–343 (2001)
22. Toscano Pulido, G., Coello Coello, C.A.: The Micro Denetic Algorithm 2: Towards Online Adaptation in Evolutionary Multiobjective Optimization. In: Fonseca, C.M., Fleming, P.J., Zitzler, E., Deb, K., Thiele, L. (eds.) EMO 2003. LNCS, vol. 2632, pp. 252–266. Springer, Heidelberg (2003)
23. Cabrera, J.C.F., Coello Coello, C.A.: Micro-MOPSO: A Multi-objective Particle Swarm Optimizer that uses a very small Population Size. In: Nedjah, N., dos Santos Coelho, L., de Macedo Mourelle, L. (eds.) Multi-Objective Swarm Intelligent Systems. SCI, vol. 261, pp. 83–104. Springer, Heidelberg (2010)
24. Li, Q., He, J.: A sophisticated architecture for evolutionary multiobjective optimization utilizing high performance dsp. In: Kang, L., Liu, Y., Zeng, S. (eds.) ICES 2007. LNCS, vol. 4684, pp. 415–425. Springer, Heidelberg (2007)
25. Larzabal, E., Cubillos, J.A., Larrea, M., Irigoyen, E., Valera, J.J.: Soft computing Testing in Real Industrial Platforms for Process Intelligent Control. In: Snasel, V., Abraham, A., Corchado, E.S. (eds.) SOCO Models in Industrial & Environmental Appl. AISC, vol. 188, pp. 221–230. Springer, Heidelberg (2013)

Solving Linear Systems of Equations from BEM Codes

Raquel González[1], Lidia Sánchez[1], José Vallepuga[2], and Javier Alfonso[1]

[1] Department of Mechanical, Computing and Aerospace Engineerings,
University of León, Campus de Vegazana s/n, 24071 León, Spain
[2] Department of Mining, Topography and Structural Technology, University of León,
Campus de Vegazana s/n, 24071 León, Spain

Abstract. In this paper, we compare different methods to solve systems of linear equations in order to determine which one allows us to reduce as much as possible the execution time. We consider contact problems that are solved by applying the Boundary Element Method that involves the resolution of different systems of linear equations. Depending on the kind of problem, thermal and elastic systems of equations have to be solved. The number of equations depends on the number of elements in which solids are discretized. This means that the more realistic are the solids defined, the more number of elements are considered. For this reason, it is really interesting to determine the most efficient method. We compare the Gauss Elimination method with and without pivoting, the Gauss-Jordan method and the LU Factorization method for several elastic and thermoelastic problems. LU Factorization provides an average reduction of a 37% in the execution times, and up to a 41% for some particular problems.

Keywords: Systems of linear equations, BEM, application performance, execution time.

1 Introduction

In the last decades, several applications that implement numerical methods have been developed, such as ANSYS [3], ABAQUS [1], ADINA [2], FEMM [7] or ParaFEM [8] working with the Finite Element Method (FEM), or BEASY [4], Concept Analyst [6] or BETL [5] working with the Boundary Element Method (BEM).

Although these applications manage many types of problems, sometimes researchers need to develop their own codes in order to solve some specific and new problems. For example, Wriggers[27] or Pantuso et al. [24] wrote FEM applications to solve contact problems, Jin [20] or Li et al. [22] for electromagnetic problems, Chung [12] for fluid dynamics, etc. We can also find BEM codes, such as those from Geng et al. [17] or Li and Huang [23] in acoustics, Ataseven et al. [10] in electro-magnetic source imaging, Andersson et al. [9], Garrido et al. [16], Giannopoulos and Anifantis [18], Keppas [21] or Vallepuga [15] in contact problems. Testing these codes usually takes a large amount of time and resources,

Á. Herrero et al. (eds.), *International Joint Conference SOCO'13-CISIS'13-ICEUTE'13*,
Advances in Intelligent Systems and Computing 239,
DOI: 10.1007/978-3-319-01854-6_9, © Springer International Publishing Switzerland 2014

so it is important to improve the efficiency of the algorithms used as long as it may reduce one of those factors.

As all these numerical methods call for the resolution of large systems of equations, and that is the most time consuming part of the code, some research have been done around the optimization of the resolution methods, mostly resorting to using parallelization techniques. Among others, it is relevant the work done by Davies [14], Cunha et al. [13], or Gueye [19]. In [25] parallel algorithms are proposed for Gauss and Gauss-Jordan methods. In this paper, we present a study on the different sequential methods to solve this type of systems, and their efficiency. We have tested the different methods in an object oriented in-house application developed to solve elastic, thermal and thermoelastic contact problems based on the algorithm provided by Vallepuga in his PhD thesis [26].

In the following sections we briefly present the problem to be solved (section 2) and the different methods that have been studied (section 3). Next, we discuss the obtained results for each method (section 4). Finally the conclusions to this work are presented (section 5).

2   Problem Description

In [15], Vallepuga presents a method to solve three dimensional thermoelastic contact problems using BEM. Considering two solids in contact, a system of static and thermal loads is applied along their boundary. Applying Vallepuga's method, the final contact zone, stresses, movements, temperatures and temperature gradients on the surface can be determined.

Being the boundary integral equations [11]:

– Heat conduction problem:

$$c(\xi)\theta(\xi) + \int_\Gamma q^*(\xi,\eta)\theta(\eta)\,d\Gamma(\eta) = \int_\Gamma \theta^*(\xi,\eta)q(\eta)\,d\Gamma(\eta) \qquad (1)$$

where $c(\xi)$ is the free term of the integral equation and $q^*(\xi,\eta)$ and $\theta^*(\xi,\eta)$ are the fundamental solutions of the potential problem.

– Three-dimensional thermoelastic problem:

$$c_{ij}(\xi)u_j(\xi) - \int_\Gamma U_{ij}^*(\xi,\eta)t_j(\eta)\,d\Gamma(\eta) + \int_\Gamma T_{ij}^*(\xi,\eta)u_j(\eta)\,d\Gamma(\eta) =$$
$$= \int_\Gamma P_i^*(\xi,\eta)\theta(\eta)\,d\Gamma(\eta) - \int_\Gamma Q_i^*(\xi,\eta)q(\eta)d\Gamma(\eta) \qquad (2)$$

where $c_{ij}(\xi)$ is the free term of the elastic problem, $T_{ij}^*(\xi,\eta)$ and $U_{ij}^*(xi,\eta)$ the Kelvin solution, and $P_i^*(\xi,\eta)$ and $Q_i^*(\xi,\eta)$ the vectors produced by considering thermal expansions in the Hooke's law.

Both solids are discretized with constant triangular elements, transforming the above equations into the following:

– Thermal problem:

$$\frac{1}{2}\theta^L + \sum_{k=1}^{N} AT^{LK}\theta^K - \sum_{k=1}^{N} BT^{LK}q^K = 0 \tag{3}$$

where θ are the temperatures and q are the thermal gradients defined by $q = -\frac{\partial\theta}{\partial n}$ and
$AT^{LK} = \int_{A_K} q^*(x^L, y^K)dA_K$
$BT^{LK} = \int_{A_K} \theta^*(x^L, y^K)dA_K$
– Thermoelastic problem:

$$\sum_{k=1}^{N} A_{ij}^{*LK}u_j^K - \sum_{k=1}^{N} B_{ij}^{LK}t_j^K = \sum_{k=1}^{N} D_i^{LK}\theta^K - \sum_{k=1}^{N} C_i^{LK}q^K \tag{4}$$

where t_j are the stresses and u_j are the movements and
$A_{ij}^{*LK} = \frac{1}{2} + A_{ij}^{LK}$ for $L = K$ and $A^{*LK} = A^{LK}$ for $L \neq K$
$A_{ij}^{LK} = \int_{A_K} T_{ij}^*(x^L, y^K)dA_K$
$B_{ij}^{LK} = \int_{A_K} U_{ij}^*(x^L, y^K)dA_K$
$D_i^{LK} = \int_{A_K} P_i^*(x^L, y^K)dA_K$
$C_i^{LK} = \int_{A_K} Q_i^*(x^L, y^K)dA_K$
$-K$ (element we integrate over)
$-L$ (node we integrate from)

Let be N^A and N^B the number of nodes of solids A and B respectively. Imposing the boundary and contact conditions [15], the equations (3) and (4) form two linear systems of equations of the form

$$Ax = b \tag{5}$$

one for the thermal problem formed by $N_A + N_B$ equations and a second one for the elastic problem formed by $3 * (N_A + N_B)$ equations.

To solve those systems of equations we follow the procedure proposed by Vallepuga [15].

3 Methods

For all the methods discussed, we consider the following system of n equations with n unknowns:

$$\begin{cases} a_{11}x_1 + a_{12}x_2 + \cdots + a_{1n}x_n = b_1 \\ a_{21}x_1 + a_{22}x_2 + \cdots + a_{2n}x_n = b_2 \\ \vdots \qquad \vdots \qquad \ddots \qquad \vdots \qquad \vdots \\ a_{n1}x_1 + a_{n2}x_2 + \cdots + a_{nn}x_n = b_n \end{cases}$$

3.1 Gauss Elimination Method

This method consists of two steps:

1. Forward elimination of the unknowns: the aim is to reduce the original system to an equivalent upper triangular one. The first step is to consider the augmented matrix and make the necessary transformations to turn the elements below the diagonal into zeros.

$$
\begin{pmatrix}
a_{11} & a_{12} & \cdots & a_{1n} & b_1 \\
a_{21} & a_{22} & \cdots & a_{2n} & b_2 \\
\vdots & \vdots & \ddots & \vdots & \vdots \\
a_{n1} & a_{n2} & \cdots & a_{nn} & b_n
\end{pmatrix}
\rightarrow
\begin{pmatrix}
a'_{11} & a'_{12} & \cdots & a'_{1n} & b'_1 \\
0 & a'_{22} & \cdots & a'_{2n} & b'_2 \\
\vdots & \vdots & \ddots & \vdots & \vdots \\
0 & 0 & \cdots & a'_{nn} & b'_n
\end{pmatrix}
$$

2. Backward substitution: once we have built the upper triangular matrix, we can solve the last equation $a'_{nn}x_n = b'_n$ to obtain x_n, substitute x_n in the equation $n-1$ and solve it to get x_{n-1} and so on. This mechanism can be generalized by the expression:

$$
x_i = \frac{b'_i - \sum_{j=i+1}^{n} a'_{ij}x_j}{a'_{ii}} \tag{6}
$$

3.2 Gauss Jordan Method

This method consists in the transformation of the original matrix into the identity matrix, getting then the solution to the system in the b' vector.

$$
\begin{pmatrix}
a_{11} & a_{12} & \cdots & a_{1n} & b_1 \\
a_{21} & a_{22} & \cdots & a_{2n} & b_2 \\
\vdots & \vdots & \ddots & \vdots & \vdots \\
a_{n1} & a_{n2} & \cdots & a_{nn} & b_n
\end{pmatrix}
\rightarrow
\begin{pmatrix}
1 & 0 & \cdots & 0 & b'_1 \\
0 & 1 & \cdots & 0 & b'_2 \\
\vdots & \vdots & \ddots & \vdots & \vdots \\
0 & 0 & \cdots & 1 & b'_n
\end{pmatrix}
$$

We can achieve this the same way as in the Gaussian elimination, but once we have the upper triangular matrix we continue with the transformations above the diagonal.

3.3 Gauss Elimination Method with Pivoting

When the A coefficient matrix has any zeros in the main diagonal, or values very close to zero, then it is necessary to make permutations between rows and columns until finding an adequate pivot that can reduce round-off error.

In partial pivoting, the entry with the largest absolute value from the column of the pivot element is selected. This pivoting is normally enough to achieve accurate results. However, sometimes it is necessary to use complete pivoting, which considers all entries in the whole matrix, interchanging rows and columns to achieve the highest accuracy. This type of pivoting is hardly ever necessary and, due to the additional computations it requires, is much more time consuming.

3.4   LU Factorization Method

If the matrix

$$
A = \begin{pmatrix}
a_{11} & a_{12} & \cdots & a_{1n} & b_1 \\
a_{21} & a_{22} & \cdots & a_{2n} & b_2 \\
\vdots & \vdots & \ddots & \vdots & \vdots \\
a_{n1} & a_{n2} & \cdots & a_{nn} & b_n
\end{pmatrix}
$$

can be factorized into the product of two matrixes of the form $A = LU$, being L a lower triangular matrix and U an upper triangular matrix, we can transform the system into the following:

$$ Ax = b \implies Ax = (LU)x \implies Ax = L(Ux) \implies L(Ux) = b $$

If we consider $y = Ux$, we can firstly solve the system $Ly = b$ by forward substitution. Knowing the y values, we can solve the $Ux = y$ system by backward substitution, obtaining the result vector x.

4   Results and Discussion

Figure 1 shows the results obtained by each method for several elastic and thermoelastic problems. The experiments have been run on a personal computer with Intel Quad Core Q8400 @ 2,66 Ghz and 4 GB of RAM. The operating system running in the machine is Ubuntu 10.04, and the application have been developed in Java as modularity is one of the goals to be achieved. Execution times presented are the average for ten runs. The characteristics of those problems are commented below.

All the problems solved consist of two solids, prisms shaped, in contact.

1. Elastic Problems:
   (a) EC0: both solids have been discretized into 396 elements. The system of equations is made up by 2376 equations. There is perfect contact between solids. The problem presents symmetries in the XZ and YZ axes.
   (b) EC1: both solids have been discretized into 396 elements. The system of equations is made up by 2376 equations. There is perfect contact between solids. The problem presents symmetries in the XZ and YZ axes.
   (c) EC2: the solids have been discretized into 328 and 424 elements respectively. The system of equations is made up by 2256 equations. There is perfect contact between solids. The problem presents symmetries in the XZ and YZ axes.
   (d) EC3: the solids have been discretized into 328 and 424 elements respectively. The system of equations is made up by 2256 equations. There is perfect contact between solids. The problem presents symmetries in the XZ and YZ axes.
   (e) ER2: the solids have been discretized into 328 and 412 elements respectively. The system of equations is made up by 2220 equations. There is perfect contact between solids. The problem presents symmetries in the XZ and YZ axes.

(f) ER1: the solids have been discretized into 182 and 218 elements respectively. The system of equations is made up by 1200 equations. There is perfect contact between solids. The problem presents symmetries in the XZ and YZ axes.

(g) Cierre: the solids have been discretized into 416 and 352 elements respectively. The system of equations is made up by 2304 equations. There is perfect contact between solids. The problem presents symmetries in the XZ and YZ axes.

(h) Cierre1: the solids have been discretized into 296 and 232 elements respectively. The system of equations is made up by 1584 equations. There is perfect contact between solids. The problem presents symmetries in the XZ and YZ axes.

(i) ERD: the solids have been discretized into 512 and 896 elements respectively. The system of equations is made up by 4224 equations. There is perfect contact between solids. The problem presents symmetries in the XZ and YZ axes.

2. Thermoelastic Problems:

(a) TE1: the solids have been discretized into 640 elements each. The thermal and the elastic systems of equations are made up by 1280 and 3840 equations respectively. There is imperfect contact between solids. There is neither conduction, nor convection in the interstitial zone. The problem presents symmetry in the YZ axe.

(b) TE2: the solids have been discretized into 640 elements each. The thermal and the elastic systems of equations are made up by 1280 and 3840 equations respectively. There is perfect contact between solids. There is neither conduction, nor convection in the interstitial zone. The problem presents symmetry in the YZ axe.

(c) TE3: the solids have been discretized into 384 and 752 elements respectively. The thermal and the elastic systems of equations are made up by 1136 and 3408 equations respectively. There is imperfect contact between solids. There is neither conduction, nor convection in the interstitial zone. The problem presents symmetry in the YZ axe.

(d) TE4: the solids have been discretized into 384 and 752 elements respectively. The thermal and the elastic systems of equations are made up by 1136 and 3408 equations respectively. There is imperfect contact between solids. There is neither conduction, nor convection in the interstitial zone. The problem presents symmetry in the YZ axe.

As it is shown in figures 1 and 2, comparing to the Gauss Elimination method with pivoting (the method used in the original code), the Gauss-Jordan method is clearly worse regarding the execution time. However, both Gauss Elimination without pivoting and LU Factorization methods reduce the execution time for all the problems tested. The first one shows an average reduction of a 27% in the execution times, and up to a 39% for some particular problems. The second one shows an average reduction of a 37% in the execution times, and up to a 41% for some particular problems. This results are obtained for any kind of problem,

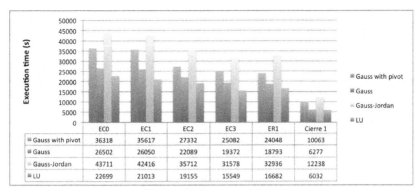

(a) Small elastic problems

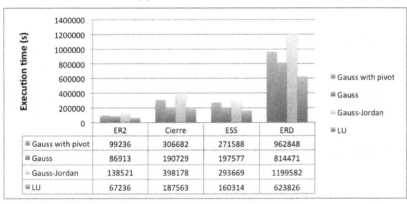

(b) Big elastic problems

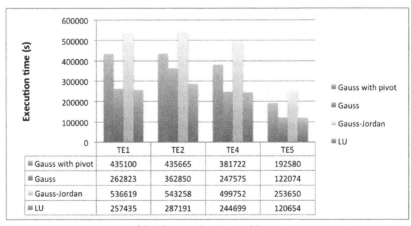

(c) Thermoelastic problems

Fig. 1. Execution times for a series of examples. We can observe how the LU factorization achieves the best reduction in every single example.

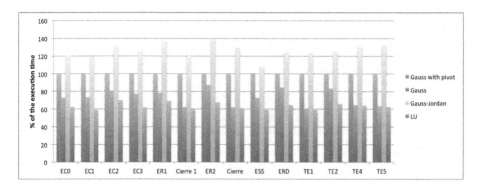

Fig. 2. Percentage of the execution times for a series of examples with respect to the Gauss with pivot method. Independently of the kind of contact problem (elastic or thermoelastic), LU factorization reduces the time execution.

no matter if it is an elastic or a thermoelastic one. It is also independent of the number of elements. Even for the different considered thermoelastic problems, with different kinds of contact, the execution time required for the resolution by using LU factorization is always lower.

It is important to say that the accuracy of all these methods for the problems solved is similar, so they can all be considered.

5   Conclusions

In this paper, we analyze how the method applied to solve a system of linear equations for contact problems that are solved by using BEM can affect the execution time. Elastic and thermoelastic contact problems involve the resolution of several systems of linear equations whose number of equations depends on the granularity of the discretization. Real problems require usually complex geometries that entail enormous systems of equations. By using the proper method to solve the systems, execution times can be drastically reduced. Comparing the performance for four methods (Gauss Elimination method with pivoting, Gauss-Jordan, Gauss Elimination without pivoting and LU factorization) we observe that LU factorization allows us to reduce a 37% the execution times (even a 41% for certain problems). Therefore, and taking into account that the accuracy of all the methods studied is similar, the recommended method among those presented in this paper for solving contact problems is this last one, since this results are obtained for elastic and thermoelastic contact problems, independently of the number of elements considered in the solid discretization.

Acknowledgements. This work has been partially supported by the research grants program from the University of León (Spain).

References

1. Sitio web de ABAQUS, http://www.simulia.com
2. Sitio web de ADINA, http://www.adina.com
3. Sitio web de ANSYS, http://www.ansys.com
4. Sitio web de BEASY, http://www.beasy.com/
5. Sitio web de BETL, http://www.sam.math.ethz.ch/betl/
6. Sitio web de Concept Analyst, http://www.conceptanalyst.com/
7. Sitio web de FEMM, http://femm.neil.williamsleesmill.me.uk
8. Sitio web de ParaFEM, http://www.parafem.org.uk/
9. Andersson, T., Fredriksson, B., Persson, B.G.A.: The boundary element method applied to two-dimensional contact problems. In: New Developments in Boundary Element Methods. Proceedings of the Second International Seminar on Recent Advances in Boundary Element Methods, Southampton, England, pp. 247–263 (1980)
10. Ataseven, Y., Akalin-Acar, Z., Acar, C.E., Gençer, N.G.: Parallel implementation of the accelerated bem approach for emsi of the human brain. Medical & Biological Engineering & Computing 46, 671–679 (2008)
11. Brebbia, C., Telles, J., Wrobel, L.: Boundary Element Techniques. Springer, Berlin (1984)
12. Chung, T.J.: Finite element analysis in fluid dynamics. McGraw Hill International Book Co. (1978)
13. Cunha, M.T.F., Telles, J.C.F., Coutinho, A.L.G.A.: A portable parallel implementation of a boundary element elastostatic code for shared and distributed memory systems. Advances in Engineering Software 35, 453–460 (2004)
14. Davies, A.: Parallel implementations of the boundary element method. Computers & Mathematics with Applications 31, 33–40 (1996)
15. Espinosa, J.V., Mediavilla, A.F.: Boundary element method applied to three dimensional thermoelastic contact. Engineering Analysis with Boundary Elements 36, 928–933 (2012)
16. Garrido, J.A., Foces, A., París, F.: Three dimensional frictional conforming contact using BEM. In: Springer-Verlag (ed.) Advances in Boundary Elements XIII, pp. 663–676. Computational Mechanics Publications (1991)
17. Geng, P., Oden, J.T., van de Geijn, R.A.: Massively parallel computation for acoustical scattering problems using boundary element methods. Journal of Sound and Vibration 191, 145–165 (1996)
18. Giannopoulos, G., Anifantis, N.: A BEM analysis for thermomechanical closure of interfacial cracks incorporating friction and thermal resistance. Computer Methods in Applied Mechanics and Engineering 196, 1018–1029 (2007)
19. Gueye, I.: Résolution de grands systèmes linéaires issus de la méthode des éléments finis sur des calculateurs massivement parallèles. Ph.D. thesis, École Nationale Supérieure des Mines de Paris (2009)
20. Jin, J.M.: The finite element method in electromagnetics. Wiley-IEEE Press (2002)
21. Keppas, L., Giannopoulos, G., Anifantis, N.: A BEM formulation to treat transient coupled thermoelastic contact problems incorporating thermal resistance. Computer Modeling in Engineering and Sciences 25, 181–196 (2008)
22. Li, J., Huang, Y., Yang, W.: Developing a time-domain finite-element method for modeling of electromagnetic cylindrical cloaks. Journal of Computational Physics 231, 2880–2891 (2012)

23. Li, S., Huang, Q.: A new fast multipole boundary element method for two dimensional acoustic problems. Computer Methods Appl. Mech. Engng. 200(9-12), 1333–1340 (2011)
24. Pantuso, D., Bathe, K.J., Bouzinov, P.A.: A finite element procedure for the analisis of termo-mechanical solids in contact. Computers and Structures 75(6), 551–573 (2000)
25. Rajalakshmi, K.: Parallel algorithm for solving large system of simultaneous linear equations. International Journal of Computer Science and Network Security 9(7), 276–279 (2009)
26. Vallepuga, J.: Análisis del problema de contacto termoelástico tridimensional sin fricción mediante el método de los elementos de contorno, aplicación en microelectrónica. Ph.D. thesis, Universidad de Valladolid (2010)
27. Wriggers, P.: Finite element algorithms for contact problems. Archives of Computational Methods in Engineering 2, 1–49 (1995)

Mean Arterial Pressure PID Control
Using a PSO-BOIDS Algorithm

Paulo B. de Moura Oliveira[1], Joana Durães[2], and Eduardo J. Solteiro Pires[1]

[1] INESC TEC – INESC Technology and Science (formerly INESC Porto, UTAD pole)
[2] Department of Engineering, School of Sciences and Technology,
5001–811 Vila Real, Portugal
oliveira@utad.pt

Abstract. A new hybrid between the particle swarm optimization (PSO) and Boids is presented to design PID controllers applied to the mean arterial pressure control problem. While both PSO and Boids have been extensively studied separately, their hybridization potential is far from fully explored. The PSO-Boids algorithm is proposed to perform both system identification and PID controller design. The advantage over a standard particle swarm optimization algorithm is the promotion of the diversity of the search procedure. Preliminary simulation results are presented.

1   Introduction

The control of mean arterial pressure (MAP) is a relevant problem in several applications such as hypertension control in the cardiac post-surgery healing process, in which the MAP has to be reduced as well as during the surgery anesthesia. The MAP lowering procedure is usually accomplished by intravenous infusion of a vasodilator such as Sodium Nitroprusside. Significant research efforts have been directed to tackle this control problem resulting in classical control approaches [1,2,3,4,7,8] and soft computing approaches such as fuzzy logic and neural networks [5,6,9]. Some of these techniques rely on the use of PID controllers, for which rather complex design methodologies are used. A swarm based optimization algorithm is proposed here as a simpler design technique for MAP PID control.

The particle swarm optimization (PSO) algorithm was originally proposed by [10] using as natural inspiration the collective behavior of bird flocks. The PSO has been successfully applied to a huge amount of search and optimization problems. One of the pioneering computational applications of a swarm inspired algorithm was developed by Reynolds [11], named Boids, to graphically simulate the coordinated movement of birds to generate computer-based animations. The boids model can be considered as a multi-agent system, as it will be reviewed in the sequel, relies on the use of three behavioral rules. Since its proposal the Boids model has been refined [15,17,18] and deployed in other applications such as data mining [12], robotics [14], flight simulation [13] and crowd behavior simulation [16]. While the PSO and Boids swarm based algorithms have been deployed separately in many applications the benefits of their hybridization has not been fully explored in terms of optimization. Indeed work in this subject is scarce [24]. This paper proposes a new algorithm based on both PSO and Boids for designing PID controllers within MAP control.

The remainder of the paper is structured as follows. Section 2 revisits Boids, presenting some introductory concepts regarding its original behavior rules. Section 3 presents the proposed PSO-BOIDS algorithm. In section 4 the mean pressure control problem is described. Section 5 presents some simulation experiments results obtained with the proposed method. Finally in section 6 some concluding remarks are discussed.

2   BOIDS: Introductory Concepts

The term Boid (bird-like object) was originally proposed by [11], to represent a flock element. Boids swarm behavior is the combined result of three rules usually referred as: cohesion, separation and alignment. Before these rules are described it is relevant to state that it is assumed in the sequel that the swarm elements position and velocity are represented by two vectors, **x** and **v**, respectively. The swarm position is updated for the *i-boid* using the following incremental equation:

$$x_i(t+1) = x_i(t) + v_i(t+1) \tag{1}$$

where: t represents the evolutionary iteration and $1 \le i \le m$, with m representing the swarm size. The velocity is the result of the combination of several components. The cohesion rule promotes the swarm centering behavior towards a perceived position centre. The entire swarm position centre, represented by x_{ctr}, can be evaluated using:

$$x_{ctr}(t) = \frac{1}{m} \sum_{j=1}^{m} x_j(t) \tag{2}$$

and the cohesion velocity component, **vc**, is governed by:

$$vc_i(t) = x_i(t) - x_{ctr}(t) \tag{3}$$

To prevent boids position abrupt changes towards the position centre, the difference expressed in (3) has to be multiplied by a factor, α_1, which regulates boids movement rate towards the perceived swarm position centre:

$$vc_i(t) = \alpha_1 [x_i(t) - x_{ctr}(t)] \tag{4}$$

The separation rule retains boids to collide and occupy the same space. This rule can be implemented using several approaches. The one adopted here consists in defining a sensor radius, S_{radius}, for each boid, and for all the other boids inside the corresponding circle, modify its velocity in order to promote their separation. The separation velocity component, **vs**, can be evaluated using:

$$vs_i(t) = \begin{cases} vs_i(t) - \alpha_2(x_i(t) - x_j(t)) \Leftarrow dist(i,j) < S_{radius} \\ vs_i(t) \Leftarrow dist(i,j) \geq S_{radius} \end{cases} \tag{5}$$

where: $dist(i,j)$ represents the Euclidian distance between boids i and j, and α_2 regulates the separation rate when boids are within a predefined exclusion radius. The alignment behavior rule promotes boids velocity alignment with neighbors. Considering a fully connected neighboring topology, the alignment velocity component, va, can be evaluated using:

$$va_i(t) = \alpha_3[v_{ctr(t)} - v_i(t)] \tag{6}$$

where α_3 defines the rate of alignment towards the swarm velocity center, v_{ctr}, evaluated with:

$$v_{ctr}(t) = \frac{1}{m-1} \sum_{(j=1)\wedge(j\neq i)}^{m} v_j(t) \tag{7}$$

The total velocity for each boid can be evaluated using:

$$v_i(t+1) = v_i(t) + \beta_1 vc_i(t) + \beta_2 vs_i(t) + \beta_3 va_i(t) \tag{8}$$

where: β_1, β_2 and β_3 represent weighting factors for each for the velocities rules components.

3   PSO-BOIDS Algorithm

The particle swarm optimization algorithm dynamics are governed by two fundamental equations, concerning the updating of swarm particles position represented by (1) and the velocity equation can be represented, for convenience of exposition, in two separated equations:

$$vpso_i(t) = c_1\varphi_c.(b_i(t) - x_i(t)) + c_2\varphi_s.(g(t) - x_i(t)) \tag{9}$$

$$v_i(t+1) = \omega v_i(t) + vpso_i(t) \tag{10}$$

In (9): b_i represents the best individual position achieved by particle i, g represents the best swarm global position achieved until iteration t. The model used considers the entire swarm as the neighborhood. In (9) c_1 and c_2 are known as the cognitive and social constants, respectively, and φ_c and φ_s represent randomly generated numbers with uniform distribution in the range [0,1]. In (10), parameter ω is referred as inertia weight. A high value assigned to the inertia weight (near one) favours exploitation of the search space, making the search global, while a small value, promotes imploitation making the search local. Thus, it is usual to linearly decrease this weight from a high initial value, ω_{nit} to a small value final value, ω_{fin}, throughout the search procedure.

Boids behavioral velocities rules can be merged with the PSO velocity equations resulting in an hybridization between both algorithms. The basic idea is to improve the particles diversity in the search and optimization procedure. The proposed PSO-Boids velocity equation is represented by:

$$v_i(t+1) = \omega v_i(t) + \alpha_1 vc_i(t) + \alpha_2 vs_i(t) + \alpha_3 va_i(t) + \alpha_4 vpso_i(t) \tag{11}$$

where α_4 represents a weighting factor for the PSO velocity component. As the information regarding the search is provided by a problem dependant objective function, the swarm is guided by the PSO velocity component. The Boids velocities components incorporate a dispersion component in the search which improves the swarm particle diversity and distribution in the search space. The proposed PSO-Boids algorithm is presented in Figure 1.

$t = 0$
initialize swarm $X(t)$ randomly
while(!(termination criterion))
  evaluate $X(t)$
  update b and g
  evaluate vc, vs, va // equations 2-6
  evaluate $vpso$ // equation 9
  update particles velocity and position // equations 10 and 1
  $t = t + 1$
end

Fig. 1. Proposed hybrid PSO-Boids algorithm

4   Mean Arterial Control: Problem Statement

Consider the general single-input single-output control structure illustrated in Figure 2, with a PID controller represented by transfer function Gc and the system to be controlled, represented here as a model of the MAP, Gp.

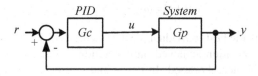

Fig. 2. PID control configuration

The model used in this study to represent the patient blood pressure was proposed by [19] as an alternative to a model proposed by [20] and has been used since in several studies [21]. This model is represented by [21]:

$$G_p(s) = \frac{Y(s)}{U(s)} = \frac{K(1+\tau_3 s)e^{-\theta s}}{[(1+\tau_3 s)(1+\tau_2 s) - \alpha](1+\tau_1 s)} \tag{12}$$

where: Y represents the drop in pressure P, due to the drug effect, U represents the rate of drug (SPN) infusion, K is the system gain, τ_1, τ_2 and τ_3 represent time constants associated respectively with the drug action, flow though pulmonary circulation and flow through a systematic circulation, θ is the system time delay and α is the fraction of recirculated drug. Considering the parameter values in (12) [21], as: K=2.5 mmHg/Cml/h; τ_1=50 s; τ_2=10 s; τ_3=30 s; L=60 s and α=0.5, results in the following model:

$$G_p(s) = \frac{Y(s)}{U(s)} = \frac{5(1+30s)e^{-60s}}{(1+130s+4600s^2+30000s^3)} \tag{13}$$

The overall objective is to design a PID controller to control the system represented by model (13) in order to achieve good time-domain step response transient characteristics. The PID controller design should also satisfy actuator clinical constraints, such as saturation limits and drug infusion rate of change.

5   PSO-BOIDS Design of MAP Control Systems

Some of the classical PID tuning methods, based on the process reaction curve, such as: Ziegler and Nichols, Cohen and Coon and the IAE [22], rely on a good approximation of the system to be controlled by a first order plus time delay model (FOPTD) as represented by:

$$G_m(s) = \frac{K}{1+sT} e^{-sL} \tag{14}$$

where: K represents the dc gain, T the dominant time constant and L the time delay. These model parameters can be estimated using a myriad of techniques. One technique which has proved well [21] is known as the two-points method proposed by [23]. This method is based on two points of the step response, corresponding to 35.3% (y_{35}) and 85.3% (y_{85}) of the final value, and respective corresponding times, (t_{35}) and (t_{85}). The time constant and time delay constants are evaluated using:

$$T = 0.67(t_{85} - t_{35}) \tag{15}$$

$$L = 1.3t_{35} - 0.29t_{85} \tag{16}$$

Expressions (15) and (16) can be used to determine good initial estimations for the time constant and time-delay, as well as to to define the parameter range to randomly generate a swarm population. This swarm is then evolved using the proposed PSO-Boids algorithm minimizing the integral of square error (ISE$_m$) between the FOPTD model output (y_{FOPTD}) and the system model (y_s), as represented by:

$$ISE_m = \int_0^{T_{sim}} e_m^2(t)\,dt, \quad e_m(t) = y_{FOPTD}(t) - y_s(t) \tag{17}$$

where: e_m represents the model error and T_{sim} represents the simulation time. The estimated FOPTD parameters can be used to obtain PID tuning parameters using the IAE method, determined by the following expressions [21,22]:

$$K_p = \frac{1.086}{K}\left(\frac{L}{T}\right)^{-0.869}; T_i = \frac{L}{0.74 - 0.13\left(\frac{L}{T}\right)}; T_d = 0.348 + T\left(\frac{L}{T}\right)^{0.914} \tag{18}$$

With the PID tuning gains determined by expressions (18) the search space can be appropriately defined to be searched by the proposed PSO-Boids. The minimization criterion used is the integral time square error (ITAE) defined by:

$$ITAE = \int_0^{T_{sim}} t\,|e(t)|\,dt, \quad e(t) = r(t) - y(t) \tag{19}$$

6   Simulation Results

The simulation experiment is organized in two parts, corresponding to an auto-tuning PID procedure: i) FOPTD model identification ii) PID tuning. Applying the two-points method described previously to the system described by (13) results in the following indentified FOPTD model (all simulations were performed using a fixed step of 0.1 s for the continuous time solver with $T_{sim}=900$ s):

$$G_{m1}(s) = \frac{5}{1+84s}e^{-79s} \tag{20}$$

With this model an $ISE_m=33.7$ is obtained indicating a good fit between model (20) and system (13). As there is no doubt about the dc gain value, the proposed PSO-Boids considers each particle encoding the time constant and time delay parameters $\{T, L\}$. The following PSO-Boids algorithm settings were used both for identification and controller tuning: $\alpha_1=0.01$, $\alpha_2=0.1$, $\alpha_3=0.125$, $\alpha_4=1$; $\beta_1=\beta_2=\beta_3=1$; Swarm size $m=30$, 100 iterations, and ω was linearly decreased between 0.9 and 0.4. The search space for the parameter identification was [75, 95] and [70, 90] for T and L respectively. The achieved optimized value was $ISE_m=15.2$ and the resulting model is represented by:

$$G_{m2}(s) = \frac{5}{1+88.9s}e^{-73.4s} \tag{21}$$

The overlap unit step responses obtained with the two-point method model, the PSO_Boids model and system (20) are presented in Figure 3a. As it can be observed in this figure both models present a good fit with the system model, however in terms of ISE_m the PSO-Boids is clearly the best one.

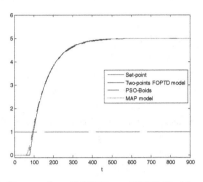
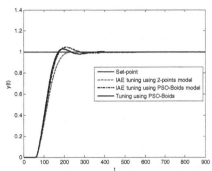

a) Two-points FOPTD and PSO-Boids
models mismatch with system (13).

b) Set-point responses IAE method
and PSO-Boids.

Fig. 3. Identification and tuning

Based on the parameter values obtained with the two-points model, Gm_1 (20), the following PID parameters were obtained using (18): $\{K_{p1}=0.22, T_{i1}=136, T_{d1}=27.6\}$, and based of the PSO-Boids identification model, Gm_2, the following parameters were obtained using (18): $\{K_{p2}=0.25, T_{i2}=141, T_{d2}=26\}$. The ITAE values obtained with both gains set are: $ITAE_1=81076$ and $ITAE_2=80057$. Using the PSO-Boids to minimize the ITAE for set-point tracking resulted in the following PID gains $\{K_{p3}=0.27, T_{i3}=148.1, T_{d3}=30.1\}$. The search interval used was [0.1 1] for K_p, [120 160] for T_i and [15 35] for T_d. The set-point tracking responses are presented in Figure 3b, and the control signals in Figure 4a. Figure 4b presents the evolution along 100 iterations for a swarm with 10 boids, in terms of the best ITAE, mean ITAE for the all swarm and individual boids ITAE. This figure clearly shows the large amplitude of variation in terms of the ITAE, corresponding to wide boids movement within the search space.

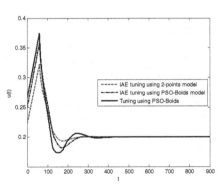
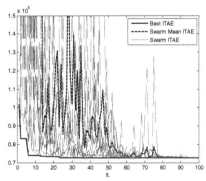

a) Control signal corresponding to set-
point responses presented in Fig 3 b).

b) Simulation for the PSO-Boids for
a swarm with $m=10$ elements.

Fig. 4. Two-points FOPTD and PSO-Boids models mismatch with system (13)

7   Conclusion

A new hybrid PSO-boids algorithm for optimization purposes was proposed. The proposed algorithm is a simple hybrid between the PSO and the original boids, using three behavior rules. The advantage of the proposed method is to promote a wider search space search. The proposed algorithm was deployed in the context of FOPTD model identification and PID controller tuning applied to the MAP control problem. Preliminary research results were presented showing the benefits of the proposed technique. Further research is necessary to consolidate the PSO-Boids algorithm, such as: i) the execution of more tests considering different swarm sizes, and weighting factors ii) the comparison with other similar PSO based methods.

References

1. Isaka, S.: Control Strategies for Arterial Blood Pressure Regulation. IEEE Trans. on Biomedical Eng. 40(4), 353–363 (1993)
2. Luginbühl, M., Bieniok, C., Leibundgut, D., Wymann, R., Gentilini, A., Schnider, T.W.: Closed-loop Control of Mean Arterial Blood Pressure during Surgery with Alfentanil. Anesthesiology 105, 462–470 (2006)
3. Yu, C., Roy, R.J., Kaufman, H., Bequette, B.W.: Multiple-Model Adaptive Predictive Control of Mean Arterial Pressure and Cardiac Output. IEEE Trans. on Biomedical Engineering 39(8), 765–777 (1992)
4. Zhu, K.Y., Zheng, H., Zhang, D.G.: A Computerized Drug Delivery Control System for Regulation of Blood Pressure. IC-MED 2(1), 1–13 (2008)
5. Furutani, E., Araki, A., Kan, S., Aung, T., Onodera, H., Imamura, M., Shirakami, G., Maetani, S.: An Automatic Control System of the Blood Pressure of Patients. International Journal of Control, Automation, and Systems 2(1), 39–53 (2004)
6. Kumar, M.L., Harikumar, R., Vasan, A.K., Sudhaman, V.K.: Fuzzy Controller for Automatic Drug Infusion in Cardiac Patients. In: Proc. of the International MultiConference of Engineers and Computer Scientists (IMECS), Hong Kong, March 18-20, vol. I (2009)
7. Ringwood, J.V., Malpas, S.C.: Slow oscillations in blood pressure via a nonlinear feedback model. American J. Physiol. Regulatory Integrative Comp. Physiology 280(4), R1105–R1115 (2001)
8. Hoeksel, S.A.A.P., Blom, J.A., Jansen, J.R.C., Maessen, J.G., Schreuder, J.J.: Computer control versus manual control of systemic hypertension during cardiac surgery. Acta Anaesthesiologica Scandinavica 45, 553–557 (2001)
9. Gao, Y., Er, M.J.: Adaptive Fuzzy Neural Control of Mean Arterial Pressure Through Sodium Nitroprusside Infusion. In: Proc. of the IEEE 42nd Conf. on Decision and Control, Hawai, vol. 3, pp. 2198–2203 (2003)
10. Kennedy, J., Eberhart, R.C.: Particle swarm optimization. In: Proc. IEEE Int'l. Conf. on Neural Networks IV, vol. 4, pp. 1942–1948. IEEE Service Center, Piscataway (1995)
11. Reynolds, C.W.: Flocks, herds and schools: A distributed behavioral model. ACM SIGGRAPH Computer Graphics 21(4), 25–34 (1987)
12. David, M.F., Castro, L.N.: A New Clustering Boids Algorithm for Data Mining. In: Proc. of the 2009 AI*IA Workshop on Complexity, Evolution and Emergent Intelligence, vol. 1, pp. 1–10, Published on CD (2009) ISBN: 978-88-903581-1-1

13. Clark, J.B., Jacques, D.R.: Flight test results for UAVs using boid guidance algorithms. Procedia Computer Science 8, 226–232 (2012)
14. Berman, S., Halász, Á.M., Kumar, V., Pratt, S.: Algorithms for the Analysis and Synthesis of a Bio-Inspired Swarm Robotic System. In: Şahin, E., Spears, W.M., Winfield, A.F.T. (eds.) SAB 2006 Ws 2007. LNCS, vol. 4433, pp. 56–70. Springer, Heidelberg (2007)
15. Reynolds, C.W.: Steering Behaviors for Autonomous Characters. In: Proc. of Game Developers Conference 1999, San Jose, California, pp. 763–782. Miller Freeman Game Group, San Francisco (1999)
16. Zaharia, M.H., Leon, F., Pal, C., Pagu, G.: Agent-Based Simulation of Crowd Evacuation Behavior. In: Proceedings of the 11th WSEAS International Conference on Automatic Control, Modelling and Simulation, pp. 529–533 (2011)
17. Chiang, C.-S., Hoffmann, C., Mittal, S.: Emergent Crowd Behavior. In: Computer-Aided Design and Applications, p. 11. CAD Solutions, LLC (2009)
18. Bajec, I.L., Zimic, N., Mraz, M.: The computational beauty of flocking: boids revisited. Mathematical and Computer Modelling of Dynamical Systems 13(4), 331–347 (2007)
19. Martin, J.F., Schneider, A.M., Smith, N.T.: Multiple-model adaptive control of blood pressure using sodium nitroprusside. IEEE Trans. Biomed. Eng. BME 34(8), 603–611 (1987)
20. Slate, J.B.: Model-based design of a controller for infusing sodium nitroprusside during postsurgical hypertension. Ph.D. dissertation, University of Wisconsin (1978)
21. Jones, R.W., Tham, M.T.: An Undergraduate CACSD Project: the Con trol of Mean Arterial Blood Pressure during Surgery. Int. J. Eng. Ed. 21(6), 1043–1049 (2005)
22. Murrill, P.W.: Automatic control of processes. International Textbook Co. (1967)
23. Sundaresan, K.R., Krishnaswamy, P.R.: Estimation of time delay, time constant parameters intime, frequency and Laplace domains. Can. J. Chem. Eng. 56, 257 (1978)
24. Cui, Z., Shi, Z.: Boid particle swarm optimisation. Int. J. Innov. Computers Applicattions 2(2), 77–85 (2009)

Empirical Study of the Sensitivity of CACLA to Sub-optimal Parameter Setting in Learning Feedback Controllers

Borja Fernandez-Gauna, Igor Ansoategui,
Ismael Etxeberria-Agiriano, and Manuel Graña

University of the Basque Country (UPV/EHU)
manuel.grana@ehu.es

Abstract. Continuous Action-Critic Learning Automaton (CACLA) offers an interesting alternative to traditional control approaches to feedback control problems. In this paper, we report results obtained on an inertial model of a feed drive with potentially sub-optimal parameter setting and designer decisions. Namely, we have tested different reward signals, different number of features to approximate value functions and policies, and different learning gains. The results show CACLA to be a very highly robust approach.

1   Introduction

Reinforcement Learning (RL) methods have been proposed [1,2,3,4] to approach feedback control problems, as an alternative to traditional control theory. The main advantage they offer is that they are able to learn from interaction with the environment requiring less input from the system designer that traditional approaches [5,6]. In principle RL methods are able to learn from experience, but there are some design aspects which must be carefully taken into account in order to obtain an accurate control policy. The most important decision is the definition of the problem itself: the reward signal to be maximized, the number of features used to approximate value functions, and the learning gain. Previous work [7] shows CACLA to be the best suited Actor-critic method for feed drive control learning. In this paper, we study the impact of the designer's decisions by comparing results obtained in computational simulations with different parameter settings.

The paper is structured as follows. Section 2 introduces the ball screw model used in the experiments. Section 3 reviews some necessary Actor-critic background. The experimental set-up is detailed in Section 4, were we also present the results. Finally, we offer our conclusions in Section 5.

2   The Ball Screw Feed Drive Model

We have used a model of a commercial *Ideko* ball screw feed drive, manufactured by *Fagor*. The components of the specific ball-screw feed drive are depicted in

Á. Herrero et al. (eds.), *International Joint Conference SOCO'13-CISIS'13-ICEUTE'13*,   101
Advances in Intelligent Systems and Computing 239,
DOI: 10.1007/978-3-319-01854-6_11, © Springer International Publishing Switzerland 2014

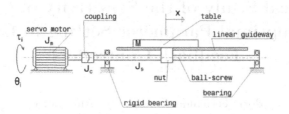

Fig. 1. Mechanical components of a ball screw feed drive

Table 1. *Ideko* ball screw feed drive model parameter settings

J_c	$170.5\ kg \cdot m^2$
J_s	$6.25\ kg \cdot m^2$
J_m	$85\ kg \cdot m^2$
p	$0.01\ m/rev$
M	$68.25\ kg$

Figure 1. To allow for extensive RL simulations, we have used a simplified inertial model:

$$\ddot{x} = \frac{\tau}{M \cdot \frac{p}{2 \cdot \Pi} + (J_c + J_s + J_m)\left(\frac{2\pi}{P}\right)}, \tag{1}$$

where τ is the torque of the motor, J_c, J_s and J_m are the coupling, ball-screw and motor inertia, p is the screw pitch, M is the mass of the table and the nut, and x is the lineal position of the supporting table. The parameter settings used in our simulations are shown in Table 1. The length of the guideways was $1\,m$ and the initial position of the table was $x\,(0) = 0.5\,m$. The setpoint $w\,(t)$ was changed to a random value in range $[0.1, 0.9]$ every T seconds, where T was randomly chosen from a range of $[0.75, 1.5]$ s. The goal of the feed drive controller is to minimize the error $e_x\,(t) = |x\,(t) - w\,(t)|$. The controller receives the position of the table $x\,(t)$ and its speed $v\,(t)$ as input, and outputs τ.

3 Actor-Critic Background

Reinforcement Learning. RL environments are Markov Decision Processes (MDP), which are defined by a tuple $\langle S, A, P, R \rangle$. In continuous domains, such as feed-back control problems, S is an n-dimensional Euclidean space $S \subseteq \mathbb{R}^n$, and A an m-dimensional Euclidean action space $A \subseteq \mathbb{R}^m$ [8]. P is the transition function $P : \ S \times A \times S \to [0, 1]$, and R is the reward function $R : \ S \to \mathbb{R}$. The goal of the learning agent is to learn a policy $\pi\,(s)$ that maximizes the expected accumulated rewards $R^\pi\,(s)$ [9].

Actor-critic architectures use two different memory structures: the *actor* which selects actions every time-step and the *critic* which assesses the quality of the policy being followed by the actor. The actor learns a policy $\pi\,(s)$ while the critic learns the state value function $V^\pi(s) = E^\pi \left\{ \sum_{k=1}^{\infty} r_{t+k} \gamma^{k-1} | s_t = s \right\}$, where s_t and r_t represent, respectively, the state and reward observed at time-step t, and γ is the discount factor.

Value Function Approximation. The size of the state space $|S|$ grows exponentially rendering tabular representation of the policy and the value function unfeasible in real-life applications. Value Function Approximators (VFA) deal with the *curse of dimensionality* using a compact approximation of the value function [10]. In this work, we have used Gaussian networks to represent both the policy and the state value function, with n_f features per state variable. The activation functions $\phi_{i,j}(x_i)$, $j = 1, \ldots, n_f$ are defined:

$$\phi_{i,j}(x_i) = \exp^{-\frac{\|x_i - c_{i,j}\|^2}{2\sigma^2}}, \tag{2}$$

where $c_{i,j}$ is the center of the activation function, x_i the value of the ith state variable, and σ^2 the shape parameter. Both the actor and the critic learn a parameter vector $\theta_t \in \Theta$. For clarity, we will denote θ_t^a and θ_t^V the parameter vector at time t being learnt by the actor and the critic, respectively.

CACLA. The Continuous Action-Critic Learning Automaton actor (CACLA) [8] algorithm updates its policy if the criticism is positive, that is, only after the policy has been improved:

$$if\ \delta_t > 0: \qquad \theta_t^a(s) \leftarrow \theta_t^a(s) + \alpha_t \cdot (a_t - \pi_a(s)) \cdot \frac{\partial \pi_a(s_{t-1})}{\partial \theta_{t-1}^\pi}, \tag{3}$$

where $\pi_a(s) = (\theta_t^a)^T \phi(s)$ represents the actual output of the policy in state s, and a_t is the action executed at time-step t. They need not be equal, because a random perturbation term $\eta(t)$ is added to the output of the policy in order to explore yet unknown policies: $a_t = \pi_a(s) + \eta(t)$. This algorithm is considered as the current state-of-the-art.

In this paper, the critic is implemented using $TD(\lambda)$ algorithm [9], which, in its most basic form $TD(0)$, updates its parameter estimates $\hat{V}_t(s) \equiv (\theta_t^V)^T \phi(s)$ using the following update-rule:

$$\theta_t^V \leftarrow \theta_{t-1}^V + \alpha_t \left(r_t + \gamma * \hat{V}_t(s_t) - \hat{V}_t(s_{t-1}) \right) \cdot \frac{\partial \hat{V}_{t-1}(s_{t-1}, a_{t-1})}{\partial \theta_{t-1}^V}, \tag{4}$$

where α_t is the learning gain.

4   Computational Experiments

We have conducted three experiments, each intended to assess the sensitivity of CACLA to a different component of the MDP, namely:

- Experiment A: the reward signal
- Experiment B: the number of features used to approximate the value function and policy
- Experiment C: the learning gain α

Common Settings. For the control of ball screw drives, RL controllers perceive the system state as a set of three variables: $x(t) \in [0.0, 1.0]$ m, $v(t) \in [-2.0, 2.0]$ $\frac{m}{s}$ and $w(t) \in [0.1, 0.9]$ m. In Experiments A and C, the domain of each state variable was approximated using 50 Gaussian Radial Basis Functions (RBF) *per* state variable. They were set so that only 3 features *per* variable were active each time.

Performance Measurement. The learning process consisted of $n_e = 5,000$ episodes of $T = 20\,s$. The performance of the controllers was measured using greedily evaluating the policies every 20 episodes [11]. Exploration was handled using a Sigmoid policy with a Gaussian disturbance signal $N(0.0, \sigma^2)$ added to the policy's output. Initially $\sigma_0^2 = 0.2$, and every episode it was decreased by $\Delta\sigma^2 = -\sigma_0^2/n_e$. The learning gain was set $\alpha = 0.05$. These values were determined empirically after running a few preliminary tests. The performance was measured using the *average absolute off-set error* of the controlled variable $x(t)$ from the setpoint $w(t)$:

$$e_T(t) = \frac{1}{T} \sum_{t=0}^{T} e_x(t).$$

All results have been averaged after three runs with different random seeds.

4.1   Experiment A: Sensitivity to Reward Function

In this first experiment, we test two different reward signals:

- $R_1^{\mu}(s) = k_1 \cdot \left[1 - \tanh^2\left(\frac{|e_x(s)|}{\mu} \right) \right] - c(s)$. (Fig. 2)
- $R_2^{\rho}(s) = k_1 \cdot \left[1 - \frac{|e_x(s)|}{\rho} \right] - c(s)$. (Fig. 3)

The first reward signal $R_1^{\mu}(s)$ is very similar to that proposed in [1] and uses a hyperbolic tangential function centered at $e_x(s) = 0$. On the other hand, the second reward signal $R_2^{\rho}(s)$ uses the absolute value of the error. In both cases, we subtract a second term $c(t)$ in order to penalize collisions with the end of the guideway. This second term is proportional to the force exerted on the border table (F_c):

$$c(t) = k_2 \cdot |F_c|. \tag{5}$$

The parameters μ and ρ set the reward function width. Several different values have been tested in order to study their influence on the learned controller. Constants k_1 and k_2 are both set to 1.000.

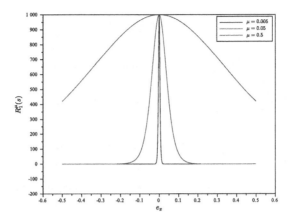

Fig. 2. Plot of the reward signal $R_1^\mu(s)$

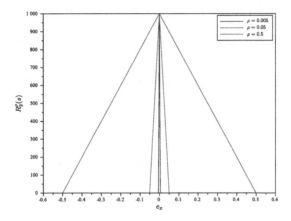

Fig. 3. Plot of the reward signal $R_2^\rho(s)$

In Fig. 4 we plot the average offset error measured obtained during the evaluation episodes with six different reward signals: $R_1^{0.005}$, $R_1^{0.05}$, $R_1^{0.5}$, $R_2^{0.005}$, $R_2^{0.05}$ and $R_2^{0.5}$. Regarding accuracy, the best results are obtained with $R_1^{0.005}$, and, after 1,000 episodes the controller is able to consistently provide an average error under $0.05m$. The use of R_1^μ with a wider bell does not improve the time needed to learn and, in fact, $R_1^{0.5}$ obtains the worst results. However, using $R_2^{0.5}$ the agent is able to learn faster than with $\rho = \{0.005, 0.05\}$ and at the end of the learning, the results are rather similar. Our results encourage the use of bell-shaped reward functions tightly delimiting the tolerance range.

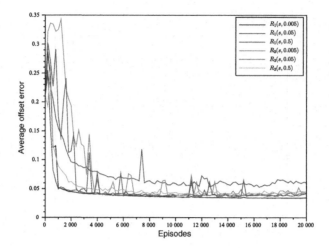

Fig. 4. Experiment A: average offset error of the controller during the learning process with different reward signals.

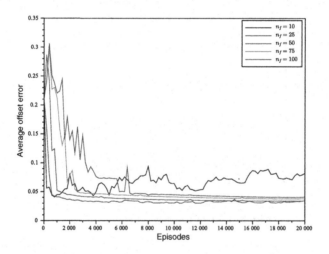

Fig. 5. Experiment B: average offset error of the controller during the learning process for different values of n_f

4.2 Experiment B: Sensitivity to Number of Features

Using $R_1^{0.005}$ as the reward function, we next test different numbers of features to represent the value function and the policy: $n_f = \{\, 10,\ 25, 50\ , 75, 100\,\}$. The number of features has direct impact on the number of times each value in vector θ is updated, the less features, the more updates. The results plotted in Fig. 5 confirm that, with less features, the learning rate is faster at earlier states of the process. With $n_f = 10$, the average offset error falls bellow $0.06m$ faster than with any other value, but as the learning process goes on, there is a damped divergence of the results. In general, the best results are obtained with $n_f = \{25, 50\}$.

4.3 Experiment C: Sensitivity to Learning Gain

Using $R_1^{0.005}$ and $n_f = 50$, we have tested different learning gains $\alpha = \{0.005, 0.025, 0.05, 0.075, 0.1\}$. Results are plotted in Fig. 6. Lower values of α result in a higher average offset error. From values 0.05 to 0.1, though, the difference is negligible. At the end of the learning process, the errors are within the range $[0.041, 0.067]$.

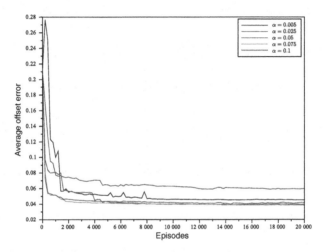

Fig. 6. Experiment C: average offset error of the controller during the learning process for learning gains

5   Conclusions

Actor-critic methods offer an alternative to traditional control approaches, such as standard PID controllers, and they are expected to require less expertise and effort from the designer than the latter in order to produce an acceptable controller in feedback control problems. In this paper, we have conducted several experiments on a feed drive computational model so as to assess the sensitivity of the state-of-the-art CACLA algorithm to sub-optimal learning parameter settings: the reward signal, the number of features used to represent both the value functions and the policy, and the learning gain. Results show that, except from some extreme parameter settings, the majority of configurations produce similar results. Future works may be addressed to compare the usability of CACLA and traditional approaches such as double-loop PID controllers, carrying more extensive experiments with groups of users with different profiles.

The use of RL methods to learn feedback controllers is a rather novel application and, because the literature lacks extensive empirical experiments, we expect more papers to be published before they become a real alternative to traditional control methods.

Acknowledgements. We acknowledge support from MICINN through project TIN2011-23823, and EU SandS project under grant agreement 317947, as well as UIF 11/07 of UPV/EHU, and University Research Group funding from the Basque Government.

References

1. Hafner, R., Riedmiller, M.: Reinforcement learning in feedback control: Challenges and benchmarks from technical process control. Machine Learning 84(1-2), 137–169 (2011)
2. Neumann, G.: The Reinforcement Learning Toolbox, Reinforcement Learning for Optimal Control Tasks. PhD thesis, Technischen Universitaet, Graz (2005)
3. Lewis, F.L., Vrabie, D., Vamvoudakis, K.G.: Reinforcement learning and feedback control. IEEE Control Systems Magazine 9, 32–50 (2012)
4. Lewis, F.L., Liu, D., et al.: Reinforcement Learning and Approximate Dynamic Programming for Feedback Control. Wiley (2013)
5. Koren, Y., Lo, C.C.: Advanced controllers for feed drives. Annals of the CIRP 41 (1992)
6. Srinivasan, K., Tsao, T.C.: Machine feed drives and their control - a survey of the state of the art. Journal of Manufacturing Science and Engineering 119, 743–748 (1997)
7. Fernandez-Gauna, B., Ansoategui, I., Etxeberria-Agiriano, I., Graña, M.: Empirical study of actor-critic methods for feedback controllers. In: Proceedings from IWINAC (2013)
8. van Hasselt, H.: Reinforcement Learning in Continuous State and Action Spaces. In: Wiering, M., van Otterlo, M. (eds.) Reinforcement Learning. ALO, vol. 12, pp. 207–251. Springer, Heidelberg (2012)
9. Sutton, R., Barto, A.: Reinforcement Learning: An Introduction. MIT Press (1998)
10. Busoniu, L., Babuska, R., De Schutter, B., Ernst, D.: Reinforcement Learning and Dynamic Programming using Function Approximation. CRC Press (2010)
11. Vamplew, P., Dazeley, R., Berry, A., Issabekov, R., Dekker, E.: Empirical evaluation methods for multiobjective reinforcement learning algorithms. Machine Learning 1, 1–30 (2011)

Optimization Design in Wind Farm Distribution Network*

Adelaide Cerveira[1,3], José Baptista[2,3], and Eduardo J. Solteiro Pires[2,3]

[1] CIO – Centro de Investigação Operacional
[2] INESC–TEC Technology and Science (formerly INESC Porto, UTAD pole)
[3] Escola de Ciências e Tecnologia, Universidade de Trás-os-Montes e Alto Douro
5000–811 Vila Real, Portugal
{cerveira,baptista,epires}@utad.pt

Abstract. Nowadays, wind energy has an important role in the challenges of clean energy supply. It is the fastest growing energy source with a increasing annual rate of 20%. This scenario motivate the development of an optimization design tool to find optimal layout for wind farms. This paper proposes a mathematical model to find the best electrical interconnection configuration of the wind farm turbines and the substation. The goal is to minimize the installation costs, that include cable cost and cable installation costs, considering technical constraints. This problem corresponds to a capacitated minimum spanning tree with additional constraints. The methodology proposed is applied in a real case study and the results are compared with the ground solution.

Keywords: Distribution networks, wind farm, optimization, capacitated minimum spanning trees, hop-indexed formulations.

1 Introduction

Most part of wind farms are constituted by a significant wind turbines number, interconnected by wired networks. Planning and design the internal distribution system configuration and finding the best location for the substation has been made, until recently, using traditional engineering solutions, with little concern for their optimization. In order to find the best solution to this problem is necessary to conduct detailed technical and economic studies, where the main objectives are the network costs and energy losses reduction. Network configuration for loss reduction is a highly complex combinatorial, non-differentiable and constrained optimization problem. This is due to the high elements number in the distribution network, and due to the non-linear constraints characteristics used to model the electrical behavior of the system [1].

Several categories of network reconfiguration techniques for loss reduction are identified in [2]. Due to the non-linear characteristics of the network behavior

---

* This work is supported by National Funding from FCT - Fundação para a Ciência e a Tecnologia, under the project: PEst-OE/MAT/UI0152.

Á. Herrero et al. (eds.), *International Joint Conference SOCO'13-CISIS'13-ICEUTE'13*, 109
Advances in Intelligent Systems and Computing 239,
DOI: 10.1007/978-3-319-01854-6_12, © Springer International Publishing Switzerland 2014

the use of heuristic techniques are often used to solve the problem addressed, as is the case of [3–6]. These models use rules based on empirical knowledge, that help the algorithm development to obtain satisfactory solutions.

Other authors, [7, 8] solve the problem using neural networks. In 1992, Nara [9] applied GA to the network reconfiguration problem. This work uses a binary codification to represent the arc branch number that identify the switch that is normally open. Other works based on evolutionary computing using genetic algorithms [1, 10–15], particle swarm optimization [16], ant colony optimization [17] and simulated annealing [18]. Evolutionary computing are efficient optimization techniques to tackle the problem of global wind farms optimization by considering the turbines layout and electrical network configuration as a whole.

Several methods have been employed on the study of substation optimization planning such as the classical optimization branch-and-bound [19], and other based on Voronoi diagrams [20].

Despite heuristic algorithms obtain good results does not guarantee that they will find the global optimum. Therefore, the paper proposes a mathematical formulation to obtain the electrical network connections in order to minimize the installation cost, considering several cables type with different cross sections. Some operating constraints described in [1] were taken into account. Radial network constraint, node voltage constraint, branch current stability constraint and Kirchoff's current and voltage laws are examples of such constraints. A cardinality constraint was included to bound the maximum number of turbines linked in a branch-line.

The paper is divided in 5 sections. Section 2 presents the case study, emphasizing some concepts related with the transmission lines, relationships which allow to calculate the voltage drop in the distribution network branches. The mathematical formulation of the problem is presented in Section 3 and the computational results are discussed in section 4. Finally, Section 5 draws the main conclusions.

2   Wind Farm Case Study

In this section a case study is presented where the location of the substation and wind turbines are previously known. The algorithm finds out the optimal configuration of the distribution network. Additionally, the obtained solution is compared with the field implementation.

2.1   Branch Current Stability Constraint

One major constraint taken into account is the maximum number of wind turbines that can be connected in parallel on the same line-branch. Therefore, the core cables type (copper or aluminum) and the largest section used must be taken into account. The current capacity that the cable can safely carry continuously (current rating of the cable: table 1) must be satisfied. Moreover, the maximum permissible voltage drop in entire path (e.g. 5%) must be satisfied.

Table 1. Characteristics of unipolar cables (LXHIOV) 12/20kV

Cross section (mm^2)	Max. Current, I_z (A)	Price (€/m)
25	122	4.50
35	144	5.30
50	170	6.80
70	209	7.12
95	249	7.98
120	283	8.70
150	316	12.77
185	357	13.23
240	413	14.89
300	463	17.50
400	526	21.09
500	592	23.77

The rated power of all wind turbines is $P_r = 2$MW and are interconnected by a $U = 20$kV grid. Usually, most of wind turbines, allows to control power factor between $0.95(c)$ to $0.95(i)$, over the entire range of power. Therefore, the power electronics is an alternative to capacitors banks, since it allows to regulate the power factor up to a certain level.

It was considered networks with unity power factor, $\tan(\varphi) = 0$, in order to check the amount of reactive power consumed, since it corresponds to the rated current drawn by each turbine that is given by equation (1).

$$I_r = \frac{P_r}{\sqrt{3} \cdot U \cdot \cos\varphi} = \frac{2 \cdot 10^6}{\sqrt{3} \cdot 20 \cdot 10^3 \cdot 1} = 57.735\text{A} \qquad (1)$$

The largest size cables used are Aluminum cables (LXHIOV) with a cross section of 500mm^2 and have a maximum permissible current in the steady state about $I_z = 592$A. This means that, under these conditions, a path can connect up to $Q = 10$ wind turbines, which corresponds to $I_{branch} = 577.35$A. Always considering the criteria defined above, where the larger currents must travel by the shortest path (see Fig. 1).

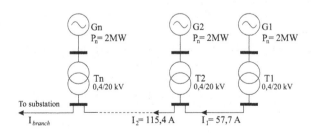

Fig. 1. Current branch drawn by wind turbines

2.2   Cost Function

The mathematical model to optimize the electrical network configuration of a wind farm aims to minimize the infrastructure cost wich takes into account the investment cost for building the network that includes the cables and their installation. The function used to evaluate the installation cost is given by (2), where L_k is the length of branch k, C_k is the total installation cost, which includes the cable and ditch cost, and n is the electrical connections number.

$$Z = \sum_{k=1}^{n} C_k L_k \qquad (2)$$

3   Mathematical Formulation

The location of turbines and substation are previously known. Let $\mathcal{N} = \{1, \ldots, n\}$ be the set of nodes corresponding to the turbines and 0 be the substation node. In order to obtain the cheapest electrical configuration, an undirected complete graph $G = (\mathcal{N}_0, \mathcal{E})$ is considered, where $\mathcal{N}_0 = \{0\} \cup \mathcal{N}$ is the set of nodes and $\mathcal{E} = \{\{i, j\}, i, j \in \mathcal{N}_0, i \neq j\}$ is the edge set, *i.e.*, the electrical branches set between the turbines and the substation. Associated with each edge $e = \{i, j\} \in \mathcal{E}$ there is a nonnegative cost c_e which depend on its length and on the number of nodes in subtree pending from this link. The goal is to find the minimum spanning tree *i.e.* the spanning tree $T = (\mathcal{N}_0, \mathcal{E}_T)$ in G, with $\mathcal{E}_T \subseteq \mathcal{E}$, of least cost $C(T) = \sum_{e \in \mathcal{E}_T} c_e$, considering an additional constraint stating that the number of nodes in every proper subtree is not greater than a given number, Q. Thus, the number of descendants of each node in $\mathcal{N}$, including itself, is at most Q and the substation, node 0, is the root node of the tree. This problem refer to the Capacitated Minimum Spanning Tree Problem (CMSTP).

It is well-known that in network design problems better formulations can be obtained by considering directed graphs (see, for instance, [21]). Therefore, it is considered the corresponding directed graph, $G' = (\mathcal{N}_0, \mathcal{A})$ with arc set $\mathcal{A} = \{(i, j), i, j \in \mathcal{N}_0, i \neq j\}$, where each edge $e = \{i, j\} \in \mathcal{E}$ is replaced with two arcs, arc (i, j) and arc (j, i) with the same cost as the original edge. It is also assumed that, in any feasible solution, the arcs are directed outward from the root and any edge $\{0, i\} \in \mathcal{E}$ is replaced by only one single arc $(0, i) \in \mathcal{A}$.

In this paper, it is proposed a mathematical formulation for the problem which is based on the Hop Index Single-flow Formulation, HSCF, for the CMSTP studied in [22].The root node, 0, serves as a source node that must send one unit of flow to every other node. In the hop-formulation models it is considered an 'hop' index. An arc (i, j) is in position t if the path between the root and node j has exactly t hops.

In order to formalize the mathematical model consider the additional notation.

Let n_c be the number of different cables sizes; $c_c \in \mathbb{R}^{n_c}$ be the array with cables price ($\text{\euro}/m$); $I_z \in \mathbb{R}^{n_c}$ the array with maximum current capacity allowed for each cable type (A); ℓ_{ij} the distance between nodes i and j, with i, $j \in \mathcal{N}_0$, (m).

The following decision variables are considered:

1. For each $t = 1, \ldots, Q$, $i \in \mathcal{N}_0$, $j \in \mathcal{N}$ with $i \neq j$, let:

   - $x_{ij}^t = \begin{cases} 1 & \text{if arc } (i,j) \text{ is in the optimal solution in position } t \\ 0 & \text{otherwise} \end{cases}$.

   - $y_{ij}^t \in \mathbb{N}$ the amount of flow on arc (i,j) if this arc is in position t.

   - $I_{b_{jit}} \in \mathbb{R}$ the current intensity through arc (j,i) in the direction of the substation, node 0, if this arc is in position t.

2. For each $i \in \mathcal{N}_0$, $j \in \mathcal{N}$, with $i \neq j$, and $k = 1, \ldots, n_c$, let:

   - $a_{ji}^k \in \{0,1\}$ the scalars in the definition of the cable price of arc (j,i).

By the definition of flow variables, if arc (i,j) is in the optimal solution in position t, then the number of nodes in the subtree pending from node $j \in \mathcal{N}$ is y_{ij}^t. For each arc (i,j) in the solution at position t ($x_{ij}^t = 1$), with $i \in \mathcal{N}_0$ and $j \in \mathcal{N}$, the current intensity through the cable from j to i in direction to the root node (substation) is given by:

$$I_{b_{jit}} = \sum_{k \in \mathcal{N}_0} y_{kj}^t \cdot I. \tag{3}$$

The value of current intensity in a arc will select the used cable type. If the current intensity in a arc, $I_{b_{jit}}$, has $e.g.$ a value between 209 and 249A then the cable with cross-section of 95mm^2 or greater should be chosen to hold the current intensity (see Table 1). Currents, $I_{b_{ji}}$, greater than 592A are not allowed thereby limiting by $Q = 10$ the turbines number linked together in a line-branch. This is assured by the following conditions:

$$I_{b_{ji}} \leq a_{ji}^1 \cdot I_z(1) + \sum_{k=1}^{n_c-1} (I_z(k+1) - I_z(k)) \cdot a_{ji}^{k+1}, \quad i \in \mathcal{N}_0, j \in \mathcal{N} \tag{4}$$

and

$$\sum_{k=1}^{n_c} a_{ji}^k \leq n_c \cdot \sum_{t=1}^{Q} x_{ij}^t, \quad a_{ji}^{k+1} \leq a_{ji}^k, \quad i \in \mathcal{N}_0, j \in \mathcal{N}, k = 1, \ldots, n_c - 1 \tag{5}$$

The problem can be formulated as follows.

$$\min \sum_{i \in \mathcal{N}_0} \sum_{j \in \mathcal{N}} 20\ell_{ij} \cdot x_{ij} + 3\ell_{ij} \cdot \left(a_{ji}^1 \cdot c_c(1) + \sum_{k=1}^{n_c-1} (c_c(k+1) - c_c(k)) \cdot a_{ji}^{k+1} \right) \tag{6}$$

subject to

$$I_{b_{jit}} \leq I_z(n_c) \cdot x_{ij}^t \qquad\qquad i \in \mathcal{N}_0, j \in \mathcal{N}, t = 1, \ldots, Q \quad (7)$$

$$I_{b_{jit}} \geq \sum_{k \in \mathcal{N}_0} y_{kj}^t \cdot I - I_z(n_c)(1 - x_{ij}^t) \qquad i \in \mathcal{N}_0, j \in \mathcal{N}, t = 1, \ldots, Q \quad (8)$$

$$I_{b_{jit}} \leq \sum_{k \in \mathcal{N}_0} y_{kj}^t \cdot I + I_z(n_c)(1 - x_{ij}^t) \qquad i \in \mathcal{N}_0, j \in \mathcal{N}, t = 1, \ldots, Q \quad (9)$$

$$I_{b_{jit}} \leq a_{ji}^1 \cdot I_z(1) + \sum_{k=1}^{n_c-1} (I_z(k+1) - I_z(k)) \cdot a_{ji}^{k+1} \quad i \in \mathcal{N}_0, j \in \mathcal{N}, t = 1, \ldots, Q$$

$$(10)$$

$$a_{ji}^{k+1} \leq a_{ji}^k \qquad\qquad i \in \mathcal{N}_0, j \in \mathcal{N}, k = 1, \ldots, n_c - 1$$

$$(11)$$

$$\sum_{k=1}^{n_c} a_{ji}^k \leq n_c \cdot \sum_{t=1}^{Q} x_{ij}^t \qquad\qquad i \in \mathcal{N}_0, j \in \mathcal{N} \quad (12)$$

$$\sum_{t=1}^{Q} \sum_{i \in \mathcal{N}_0} x_{ij}^t = 1 \qquad\qquad j \in \mathcal{N} \quad (13)$$

$$\sum_{i \in \mathcal{N}_0} y_{ij}^t - \sum_{i \in \mathcal{N}} y_{ji}^{t+1} = \sum_{i \in \mathcal{N}_0} x_{ij}^t \qquad\qquad j \in \mathcal{N}, t = 1, \ldots, Q - 2 \quad (14)$$

$$\sum_{i \in \mathcal{N}_0} y_{ij}^{Q-1} + \sum_{i \in \mathcal{N}} y_{ij}^Q - \sum_{i \in \mathcal{N}} y_{ji}^Q = \sum_{i \in \mathcal{N}_0} x_{ij}^{Q-1} + \sum_{i \in \mathcal{N}_0} x_{ij}^Q \quad j \in \mathcal{N} \quad (15)$$

$$x_{ij}^t \leq y_{ij}^t \leq (Q - t + 1) \cdot x_{ij}^t \qquad\qquad i \in \mathcal{N}_0, j \in \mathcal{N}, t = 1, \ldots, Q \quad (16)$$

$$x_{ij}^t \in \{0, 1\} \qquad\qquad i \in \mathcal{N}_0, j \in \mathcal{N}, t = 1, \ldots, Q \quad (17)$$

The objective function (6) corresponds to minimization of the total costs. The first term $20\ell_{ij} \cdot x_{ij}$ represents the ditch cost, which is 20€ per meter, and the second term represents the cost of three-phase cables.

Constraints (7)-(9) assure that if an arc (i, j) is in the solution in position t, the current intensity on (j, i) is given by equation (3), otherwise is zero. Constraints (10)-(11) set values for the scalars a_{ij}^k, to guarantee that the selected cable type satisfies the maximum limits on current intensity. Constraints (12) assure that the scalars will be zero for a cable not present in the optimal solution. These constraints correspond to (4) and (5).

Constraints (13)- (17) refer to the CMSTP. Constraints (13) assure that each node is in the solution. Constraints (14) and (15) model flow balance at nodes. Constraints (14) state that if a node j is in the position $t = 1, \ldots, Q - 2$

(*i.e.*, there exist an arc in position t converging into that node), then the flow comes into that node through an arc in position t and the flow leaving that node follows along arcs in position $t+1$. Constraints (15) are the previous set of constraints but refer to position $Q-1$. The coupling constraints (16) ensure that it is possible to send flow throughout the arc only if the arc is in the solution and state that the maximum admissible value of the flow in each arc decreases with its position in the solution. The biggest value is Q for the arcs that come from the root. Constraint (17) establish binary value for the design variables.

4   Case Study and Results

The proposed model was applied to the Montalegre wind farm, located in the north of Portugal, constituted by $n = 25$ wind turbines and one substation. The obtained optimal solution is therefore compared with the ground solution. In computational experiments was considered $n_c = 12$ cable sizes (see Table 1).

To solve the model, a branch and bound method is applied using the Xpress 7.2 software (available on http://www.fico.com/Xpress). Computations were performed on a computer i5-2500 3.3GHZ/3.7 Turbo with 8GB RAM.

The model has 23175 variables and 34525 constraints and the algorithm stopped 14 hours later with a gap of 5.6%. Should notice that although the execution time is large, this effort is made only once, during the construction phase of the wind farm.

Table 2. Parameters of Montalegre wind farm layout: (a) Optimization design; (b) Ground design

Branch	I (A)	Section (mm²)	Length (m)	Branch	I (A)	Section (mm²)	Length (m)
(7,0)	404.145	240	1415.900	(7,0)	404.145	300	1415.900
(15,0)	461.880	300	118.418	(13,0)	173.205	240	928.313
(16,0)	577.350	500	23.770	(15,0)	288.675	240	118.418
(6,7)	346.410	185	432.312	(16,0)	115.470	120	369.770
(14,15)	404.145	240	584.520	(18,0)	461.880	300	2031.490
(17,16)	57.735	25	711.113	(6,7)	346.410	300	432.312
(18,16)	461.880	300	1733.09	(11,13)	115.470	120	495.262
(13,14)	346.410	185	252.189	(14,15)	230.940	240	584.520
(19,18)	404.145	240	595.856	(17,16)	57.735	120	711.113
(5,6)	288.675	150	637.157	(19,18)	404.145	300	595.856
(12,13)	288.675	150	248.584	(5,6)	288.675	300	637.157
(20,19)	346.410	185	589.028	(10,11)	57.735	120	248.452
(4,5)	230.940	95	282.529	(12,14)	173.205	240	500.34
(11,12)	230.940	95	246.702	(20,19)	346.410	300	589.028
(21,20)	288.675	150	252.085	(4,5)	230.940	240	282.529
(3,4)	173.205	70	254.686	(9,12)	115.470	120	745.544
(10,11)	173.205	70	248.452	(21,20)	288.675	240	252.085
(22,21)	230.940	95	251.863	(3,4)	173.205	240	254.686
(1,3)	115.470	25	249.154	(8,9)	57.735	120	253.217
(9,10)	115.470	25	310.728	(22,21)	230.940	240	251.863
(23,22)	173.205	70	804.568	(1,3)	115.470	120	249.154
(2,1)	57.735	25	415.041	(23,22)	173.205	120	804.568
(8,9)	57.735	25	253.217	(2,1)	57.735	120	415.041
(24,23)	115.470	25	606.158	(24,23)	115.470	120	606.158
(25,24)	57.735	25	247.838	(25,24)	57.735	120	247.838

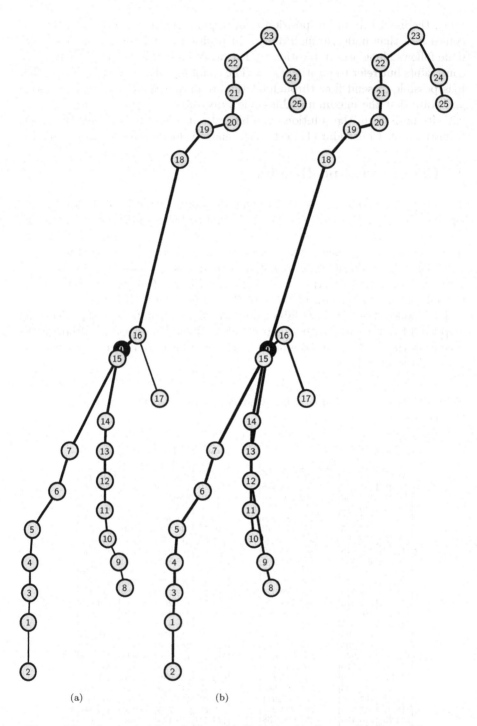

(a)   (b)

Fig. 2. Montalegre wind farm layout: (a) Optimization layout; (b) Ground layout

The wind farm layout obtained is illustrated in Fig. 2(a). The current, the cross-section and the length for all branches presented in the optimal solution are shown in Table 2(a). In order to compare with the ground solution, Fig. 2(b) and Table 2(b) present the correspondent layout and parameters. The main configuration difference, between both solutions, is the replacement of two branch-lines $10 - 11 - 13 - 0$ and $8 - 9 - 12 - 14 - 15 - 0$, in the ground solution, by only one branch-line $8 - 9 - 10 - 11 - 12 - 13 - 14 - 15 - 0$. Moreover, in the obtained optimal solution the peripheral wind turbines are connected by thinner cables. The total cost was 833875€ and 660245€ for the the ground solution and the obtained optimal solution, respectively. The optimization model leads to a safely network design with 30% cost reduction.

5   Conclusion

In the present study, a mathematical model to optimize the design of electrical connections between turbines and the substation in wind farms was presented. The goal is to minimize the investment costs that include cable cost and cable installation costs. The cable cost depend on its cross-section area which must safely carry the current intensity. This problem could be modeled as a minimum spanning tree problem.

One of the major constraint taken into account is the bound of the maximum number of wind turbines that can be connected in parallel on the same line-branch due to the constraints of the physical cable available. These additional constraints lead to a capacitated minimum spanning tree problem. An Hop Index Single-flow formulation was considered, once by the computational experiments made with other approaches, this formulation was the one that achieved better results.

The method was applied to a real wind farm, Montalegre, located in the north of Portugal. The results show that the use of the optimization method proposed is a very useful tool. In the case study, the method leads to a 30% reduction on the total costs, when compared with the ground solution. Other studies applied heuristics methods to the problem of turbine placement in wind farm. However, it should be noted that the presented approach reaches optimal solutions.

References

1. Carreno, E.M., Moreira, N., Romero, R.: Distribution network reconfiguration using an efficient evolutionary algorithm. In: IEEE Power Engineering Society General Meeting, Tampa, Florida, USA (June 2007)
2. Fan, J.Y., Zhang, L., McDonald, J.D.: Distribution network reconfiguration: single loop optimization. IEEE Trans. Power Systems 11(3), 1643–1647 (1996)
3. Borozan, V., Rajicic, D., Ackovski, R.: Improved method for loss minimization in distribution networks. IEEE Trans. Power Systems 10(4), 1420–1425 (1995)

4. Li, K., Chen, G., Chung, T., Tang, G.: Distribution planning using a rule-based expert system approach. In: IEEE International Conference on Electric Utility Deregulation, Restructuring and Power Technologies, Hong Kong (April 2004)
5. McDermott, T.E., Drezga, I., Broadwater, R.P.: A heuristic nonlinear construtive method for distribution system reconfiguration. IEEE Trans. Power Systems 14(2), 478–483 (1999)
6. Shirmohammadi, D., Hong, H.: Reconfiguration of electric distribution networks for resistive line losses reduction. IEEE Trans. Power Delivery 4(2), 1492–1498 (1989)
7. Bouchard, D., Chikhani, A., John, V.I., Salama, M.M.A.: Applications of hopfield neural-networks to distribution feeder reconfiguration. In: Applications of Neural Networks to Power Systems, IEEE ANNPS, Japan, pp. 311–316 (1993)
8. Kim, H., Ko, Y., Jung, K.H.: Artificial neural-network based feeder reconfiguration for loss reduction in distribution systems. IEEE Trans. Power Delivery 8(3), 1356–1366 (1993)
9. Nara, K., Shiose, A., Kitagawa, M., Ishihara, T.: Implementation of genetic algorithm for distribution systems loss minimum re-configuration. IEEE Transactions on Power Systems 7(3), 1044–1051 (1992)
10. Mendoza, J., Lopez, R., Morales, D., Lopez, E., Dessante, P., Moraga, R.: Minimal loss reconfiguration using genetic algorithms with restricted population and addressed operators: real application. IEEE Transactions on Power Systems 21(2), 948–954 (2006)
11. Zhu, J.Z.: Optimal reconfiguration of electrical distribution network using the refined genetic algorithm. Electric Power Systems Research 62(1), 37–42 (2002)
12. Miranda, V., Ranito, J., Proenca, L.: Genetic algorithms in optimal multistage distribution network planning. IEEE Trans. Power Systems 9(4), 1927–(1933)
13. Grady, S., Hussaini, M., Abdullah, M.: Placement of wind turbines using genetic algorithms. Renewable Energy 30(2), 259–270 (2005)
14. González, J.S., Rodriguez, A.G.G., Mora, J.C., Santos, J.R., Payan, M.B.: Optimization of wind farm turbines layout using an evolutive algorithm. Renewable Energy 35(8), 1671–1681 (2010)
15. Braz, H.D.M., Melo, G.H.S.V., Souza, B.A., de Souza, A.C.Z.: Planejamento da rede coletora de um parque de geração eólica usando algoritmos genéticos. In: Simpósio Brasileiro de Sistemas Elétricos, UFCG, Brasil, July 17-19, pp. 1–6 (2006)
16. Wang, C., Yao, G., Wang, X., Zheng, Y., Zhou, L., Xu, Q., Liang, X.: Reactive power optimization based on particle swarm optimization algorithm in 10kv distribution network. In: Tan, Y., Shi, Y., Chai, Y., Wang, G. (eds.) ICSI 2011, Part I. LNCS, vol. 6728, pp. 157–164. Springer, Heidelberg (2011)
17. Abdelaziz, A., Osama, R., El-Khodary, S., Panigrahi, B.: Distribution systems reconfiguration using the hyper-cube ant colony optimization algorithm. In: Panigrahi, B.K., Suganthan, P.N., Das, S., Satapathy, S.C. (eds.) SEMCCO 2011, Part II. LNCS, vol. 7077, pp. 257–266. Springer, Heidelberg (2011)
18. Jeon, Y., Kim, J., Kim, J., Shin, J., Lee, K.Y.: An efficient simulated annealing algorithm for network reconfiguration in large-scale distribution systems. IEEE Trans. Power Delivery 17(4), 1070–1078 (2002)
19. Thompson, G.L., Wall, D.L.: A branch and bound model for choosing optimal substation locations. IEEE Power Engineering PER-1(6), 69–70 (1981)

20. Fangdong, W., Han, L., Fangdong, W., Buying, W.: Substation optimization planning based on the improved orientation strategy of voronoi diagram. In: 2010 2nd International Conference on Information Science and Engineering (ICISE), pp. 1563–1566 (2010)
21. Magnanti, T.L., Wolsey, L.A.: Chapter 9 optimal trees. In: Ball, M., Magnanti, T., Monma, C., Nemhauser, G. (eds.) Network Models. Handbooks in Operations Research and Management Science, vol. 7, pp. 503–615. Elsevier (1995)
22. Gouveia, L., Martins, P.: The capacitated minimum spanning tree problem: revisiting hop-indexed formulations. Computers & OR 32, 2435–2452 (2005)

Intelligent Model to Obtain Initial and Final Conduction Angle of a Diode in a Half Wave Rectifier with a Capacitor Filter

José Luis Casteleiro-Roca[1], Héctor Quintián[1], José Luis Calvo-Rolle[1], Emilio Corchado[2], and María del Carmen Meizoso-López[1]

[1] Universidad de A Coruña,
Departamento de Ingeniería Industrial,
Avda. 19 de febrero s/n, 15.495, Ferrol, A Coruña, España
[2] Universidad de Salamanca,
Departamento de Informática y Automática,
Plaza de la Merced s/n, 37.008, Salamanca, Salamanca, España

Abstract. The half wave rectifier with a capacitor filter circuit is a typically non-linear case of study. It requires a hard work to solve it on analytic form. The main reason is due to the fact that the output voltage comes alternatively from the source and from the capacitor. This study describes a novel intelligent model to obtain the time when the changes of the sources occur. For the operation range, a large set of work points are calculated to create the dataset. To achieve the final solution, several simple regression methods have been tested. The novel model is verified empirically by using CAD software to simulate electronic circuits and by analytical methods. The novel model allows to obtain good results in all the operating range.

Keywords: Single phase wave rectifier, capacitance filter, neural networks.

1   Introduction

Since semiconductors emerge in electronic field, one of the most researched circuits has been the rectifier. It is well-known that the function of the rectifiers, in general terms, is to obtain a continuous signal from an alternate signal [1]. These circuits are very common in applications like DC power supplies [2], peak signal detectors [3], and so on.

The basic rectifier is one of the most traditional circuits to learn electronic in the first courses [4]. Despite this, this type of schema could be not easy to solve and obtain its parameters. Fundamentally, the difficulty depends on: the number of phases of the source, the characteristics of the load and if it is controlled or a not controlled type [1].

A more specific example of this fact is the half wave rectifier with a capacitor filter. Due to the capacitive component added to a resistive load, the voltage has a less ripple wave [1]. Then the output voltage comes alternatively from

Á. Herrero et al. (eds.), *International Joint Conference SOCO'13-CISIS'13-ICEUTE'13*,   121
Advances in Intelligent Systems and Computing 239,
DOI: 10.1007/978-3-319-01854-6_13, © Springer International Publishing Switzerland 2014

the source and from the capacitor [2]. The main problem is to know the time where the diode is in conduction (source provides the power to the load) or not (capacitor provides the power) [2]. As would be seen in the 'Case of study' section, this type of rectifier is not easy to solve in analytical form, due to the non-linear nature of its topology [3].

The knowledge of experts is used to create rule based systems models [5, 6]. Human experts extract rules from a system operation and then they structure it according to the system performance [5]. These methods allow to develop models to emulate the expert's behavior in a certain field [5, 7], and have been one of the most used methods in both research and operation [7].

The traditional regression methods are based on Multiple Regression Analysis (MRA) methods [8]. MRA-based methods are very popular among others because of their applications in many different fields [9, 10]. It is well-known that these methods have limitations [8, 11]. With the aim of solving these limitations, regression methods based on Soft Computing techniques have been proposed [12–17]. Taking into account this fact, a novel approach is proposed in this research. Therefore a novel Intelligent Model has been developed based on rules and Soft Computing techniques. The model allows obtaining the angles where a diode is conducting or not on the half wave rectifier with a capacitor filter circuit. Many Soft Computing techniques [18] have been tested in order to obtain the best fitness of the created model.

The rest of the paper is organized as follows: section 2 presents a brief description of the half wave rectifier with a capacitor filter is exposed. Then, the novel approach is presented in section 2.2. Section 3 describes the data set to achieve the model, and the intelligent regression techniques tested in this study to complement the experimental rule-based system. The results for the tested methods are described in section 4; and conclusions and future works are exposed at the end of the paper.

2   Case of Study

2.1   Half Wave Rectifier with a Capacitor Filter Circuit

The half wave rectifier with a capacitor filter circuit is shown in figure 1. The resistance R in parallel with capacitor filter C is fed for the voltage source V via diode D. The purpose of the capacitor is to reduce the variation in the output voltage, making it more like *DC Power*. The resistance represents a load and the capacitor represents a filter which is part of the rectifier circuit.

As the source decreases after the sinusoid peak, the capacitor discharges into the load resistor. Just at the same point, the voltage of the source becomes less than the output voltage, then the diode turns off, and the output voltage is a decaying exponential. The point where the diode turns off ($\omega t = \theta$) is determined by comparing the rates of change of the source and the capacitor voltages (figure 2). The angle, at which the diode turns on in the second period ($\omega t = 2\pi + \alpha$), is the point where the sinusoidal source reaches the same value as the decaying

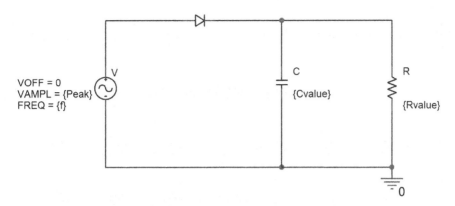

Fig. 1. Half wave rectifier with a capacitor filter circuit

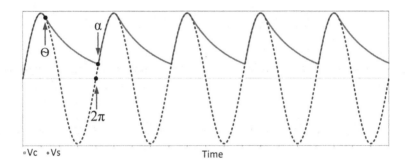

Fig. 2. Source and load voltage

exponential output. θ is obtained by equation 1, where ω is the frequency, R is the resistive value of the load and C is the capacitive value of the filter.

Two are the possibilities to obtain α value measured since 2π instant:

The first one is when the capacitor discharge totally before the next positive period of the input source, the way to calculate the angle α is to use the equation 2. In this case the sign angle is negative.

The second one is obtained by solving the nonlinear equation 3.In this case the sign angle is positive.

$$\theta = -tan^{-1}\left(\omega RC\right) + \pi \tag{1}$$

$$\alpha = 2\pi - \theta + \omega RC ln \frac{1}{1000 sin\left(\theta\right)} \tag{2}$$

$$sin\left(\alpha\right) - \left(sin\left(\theta\right)\right) e^{\frac{-\left(2\pi + \alpha - \theta\right)}{\omega RC}} = 0 \tag{3}$$

Figure 3 shows partially the operation range. On it, it is possible to appreciate two different regions depending of the α sign angle.

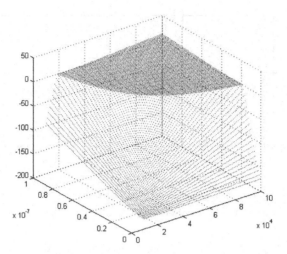

Fig. 3. Dataset to calculate α's model

Figure 2 shows the source voltage (dots in red) and voltage in the load (continuous line in green). As can be shown in figure 2, the output voltage in the load is like *DC Power* but with a ripple voltage. θ and α instants are specified at figure 2.

As there is not any analytical solution for the equation 3, it is necessary to use numeric methods to solve the equation as the one mentioned before [1, 2].

2.2   Novel Approach

The general schema of the proposed topology where the novel Intelligent Model is used to obtain the value of angles is illustrated in figure 4.

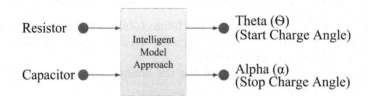

Fig. 4. General schema

As can be seen in figure 4, the model has two inputs: the resistor value and the capacitor value. The model provides the start discharge angle when diode turns off (θ) and the stop discharge angle when diode turns on (α).

The internal layout of the general schema is shown on figure 5. In figure 5 are identified the different parts of the proposal in a diagram block form. The novel approach has four blocks whose inputs are the same of the general schema (resistive value and the capacitive value of case of study). One of them is an Artificial Neural Networks (ANN) to obtain the α angle when it has a positive value (like non linear equation 3). Two of them are functions: the first one to obtain the θ angle according with equation 1, and the second one is to obtain α angle when it has a negative value according with equation 2. The fourth block, with the help of the multiplexer, that makes possible to choose between the blocks to obtain the α angle depending of its sign.

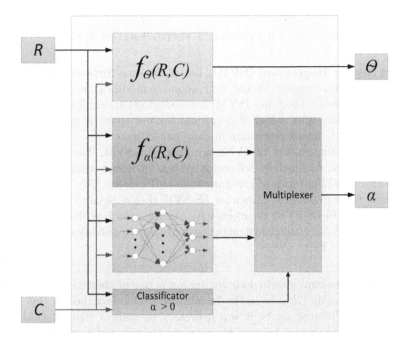

Fig. 5. Internal layout of the general schema

3   Model Approach

3.1   Used Techniques

The techniques tested in the study to select the best one, are described as follows.

Artificial Neural Networks (ANN): Multilayer Perceptron (MLP). A multilayer perceptron is a feedforward artificial neural network [18]. It is one of the most typical ANNs due to its robustness and relatively simple structure. However the ANN architecture must be well selected to obtain good results.

The MLP is composed by one input layer, one or more hidden layers and one output layer, all of them made of neurons and with pondered connections between neurons of each layer.

Polynomial Regression. Generally, a polynomial regression model [18] may also be defined as a linear summation of basis functions. The number of basis functions depends of the number of inputs of the model, and the degree of the polynomial used.

With a degree 1, the linear summation could be defined as the one shown in equation 4. The model becomes more complex as the degree increases (equation 5 shows a second degree polynomial for the model).

$$F(x) = a_0 + a_1 x_1 + a_2 x_2 \tag{4}$$

$$F(x) = a_0 + a_1 x_1 + a_2 x_2 + a_3 x_1 x_2 + a_4 x_1^2 + a_5 x_2^2 \tag{5}$$

Support Vector Regression (SVR), Least Square Support Vector Machine (LS-SVM). Support Vector Regression is a modification of the algorithm of the Support Vector Machines (SVM) for classification. In SVR the basic idea is to map the data into a high-dimensional feature space F via a non linear mapping and to do linear regression in this space [19].

Least Square formulation of SVM, are called LS-SVM. The approximation of the solution is obtained by solving a system of linear equations, and it is comparable to SVM in terms of generalization performance [20]. The application of LS-SVM to regression is known as LS-SVR (Least Square Support Vector Regression) [21]. In LS-SVR, the insensitive loss function is replace by a classical squared loss function, which constructs the Lagragian by solving a linear Karush-Kuhn-Tucker (KKT).

Logistic Regression. The logistic regression is one of the most simplex algorithm used to classify [22]. As in the Intelligent Model described in this contribution has two different parts, it was necessary to make a classification of the samples to decided the model used to calculate the value of α angle.

The results of this classification is showed in table 4 in section 'Results'. The decision to used this method respond a principle of simplification and less computing requirements.

3.2   Case of Study

The values of the resistor load went from $0.1\,\Omega$ to $1M\,\Omega$. As this range is quite big, the values were not taking in a linear distribution; instead of that, 5 subranges were created with deferents steps each one (table 1).

In the same way, the values of the capacitor filter went from $1nF$ up to $1mF$. This range was not as large as the resistor's one, but it was also sub-divided in 3 subranges (table 2).

Whit this ranges of values, all typical working points of a rectifier are covered. Several scripts were implemented and used to calculate the starting point of the capacitor charge (angle θ), and the final angle of the capacitor discharge (angle α).

Table 1. Ranges used for the load value

Initial value	Final value	Step size
0.1Ω	100Ω	1Ω
100Ω	$1k\Omega$	10Ω
$1k\Omega$	$10k\Omega$	100Ω
$10k\Omega$	$100k\Omega$	$1k\Omega$
$100k\Omega$	$1M\Omega$	$10k\Omega$

Table 2. Ranges used for the capacitor filter value

Initial value	Final value	Step size
$1nF$	$100nF$	$1nF$
$100nF$	$1\mu F$	$100nF$
$10\mu F$	$1mF$	$10\mu F$

The calculation of the starting point is easy, because it's only necessary to apply the formula showed in equation 1.

Instead of the first angle, the stop discharging angle is calculated with two different formulas depending on the predicted sign of this angle (equations 3 and 2). One of these formulas is a non linear equation, and its solution is achieved by an iteration method. The number of iterations necessaries depends on the desired precision. In this study, the precision is set to ± 0.01 *rad*, enough for the propose of this study, and better than the normal tolerance of the typical components.

3.3   Dataset Conditioning

For obtaining the proposed model, it has been used a dataset which has 2 inputs variables (resistor load, and capacitor filter) and 2 outputs variables (θ and α angles). In figure 3 it is possible to see the difference depending on the sign of the angle α. That is the reason why it is necessary to use a classification system to calculate the value of this angle.

The dataset with a total of 139200 *samples*, has been divided into two dataset, training dataset (92800 *samples*, two thirds of the original dataset), and test dataset (46400 *samples*, one third of the original dataset).

Each model has been trained and later evaluated with the test dataset, in terms of the MSE (Mean Square Error).

4   Results

Once the models were trained, they were tested getting the results shown on tables 3 and 4 in terms of MSE.

The best result with ANN were achieved using only one hidden layer in each neural network, with an hyperbolic tangent, with 9 neurons in this hidden layer.

With polynomial regression, best results were obtained with a 6 degree polynomial model. The best result with LS-SVR was achieved with parameters: $\gamma = 1.8324\ e^7$ and $\sigma^2 = 2.0342$.

Based on the obtained results (tables. 3), it is possible to conclude that the best models are ANN using 9 and 10 neurons in the hidden layer. LS-SVR is also getting good results (see table 3). The main reason to choose ANN is due to the required optimization process of LS-SVR for getting the best parameters (γ, σ) necessary for the training process, but once it has been trained, the computational cost is the same as for ANN.

Table 3. MSE of the different tested regression models

	$\alpha > 0$
ANN_12	0.0097
ANN_11	0.0086
ANN_10	0.0080
ANN_09	**0.0014**
ANN_08	0.0916
Poly_08	0.0422
Poly_07	0.0792
Poly_06	0.0036
Poly_05	0.0959
LS-SVM	0.0049

The best results to classification technique are obtained using a regularization constant with a value of 0.8. The confusion matrix for the training is shown in table 4.

Table 4. Confusion matrix of the classification

	$\alpha < 0$ (real)	$\alpha > 0$ (real)
$\alpha < 0$	15342	118
(test)	66.11%	0.56%
$\alpha > 0$	262	30678
(test)	0.25%	33.06%

5   Conclusions

With the novel intelligent model approach achieving very good results in general terms, getting the instants when the output voltage comes from the source or from de capacitor. With this proposal created based on intelligent techniques and rules is not necessary to solve the non-linear equation, both analytically or simulation.

As can be seen in 'Results' section, the best graphical adjustment of the objective data is obtained with ANN, where the MSE has a value of 0.0014. Other techniques allow good results like LS-SVM or Polynomial regression. With any of them, the time would be obtained, among others because the real components have normally tolerances bigger than 1%.

Acknnowledgement. This research is partially supported by the Spanish Ministry of Economy and Competitiveness under project TIN2010-21272- C02-01 (funded by the European Regional Development Fund), SA405A12-2 from Junta de Castilla y León. This work was also supported by the European Regional Development Fund in the IT4Innovations Centre of Excellence project (CZ.1.05/1.1.00/02.0070).

References

1. Hart, D.W.: Power electronics. McGraw-Hill, New York (2011)
2. Rashid, M.H.: Power electronics handbook: devices, circuits, and applications. Butterworth-Heinemann (2011)
3. Luo, F., Ye, H.: Power electronics. CRC Press, Singapure (2011)
4. Malvino, A.P., Bates, D.J.: Electronic principles. Recording for Blind & Dyslexic (2008)
5. Hayes-Roth, F., Waterman, D., Lenat, D.: Building expert systems. Addison-Wesley Pub. Co. (1983)
6. Cimino, M.G.C.A., Lazzerini, B., Marcelloni, F., Ciaramella, A.: An adaptive rule-based approach for managing situation-awareness. Expert Syst. Appl. 39, 10796–10811 (2012)
7. Hayes-Roth, F.: Rule-based systems. Commun. ACM 28, 921–932 (1985)
8. Mark, J., Goldberg, M.: Multiple regression analysis and mass assessment: A review of the Issues. Appraisal Journal 56(1), 89–109 (1988)
9. Yankun, L., Xueguang, S., Wensheng, C.: MRA-based revised CBR model for cost prediction in the early stage of construction projects. Expert Systems with Applications 39, 5214–5222 (2012)
10. Ho, L.H., Feng, S.Y., Lee, Y.C., Yen, T.M.: Using modified IPA to evaluate supplier's performance: Multiple regression analysis and DEMATEL approach. Expert Systems with Applications 39, 7102–7109 (2012)
11. Do, A.Q., Grudnitski, G.: A neural network approach to residential property appraisal. The Real Estate Appraiser 58(3), 38–45 (1992)
12. Guan, J., Zurada, J., Levitan, A.S.: An adaptive neuro-fuzzy inference system based approach to real estate property assessment. Journal of Real Estate Research 30(4), 395–420 (2008)
13. Peterson, S., Flanagan, A.B.: Neural network hedonic pricing models in mass real estate appraisal. Journal of Real Estate Research 31(2), 147–164 (2009)
14. Calvo-Rolle, J., Manchón-González, I., López-García, H.: Neuro-robust controller for non-linear systems. DYNA 86(3), 308–317 (2011)
15. Taffese, W.Z.: Case-based reasoning and neural networks for real estate valuation. In: Proceedings of 25th International Multi-conference: Artificial Intelligence and Applications, Innsbruck, Austria, vol. 84(9) (2007)

16. Nieves-Acedo, J., Santos-Grueiro, I., Garcia-Bringas, P., et al.: Enhancing the prediction stage of a model predictive control systems through meta-classifiers. DYNA 88(3), 290–298 (2013)
17. Alvarez-Huerta, A., Gonzalez-Miguelez, R., Garcia-Metola, D., et al.: Drywell temperature prediction of a nuclear power plant by means of artificial neural networks. DYNA 86(4), 467–473 (2011)
18. Bishop, C.M.: Pattern recognition and machine learning. Springer, New York (2006)
19. Vapnik, V.: The nature of statistical learning theory. Springer (1995)
20. Ye, J., Xiong, T.: SVM versus least squares SVM. In: 11th International Conference on Artificial Intelligence and Statistics (AISTATS), pp. 640–647 (2007)
21. Yankun, L., Xueguang, S., Wensheng, C.: A consensus least support vector regression (LS-SVR) for analysis of near-infrared spectra of plant samples. Talanta 72, 217–222 (2007)
22. O'Connell, A.A.: Logistic regression models for ordinal response variables. Sage Publications (2006)

Prosodic Features and Formant Contribution for Speech Recognition System over Mobile Network

Lallouani Bouchakour and Mohamed Debyeche

Faculty of Electronics and Computer Sciences (FEI), USTHB
Speech Communication and Signal Processing Laboratory (LCPTS)
P.O. Box 32, Bab Ezzouar, Algiers, Algeria
{lbouchakour,mdebyeche}@usthb.dz, mdebyeche@gmail.com

Abstract. This paper investigates the contribution of formants and prosodic features like pitch and energy on automatic speech recognition system performance in mobile networks especially the GSMEFR (Global System for Mobile Enhanced Full Rate) codec.The front-end of the speech recognition system combines feature extracted by converting the quantized spectral information of speech coder, prosodic information and formant frequencies. The quantized spectral information is represented by the LPC (Linear Predictive Coding) coefficients, the LSF (Line Spectral Frequencies) coefficients, the approximation of the LSF's to the LPC Cepstral Coefficients (LPCC's) that are the Pseudo Cepstral Coefficients (PCC) and the Pseudo-Cepstrum (PCEP) coefficients. The achieved speaker-independent speech recognition system is based on Continuous Hidden Markov Model (CHMMs) classifier. The obtained results show that the resulting multivariate feature vectors lead to a significant improvement of the speech recognition system performance in mobile environment, compared to speech coder bit-stream system alone.

Keywords: ASR,GSMEFR, CHMM, ARADIGIT, bit-stream, Formant, Pitch.

1 Introduction

Automatic Speech Recognition (ASR) over Mobile networks has attracted various research activities since the latter decades. Speech is the most important means of interpersonal communication, machines and computers can be operated more conveniently with the help of automated speech recognition and understanding. However, there are several   degradation sources on the performances of speech recognition system over mobile communication networks. The ASR performance degradation include distortion from low-bit-rate speech coders employed in the networks and the distortions arising from transmission errors occurring over the associated communication channels.The speech coder can compress speech signal with near transparent quality from a perceptual point of view; the performance of an ASR system using the decoded speech can degrade relative to the performance obtained for the original speech. This paper addresses the effect of GSMEFR speech coder on speech recognition performance. We have used two variants of extraction parameters, the first based on the quantized spectral information bit-stream-based technique. Its parameters are

Á. Herrero et al. (eds.), *International Joint Conference SOCO'13-CISIS'13-ICEUTE'13*,   131
Advances in Intelligent Systems and Computing 239,
DOI: 10.1007/978-3-319-01854-6_14, © Springer International Publishing Switzerland 2014

LPC, LSF, the approximation of the LSF coefficient to the LPC Cepstral Coefficients (LPCC's) that are PCC (Pseudo-Cepstral Coefficients) and Pseudo-Cepstrum (PCEP) coefficients. The second variant combines the auxiliary parameters and the quantized spectral information (bit-stream-based technique). We can refer to many works that tried to improve the robustness of ASR system by using several streams of features that rely on different underlying assumptions and exhibit different properties. Formant and auditory-based acoustic cues are used together with MFCC in [1]. In [2], [3], a multi-stream approach is used to combine MFCC features with formant estimates and a selection of acoustic cues such as acute/grave, open/close, tense/lax, etc. Pitch has been also taken into account in many works for the recognition of tonal languages[4], [5]. For the same purpose, many works in audio-visual domain have investigated the contribution of the visual information on the acoustic recognition system in noisy environments [6], [7].This work aims to improve ASR system over mobile networks by using a combination of auxiliary parameters that are the first three formant frequencies, frames energy, the pitch and the quantized spectral information (bit-stream-based technique). The implemented speech recognition system is based on the Continuous Hidden Model Markov (CHMM) probabilistic classification model. Our experiments are done on the ARADIGIT database [8]. This paper is organized as follows: after this introductive section 1; the GSMEFR codec is presented in section 2; the developed speech recognition system is presented in section 3; the features extraction in section 4 ; the experimental results are given in section 5. Finally, conclusion and future work are summarized in section 6.

2   GSM Speech Coders

There exist three different speech coders standardized for use in GSM communication networks. They are referred to as the Full Rate, Half Rate and Enhanced Full Rate GSM coders. Their corresponding European Telecommunications Standards are the GSM 06.10, GSM 06.20 and GSM 06.60, respectively. These coders work on a 13 bits uniform PCM speech input signal, sampled at 8 kHz. The GSM EFR speech coder is based on the ACELP algorithm (Algebraic Code Excited Linear Prediction) [9]. In the ACELP speech synthesis model, the excitation signal at the input of the short-term LP synthesis filter is constructed by adding two excitation vectors from adaptive and fixed (innovative) codebooks [10]. The speech is synthesized by feeding the two properly chosen vectors from these codebooks through the short-term synthesis filter. The optimum excitation sequence in a codebook is chosen using an analysis-by-synthesis search procedure in which the error between the original and synthesized speech is minimized according to a perceptually weighted distortion measure. The speech coding (source coding) bit-rate is 12.2 Kbit/s and for channel coding (error protection) 10.6kbit/s bit-rate is used, resulting in 22.8 Kbit/s channel bit-rate [11][12]. In each 20 msec speech frame, 244 bits are produced, corresponding to a bit rate of 12.2 Kbit/s.The function of the decoder consists of decoding the transmitted parameters (LSF parameters, adaptive codebook vector, adaptive codebook gain, fixed codebook vector, fixed codebook gain) and performing synthesis to obtain the reconstructed speech.

3   Speech Recognition System

Nowadays, ASR systems are primarily based on the principles of statistical pattern recognition, in particular the use of Hidden Markov Models (HMMs). The HMM is a powerful statistical method for characterizing the observed data samples of a discrete-time series. The underlying assumptions for applying HMMs to ASR are that the speech signal can be conveniently characterized as a parametric random process and that the parameters of the process can be precisely estimated. The Markov model concept is that, we have two superposed processes. One of them is observable (the sequence of observations), but the other one is "hidden" (sequence of states). In the same way as for Markov concept processes, it is possible to calculate the probability of a sequence of observations $O = o_1, o_2 ....o_T$ of length T generated by an HMM, although we will formally define them before the concept of HMM[13][14]. The architecture of a typical ASR system, depicted in Fig.1, shows a sequential structure of ASR including such components as speech signal capturing, front-end feature extraction and back-end recognition decoding. Feature vectors are first extracted from the captured speech signal and then delivered to the ASR decoder. The decoder searches for the most likely word sequence that matches the feature vectors on the basis of the acoustic model. The output word sequence is then forwarded to a specific application [15].

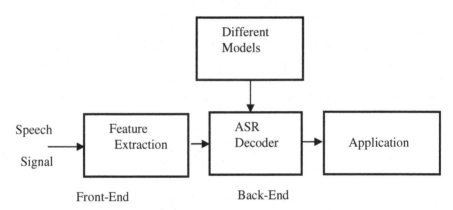

Fig. 1. Architecture of an ASR system

4   Parameter Extraction

Various features are used to characterize the speech word, and perform a speech recognition system, such as: Mel Frequency Cepstral Coefficients (MFCC)[16], LSF, LPC, Linear Predictive CepstralCoefficients LPCC, Pitch, Energy and the Formants. In this work we have used two variants of parameterization. In the first variant we have used the parameters of the coder (LPC, LSF) and the approximation of the LSF's to the LPCC's that are PCC and PCEP. In the second variant,   we added the prosodic parameters and formant.

4.1   Pitch

There is actually a very wide variety of pitch extraction methods in published literature, although we will here describe one of the simpler, and more common methods
used in speech coders. This method relies upon minimizing the mean-squared error
between an LPC residual (containing pitch), and the reconstructed pitch signal resulting from the analysis. If E is the mean-squared error, e is the residual, then [17]:

$$E(M, \beta) = \sum_{n=0}^{N-1} \left\{ S(n) - \beta S(n - M) \right\}^2 \tag{1}$$

Where:

β: is the pitch gain.
M:is the pitch.
N: is the analysis window size.
S(n): is the speech signal.

4.2   Formant Frequencies

The maximum of the spectral envelop represents the resonant frequencies of the vocal
tract called: "formants". Which are the peaks in the frequency spectrum of speech that
characterize vowels, vary considerably depending on the speaker and on the context
of speech. The formant values change with different sounds; however, we can identify
ranges in which they occur. For example, the first formant occurs in the range 200-
800 Hz [18] for a male speaker and in the range 250-1000 Hz for a female speaker.

The poles of the $H(z)$ are just the formants resonances.

$$H(z) = \frac{1}{A(z)} = \frac{1}{1 + \sum_{i=1}^{m} a_i z^i} \tag{2}$$

Where a_i are the Linear Prediction (LP) coefficients.

4.3   Energy

Energy is defined as the variation of the signal amplitude caused by the force coming
from the pharynx. The energy is given by [18]:

$$E = \frac{1}{T} \sum_{t=1}^{T} S(t)^2 \tag{3}$$

Where:

$S(t)$: is the speech signal.
T:is the window signal size.

4.4   Mel-Frequency Cepstral Coefficient (MFCC)

The power spectrum (that is, the magnitude spectrum squared) of the speech frame is found. Then, the power spectrum is filtered by a series of triangular shaped band pass filters. These filters are spaced uniformly on a non-linear frequency scale called the Mel scale, which is a psychoacoustic measure of pitches judged by human. The Mel scale is approximately linear to the linear frequency scale below 1 kHz and logarithmic above 1 kHz. The Mel scale can be calculated by equation (4) [19].

$$\text{Mel} \ \ (f) = 2595 \times \log_{10} (1 + \frac{f}{700}). \tag{4}$$

where f is the frequency in the linear scale and Mel (f) is the resulting frequency in Mel-Scale.

After the Mel filter bank, the filter outputs are compressed by the logarithmic function. Finally, the log filter outputs are decorrelated by discrete cosine transform (DCT) [19].

$$C_n = \sqrt{\frac{2}{N}} \sum_{j=1}^{N} A_j \cos\left(\frac{\pi n}{N} (j - 0.5) \right). \quad n = 1, \ldots, P. \tag{5}$$

Where:

N: is the total number of filter banks.
A_j is theenergy output of j filter.
P: is the number of coefficients MFCC.
C_n are the n-th MFCC coefficients.

This allows obtaining speech recognition parameters directly from the bit-stream transmitted to the receiver over digital mobile networks. This technique is based on the decoded speech to avoid the stage of reconstructing speech from the coded speech parameters. In this scenario, we are transforming the speech coding parameters to speech recognition parameters as introduced to the ASR systems [19].

4.5   Pseudo-Cepstral (PCC)

The PCC parameters are computed directly from the LSF parameters. Mathematical manipulations and approximations allow it to be expressed in terms of the LSFs. The n-th PCC is given by the equation [20][21]:

$$C_n^{PCC} = \frac{(1 + (-1)^n)}{2n} + \frac{\sum_{i=1}^{p} \cos(nw_i)}{n} \tag{6}$$

Where wi is the i-th LSF parameters.

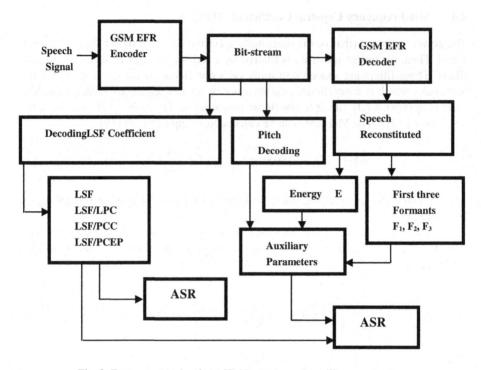

Fig. 2. Feature extraction for ASR bit-stream and auxiliary parameters

4.6   Pseudo-Cepstrum (PCEP)

Using the mathematical expression of the PCC parameters, it is somewhat trivial to obtain the PCEP. They are derived from the PCC parameters by eliminating the term

$$\frac{(1 + (-1)^n)}{2n} \tag{7}$$

The n-th PCEP parameters expression is given by [20][21]:

$$C_n^{PCEP} = \frac{\sum_{i=1}^{P} \cos(nw_i)}{n} \tag{8}$$

5   Experimental Results

5.1   Databases

The database used in this work is the ARADIGIT database. It consists of the 10 Arabic digits spoken by 60 speakers of both sexes. This database was recorded by Algerian speakers aged between 18 and 50 years in a quiet environment. With an ambient noise level below 35 dB, in ".wav" file format, with a sampling frequency of 8 KHz. This database is divided into two corpuses, in speaker-independent mode:

Train corpus: Consisting of 1200 speech files (words) pronounced by 40 speakers including the two sexes, where, each speaker repeats the 10 Arabic digits three times.

Test corpus: Consisting of 600 speech files (words) pronounced by 20 speakers including the two sexes, where each speaker repeats the 10 Arabic digits three times.

5.2   Recognition Accuracy Using Quantized Spectral Information (Bit-Stream)

These experiments were conducted using CHMM of N=3 states the number of Gaussian components is varied from 1 until 12, a set of HTK Hidden Markov Model Toolkit (ver. 3.3) [22]. The basic test consists of 600 words. The frame length is 240 samples. These experiences aim to study the influence of the GSM EFR speech coder on the performance of the speech recognition system. These experiences are applied to transcoded   database represented by 12 MFCC coefficients and database coding bit-stream parameters, as produced by the channel decoder in a mobile cellular communications network represented by 10 LSF coefficients and its approximation 10 for LPC, 12 for PCCand 13 for PCEP coefficients. The recognition results are given by the Recognition Accuracy (RA).

$$ RA = \frac{N - D - S}{N} \times 100 \qquad (9) $$

Where N is the total number of units (words), D is the number of deletion errors and S is the number of substitution errors.The obtained results are summarized in Table 1 and Table 2.

Table 1. Recognition results in (%) of transcoded speech GSM EFR

Number of Gaussian	1	2	4	8	10	12
RA (%) MFCC	66	77,50	77,83	75,83	76,67	74,67

Table 2. Recognition results of quantized spectral information (bit-stream)

Number of Gaussian	1	2	4	8	10	12
RA (%)LPC	66,17	72,33	68,50	73,17	70,83	70,33
RA(%)LSF	71,17	79,17	82,50	82,33	81,83	82,50
RA(%)PCC	80,33	86,33	87,83	89,17	88,50	89,67
RA(%)PCEP	81,83	87,33	88,17	87,50	87,67	88,50

Table 1 and Table 2 show that, the speech recognition rate results respectively obtained on GSMEFR ARADIGIT transcoded, and the GSMEFR ARADIGIT coding bit-stream. The results of Table 1 can be explained by the degradation of signal speech quality caused by the GSMEFR coder. The Table 2 represented the result of speech coding parameters called bit-stream. Its parameters are the LFS, LPC and the approximation of the LSF's to the LPC Cepstral Coefficients (LPCC's), that are PCC and PCEP. The transformation of speech coding parameters to speech recognition parameters motivated by the fact that ASR feature parameters are based on the speech spectral envelope and not on the excitation. The bit-stream parameters improved the performance of the system ASR comparison by the MFCC parameters.

5.3   Recognition Accuracy Using Quantized Spectral Information (It-Stream) + Auxiliary Parameters

In this variant, we concatenate the quantized spectral information that are, the LSF, LPC, PCC and the PCEP coefficients with prosodic parametersthat are, theenergy frame, pitch and the first three formants (3Fi) Fig. 1. This parameters are referred by (LSF+ auxiliary parameters) is composed of 15 dimensional observation vectors. The LPC parameters are referred by (LPC+ auxiliary parameters) is composed of 15 dimensional observation vectors. The approximation of the LSF's to the LPC Cepstral Coefficients (LPCC's) that are PCC (Pseudo-Cepstral Coefficients) and PCEP (Pseudo-Cepstrum) are referred by (PCC+ auxiliary parameters) and   (PCEP + auxiliary parameters) their dimensions are respectively17 and 18. Table 3 summarizes the comparative results.

Table 3. Recognition results of quantized spectral information (bit-stream) + auxiliary parameters

Number of Gaussian	1	2	4	8	10	12
RA LPC + auxiliary parameters	83.5 %	88.17%	90.17%	91.17%	91.17%	92.00%
RA LSF + auxiliary parameters	83.67%	91.33%	93.33%	94.17%	94.17%	93.83%
RA PCC + auxiliary parameters	87.83%	91.67%	94.17%	94.5%	95.33%	95.33%
RA PCEP + auxiliary parameters	87.67%	91.17%	93.17%	94.5%	95.5%	94.17%

The results illustrate in Table 3 for the second variant (Bit-stream parameters + the auxiliary parameters) has performed a better recognition rate than the first variant (Bit-stream parameters). The results shown in Table 3 prove that the fact of mixing different pertinent parameters to be used as discriminative features improve the speech recognition accuracy rate.

It's clear that in the second variant, when we used the auxiliary parameters the recognition rates are increased comparing than the first variant, because the auxiliary parameters (pitch + energy + 3Fi) are the distinct parameters for   speech characterization process. The proposed system that includes additional features (bit-stream + auxiliary parameters) clearly outperforms the bit-stream method.

6   Conclusion

In this paper, we have studied the contribution of various speech parameters that must be addressed to facilitate robust automatic speech recognition over mobile communication networks, with the aim of minimizing the impact of speech coder on the performance of text-independent speech recognition system. Two variants of speech parameters extraction are presented. The first one is the quantized spectral information (bit-stream) and the second one is a combination of the quantized spectral information and the auxiliary parameters. The obtained results suggest that bit-stream parameters added to the auxiliary features could improve the ASR performance in mobile communication. In fact, this inclusion yields an improvement of more than 7% of correct recognition rate.

References

1. Holmes, J.N., Holmes, W.J.: Using formant frequencies in speech recognition. In: Proc. Eurospeech, Rhodes, pp. 2083–2086 (1997)
2. Selouani, S.A., Tolba, H., O'Shaughnessy, D.: Auditory-based acoustic distinctive feature-sand spectral cues for automatic speech recognition using a multi-stream paradigm. In: Proc. of ICASSP, pp. 837–840 (2002)
3. Tolba, H., Selouani, S.A., O'Shaughnessy, D.: Comparative experiments to evaluate theuse of auditory-based acoustic distinctive features and formant cues for robust automaticspeech recognition in low snr car environments. In: Proc. of Eurospeech, pp. 3085–3088 (2003)
4. Chongjia, N.I., Wenju, L., Xu, B.: Improved Large Vocabulary Mandarin Speech Recognition Using Prosodic and Lexical Information in Maximum Entropy Framework. In: Proc. CCPR 2009, pp. 1–4 (2009)
5. Ma, B., Zhu, D., Tong, R.: Chinese Dialect Identification Using Tone Features Based on Pitch Flux. In: Proc. ICASSP, p. I (2006)
6. Gurbuz, S., Tufekci, Z., Patterson, E., Gowdy, N.J.: Multi-stream product modalaudiovisual integration strategy for robust adaptive speech recognition. In: Proc. ICASSP, pp. II-2021–II-2024 (2002)
7. Guoyun, L.V., Dongmei, J., Rongchun, Z., Yunshu, H.: Multi-stream AsynchronyModeling for Audio-Visual Speech Recognition. In: Proc. ISM, pp. 37–44 (2007)
8. Amrouche, A., Debyeche, M., Taleb-Ahmed, A., Rouvean, J.M., Yagoub, M.C.E.: An efficient speech recognition system in adverse conditions using the nonparametric regression. International Journal of Engineering Applications of Artificial Intelligence, IJEAAI 2010 23(1), 85–94 (2010)
9. Jarvinen, K., Vainio, J., Kapanen, P., Honkanen, T.: GSM Enhanced Full Rate Speech Codec. In: ICC 1997, Montreal, pp. 721–724. IEEE (1997)
10. Honkanen, T., Vainoi, J., Jarvinen, K., Haavisto, P., Salami, R.: Enhanced Full Rate Speech Code For Is-136 Digital Cellular System. In: ICASSP 1997, vol. 2, pp. 731–734. IEEE (1997)
11. Salami, R., Laflamme, C.: Description of GSM Enhanced Full Rate Speech Codec. In: ICC 1997, Montreal, pp. 725–729. IEEE (1997)
12. Digital Cellular Telecommunications System Enhanced Full Rate (EFR) SpeechTranscoding GSM 06.60 ETSI Technical Report version 8.0.1. Release (1999)

13. Fink, G.A.: Markov Models for Pattern Recognition, pp. 61–92. Springer (2008)
14. Furui, S.: Digital Speech Processing, Synthesis and Recognition, 2nd edn., pp. 243–328 (2001)
15. Peinado, A.M., Segura, J.C.: Speech Recognition Over Digital Channels, pp. 7–29. John Wiley & Sons Ltd. (2006)
16. Holmes, J., Holmes, W.: Speech Synthesis and Recognition, 2nd edn., pp. 161–164. Taylor & Francis e-Library (2003)
17. Chu, W.C.: Speech Coding Algorithms, pp. 33–44. John Wiley (2003)
18. Sayoud, K.: Introduction To Data Compression, 3rd edn., pp. 540–542. Elsevier (2006)
19. Tan, Z.H., Lindberg, B.: Automatic Speech Recognition On Mobile Devices And Over Communication Networks, pp. 41–117. Springer (2008)
20. Fabregas Surigué de Alencar, V., Alcaim, A.: Transformations of LPC and LSF Parameters to Speech Recognition Features, pp. 522–528. Springer (2005)
21. Kim, H.K., Choi, S.H., Lee, H.S.: On Approximating Line Spectral Frequencies To Lpc Cepstral Coefficients. IEEE 8(2) (March 2000)
22. Young, S., Odell, J.: The HTK Book Version 3.4 (December 2006), http://htk.eng.cam.ac.uk/

The Fuzzy WOD Model with Application to Biogas Plant Location*

Camilo Franco, Mikkel Bojesen, Jens Leth Hougaard, and Kurt Nielsen

Department of Food and Resource Economics, Faculty of Science,
University of Copenhagen, Denmark

Abstract. The decision of choosing a facility location among possible alternatives can be understood as a multi-criteria problem where the solution depends on the available knowledge and the means of exploiting it. In this sense, knowledge can take various forms, where the imprecise nature of information can be expressed by degrees of intensity in which the alternatives satisfy the given criteria. Hence, such degrees can be gradually expressed either by unique values or by intervals, in order to fully represent the characteristics of each alternative. This paper examines the selection of biogas plant location based on a decision support model capable of handling and exploiting both interval and non-interval forms of knowledge. Such model is built on a fuzzy approach to weighted overlap dominance, where an interactive procedure is developed allowing the individuals to explore and put into perspective how their different attitudes affect the final ranking of alternatives.

Keywords: Decision support, fuzzy data, weighted overlap dominance, interval degrees, biogas plant location.

1 Introduction

The selection of facility location is a decision problem that can be analyzed from many perspectives and using many different criteria, following some suitable methodology that allows aggregating and exploiting the different pieces of available information (see e.g., [5,8,10]). In particular, such information can be expressed in precise or imprecise form, i.e., by a unique value or by an interval set of values, according to the type of knowledge that exists on how the alternatives satisfy some property of interest. In this way, the type of knowledge that can be expressed by means of one unique degree is taken as being precise, and the one that can be expressed by an interval degree is taken to be imprecise. Hence, by dealing with interval data, we deal with *uncertainty due to imprecision* (on the contrary to other approaches where intervals represent lack of knowledge, as in [2,12]).

* Financial support from the Center for Research in the Foundations of Electronic Markets (CFEM), funded by the Danish Council for Strategic Research, is gratefully acknowledged.

Á. Herrero et al. (eds.), *International Joint Conference SOCO'13-CISIS'13-ICEUTE'13*, 141
Advances in Intelligent Systems and Computing 239,
DOI: 10.1007/978-3-319-01854-6_15, © Springer International Publishing Switzerland 2014

Recently a decision support model has been developed for treating multi-dimensional interval data in an interactive way, named the Weighted Overlap Dominance (WOD) Model [9]. This model places itself within the existing outranking literature for decision aid (see e.g., [3,4,16,17]), with the advantage of being built for the direct handling of intervals, addressing the particular challenges of interval pairwise comparisons. Therefore, examining the WOD Model from the perspective of fuzzy set theory [6,20], information can be gradually verified and aggregated taking into consideration the imprecision associated to each alternative. Based on this approach, we are able to use all the available knowledge, without requiring data to take the form of a unique value. Furthermore, we are able to identify a ranking of the alternatives according to their relevance and degree of satisfaction of the most desired and non-rejected properties. Hence, the fuzzy-WOD Model allows the decision-maker (DM) to take into account not only the satisfaction and non-rejection of criteria but also, the magnitude of uncertainty due to the imprecision associated with each alternative.

The objective of this paper is to illustrate the fuzzy-WOD methodology and its application as a decision support tool. We apply the model to a case study on selecting locations for slurry based biogas plants in the municipality of Ringkøbing-Skjern, Denmark. This is a problem that rises from national policies regarding the construction of new biogas plants that use animal slurry as main resource input for producing combined heat and electricity, where different location criteria have to be satisfied to varying degrees. Such criteria refer to economic, social, political and environmental dimensions, where the DM has to consider all of them for arriving at a balanced and sustainable conclusion. Then, following the fuzzy-WOD methodology and given a set of alternatives, we exploit the existing knowledge for recommending which are the better locations for building the biogas plants.

This paper is organized as follows. First, the fuzzy-WOD Model is presented, introducing its basic definitions and explaining its methodology. Second, this methodology is illustrated by implementing the fuzzy-WOD procedure on the location selection problem for biogas plants. Lastly, results are presented for discussion and sum up with directions for future research.

2   The Fuzzy-WOD Model

The WOD methodology has been initially presented in [9], as an outranking procedure proposed for handling multidimensional interval data. Here we present a fuzzy approach on this methodology for exploring the meaning of interval aggregation and achieving a ranking of alternatives, which can be directly weighed by the DM.

In this way, given a set of alternatives $N = \{1, ..., n\}$ and a set of criteria $M = \{1, ..., m\}$, we want to verify up to which extent some alternative $a \in N$ satisfies a given criterium $j \in M$. Then, we can use fuzzy sets [6,20] in order to explore how this verification occurs.

Definition 1. *A fuzzy set A is given by $A = \{\langle a, \mu_A(a) \rangle \,|\, a \in N\}$, where $\mu_A(a):$ $N \rightarrow [0,1]$ is its characteristic function and $\mu_A(a) \in [0,1]$ is its membership* *degree, expressing up to which extent the object $a \in N$ satisfies the property represented by A.*

Under this definition, a unique value is used for representing the degree in which A is verified by some $a \in N$. Nonetheless, such value may not be unique, and intervals may be necessary for representing the membership degree (see e.g., [2,12,13,15]). Then interval-valued fuzzy sets can be used instead [7].

Definition 2. *An interval-valued fuzzy set Z is given by $\mu_Z(a) : N \to L([0,1])$, where $L([0,1]) = \left\{ x = [x^L, x^U] \mid (x^L, x^U) \in [0,1]^2 \text{ and } x^L \leq x^U \right\}$ denotes the set of all closed sub-intervals in $[0,1]$ and $\mu_Z(a) \in L([0,1])$ is the membership degree of Z.*

In this sense, let Z_j represent the property characterizing some criterium $j \in M$, such that $\mu_{Z_j}(a) = \left[\mu_{Z_j}^L(a), \mu_{Z_j}^U(a) \right]$ is the interval degree in which the predicate "a satisfies the property Z_j" is verified. Then, considering all of the m criteria $j \in M$, we have that the input data for the fuzzy-WOD model is given by the m-dimensional cubes $c_a = \left[\mu_{Z_j}^L(a), \mu_{Z_j}^U(a) \right]^m$, containing the interval degrees for every alternative in N.

Now, this model develops its procedure by means of the imprecission-volume operator V, which measures the magnitude of imprecision associated to every alternative. Such operator is characterized as an m-dimensional imprecision measure in the following sense.

Definition 3. *Let μ_{Z_j} be the membership degree of Z_j, given by the respective lower and upper bounds $\mu_{Z_j}^L$ and $\mu_{Z_j}^U$, and let $\varepsilon \in (0,1)$. Then, the function $\delta : [0,1]^2 \to [\varepsilon, 1 + \varepsilon]$ is an imprecision measure if and only if it fulfills the following conditions:*

1. $\delta\left(\mu_{Z_j}\right) = \varepsilon$ if and only if $\mu_{Z_j}^U = \mu_{Z_j}^L$.
2. $\delta\left(\mu_{Z_j}\right) = 1 + \varepsilon$ if and only if $\mu_{Z_j}^L = 0$ and $\mu_{Z_j}^U = 1$.
3. If $\mu_{Z_j}^U > \mu_{Z_j}^L$ then $\delta\left(\mu_{Z_j}\right) > \varepsilon$.
4. For two memebership degrees μ_{Z_j} and ν_{Z_j}, $\delta\left(\mu_{Z_j}\right) > \delta\left(\nu_{Z_j}\right)$ if and only if $\left(\mu_{Z_j}^U - \mu_{Z_j}^L \right) > \left(\nu_{Z_j}^U - \nu_{Z_j}^L \right)$.

In this way, Definition 3 states that (1) imprecision is minimum if and only if the lower and upper bounds of the membership degree are the same, (2) imprecision is maximum if and only if the lower bound is minimum and the upper bound is maximum, (3) imprecision is greater than ε every time the upper bound of the interval is greater than its lower bound, and (4) imprecision increases monotonically with the difference between the lower and upper bounds.

Hence, the function

$$\delta'\left(\mu_{Z_j}\right) = \mu_{Z_j}^U - \mu_{Z_j}^L + \varepsilon \tag{1}$$

is an imprecision measure for the cases of one unique degree or an interval degree, and its respective m-dimensional formulation is given by,

$$V(\mu) = \delta'\left(\mu_{Z_1}\right) \cdot \delta'\left(\mu_{Z_2}\right) \cdot \ldots \cdot \delta'\left(\mu_{Z_m}\right) = \delta'(\mu)^m. \tag{2}$$

Now, we need the following definitions for obtaining the *outranking* and *indifference* relations for any pair of alternatives $a, b \in N$. First, define a vector of weights $w \in \mathbb{R}_+^m$ according to the importance of the criteria, and label the alternatives such that $w \cdot \mu^U(a) > w \cdot \mu^U(b)$ holds, where $\mu^U(a) = \left(\mu_{Z_1}^U(a), ..., \mu_{Z_m}^U(a) \right)$. In case the equality $w \cdot \mu^U(a) = w \cdot \mu^U(b)$ holds, label the alternatives according to their imprecision, such that $V(w \cdot \mu(a)) > V(w \cdot \mu(b))$ holds. In any other case, if $w \cdot \mu^U(a) = w \cdot \mu^U(a)$ and $V(w \cdot \mu(a)) = V(w \cdot \mu(b))$ hold, then a and b are indifferent, i.e., it is true that $a \sim b$.

Second, define the three following sets,

$$\hat{Z}(a,b) = \left\{ i \in c_a | w \cdot i > w \cdot \mu^U(b) \right\}, \tag{3}$$

$$\check{Z}(a,b) = \left\{ i \in c_a | w \cdot i < w \cdot \mu^L(b) \right\}, \tag{4}$$

$$\tilde{Z}(a,b) = \left\{ i \in c_b | w \cdot i > w \cdot \mu^L(a) \right\}. \tag{5}$$

In this way, we have three different cases that require our attention.

Case 1. Strict domination. This is the case where $\tilde{Z} = \varnothing$ holds true. Here, if it is verified that $\tilde{Z} = \varnothing$, we say that a *strictly dominates* b, and it holds that a outranks b ($a \succ b$).

Case 2. Partial overlap. This is the case where $\check{Z} = \varnothing$ holds true. Then, it is said that a *outranks* b, i.e., $a \succ b$ holds, if and only if the following expression holds,

$$P(a,b) = \frac{V(\hat{Z})}{V(c_a)} + \frac{V(c_a \setminus \hat{Z})}{V(c_a)} \frac{V(c_b \setminus \tilde{Z})}{V(c_b)} > \beta, \tag{6}$$

where $\beta \in [0,1]$. Here, both alternatives are indifferent ($a \sim b$) if and only if it is true that $P(a,b) \le \beta$.

Case 3. Complete overlap. This is the case where $\check{Z} \ne \varnothing$ holds true. Here it is said that for some $\gamma \in \mathbb{R}_+$, $a \succ b$ holds if and only if it is verified that

$$\frac{V(\hat{Z})}{V(\check{Z})} > \gamma, \tag{7}$$

$a \sim b$ holds if and only if it is verified that

$$\frac{V(\hat{Z})}{V(\check{Z})} = \gamma, \tag{8}$$

and $b \succ a$ holds if and only if it is verified that

$$\frac{V(\hat{Z})}{V(\check{Z})} < \gamma. \tag{9}$$

Hence, the alternatives can be ordered according to their outranking and indifference relations, and the DM is allowed to interact with the model by updating the free parameters β and γ. The specification of these parameters allow exploiting the flexibility of the interval data, obtaining different relational situations depending on their values. Besides β and γ, a third parameter α is defined in order to allow the DM to reduce the initial set of alternatives, focusing on the ones that obtain a higher position in the outranking order. This parameter is defined as follows. Let alternative $a_0 \in N$ be the alternative with the maximal weighted value, such that $a_0 = \arg\max_{a \in N} w \cdot \mu^U(a)$,and for every alternative $a \in N$, define the set $c_a|a_0 = \left\{ i \in c_a | w \cdot i \geq w \cdot \mu^L(a_0) \right\}$,such that the ratio between $c_a|a_0$ and c_a refers to the weighted overlap of a relative to a_0. Then, the DM can choose to consider only the alternatives with a weighted overlap relative to a_0 greater than $\alpha \in [0, 1]$.

Notice that up to this moment, the fuzzy-WOD approach allows the interaction of the individual with the ordering process by means of user-defined parameters, obtaining an outranking order over the alternatives. But due to the semi-tranistive character of WOD's outranking relation (see [9]), and the consequent intransitivity of its indifference relation, it still remains the task of building a ranking over the set of alternatives. Hence, the fuzzy-WOD procedure ranks alternatives according to their relevance in the following way. Let the importance of an object $a \in N$ be given by the number of objects that it outranks, denoted by d_a, and let D_a be the set of all objects that are outranked by a. Then, relevance degrees are defined for every alternative $a \in N$, by,

$$\sigma_a = d_a + \sum_{b \in D_a} d_b. \tag{10}$$

As a result, based on the fuzzy-WOD outranking procedure, alternatives are ranked according to their relevance degrees, and the DM can find support for understanding the decision problem and the possible solutions that may exist. In the following Section we apply this procedure over the location selection for biogas plants, and in that way, not only do we offer support for the DM, but also, we illustrate the fuzzy-WOD methodology for its better understanding.

3   Application to Location Selection of Biogas Plants

Site selection for biogas plants is a multi-criteria problem that policy authorities and municipal planners face in the effort of using the available natural resources and agricultural by-products in a *sustainable* way, i.e., being the most efficient and least damaging as possible. Consequently, the municipal planners have to consider the multiple attributes and restrictions associated with such location problem. Such a task is complex and often very time demanding, which is why some kind of decision support is appropriate and strongly needed. Hence, the mission of the decision support model is to first, serve as analytical tool for understanding the problem, and second, identify possible courses of action that satisfy the given set of criteria. Here we propose the use of the fuzzy-WOD Model for providing decision support on choosing between suitable locations for biogas plants in the municipality of Ringkøbing-Skjern, Denmark.

Ringkøbing-Skjern municipality is Denmark's largest municipality, with an area of 1470 km^2, and at the same time one of the least populated municipalities, with a population of 39 inhabitants per km^2 [19]. The local authorities have the ambition that the municipality becomes self-sufficient with respect to energy consumption by 2020, producing all of its electricity in the form of renewable energy. As part of fulfilling this policy goal, biogas production based on the exploitation of pork and cattle manure is thought to play an important role, taking into account that Ringkøbing-Skjern is intensively farmed, with a stock of 566.000 animal units of pigs and 484.000 animal units of dairy cattle within the municipal boundaries [19].

Due to the bioenergy ambitions in Ringkøbing-Skjern, the municipal authorities have been working with biogas planning for years, gaining experience and knowledge about the many factors that need to be considered and balanced. This, combined with the production opportunities and policy objectives in Ringkøbing-Skjern, provides a good basis for a case study on how the fuzzy-WOD procedure offers decision support to real world situations.

Therefore, the available knowledge on the suitability of an alternative, i.e., a biogas plant location of 1 km^2, for being accepted or rejected, can be given by a geographical map (see e.g. [8]), where regions are colored according to their acceptability or rejectability (see Fig. 1). There are four classes of regions, given by *green* (*G*), *yellow* (*Y*), *orange* (*O*) and *red* (*R*), where their meaning is to be taken as "green is acceptable", "yellow is weakly acceptable", "orange is weakly rejectable" and "red is rejectable".

Then, given the set of alternatives/locations $N = \{1, ..., n\}$, the four functions $g', y', o', r' : N \to [0, 1]$ are assumed to exist, respectively assigning to each alternative a value according to their classification under G, Y, O and R. In this way, alternatives agree with not just one of the classes $\{G, Y, O, R\}$, but several of them (being this a main attribute of *fuzzy data*), due to the fact that an alternative may share parts of the different colored regions. In consequence, there is information on the acceptability of the alternatives, where a completely green alternative is acceptable with a maximum degree but also, on their rejectability, where a completely red alternative is rejectable with a maximum degree. Hence, there is a positive and a separate negative dimension, grouping all the information into two families of criteria for respectively accepting and rejecting some alternative $a \in N$.

Now, membership degrees can be built from the available knowledge, so they can be used as input for the fuzzy-WOD Model. Let $\mu_{Z_1}(a)$ represent the degree of verification for the predicate "a satisfies the property of being accepted", and let $\mu_{Z_2}(a)$ represent the degree of verification for the predicate "a satisfies the property of being rejected". While the former refers to a property that has to be satisfied, the latter refers to a property that has to be avoided, or not-satisfied. Hence, the information on rejectability has to be negated (by means of an involutive operator) so the fuzzy-WOD Model is able to weigh it accordingly.

Fig. 1. Alternative locations for biogas plants over acceptability-rejectability zones

Then, the lower and upper bounds for these membership degrees can be obtained from standard fuzzy aggregation operators [18], such that for some t-conorm S, like e.g. the Lukasiewicz t-conorm S^L, where for any $x, x' \in [0, 1]$, $S^L(x, x') = \min(x + x', 1)$, and some strong negation n, like e.g. the complement $n(x) = 1 - x$, we have that,

$$\mu_{Z_1}(a) = [g'(a), S(g'(a), y'(a))], \tag{11}$$

$$\mu_{Z_2}(a) = [n(S(r'(a), o'(a))), n(r'(a))]. \tag{12}$$

Notice that accceptability will only coincide with non-rejectability if the equality $\mu_{Z_1} = \mu_{Z_2}$ holds, which will occur only if $g'(a) = n(r'(a))$ holds true (for more details on the particular relation between interval and positive-negative knowledge, see e.g. [1,15]).

Now, we can apply the fuzzy-WOD procedure in order to rank the alternatives in N acccording to their acceptability and non-rejectability degrees, where for every alternative $a \in N$ and criteria $j \in M$, the input data for the fuzzy-WOD model is given by the 2-dimensional cubes $c_a = \left[\mu_{Z_j}^L(a), \mu_{Z_j}^U(a) \right]^2$.

4   Results and Discussion

The decision support system that is built on the fuzzy-WOD Model is conceived as a dynamic mechanism, representing decision as a learning process, where the individual is able to *revise* the values of the parameters and of the criteria weights (for more details see [9]). In this way, the input data that is introduced into the system consists of: (1) the m-dimensional cubes c_a for every alternative $a \in N$; (2) the vector of weights w according to the importance of each criterium $j \in M$; and (3) the vector of free parameters p containing the initial values for α, β and γ. Then, an initial outcome is produced, under the first iteration $t \in T$, offering a ranking of alternatives dependent of w and p, denoted by $Rank(t)$. Once $Rank(t)$ is obtained for the first iteration, the DM is allowed to revise any of the values either for the criteria weights w_j or for the free parameters in p. If he revises, the new input data (c_a, w', p') goes under the fuzzy-WOD procedure for a new ranking $Rank(t+1)$to be obtained. This procedure can be repeated as long as the DM desires to.

Hence, introducing the input data consisting of 365 different alternatives, represented each one by a cube c_a for every $a, b, c, ... \in N$, we iterate $t = 8$ times using different values for w and p. The different combination of weights and parameters are shown in Table 1.

Table 1. Different combinations of parameters and weights for each iteration t

Iteration t	α	β	γ	w_1	w_2
1	1	0.3	1	1	1
2	0.8	0.3	1	1	1
3	1	0	1	1	1
4	1	0.5	1	1	1
5	1	0.5	1.3	1	1
6	1	0.5	0.7	1	1
7	1	0.5	1	1	2
8	1	0.5	1	2	1

As a result, according to each $Rank(t)$, we obtain a certain ranking for the most relevant alternatives, as shown in Table 2. First, the system obtains $Rank(1)$. Then, in $Rank(2)$, due to the revision of α, alternatives g and i are left out from the top 9 and alternative m is ranked 7th. Notice that when the α parameter is given the value of 1, we are considering only the alternatives that have a maximum overlap with a_0. Hence, if we lower its value, more alternatives will come into play. In this sense, the more alternatives are considered, the more *confident* the DM can be that the system is not leaving behind any relevant alternatives. Then, in $Rank(3)$ and $Rank(4)$, due to changes in β, the indifference between d and e is refuted due to the outranking of f over d. In this way, β refers to the estimation of the outranking threshold for the *partial overlap* case. Next, due to changes in γ, alternative c is relegated to the 6th position in $Rank(5)$

while in $Rank\,(6)$, c recovers the 3rd place but now alternative k, which used to be ranked last or simply ignored, is now 5th, being indifferent with f. Hence, the specification of the parameter γ has to be done carefully (due to the sensitivity of the final ranking to its value), mainly because it determines, by one unique value, if for some $a, b \in N$, any of the three situations $a \succ b$, $b \succ a$ or $a \sim b$ holds. Following a neutral attitude, this parameter is fixed to 1. Finally, giving more weight to the negative information, $Rank\,(7)$ is obtained, while assigning more weight to the positive information, $Rank\,(8)$ is obtained instead. Here, notice that alternative e is ranked third over c when more strength is assigned to the negative information. Therefore, alternative e is less harmful or damaging than c. On the contrary, assinging more strength to the positive information, alternatives c, d and f appear first than e.

Table 2. Different rankings for each iteration t

Ranking	$t = 1$	$t = 2$	$t = 3$	$t = 4$	$t = 5$	$t = 6$	$t = 7$	$t = 8$
1	a	a	a	a	a	a	a	a
2	b	b	b	b	b	b	b	b
3	c	c	c	c	e	c	e	c
4	$d \sim e$	$d \sim e$	$d \sim e$	e	$g \sim h$	e	c	d
5	f	f	f	f	i	$k \sim f$	d	f
6	$g \sim h$	j	$g \sim h$	d	c	d	$g \sim h$	e
7	i	m	i	$g \sim h$	d	l	i	$g \sim h$
8	j	h	j	i	f	j	f	i
9	k	k	k	j	j	g	l	j

In consequence, alternatives a and b are recommended for building the new biogas plants, and if a third plant is required, the DM may decide either to avoid the negative aspects of c by choosing e, or to select c based on the possible better outcomes that is can provoke.

Notes and Comments. As future lines of research, there is a certain difficulty in forcing the DM to find a unique value in order to express his attitudes in the form of the parameters β and γ. Hence, such parameters could be expressed in the form of words, extending the fuzzy-WOD procedure into a Computing with Words environment (see e.g., [11]). Also, notice that there may be different experts with different beliefs, such that a social approach to the fuzzy-WOD model could explore how distinct rankings can be aggregated for identifying a single choice (as in the case of placing a biogas plant). From an applied perspective, it is pointed out that the positive and negative family of criteria can be explored in more detail, directly addressing distinct criteria for selecting the most sustainable biogas plant locations. Some of these criteria refer to production potentials, transport economics, distribution of district heating or population density. As it is an ongoing investigation, we have to deal not only with precise criteria, but also with the direct elicitation of their weights from farmers or experts in the field. Besides, the fuzzy-WOD results should be compared to the ones of other outranking approaches found in multi-criteria decision making (see e.g., [3,4,16,14,17]).

References

1. Atanassov, K., Gargov, G.: Interval valued intuitionistic fuzzy sets. Fuzzy Sets and Systems 31, 343–349 (1989)
2. Barrenechea, E., Fernández, A., Herrera, F., Bustince, H.: Construction of interval-valued fuzzy preference relations using ignorance functions: Interval-valued non dominance criterion. In: Melo-Pinto, P., Couto, P., Serôdio, C., Fodor, J., De Baets, B. (eds.) Eurofuse 2011. AISC, vol. 107, pp. 243–255. Springer, Heidelberg (2011)
3. Behzadian, M., Kazemzadeh, R., Albadvi, A., Aghdasi, M.: PROMETHEE: A comprehensive literature review on methodologies and applications. European Journal of Operational Research 200, 198–215 (2010)
4. Chen, T., Tsao, C.: The interval-valued fuzzy TOPSIS method and experimental analysis. Fuzzy Sets and Systems 159, 1410–1428 (2008)
5. Farahani, R., SteadieSeifi, M., Asgari, N.: Multiple criteria facility location problems: A survey. Applied Mathematical Modelling 34, 1689–1709 (2010)
6. Goguen, J.: The logic of inexact concepts. Synthese 19, 325–373 (1969)
7. Grattan-Guinness, I.: Fuzzy membership mapped onto interval and many-valued quantities. Mathematical Logic Quarterly 22, 149–160 (1976)
8. Higgs, G.: Integrating multi-criteria techniques with geographical information systems in waste facility location to enhance public transportation. Waste Management & Research 24, 105–117 (2006)
9. Hougaard, J.L., Nielsen, K.: Weighted Overlap Dominance - A procedure for interactive selection on multidimensional interval data. Applied Mathematical Modelling 35, 3958–3969 (2011)
10. Kahraman, C., Ruan, D., Dogan, I.: Fuzzy group decision-making for facility location selection. Information Sciences 157, 135–153 (2003)
11. Martínez, L., Ruan, D., Herrera, F.: Computing with words in decision support systems: an overview on models and applications. International Journal of Computational Intelligence Systems 3, 382–395 (2010)
12. Montero, J., Gómez, D., Bustince, H.: On the relevance of some families of fuzzy sets. Fuzzy Sets and Systems 158, 2429–2442 (2007)
13. Özturk, M., Pirlot, M., Tsoukias, A.: Representing preferences using intervals. Artificial Intelligence 175, 1194–1222 (2011)
14. Özturk, M., Tsoukias, A.: Bipolar preference modeling and aggregation support. International Journal of Intelligent Systems 23, 970–984 (2008)
15. Rodríguez, J.T., Franco, C.A., Montero, J.: On the semantics of bipolarity and fuzziness. In: Melo-Pinto, P., Couto, P., Serôdio, C., Fodor, J., De Baets, B. (eds.) Eurofuse 2011. AISC, vol. 107, pp. 193–205. Springer, Heidelberg (2011)
16. Roy, B.: The outranking approach and the foundations of ELECTRE methods. Theory and Decision 31, 49–73 (1991)
17. Sayadi, M., Heydari, M., Shahanaghi, K.: Extension of VIKOR method for decision making problem with interval numbers. Applied Mathematical Modelling 33, 2257–2262 (2009)
18. Schweizer, B., Sklar, A.: Probabilistic Metric Spaces. Elseiver Science Publishing Co., New York (1983)
19. Statistikbanken (2013), http://www.statistikbanken.dk
20. Zadeh, L.: The concept of a linguistic variable and its application to approximate reasoning-I. Information Sciences 8, 199–249 (1975)

The Application of Metamodels Based on Soft Computing to Reproduce the Behaviour of Bolted Lap Joints in Steel Structures

Julio Fernández-Ceniceros, Ruben Urraca-Valle,
Javier Antoñanzas-Torres, and Andres Sanz-Garcia

EDMANS Research Group, University of La Rioja, Logroño, Spain
http://www.mineriadatos.com

Abstract. A promising field of research in steel structures regarding their preliminary design and optimization is the replacement of expensive computational finite element models with more efficient techniques. Without a significant loss of accuracy, new proposals should be able to consider not only the ideal load-displacement response but also relevant failure mechanisms and imprecisions in material properties. The article proposes the use of metamodels based on soft computing as an overall approximation system for structures analysis. This approach has been applied in several fields but, till nowadays, its implementation on structural analysis in early esign seems quite limited to a few theoretical cases. Taking advantage of artificial neural network as global approximation technique, the parameters for more realistic and informative load-displacement curve including nonlinear effects (damage mechanics) are estimated for bolted steel lap joints. Our results demonstrate the accuracy of the metamodel implemented can be close to simulations and also real experimental tests.

Keywords: Artificial Neural Network, Metamodel, Finite Element Analysis, Steel Structure, Bolted Lap Joint.

1 Introduction

In the existing literature, many approaches to early structures design aim primarily at the development of strategies for rapidly finding an optimal design. It is well known that expensive computational finite element models (FEMs) do not fit correctly with the requirements in computation speed of early design applications. Therefore, a promising field of research is the replacement of these by more efficient systems. For instance, metamodel driven design optimization (MDDO) is focused on finding functions that estimate the behavior on an entire design space with reasonable accuracy. The resulting metamodels, also named response-surface models, are subsequently used for searching the optimal design of the structure. These are composed of several single models that are frequently based on soft computing (SC) due to their high generalization capacity [6].

Á. Herrero et al. (eds.), *International Joint Conference SOCO'13-CISIS'13-ICEUTE'13*, 151
Advances in Intelligent Systems and Computing 239,
DOI: 10.1007/978-3-319-01854-6_16, © Springer International Publishing Switzerland 2014

Traditional analyses of steel structures are usually carried out without taking into account interactions between their members and joints. However, recent studies together with modern design codes have shown the necessity of design structures assuming that the joints are transferring the loads from one member to the others. This assumption requires that additional variables have to be included in the design of structures, increasing the complexity so much that analytical evaluation is not suitable. The alternative is usually based on trial and error approach combining finite element analysis (FEA) and validation tests. FEA is a nonlinear modeling technique that allows to solve these problems with a large number of design parameters provideing a more realistic behavior of steel structures. The main requirements are the availability of experimental data to validate the FEM and enough computational time to achieve the convergence of the solution. However, as many other parameters are considered in the calculation, such as failure criteria, crack propagation and imprecision in materials, the FEA becomes complex and inefficient.

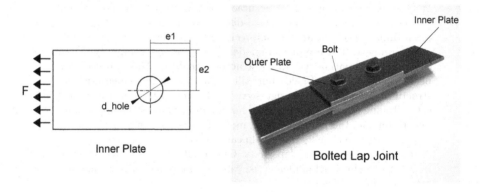

Fig. 1. Geometrical parameters of bolted lap joints

In this study, the FEA is replaced by other based on metamodeling using SC models that show higher computational efficiency but less accuracy. In recent years, the use of metamodels has been explored in several fields. Till nowadays, no efficient implementations of metamodels on the basis of SC have been reported for solving highly nonlinear structural problems. Therefore, we propose a solution based on the use of artificial neural networks (ANNs) as global estimators to create a metamodel that describes the realistic behavior of steel joints.

2   Methodology

2.1   Basic Metamodeling Techniques

Metamodels, also known as approximation models or response surface models, imitate the behavior of complex computational models. There are many types such Universal Kriging, Kriging with External Drift (KED), ANN, etc. and they have become relatively common nowadays. Generally speaking, there are mainly two applications of metamodels. First, the most widely known type encompasses the construction of a simple system for helping in optimization processes. Its goal is to direct the search of the global optimum of a specific problem. The second type is not focused on previous search, but rather on accurately predict the behavior of a complete system. The main goal is to tune the metamodel to imitate the underlying native model as closely as possible on a specific region of the domain. In this article, authors are concerned with the latter.

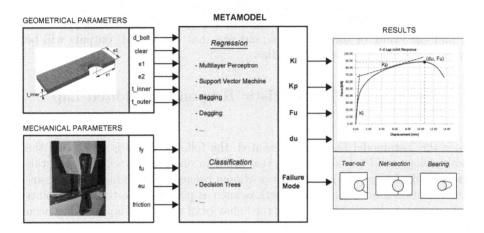

Fig. 2. Basic scheme of the proposed metamodel for predicting a bolted lap joint response

2.2   Basic Soft Computing Models

Through a wide list of SC models available we have selected one machine learning (ML) algorithms that provides the best generalization capacity, the multilayer-perceptron (MLP) neural network. This is a type of ANN considered one of the best-known biological-based information processing methods. MLPs can solve challenging problems after they are correctly trained. Their generalization ability is their strong point because they are able to tackle different cases with similar characteristics [9,7]. The information is moved forward in the MLP without loops from the input nodes through the hidden nodes to the output node or nodes [5]. It is also widely known only one hidden layer is required to approximate any continuous function if the number of connection weights is enough [8]. Theoretically, MLP model with only one hidden layer is able to predict the majority of continuous complex functions and also solve any high dimensional classification problem. Mathematically, we can express the output of the MLP with one hidden layer as follows:

$$\hat{y}\left(k\right) = \alpha_0 + \sum_{i=1}^{n} \alpha_i \cdot \phi_i \left(\sum_{j=1}^{l} w_{ij} \cdot x_j\left(k\right) + w_{i0}\right) \tag{1}$$

where $x_j\left(k\right) = \left[x_1\left(k\right), x_2\left(k\right), \ldots, x_l\left(k\right)\right]^T$ is the input vector at instant k with length l, α_i and w_{ji} are the weighting coefficients of the network (α_0 and w_{i0} are the output and input layer bias respectively, ϕ_i is the nonlinear activate function, n is the number of hidden neurons, and $\hat{y}\left(k\right)$ represents the estimation of the output variable y. In regression problems with a single output y there is only one neuron in the output layer. For the learning process, the error back-propagation (BP) algorithm is usually the most applied. The goal of this training process is to find the values of the weights α_i and w_{ji} that cause MLP outputs will be equal or very close to the target values.

3   Case Study: Characteristic Response of Bolted Lap Joint

Once the metamodel has been presented, the following step is to explain the methodology used to validate in a real problem on steel connections. For this purpose, the bolted lap joint has been chosen because it is a fundamental component within more complex connections such as pinned beam-to-column joints or beam splices joints. Moreover, as the behavior of the bolted lap joint has been widely investigated and is well-documented, a great amount of experimental tests in the literature can be useful to validate our methodology.

Regarding the configuration of a bolted lap joint, it is composed of the bolts and the connected elements (column flanges, beam webs, cover plates), which will be treated as plates. According to the Fig. 1 when the lap joint is loaded in shear, forces are transferred from the inner plates to the outer plates by plate-to-bolt contact [2]. Despite the fact that there are analytical models which accomplish the maximum force supported by the joint, there is a lack of knowledge about their ductility and the assessment of the whole force-displacement characteristic response beyond the plastic zone. The main reason of this issue is the nonlinearities involved in the deformation process, i.e. contacts between elements, stress concentrations, and material plasticity. Moreover, there is another difficulty associated to the onset of the failure and how this failure occurs. In fact, the literature reports four main failure modes for the bolted lap joint:

- *Bearing failure in the plates*: it is considered to constitute failure by the excessive deformation of material behind the bolt, regardless of whether the connection has reserve strength.
- *Tear-out failure in the plates*: it is associated with connections which have a relatively small end distance and pitch, while having a comparatively large edge distance and gauge to avoid net section failure mode.
- *Net-section failure in the plates*: a critical fracture in connections with relatively narrow plate widths.

– *Shear failure in the bolt*: a brittle fracture occurs in the bolts when the shear
load exceeds its capacity.

Consequently, we consider of interest to develop a refined FEM which takes into
account not only the elastic and plastic response, but also damage mechanics to
detect the onset of the failure and the failure mode. However, the complexity of
this model leads to extremely high computational costs many times, becoming
the FEA an ineffective tool for structural engineers in their daily routine. For
that reason, we propose a methodology which takes advantage of the accuracy of
the FEA to generate a useful training database. The final goal is to train a soft-
computing-based metamodel able to predict the output parameters needed to
reproduce the characteristic response and the failure mode of the lap joint. The
overall scheme of the methodology proposed to predict the lap joint response is
shown in Fig. 3.

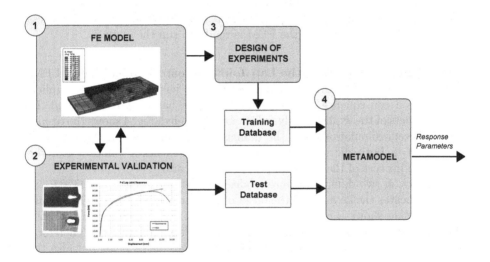

Fig. 3. Methodology for designing a metamodel based on data collected from FEA

3.1   Development of the Refined FEM

An improvement in the original FEM of the bolted lap joint [3], has been carried
out in order to include a model of material damage which enables to predict the
onset of the damage and estimate the ductility of the joint. The numerical model
has been created using ABAQUS v.6.11, a general purpose FEA software. The
main features of the new refined model are described below.

Progressive Damage for Ductile Materials. In order to detect the point
in which the damage is initiated and estimate the ductility of the joint, a failure
criterion has been included to the original FEM. For this purpose, ABAQUS
provides models to predict progressive damage and failure in ductile materials.

Progressive damage refers to the stiffness degradation of the material which leads to the loss of load-carrying capacity. This process can be divided into a criterion for the onset of damage and the law which governs the damage evolution [1]. At this stage of the work, only the onset of damage has been taken into account.

The ABAQUS capabilities for predicting the onset of damage due to nucleation, growth, and coalescence of voids assume that the equivalent plastic strain at the onset of damage is a function of stress triaxiality. Moreover, according to the research work carried out by Lemaitre in the continuum damage mechanics [11,10], the equivalent strain at failure can be calculated as follow:

$$\varepsilon_f^* = \varepsilon_f \left[\frac{2}{3}\left(1+\nu\right) + 3\left(1-2\nu\right) \left(\frac{\sigma_m}{\sigma_{eq}} \right)^2 \right]^{-1} \tag{2}$$

where ε_f is the strain at rupture from uniaxial tension tests, ν is the Poisson ratio, σ_m is the mean normal stress and σ_{eq} is the effective flow stress [4].

Therefore, once the relationship between plastic strain at failure and stress triaxility has been established, the FEM is ready to run the analysis.

Finite Element Analysis of the Lap Joint. In contrast to the original FEM, an improvement in the initial boundary conditions has allowed to use implicit analysis. As a consequence, the computational cost has been reduced and the dynamic effects of the explicit analysis have been avoided. Fig. 4 shows the results of two different simulations and the damage initiation criterion (DUCTCRT) is displayed in both plots. When this parameter reaches a value of 1.0, the degradation process of the material properties appears.

In the Fig. 4, two different failure modes are identified. The DUCTCRT parameter indicates the area where the onset of the damage is located. The picture on the left side shows a bearing failure whereas in the second picture (on the right side) the plates fails according to the net-section failure mode. Therefore, the refined FEM provides not only forces and displacements fields but also the failure mode of the plates.

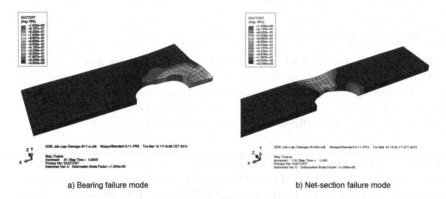

a) Bearing failure mode   b) Net-section failure mode

Fig. 4. Two examples of damage initiation criterion

3.2   Experimental Validation

Experimental tests from different sources in the literature have been used to validate the refined FEM. The experimental programs made at Delft University of Technology and Ljubljana University are summarized by [2] and provide detailed information about the specimens and material properties. High strength steel (S690) was used in all the tests with the scope of studying the bearing and the net-section failure modes.

A total of 19 specimens were compared against our numerical model. Three characteristic parameters for each test were checked in order to assess the accuracy of the refined FEM. These parameters were the maximum force supported by the joint (F_u), the ultimate displacement (d_u), and the failure mode (FM). The results are summarized in Table 2, where the ratios between experimental tests and FEM are provided. An excellent agreement has been achieved in the assessment of the maximum force. However, some discrepancies have been found in the estimation of the ultimate displacement and the failure mode due to the high sensitivity of the material properties in the plastic domain. On the whole, we state that the refined FEM is a reliable tool to simulate the response of bolted lap joints.

3.3   Design of Experiments

A FEM is created and validated against experimental tests. Taking advantages of the accuracy of this kind of models, a representative dataset is possible to create with them. This data should include the conventional range of geometrical and mechanical parameters in bolted lap joints. To this end, one of the most popular space-filling sampling methods is the Latin Hypercube Sampling (LHS) introduced by [12] in 1979. The method divides each input variable into n equally probable intervals and then samples random values for each interval. A LHS thus consists of k input variables and n experiments which can be defined as *LHS (n,k)*. Its principal advantage is that each input variable is represented in all divisions of its range [14].

The experiments are generated combining ten input parameters by the LHS method. All the variables are continuous with the exception of the bolt diameter *(d_ bolt)*, which only takes the nominal values of the ISO metric screw thread. Table 1 shows the description of each attribute, the range of its values, the mean, and the standard deviation. For each possible combination of inputs, a FEA is carried out to obtain the same output parameters mentioned above. The training dataset is finally composed of 10 input variables, 3 outputs and 600 instances.

3.4   Results

The generation of the models was developed on a Linux operating system SUSE 10.3 OS running in a Quad-Core Opteron server. The statistical software R-project 3.0 [13] and particularly the packages CLASS and AMORE are the tools used to preprocessing the data and training the classifiers and predictors.

Table 1. Statistics of input parameters in the dataset

Attributes	Description [units]	Range	Mean	Sd.
d_bolt	Bolt diameter [mm]	$M12 - M27$	-	-
clearance	Difference between d_bolt and bolt hole	$0.5 - 2.0$	1.25	0.43
e1	End distance [mm] (Fig. 1)	$10 - 58$	28.41	9.94
e2	Edge distance [mm] (Fig. 1)	$10 - 58$	28.37	9.90
t_inner	Inner plate thickness [mm]	$4 - 16$	10.00	3.47
t_outer	Outer plate thickness [mm]	$2.8 - 19.2$	9.52	3.70
friction	Friction coefficient between elements	$0.1 - 0.5$	0.3	0.12
fy	Yield strength of steel plates [MPa]	$235 - 355$	295.00	34.67
fu	Ultimate tensile strength [MPa]	$355 - 887.5$	590.38	112.53
eu	Deformation at ultimate strength	$0.15 - 0.40$	0.28	0.07

The dataset for training the metamodel is based on the results of the FEA. The first step carried out is a data normalization into a range of [0, 1] instead of a statistical standardization. In our opinion, the normalization will work well because the data are positive or zero. Then, the normalized dataset is used for training and validating the models using 10-fold cross validation (CV) and repeating the operation for 100 times. The training data are directly used to the modeling, and the validation data estimate how accurately a predictive model will perform in practice. The parameters for training the models are initially adjusted by trial and error. During training process, an additional validation dataset (20% of training dataset) was used for stopping this process when validation error is constant or starts to increase its value for a number of 10 iterations. This control prevents model overfitting.

The errors predicting the maximum force (F_u), the ultimate displacement (d_u) and the failure mode (FM) in bolted lap joints are summarized in following paragraph. These correspond to the mean and the standard deviation (sd) of the best estimator for each output variable. For each model, we obtain training and validation root mean squared errors ($RMSE_{tr}$ and $RMSE_{val}$ respectively). Finally, the experimental errors are computed in for the experimental test explained before. Additionally, we directly compare the prediction of the metamodel with data obtain from previous experimental tests to check the generalization capacity of the model.

According to the results, the $RMSE_{val}$ of the best MLP model for both outputs, which corresponds to a MLP with 9 neurons in the hidden layer, are as follows: $RMSE_{val}^{Fu} = 0.2781$, $RMSE_{val}^{du} = 0.2157$. Different MLP settings have been used in order to find the most accurate model, but no significant differences are observed using from five to nine neurons in the hidden layer. The rest of the MLP parameters were maintained at their same values after an initial trial and error adjustment. However, future experiments will focus on varying those parameters to obtain additional improvements.

On the other hand, the testing errors for the 19 experimental test are as follows: $RMSE_{tst}^{Fu} = 0.3581$, $RMSE_{tst}^{du} = 0.4303$. These are achieved in the prediction of the maximum force and the ultimate displacement. These are not a really low error considering those obtained by FEA, but it is useful to show the possibilities that the metamodel is able to provide. Moreover, a high precision is obtained in the prediction of the failure mode as (see Table 2). However the ratios between FEA and metamodel prediction are not so different. As Table 2 shows, the majority of the models studied present a ratio $Test/MM$ no much worse than the ratio $Test/EC$. For that reason, we still believe that the metamodel can be improved by using a better training procedure or increasing the number of instances in the training dataset.

Finally, it is worthy to mention the significant time reduction in the use of metamodels. Each of the FEA runs took between 24-41 min, depending on the size of the model and the number of elements. On the other hand, the predictive model took between 15-20 s during the training and cross validation process. However, once the predictive model was trained, the lap-joint response can be obtained immediately, without computational cost.

Table 2. Comparison of the results obtained in experimental tests, Eurocode 3 formulae, FEA and metamodel predictions

Id	Maximum Force (kN)					Ultimate Disp (mm)				Failure Mode		
	Test	$\frac{Test}{FEM}$	$\frac{Test}{EC}$	$\frac{Test}{MM}$	$\frac{FEM}{MM}$	Test	$\frac{Test}{FEM}$	$\frac{Test}{MM}$	$\frac{FEM}{MM}$	Test	FEM	MM
B102	273	0.99	1.19	0.66	0.67	17.9	0.96	0.4	2.4	B	B	B
B103	342	0.96	1.49	0.49	0.51	23.8	1.15	0.45	2.56	B/N	B	B
B104	360	0.94	1.57	0.46	0.48	8.4	0.98	1.86	0.53	B	N	B
B105	355	0.93	1.55	0.56	0.6	6.9	0.81	1.59	0.51	N	N	N
B106	445	0.99	1.94	0.7	0.7	10.2	1.23	1.3	0.95	N	B	N
B107	440	0.97	1.92	0.79	0.81	9.8	1.16	1.33	0.87	N	B	B
B109	228	0.98	0.99	0.89	0.91	15.9	0.90	1.75	0.51	B	B	B
B110	286	1.00	1.25	0.31	0.45	23.2	1.03	0.77	1.34	B	B	B
B111	363	1.00	1.58	0.34	0.34	22.1	1.03	0.58	1.78	B	B	B
B112	483	1.01	2.1	0.91	0.9	21.3	1.90	0.6	1.67	B/N	B	N
B113	516	1.03	2.25	1.08	1.05	12.4	1.59	0.51	3.12	N	B	B
B114	510	1.02	2.22	1.18	1.15	12.8	1.62	0.3	5.4	N	B	B
B116	371	1.03	1.62	0.27	0.26	12.2	0.56	0.1	5.6	B	B	B
B117	362	1.00	1.58	0.34	0.34	15.4	0.72	0.77	0.94	B	B	B
B118	392	1.05	1.71	0.5	0.47	21.9	1.03	0.65	1.56	B	B	B
B119	530	1.10	1.6	1.21	1.1	18.9	1.65	0.56	2.9	B	B	B
B120	629	1.12	1.9	1.34	1.2	25.2	3.22	2.1	1.53	B	B	B
B123	483	1.06	2.21	0.9	0.85	17.1	1.74	1.8	0.97	N	B	B
B124	400	1.04	1.83	0.65	0.63	13.8	1.38	1.9	0.73	N	B	B
mean		1.01	1.71	0.65	0.71		1.29	1.01	1.89	–	–	–
SD		0.05	0.35	0.4	0.29		0.59	0.64	1.5	–	–	–

4   Conclusions and Future Work

In the article, we present the possibilities that SC metamodeling offers to engineers for structural early design. Taking advantage of the accuracy achieved using FEM, metamodels can be developed providing shorter times with low errors when predicting the behaviour of steel structures. Indeed, both techniques present the same shortcomings in case of some inherent nonlinearities, e.g. failure mechanism or plastic behaviour, are taken into consideration. So, determining the ultimate displacement and the failure mode is still far from being reliable. It seems that the high variability of these outputs in the original dataset should be reduced before the training of the models. Future works will focus on reducing these errors to assess the performance of joints. This is a first step before including the metamodel into comercial software for steel frames design or other complex connections. Moreover, metamodeling should be also explored in other fields like connections of composite materials or subjected under fire conditions.

References

1. ABAQUS v.6.11. Analysis User's Manual
2. Ductility Requirements in Shear Bolted Connections. Master's thesis, University of Coimbra (2007)
3. Fernández-Ceniceros, J., Sanz-García, A., Antoñanzas-Torres, F., Martínez-de-Pisón-Ascacibar, F.J.: Multilayer-perceptron network ensemble modeling with genetic algorithms for the capacity of bolted lap joint. In: Corchado, E., Snášel, V., Abraham, A., Woźniak, M., Graña, M., Cho, S.-B. (eds.) HAIS 2012, Part III. LNCS, vol. 7208, pp. 545–556. Springer, Heidelberg (2012)
4. Rotation capacity of partial strength steel joints with three-dimensional finite element approach. Computers & Structures 116, 88–97 (2013)
5. Bishop, C.: Neural Networks for Pattern Recognition. Oxford University Press, Oxford (1995)
6. Corchado, E., Herrero, A.: Neural visualization of network traffic data for intrusion detection. Applied Soft Computing (2010)
7. Haykin, S.: Neural networks: a comprehensive foundation. Prentice Hall (1999)
8. Hornik, K., Stinchcombe, M.B., White, H.: Multilayer feedforward networks are universal approximators. Neural Networks 2(5), 359–366 (1989)
9. Jones, M.T.: Artificial Intelligence: A Systems Approach. Infinity Science Press LLC (2008)
10. Lemaitre, J.A.: A course on damage mechanics. Springer (1992)
11. Lemaitre, J.A.: A continuous damage mechanics model for ductile fracture. J. Eng. Mater. Technol. 107, 83–90 (1985)
12. Mckay, M., Beckman, R., Conover, W.: A comparison of three method for selecting values of input variables in the analysis of output from a computer code. Technometrics 21, 239–245 (1979)
13. R Core Team: R: A Language and Environment for Statistical Computing. R Foundation for Statistical Computing, Vienna, Austria (2013)
14. Sacks, J., Welch, W., Mitchell, T., Wynn, H.: Design and analysis od computer experiments. Statistical Science 4, 409–423 (1989)

Bayesian Networks for Greenhouse Temperature Control

José del Sagrado, Francisco Rodríguez, Manuel Berenguel, and Rafael Mena

Departamento de Informática,
Universidad de Almería,
Crtra. de la Playa s/n,
04120 Almería, Spain
{jsagrado,frrodrig,beren,rafael.mena}@ual.es

Abstract. Greenhouse production processes are heavily influenced by greenhouse climate conditions, as crop growth performance is directly influenced by these conditions. A solution to the problem of controlling the temperature in greenhouses using an open–loop control system based on Bayesian networks is presented in this paper. The system is built and tested using data gathered from a real greenhouse. The results show the performance and applicability of this type of systems.

Keywords: Bayesian networks, Greenhouse climate control, Decision support.

1 Introduction

Nowadays, agriculture must meet an increasing number of regulations on quality and environmental impact. The incorporation of new technologies in the agro-alimentary sector can serve as an aid for the fulfillment of these requirements. This fact makes it a propitious field for the application of automatic control techniques [21], [5], [20].

In greenhouses, fertirrigation and climate systems are the focus of control problems. The fertirigation control problem consists in providing the amount of water and fertilizers that the crop requires, whereas the climate control problem has to do with maintaining the greenhouse temperature within appropriate range despite of disturbances. This paper will be focused on controlling the air temperature inside the greenhouse by acting on ventilations aperture. As ventilations aperture and inside temperature follow a non linear relationship, it is suitable the use of Bayesian networks.

This work presents preliminary ideas and simulations about the combination of Bayesian networks and control systems to be applied in greenhouses. As a first approximation, an open-loop Bayesian network-based control has been applied for the temperature control problem using the ventilation system. The paper is organized as follows. Section 2 is devoted to describe the greenhouse climate control problem. Afterwards, we briefly introduce Bayesian networks and describe its use in Section 3. The Bayesian network-based control system is discussed and simulation results are presented in Section 4. Finally, some conclusions are given in Section 5.

Á. Herrero et al. (eds.), *International Joint Conference SOCO'13-CISIS'13-ICEUTE'13*,
Advances in Intelligent Systems and Computing 239,
DOI: 10.1007/978-3-319-01854-6_17, © Springer International Publishing Switzerland 2014

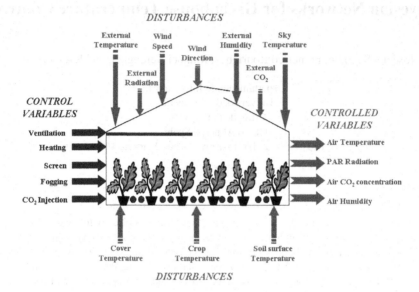

Fig. 1. Greenhouse climate control problem

2   Greenhouse Climate Control Problem

A greenhouse is an enclosure which can maintain constant temperature, humidity and other environmental factors to promote the cultivation of plants. Therefore its main aim is to provide near-optimal environmental conditions for a crop at different stages of crop development and even obtaining crop production in other seasons different from normal seasons. Inside this closed environment climatic and fertirrigation variables can be controlled to allow an optimal growth and development of the crop. Within the objectives of this problem are the maximization of profit, fruit quality, and water-use efficiency, these being currently fostered by international rules. The complex problem of greenhouse crop production, that involves climate and fertirrigation systems, can be simplified by assuming that plants receive, in every moment, the amount of water and fertilizers they require. As result of this simplification, the problem reduces to control the crop growth as a function of climate environmental conditions [17], [18], [14].

In this framework, the term "greenhouse climate" is used to denote the set of environmental physical quantities produced in a greenhouse that affect the growth and development of a crop [1], [2], [3], [22]. And the main purpose of a greenhouse is to help to achieve the optimum conditions for a particular crop and season, based on the needs of the plants.

The dynamic behaviour of the greenhouse climate is a combination of physical processes involving energy transfer (radiation and heat) and mass balance (water vapour fluxes and CO_2 concentration). These processes are affected by the outlet environmental conditions, structure of the greenhouse, type and state of the crop, and on the effect of the control actuators [1], [2]. Thus, greenhouse climate (see Figure 1) is mainly controlled by using ventilation and heating (to modify inside temperature and humidity conditions), shading and artificial light (to change internal radiation), CO_2 injection (to influence photosynthesis), and fogging/misting (for humidity enrichment). A deeper study about the features of the climate control problem can be found in [1], [17].

The approach presented in this paper is focused on the climatic conditions of mild winter in Southern Europe, where the production in greenhouses is made without CO_2 enrichment and there is an increasing demand of quality products. In this geographical area the main variables to control are temperature and humidity, taking into account the greenhouse structures, the commonest actuators, the crop types, and the commercial conditions. The PAR (*Photosynthetically Active Radiation* is the spectral range from 400 to 700 Wm^2, which is used by the plants as energy source in the photosynthesis process) radiation is controlled with shade screens, but its use is not much extended. So, this paper is focused on the temperature control problem.

2.1 Air Temperature Control

Inside the greenhouse, the crop growth is influenced by PAR radiation, temperature and CO_2 level. Under diurnal conditions PAR radiation and temperature influence the process of plant photosynthesis. In particular, temperature influences the speed of sugar production by photosynthesis and a higher radiation level implies a higher temperature. Thus radiation and temperature have to be in balance and it is necessary to maintain the temperature in a high level, being optimal for the photosynthesis process. On the other hand, under nocturnal conditions, the plants are not active (the crop does not grow) and, therefore, it is not necessary maintain such a high temperature. These are the reasons why two temperature set-points (diurnal and nocturnal) are usually considered [11], [18].

Due to the favorable climate conditions of the South-East of Spain, during the diurnal intervals the sun provides the energy required to reach the optimal temperature. In fact, the diurnal temperature control problem is the refrigeration of the greenhouse (when temperature is higher than the diurnal set-point) using natural ventilation to reach the diurnal optimal temperature. Whilst the nocturnal temperature control problem is the heating of the greenhouse (when temperature is lower than the nocturnal set-point) using heating systems to reach the nocturnal optimal temperature.

The air exchange and flow inside the greenhouse is determined by natural ventilation as a consequence of the difference between inside and outside temperatures. The objective of the control system is to maintain the inside temperature in an optimal level. It is known that vents aperture and inside temperature follow a non linear relationship [16], so a first approach treated in this paper is to implement an open–loop control system based on Bayesian networks (see Figure 2).

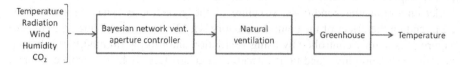

Fig. 2. Open-loop controller based on Bayesian network

The temperature control has the main priority because it affects to the crop growth directly, even though controlling temperature affects on humidity. Humidity inside the greenhouse is another important aspect, because high humidity may produce the appearance of diseases and decrease transpiration of the crop, whereas low humidity may cause hydric stress, closing the stomata and reducing the photosynthesis due to a decrease in the CO_2 assimilation. Humidity control presents two problems: (1) the greenhouse inside temperature and the relative humidity are inversely related when the greenhouse air is not mixed with the external air, generally colder and drier (when one of them increases the other one decreases and vice versa); (2) the same actuators are used for controlling temperature and humidity. The temperature control has the main priority because it affects to the crop growth directly and this is why we have focused our attention on it.

Therefore, modeling dependency among the variables involved in the greenhouse climate control problem is an important unsolved topic. Bayesian networks provide a simple and sound framework for this task.

3   Bayesian Networks

In this section we briefly introduce Bayesian networks and describe its use, before building a model for the greenhouse climate control problem.

3.1   Bayesian Networks Basics

The idea underlying Bayesian networks [15], [9] is to represent statistical dependences between a set of variables $X = \{X_1, X_2, \cdots, X_n\}$ in qualitative and quantitative ways. A directed acyclic graph (DAG) is used to represent graphically (qualitatively) dependences between variables, so that each variable is represented as a vertex and each link indicates the presence of statistical dependence between variables. Also a joint probability distribution of the variables $P(X_1, X_2, \cdots, X_n)$ can be defined (quantitatively), according to the graph structure, as the product of the conditional probability distributions of each variable given its parents in the graph

$$P(x_1, x_2, \cdots, x_n) = \prod P(x_i \,|\, pa(x_i)), \; i = 1, \cdots, n \tag{1}$$

Thus, the dependences in the graph are translated (and quantified) to the joint probability distribution in a sound form.

3.2 Reasoning with Bayesian Networks

One of the most important features of Bayesian networks is that it is possible to find out, without the need of performing any numerical computation, which variables are relevant or irrelevant for some other variable of interest on the structure of the directed acyclic graph. The information flow (regardless of the direction) over a Bayesian network structure can be modeled using the following rules: (1) information may flow through a *serial* ($X{\rightarrow}Y{\rightarrow}Z$)/*diverging* ($X{\leftarrow}Y{\rightarrow}Z$) connection unless the state of the intermediate variable Y is known; (2) information may flow through a *converging* ($X{\rightarrow}Y{\leftarrow}Z$) connection whenever the state of the intermediate variable Y or one of its descendants is known. With these rules is possible to determine which are the relevant variables with respect to a given goal variable. For instance, we could determine the variables over which we have to operate in order to change the value of the temperature inside the greenhouse.

Now that it has been described how to reasoning over the structure of a Bayesian network, let's look at the numbers. Reasoning is done by computing the posterior probability distribution given new information on some variables (i.e. evidence). Assume X_i is the goal variable and $\mathbf{X}_E$ is a set of variables for which we can gather information (e.g. based on sensors). The likelihood of each possible value of X_i given each possible configuration of $\mathbf{X}_E$ can be obtained by computing the distribution $P(X_i \mid \mathbf{X}_E)$. This distribution can be computed efficiently taking advantage of the factorization of the joint probability distribution (see Eq. (1)) imposed by the network structure [19], [13] without the need of computing the joint distribution.

3.3 Learning Bayesian Networks from Data

The next question that arises for Bayesian networks is how to obtain them. The problem of learning a Bayesian network from data consists on finding the network that best approximates the data. This problem can be tackled as a search guided by a score function g, as in K2 [4], BDe [8] or MDL [12]. The score function g is used to evaluate each candidate network against the data set and the search tries to find the best structure with respect to the score function that it is being applied. A feature of these methods is that they work on qualitative variables, so continuous variables have to be discretized previously. A review of discretization methods can be found in [10]. In this work, the K2 algorithm has been used in order to learn the Bayesian network from data.

The K2 learning algorithm [4] is based on a series of assumptions (independence of cases in the data set $\mathbf{D}$, absence of missing data, uniformity of the probability distributions of the parameters of a given network) to derive the joint probability distribution of G network structure and a data set $\mathbf{D}$ as

$$P(G, \mathbf{D}) = P(G) \prod_{i=1}^{n} g(V_i, pa(V_i)) = P(G) \prod_{i=1}^{n} \prod_{j=1}^{c_{pa_i}} \frac{(c_i - 1)!}{(N_{ij} + c_i - 1)!} \prod_{k=1}^{c_i} N_{ijk}! \qquad (2)$$

where c_i is the number of cases of the variable X_i; c_{pai} is the number of cases of $pa(X_i)$; N_{ijk} is the number of cases in which X_i takes its k-th value, $X_i = v_{ik}$ and $pa(X_i)$ takes its j-th value, $pa(X_i) = w_{ij}$; and $N_{ij} = \sum_{k=1}^{c_i} N_{ijk}$.

To make the search space treatable it is also required that variables are ordered σ_X and a uniform prior distribution on the different structures (this allows the term $P(G)$ disappears). By introducing an order σ_X among the variables, the previous metric (Eq. (2)) can be maximized by working separately with each node X_i and its set of parents $pa(X_i)$. For this, the algorithm proceeds by considering the variables in the preset order and for each of them shapes the set of parents (incrementally from the empty set) which provides the greatest value of the function $g(V_i, pa(V_i))$, that measures the likelihood of the resultant structure, until there is no improvement, and then returns the set of parents that is being considered at the time.

Once we have learned the structure of the network, we learn the parameters, $P(x_i \mid pa(x_i))$, from the data set $\mathbf{D}$, using *m-estimation* [7].

Table 1. Variables considered in the air temperature control problem

Variable	Description	Variable	Description
G_R_OUT	Global radiation outside	G_R_IN	Global radiation inside
P_R_OUT	PAR radiation outside	P_R_IN	PAR radiation inside
T_OUT	Outside temperature	T_IN	Inside temperature
T_WC_OUT	West side outer cuvette temperature	T_WC_IN	West side inner cuvette temperature
T_EC_OUT	East side outer cuvette temperature	T_EC_IN	East side inner cuvette temperature
H_OUT	Outdoor relative humidity	H_IN	Indoor relative humidity
W_V	Outdoor wind velocity	T_R1	Roofing temperature
W_D	Outdoor wind direction	T_R2	Roofing temperature
RAIN	Rainfall	T_SOIL_3	Soil temperature at 3 cm
CO2_OUT	Outdoor CO_2	T_SOIL_40	Soil temperature at 40 cm
CO2_IN	Indoor CO_2	VENT_N	North side vent. aperture
T_LEAF	Leaf temperature	VENT_R	Roof ventilation aperture
T_SUBS	Temperature of the netting that covers the soil	VENT_AP	Ventilation aperture (class)

4   Case Study

The data set used (see http://aer.ual.es/CJPROS/engindex.php) was obtained at a greenhouse belonging to "Las Palmerillas Experimental Station" of Cajamar Foundation, which is located at El Ejido (36°48'N, 2°43'W and a height above sea level of 63 m). The structure of the greenhouse is symmetric in area and the roof runs from East to West. The surface area is 877 m² with a variable height (between 2.8 and 4.4 m). The available control systems are as follows: natural ventilation (in roof and sides) and shades. The climatic conditions of this greenhouse were controlled during a season with the objective of regulate diurnal temperature through the control of ventilation. During the season several internal and external measurements (see Table 1) were taken using a data acquisition system made up of an Ethernet network with Compact Field Point modules from National Instruments. The internal temperature was measured with sensors situated 2.5 m above the crop. Besides, to have a record of the internal state of the greenhouse we took measurements such as, relative humidity, radiation, soil temperatures at 40 and 3 cm, and temperatures of the netting that covers the soil, of the interior part of the roofing, and of the leaves of the plants. In addition, the following exterior measurements were taken: air temperature, relative humidity, global radiation, PAR radiation, rainfall and the direction and velocity of the wind. Data were recorded daily at 5 minutes intervals (i.e., 288 records per day) during the 140 days of the campaign (i.e., 40320 records in total).

4.1 Ventilation Forecasts

A Bayesian network can be used as a predictor simply by considering one of the variables as the class and the others as predicting variables (characteristics or features that describe the object that has to be classified). The posterior probability of the class is computed given the characteristics observed. The value assigned to the class is the one that reaches the highest posterior probability value. A predictor based on a Bayesian network model provides more benefits, in terms of decision support, than traditional predictors, because it can perform powerful *what-if* problem analyses.

All variables in the Bayesian network are discrete finite, thus the initial values in the data set were discretized. First, we discretized manually the ventilation aperture values (i.e. the class VENT_AP) using percentiles, obtaining 12 states as intervals for the percentage of ventilation aperture (see Class column in Table 2). Then the rest of the variables were discretized, based on class values and the information in the data set, using the supervised discretization algorithm proposed by [6], which tries to obtain, based on the class, the optimal number of states of each variable to discretize. Finally, two Bayesian networks for controlling greenhouse ventilation were learned, based on a naïve Bayes and TAN-Tree Augmented Network (applying the K2 algorithm with 2 parents) structures, using 10-fold cross-validation. The TAN obtained following this procedure is shown in Figure 3.

4.2 Evaluation

The features independence assumption penalizes the performance of naïve Bayes, as shows its 77.2% of correctly classified instances versus the 91.6% achieved by the TAN predictor. In order to evaluate performance, we use several measures (see Table 2), such as: precision (i.e. ratio of correctly classified instances), sensitivity, specificity and AUC (Area Under ROC Curve). As it can be seen the TAN achieves high specificity and AUC on all classes with good sensitivity and precision on extreme classes (i.e. those representing ventilation apertures below 8.53% or above 69.25%). Note that those classes represent the 92.99% of the data cases and the significant influence that discretization has on the effectiveness of classification.

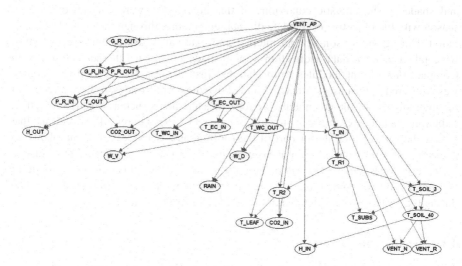

Fig. 3. Bayesian network used for controlling greenhouse ventilations aperture

Table 2. Bayesian networks performance measures

Class	Naïve Bayes				TAN			
	Prec.	*Sens.*	*Spec.*	*AUC*	*Prec.*	*Sens.*	*Spec.*	*AUC*
0	0.995	0.809	0.996	0.981	0.976	0.964	0.980	0.997
(0, 8.53]	0.114	0.502	0.952	0.930	0.325	0.516	0.987	0.973
(8.53, 16.25]	0.125	0.264	0.981	0.919	0.281	0.267	0.993	0.961
(16.25, 23.53]	0.235	0.154	0.995	0.929	0.327	0.211	0.996	0.966
(23.53, 31.8]	0.454	0.220	0.997	0.934	0.372	0.335	0.994	0.963
(31.8, 38.81]	0.159	0.535	0.973	0.925	0.238	0.249	0.992	0.964
(38.81, 47.65]	0.175	0.075	0.996	0.934	0.365	0.387	0.992	0.971
(47.65, 53.602]	0.358	0.353	0.994	0.937	0.667	0.401	0.998	0.972
(53.602, 69.25]	0.231	0.237	0.992	0.904	0.376	0.314	0.995	0.969
(69.25, 78.33]	0.116	0.602	0.954	0.923	0.662	0.580	0.997	0.973
(78.33, 98.97]	0.097	0.566	0.947	0.913	0.465	0.553	0.994	0.975
(98.97, 100]	0.973	0.830	0.982	0.988	0.975	0.990	0.980	0.998

As example, to show how the proposed controller works, the reference signal for controlling greenhouse ventilation during a whole day is depicted in top of Fig.4. The average external temperature of the selected day was of 16.9°C with a deviation of 2.3°C, whereas the average controlled temperature achieved inside of the greenhouse was 20.9°C with a deviation of 4.3°C. Data from sensors were sampled at 5 minutes intervals and sent, as evidences, to the Bayesian network that propagates them in order to obtain the state of the class variable (i.e. ventilation aperture) reaching the maximum a posteriori probability value. The control signal computed, in this way, by the proposed controller is depicted in Fig. 4. The differences between the two control signals depicted coincide with the sensitivity and precision on extreme classes exhibited by the Bayesian network predictor, and the overall performance of the open–loop control system based on Bayesian networks appears to be more than acceptable for this type of system.

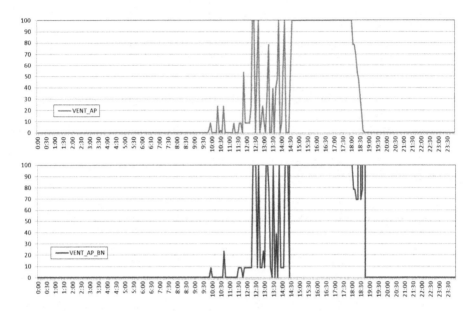

Fig. 4. Reference (top) and obtained control signal for ventilations aperture

5   Conclusions

The aim of this work was to develop a control system based on Bayesian networks for maintaining the greenhouse inside temperature by acting on vents aperture. To achieve this goal, we have shown how Bayesian networks can be learned from data using a widely applied score plus search approach. Once the model was constructed, the proposed technique was applied to a case study, showing that the designed controller exhibits a more than acceptable behavior.

As future work is planned to implement the controller in a real greenhouse environment in order to verify the results obtained in this study. Also we want to compare it with other type of controllers previously used in the greenhouse air control problem. Furthermore is planned to apply these ideas to other greenhouse control problems, e.g. fertirrigation control, and to evaluate the results from an economical point of view.

Acknowledgments. This research has been funded by the Spanish Ministry of Science and Innovation and EU-ERDF funds under grants DPI2010-21589-C05-04 and DPI2011-27818-C02-01, by the Spanish Ministry of Science and Innovation under project TIN2010-20900-C04-02, by Fundación Cajamar, and by the Consejería de Economía, Innovación y Ciencia de la Junta de Andalucía under grant Controlcrop (PIO-TEP-6174).

References

1. Bot, G.P.A.: Greenhouse climate from physical processes to a dynamic model. PhD thesis, Agricultural University of Wageningen: The Netherlands (1983)
2. Bot, G.P.A.: Physical modelling of greenhouse climate. In: Proc. of the IFAC/ISHS Workshop, pp. 7–12 (1991)
3. Boulard, T., Baille, A.: A simple greenhouse climate control model incorporating effects on ventilation and evaporative cooling. Agricultural and Forest Meteorology 65, 145–157 (1993)
4. Cooper, G.F., Herskovits, E.: A Bayesian method for the induction of probabilistic networks from data. Machine Learning 9, 309–348 (1992)
5. Farkas, I.: Modelling and control in agricultural processes. Computers and Electronics in Agriculture 49, 315–316 (2005)
6. Fayyad, U.M., Irani, K.B.: Multi-Interval Discretization of Continuous-Valued Attributes for Classification Learning. In: Proc. IJCAI 1993, Chambéry, France, pp. 1022–1029 (1993)
7. Friedman, N., Geiger, D., Goldszmidt, M.: Bayesian network classifiers. Machine Learning 29, 131–163 (1997)
8. Heckerman, D., Geiger, D., Chickering, D.M.: Learning Bayesian networks: the combination of knowledge and statistical data. Machine Learning 20, 197–243 (1995)
9. Jensen, F.V., Nielsen, T.D.: Bayesian networks and decision graphs, 2nd edn. Springer, New York (2007)
10. Jin, R., Breitbart, Y., Muoh, C.: Data discretization unification. Knowledge Information Systems 19, 1–29 (2009)
11. Kamp, P.G.H., Timmerman, G.J.: Computerized environmental control in greenhouses. A step by step approach. IPC Plant, The Netherlands (1996)
12. Lam, W., Bacchus, F.: Learning Bayesian belief networks. An approach based on the MDL principle. Computational Intelligence 10, 269–293 (1994)
13. Madsen, A., Jensen, F.V.: Lazy propagation: a junction tree inference algorithm based on lazy evaluation. Artificial Intelligence 113, 203–245
14. Pawlowski, A., Guzman, J.L., Rodríguez, F., Berenguel, M., Sánchez, J., Dormido, S.: Simulation of Greenhouse Climate Monitoring and Control with Wireless Sensor Network and Event-Based Control. Sensors 9, 232–252 (2009), doi:10.3390/s90100232
15. Pearl, J.: Probabilistic Reasoning in Intelligent Systems: Networks of Plausible Inference. Morgan Kaufmann, San Mateo (1988)
16. Rodríguez, F., Berenguel, M., Arahal, M.R.: Feedforward controllers for greenhouse climate control based on physical models. In: Proc. ECC 2001, Oporto, Portugal (2001)
17. Rodríguez, F.: Modeling and hierarchical control of greenhouse crop production. PhD thesis, University of Almería, Spain (2002) (in Spanish), http://aer.ual.es/TesisPaco/TesisCompleta.pdf
18. Rodríguez, F., Guzmán, J.L., Berenguel, M., Arahal, M.R.: Adaptive hierarchical control of greenhouse crop production. Int. J. Adap. Cont. Signal Process. 22, 180–197 (2008)
19. Shafer, G., Shenoy, P.: Probability propagation. Annals of Mathematics and Artificial Intelligence 2 (1990)
20. Sigrimis, N., Antsaklis, P., Groumpos, P.P.: Advances in control of agriculture and the environment. IEEE Control Systems 21(5), 8–12 (2001), doi:10.1109/37.954516
21. van Straten, G.: What can systems and control theory do for agriculture? Automatika 49(3-4), 105–107 (2008)
22. van Straten, G., van Willigenburg, G., van Henten, E., van Ooteghem, R.: Optimal control of greenhouse cultivation, p. 305. CRC Press, USA (2010)

Towards Improving the Applicability of Non-parametric Multiple Comparisons to Select the Best Soft Computing Models in Rubber Extrusion Industry

Ruben Urraca-Valle, Enrique Sodupe-Ortega,
Alpha Pernía-Espinoza, and Andres Sanz-Garcia

EDMANS Research Group, University of La Rioja, Logroño, Spain
ruben.urraca@alum.unirioja.es,
{enrique.sodupeo,alpha.pernia,andres.sanz}@unirioja.es
http://www.mineriadatos.com

Abstract. In this paper we propose different strategies to apply non-parametric multiple comparisons in industrial environments. These techniques have been widely used in theoretical studies and research to evaluate the performance of models, but they are still far from being implemented in real applications. So, we develop three new automatized strategies to ease the selection of soft computing models using data from industrial processes. A rubber products manufacturer was selected as a real industry to conduct the experiments. More specifically, we focus our study on the mixing phase. The rheology curve of rubber compounds is predicted to anticipate possible failures in the vulcanization process. More accurate predictions are needed to provide set points to enhance the control the process, particularly working in this rapidly changing environment. Selecting among a wide range of models increases the probability of achieving the best predictions. The main goal of our methodology is therefore to automatize the selection process when many choices are available. The models based on soft computing used to validate our proposal are neural networks and support vector machines and also other alternatives such as linear and rule-based models.

Keywords: Support Vector Machine, Multilayer Perceptron, Non-parametric comparison, Friedman, Rubber Mixing Process.

1 Introduction

Rubber manufacturers are demanding more homogeneous rubber mixtures to achieve higher quality in their extrusion processes (see Fig 1.a). This is particularly critical in the products generated for automotive industry such as rubber door seals. The complexity and diversity of these rubber profiles can create great problems to companies for continue being competitive. So, companies have increased their investments for improving the design and control of rubber mixers. One key strategy is to develop new soft computing (SC) models [5,3,4] that accurately predict the final quality of rubber compounds after the mixing phase. This improvement can reduce the high variability inherent in the mixing process and provide useful information about important quality reductions.

Á. Herrero et al. (eds.), *International Joint Conference SOCO'13-CISIS'13-ICEUTE'13*,   171
Advances in Intelligent Systems and Computing 239,
DOI: 10.1007/978-3-319-01854-6_18, © Springer International Publishing Switzerland 2014

Models based on SC yield optimal results in problems where large volumes of data are available such as the estimation of the set points for control systems. However, selecting the best performance model is not an easy task and an iterative procedure is always needed to find the best solution, particularly in real industrial problems. The main article's issue is to design new strategies based on multiple statistical comparison to select the best models in real engineering problems. In the article, SC models will be used to estimate the parameters that define the final quality of rubber compounds after the mixing phase. First, several types of models, including those based on SC aproach, are tested. Then, an automatic method using multiple statistical comparison is proposed to select which of the algorithms trained gives significant better results.

The dataset has already been described by [7], who preprocessed the complete set of experimental data. It contains 1240 samples of 6 different compound formulas but an unequal number of samples is available per each formula (627, 122, 228, 91, 82, 90). Fig 1.b summarizes the main scope of the models developed in this article. The rheology curve for rubber compounds that characterizes the behaviour of the rubber during the course of a vulcanization.

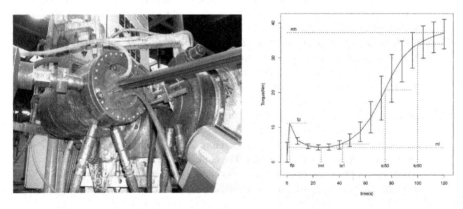

Fig. 1. a) Rubber extrusion process. b) Rheology curve of rubber compound

2   Problem Statement

The main goal is to enhance the automation of the selection process for prediction models used in industrial production lines. In model selection, the naivest and widely used procedure is the straight comparison of prediction errors. The reliability of this comparison is often increased by using more complex validation techniques such as bootstrap or cross-validation (CV). In addition, these straightforward procedures can be also improved by including statistics to measure the probability that two or more models yield different predictions. Statistical measures have already been successfully implemented in some engineering tasks such as evaluating two prediction models with just one dataset [9]. However, we distinguish two outstanding cases: when plant engineers need to compare more than a pair of prediction models and second, a situation in which not only a single output has to be estimated. Besides, results obtained using data mining rarely follow the assumptions required by a parametric procedures (independence, normality and homoscedasticity). Non-parametric statistics are thus required for multiple comparisons in real engineering problems.

In this work, we deal with the problem of comparing K different algorithms to evaluate their performance using N independent datasets. It is known that the number of datasets has to be higher $(N > K)$ than the algorithms compared to obtain confident results [6]. Traditionally, these techniques have been implemented in theoretical fields of Computer Science. The basic idea is to compare the performance of a new algorithm with others $(K - 1)$. The N independent datasets required are independent datasets selected from different problems (or public repositories) [6]. This is an algorithm-focused problem, in which datasets are just one component of algorithm evaluation process.

Working on engineering problems the aim is to solve a particular problem, which often consists on predicting several N_y outputs. The outputs are predicted by different types of models trying to minimize the errors but often using just one dataset. This is a database-focused problem, in which algorithms are the components selected depending of the properties of a single database. The main issue is to obtain N independent datasets from just one database higher than the number of prediction models K to ensure confident results.

3 Methodology

The methodology proposes different strategies for selecting the datasets that can allow the implementation of multiple statistical comparisons.

3.1 Basic Statistical Comparative Methods

The aim is to find the best model predicting a set of outputs. The starting point is to train several types of prediction methods with different setting parameters. Several error measures can be computed to know model performance but the validation method is usually the k-fold CV, obtaining k independent errors that estimate how accurately a model will perform in practice. Multiple statistical comparisons are thus proposed to carry out this decision with higher support, which should be based on selecting the best method and also finding its optimal parameters. Some of these techniques are described as follows:

1. Non-parametric Friedman test is used to determine whether there is any significant difference within the whole group of regression models. Friedman evaluates the K different algorithms in N different datasets and ranks them based on the error results. The final Friedman's statistic is computed based on the average rank of each model.

2. Exploratory comparison $(N \times N)$ using Holm post hoc is sometimes applied to have an initial preview. The difference between models is quantify in terms of the p-value where a value of 0% means totally different algorithms, while 100% means statistically equivalent algorithms. This step is useful when the control method selection for step 3 is not clear in previous steps.

3. Comparison $(1 \times N)$ using Finner post hoc contrasts a control method against $(K - 1)$ left algorithms. The control is the model that seems to perform better against the rest. The advantage of isolating a control method is lower p-values are obtained better establishing differences between algorithms.

3.2 Description of the Comparative Methodology

Three strategies are proposed to obtain a number of N datasets higher than the K models compared having just one database composed of N_y outputs [6]. The strategies are described as following:

1. $N_y > K$. Multiple statistical comparison can be implemented by considering each output variable (prediction error) as an independent dataset. The N datasets needed are the N_y output variables predicted by the K models trained $(N = N_y)$. The average error per output from the k-fold CV is chosen as the measure to make comparisons.
2. $N_y \leq K$. More datasets than the N_y outputs are needed. Due to the independency of the performance measurmentes obtained in k-fold CV, each fold result is considered as an inpenedent dataset, making a total amount of N of datasets $(N = N_y \times k)$
3. $N_y = 1$. This is the situation where the comparison is focused on one output. In order to generate datasets, only folds results are available to make statistical comparison. Consequently, the k-folds from CV are the N independent datasets where the algorihtm are tested $(N = k)$.

To select one of these strategies, in addition to dataset availability, the nature of the predicted outputs has to be considered. On one hand, strategies 1 and 2 should be used when just one model is required to predict several output variables. This is recommended in situations where all outputs have similar meaning. Therefore the use of different models for each output may increase the complexity of the solution and may also give importance to spurious outputs, i.e. sensor outputs with calibration problems. On the other hand, strategy 3 should be used when one regression model is needed for predicting each output. This clearly minimizes the prediction errors for each output and it is useful when outputs have not uniform behaviour or we study a specific variable of interest.

As a result, an expert decision is required prior to perform multiple statistical comparison to decide how many models are needed for a given set of outputs.

4   Results and Discussion

In this section, three strategies proposed are compared by using two different experiments with the same data from a real mixing process of rubber compounds and four types of regression methods.

4.1   General Description of the Experimental Evaluation

The four regression methods implemented varying their settings are following:

- Support Vector Machines (SVM) with different settings but only using two kernel functions: linear and nonlinear [2].
- Multilayer Perceptron Neural Networks (MLP) using seven different setting parameters and varying the number of neurons in the hidden layer between 3, 5, 7, 9 and 11. One MLP includes *linear* activation functions and the other six MLPs use *tansig* functions in the neurons of the hidden layer [1].
- Rule-based models (Cubist) with only one configuration [8].
- Linear regression models (LM) [10].

The performance of models is computed by using MAE errors and the validation procedure is the 10-fold CV $(k = 10)$ repeated for the $N_y = 5$ output variables (*ml*, *mh*, *tc50*, *tc90*, *ts1*). The results are finally compared using nonparametric statistical metrics. To show all the possible variants, two different experiments are carried out, which are described as following:

- Experiment 1, dividied in two parts, contrasts the prediction of all outputs with one type of regression method against the prediction of a single output with a specific one. Strategies 1 and 3 are applied in this experiment, in which $K = 4$ different regression models are compared.
- Experiment 2 analyzes the prediction of several outputs with one method in case of having less output variables than the number of models to be compared $(N_y < K)$. Strategy 2 is applied in this experiment and $K = 7$ different settings for the MLP are developed instead of using different methods.

4.2   Experiment 1

The main issue in this experiment is to evaluate the effects of how many output variables are predicted by the same regression model. This is divided in two parts where different strategies are implemented. First, $N_y = 5$ output variables strongly related are selected. Thus, the best choice for predictig all outputs may be to consider only one method. Second, variable *tc50* is strongly believed an output which entails critical information. One model is therefore trained just to predict it but $K = 4$ soft computing methods are compared. Particularly, SVM with non-linear kernel and MLP with 3 neurons in the hidden layer (MLP-3) including *tansig* activation functions are selected as the best configurations for SVM and MLP models.

Experiment 1 Following Strategy 1. In this first part, one general model is required for predicting the $N_y = 5$ outputs. $K = 4$ regression models are compared, thus $N > 4$ different datasets are required. Strategy 1 is applied. The $N = 5$ independent datasets are the average error of the $N_y = 5$ outputs prediction. The MAE average results of $K = 4$ regression models in $N = 5$ outputs are shown in Table 1.

The Friedman Ranked Test ranks of the regression models are shown in Table 2. A Friedman statistic of 12.6 with a p-value of 0.0055886 is obtained, showing that there is a significant difference between the group of algorithms.

Table 1. MAE of testing dataset and standard deviation in 10-fold CV

	SVM	MLP 3	LR	CUBIST
ml	0.1706 (0.0127)	0.1834 (0.0143)	0.1796 (0.0122)	0.1646 (0.0131)
ts1	1.4498 (0.1057)	1.5799 (0.1184)	1.5593 (0.1078)	1.4287 (0.1222)
tc50	1.6038 (0.1193)	1.7400 (0.1404)	1.7575 (0.1185)	1.6359 (0.1459)
tc90	1.0925 (0.0790)	1.1461 (0.0890)	1.1563 (0.0800)	1.0646 (0.0899)
mh	1.1531 (0.0904)	1.2498 (0.0911)	1.2368 (0.0837)	1.1380 (0.0963)

Table 2. Friedman ranked test results for $N_y=5$ outputs

	CUBIST	MLP 3	LR	SVM
dataset1	1.00	4.00	3.00	2.00
dataset2	1.00	4.00	3.00	2.00
dataset3	2.00	3.00	4.00	1.00
dataset4	1.00	3.00	4.00	2.00
dataset5	1.00	4.00	3.00	2.00
avg. rank	1.20	3.60	3.40	1.80

Table 3. Results of exploratory NxN comparison process

	z value	unadj. p	Holm
CUBIST vs MLP 3	2.939388	0.003289	0.019732
CUBIST vs LR	2.694439	0.007051	0.035254
MLP 3 vs SVM	2.204541	0.027486	0.109945
LR vs SVM	1.959592	0.050044	0.150131
CUBIST vs SVM	0.734847	0.462433	0.924865
MLP 3 vs LR	0.244949	0.806496	0.924865

Table 4. Results of the 1xN comparison process

	z value	unadj_P	APV_Finner
MLP 3	2.939388	0.003289	0.009833
LR	2.694439	0.007051	0.010557
SVM	0.734847	0.462433	0.462433

The exploratory NxN comparison in Table 3 shows that Cubist and SVM results are statistically better than those obtained with LM and MLP. SVM or Cubist have to be chosen as control method. Cubist has slightly smaller errors and it lower ranked in Friedman test, thus, Cubist is chosen as control method. In the $1 \times N$ comparison in Table 4, the Finner post hoc shows that Cubist is significantly better than MLP and LM but there is only a 46% probability of obtaining better results using Cubist than using SVM. If the threshold is set in 10%, both regression models cannot be considered significantly different despite of Cubist smaller errors. When two models statistically yield the same results, the most parsimonious model is usually chosen based on Occam's razor criterion because it is easier to implement and extra information may be obtained. In this situation, despite of both being non parametric techniques, Cubist should be chosen. A tree based model can be better interpreted when the number of variables is not to high. However, SVM are black boxes where not information can be obtained. Instead of setting a threshold, the following step would be to use the p-value obtained with a cost parameter based on complexity to create a weighted function that automatically discern between the models compared.

Experiment 1 Following Strategy 3. In a second part, one model is trained to predict just one output ($tc50$). $K = 4$ regression models still have to be compared. Consequently, N>4 independent datasets are needed, but only prediction results for one output variable are available. Strategy 3 is used to increase the number of datasets. The N independent datasets are the k errors computed in the k folds of 10-fold CV process ($N = k = 10$).

Table 5. MAE test error in Experiment 1 - Strategy 3

	SVM	MLP 3	LR	CUBIST
fold 1 tc50	1.543	1.599	1.857	1.664
fold 2 tc50	1.684	2.016	1.654	1.672
fold 3 tc50	1.574	1.746	1.633	1.512
fold 4 tc50	1.457	1.585	1.776	1.898
fold 5 tc50	1.471	1.639	1.625	1.795
fold 6 tc50	1.601	1.724	1.743	1.800
fold 7 tc50	1.594	1.922	1.746	1.465
fold 8 tc50	1.763	1.846	1.663	1.713
fold 9 tc50	1.760	2.022	1.761	1.919
fold 10 tc50	1.527	1.789	1.725	1.536

Table 6. Friedman ranked Test in Experiment 1 - Strategy 3

	SVM	MLP 3	LR	CUBIST
avg. rank	1.60	3.30	2.50	2.60

Table 7. Nx1 comparison in Experiment 1 - Strategy 3

	z value	unadj_P	APV_Finner
MLP 3	2.944486	0.003235	0.009673
CUBIST	1.732051	0.083265	0.122260
LR	1.558846	0.119033	0.122260

The prediction MAE results for $tc50$ divided by the 10 folds are shown in Table 5 and the average rank in Table 6. SVM is the technique with higher accuracy predicting $tc50$. No $N \times N$ comparison is needed to set the control method. Looking to $1 \times N$ comparison results in Table 7 there is only a probability of 12.26 that Cubist and LM perform predicting $tc50$ with the same accuracy as SVM. Setting the p-value threshold in 10%, LM should be implemented as the most parsimonious model. Onthe other hand, setting it in 15% or a higher percentage, SVM is the method that yields statistically better results than the others.

This experiment shows how model selection can vary from Cubist, SVM or LM depending which strategy is followed and the significance threshold set. For this problem, if one the same regression model has to predict all outputs, with a 10% p-value threshold, cubist is the chosen model . However, when the prediction problem focuses on a variable of special interest ($tc50$), LM is chosen when the threshold in 10% and SVM should be implemented if the threshold is 15% or higher. These results show how the model obtained is strongly dependent on the strategy selected to predict the outputs.

4.3   Experiment 2

In experiment 1, the comparion methods have been applied to contrast different regression models. Experiment 2 shows that the same procedure can be applied to compare different settings for the same regression model. $K = 7$ different settings of the MLP are compared in the experiment. $N > 7$ different subsets are needed. Experimt 2 considers a situation in between part 1 and part 2 of the previous experiment. It is supposed that torque outputs (mh, ml) are strongly related variables. So, a single configuration is only used to predict mh and ml. When the one model is used to predict different outputs but the number of outputs is smaller than the number of compared ($N_y < K$), strategy 2 is used combining the basics of both, strategy 1 and strategy 2. The N independent datasets corresponds to the k errors computed in the folds of the 10-CV carried out for both outputs ($N = 2 \times 10 = 20$). The MAE results of the MLP $K = 7$ different settings for predicting ml and mh are depicted in Table 8.

Friedman average rank of Table 9 yields a Friedman statistic of 9.792857 and a p-value of 0.1336507. This p-value shows that there is not a strong difference between the settings compared. However, despite of Friedman p-value advises that differences are not so high, post hoc procedures are still used to quantify the difference between the settings. From the average rank it is observed that MLP 3 yields the most accurate results. Besides, MLP 3 is the most parsimonious methods of those one with a *tansig* activation function. Thus MLP 3 is choses as control method without the necessity of an exploratory $N \times N$ comparison.

In Table 10 MLP 3 is proved to be significantly better that all the settings except of MLP 5 with a p-value threshold of 15%. Statistically, only MLP 5 predict the torque outputs (mh, ml) with the same accuracy as MLP 3, but MLP 5 presents higher complexity. Consequently, a *tansig* activation function with 3 neurons in the hidden layer is the best setting for the MLP to predict torque.

This experiment shows how statistical comparison may have future applications in the existing auto-tuning functions to select the best settings for a model, which make a balance between simplicity and accuracy. For instance, the p-values obtained from the comparison can better establish the accuracy difference between the settings, improving the quality of the final model.

Table 8. MAE test errors in Experiment 2

	MLP 3	MLP 5	MLP 7	MLP 9	MLP 11	MLP 13	MLP linear
1	0.184	0.179	0.182	0.186	0.186	0.185	0.186
2	0.166	0.166	0.170	0.170	0.168	0.170	0.170
3	0.191	0.197	0.203	0.198	0.201	0.203	0.194
4	0.192	0.189	0.184	0.184	0.183	0.183	0.189
5	0.195	0.190	0.190	0.188	0.188	0.189	0.199
6	0.170	0.170	0.172	0.173	0.175	0.175	0.175
7	0.188	0.184	0.192	0.189	0.196	0.190	0.184
8	0.189	0.192	0.192	0.192	0.190	0.191	0.189
9	0.171	0.176	0.175	0.178	0.176	0.175	0.167
10	0.176	0.180	0.176	0.179	0.181	0.177	0.170
11	1.189	1.204	1.194	1.195	1.189	1.204	1.222
12	1.293	1.256	1.273	1.275	1.263	1.275	1.287
13	1.394	1.380	1.384	1.387	1.378	1.390	1.376
14	1.205	1.210	1.215	1.212	1.214	1.211	1.216
15	1.034	1.036	1.041	1.042	1.044	1.060	1.077
16	1.307	1.315	1.307	1.309	1.297	1.285	1.313
17	1.342	1.357	1.346	1.339	1.360	1.344	1.302
18	1.205	1.215	1.218	1.221	1.219	1.216	1.237
19	1.236	1.236	1.237	1.228	1.231	1.234	1.251
20	1.190	1.198	1.202	1.200	1.199	1.208	1.264

Table 9. Friedman ranked Test in Experiment 2

	MLP 3	MLP 5	MLP 7	MLP 9	MLP 11	MLP 13	MLP linear
avg. rank	2.95	3.55	4.15	4.35	4.10	4.20	4.70

Table 10. Nx1 comparison in Experiment 2

	z value	unadj_P	APV_Finner
MLP linear	2.561738	0.010415	0.060885
MLP 9	2.049390	0.040424	0.116436
MLP 13	1.829813	0.067278	0.130030
MLP 7	1.756620	0.078983	0.130030
MLP 11	1.683428	0.092292	0.130030
MLP 5	0.878310	0.379775	0.379775

5 Conclusions and Future Work

The paper shows that multiple nonparametric comparisons are useful to select the best performing prediction models in real engineering problems. Additionally, we also demonstrate that these techniques can be useful for obtaining the proper model settings. The use of our three strategies is much more sophisticated than a visual comparison of the magnitude of the prediction errors, specially when one set of data is only available and several outputs have to be predicted. The three strategies proposed not only make the comparison feasible but also they provide more confidence because models are evaluated in multiple scenarios. The results of experiment 1 show that the selection of the best model varies depending on the number of variables predicted and their relationships. In authorsÂŽ opinon, the expert judgment is still needed when choosing between the strategies. In case of experiment 2, the results demonstrate that the p-values obtained after the statistical comparison of the final models can quantify better their accuracy.

The p-values could be in the future included to an auto-tuning system based on a balance between accuracy and complexity.

Acknowledgments. First, we would like to convey our gratitude to *Standard Profil* for their support. On the same line, we would also like to thank to the *Autonomous Government of La Rioja* for the continuous encouragement by the means of the *"Tercer Plan Riojano de Investigación y Desarrollo de la Rioja"* on the project FOMENTA 2010/13, and to the *University of La Rioja* and *Santander Bank* for the project API11/13.

References

1. Bradley, J.B.: Neural networks: A comprehensive foundation. Information Processing & Management 31(5), 786–794 (1995) 4.1
2. Burges, C.J.C.: A tutorial on support vector machines for pattern recognition. Data Mining and Knowledge Discovery 2, 121–167 (1998) 4.1
3. Corchado, E., Abraham, A., Carvalho, A.: Hybrid intelligent algorithms and applications. Information Sciences 180(14), 2633–2634 (2010) 1
4. Corchado, E., Graña, M., Woźniak, M.: New trends and applications on hybrid artificial intelligence systems. Neurocomputing 75(1), 61–63 (2012) 1
5. Corchado, E., Herrero, Á.: Neural visualization of network traffic data for intrusion detection. Applied Soft Computing 11(2), 2042–2056 (2011) 1
6. Derrac, J., García, S., Molina, D., Herrera, F.: A practical tutorial on the use of nonparametric statistical tests as a methodology for comparing evolutionary and swarm intelligence algorithms. Swarm and Evolutionary Computation 1(1), 3–18 (2011) 2, 3.2
7. Marcos, A., Espinoza, A., Elas, F., Forcada, A.: A neural network-based approach for optimising rubber extrusion lines. International Journal of Computer Integrated Manufacturing 20(8), 828–837 (2007) 1
8. Quinlan: Combining instance-based and model-based learning. In: Proceedings of the Tenth International Conference on Machine Learning, pp. 236–243 (1993) 4.1
9. Wilcoxon, F.: Individual comparisons by ranking methods. Biometrics Bulletin 1(6), 80–83 (1945) 2
10. Wilkinson, G.N., Rogers, C.E.: Symbolic description of factorial models for analysis of variance. Journal of the Royal Statistical Society. Series C (Applied Statistics) 22, 392–399 (1973) 4.1

Ear Biometrics: A Small Look at the Process of Ear Recognition

Pedro Luis Galdámez and María Angélica González Arrieta

University of Salamanca, Plaza de los Caídos, 37008 Salamanca, Spain
{peter.galdamez,angelica}@usal.es

Abstract. This document provides an approach to biometrics analysis which consists in the location and identification of ears in real time. Ear features, which is a stable biometric approach that does not vary with age, have been used for many years in the forensic science of recognition. The ear has all the properties that a biometric trait should have, i.e. uniqueness, permanence, universality and collectability. Because it is a field of study with potential growth, in this paper, we summarize some of the approaches to the detection and recognition in existing 2D images in order to provide a perspective on the possible future research and the develop of a practical application of some of these methodologies to create finally a functional application for identification and recognition of individuals from an image of the ear, the above in the context of intelligent surveillance and criminal identification, one of the most important areas in the processes of identification.

Keywords: Neural Network, System Identification, Ear Recognition.

1 Introduction

The Ear does not have a completely random structure. It has standard parts as other biometric traits like face. Unlike human face, ear has no expression changes, make-up effects and more over the color is constant throughout the ear. Although the use of information from ear identification of individuals has been studied, is still an open question by specifying and determining whether or not the ear can be considered unique or unique enough to be used as a biometric. Accordingly, any physical or behavioural trait can be used as biometric identification mechanism provided which is universal, that every human being possesses the identifier, being distinctive and unique to each individual, invariant in time, finally measurable automatically or manually, the ear accomplish all these characteristics. From this reading we can conclude that the detailed structure of the ear is not only unique, but also permanent, the ear does not change during the human life. Furthermore, capturing images of the ears do not necessarily requires the cooperation of a person, so it can be considered non-intrusive. Because of these qualities, the interest in recognition systems through the ear has grown significantly in recent years and generally the increasing need to automatically authenticate people.

Á. Herrero et al. (eds.), *International Joint Conference SOCO'13-CISIS'13-ICEUTE'13*,   181
Advances in Intelligent Systems and Computing 239,
DOI: 10.1007/978-3-319-01854-6_19, © Springer International Publishing Switzerland 2014

Traditional means of automatic recognition such as passwords or ID cards can be stolen, counterfeit, or forgotten, however, the biometric characteristics such as the ear, are universal, unique, permanent, and measurable, leaving the field open for further continuously research. The figure 1 shows the major steps of a traditional recognition system.

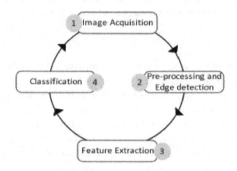

Fig. 1. Major Steps of ear recognition system

The Diagram describes task flow of a traditional recognition system, this flow is expose in the next sections, where we will see preprocessing methods and the flow catches ranging from the acquisition of the images to their identification, subsequently we will make emphasis in techniques applied in the heart system to achieve the goal of ear recognition.

2   Image Acquisition

In this section we need to differentiate from two tasks, the first one is the creation of the Ears Database, and second one is the image acquisition of the person to identify.

2.1   Ear Database

When we talk about an ear database is undisputed mentioned that exist a large group of databases for ear detection and recognition, this information is vital to test and compare the detection or recognition performance of a computer vision system, in general. The University of Science and technology in Beijing offers four collections of images that we could use to test taking any ear recognition system as they. The University of Notre Dame (UND) offers a large variety of different image databases and the Hong Kong Polytechnic University IIT Delhi has a database that consists in 421 images.

All this set of images could be used to test the performance of a ear recognition system by computer, but in our research we created our own set of images with the help of the police school of Ávila, which is a city located in Spain. This database consist in two hundred images two for each person which does not means that we do not use the other datasets, in fact, for this research we also are using the IIT Delhi dataset. [10] The figure 2 shows a sample of the dataset created.

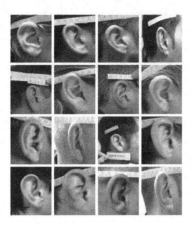

Fig. 2. Ear dataset. Police School of Ávila.

In the images obtained we tried to prevent the occlusion that hair and ear rings can make. The goal of create this dataset is to test our system with images that we already know to whom it belongs, in short words, this is our test set and sometimes we could use these images to train the system, in fact, we have the profile face of each person which we use to test algorithms to identify ears. This brings us to the second point to be analyzed in this section.

2.2   Image to Identify

Now, we need to capture the image to be analyzed, for that we use a simple web camera. The figure 3 shows a screen capture from the application that we are developing in order to achieve an application able to recognize a person in real time from a snapshot of its ear. For now it is a small system that can capture the ear from a web cam video using EmguCV [4] which is a wrapper from OpenCV that allows to develop in Visual Studio .Net and Java.

The application allows the user to create and manipulate images applying image filter like color filters and edge detectors. It is an image lab processing that also permits to introduce video documents as shown in figure 3, also we can edit the configuration to select what object we want to detect, this can vary, for example we could detect frontal and profile faces, right and left ear and eyes. But in this project we are focussing in ears, how we use the emgucv framework, we have access to the viola-jones classifier included in that library, to detect the ear in video, this application is using the haar-cascade classifier developed by

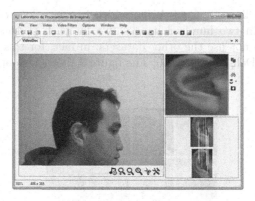

Fig. 3. Ear capture in the application

Modesto Castrillón-Santana [2] this allow us to capture ears on a easy way in real-time video. With the acquisition of the images defined, we must move into the second stage of the recognition system which is the pre-processing of the images.

3   Image Pre-processing and Feature Extraction

With each image in the database and the captured image, we begin the normalization process, first we perform the segmentation of the image applying a mask to extract only the ear, then the image is converted to an edge map using the canny edge filter. The Figure 4 is a summary of the pre-processing activities.

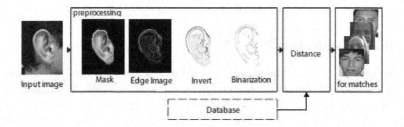

Fig. 4. Image pre-processing

If w is the width of the image in pixel and h is the height of the image in pixel, the canny edge detector takes as input an array $w \times h$ of gray values and sigma. The output is a binary image with a value 1 for edge pixels, i.e., the pixel which constitute an edge and a value 0 for all other pixels. We calculate a line between major and minor y value in the edge image to rotate and normalize each image, trying to put the lobule of the ear in the centre. This process is to try to get all the images whose shape is similar to the image to identify. We identify some points on the external shape of the ear and the angle created by the center of the line drawn before and the section in the ear's tragus with the major x value. All these values will be used like complement of an input in a neural network. Once we have all data processed we proceed to extract the features for the recognition.

4   Classification

4.1   Principal Component Analysis (PCA)

The ear recognition algorithm with eigenears is described basically in the figure 6. First, the original images of the training set are transformed into a set of eigenears E, Then, weights are calculated for each image on the (E) set, and then are stored in the (W) set. Observing an image X unknown, weights are calculated for that particular image, and stored in the vector W_X. Subsequently, W_X compared to the weights of images, which is known for sure that they are ears (the weights of the training set W) [8].

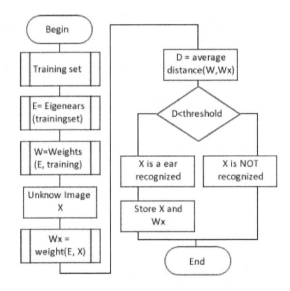

Fig. 5. Ear recognition algorithm based on eigen vectors

Classification of a New Ear. The process of classifying a new ear in the Γ_{new} to another category (known ears) is the result of two steps. First of all, the new image is transformed into its eigenear components. The resulting weights forms the weight vector Ω_{new}^T.

$$\omega_k = u_k^T(\Gamma_{new} - \Psi) \quad k = 1, ..., M'$$
$$\Omega_{new}^T = [\omega_1 \ \omega_2 \ ... \ \omega_{M'}] \tag{1}$$

The Euclidean distance between two vectors of weights $d(\Omega_i, \Omega_j)$ provides a measure of similarity between the corresponding images i and j. If the Euclidean distance between Γ_{new} and the rest of images on average exceeds a certain threshold value, through this can be assumed that Γ_{new} is not a recognizable ear [8].

4.2 Fisher Algorithm

The PCA algorithm is a linear combination of functions that maximizes the variance of the information. This can result in poor performance, especially when we are working with image noise such as changes in the background, light and perspective. So the PCA algorithm can find faulty components for classifying. To prevent this problems, we implement the Fisher algorithm to compare results in the ear recognition process. The Fisher algorithm that we implement basically goes like this [1,13]:

Construct the Image matrix x with each column representing an image. Each image is assigned to a class in the corresponding class vector c. Project x into the $(N - c)$ dimensional subspace as P with the rotation matrix $WPca$ identified by a Principal Component Analysis, where N is the number of samples in x.

c is unique number of classes $(length(unique(C)))$ Calculate the between-classes scatter of the projection P as:

$$Sb = \sum_{i=1}^{c} N_i * (mean_i - mean) * (mean_i - mean)^T \qquad (2)$$

Where $mean$ is the total mean of P $mean_i$ is the mean of class i in P, N_i is the number of samples for class i. Calculate the within-classes scatter of P as:

$$Sw = \sum_{i=1}^{c} \sum_{x_k \in X_i} (x_k - mean_i) * (x_k - mean_i)^T \qquad (3)$$

Where x_i are the samples of class i x_k is a sample of x_i $mean_i$ is the mean of class i in P. Apply a standard Linear Discriminant Analysis and maximize the ratio of the determinant of between-class scatter and within-class scatter. The solution is given by the set of generalized eigenvectors $Wfld$ of Sb and Sw corresponding to their eigenvalue. The rank of Sb is atmost $(c-1)$, so there are only $(c-1)$ non-zero eigenvalues, cut off the rest. Finally obtain the Fisherears by $W = WPca * Wfld$ [13].

4.3 Ear Classification

With all the pictures processed, an array is created with labels indicating to whom belongs each image, the vector values are calculated with the PCA and Fisher algorithm. Subsequently, it obtains the Euclidean distance of the comparison of the weight vectors obtained. Defining a threshold value of 1000 representing a similarity between two images of 90%, under the assumption that a smaller distance means a greater similarity between the sets.

In other words, the eigen and fisher vectors are obtained from the set of images stored in the database, the vectors values and the unknown image are compared both with the distance measurement, resulting in a distance vector from the input image regarding the collection of the database. The lowest value is obtained and compared with the threshold, being the label with the lower value the identified user.

5 Neural Network for Ear Recognition

Neural networks have been trained to perform complex functions in various fields of application including pattern recognition, speech, vision and control systems. In this project, there is a neural network that identifies each person in the database. After calculating the eigen and fisher ears, the feature vectors of the ears are stored in the database. These vectors are used as inputs to train the network. In the training algorithm, the vectors of values belonging to a person, are used as positive for returning said individual neuron 1 as the neuron output assigned to that user and 0 in other neurons.

When the new image has been captured, the feature vectors are calculated from the eigenears obtained before, we compute new descriptors of the unknown image. These descriptors are entered into the neural network, the outputs of individual neurons are compared, and if the maximum output level exceeds the predefined threshold, then it is determined that the user belongs to the ear assigned to the neuron with the index activated. The algorithm implementation in the approach of ear recognition using eigen, fisher and geometric image preprocessing with canny edge filter basically is summarized building a library of ears, choose a training set M that includes if it is possible more than one image per person, these images can have variation in lighting and perspective, use the eigen and fisher values as input in a neural network, in our case we add at the input two values previously calculated.

These values are the average of the points detected in the geometric normalization, and the angle that forms the center and the major x value on the edge map in the tragus area of the ear. Create one output neuron per person in the database and finally after the training select the neuron with the maximum value. If the output of the selected neuron passes a predefined threshold, is presented as the recognized person.

6 Experimental Results

In this section we present each of the mentioned ear recognition techniques applied over the new database created, our first hypothesis says that variable illumination could affect the performance of the algorithms. In the process, all images were cropped within the ear therefore the contour around the ear was excluded. classification was performed using a nearest neighbour classifier. All training images of an individual were projected into the feature space. Each image in the database was taken while the subject ear was being illuminated by enough light. The methods reveal a number of interesting points:

1. Both algorithms perform perfectly with frontal light, however, as we are using a real-time video, the change of perspective, and when the lighting became darkness made a significant performance difference between the two methods.
2. The algorithm with less errors classifying the ear when we change the illumination and perspective was Fisher method.
3. Neural network using as input the eigenvalues computed have accomplished better performance in ear recognition process with change on illumination.

Table 1. Normal Conditions

	PCA		Fisher		NeuralNetwork	
	Positive	Negative	Positive	Negative	Positive	Negative
Positive	101	19	177	17	163	18
Negative	11	88	18	21	8	33

Under normal conditions we obtain the previous confusion matrix, assuming that true positive values are people classified correctly. True negatives are the users that the system should not recognize because they do not exist in the database, and indeed does not recognize, a false negative is when the system predicts that the user does not exist but it does exist, finally a false positive is when the system make a mistake to identify a person, but really it is not in the database.

Table 2. Changing Illumination and Perspective Conditions

	PCA		Fisher		NeuralNetwork	
	Positive	Negative	Positive	Negative	Positive	Negative
Positive	64	72	93	32	84	44
Negative	35	67	23	46	23	47

With perspective and illumination in normal conditions, we get 87% of succeed in recognition with PCA, 91% with fisher algorithm, using the neural network, the percentage increased to 93%, over more than 200 attempts of different individuals. When we change illumination and perspective conditions the error rate increase, leaving the PCA algorithm with a success rate of only 54%, neural network with 71% and Fisher Algorithm in 77%. This percentages are calculated using the $F_1 Score$ (over tables 1 and 2) which associate recall and precision measures to give us a value that represent, how well the system makes the recognition.

The method that has being used in this research is to try to put together some of the most common approaches in the recognition process, the project is not presented as unique and exceptional, but upon the approaches that other researchers have proposed, combining and comparing them, and trying to select a combination of these approaches to successfully implement a fully functional system capable of recognizing a person across its ear and use this system to identify criminals, using the database created in the police academy of Ávila. As the title of this paper indicates this is a small look at the process of recognition of people from captured images of their ears in real-time video. Throughout the article presents the first techniques studied, therefore the quantitative results of the project can be considered preliminary, but they provide a clearer picture to where should point this research in the future, observing some of the strengths and weaknesses of the algorithms studied in order to strengthen pre-processing tasks and / or implementation of more robust algorithms.

7   Conclusion and Future Work

The algorithms perform a good ear recognition process if the video captures an image very similar with one in the training set. Fisher method appears to be better over variation in lighting. Use a neural network with the eigenvalues calculated as input makes a better performance than the eigenear traditional method over changes on illumination and perspective. Changes in pre-processing process allows better results if all images have the same angle and illumination, other techniques of pre-processing images may improve the ear recognition process. If these techniques allow recognize a person through the ear, exist other methods like Ray Image Transform, Histograms of Categorized Shapes, Edge orientation pattern that can obtain better results.

Our goal is to create an application that identify one person in a real-time video for that we are interested in the study of these techniques. This paper is our first look to the ear recognition process with encouraging results in real-time video identify process. Our future work will be modify the application to use more complex algorithms like Ray Image Transform and improve the pre-processing step.

References

1. Belhumeur, P.N., Hespanha, J.P., Kriegman, D.J.: Eigenfaces vs. Fisherfaces: Recognition Using Class Specific Linear Projection. IEEE Transactions on Pattern Analysis and Machine Intelligence (1997)
2. Castrillón-Santana, M., Lorenzo-Navarro, J., Hernández-Sosa, D.: An Study on Ear Detection and Its Applications to Face Detection. In: 14th Conference of the Spanish Association for Artificial Intelligence, La Laguna, Spain, pp. 313–322 (2011)
3. ElBakry, H.M., Mastorakis, N.: Ear Recognition by using Neural Networks. Technical University of Sofia Faculty of Computer Science and Information, Sofía (2007)
4. Intel OpenCV wrapper EmguCV (April 08, 2013),
   http://www.emgu.com/wiki/index.php/Main_Page
5. Islam, S., Davies, R., Mian, A., Bennamoun, M.: A Fast and Fully Automatic Ear Recognition Approach Based on 3D Local Surface Features, The University of Western Australia, Crawley, WA 6009, pp. 1081–1092. ACIVS, Australia (2008)

6. Lammi, H.-K.: Ear Biometrics. Lappeenranta, Lappeenranta University of Technology, Department of Information Technology, Laboratory of Information Processing, Finland (2004)
7. Marti-Puig, P., Rodríguez, S., De Paz, J.F., Reig-Bolaño, R., Rubio, M.P., Bajo, J.: Stereo Video Surveillance Multi-Agent System: New Solutions for Human. Journal of Mathematical Imaging and Vision (2011), ISSN: 0924-9907, doi:10.1007/s10851-011-0290-2
8. Dimitri, P.: Eigenface-based facial recognition (2002)
9. Prakash, S., Gupta, P.: An Efficient Ear Recognition Technique Invariant to Illumination and Pose. Indian Institute of Technology Kanpur, Kanpur-208016, India.: Department of Computer Science and Engineering (2010)
10. Pug, A., Busch, C.: Ear Biometrics: A Survey of Detection, Feature Extraction and Recognition Methods. IET Biometrics, Darmstadt (2012)
11. Islam, S.M.S., Davies, R., Bennamoun, M., Mian, A.S.: Efficient Detection and Recognition of 3D Ears. International Journal of Computer Vision, 52–73 (2011)
12. Narendira Kumar, V.K., Srinivasan, B.: Ear Biometrics in Human Identification System. Information Technology and Computer Science, 41–47 (2012)
13. Wagner, P.: Fisherfaces (January 13, 2013), http://www.bytefish.de/blog/fisherfaces/
14. Xin Dong, Y.G.: 3D Ear Recognition Using SIFT Keypoint Matching. Journal of Theoretical and Applied Information Technology (2013)
15. Yuan, L., Mu, Z., Xu, Z.: Using Ear Biometrics for Personal Recognition. In: Advances in Biometric Person Authentication, pp. 221–228 (2005)

Enhanced Image Segmentation Using Quality Threshold Clustering for Surface Defect Categorisation in High Precision Automotive Castings

Iker Pastor-López, Igor Santos, Jorge de-la-Peña-Sordo, Iván García-Ferreira, Asier G. Zabala, and Pablo García Bringas

S³Lab, DeustoTech - Computing, University of Deusto, Bilbao, Spain
{iker.pastor,isantos,jorge.delapenya,ivan.garcia.ferreira,
asier.gonzalez,pablo.garcia.bringas}@deusto.es

Abstract. Foundry is an important industry that supplies key products to other important sectors of the society. In order to assure the quality of the final product, the castings are subject to strict safety controls. One of the most important test in these controls is surface quality inspection. In particular, our work focuses on three of the most typical surface defects in iron foundries: inclusions, cold laps and misruns. In order to automatise this process, we introduce the QT Clustering approach to increase the perfomance of a segmentation method. Finally, we categorise resulting areas using machine-learning algorithms. We show that with this addition our segmentation method increases its coverage.

Keywords: Computer Vision, Machine-Learning, Defect Categorisation, Foundry.

1 Introduction

Foundry process is one of the most relevant indicatives of the progress of a society. Generally, it consists on melting a material and pouring it into a mould where it solidifies into the desired shape. Later, the resulting castings are used in other industries like aeronautic, automotive, weaponry or naval, where they are critical and any defect may be critical. For this reason, the manufactured castings must success very strict safety controls to ensure their quality.

In this context, there are many defects that may appear on the surface of the casting. In this paper, we focus on three of the most common surface defects: (i) inclusions, which are little perforations caused by an excess of sand in the mould; (ii) misruns, that appear when not enough material is poured into the mould; and finally, (iii) cold laps, which are produced when part of the melted material is cooled down before the melting is completed.

Currently, the visual inspection and quality assessment are performed by human operators [1,2]. Although, people can perform some tasks better than machines, they are slower and can get easily exhausted. In addition, qualified operators are hard to find and to maintain in the industry since they require capabilities and learning skills that usually take them long to acquire.

Á. Herrero et al. (eds.), *International Joint Conference SOCO'13-CISIS'13-ICEUTE'13*, 191
Advances in Intelligent Systems and Computing 239,
DOI: 10.1007/978-3-319-01854-6_20, © Springer International Publishing Switzerland 2014

2 Proposed Machine Vision System

For the casting surface information retrieval, a simple computer-vision system
was developed, composed by [3]: (i) image device, (ii) processing device and (iii)
robotic arm.

1. **Image device:** We obtain the three-dimensional data through a laser-based
   triangulation camera. By taking advantage of the high-power (3-B class)
   laser, we are able to scan the casting even though their surface tends to be
   dark.
2. **Processing device:** We utilise a high-speed workstation. In particular, we
   use a workstation with a XENON E5506 processor working with 6GB of
   RAM memory and a QUADRO FX1800 graphic processing unit. This component controls the camera and the robotic arm. Besides, it processes the
   information retrieved by the image capturing device and transforms it into
   segments.
3. **Robotic arm:** The function of the robot is to automate the gathering phase
   of the system, making every necessary move to successfully acquire the data.
   There are two working options [4]: (i) to use the arm in order to handle
   the tested castings, leaving the image device in a fixed position or (ii) to
   attach the camera to the robotic arm. We selected the second one due to the
   diversity of the castings.

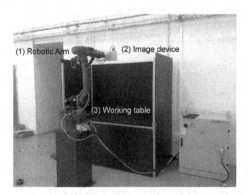

Fig. 1. The architecture of the machine vision system. (1) is the robotic arm, (2) is
the image device and (3) is the working table where the castings are put for analysis.

The casting is positioned on a working table using a manually adjusted foundry
mould. The mould is built with a material similar to common silicone, which is
easily malleable. In the case that we decide to change the casting type, we will
only have to change the mould. In this way, we ensure that the vision system
allows us to analyse every type of casting in the same position.

With this system, we capture the information of the casting surface. The process starts by putting the casting manually on a working table. Then, we use a mould that is built with a material similar to common silicone. In this way, we ensure that the vision system allows us to analyse every type of castings in the same position.

When the casting is on the working table, the robotic arm makes a linear movement, retrieving a set of profiles based on the generated triangulation of the laser and the optical sensor. In other words, a foundry casting $\mathcal{C}$ is composed of profiles $\mathcal{P}$ such as $\mathcal{C} = \{\mathcal{P}_1, \mathcal{P}_2, ..., \mathcal{P}_{n-1}, \mathcal{P}_n\}$. Each profile is retrieved with a thickness of 0.2mm. These profiles are vectors p composed of the heights of each point $p_{x,y}$. Joining these profiles, we represent the casting $\mathcal{C}$ as a height matrix $\mathcal{H}$

$$\mathcal{H} = \begin{pmatrix} h_{1,1} & h_{1,2} & ... & h_{1,m-1} & h_{1,m} \\ h_{2,1} & h_{2,2} & ... & h_{2,m-1} & h_{2,m} \\ & & ... & & \\ h_{\ell-1,1} & h_{\ell-1,2} & ... & h_{\ell-1,m-1} & h_{\ell-1,m} \\ h_{\ell,1} & h_{\ell,2} & ... & h_{\ell,m-1} & h_{\ell,m} \end{pmatrix} \tag{1}$$

where each $h_{x,y}$ represent the height of the point in the space (x, y). Therefore, the number of profiles of each casting depends on its size.

Once the system has computed the matrix $\mathcal{H}$, we have to remove the possible existing noise, as well as the data unrelated to the casting surface. To this end, we establish a height threshold empirically.

Finally, we generate the following representations of the information, besides from the heigh matrix:(i) normals matrix and (ii) normals map coded in RGB.

- **Normals Matrix.** This representation is generated by means of the height matrix, but shows the direction of the normal vector of the surface for each point in the matrix. Each vector for each point have three components (x, y, z).
- **Normals Map coded in RGB.** This image represents the information of the Normals Matrix corresponding red component to the x value, green component to the y value and blue component to the z value.

3   Enhanced Segmentation Method

Image segmentation consists on the subdivision of an image into disjointed regions [5]. Usually, these regions represent areas of the original image that contain at least one irregularity (defect or regular structure). In [6], we presented a model based approach for image segmentation. This method uses correct castings to compare with the normals map coded in RGB of the potentially defective surfaces. Then, we employed 176 correct castings to build the model, and we confirmed that if the number of correct castings increases, the performance of the segmentation method decreases.

In this paper we optimise this segmentation method, using a combination of: (i) image filters (to emphasise the defective areas); and (ii) the use of QT clustering algorithm (to reduce the information of the good castings used like a models).

Quality threshold algorithm was proposed by Heyer et al. [7] to extract relevant information of big datasets. Specifically, we use the implementation of this algorithm proposed by Ugarte-Pedrero et al. [8]. The Figure 2 shows the pseudo-code of this implementation and how are centroid vectors generated.

```
input  : The original dataset V, the distance threshold for each cluster threshold, and the
         minimum number of vectors in each cluster minimumvectors
output: The reduced dataset R
// Calculate the distance from each vector (set of executable features) to the rest of
   vectors in the dataset.
foreach {v_i|v_i ∈ V} do
    foreach {v_j|v_j ∈ V} do
        // If a vector v_j's distance to v_i is lower than the specified threshold, then
           v_j is added to the potential cluster A_i, associated to the v_i vector
        if distance(v_i,v_j) ≥ threshold then
        |   A_i.add(v_j)
        end
    end
end
// In each loop, select the potential cluster with the highest number of vectors.
while ∃A_i ∈ A : |A_i| ≥ minimumvectors and ∀A_j ∈ A : |A_i| ≥ |A_j| and i ≠ j do
    // Add the centroid vector for the cluster to the result set R.add(centroid(Ai))
    R.add(centroid(A_i))
    // Discard potential clusters associated to vectors v_j ∈ A_i
    foreach {v_j|v_j ∈ A_i} do
    |   A.remove(A_j) V.remove(v_j)
    end
    // Remove vectors v_j ∈ A_i from the clusters A_k remaining in A
    foreach {A_k|A_k ∈ A} do
        foreach {v_j|v_j ∈ A_k and v_j ∈ A_i} do
        |   A_k.remove(v_j)
        end
    end
end
// Add the remaining vectors to the final reduced dataset
foreach {v_j|v_j ∈ V} do
|   R.add(v_j)
end
```

Fig. 2. Pseudo-code of the implementation of QT Clustering based model reduction algorithm proposed by Ugarte-Pedrero et al. [8]

The input vectors are composed of the heigh matrix of the correct casting surfaces, concatenating each row in a sole row. In other words, if we have a height matrix $\mathcal{H}$, the input vector v that represents $\mathcal{H}$ is the following

$$v = \{h_{1,1}, h_{1,2}, ..., h_{1,n-1}, h_{1,n}, h_{2,1}, h_{2,2}, ..., h_{m,n-1}, h_{m,n}\} \tag{2}$$

QT clustering algorithm requires to fix a threshold to determine the maximum distance between two vector of the same cluster. Specifically, we use Euclidean distance and we set different values for the threshold to optimise the performance of the segmentation method.

Next, the centroids are generated, replacing the original model castings and the segmentation process continues with the following steps:

1. The process starts converting to grey-scale the normals map coded in RGB of the casting and of the correct models. This step is necessary to remove any noise of the rugosity of the surface.
2. The Gaussian Blur [9] filter is applied.
3. The process continues applying the difference filter between the result image of the previous steps and each model image.
4. The system applies a intersection filter between the differences computed in the previous step.
5. The result image is binarized.
6. The process ends with an algorithm that extracts the areas potentially faulty, removing the ones which are excessively small.

For each extracted area, several features are computed. These features can be divided into the following categories:

- **Features of the segmented image:** The segmented image is the result of the segmentation process applied to the normal map. We use: (i) the width, height and perimeter of the area; (ii) the euclidean distance of the center of gravity of the area to origin of coordinate axes; and (iii) the fullness, which is computed as $Area/(Width * Height)$.
- **Features of the integral image of segmented binary image:** These features are obtained from the conversion to the integral image of the segmented version of the image. An integral image is defined as the image in which the intensity at a pixel position is equal to the sum of the intensities of all the pixels above and to the left of that position in the original image [10]. We use: (i) mean value of pixels in the integral image and (ii) the result of addition of the pixels values in the integral image.
- **Features of the height matrix:** They are extracted from the computed segments in the original grey-scale height map. We use: (i) summation, mean, variance, standard deviation, standard error, min, max, range, median, entropy, skewness and kurtosis of the height matrix values; and (ii) summation, mean, variance, standard deviation, standard error, min, max, range, median, entropy, skewness and kurtosis of the height matrix without zero pixels values.
- **Features of the normals matrix:** These features are extracted from the computed segments in the original normals matrix. We use: (i) summation, mean, variance, standard deviation, standard error, min, max, range, median, entropy, skewness and kurtosis of the x component; (ii) summation, mean, variance, standard deviation, standard error, min, max, range, median, entropy, skewness and kurtosis of the x component without zero pixels values; (iii) summation, mean, variance, standard deviation, standard error, min, max, range, median, entropy, skewness and kurtosis of the y component; (iv) summation, mean, variance, standard deviation, standard error,

min, max, range, median, entropy, skewness and kurtosis of the y component without zero pixels values; (v) summation, mean, variance, standard deviation, standard error, min, max, range, median, entropy, skewness and kurtosis of the z component; and (vi) summation, mean, variance, standard deviation, standard error, min, max, range, median, entropy, skewness and kurtosis of the z component without zero pixels values.

4   Empirical Validation

To evaluate our casting defect detector and categoriser, we collected a dataset from a foundry, which is specialised in safety and precisions components for the automotive industry (principally, in disk-brake support with a production over 45,000 tons per year). Three different types of defect (i.e., inclusion, cold lap and misrun) were present in the faulty castings.

To construct the dataset, we analysed 639 foundry castings with the segmentation machine-vision system described in Section 2 in order to retrieve the different segments and their features. In particular, we used 236 correct castings as input for the clustering algorithm and the remainder for testing.

The acceptance/rejection criterion of the studied models resembles the one applied by the final requirements of the customer. Pieces flawed with defects must be rejected due to the very restrictive quality standards (which is a requirement of the automotive industry). We labelled each possible segment within the castings with its defects.

First, we evaluate the coverage of our segmentation method using different values for QT Clustering threshold. To this end, we define the metric 'Coverage' as:

$$Coverage = \frac{S_{s \to s}}{S_{s \to s} + S_{c \to s}} \cdot 100 \tag{3}$$

where $S_{s \to s}$ is the number of segments retrieved by the segmentation system which are defects and $S_{c \to s}$ are the number of defects that our segmentation method does not gather.

The Table 1 shows the evolution of the coverage of the segmentation method using different values for the clustering threshold.

We can notice that the coverage increases with higher values of the threshold. Besides, when we use a threshold higher than 450, the segmented areas are too big and the method loses precision. For this reason, we compute the segmentation process using the 51 centroids vectors.

By means of this analysis, we constructed a dataset of 6,150 segments to train machine-learning models and determine when a segment is defective. Besides, we added a second category to identify the noise that our machine vision system retrieves called 'Correct', which represents the segments gathered by the segmentation method that are correct even though the method has marked them as potentially faulty. In particular, 5,686 were correct and 464 were faulty.

Table 1. Coverage results and generated centroids for different threshold values

Threshold	Number of centroids	Coverage(%)
300	236	59.71
325	234	59.87
350	218	60.69
375	182	62.97
400	128	66.72
425	68	77.00
450	51	78.79

Table 2. Number of samples for each category

Category	Number of samples
Correct	33,216
Inclusion	553
Cold Lap	20
Misrun	60

Next, we evaluate the precision of the machine-learning methods to categorise the segments. To this extent, by means of the dataset, we conducted the following methodology to evaluate the proposed method:

- **Cross validation:** This method is generally applied in machine-learning evaluation [11]. In our experiments, we performed a K-fold cross validation with $k = 10$. In this way, our dataset is split 10 times into 10 different sets of learning (90% of the total dataset) and testing (10% of the total dataset).
- **SMOTE:** The dataset was not balanced for the different classes. To address unbalanced data, we applied Synthetic Minority Over-sampling TEchnique (SMOTE) [12], which is a combination of over-sampling the less populated classes and under-sampling the more populated ones. The over-sampling is performed by creating synthetic minority class examples from each training set. In this way, the classes became more balanced.
- **Learning the model:** For each fold, we accomplished the learning step using different learning algorithms depending on the specific model. Particularly, we used the following models:
  - *Bayesian Networks (BN):* With regards to Bayesian networks, we utilize different structural learning algorithms: K2 [13] and Tree Augmented Naïve (TAN) [14]. Moreover, we also performed experiments with a Naïve Bayes Classifier [11].
  - *Support Vector Machines (SVM):* We performed experiments with a polynomial kernel [15], a normalized polynomial Kernel [16], a Pearson VII function-based universal kernel [17] and a radial basis function (RBF) based kernel [18].

- *K-Nearest Neighbour (KNN):* We performed experiments with $k = 1$, $k = 2$, $k = 3$, $k = 4$, and $k = 5$.
- *Decision Trees (DT):* We performed experiments with J48(the *Weka* [19] implementation of the *C4.5* algorithm [20]) and Random Forest [21], an ensemble of randomly constructed decision trees. In particular, we tested random forest with a variable number of random trees N, $N = 10$, $N = 25$, $N = 50$, $N = 75$, and $N = 100$.

– **Testing the model:** To test the approach, we evaluated the percent of correctly classified instances and the area under the ROC curve, which establishes the relation between false negatives and false positives [22].

Regarding the coverage results, our segmentation method is able to detect 78.79% of the surface defects. This coverage value is higher than we obtained without clustering. In particular, the coverage increases in 19.09 points.

Table 3. Results of the categorisation in terms of accuracy and AUC

Model	Accuracy(%)	AUC
Bayes K2	97.21	0.8364
Bayes TAN	98.40	0.7229
Naïve Bayes	81.83	0.8938
SVM: Polynomial Kernel	93.22	0.9543
SVM: Normalised Polynomial Kernel	96.85	0.9611
SVM: Pearson VII Kernel	98.81	0.9516
SVM: Radial Basis Function Kernel	93.98	0.9578
KNN K = 1	98.03	0.5584
KNN K = 2	98.17	0.5860
KNN K = 3	98.11	0.6104
KNN K = 4	98.16	0.6277
KNN K = 5	98.09	0.6450
J48	97.58	0.7911
Random Forest N = 10	98.63	0.9497
Random Forest N = 25	98.62	0.9621
Random Forest N = 50	98.66	0.9680
Random Forest N = 75	98.67	0.9692
Random Forest N = 100	98.70	0.9689

If we focus in the precision of the categorisation of the segments, Table 3 shows the results of the categorisation phase. In particular, the best results were obtained by the Random Forest trained with more than 50 trees with an accuracy of more than 98% and an AUC of 0.96. SVM trained with a Radial Basis Function kernel and trained with Polynomial Kernel obtained poor results, implying that a radial division of the space is not as feasible as others, because the rest of the SVMs behaved with accuracies higher 98% in the case of

Pearson VII and near 97% in the case of the Normalised Polynomial kernel. Surprisingly, the lazy classifier KNN achieved high results, ranging from 98.03% to 98.17% of accuracy and from 0.55 to 0.64 or AUC. J48 was an average classifier that achieved an AUC of 0.79.

5   Conclusions and Future Work

In this paper, we proposed an improvement for a machine vision system. Concretely, we used Quality Threshold Clustering to reduce the data of the correct castings used in the segmentation methods. Also, with this enhancement we have increased the coverage of the method. Then we evaluated our new segmentation method using machine learning models to categorise the detected areas into correct, inclusion, cold lap or misrun. For this classification, we proposed new features, using different representations. The experimental results showed that, albeit our precision in categorisation is very high, the coverage of the segmentation method had increased.

Future work is oriented in 2 main ways. First, we are going to develop new segmentation methods in order to enhance the coverage results and the system performance. Second, we will evaluate different features and approaches in order to improve the categorisation process.

References

1. Mital, A., Govindaraju, M., Subramani, B.: A comparison between manual and hybrid methods in parts inspection. Integrated Manufacturing Systems 9(6), 344–349 (1998)
2. Watts, K.P.: The effect of visual search strategy and overlays on visual inspection of castings. Master's thesis, Iowa State University (2011)
3. Pernkopf, F., O'Leary, P.: Image acquisition techniques for automatic visual inspection of metallic surfaces. NDT & E International 36(8), 609–617 (2003)
4. vom Stein, D.: Automatic visual 3-d inspection of castings. Foundry Trade Journal 180(3641), 24–27 (2007)
5. Castleman, K.: 2nd edn. Prentice-Hall, Englewood Clliffs, New Jersey (1996)
6. Pastor-Lopez, I., Santos, I., Santamaria-Ibirika, A., Salazar, M., de-la Pena-Sordo, J., Bringas, P.: Machine-learning-based surface defect detection and categorisation in high-precision foundry. In: 2012 7th IEEE Conference on Industrial Electronics and Applications (ICIEA), pp. 1359–1364 (2012)
7. Heyer, L.J., Kruglyak, S., Yooseph, S.: Exploring expression data: identification and analysis of coexpressed genes. Genome Research 9(11), 1106–1115 (1999)
8. Ugarte-Pedrero, X., Santos, I., Bringas, P., Gastesi, M., Esparza, J.: Semi-supervised learning for packed executable detection. In: Proceedings of the 5th International Conference on Network and System Security (NSS), pp. 342–346 (2011)
9. Gonzalez, R., Woods, R.: Digital image processing, vol. 16(716). Addison-Wesley, Reading (1992)
10. Viola, P., Jones, M.: Robust real-time face detection. International Journal of Computer Vision 57(2), 137–154 (2004)

11. Bishop, C.M.: Neural Networks for Pattern Recognition. Oxford University Press (1995)
12. Chawla, N., Bowyer, K., Hall, L., Kegelmeyer, W.: SMOTE: synthetic minority over-sampling technique. Journal of Artificial Intelligence Research 16(3), 321–357 (2002)
13. Cooper, G.F., Herskovits, E.: A bayesian method for constructing bayesian belief networks from databases. In: Proceedings of the 1991 Conference on Uncertainty in Artificial Intelligence (1991)
14. Geiger, D., Goldszmidt, M., Provan, G., Langley, P., Smyth, P.: Bayesian network classifiers. Machine Learning, 131–163 (1997)
15. Amari, S., Wu, S.: Improving support vector machine classifiers by modifying kernel functions. Neural Networks 12(6), 783–789 (1999)
16. Maji, S., Berg, A., Malik, J.: Classification using intersection kernel support vector machines is efficient. In: Proc. CVPR, vol. 1, p. 4 (2008)
17. Üstün, B., Melssen, W., Buydens, L.: Visualisation and interpretation of support vector regression models. Analytica Chimica Acta 595(1-2), 299–309 (2007)
18. Cho, B., Yu, H., Lee, J., Chee, Y., Kim, I., Kim, S.: Nonlinear support vector machine visualization for risk factor analysis using nomograms and localized radial basis function kernels. IEEE Transactions on Information Technology in Biomedicine 12(2), 247 (2008)
19. Garner, S.: Weka: The Waikato environment for knowledge analysis. In: Proceedings of the 1995 New Zealand Computer Science Research Students Conference, pp. 57–64 (1995)
20. Quinlan, J.: C4. 5 programs for machine learning. Morgan Kaufmann Publishers (1993)
21. Breiman, L.: Random forests. Machine learning 45(1), 5–32 (2001)
22. Singh, Y., Kaur, A., Malhotra, R.: Comparative analysis of regression and machine learning methods for predicting fault proneness models. International Journal of Computer Applications in Technology 35(2), 183–193 (2009)

Orthogonal Matching Pursuit Based Classifier for Premature Ventricular Contraction Detection

Pavel Dohnálek[1,2], Petr Gajdoš[1,2], Tomáš Peterek[2], and Lukáš Zaorálek[1,2]

[1] Department of Computer Science
[2] IT4 Innovations, Centre of Excellence
VŠB - Technical University of Ostrava
17. listopadu 15, 708 33 Ostrava, Czech Republic
{pavel.dohnalek,petr.gajdos,tomas.peterek,lukas.zaoralek}@vsb.cz

Abstract. Premature Ventricular Contractions (PVCs) are a common topic of discussion among cardiologists as this type of heart arrhythmia is very frequent among the general population, often endangering people's health. In this paper, a software system is proposed that differentiates PVCs from normal, healthy heartbeats collected in the MIT-BIH Arrhythmia Database. During classification, training data were recorded from subjects different than those from which testing data were measured, making the classifiers attempt to recognize patterns they were not trained for. A modification of the Orthogonal Matching Pursuit (OMP) based classifier is described and used for comparison with other, well-established classifiers. The absolute accuracy of the described algorithm is 87.58%. More elaboration on the results based on cross-reference is also given.

Keywords: orthogonal matching pursuit, electrocardiogram, heart arrhythmia, pattern matching, sparse approximation.

1 Introduction

Ever since their invention, electrocardiogram (ECG) devices have been providing cardiologists with vast amounts of data describing functions and malfunctions of the human heart. Processing and labeling this data can be difficult and time consuming, leading many researchers to ideas on how the process could be automated. Given the diversity of ECG signals in different patients and the amount of noise collected along with useful measurements, the task of building an autonomous recognition device, be it a small wearable apparatus or a high performance system, is challenging but greatly rewarding as timely recognition of premature ventricular contractions (PVCs) can prevent many health-endangering situations.

In this paper, we explore the possibility of differentiating PVC heartbeats from healthy heart waveforms with minimal sensor data preprocessing. While it is a common practice to combine and perform several data preprocessing techniques before classification and recognition stages of a pattern matching

Á. Herrero et al. (eds.), *International Joint Conference SOCO'13-CISIS'13-ICEUTE'13*, 201
Advances in Intelligent Systems and Computing 239,
DOI: 10.1007/978-3-319-01854-6_21, © Springer International Publishing Switzerland 2014

system, preprocessing can be computationally expensive and thus unsuitable for small, low-power or low-cost devices. That is why a minimalistic, simple preprocessing was chosen for this research and its effect on PVC from healthy heartbeat differentiation accuracy measured. The classifiers were chosen so that they represent both common and well-established approaches and less common, rather novel techniques. Each of the compared classifier is briefly described in the following section.

For a brief description of electroencephalography, it is a non-invasive examination method that takes readings of electric potentials in the human heart. An action potential originates in the sinoatrial node from which it spreads through atrioventricular node and bundle branches to Purkinje fibers. Purkinje fibers, in turn, spread this excitation throughout the whole ventricular muscle. The excitation causes a mechanical contraction of the heart tissue, ejecting blood into the bloodstream. An electrocardiogram is a recording of this process. One period of an ECG record consists of three phases: a P-wave (atrium depolarization), a QRS complex (ventricle depolarization) and T-wave (ventricle repolarization). Particular heart disorders manifest themselves as changes in the shape of particular heart muscle segments. From ECG, a cardiologist can recognize pathological states like ischaemic heart disease, myocardial infarction or different types of arrhythmia, one of which is PVC.

Premature ventricular contraction is one of the most common and frequently occurring heart arrhythmias. Long-term researches indicate that at least one PVC beat occurs every hour in up to 33% healthy males without any heart disorders. This percentage increases with the presence of circulatory system defects to up to 58% of females and 49% of males, topped by up to 84% of pensioners. PVCs can occur due to psychological stress, fatigue, alcohol and caffeine consumption. Physiologically it can be caused by low levels of oxygen or potassium in the body. During this type of arrhythmia, the action potential originates in Purkinje fibres instead of the sinoatrial node. In contrast with the standard QRS comples, a PVC beat is not preceded by a P-wave, is morphologically different and takes longer than 120 milliseconds. PVCs put the patients in danger of ventricular fibrillation, myocardial infraction or sudden death, greatly motivating researches in this area.

When addressing the PVC detection problem, research authors currently focus on preprocessing ECG data, applying various feature extraction and domain transformation algorithms. Wavelet transformations are a reoccurring theme in this area [1–3]. Classification methods currently widely used in the area are based both on classic algorithms like the Classification And Regression Tree (CART) [4] or k-Nearest Neighbor (kNN) [5] and more advanced techniques like the Adaptive Neuro Fuzzy Interference System (ANFIS) [6, 7] or Iterative Dichotomiser 3 (ID3) [8] alike. In this paper, the idea behind the proposed classifier is sparse approximation.

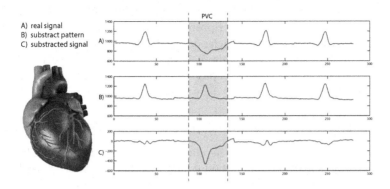

Fig. 1. An example of the preprocessing method and its resulting signal. There are four beats, the PVC beat is highlighted, the rest are usual heartbeat waveforms.

2   Methods Evaluated for PVC Detection

This section provides a brief description for each of the classifiers used in the comparison in this paper. The OMP classifier and its modification are left out here as they are more elaborately described in their own section.

2.1   k-Nearest Neighbors

kNN is a non-parametric algorithm, meaning that it makes no assumptions about the structure or distribution of the underlying data, thus being suitable for real-world problems that usually do not follow the theoretical models exactly. The method is also considered to be a lazy learning algorithm as it performs little to no training during computation. As a result, the method uses the whole training dataset during classification. kNN is well known for its simplicity, speed and generally good classification results.

2.2   Nearest Centroid Classifier

An extremely fast classifier. The approach is similar to that of kNN, but instead of k closest training samples the method picks the label of the class whose training samples' mean (centroid) is closest to the signal query. The speed and simplicity of the algorithm is compensated by low classification performance. Therefore the classifier is usually coupled with one or more data preprocessing techniques. In many implementations the method has been successfully used to create pattern recognition systems in bioinformatics [9].

2.3   Classification and Regression Tree

This algorithm classifies a sample based on groups of other samples with similar properties. During training, the training data is continuously divided into smaller subsets (tree nodes). When the divisions are finished, the samples are clustered together according to their properties. Testing samples are then evaluated against certain conditions in each node and propagated throughout the tree. When the sample reaches a leaf node, it is then assigned the class to which the samples in that node belong. In this paper, a binary tree with logical conditions was used.

2.4 Singular Value Decomposition

SVD is an algebraic extension of the classic vector model. It is similar to the Principle Component Analysis method which was the first method used to generate eigenfaces. Informally, SVD discovers significant properties and represents the EEG signals as linear combinations of the base vectors. Formally, an ECG signal matrix A is decomposed with SVD, calculating singular values and singular vectors of A [10]. Since SVD is not a classification algorithm on its own, it needs to be followed by a classifier. Given the complexity of the SVD algorithm itself, a simple Euclidean distance based classifier was used to assign a class to a given test sample in order to avoid creating a computationally expensive workflow.

2.5 Common Tensor Discriminant Analysis

As the name of the method suggests, CTDA is a feature extraction algorithm that processes tensors formed from a signal. An input signal is split into epochs of equal lengths. Each epoch is then transformed into a tensor whose covariance tensor is subsequently computed. All the covariance tensors are then used to compute a single, representative covariance tensor for a given class. The goal of the CTDA training process is to find the transformation matrices, two for each class, that project a query signal corresponding to a class on the first and second mode of the tensor [11]. Once the matrices are computed, a feature extraction takes place. As with SVD, CTDA is not a self-contained classifier. Features extracted from the test samples are passed through the same Euclidean distance based classification.

To conform to the workflow set by the authors of [11], CTDA is the only method for which more data preprocessing took place. Train and test samples alike were, in addition to above mentioned sample substraction, preprocessed with wavelet transformation. The transformation was performed by MATLAB function CWT using the mexican hat wavelet and 17 scales corresponding to 5-60Hz frequencies. The mexican hat wavelet is commonly used in heartbeat recognition as its waveform often fits a heartwave the best, while the frequencies were limited to those that are the most probable to occur during PVC.

3   OMP in PVC Detection

In order to use the sparse approximation of a signal for classification, sparse coefficients must first be computed. To do this, the OMP algorithm defined in [12] is used. It is a greedy algorithm that selects vectors from the entire training set that are the closest match to the signal being classified (the query signal). The training set can be represented as an $m \times n$ real-valued matrix A, where m is the length of both training and query signal and n is the number of training samples. Each column of every training matrix is normalized to unit length. The iterative nature of the algorithm allows for sparse coefficient number to be chosen in advance. It stands to reason to limit the number of sparse coefficients s such that $s \leq m$, although the number can truly be limited only by the lowest number of training samples across classes.

3.1 Training Matrix Preparation

Originally, the classifier proposed in [13] requires $n = t \times c$, where t is the number of training samples for a given class and c is the number of classes. The classification algorithm requires the training set to contain the same number of training samples for each class. It is also necessary to keep the samples of a given class grouped together. Therefore, the training matrix has the form of $A = [\mathbf{a}_{11}, \mathbf{a}_{21}, ..., \mathbf{a}_{t1}, \mathbf{a}_{12}, ..., \mathbf{a}_{tc}]$, where $\mathbf{a}_{ij}, i = 1..t, j = 1..c$ is the ith training sample of class j and length m.

The proposed modification changes the meaning of t and the resulting number of samples in the training set. Here,

$$n = \|\mathbf{t}\|_1 = \sum_{x=1}^{c} \mathbf{t}_x, \tag{1}$$

where $\mathbf{t}$ is the c-dimensional vector consisting of numbers of training samples for a given class. By this, the limitation imposed on the number of training samples in the original classification approach is lifted, yielding a training matrix in the form of $A = [\mathbf{a}_{11}, \mathbf{a}_{21}, ..., \mathbf{a}_{t_11}, \mathbf{a}_{12}, ..., \mathbf{a}_{t_cc}]$, where $\mathbf{a}_{ij}, i = 1..\mathbf{t}_x, x = 1..c, j = 1..c$ is the ith training sample of class j and length m.

The sparse coefficients are obtained by finding the sparse solution to the equation

$$\mathbf{y} = A\mathbf{s}, \tag{2}$$

where $\mathbf{y} \in \mathbb{R}^m$ is the query vector, $A \in \mathbb{R}^{m \times n}$ is the training matrix and $\mathbf{s} \in \mathbb{R}^n$ is the sparse coefficient vector. The stopping criterion in the implementation is reaching the sparse coefficient vector with the desired number of non-zero values.

3.2 Classification

To classify the query signal vector, a strategy of computing the residual value from the difference between the query vector and its sparse representation converted into the vector space of the training matrix vectors is employed. This is performed for each class resulting in c residuals. The classification is then based on the minimum residual. Formally, the classification problem can be stated as follows:

$$arg\, min\, r_k(\mathbf{y}) = \|\mathbf{y} - A\mathbf{s}_k\|_2. \tag{3}$$

Here, $\mathbf{s}_k$ is an n-dimensional vector with non-zero elements located only on indices corresponding to the kth class in the training matrix, hence the need for the training samples of a given classes to be grouped together in the matrix. The algorithm could be described by the following steps:

- Set the iteration variable i to 1
- Replace all sparse coefficients not belonging to class i with zeros
- Multiply the training matrix with the modified vector $\mathbf{s}$
- Compute the ℓ^2-norm of the resulting vector
- Increase i by 1 and repeat for all classes
- Output the class whose ℓ^2-norm is the lowest

Calculating the residuals is generally not computationally expensive and can be performed in real time, depending on the size of the training matrices. Only very large training matrices can slow the process down significantly.

4   The MIT-BIH Arrhythmia Database

The distribution of the MIT-BIH Arrhythmia Database started in 1980, the same year it was compiled into its final form, with half of it freely accessible after being released by PhysioNet [14] in 1999. Its data consist of 48 half-hour dual-channel ECG recorded from 47 subjects. The signals were discretized at 360Hz sampling frequency with 11-bit resolution and 10mV range. The recordings contain healthy heart beats as well as several different types of arrhythmic beats occurring at various rates. Detailed information about the signal properties and arrhythmias included in the database can be found in [15] and [16].

4.1   Dataset Preprocessing

In order to improve the accuracy of distinguishing the two beats from each other, the input data has been passed through a very quick preprocessing. First, a number of healthy beat samples of the training set were separated. These samples were then substracted from the remaining training set. In ideal conditions, all healthy beats in the training set would be reduced to a zero flat line, leaving only the PVC beat signal distorted. In reality, however, every one of the heartbeats, even of the same type, has a slightly different waveform, even more so if measured from a different subject. It can be seen that even this extremely simple preprocessing technique can have a significant positive impact on recognition accuracy, making it suitable for low-performance devices. Example of the preprocessed signal can be seen in Figure 1.

5   Experiments

The following section summarizes the experiments performed and the resulting recognition accuracy, pointing where exactly is the proposed method more accurate then the original OMP based classifier and vice versa. A comparison between all used methods is also shown. Classifier results were cross-referenced and evaluated in more detail.

Table 1. Overall recognition accuracy of all compared methods

Dataset	3-NN	OMP	OMP(mod)	CART	NCC	CTDA	SVD
Unprocessed	89.16	58.11	74.39	84.64	**79.92**	**96.26**	88.04
Preprocessed	**93.53**	**68.83**	**87.58**	**88.70**	65.86	83.38	**92.18**

5.1 Experimental Setup

The database was divided into a training set consisting of 20 subjects (subjects 101, 104, 105, 106, 109, 112, 113, 117, 119, 124, 200, 201, 208, 210, 215, 219, 223, 228, 230, 233) and a testing set containing data from 24 subjects different from the training set (subjects 100, 102, 103, 107, 108, 111, 114, 115, 116, 118, 121, 122, 123, 202, 203, 205, 212, 213, 214, 217, 220, 221, 231 and 234). All the signal data of each subject were analyzed and heartbeat waves extracted. In order to extract the waves, the index of a QRS complex is found in the signal description data of the dataset and then 70 values (35 on the left of the complex, 34 on the right + the complex peak itself) are taken to form one sample of a heartbeat. In total, 39510 of such samples of healthy and PVC beats were allocated for the training phase of the algorithms and 37878 samples for testing.

For preprocessing, 19755 healthy heartbeat samples were separated from the training set and substracted from the remaining training samples. Since the testing set is larger than 19755 samples, the separated samples were used repeatedly to accommodate this difference in size. The settings for each algorithm was as follows: both OMP-based classifiers computed 10 sparse coefficients for each testing sample. The k parameter for k-NN was set to 3.

5.2 Results

During the evaluation of the methods, it became clear that substracting healthy heartbeats from PVC beats can both benefit and hurt the recognition accuracy, depending on the method used as the classifier. While for the k-NN, OMP, proposed OMP modification, CART and SVD the preprocessing did indeed bring positive results, boosting the recognition accuracy by up to 13.19%, some classifiers, namely NCC and CTDA, prove that the suitability of the presented preprocessing technique is dependent on what classification technology follows. Table 1 sums up the absolute recognition accuracies for each of the tested classifier.

Tables 2 and 3 elaborate on the results in terms of cross-reference evaluation. The abbreviations stand for the following:

- **TP:** True/Positive - PVC beat was supposed to be and was recognized.
- **TN:** True/Negative - Healthy beat was supposed to be and was recognized.
- **FP:** False/Positive - Healthy beat was supposed to be recognized, but PVC was detected.
- **FN:** False/Negative - PVC beat was supposed to be recognized, but healthy beat was detected.

Table 2. Cross-reference data for the preprocessed training set

	3-NN	OMP	OMP(mod)	CART	NCC	CTDA	SVD
TP	1461	1670	1490	1429	923	757	1464
TN	33965	24401	31684	32168	24023	30825	33452
FP	2086	11650	4367	3883	12028	5226	2599
FN	366	157	337	398	904	1078	363
Sensitivity	0.7997	0.9141	0.8155	0.7822	0.5052	0.4143	0.8013
Specificity	0.9421	0.6768	0.8789	0.8923	0.6664	0.8550	0.9279

Table 3. Cross-reference data for the unprocessed training set

	3-NN	OMP	OMP(mod)	CART	NCC	CTDA	SVD
TP	1591	1756	1701	1576	798	417	1618
TN	32179	20255	26477	30484	29473	36042	31731
FP	3872	15796	9574	5567	6578	5	4320
FN	236	71	126	251	1029	1410	209
Sensitivity	0.8708	0.9611	0.9310	0.8626	0.4368	0.2282	0.8856
Specificity	0.8926	0.5618	0.7344	0.8456	0.8175	0.9999	0.8802

Since the goal is to detect PVC arrhythmias, it stands to reason to also evaluate the methods in terms of error rates in PVC detection. From this point it can be seen that OMP combined with unprocessed data detects the PVC beats better than the other methods. It correctly detected 1756 PVC beats and missed only 71 of them, resulting in the accuracy of 96.11%. Following is the suggested modification of the OMP classifier with slightly worse results in PVC detection (93.10%), but significantly improving detection reliability for healthy heartbeats (73.44% against 56.18%), explaining the increase in the absolute recognition accuracy. For the preprocessed dataset, PVC recognition was generally worse for OMP, dropping to 91.41% for the original classifier and 81.56% for the modified version, but the differentiation for healthy beats was boosted significantly (67.69% original OMP, 87.89% modified OMP). Interesting results are shown by CTDA. For unprocessed data (processed only by the above-mentioned wavelet transformation), the method performed almost flawlessly when recognizing healthy heartbeats, misclassifying only five. However, its performance in terms of PVC detection was the worst out of all evaluated methods, no matter the substraction preprocessing. Curiously, NCC and CTDA were the only two classifiers for which substraction improved the PVC detection rate, albeit with significant loses in healthy beat recognition. In the scope of the tested methods, k-NN, CART and SVD provided average recognition accuracies.

6   Conclusions and Future Research

This paper reviewed several classification algorithms and their performance in terms of PVC detection in human subjects with minimalistic prior data preprocessing, training the classifiers on different subjects than they were tested on. While the overall performance of some of the classifiers, namely k-NN, CTDA and SVD based, was satisfactory, none of the methods have an obvious advantage over the other, leaving space for significant improvements to achieve greater accuracy in PVC detection, healthy beat detection and overall accuracy alike. One such improvement was suggested and shown to significantly boost the algorithm's overall performance and although still not being able to surpass all of its competitors, it retains its original variant's superior PVC detection rate. The presented results motivate for further research of the OMP-based classifier and suggest that the approach, with further modifications, might have the potential to outperform both simple and more sophisticated methods and possibly become viable to be practically implemented in wearable monitoring devices. The focus of future research lies in more elaborate, yet still computationally inexpensive preprocessing as well as modifying the classifiers for greater recognition accuracy in the PVC detection area. Combining multiple preprocessing techniques or inserting a few beat samples relevant to the test subjects into the training set is also a viable option for further studies.

Acknowledgements. This article has been elaborated in the framework of the IT4Innovations Centre of Excellence project, reg. no. CZ.1.05/1.1.00/02.0070 funded by Structural Funds of the European Union and state budget of the Czech Republic. The work is partially supported by Grant of SGS No. SP2013/70, VB - Technical University of Ostrava, Czech Republic. This work was also supported by the Bio-Inspired Methods: research, development and knowledge transfer project, reg. no. CZ.1.07/2.3.00/20.0073 funded by Operational Programme Education for Competitiveness, co-financed by ESF and state budget of the Czech Republic.

References

1. Haibing, Q., Xiongfei, L., Chao, P.: A method of continuous wavelet transform for qrs wave detection in ecg signal. In: 2010 International Conference on Intelligent Computation Technology and Automation (ICICTA), vol. 1, pp. 22–25 (2010)
2. Huptych, M., Lhotsk, L.: Proposal of feature extraction from wavelet packets decomposition of qrs complex for normal and ventricular ecg beats classification. In: Vander Sloten, J., Verdonck, P., Nyssen, M., Haueisen, J. (eds.) ECIFMBE 2008. IFMBE Proceedings, vol. 22, pp. 402–405. Springer, Heidelberg (2009)
3. Inan, O., Giovangrandi, L., Kovacs, G.T.A.: Robust neural-network-based classification of premature ventricular contractions using wavelet transform and timing interval features. IEEE Transactions on Biomedical Engineering 53(12), 2507–2515 (2006)

4. Loh, W.-Y.: Classification and regression trees. Wiley Interdisc. Rew.: Data Mining and Knowledge Discovery 1(1), 14–23 (2011)
5. Bortolan, G., Jekova, I., Christov, I.: Comparison of four methods for premature ventricular contraction and normal beat clustering. In: Computers in Cardiology, pp. 921–924 (2005)
6. Jang, J.-S.R., Sun, C.-T.: Neuro-fuzzy and soft computing: a computational approach to learning and machine intelligence. Prentice-Hall, Inc., Upper Saddle River (1997)
7. Gharaviri, A., Dehghan, F., Teshnelab, M., Moghaddam, H.: Comparison of neural network, anfis, and svm classifiers for pvc arrhythmia detection. In: 2008 International Conference on Machine Learning and Cybernetics, vol. 2, pp. 750–755 (2008)
8. Lavanya, D., Rani, D.K.: Performance evaluation of decision tree classifiers on medical datasets. International Journal of Computer Applications 26(4), 1–4 (2011)
9. Dabney, A.R., Storey, J.D.: Optimality driven nearest centroid classification from genomic data. PloS One 2(10) (2007)
10. Gajdos, P., Moravec, P., Snasel, V.: Preprocessing methods for svd-based iris recognition. In: 2010 International Conference on Computer Information Systems and Industrial Management Applications (CISIM), pp. 48–53 (October 2010)
11. Frolov, A., Husek, D., Bobrov, P.: Brain-computer interface: Common tensor discriminant analysis classifier evaluation. In: 2011 Third World Congress on Nature and Biologically Inspired Computing (NaBIC), pp. 614–620 (2011)
12. Blumensath, T., Davies, M.E.: On the difference between Orthogonal Matching Pursuit and Orthogonal Least Squares. University of Edinburgh. Tech. Rep. (March 2007)
13. Wright, J., Yang, A.Y., Ganesh, A., Sastry, S.S., Ma, Y.: Robust face recognition via sparse representation. IEEE Transactions on Pattern Analysis and Machine Intelligence 31, 210–227 (2009)
14. Goldberger, A.L., Amaral, L.A.N., Glass, L., Hausdorff, J.M., Ivanov, P.C., Mark, R.G., Mietus, J.E., Moody, G.B., Peng, C.-K., Stanley, H.E.: PhysioBank, PhysioToolkit, and PhysioNet: Components of a new research resource for complex physiologic signals. Circulation 101(23), 215–220 (2000)
15. Moody, G., Mark, R.: The impact of the mit-bih arrhythmia database. IEEE Engineering in Medicine and Biology Magazine 20(3), 45–50 (2001)
16. Moody, G., Mark, R.: The mit-bih arrhythmia database on cd-rom and software for use with it. In: Proceedings of the Computers in Cardiology 1990, pp. 185–188 (1990)

Text Classification Techniques in Oil Industry Applications

Nayat Sanchez-Pi[1], Luis Martí[2], and Ana Cristina Bicharra Garcia[1]

[1] ADDLabs, Fluminense Federal University
Rua Passo da Pátria, 156. 24210-240 Niterói (RJ) Brazil
`nayat,cristina@addlabs.uff.br`
[2] Dept. of Electrical Engineering, Pontifícia Universidade Católica do Rio de Janeiro,
R. Marquês de São Vicente, 225. Rio de Janeiro (RJ) Brazil
`lmarti@ele-puc-rio.br`

Abstract. The development of automatic methods to produce usable structured information from unstructured text sources is extremely valuable to the oil and gas industry. A structured resource would allow researches and industry professionals to write relatively simple queries to retrieve all the information regards transcriptions of any accident. Instead of the thousands of abstracts provided by querying the unstructured corpus, the queries on structured corpus would result in a few hundred well-formed results.

On this paper we propose and evaluate information extraction techniques in occupational health control process, particularly, for the case of automatic detection of accidents from unstructured texts. Our proposal divides the problem in subtasks such as text analysis, recognition and classification of failed occupational health control, resolving accidents.

Keywords: text classification, ontology, oil and gas industry.

1 Introduction

Health, safety and environment (HSE) issues are priority matter for the offshore oil and gas industry. This industry is frequently in the news. Much of the time it is because of changes in prices of oil and gas. Other —less frequent but perhaps more important— subject of media attention is when disasters strike, as is the case of offshore oil drilling platform explosions, spills or fires. These incidents have a high impact on lives, environment and public opinion regarding this sector. That is why a correct handling of HSE is a determining factor in this industry long–term success.

Today, with the advances of new technologies, accidents, incidents and occupational health records are stored in heterogeneous repositories. Similarly, the amount of information of HSE that is daily generated has become increasingly large. Furthermore, most of this information is stored as unstructured or poorly-structured data and. This poses a challenge is a top priority for industries that are looking for ways to search, sort, analyze and extract knowledge from masses of data.

Á. Herrero et al. (eds.), *International Joint Conference SOCO'13-CISIS'13-ICEUTE'13*, 211
Advances in Intelligent Systems and Computing 239,
DOI: 10.1007/978-3-319-01854-6_22, © Springer International Publishing Switzerland 2014

The development of automatic methods to produce usable structured information from unstructured text sources would be extremely valuable to the oil and gas industry. A structured resource would allow researches and industry professionals to write a relatively simple queries to retrieve all the information regards transcriptions of any accident. Instead of the thousands of abstracts provided by querying the unstructured corpus, the queries on structured corpus would result in a few hundred well-formed results. This would obviously save time and resources while also empowering the decision maker with consistent and valuable information.

On this paper we propose and evaluate information extraction techniques in occupational health control process, particularly, for the case of automatic detection of accidents from unstructured texts. Our proposal divides the problem in subtasks such as: (i) text analysis, (ii) recognition and (iii) classification of failed occupational health control, resolving accidents. We present an ontology-based approach to the automatic text categorization.

An important and novel aspect of this approach is that our categorization method does not require a training set, which is in contrast to the traditional statistical and probabilistic methods that require a set of pre-classified documents in order to train the classifier. Also relevant is the use of a word thesaurus for finding non-explicit relations between classification text and ontology terms. This feature widens the domain of the classifier allowing it to respond to complex real-life text and more resilient to ontology incompleteness.

The rest of this work goes on by describing the theoretical foundations that support it. After that, in Section 3, we describe the elements that are involved in our proposal, that is: (i) the elaboration of the ontology, (ii) the use of a thesaurus as a crawling tool and (iii) the use of the ontology as a classifier. Subsequently, in Section 4 we present the prototype that has been developed under the scope of this work. Finally, some conclusive remarks are put forward.

2   Foundations

Automatic text categorization is a task of assigning one or more pre-specified categories to an electronic document, based on its content. Nowadays, automatic text classification is extensively used in many contexts. A typical examples is the automatic classification of incoming electronic news into categories, such as entertainment, politics, business, sports, etc.

Standard categorization approaches utilize statistical or machine learning methods to perform the task. Such methods include Naïve Bayes [1], Support Vector Machines [2], Latent Semantic Analysis [3] and many others. A good overview of the traditional text categorization methods is presented in [4]. All of these methods require a training set of pre-classified documents that is used for classifier training; later, the classifier can correctly assign categories to other, previously unseen documents.

However, it is often the case that a suitable set of well categorized (typically by humans) training documents is not available. Even if one is available, the set may be too small, or a significant portion of the documents in the training set may not have been classified properly. This creates a serious limitation for the usefulness of the traditional text categorization methods.

Ontologies [13] offer knowledge that is organized in a more structural and semantic way. Their use in text categorization and topic identification has lately become an intensive research topic. As ontologies provide named entities and relationship between them, an intermediate categorization step requires matching terms to ontological entities. Afterwards, an ontology can be successfully used for term disambiguating and vocabulary unification, as presented in [5]. Another approach, presented in [6], reinforces co-occurrence of certain pairs of words or entities in the term vector that are related in the ontology.

The use of descriptions of neighboring entities to enrich the information about a classified document is described in [9]. Interesting approach, although very different, is presented in [10]; where authors automatically build partial ontology from the training set to improve keyword-based categorization method. Other categorization approaches based on using recognized named entities are described in [11] and [12].

The knowledge represented in a comprehensive ontology can be used to identify topics (concepts) in a text document, provided the document thematically belongs to the domain represented in the ontology. Furthermore, if the concepts in the ontology are organized into hierarchies of higher-level categories, it should be possible to identify the category (or a few categories) that best classify the content of the document.

3   Proposal

In this section we introduce a novel text categorization method based on leveraging the existing knowledge represented in a domain ontology. The novelty of this approach lays in that it is not dependent on the existence of a training set, as it relies solely on the entities, their relationships, and the taxonomy of categories represented in the ontology.

In the proposed approach, the ontology effectively becomes the classifier. Consequently, classifier training with a set of pre-classified documents is not needed, as the ontology already includes all important facts. The proposed approach requires a transformation of the document text into a graph structure, which employs entity matching and relationship identification.

The categorization is based on measuring the semantic similarity between the created graph and categories defined in the ontology require a training set of pre-classified documents that is used for classifier training; later, the classifier can correctly assign categories to other, previously unseen documents.

Our strategy is to use an ontology as the key component of our text classification heuristic algorithm. Besides the ontology itself, the algorithm is composed of the following set of modules:

1. A lemmatization, stemming and stop-word removing preprocessing. In this work we applied for this task the functionality provided by the Apache Lucene framework [15].

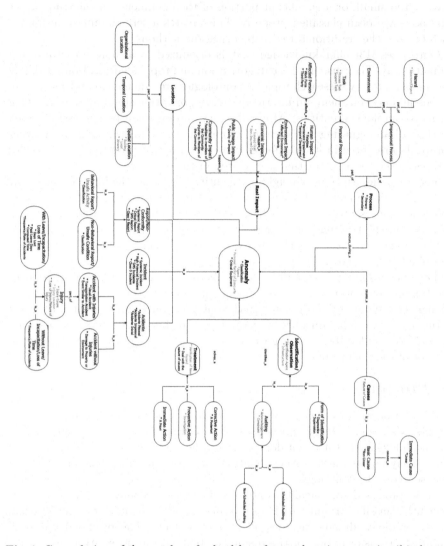

Fig. 1. General view of the ontology for health, safety and environment in oil industry

2. A thesaurus for locating words appearing in the text in the ontology. In our case we used a customized version of OpenOffice Brazilian Portuguese thesaurus [16].
3. Set of ontology elements tagged with its corresponding classification label.
4. A thesaurus crawling algorithm that takes care of determining the matching degree of text words with a corresponding ontology term.

There are some other proposals that also employ mechanisms that rely on ontologies, for example, [7, 8], and many more. However, in our case, the use of the thesaurus makes the approach more flexible and capable or handling real-world

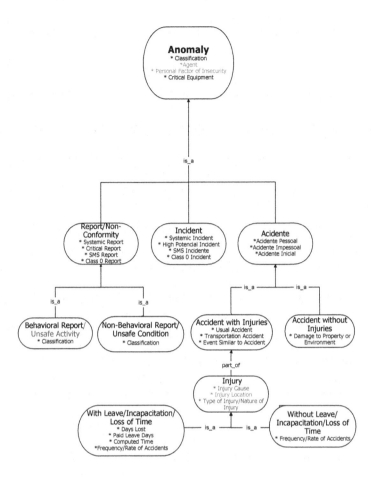

Fig. 2. The anomaly concept of the ontology for health, safety and environment in oil industry

applications. An ontology describes the application domain in a comprehensive but inflexible way. Natural language is, on the other hand, highly-irregular something rather impossible to grasp by directly using an ontology. The use of the thesaurus along with lemmatization and text processing bridges this gap effectively.

3.1   Health, Safety and Environment ontology

As part of this work, we devised a domain ontology for Health, Safety and Environment (HSE) for oil and gas application contexts (see Figure 1). This ontology was elaborated after interviewing field experts, an extensive reviewing of related literature and the analysis of the existing data sources.

We also obtained the inferences that describe the dynamic side and finally we group the inferences sequentially to form tasks. Principal concepts of the ontology, see Figure 2, are the following:

- **Anomaly:** Undesirable event or situation which results or may result in damage or faults that affect people, the environment, equity (own or third party), the image of the Petrobras System, products or production processes. This concept includes accidents, illnesses, incidents, deviations and non-conformances.
  - **Neglect:** Any action or condition that has the potential to lead to, directly or indirectly, damage to people, to property (own or third party) or environmental impact, which is inconsistent with labor standards, procedures, legal or regulatory requirements, requirements management system or practice.
    * **Behavioral neglect:** Act or omission which, contrary provision of security, may cause or contribute to the occurrence of accidents.
    * **Non-behavioral neglect:** Environmental condition that can cause an accident or contribute to its occurrence. The environment includes adjective here, everything that relates to the environment, from the atmosphere of the workplace to the facilities, equipment, materials used and methods of working employees who is inconsistent with labor standards, procedures, legal requirements or normative requirements of the management system or practice.
  - **Incident:** Any evidence, personal occurrence or condition that relates to the environment and/or working conditions, can lead to damage to physical and/or mental.
  - **Accident:** Occurrence of unexpected and unwelcome, instant or otherwise, related to the exercise of the job, which results or may result in personal injury. The accident includes both events that may be identified in relation to a particular time or occurrences as continuous or intermittent exposure, which can only be identified in terms of time period probable. A personal injury includes both traumatic injuries and illnesses, as damaging effects mental, neurological or systemic, resulting from exposures or circumstances prevailing at the year's work force. In the period for meal or rest, or upon satisfaction of other physiological needs at the workplace or during this, the employee is considered in carrying out the work.
    * **Accident with injury:** It's all an accident in which the employee suffers some kind of injury. Any damage suffered by a part of the human organism as a consequence of an accident at work.
      · **With leave:** Personal injury that prevents the injured from returning to work the day after the accident or resulting in permanent disability. This injury can cause total permanent disability, permanent partial disability, total temporary disability or death.

```
function:   ClassifyText(t): ⟨Γ, l⟩
   input:   t: string – the text to be classified.
  output:   Γ := {w_0, ..., w_n} – set of ontology terms that can classify the current text.
            l: integer – level of similarity of the terms.

       1    Ω = PreprocessText(t) – yields a set of words after lemmatization, stemming
            and stop-word removal.
       2    l = ∞  – best similarity level.
       3    Γ = ∅.
       4    for each w_k ∈ Ω do
       5        Θ_k = ComputeSimilarityLevels(w_k, 0).
       6        for each ⟨w_j, l_j⟩ ∈ Θ_k
       7            if l_j < l then
       8                Γ = {w_j}.
       9                l = l_j.
      10            else if l_j = l then
      11                Γ = Γ ∪ {w_j}.
      12            end-if
      13        end-for each
      14    end-for each
      15    return ⟨Γ, l⟩.
```

Fig. 3. Pseudo-code description of the algorithm used to compute the levels of similarity between a given word found in text and ontology terms

· **Without leave:** Personal injury that does not prevent the injured to return to work the day after the accident, since there is no permanent disability. This injury, not resulting in death, permanent total or partial disability or total temporary disability, requires, however, first aid or emergency medical aid. Expressions should be avoided "lost-time accident" and "accident without leave", used improperly to mean, respectively, "with leave injury" and "injury without leave."

* **Accident without injury:** Accident causes no personal injury.

3.2   Classification Algorithm

As already mentioned, the classification algorithm proposed in this work relies on the previous ontology, a thesaurus to establish the degree of matching between a given text fragment and some terms of interest that are present in the ontology.

The algorithm is presented as pseudo-code in Figure 3 as function **ClassifyText()**. It proceeds by first filtering and rearranging the input sentence in order to render it in a format suitable for processing (**PreprocessText()** method in step 1 of Figure 4). We have employed Apache Lucene text processing tools for stemming, lemmatization and stop-word removal.

Having the filtered text represented as a set of words, the algorithm proceeds to identify which terms of the ontology are most closely related to that set. It carries that out by invoking for each word the function **ComputeSimilarityLevels()**. This function —which is described in Figure 4— returns the set

```
function:  ComputeSimilarityLevels(w_0, l_0): Θ
   input:  w_0: string – the word to be processed.
           l_0: integer – initial level.
  global:  l_max: integer – maximum number of levels to be reached.
  output:  Θ := {⟨w_1, l_1⟩, ..., ⟨w_n, l_n⟩} – set of pairs of ontology term and level of similarity.
           Each pair represents how "close" is word w_n to w.

      1    if l_0 = l_max then
      2        return Θ = ∅ – max. number of levels reached.
      3    end-if
      4    if OntologyContains(w_0) then
      5        return Θ = {⟨w_0, l_0⟩} – the word is a term of the ontology, no further search
           is necessary.
      6    end-if
      7    Υ = Thesaurus(w_0) – determine the synonyms of w_0.
      8    Θ = ∅.
      9    for each w_k ∈ Υ do
     10        Θ_k = ComputeSimilarityLevels(w_k, l_0 + 1).
     11        Θ = Θ ∪ Θ_k.
     12    end-for each
     13    return Θ.
```

Fig. 4. Pseudo-code description of the algorithm used to compute the levels of similarity between a given word found in text and ontology terms

of ontology terms that are related with a given word by recursively traversing a thesaurus up to a given number of levels. If a connection between a word and a term is established that term is included, along with its level of similarity in the set of related terms Θ. The level of similarity is defined as the number of jumps needed to get from to word to the term using the thesaurus. A lower level implies higher similarity.

The result of the classification is one or more ontology terms that are most closely related to the text, or, posed in other words, the terms with minimal level of similarity. It should be beard in mind that the two functions presented here have been simplified for didactical reasons, and in practice some a harder to read but more efficient option is used.

4 Results

The classification algorithm proposed herein has an adequate computational performance. However it has some clear drawbacks when confronted to complex and contradictory texts. This is not an issue for our application domain. In spite of the texts are written in a natural language, for this particular domain, unstructured texts are written in a very direct discourse and there was no a large variation in the amount of information in each text, issues that were good for the step 1 (see Figure 5).

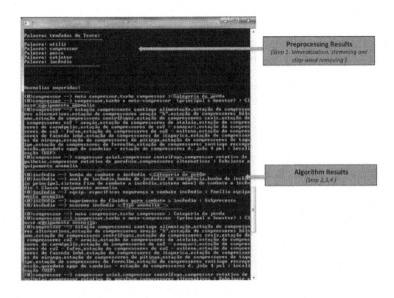

Fig. 5. Preprocessing and algorithm results

5   Final Remarks

In this paper we introduced a novel text classification method based on leveraging the existing knowledge represented in domain ontology. We have focused our approach on a real-life high-relevance problem: the health, safety and environment issues in the oil and gas industry.

The novelty of this approach is that it is not dependent on the existence of a training set, as it relies solely on the entities, their relationships, and the taxonomy of categories represented as an ontology. It might be argued that the synthesis of such ontology is comparable at some degree with the preparation of an annotated training set. However, when analyzing this issue at a deeper level it may be realized that an ontology-based solution is better mainly because an ontology can be easily contrasted and verified, both by formal means and by members of the research team. Therefore, this approach is less prone to bias, inconsistency and prejudice. Similarly, the resulting ontology is a relevant asset on its own right.

This approach has the additional novelty of incorporating a thesaurus for overcoming the possible narrow classification domain imposed by the limited set of terms that are present in the ontology. This feature makes the method more flexible and resilient to real-life texts that are hardly written in a homogeneous or exact form.

It must be said that this paper presents a set of results that is susceptible of being improved. In particular, we are interested on using the ontology to provide a more granular classification.

References

[1] Lewis, D.D.: Naive (Bayes) at forty: The independence assumption in information retrieval. In: Nédellec, C., Rouveirol, C. (eds.) ECML 1998. LNCS, vol. 1398, Springer, Heidelberg (1998)

[2] Vapnik, V.: The nature of statistical learning theory. Springer (1995)

[3] Deerwester, S., Dumais, S., Furnas, G.W., Landauer, T.K., Harshman, R.: Indexing by Latent Semantic Analysis. Journal of the Society for Information Science 41, 391–407 (1990)

[4] Sebastiani, F.: Machine learning in automated text categorization. ACM Computing Surveys (CSUR) 34, 1–47 (2002)

[5] Bloehdorn, S., Hotho, A.: Text Classification by Boosting Weak Learners based on Terms and Concepts. In: 4th IEEE International Conference on Data Mining, ICDM 2004 (2004)

[6] Nagarajan, M., Sheth, A.P., Aguilera, M., Keeton, K., Merchant, A., Uysal, M.: Altering Document Term Vectors for Classification - Ontologies as Expectations of Co-occurrence. LSDIS Technical Report (November 2006)

[7] Fang, J., Guo, L., Wang, X., Yang, N.: Ontology-Based Automatic Classification and Ranking for Web Documents. In: Fourth International Conference on Fuzzy Systems and Knowledge Discovery (FSKD 2007), pp. 627–631 (2007)

[8] Camous, F., Blott, S., Smeaton, A.F.: Ontology-based MEDLINE document classification. In: Hochreiter, S., Wagner, R. (eds.) BIRD 2007. LNCS (LNBI), vol. 4414, pp. 439–452. Springer, Heidelberg (2007)

[9] Gabrilovich, E., Markovitch, S.: Overcomingthe Brittleness Bottleneck using Wikipedia: Enhancing Text Categorization with Encyclopedic Knowledge. In: 21st National Conference on Artificial Intelligence, Boston, MA, USA (2006)

[10] Wu, S.-H., Tsai, T.-H., Hsu, W.-L.: Text categorization using automatically acquired domain ontology. In: 6th International Workshop on Information Retrieval with Asian Languages, Sapporo, Japan, vol. 11 (2003)

[11] Sheth, A.P., Bertram, C., Avant, D., Hammond, B., Kochut, K.J., Warke, Y.: Semantic Content Management for Enterprises and the Web. IEEE Internet Computing (July/August 2002)

[12] Hammond, B., Sheth, A.P., Kochut, K.J.: Semantic Enhancement Engine: A Modular Document Enhancement Platform for Semantic Applications over Heterogeneous Content. In: Real World Semantic Web Applications. IOS Press (2002)

[13] Gruber, T.: A Translation Approach to Portable Ontology Specifications. Knowledge Acquisition 5, 199–220 (1993)

[14] Sheth, A.P., Arpinar, I.B., Kashyap, V.: Relationships at the Heart of Semantic Web: Modeling, Discovering, and Exploiting Complex Semantic Relationships. In: Nikravesh, M., Azvin, B., Yager, R., Zadeh, L. (eds.) Enhancing the Power of the Internet. Stud Fuzz. Springer (2003)

[15] Gospodnetic, O., Hatcher, E., McCandless, M.: Lucene in Action, 2nd edn. Manning Publications (2009) ISBN 1-9339-8817-7

[16] DicSin: Dicionário de Sinônimos Português Brasil. Apache OpenOffice.org (2013), http://extensions.openoffice.org/en/project/DicSin-Brasil

A Domestic Application of Intelligent Social Computing: The SandS Project

Manuel Graña[1], Ion Marqués[1], Alexandre Savio[1], and Bruno Apolloni[2]

[1] Grupo de Inteligencia Computacional (GIC), Universidad del País Vasco, Spain
manuel.grana@ehu.es
[2] Dept. of Computer Science, University of Milano, Milano, Italy

Abstract. This paper introduces principal ideas of new ways to mediate the interaction between users and their domestic environment, namely the set of household appliances owned by the user. These ideas are being developed in the framework of the Social and Smart (SandS) project, which elaborates on the idea of a social network of home appliance users that exchange information and insights about the use of their appliances. This interaction constitutes the conscious social computing layer of the system. The system has a subconscious computing layer consisting of a networked intelligence that strives to provide innovative solutions to user problems, so that the system goes beyond being a recollection of appliance recipes. This paper discusses the structure of the system, as well as some data representation issues that may be instrumental to its development, as part of the development work leading to the final implementation of the project ideas.

1 Introduction

Social networks may act as a repository of information and knowledge that can be probed by the social agents to solve specific problems or to learn procedures relative to a common knowledge domain. This is well known in the social sciences, where social networks have been useful to spread educational innovations in health care [4], manage product development programs [6], engagement in agricultural innovations by farmers [7].

Social computing was defined by Vannoy and Palvia [8] as "intra-group social and business actions practiced through group consensus, group cooperation, and group authority, where such actions are made possible through the mediation of information technologies, and where group interaction causes members to conform and influences others to join the group" in the context of a study of social models for technology adoption. We qualify the term social computing with conscious levels as follows:

- Conscious social computing: the information processing done under the complete user awareness and participation.
- Unconscious social computing: the information processing done on the social data without user awareness, usually kept hidden from the user.

Á. Herrero et al. (eds.), *International Joint Conference SOCO'13-CISIS'13-ICEUTE'13*,   221
Advances in Intelligent Systems and Computing 239,
DOI: 10.1007/978-3-319-01854-6_23, © Springer International Publishing Switzerland 2014

– Subconscious social computing: the information processing by delegation from the users but of which the user is aware, but not in detailed control.

We would say that Vannoy and Palvia's definition corresponds to Conscious social computing. Social subconscious computing can be termed *intelligent* when new solutions to new or old problems are generated when posed to it. Let us clarify ideas by proposing a short taxonomy of tasks that can be done by the social subconscious computing:

– Crowd-sourcing: the social players explicitly cooperate to build a knowledge object following some explicit and acknowledged rules. The foremost example: wikipedia. Sometimes the user contribution is stimulated by games and competition. For instance, the Lego design competition is a strong source of innovation obtained from the conscious social computing carried out by the players. Other example are the image labeling games proposed to obtain ground truth for the design of automatic content based image retrieval algorithms.
– Information gathering: The social player asks for a specific data, i.e. the restaurant closest to a specific landmark, and the social framework searches for it. In the internet of things framework [2], this search may mean that "things" are chatting between them sharing bits of information that lead to the appropriate information source.
– Solution recommendation: the social player asks for the solution of a problem, i.e. the best dating place for a first date, and the social framework broadcasts the question searching for answers in the form of recommendations by other social players. Answers can be tagged by trust values. The recommendations can be also produced by automatic recommender systems which need the precise specification of questions and answers in a repository while users may process natural language uncertain and vague issues directly.
– Solution generation: the social player asks for the solution of a problem, i.e. how to cook a 5 kg turkey?, and the social framework provides solutions based on previous reported experience from other social players. This experience can be explicitly or implicitly provided by the social players, but the hidden intelligence layer effectively strive to provide the best fit solution, even if nobody in the social network has ever cooked a turkey.

Social and Smart (SandS) project aims to lay the foundations of social subconscious intelligent systems in the realm of household appliances and domestic services. The contents of the paper are as follows: Section 2 gives an intuitive definition of Sands Network.

2 Intuitive Definition of SandS

Figure 1 gives an intuitive representation of the architecture and interactions between the system elements. The SandS Social Network mediates the interaction between a population of users (called eahoukers in SandS [1]) each with its own set of appliances. The SandS Social Network has a repository of tasks that have been posed by the eahoukers and a repository of recipes for the use of appliances. These two repositories are related by a map between (to and from) task and recipes. This map needs not to be one-to-one. Blue arrows correspond to the path followed by the eahouker queries, which are used to interrogate the database of known/solved tasks. If the task is already known, then the corresponding recipe can be returned to the eahouker appliance (black arrows). The eahouker can express its satisfaction with the results (blue arrows). When the queried task is unknown and unsolved then the social network will request a solution from the SandS Networked Intelligence that will consists in a new recipe deduced from the past knowledge stored in the recipe repository. This new solution will be generated by intelligent system reasoning. The eahouker would appreciate some explanation of the sources and how it has been reasoned to be generated, therefore explicative systems may be of interest for this application.

The repository of recipes solving specific tasks can be loaded by the eahoukers while commenting among themselves specific cases, or by the appliance manufacturing companies as an example of use to foster their sales by appealing their clients with additional services. These situations correspond to the conscious computing done on the social web service by human agents. The role of the Networked Intelligence is to provide the subconscious computing that generates innovation without eahouker involvement.

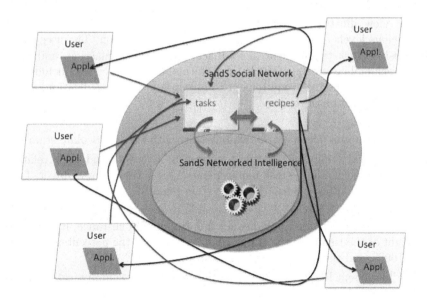

Fig. 1. Social and Smart system prototypical architecture

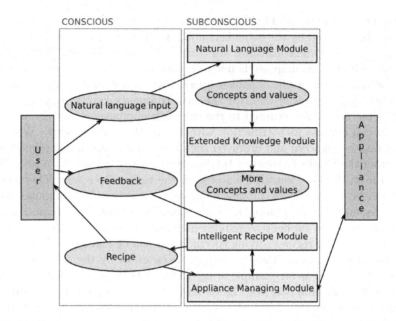

Fig. 2. A visualization of the SandS task and recipe processing flow chart

We can view the main components of the intelligent system from another point of view illustrated in figure 2. The figure separates the conscious and subconscious levels of information transfer and computation. The user can fed consciously a task requirement, a feedback about his satisfaction, and an appliance recipe. The subconscious layer performs several processes, including the interpretation of the task requirement to obtain a precise specification, and the intelligent generation of recipes:

1. The user inputs "clean cotton white t-shirt with small chocolate stain".
2. The Natural Language Module extracts the concepts and their relational structure specifying the task requirements, i.e. "Action: clean. Object: Tshirt. Attributes: Cotton made. Chocolate stain (small)".
3. The Extended Knowledge Module is able to deduce more information that may be needed by the appliance for fine tuning. For instance, it can set the weight of the t-shirt to the standard 200 grams, set the percentage of cotton to 100%, and the size of the stain to around 3 cm^2.
4. The Intelligent Recipe Module generates a recipe. It communicates with the Appliance Managing Module and the user, in order to execute the task and learn. The appliance and the user can give useful feedback information that ends in the Intelligent Recipe Module. For instance, if the user says "OK", then the recipe is known to be good. If the washing machine senses low weight, it could learn to reduce the water quantity accordingly.

3   Task and Recipe Domains

The mapping between task and recipe semantic domains is illustrated in figure 3 for a washing machine case. The task domain corresponds to graph on the left side of the figure, and it is a representation of the conscious aspect of the conscious interaction of the user and his appliance mediated by the social network service. If describes the concepts relating to the clothes and stains that can appear on them, so that a description of a washing task would be composed of some keywords extracted from this domain specification. They conform the basic vocabulary that the user wants to use to communicate, the natural language of the events related to her needs. The graph on the right side of the figure corresponds to the semantic domain of the appliance recipes. Recipes will be composed by sequences of actions and perceptually conditioned decisions whose atomic elements would be extracted from the ontology. Mapping tasks into appliance recipes may not be performed by ontology matching or graph homomorphism computing. The formalization of the required mapping must follow the mapping between sequences of terms that can be affected by the restrictions imposed by the semantic constraints. Besides the importance and scientific interest that the mathematical formalization may have by itself, the important point here is to notice that we can relate the above said social computing qualifiers to these two semantic domains:

- The conscious social computing of the user goes in terms of the specification of task to be performed (washing a cloth with some stains), as well as the informal reasoning about the most adequate washing recipe (temperature setting, kind of detergent). However, not all users are willing or capable of discussing washing recipes, so they can also reason in terms of the similitude of tasks (same or different materials, etc.).
- The subconscious social computing aspect of the system consists in the formal translation of tasks into recipes and vice versa, as well as the formal manipulation of the recipes to obtain new recipes that may provide enhanced solutions to the tasks that are presented by the user. The aim of the subconscious computing is to solve some vague optimization problem, which could be posed as a learning problem. It is in fact the function of the networked intelligence level of the proposal main structure.

The subconscious social computing presents, therefore, a strong learning and validation problem, as far as the there are no actually gathered data to perform them. In fact, the project is in the stage of setting the formal descriptions and computational solutions adopted to carry out the learning problem. Validation requires the user satisfaction feedback as the gold standard or ground truth, which involves real life experimentation and some extra personal costs. However, user satisfaction does not give enough information to derive rules that may perform the recipe adaptation in terms of the approximation to the desired task performance. Gradient operators in the task and recipe spaces, and the relation between them are crucial steps to develop the subconscious social intelligence in the SandS system. Recipes may be specified as graphs of processes, so that recipe gradient operators would be some kind of graph matching procedures. Alternatively they can be represented as empirical discrete time function specifications, which amount to some high dimensional vector time sequences, so that recipe gradient operators would be some operator in a high dimensional Euclidean space.

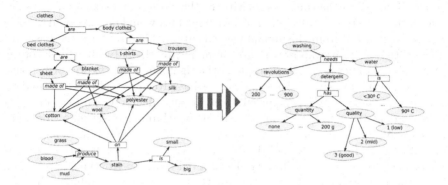

Fig. 3. Mapping the task (left) and appliance recipe (right) semantic domains

4   Knowledge Storage

Knowledge data storage and retrieval is not a trivial problem in the SandS system. Besides the already difficult problem of storing and retrieving social information data, which is involves many recursive call explorations , we have the problem of storing and retrieving the task and recipe data, which may be more efficiently considered as graphs in some sense. It would be desirable to have all these data stored in an integrated and unique database.

Graph databases have been developed since more than twenty years ago [5]. Lately, they have made place into a knowledge domain known as NOSQL (http://nosql-database.org/) whose intention is modern databases, i.e., non-relational, distributed, open-source and horizontally scalable. Graph databases leverage complex and dynamic relationships in highly-connected data to generate a different insight [3]. Nowadays, general-purpose graph databases are widely available (http://www.neo4j.org/). There are three popular graph models: the property graph, hypergraphs and RDF triples, although the former is the most implemented. A property graph contains vertices and edges where both of them contain properties (key-value pairs), edges are named and directed.

Figure 4 shows an snapshot of the possible database structure containing the knowledge of the system. Users and appliances are kind of terminal (blue and green) nodes of the system, storing their respective parameters. Tasks are also represented as single magenta nodes, connected to the user(s) that has requested them, and the recipes that may provide a solution to them. Recipes are represented by graphs with an special connection node, which the one used to connect them to the other kinds of nodes. We have not represented relations

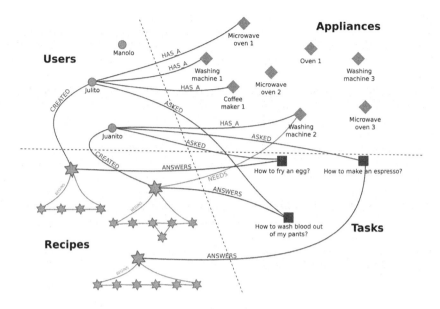

Fig. 4. Graph database snapshot

among recipes that may lead to think of any learning process between recipes. Though this representation is improving and evolving, it already shows the power of the graph database for our problem. A key problem is the definition of the storage and retrieval operators corresponding to the gradient operators in the recipe space.

5 Conclusions

We have discussed some salient features of the SandS project going beyond the accepted definition of social systems, introducing a subconscious intelligent computing layer that would be a step ahead in the social computing environment. Though the project is focused on household appliances, its philosophy may be exported to many other domains. The main difference with other approaches is that the system will autonomously elaborate on the knowledge provided by the social players to innovate and obtain solutions to new problems, and to increase the satisfaction of the eahouker by solving better old problems by underground reinforcement learning, obtaining thus a personalization of the appliances to the user and its conditions.

References

1. Apolloni, B., Fiasche, M., Galliani, G., Zizzo, C., Caridakis, G., Siolas, G., Kollias, S., Grana Romay, M., Barriento, F., San Jose, S.: Social things - the sands instantiation. In: IoT-SoS 2013. IEEE (2013)

2. Atzori, L., Iera, A., Morabito, G., Nitti, M.: The social internet of things (siot) – when social networks meet the internet of things: Concept, architecture and network characterization. Computer Networks 56, 3594–3608 (2012)
3. Robinson, J.W.I., Eifrem, E.: Graph Databases, 1st edn. O'Reilly (April 11, 2013) (early release revision)
4. Jippes, E., Achterkamp, M.C., Brand, P.L.P., Kiewiet, D.J., Pols, J., van Enge-len, J.M.L.: Disseminating educational innovations in health care practice: Training versus social networks. Social Science & Medicine 70(10), 1509–1517 (2010)
5. Kiesel, N., Schuerr, A., Westfechtel, B.: Gras, a graph-oriented (software) engineer-ing database system. Information Systems 20(1), 21–51 (1995)
6. Kratzer, J., Leenders, R.T.A.J., van Engelen, J.M.L.: A social network perspec-tive on the management of product development programs. The Journal of High Technology Management Research 20(2), 169–181 (2009)
7. Oreszczyn, S., Lane, A., Carr, S.: The role of networks of practice and webs of influencers on farmers' engagement with and learning about agricultural innovations. Journal of Rural Studies 26(4), 404–417 (2010)
8. Vannoy, S.A., Palvia, P.: The social influence model of technology adoption. Com-mun. ACM 53(6), 149–153 (2010)

Clustering of Anonymized Users
in an Enterprise Intranet Social Network

Israel Rebollo Ruiz[1] and Manuel Graña[2]

[1] Unidad I+D empresarial Grupo I68, Computational Intelligence Group,
University of the Basque Country
beca98@gmail.com
[2] Computational Intelligence Group, University of the Basque Country
ccpgrrom@gmail.com

Abstract. Modern enterprises aim to implement increasingly efficient working methods. One of the trends is the use of process knowledge to guide users about the proper way of performing specific tasks, getting rid of the misleading influence of inexperienced users or veteran users with incorrect habits. But implementation of this kind of working methods can give rise to conflicts and general stressful interdepartmental atmosphere within the company. To this end we are designing a system that allows user ordering based on the amount of daily work done so that more proficient users may have a greater influence in their group, helping all the members of the group to a reach higher levels of performance. A key feature of the proposed system is that the users are anonymized at all times. This measures seeks to avoid personal feelings to interfere on the acceptance and generation of working recommendations. Even the number of user groups and their components are unknown at all times, so that the recommendation system is intended to be fully autonomous and self-managing, increasing overall work efficiency by using the experience of some workers but without disclosing the identity of those employees. The aim is to obtain the personal alignment of the user with the proposed recommendations. We report evaluation of the approach on two test cases in a real business environment, where it has been observed that the proposed system is capable of correctly clustering users and identify the most proficient within each group.

Keywords: Anonimization, ERP recommendations, Social Networks, user clustering.

1   Introduction

Even in companies where manufacturing labor is the predominant activity, bureaucratic management is becoming more important, needing tools capable of carrying out all work in the shortest time possible and with the least number of incidents [1]. To do this, the Enterprise Resource Planning (ERP) software packages offer a multitude of functionalities to perform all necessary management tasks. However, these tools are increasingly complex, the amount of data

Á. Herrero et al. (eds.), *International Joint Conference SOCO'13-CISIS'13-ICEUTE'13*,   229
Advances in Intelligent Systems and Computing 239,
DOI: 10.1007/978-3-319-01854-6_24, © Springer International Publishing Switzerland 2014

handled by the user, and the interaction with other processes makes it sometimes difficult to devise an efficient flow of actions to perform various tasks [2].

The users of these ERPs perform tasks, based on their training and their own personal experience, in patterns that are not always correct [3]. Some users even refuse to use the new tools and try to continue using as far as possible the familiar old methods [4]. There are many factors that lead users to use the ERP tools in one form or another, such as age, education level or even gender [5]. However, in many cases these usage guidelines are incorrect and may lead new employees to adopt them as working habit regardless of its effectiveness and alignment with the guidelines set by the company.

To correct this kind of behaviors, the only effective measure is to provide the correct formation and convince the users that the correct way to perform the tasks is the one set by the company [6]. Training is expensive and has a strong financial impact on ERP implementation [7]. To minimize high training costs, Grupo I68 is designing a system in which the ERP is able to identify the most common actions that a user employs to perform its assigned tasks, based on expert system recommendations and on an ERP function usage log [8]. In the previous work the most active users exert a big influence in other users without taking into account the user skill, because the system learns from the user and suggests the most suitable applications at any time. However this does not solve the problem of mimicking user bad practices. Additional user clustering creates profiles based on tasks performed so that ERP programs suggested by the system are related to the overall performance of the group, and not just the user. In this article we add the user proficiency as a further factor for recommendation generation. Therefore, a skilled user will have greater influence on the group leading other users to mimic the expert actions, also diminishing the influence of less skillful users. It is very difficult to identify an skilled. The aim of this paper is also to establish indicators that allow us to identify these "good" users, that we keep for future work.

This system can be met with strong opposition from some company employees who despite having extensive experience can not be a role model for the rest of the ERP user group. Thus the members of each user groups, and the expert(s) included in them, are anonymous so that the user does not know if the recommendation received comes from his/her own experience or that of others, evaluating it only by increased work satisfaction [9]. The number of user groups is unknown, and are spontaneously formed according to user behavior, so that the number of groups varies from one moment to another. For experimental purposes, the identification of the experts has been done manually in the works reported in this paper, assigning a high experience value to a user by a high rank officer of the company, but always preserving anonymity or automatically using a marker with which to measure the efficiency of each user.

The rest of the paper is organized as follow: Section 2 presents the user clustering. In Section 3 we discuss the anonymization of user groups. Finally, 4 present experimental results. The last section is for our conclusions.

2   User Clustering

In order to perform user clustering, we must correctly define the common characteristics shared by two users identifying them as belonging to the same group. We first define the affinity A between two users as the sum of the Euclidean distance between their working characteristics so that if this sum is less than a threshold U, then it will establish an affinity rapport between the two users.

$$A = \sum_{1}^{n} ||fa_i - fb_i||,$$

$$(A < U) \Rightarrow e(a, b) = 1,$$

where fa_i and fb_i are the i-th features of users a and b, and where $e(a, b) = 1$ means that there is an edge between both users in the affinity graph.

Calculating the affinity pairs, we obtain an affinity graph [10] which defines the different groups, but it may be the case that a user is related to two groups without that these groups are related to each other. For these cases, using a smaller threshold can broke the tie between groups. If lowering the threshold too much dissolves the groups, the user should choose randomly a group. A threshold too low may cause excessive fragmentation of the groups while a very large threshold can create groups in which there is too much difference between any pair of members. The number of groups in which users are divided is arbitrary, but to make a correct tuning of the system, we need to establish upper and lower

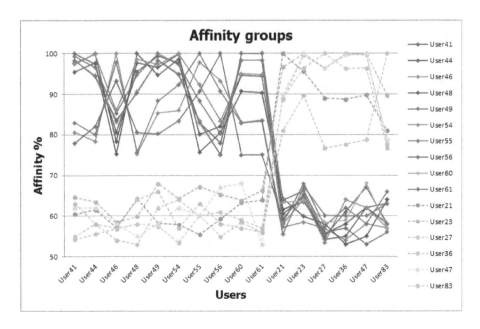

Fig. 1. Affinity Groups

bounds on the number of groups. These parameters depend on the number of users of the company. User clustering can be achieved by an algorithm for graph coloring, as we have done previously by a Swarm algorithm [11] that offers very good results compared to others in the literature [12,13].

Because each company is different respect another company, the threshold must be accurate for each company. The company economic area, number of employees or internal hierarchy, define differences so great that it is impossible to determine some common parameters.

For illustration purposes, in Figure 1 we plot the affinities between some users in one of the industrial studies performed. We can observe that users {User41, User44, User46, User48, User49, User54, User55, User56, User60} have affinity values between them in the order 75% and 100%, allowing to create an affinity group. On the other hand, users {User21, User23, User27, User36, User83, User47} also have a great affinity between them, creating another group. Moreover, this grouping is confirmed by the low affinity between the users from both groups, so that no users can belong to both groups. The data have been obtained from a company with over 400 user, and having anonymized user names for confidentiality issues, a human resources expert from the company has confirmed that the groups generated by the algorithm greatly resembled the daily work members of the same in most cases.

3   Anonymous User Groups

In the previous section we have grouped the users, taking into account various factors. These groups have been reviewed by a human resources expert to check that represent coherent user groups [14]. Whether the expert would have to make the groups would have been very similar. But the problem remains of finding the best performing user. The first approach to identify it would be to check who has the higher average affinity with other members of the group, which means that in a group of users with similar tasks, we select the user that matches in carrying out the tasks with more group members assuming that the expert will be mimicked by others. However, popular users with bad practices may adversely affect the rest of the group. This approach helps to make experiments, but is very poor identifying expert users so we will need a more accurate way to identify the expert.

Figure 2 shows different user states, and how they can evolve from one to another [15]. Using these relationships we are able to characterize the expert user, beyond routinely performed tasks. The experience is achieved with training supported by appropriate studies. Studies and training require an effort that can sometimes be limited by age. Age is an important factor in the behavior of people and the influence of society and, in the case of a company, its working peers. Happiness is another factor that affects the behavior, or that, vice verse, is a consequence of the behavior. We must differentiate between personal and impersonal social influence, where the latter is exerted regardless of personal traits, only on the basis of working performance. Then, to ensure that system experts providing recommendations to the other users are selected on the basis of impersonal value, the system needs to emphasize the need for privacy and anonymous clustering and expert selection. Our system will hide the experience so that it is an imposition that may cause rejection of the employees but will be anonymized. This anonymization will result in increasing working satisfaction of ERP users by excellency in user recommendations.

Human Features

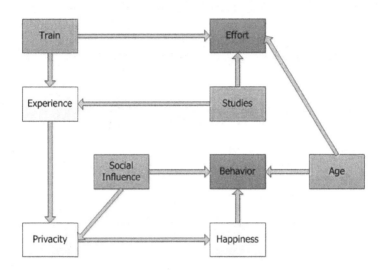

Fig. 2. Human Features relationship

4   Experiment Results

After consulting with the company human resources expert we have performed the following experiment. We have selected five different tasks consisting of six steps. The work had already been done by the users in the group, so that we known *a priory* their behavior. The system has recommended certain actions taking into account the experience of each user and group ownership, taking into account the experience of the group is closely tied to that of its expert. The first steps, the overlap between what the system offers and what users actually did was low, but as the task progresses users tend to make the expert's recommendations.

To prove these kind of systems is very difficult because is very complicated to identify if recommendations improve the productivity of the company or not. Only, when this system is installed in a real company, and used for a long period of time, we can extract conclusions about goodness of the system. But even in that case we can not ensure that an improvement or worsening is due to the system. The human opinion is the most valuable conclusion of our experiments.

In figure 3 we plot the percentage of success of the expert on each task, measured as the percentage of users that perform the same action as the expert which performing a given task. We observe that users tend to perform the same tasks as the expert. The figure indicates the percentage of success as it seems the action of the expert user. In the first action, the users take their own decision about what to do, but then more users performs the same action as the expert. These results are encouraging to the implementation of the anonymous user clustering and expert detection system, showing that this anonymous "follow the leader" process may improve performance of the overall system by levels above 80 % adoption of expert recommendations in some task performing sequences of actions.

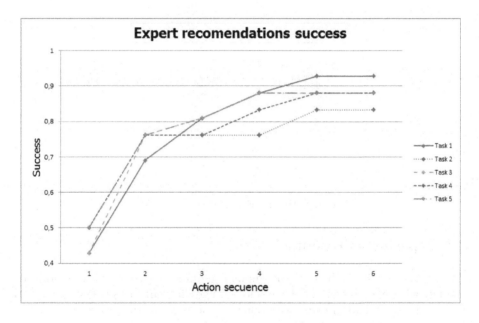

Fig. 3. Expert recommendation success for different tasks

5   Conclusions and Future Work

We have already managed to create groups of users within a company on the basis of ERP logs. Evaluation by company human resource expert, these groups have been found to correspond to the consistent user categories. In this paper we argue that the user grouping information is to be kept anonymous. We have found that even though you may give upper and lower bounds on the number of groups to generate, they are spontaneously formed without having any kind of outside interference. In this paper we have required expert opinion to validate the composition of the user group, however the system will be capable of generating autonomously user groups following the dynamic behavior variations of each user.

In addition we found that if we identify a hierarchy of experience within each group by assigning more value to their recommendations to those of others and not leaving you that will influence others, actions taken to solve various tasks are similar to those of the expert.

Future work is to define the requirements a user must meet in order to be considered an expert. In this way the system will be able to locate the experts in each group autonomously.

Acknowledgments. The Unidad I+D empresarial Grupo I68 company encourages participation in academic meetings, and academic production, within the limits of business confidentiality. The work reported here has been supported by project Gaitek IG-2012/00989 funded by the Basque Government, and also by project of the CDTI IDI-20130070 funded by Spanish Government.

References

1. Kanellou, A., Spathis, C.: Accounting benefits and satisfaction in an ERP environment. International Journal of Accounting Information Systems (2013)
2. Ram, J., Corkindale, D., Wu, M.L.: Implementation critical success factors (csfs) for erp: Do they contribute to implementation success and post-implementation performance? International Journal of Production Economics (2013)
3. Khan, R.Q., Corney, M.W., Clark, A.J., Mohay, G.M.: A role mining inspired approach to representing user behaviour in erp systems. In: Oyabu, T., Gen, M. (eds.) The 10th Asia Pacific Industrial Engineering and Management Systems Conference, pp. 2541–2552 (December 2009)
4. Gumussoy, C., Calisir, F., Bayram, A.: Understanding the behavioral intention to use erp systems: An extended technology acceptance model. In: 2007 IEEE International Conference on Industrial Engineering and Engineering Management, pp. 2024–2028 (2007)
5. Kanwal, S., Manarvi, I.A.: Evaluating erp usage behavior of employees and its impact on their performance: A case of telecom sector. Global Journal of Computer Science and Technology 10(9), 34–41 (2010)
6. Holland, C.P., Light, B.: A critical success factors model for erp implementation. IEEE Softw. 16(3), 30–36 (1999)
7. Poston, R., Grabski, S.: Financial impacts of enterprise resource planning implementations. International Journal of Accounting Information Systems 2(4), 271–294 (2001)
8. Ruiz, I.C.R., Romay, M.G.: User assistance tool for a webservice erp. In: Pérez, J.B., et al. (eds.) Trends in Pract. Appl. of Agents & Multiagent Syst. AISC, vol. 221, pp. 193–200. Springer, Heidelberg (2013)
9. Phelps, E., Zoega, G.: Corporatism and job satisfaction. Journal of Comparative Economics 41(1), 35–47 (2013)
10. Opsahl, T., Panzarasa, P.: Clustering in weighted networks. Social Networks 31(2), 155–163 (2009)

11. Ruiz, I.R., Romay, M.G.: Gravitational Swarm Approach for Graph Coloring. In: Pelta, D.A., Krasnogor, N., Dumitrescu, D., Chira, C., Lung, R. (eds.) NICSO 2011. SCI, vol. 387, pp. 159–168. Springer, Heidelberg (2011)
12. Rebollo, I., Graña, M.: Further results of gravitational swarm intelligence for graph coloring. In: Nature and Biologically Inspired Computing, pp. 183–188 (2011)
13. Folino, G., Forestiero, A., Spezzano, G.: An adaptive flocking algorithm for performing approximate clustering. Information Sciences 179(18), 3059–3078 (2009)
14. Rebollo-Ruiz, I., Graña-Romay, M.: Swarm graph coloring for the identification of user groups on erp logs. Cybernetics and Systems (in press, 2013)
15. Destre, G., Levy-Garboua, L., Sollogoub, M.: Learning from experience or learning from others?: Inferring informal training from a human capital earnings function with matched employer-employee data. The Journal of Socio-Economics 37(3), 919–938 (2008)

Intelligent Model for Fault Detection on Geothermal Exchanger of a Heat Pump

José Luis Casteleiro-Roca[1], Héctor Quintián[1], José Luis Calvo-Rolle[1],
Emilio Corchado[2], and María del Carmen Meizoso-López[1]

[1] Universidad de A Coruña,
Departamento de Ingeniería Industrial,
Avda. 19 de febrero s/n, 15.495, Ferrol, A Coruña, España
[2] Universidad de Salamanca,
Departamento de Informática y Automática,
Plaza de la Merced s/n, 37.008, Salamanca, Salamanca, España

Abstract. The Heat Pump with geothermal exchanger is one of the best methods to heat a building. The heat exchanger is an element with probabilities of failure due its size and due it is outside construction. The present study shows a novel intelligent system design to detect faults on this type of heating equipment. The novel approach has been successfully empirically tested under a real dataset obtained during measurements along one year. It is based on classification techniques with the aim to detect failures in real time. Then the model is validated and verified over the building; it allows to obtain good results in all the operating conditions ranges.

Keywords: MLP, J48, FLDA, Heat Exchanger, Heat Pump, Geothermal Exchanger.

1 Introduction

A Heat Pump provides heat by taking out it from a source, and then transferring it into a house [1]. Heat can be obtained from any source, whether it is cold of heat. But if this source is warm then it is possible to achieve higher efficiency [2]. The ground can be a source for the Heat Pump and, the heat exchangers topology can be vertical or horizontal [1,3]. Usually horizontal exchanger is more economical than the vertical configuration [3]; however this configuration has less efficiency than the other one. With the aim to increase the performance of the horizontal exchanger, frequently, installers place the exchanger more deeply in the ground [4].

Both configurations, vertical and horizontal have their own operation problems, but the horizontal has more inconvenient than the other, among others because the exchanger is closer to the ground surface [5,6]. Due to the proximity to the surface, the weather has influence over the exchanger and the efficiency could be lower [7,8]. For the same reason the installation may be damaged due

Á. Herrero et al. (eds.), *International Joint Conference SOCO'13-CISIS'13-ICEUTE'13*,   237
Advances in Intelligent Systems and Computing 239,
DOI: 10.1007/978-3-319-01854-6_25, © Springer International Publishing Switzerland 2014

to different reasons like crushing, perforations, and so on [9]. In normal, the performance is the same during throughout the year, but if any problem appears, then, the efficiency could drop significantly or even stop working [5].

Fault detection involves the monitoring of a system and the detection when a fault has occurred [10]. The system must be modeled or a knowledge based system must be created with the aim to detect deviations of the correct performance [11]. There are typical systems where fault detection has been implemented with satisfactory results. For instance, [12] propose a two-stage recognition system for continuous analysis of ElectroEncephaloGram (EEG) signals. [13] proposes a model-based Robust Fault Detection and Isolation (RFDI) method with hybrid structure. In [14] is presented a fault detection strategy for wireless sensor networks. In [15] a hybrid two stage one-against-all Support Vector Machine (SVM) approach is proposed for the automated diagnosis of defective rolling element bearings, [16] shows the robust fault detection problem for non-linear systems considering both bounded parametric modeling errors and measurement noises. As can be seen on the mentioned examples of fault detection, different soft computing techniques have been used to solve the problem.

This research presents a new intelligent method to make fault detection on buildings for a Heat Pump system using geothermal exchanger. The way to detect wrong work-points of operation is creating a model based on classification. This model has been trained with a big dataset of a year of operation, and consequently all weather seasons are taking into account. The model was tested with a real dataset of failures samples obtained for this purpose.

This paper is organized as follows. It begins with a brief description of the case of study followed by an explanation of the model approach and the dataset conditioning to perform fault detection through classification techniques. In the next section results are presented, and finally the conclusions and future work are exposed.

2   Case of Study

The novel model has been applied to do fault detection in a Geothermical Heat Pump. It is part of the systems installed within a real bioclimatic house. The physical system is described in detail as follows.

2.1   Sotavento Bioclimatic House

Sotavento bioclimatic house is a project of *demonstrative* bioclimatic house of Sotavento Galicia Foundation. The house is located within the *Sotavento Experimental Wind Farm*, which is a center of dissemination of renewable energy and energy saving. The farm is located between the councils of Xermade (Lugo) and Monfero (A Coruña), in the autonomous community of Galicia (Spain). It is at coordinates 43°21′ North, 7°52′ West, at an elevation of 640 m above sea level and at a distance of 30 Km from the sea. Figure 1 shows geographic location in the Spanish territory (left), and an external view of the bioclimatic house (right).

Fig. 1. Left: Geographic location of the bioclimatic house. Right: External view of the bioclimatic house

2.2   Installations of the Bioclimatic House

Thermal and electrical installations of the bioclimatic house have various renewable energy systems to complement these installations. Figure 2 describes through a schema, the main systems and component of thermal and electrical installations. The thermal installation consists of 3 renewable energy systems (solar, biomass and geothermal) that serve the DHW (Domestic Hot Water) system and the heating system. The electrical installation consists of two renewable energy systems (wind and photovoltaic) and one connection to the grid power, getting supply to the lighting and power systems of the house.

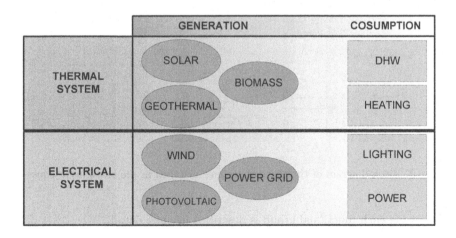

Fig. 2. Thermal and electrical installations of the house

2.3   Description of the Thermal Installation

Figure 3 shows through a block diagram, the different components of the thermal installation and their interconnections. Overall, the heating installation can be divided into 3 functional groups:

- Generation group: Solar thermal (1), biomass boiler (2) and geothermal (3).
- Energy accumulation group: Inertial accumulator (5), solar accumulator (4) and preheating (8).
- Consumption group: Underfloor heating (6) and DHW (7).

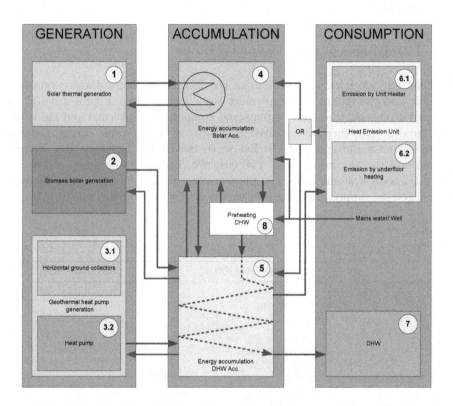

Fig. 3. Block diagram of the thermal systems installed in the bioclimatic house

Following each functional group is described in more detail.

Generation. The generation group has three collection systems of renewable energy:

- Solar thermal system: It consists of eight solar panels that capture energy from the solar radiation and use it to heat a fluid (ethyleneglycol) that flows inside the panels. The heated fluid is transported to the accumulation zone, where through the heat exchanger of the solar accumulator (4) gives up its heat to water stored inside the accumulator, getting to raise the water temperature for later use.

- Biomass boiler system: It has a biomass boiler type Ökofen, model Pallematic 20, with adjustable power from 7 kW to 20 kW, with a yield of pellets of 90%. It gives the hot water directly to the inertial accumulator to $63°C$.
- Geothermal system: The system consists of a horizontal collector with 5 loops of 100 meters, each one buried at a depth of 2 meter and a Heat Pump with a nominal heating power of 8.2 kW with a nominal electrical power consumption of 1.7 kW. It is possible to extract energy of the ground that is used to heat a fluid (water with glycol) that heats the water by geothermal pump and is driven directly to the inertial accumulator.

Accumulation. The system has a solar accumulator with storage capacity of 1000 liters which receives energy contribution of solar thermal system. This accumulator is connected in series with the inertia accumulator that has storage capacity of 800 liters, also it receives energy from the biomass boiler and geothermal system. The use of inertial accumulators is generally recommended in all types of heating systems and its function is to work as thermal energy storage to minimize starts and stops of the boiler.

Consumption. The house is equipped with DHW and underfloor heating systems, both supplied through the inertial accumulator. The DHW system has been sized based on the Spanish Technical Building Code, taking into account that the house only has a bathroom and a kitchen for demonstration purposes. Thus, the DHW system obtained was sized for 240 liters per day. The underfloor heating system consists of a network of crosslinked polyethylene with barrier etilvinil-alcohol (EVAL) pipes at a distance of 5 cm below the floor of the rooms of the house. It has 4 distribution collectors above the floor to enable the purging. This system is able to maintain the temperature inside the house between $18°C$ and $22°C$ for which the water circuit should only be between $35°C$ y $40°C$.

2.4   Geothermal System under Study

This section gives a detailed description of the operation of the real geothermal system and its components.

System Description. The Heat Pump is a MAMY Genius - 10.3 kW with a horizontal heat exchanger as is shown in figure 4.

This study is only focussed on the two sensors shown in figure 4 (thick arrows), in the primary circuit of the Heat Pump. The Heat Pump have two different circuits; the primary one provides the heat from the ground (the exchanger), and the other one is connected to the inertial accumulator.

Geothermal Exchanger. The horizontal exchanger has four different circuits, with the aim to isolate parts in case of discover a failure in one circuit. The installation has several temperature sensors installed along the heat exchanger

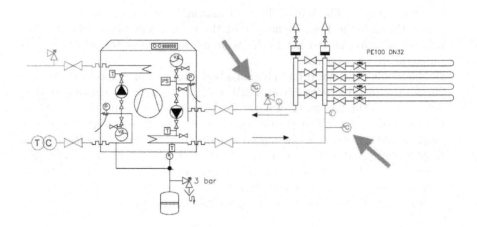

Fig. 4. Heat Pump and horizontal exchanger layout

to study the ground temperature while the system is running (figure 5 shows the exchanger, and its connection to the house). The four circuits are connected all together outside the house and only one pipe goes to the Heat Pump (one output and one input of the exchanger).

3   Model Approach

The fault detection model is based on the temperature difference at the output of the heat exchanger. These temperatures are measured at the Heat Pump. This is an improvement of the method, because it is not necessary to install sensors outside the building.

The model have three inputs. The first one is to indicate the state of the Heat Pump (*on* or *off*). The other two inputs are the temperatures at the input and at the output of the heat exchanger (figure 6).

3.1   Obtaining and Conditioning Dataset

The real dataset has been obtained by measurements taken along one year. Each measure has been taken every ten minutes.

Figure 7 shows the 144 measures of a day (one every 10 minutes). In this figure, Frame A shows a *complete* running cycle. Frame B is the part of the cycle when the Heat Pump is *on*, and Frame C is the restoration of the temperatures after the system turn *off*. In Frame D appears a cycle when the Heat Pump is *off* more time, and it is possible to see that the difference of the temperatures achieved more or less is the same.

$$\Delta T = T_{output} - T_{input} \tag{1}$$

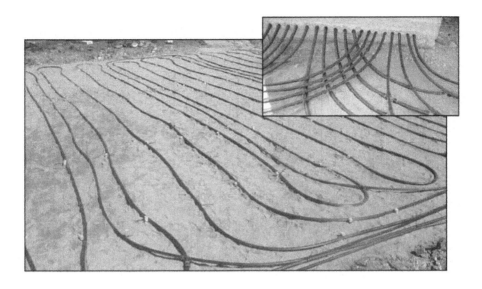

Fig. 5. Installation of the horizontal exchanger

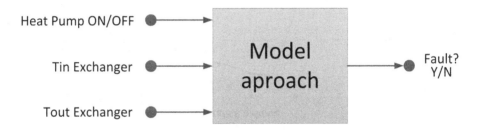

Fig. 6. Model approach

The model studies the variation of the temperature ratio from the input to the output of the Heat Pump (equation 1). If the Heat Pump is running or not, the model tolerate more or less changes, then:

- If the Heat Pump is *on*, the difference between temperatures is going to increase at first time (see figure 8, into the frame).
- At certain time when the Heat Pump turn *off*, and between 30 and 40 minutes after that, the difference between temperatures decreases (see figure 8, out the frame).

When the Heat Pump is *off* during a long time, the corresponding data are not taking into account. The dataset is obtained when the Heat Pump is in operation, and it turns *off* while temperatures do not achieved the steady state.

The initial dataset contains 52705 samples, and after discarding the non representative data, with the rules explained before, the data set contains 13612 samples.

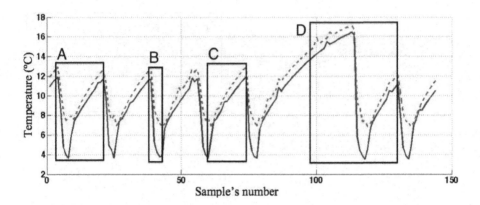

Fig. 7. One day of running cycles

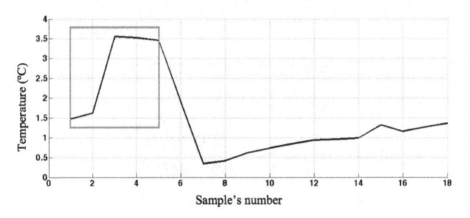

Fig. 8. Ratio between input and output of the Heat Pump

3.2 Classification Techniques Considered to Create the Model

This section describes the classification techniques used to create the model.

Fishers Linear Discriminant Analysis FLDA. This technique is used for reduction the dimension on a high-dimensional data project. The reduction achieves a low-dimensional space, where all the classes are separated [17,18]. FLDA create hyperplanes to discriminate the class as a result of analyzing the training data.

J48 Learning Algorithm. A decision tree is approach with this method, and there are the most common approaches in machine learning [19,20,21]. The decision trees are used to classify the data into different groups, depending of the variables used in training [19]. The decision trees were obtained by using the J48 algorithm [19,22,23]. J48 algorithm has better performance in most circumstances than other algorithms [23] This algorithm in based on an entropy calculation to develop the decision tree. Entropy is the probable information based on the partitioning into subcategories according to attribute. The clarity on the subcategory partitions achieve the greatest advantage. The feature with the greatest entropy reduction is chosen as the test attribute for the present node.

A Multilayer Perceptron (MLP) [19,24]. The architecture on the ANN must be carefully selected to achieve the best results. In this study several activation function have been tested.

4   Results

The results of the classification is shown in table 1. The performance of the classification techniques is obtained using the following parameters: Sensitivity (SE), Specificity (SPC), Positive Prediction Value (PPV), Negative Prediction Value (NPV) and Accuracy (ACC). The equations to calculate these parameters are shown in equations 2 to 6 respectively. Where, TP are the samples where there are no failures, TN are the samples where there are failures, and the model classify the samples in a good way. And, FP are the samples where there are failures, FN are the samples where there are no failures, but the model classifies the samples in a wrong way.

$$SE = \frac{TP}{TP + FN} \tag{2}$$

$$SPC = \frac{TN}{TN + FP} \tag{3}$$

$$PPV = \frac{TP}{TP + FP} \tag{4}$$

$$NPV = \frac{TN}{TN + FN} \tag{5}$$

$$ACC = \frac{TP + TN}{TP + TN + FP + FN} \tag{6}$$

The dataset has been increased by including one hour of real time measurements. When the system acquires samples in real time, it saves the parameters every second. During this time of fast acquisition, it has been generated some typical anomalies in this type of systems. It was simulated a fluid leak in the system, an obstruction in a pipe, a partial failure in the exchanger...

The dataset has been divided in two groups; one for training (70%, 12019 samples), and other to test the classification techniques (30%, 5151 samples). Table 1 shows the confusion matrix, and the performance parameters that were explained before (equations 2 to 6). The actual data are the real classification of the data, and the predicted data are the classification output of the algorithms.

Table 1. Confusion matrix

	FLDA		J48		MLP	
	No Failure (actual)	Failure (actual)	No Failure (actual)	Failure (actual)	No Failure (actual)	Failure (actual)
No Failure (predicted)	2431 47.19%	467 9.07%	3572 69.35%	163 3.16%	3815 74.06%	50 0.97%
Failure (predicted)	1621 31.47%	632 12.27%	479 9.30%	937 18.19%	269 5.22%	1017 19.74%
SE	0.6000		0.8818		0.9341	
SPC	0.5751		0.8518		0.9531	
PPV	0.8389		0.9564		0.9871	
NPV	0.2805		0.6617		0.7908	
ACC	0.5946		0.8754		0.9381	

5 Conclusions

With the novel model approach, it is possible to know when the geothermal exchanger of an installation based on a Heat Pump is failing. Very good results have been obtained in general terms. With this approach based on intelligent techniques working as a classification system, it is possible to predict malfunction states.

As can be seen in the results section, the best classification is achieved with MLP, where the percentage of classification accuracy is near to 94%. Models created with other classification techniques also achieve good results as J48 algorithm with an accuracy of 87.5%, however FLDA algorithm do not allow to achieve good result. The best detection failures only has an accuracy of 59.5 %.

Future work will be based on the application of on-line anomaly detection, with the aim to adapt the model to the progressive changes of the system. These changes can occur due to several reasons like the wear or the dirt.

Acknowledgments. This research is partially supported by the Spanish Ministry of Economy and Competitiveness under project TIN2010-21272- C02-01 (funded by the European Regional Development Fund), SA405A12-2 from Junta de Castilla y León. This work was also supported by the European Regional Development Fund in the IT4Innovations Centre of Excellence project (CZ.1.05/1.1.00/02.0070).

References

1. Langley, B.C.: Heat Pump Technology, 3rd edn. Prentice Hall, Upper Saddle River (2002)
2. Sauer, H.J., Howell, R.H.: Heat Pump Systems. Wiley-Interscience, New York (1983)
3. Kakac, S., Liu, H.: Heat Exchangers: Selection, Rating and Thermal Design, 2nd edn. CRC Press, Boca Raton (2002)

4. Rezaei, A.B., Kolahdouz, E.M., Dargush, G.F., Weber, A.S.: Ground Source Heat Pump Pipe Performance with Tire Derived Aggregate. International Journal of Heat and Mass Transfer 55(11-12), 2844–2853 (2012)
5. Lee, C.K.: Effects of multiple ground layers on thermal response test analysis and ground-source heat pump simulation. Applied Energy 88(12), 4405–4410 (2011)
6. Sanner, B., Karytsas, C., Mendrinos, D., Rybach, L.: Current Status of Ground Source Heat Pumps and Underground Thermal Energy Storage in Europe. Geothermics 32(4-6), 579–588 (2003)
7. Tarnawski, V.R., Leong, W.H., Momose, T., Hamada, Y.: Analysis of Ground Source Heat Pumps with Horizontal Ground Heat Exchangers for Northern Japan. Renewable Energy 34(1), 127–134 (2009)
8. Omer, A.M.: Ground-Source Heat Pumps Systems and Applications. Renewable and Sustainable Energy Reviews 12(2), 344–371 (2008)
9. Banks, D.: An Introduction to Thermogeology: Ground-Source Heating and Cooling. Wiley/Blackwell, Oxford (2008)
10. Isermann, R.: Fault-Diagnosis Systems: An Introduction from Fault Detection to Fault Tolerance. Springer, Germany (2006)
11. Isermann, R.: Fault-Diagnosis Applications: Model-Based Condition Monitoring: Actuators, Drives, Machinery, Plants, Sensors, and Fault-tolerant Systems. Springer, Germany (2011)
12. Hsu, W.Y.: Continuous EEG Signal Analysis for Asynchronous BCI Application. International Journal of Neural Systems 21(04), 335–350 (2011)
13. Nozari, H.A., Shoorehdeli, M.A., Simani, S., Banadaki, H.D.: Model-based robust fault detection and isolation of an industrial gas turbine prototype using soft computing techniques. Neurocomputing 91, 29–47 (2012)
14. Khan, S.A., Daachi, B., Djouani, K.: Application of fuzzy inference systems to detection of faults in wireless sensor networks. Neurocomputing 94, 111–120 (2012)
15. Gryllias, K.C., Antoniadis, I.A.: A Support Vector Machine Approach Based on Physical Model Training for Rolling Element Bearing Fault Detection in Industrial Environments. Engineering Applications of Artificial Intelligence 25(2), 326–344 (2012)
16. Tornil-Sin, S., Ocampo-Martinez, C., Puig, V., Escobet, T.: Robust Fault Detection of Non-Linear Systems Using Set-Membership State Estimation Based on Constraint Satisfaction. Engineering Applications of Artificial Intelligence 25(1), 1–10 (2012)
17. Koç, M., Barkana, A.: A new solution to one sample problem in face recognition using FLDA. Applied Mathematics and Computation 217(24), 10368–10376 (2011)
18. Calvo-Rolle, J.L., Casteleiro-Roca, J.L., Quintián, H., Meizoso-Lopez, M.C.: A hybrid intelligent system for PID controller using in a steel rolling process. Expert Systems with Applications 40(13), 5188–5196 (2013)
19. Parr, O.: Data Mining Cookbook. Modeling Data for Marketing, Risk, and Customer Relationship Management. John Wiley & Sons, Inc., New York (2001)
20. Duda, R.O., Hart, P.E., Stork, D.G.: Pattern Classification. John Wiley & Sons, Inc., Canada (2001)
21. Khan, R., Hanbury, A., Stöttinger, J., Bais, A.: Color based skin classification. Pattern Recognition Letters 33(2), 157–163 (2012)
22. Frank, E., Witten, I.: Data Mining: Practical Machine Learning Tools and Techniques, 2nd edn. Morgan Kaufmann (2005)
23. Rokach, L., Maimon, O.: Data Mining with Decision Trees: Theory and Applications. World Scientific Publishing, USA (2008)
24. Alpaydin, E.: Introduction to Machine Learning. The MIT Press, Oxford (2009)

A Comparative Study of Machine Learning Regression Methods on LiDAR Data: A Case Study

Jorge Garcia-Gutierrez[1], Francisco Martínez-Álvarez[2],
Alicia Troncoso[2], and Jose C. Riquelme[1]

[1] Department of Computer Science, University of Seville, Spain
{jgarcia,riquelme}@lsi.us.es
[2] Department of Computer Science, Pablo de Olavide University, Spain
{fmaralv,ali}@upo.es

Abstract. Light Detection and Ranging (LiDAR) is a remote sensor able to extract vertical information from sensed objects. LiDAR-derived information is nowadays used to develop environmental models for describing fire behaviour or quantifying biomass stocks in forest areas. A multiple linear regression (MLR) with previous stepwise feature selection is the most common method in the literature to develop LiDAR-derived models. MLR defines the relation between the set of field measurements and the statistics extracted from a LiDAR flight. Machine learning has recently been paid an increasing attention to improve classic MLR results. Unfortunately, few studies have been proposed to compare the quality of the multiple machine learning approaches. This paper presents a comparison between the classic MLR-based methodology and common regression techniques in machine learning (neural networks, regression trees, support vector machines, nearest neighbour, and ensembles such as random forests). The selected techniques are applied to real LiDAR data from two areas in the province of Lugo (Galizia, Spain). The results show that support vector regression statistically outperforms the rest of techniques when feature selection is applied. However, its performance cannot be said statistically different from that of Random Forests when previous feature selection is skipped.

Keywords: LiDAR, regression, remote sensing, soft computing.

1 Introduction

Light Detection and Ranging (LiDAR) is a remote laser-based technology which differs from optic sensors in its ability to determine heights of objects. LiDAR is able to measure the distance from the source to an object or surface providing not only x-y position, but also the coordinate z for every impact. The distance to the object is determined by measuring the time between the pulse emission and detection of the reflected signal taking into account the position of the emitter.

LiDAR sensors have transformed the way to perform many important tasks for the natural environment. The work previously done with expensive or not

Á. Herrero et al. (eds.), *International Joint Conference SOCO'13-CISIS'13-ICEUTE'13*, 249
Advances in Intelligent Systems and Computing 239,
DOI: 10.1007/978-3-319-01854-6_26, © Springer International Publishing Switzerland 2014

always-feasible fieldwork has partially been replaced by the processing of airborne LiDAR point cloud (initial product obtained from a LiDAR flight). In this context, research work focuses on the extraction of descriptive variables from LiDAR and their relation with field measurements. Following this philosophy, LiDAR is currently used to develop forest inventories [1] or fuel models [2] and to estimate biomass in forest areas [3], among other applications.

LiDAR-derived models are usually based on the estimation of parameters regressed from LiDAR statistics through multiple linear regression (MLR). The main advantage of using this type of methodology is the simplicity of the resulting model. In contrast, the selected method also has some drawbacks: this process results a set of highly correlated predictors with little physical justification [4] and, as a parametric technique, it is only recommended when assumptions such as normality, homoscedasticity, independence and linearity are met [5].

With the previous in mind, it is important to outline that methodologies to develop regression models between field-work data and LiDAR are being reviewed [6]. As a consequence, machine learning non-parametric regression techniques have recently started to be applied with success. For example, Hudak et al. [7] applied nearest neighbour to extract relations between LiDAR and fieldwork for several vegetation species at plot level. Chen and Hay [8] used support vector regression to estimate biophysical features of vegetation using data fusion (LiDAR + multiespectral). In the same line, Zhao et al. [9] provided a comparison between Gaussian processes and stepwise MLR where the first clearly improved the results after a set of composite features were extracted from a LiDAR point cloud. Decision trees in the form of random forests have also been applied with good results. Thus, Latifi et al. showed [10] how random forests could be used for biomass estimation and outperform classical stepwise regression after evolutionary feature selection.

Although machine learning seems to provide a suitable tool to extract meaningful information from LiDAR, few studies have been provided to compare the quality of the regressions obtained by different sets of techniques. For instance, Gleason and Im [11] showed a partial comparison of methods where support vector regression outperformed random forests. Unfortunately, no statistical validation was performed which is necessary to generalize their conclusions.

Our aim in this work was to compare the most well-known regression techniques of machine learning in a common framework. We established a ranking when they were applied to forest variable estimation to help environmental researchers the selection of the most suitable technique for their needs. The different techniques were tested and statistically validated using their results on two LiDAR datasets from two different areas of the province of Lugo (Galizia, Spain).

The rest of the paper is organized as follows. Section 2 provides a description of the LiDAR data used in this work as well as the methodology used. The results achieved, their statistical validation and the main findings in this work are shown in Section 3. Finally, Section 4 is devoted to summarize the conclusions and to discuss future lines of work.

2   Materials and Method

2.1   Study Sites

Aerial LiDAR data in two forest areas in the northwest part of the Iberian Peninsula (Fig. 1) were used for this study (more details about both areas can be found in Goncalves-Seco et al. [12] and Gonzalez-Ferreiro et al. [13], respectively).

The first study area (hereafter site A) was located in Trabada, concretely in the municipality of Vilapena (Galicia, NW Spain; boundaries 644800; 4806600 and 645800; 4810600 UTM). *E. globulus* stands, with low intensity silvicultural treatments and the presence of tall shrubs, dominated the forest type.

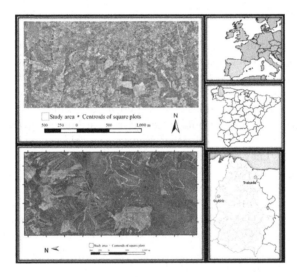

Fig. 1. Study sites located in the province of Lugo (NW of Spain). Top: study site of Guitiriz. Bottom: study site of Trabada.

The second study area (hereafter site B) was also located in Galicia (NW Spain), in the municipality of Guitiriz, and covered about 36 km^2 of *P. radiata* forests (boundaries 586315; 4783000 and 595102; 4787130 UTM). *P. radiata* was the main forest type in this area and its stands were also characterized by low-intensity silvicultural treatments and by the presence of tall shrubs.

2.2   Field Data

Field data from the two study sites were collected to obtain the dependent variables for the regressions in this work. Thus, 39 instances (one per training plot in the study site) were located and measured on site A. On site B, a similar process was carried out for a total of 54 plots. The plots were selected to represent the existing range of ages, stand sizes, and densities in the studied forests.

For site A and B, the dry weight of the biomass fractions of each tree was estimated using the equations for *E. globulus* in Galicia reported by Dieguez-Aranda et al. [14]. In order to define the dependent variables, the field measurements (heights and diameters) and the estimated dry weight of the biomass fractions were used to calculate the following stand variables in each plot: stand crown biomass (W_{cr}), stand stem biomass (W_{st}), and stand aboveground biomass (W_{abg}).

In the case of site B, the field measurements (heights and diameters) and the estimated volumes and dry weight of the biomass fractions helped to estimate the following additional stand variables in each plot: stand basal area (G), dominant height (H_d), mean height (H_m), and stand volume (V).

2.3 LiDAR Data

The LiDAR data from site A were acquired in November 2004. The first and last return pulses were registered. The whole study area was flown over 18 strips and each strip was flown over three times, which gave an average measurement density of about 4 pulses m^{-2}. The LiDAR data for site B were acquired in September 2007. A theoretical laser pulse density of 8 pulses m^{-2} was obtained. In order to obtain two additional different resolutions, an artificial reduction based on a random selection of LiDAR returns in a grid cell of 1 m^2 was carried out for each flight. They resulted in two new LiDAR datasets with a pulse density of 0.5 pulses [13].

Intensity values in both study sites were normalized to eliminate the influence of path height variations [1]. Filtering, interpolation, and the development of Digital Terrain and Canopy Models (DTM/DCM) were performed by FUSION software [15]. This software also provided the variables related to the height and return intensity distributions within the limits of the field plots in the four datasets (original and reduced data from study sites A and B). Table 1 shows the complete set of metrics and the corresponding abbreviations used in this article.

After the LiDAR data processing, we obtained 60 databases with 48 independent variables ($cover_{FP}$ and *returns* in Table 1 plus the rest calculated for intensity and heights). The first 20 datasets were composed of the previous statistics and each fieldwork variable as dependent variable (for each study site and resolution). The rest were obtained using two types of feature transformation (allometric and exponential, [13]), respectively.

2.4 Regression Techniques Comparison

The goal of this paper was to compare the results of several families of machine learning techniques when applied to LiDAR data for estimation of forest variables. For comparison, we selected the most extended machine learning algorithms in the literature from the software WEKA [16]: M5P (regression tree), SMOreg (support vector machine for regression), LinearRegression (classic multiple linear regression), MultilayerPerceptron (artificial neural network),

Table 1. Statistics extracted from the LiDAR flights' heights and intensities used as independent variables for the regression models

Description	Abbreviation	Description	Abbreviation
Percentage of first		25th percentile	P25
returns over 2m	cover_FP	50th percentile	P50
Number of returns above 2 m	returns	75th percentile	P75
Minimum	min	5th percentile	P05
Maximum	max	10th percentile	P10
Mean	mean	20th percentile	P20
Mode	mode	30th percentile	P30
Standard deviation	SD	40th percentile	P40
Variance	V	60th percentile	P60
Interquartile distance	ID	70th percentile	P70
Skewness	Skw	80th percentile	P80
Kurtosis	Kurt	90th percentile	P90
Average absolute deviation	AAD	95th percentile	P95

IBk (nearest neighbor). We also developed an ad-hoc Random Forest (ensemble of regression trees) based on the original implementation in Weka but replacing its random trees by M5P trees. This change was necessary because the original implementation in Weka only allows its use for classification and not for regression. In any case, all algorithms were used with default parameters after applying a preprocessing phase of normalization, elimination of missing values, and feature selection (to avoid the Hughes phenomenon [17]) based on the Correlation Feature Selection (CFS) filter of Weka. The comparison was defined from the coefficients of determination (R^2) obtained in a process of 5-Fold Cross-Validation (5FCV).

A key factor for the performance of the techniques is the set of selected attributes in the preprocessing step. In certain cases, such as SVM and Random Forest, techniques perform their own selection of best attributes. A previous selection could therefore affect the quality of the predictions. To study feature selection's influence, we repeated the 5FCV in a second level of experimentation for the best two techniques without applying previous feature selection.

2.5   Statistical Analysis

After the generation of the quality results for the different models, a statistical analysis was used (using the open-source platform StatService [18]) to check the significance in the differences among multiple methods in terms of R^2. ANOVA is usually used for multiple comparison of results if parametric conditions (homoscedasticity, independence, normality) are met [19]. Parametric conditions were checked using the Shapiro-Wilk and Lilliefors tests for normality and the Levene test for homoscedasticity. If parametric conditions were not met, a nonparametric procedure would be selected. This procedure, firstly, would obtain the average ranks taking into account the position of the compared results with respect to each other. Thus, a value of 1 for a rank would mean that the method would be the best for a test case, while a rank of n would mean it was the worst of the n compared methods. Finally the chosen procedure would use the Friedman test and the Holm post-hoc procedure (see [20] for a complete description of both non-parametric methods) to statistically validate the differences in the mean ranks.

In addition, for the second level comparison (between the two best methods) a similar procedure was done using a Student's T or a Wilcoxon test which are the corresponding parametric and non-parametric statistical test for pairwise comparisons, respectively [19].

3   Results

Due to the high number of datasets studied, we provide Table 2 which sums up the main statistics for every technique besides their mean ranking throughout the 60 datasets. The whole set of results are also depicted in Fig. 2 and 3. Figures show the results obtained by nearest neigbour (NN), support vector machines (SVM), artificial neural networks (ANN), multiple linear regression (MLR), regression trees (RT) and random forests (RF) for the 60 datasets separated in two subgroups for clarity. They both show the results of the globally best technique (SVM obtained the best mean ranking) when the complete data mining framework (including feature selection) was applied.

Table 2. Mean ranking and main statistics from the results obtained for every regression technique in terms of R^2 throughout the 60 datasets when preprocessing included feature selection

Technique	Mean ranking	R^2 Mean	R^2 Standard deviation
SVM	1.700	0.844	0.062
RF	2.783	0.827	0.071
RT	2.967	0.820	0.079
MLR	3.133	0.815	0.083
ANN	5.133	0.710	0.148
NN	5.283	0.723	0,098

Table 3. P-values and α values for each pairwise comparison in the Holm's procedure

DataSet	p	Holm
NN	0.000	0.010
ANN	0.000	0.013
MLR	0.000	0.017
RT	0.000	0.025
RF	0.002	0.050

Table 4. Mean ranking and main statistics from the results obtained for the two best regression techniques in terms of R^2 when preprocessing did not include feature selection

Technique	Ranking	R^2 Mean	R^2 Standard deviation
RandomForest	1.45	0.772	0.011
SVM	1.55	0.774	0.006

Rankings were used to assess the statistical significance of the study since Liliefors test rejected the normality hypothesis of the results with a p-value of 0.10 for an $\alpha = 0.05$. In this case, the p-value for the Friedman test was less than 0.0001 so it rejected the null hypothesis (all the techniques behave in a similar way) with a level of significance of $\alpha = 0.05$. Then, the Holm post-hoc procedure was applied. The p-values for the several pairwise comparisons can be found in Table 3. As can be seen, every p-value was lower than the α required by Holm (Holm column in the table) so the procedure concluded that pairwise differences between SVM and the rest of the regression techniques were also statistically significant.

The top two regressors in the previous test were SVM and RF. Both were selected for a subsequent pairwise comparison where the experiment was replicated without performing previous feature selection. Their results are visually presented in Fig. 4 and summarized regarding the ranking and main statistics in Table 4. In this case, the use of the Wilcoxon test with a p-value of 0.9325 could not reject the null hypothesis (i.e., there were no significant differences between them) with a level of significance of $\alpha = 0.05$.

Through the analysis of the results of experimentation, it is possible to draw some important findings. First, our experiments confirmed that both SVM and RF are suitable tools to improve the performance of classical predictions (e.g., MLR or NN) for estimation of forest variables from LiDAR statistics although there is still room to improve the predictions.

Regarding the application of SVM or RF, our experiments showed that although feature selection outperformed every technique, SVM is more sensitive to the feature selection method since its performance decreased more in the second part of the experimentation. Moreover, if the feature selection is not adequate or does not exist, there would be little difference between RF and SVM (in our case, RF even globally outperformed SVM in most cases) as was shown in Table 4. This finding could justify the results of Gleason and Im [11] (where SVM outperformed RF) since the authors made the feature selection manually.

The fact that SVM outperformed RF can be also attributed to CFS feature selection provided a better set of attributes for SVM than for RF. More experimentation is needed to check if the same results can be obtain with other automatic feature selection techniques. In addition, it was also possible that the random nature of RF did not optimally combine the selected features. In any case, this study confirmed the well-known importance of feature selection for the performance of machine learning also in the LiDAR-regression context.

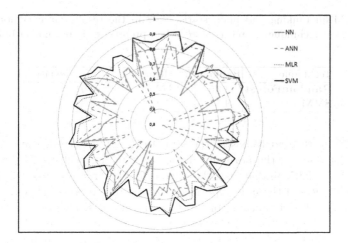

Fig. 2. Results of MLR, NN, ANN, and the averaged best technique (SVM) in terms of R^2 for the 60 datasets

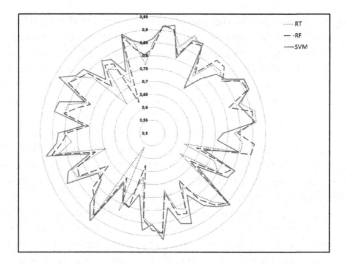

Fig. 3. Results of RT, RF, and the averaged best technique (SVM) in terms of R^2 for the 60 datasets

Finally, an issue not covered in this study and that should be considered in future studies is the influence of the parameters on the results. This point as feature selection will be addressed in future work.

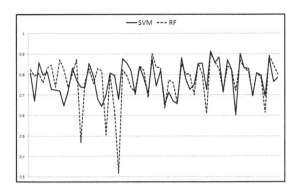

Fig. 4. Results for every dataset in terms of R^2 of Random Forests and SVM when no feature selection was applied

4 Conclusions

This paper presented a comparison between common regression techniques in machine learning (ANN, RT, SVM, NN, and ensembles such as RF) and the classic MLR-based methodology. The selected techniques were applied to real LiDAR data from two areas in the province of Lugo (Galizia, Spain). The results showed that support vector regression statistically outperformed the rest of techniques when feature selection is applied but its performance could not be said statistically different from that of Random Forests when feature selection was skipped. Nevertheless, results confirmed recent bibliography since SVM and RF behaved the best for the 60 experimental datasets.

Future work should address gaps not covered in this work. Thus, we must complete the framework with an ad-hoc feature selection for each specific method. In the same line, parametrization will have to be addressed as another important issue which can change the results of the predictors. Both problems can be solved at the same time with the application of evolutionary computation although a trade-off between optimization and run time should be reached for industrial uses.

References

1. Garcia, M., Riano, D., Chuvieco, E., Danson, F.M.: Estimating biomass carbon stocks for a mediterranean forest in central spain using LiDAR height and intensity data. Remote Sensing of Environment 114(4), 816–830 (2010)
2. Mutlu, M., Popescu, S.C., Stripling, C., Spencer, T.: Mapping surface fuel models using LiDAR and multispectral data fusion for fire behavior. Remote Sensing of Environment 112(1), 274–285 (2008)
3. Gonzalez-Ferreiro, E., Dieguez-Aranda, U., Gonçalves-Seco, L., Crecente, R., Miranda, D.: Estimation of biomass in eucalyptus globulus labill. forests using different LiDAR sampling densities. In: Proceedings of ForestSat (2010)

4. Muss, J.D., Mladenoff, D.J., Townsend, P.A.: A pseudo-waveform technique to assess forest structure using discrete LiDAR data. Remote Sensing of Environment 115(3), 824–835 (2010)
5. Osborne, J., Waters, E.: Four assumptions of multiple regression that researchers should always test. Practical Assessment, Research and Evaluation 8(2) (2002)
6. Salas, C., Ene, L., Gregoire, T.G., Næsset, E., Gobakken, T.: Modelling tree diameter from airborne laser scanning derived variables: A comparison of spatial statistical models. Remote Sensing of Environment 114(6), 1277–1285 (2010)
7. Hudak, A.T., Crookston, N.L., Evans, J.S., Halls, D.E., Falkowski, M.J.: Nearest neighbor imputation of species-level, plot-scale forest structure attributes from LIDAR data. Remote Sensing of Environment 112, 2232–2245 (2008)
8. Chen, G., Hay, G.J.: A support vector regression approach to estimate forest biophysical parameters at the object level using airborne lidar transects and quickbird data. Photogrammetric Engineering and Remote Sensing 77(7), 733–741 (2011)
9. Zhao, K., Popescu, S., Meng, X., Pang, Y., Agca, M.: Characterizing forest canopy structure with lidar composite metrics and machine learning. Remote Sensing of Environment 115(8), 1978–1996 (2011)
10. Latifi, H., Nothdurft, A., Koch, B.: Non-parametric prediction and mapping of standing timber volume and biomass in a temperate forest: Application of multiple optical/LiDAR-derived predictors. Forestry 83(4), 395–407 (2010)
11. Gleason, C.J., Im, J.: Forest biomass estimation from airborne LiDAR data using machine learning approaches. Remote Sensing of Environment 125, 80–91 (2012)
12. Goncalves-Seco, L., Gonzalez-Ferreiro, E., Dieguez-Aranda, U., Fraga-Bugallo, B., Crecente, R., Miranda, D.: Assessing attributes of high density eucalyptus globulus stands using airborne laser scanner data. International Journal of Remote Sensing 32(24), 9821–9841 (2011)
13. Gonzalez-Ferreiro, E., Dieguez-Aranda, U., Miranda, D.: Estimation of stand variables in pinus radiata d. don plantations using different lidar pulse densities. Forestry 85(2), 281–292 (2012)
14. Dieguez-Aranda, U., et al.: Herramientas selvicolas para la gestion forestal sostenible en Galicia. Xunta de Galicia (2009)
15. McGaughey, R.: FUSION/LDV: Software for LIDAR Data Analysis and Visualization. US Department of Agriculture, Forest Service, Pacific Northwest Research Station, Seattle (2009)
16. Hall, M., Frank, E., Holmes, G., Pfahringer, B., Reutemann, P., Witten, I.H.: The WEKA data mining software: An update. SIGKDD Explorations 11(1) (2009)
17. Hughes, G.F.: On the mean accuracy of statistical pattern recognizers. IEEE Transactions on Information Theory 14, 55–63 (1968)
18. Parejo, J.A., García, J., Ruiz-Cortés, A., Riquelme, J.C.: Statservice: Herramienta de análisis estadístico como soporte para la investigación con metaheurísticas. In: Actas del VIII Congreso Expañol sobre Metaheurísticas, Algoritmos Evolutivos y Bio-inspirados (2012)
19. Demsar, J.: Statistical comparisons of classifiers over multiple data sets. Journal of Machine Learning Research 7, 1–30 (2006)
20. Luengo, J., Garcia, S., Herrera, F.: A study on the use of statistical tests for experimentation with neural networks: Analysis of parametric test conditions and non-parametric tests. Expert Systems with Applications 36, 7798–7808 (2009)

Principal Component Analysis on a LES
of a Squared Ribbed Channel

Soledad Le Clainche Martínez[1], Carlo Benocci[2], and Alessandro Parente[3]

[1] School of Aeronautics, Universidad Politécnica de Madrid, Spain
soledad.leclainche@upm.es
[2] Department of Environmental and Applied Fluid Dynamics,
von Karman Institute for Fluid Dynamics, Belgium
benocci@vki.ac.be
[3] Service d'Aéro-Thermo-Mécanique, Université Libre de Bruxelles, Belgium
Alessandro.Parente@ulb.ac.be

Abstract. The present paper reports on the application of Principal Component Analysis (PCA) on the flow and thermal fields generated by the large-eddy simulation (LES) of a square ribbed duct heated by a constant heat flux applied over the bottom surface of the duct. PCA allows to understand the complexity of the resulting turbulent heat transfer process, identifying the flow and thermal quantities which are most relevant to the process. Different algorithms have been employed to perform this analysis, showing high correlation between turbulent coherent structures, identified by $Q - criterion$, and the heat transfer quantified by the non-dimensional magnitude Enhancement Factor (EF), both identified as Principal Variables (PV) of the process.

Keywords: Conjugate heat transfer, Large Eddy Simulation, Principal Component Analysis, Flow topology.

1 Introduction

Most physical and engineering processes depend on the interaction of a high number of different physical parameters. A such example is represented by the internal cooling of turbines blades. The turbine efficiency of a jet engine can be improved increasing the turbine inlet temperature. However, temperatures higher than the melting point temperature of the blade material could be reached, as they would cause catastrophic failure. Thus, an efficient blade cooling must be introduced to avoid overheating and internal ducts, transporting a cooling fluid, are typically employed within blades. The efficiency of the cooling process can be improved by means of turbulence generators, which increase the turbulent mixing and turbulent heat transfer. A typical generator is a set of ribs installed on the walls of the cooling ducts. The obvious drawback associated with the use of ribs is the increase of pressure drop; therefore, the advantages and shortcomings of this approach must be accurately assessed. A deeper discussion about the topic can be found in [1], which provides an overview of heat transfer in gas turbines.

Á. Herrero et al. (eds.), *International Joint Conference SOCO'13-CISIS'13-ICEUTE'13*, 259
Advances in Intelligent Systems and Computing 239,
DOI: 10.1007/978-3-319-01854-6_27, © Springer International Publishing Switzerland 2014

The number of parameters that can potentially affect the heat transfer process is very high and it is therefore very complex to identify the most relevant ones. In the present work, Principal Component Analysis (PCA) is applied to the matrix containing all the parameters potentially influencing the process, to identify the Principal Variables (PV) and quantify their effect on the system quantify of interest, i.e. the heat removal rate. PCA is a well-known statistical technique that transforms a set of correlated variables into a smaller number of uncorrelated variables, called Principal Components (PCs) [3,9]. The PC are determined to successively maximize the amount of variance contained in the original data sample. The approach performs a rotation within the original data space to reveal the internal structure of the sample and to maximize the amount of information accounted for by a smaller number of parameters. The PCs are, by definition, linear combinations of all variables available. However, in some applications, it can be useful working in terms of the original variables, to help the physical interpretation of the results. Therefore, the PCA analysis of the thermo-fluid dynamic state is completed with the determination of the Principal Variables [3].

The present paper presents the application of PCA to the numerical simulations carried out in [8], with the main objective of identifying the parameters which control the heat transfer process in proximity of the rib and clarifying the effect of the rib on heat transfer enhancement. Additionally, the study aims to investigate the relevance of turbulent coherent structure on the heat transfer process, as recent studies [6] pointed out how heat transfer enhancement always occurs in regions of the flow characterized by the existence of presence of organized turbulent flow structures.

2   Physical Principles and Numerical Simulations

The investigated case is presented in Fig. 1. It represents conjugate heat transfer in a squared cross-section ribbed duct with five consecutive ribs [8]. The ribs are characterized by a squared cross section positioned perpendicularly to the channel axis. The rib hydraulic diameter is $D = 0.075$ m, the rib height is $h/D = 0.3$, the pitch length, $p/h = 10$ and the thickness of the coupled solid domain is $s/h = 1.1$. The Reynolds number based on the hydraulic diameter and bulk velocity U_b is $4 \cdot 10^4$ and the Prandtl number is equal to 0.7. The simulations were solved using the Large Eddy Simulations (LES) module of the commercial software FLUENT 6.3. Numerical results have been validated with some experimental measurements carried out at von Karman Institute employing Particle Image Velocimetry (PIV) and Infrared thermography (ICR). In all cases, a satisfactory agreement between simulations and experimental data was found.

Fig. 2 presents an overview of the high complexity fluid dynamic problem, showing how the flow is dominated by regions of separated flow. Fig. 2a shows the the vortex structures around the 4th rib, on the symmetry plane (up) and the thermal field in the fluid and solid part of the channel on the same plane (down).

Temperature decreases in presence of vortex structures. Fig. 2b shows the heat transfer flux (up), defined as the non-dimensional magnitude Enhancement Factor (EF) and coherent structures (down) defined with the $Q - criterion$. The presence of coherent structures enhances the heat transfer flux on the bottom surface.

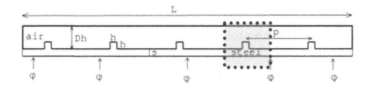

Fig. 1. Computational domain for the ribbed channel geometry

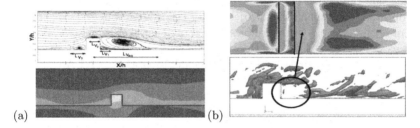

(a)   (b)

Fig. 2. Left: (up) Vortex structures and (down) thermal field on the mid plane around the 4thrib. Right: (up) Enhancenment Factor and (down) Coherent Structures isosurfaces $(Q/(U_b^2/h^2) = 24)$ on the 4th rib. Legend: blue colors: negative value, green colors: zero value, red colors: positive value.

3   Theory

3.1   Coherent Structures

Coherent structures (CS) are large and strongly organized turbulent structures associated with the process of production of turbulent kinetic energy. Within the present context, the role of CS with respect to the heat transfer mechanism is of interest. A common criterion for the detection of coherent structures in a turbulent flow is the so-called Q-criterion [2], $Q = \frac{1}{2}(\Omega_{ij}\Omega_{ij} - S_{ij}S_{ij})$, where S_{ij} is the strain tensor and Ω_{ij} is the rotation tensor. When rotational effects are dominant over strain effects ($Q > 0$), a coherent structure identified by Q exists. Fig. 2b down shows the iso-surface of $Q/(U_b^2/h^2) = 24$, where U_b is the bulk velocity and h is the height of the rib.

3.2 Enhancement Factor

The effect of the rib on heat transfer is assessed by means of the so-called Enhancement Factor (EF), which represents the ratio of the Nusselt number along the actual (ribbed) duct, Nu, and that on a smooth surface, Nu_0, $EF = Nu/Nu_0 = h \cdot D_h/k_{air} \cdot Nu_0$, where h is the heat transfer coefficient, D_h is the hydraulic diameter, k_{air} is the air thermal conductivity, Re and Pr are the Reynolds and Prandtl number, respectively. Fig. 2b, shows in the upper part the EF distribution on the bottom solid surface of the body.

3.3 Principal Component Analysis

PCA is a well-known statistical technique that transforms a set of possibly correlated variables into a smaller number of uncorrelated variables, called Principal Components (PCs) [3,9]. For a multivariate data set, PCA is usually employed for detecting the directions that carry most of the data variability, thus providing an optimal low-dimensional projection of the system. For a data set, $\mathbf{X}$, consisting of n observations of p variables, the sample covariance matrix, $\mathbf{S}$, can be defined as $\mathbf{S} = 1/(n-1)\mathbf{X}^T\mathbf{X}$. Recalling the eigenvector decomposition of a symmetric, non singular matrix, $\mathbf{S}$ can be decomposed as $\mathbf{S} = \mathbf{ALA}^T$, where $\mathbf{A}$ is the (p x p) matrix whose columns are the eigenvectors of $\mathbf{S}$, and $\mathbf{L}$ is a (p x p) diagonal matrix containing the eigenvalues of $\mathbf{S}$ in descending order, $l_1 > l_2 > \ldots > l_p$. Once the decomposition of the covariance matrix is performed, the Principal Components (PC), $\mathbf{Z}$, are defined by the projection of the original data onto the eigenvectors, $\mathbf{A}$, of $\mathbf{S}$, $\mathbf{Z} = \mathbf{XA}$. Then, the original variables can be stated as a function of the PC as $\mathbf{X} = \mathbf{ZA}^T$. The main objective of PCA is to replace the p elements of $\mathbf{X}$ with a much smaller number, q, of PC, preserving at the same time the amount of information originally contained in the data. If a subset of size $q << p$ is used, the truncated subset of PC is $\mathbf{Z_q} = \mathbf{XA}_q$. This relation can be inverted to obtain an approximation of the original state space, $\widetilde{\mathbf{X}}_q = \mathbf{Z}_q\mathbf{A}_q^T$. A representation of the PCA reduction process is shown in Fig. 3.

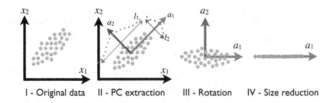

Fig. 3. PCA reduction process

3.4 Principal Component Rotation and Principal Variables

The physical interpretation of Principal Components is generally not straightforward as they are in principle linear combination of all the original variables.

Rotation represents an attempt to overcome such difficulty: with rotation, the weights can be redefined to attain a simple structure for $\mathbf{A_q}$, so that weights on a PC are either close to unity or close to zero and, thus, variables have large weights on only few or (ideally) one PC. The most common orthogonal rotation is based on the maximization of the VARIMAX criterion [4]. When rotation is applied, the total variance within the rotated q-dimensional subspace is re-distributed amongst the rotated components more evenly than before rotation; however it remains the maximum that can be achieved.

Principal variable (PV) represents a further attempt to help the physical understanding of Principal Components. Differently from PC rotation, PV algorithms try to link the PC back to a subset of the original variables, which satisfy one or more optimal properties of PCA. A number of methods exist for selecting a subset of m original variables. Within the present investigation, the M2 method by Krzanowski [5] and the approaches proposed by McCabe [7] were selected. In the M2 method, PCA is performed on the original data matrix and the scores $\mathbf{Z_q}$ are then then evaluated. Assuming that q is the true data dimensionality, the approximation, $\widetilde{\mathbf{Z_q}}$, of $\mathbf{Z_q}$ obtained by keeping only q variables of the original data set is evaluated, to find the subset of variables yealding the least difference between $\mathbf{Z_q}$ and $\widetilde{\mathbf{Z_q}}$. The approaches by McCabe [7], MC1-MC3, originates from the observation that the principal components satisfy a certain number of optimality criteria, i.e. maximal variance and minimum reconstruction error. Therefore, a subset of the original variables optimizing one of these criteria defines a set of principal variables.

4   Results

PCA analysis has been performed in order to determine the most relevant quantities for the heat transfer process. The variables selected for the analysis are the mean stream-wise (U), chord-wise (V) and span-wise (W) velocities and their variances, the mean temperature (T) and its variance, the turbulent kinetic energy (k), the mean vorticity (Ω) and mean strain (S) tensors, the enhancement factor (EF), the mean heat flux (q) and the mean velocity and temperature correlations UV, UW, VW, UT, VT, WT. The matrix to be processed with PCA is built collecting data on different planes in the computational domain. In particular, the heat transfer related quantities (heat flux, EF, T) are taken at the first grid point above the ribbed wall and around the rib itself. On the other hand, the flow-related quantities (velocities, strain, vorticity, Q) are more meaningful at a certain distance from the wall. Fig. 4a clearly indicates that the coherent structures are present at a certain distance from the wall, they interact with the bottom surface, leading to a reduction of temperature, and then they mix back with the bulk flow. Fig. 4b shows the different sampling planes employed for the PCA analysis. Since the computational grid is unstructured, the data extracted for the PCA analysis were interpolated over a structured grid.

Two different data set have been analyzed. The resulting data set for PCA and PV analysis, indicated as CHT1, consists of 20 state variables and about

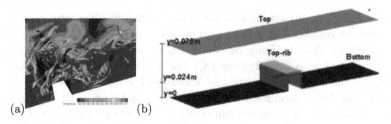

Fig. 4. Iso-surfaces at instantaneous $Q/(U_b^2/h^2) = 24$ colored by temperature (a) and sampling planes for the application of PCA and PV (b)

6000 observations. The second data set, indicated as CHT2, is then built to focus the statistical analysis on the observations corresponding to positive values of Q, to remove the effect of strain dominated regions on the results and point out the role of coherent structures on the heat transfer process. The number of observations for the reduced CHT2 data set is about 2000.

4.1 PCA Analysis

Fig. 5 shows the magnitude of the eigenvalues associated with the PCA reduction (left), and the contribution of the largest eigenvalues to the amount of variance explained by the new basis vectors for the CHT1 (a) and CHT2 (b) data sets. The first five eingenvalues (representing 25% of the total number of variables) account for more than 75% and 80% of the total variance in the original CHT1 and CHT2 data sets, respectively. On the other hand, the last six smallest eigenvalues contain no useful information and only explain linear dependencies among the original variables. Therefore, a non-negligible size reduction can be accomplished with PCA, through the identification of the most active directions in the data. Table 1 shows a selection of the correlation matrix, built for the extraction of the PC and PV, showing the correlations between k, EF and Q and a subset of variables, for the CHT1 and CHT2 data sets. For CHT1, it is observed that Q is mostly correlated to T', V, and the correlations UV and VT (beside the obvious correlation with strain and vorticity). This is in agreement with the qualitative observation (Fig. 4a) that the vortex structures present in the flow are driven by the temperature gradient determined by the heat exchange process. Their movement is mostly perpendicular to the rib plane (thus explaining the V component of velocity), allowing the structures to reach the ribbed surface and to mix back into the flow bulk. On the other hand, EF is negatively correlated with T (EF is larger when the temperature decreases) and it shows large positive correlations with T' and VT, confirming the interaction between coherent structures and the heat transfer process. As far as turbulent kinetic energy is concerned, k shows a positive correlation with the variables representing the heat transfer process, indicating that an increase in turbulence levels could be effectively used to reduce rib temperatures (if compatible with pressure drops). The same analysis for the CHT2 data set confirms the correlations pointed out for CHT1 and supports the phenomenological explanation provided above, with

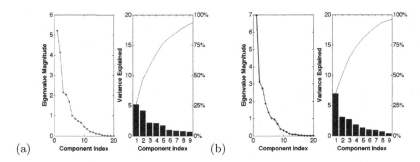

Fig. 5. Scree-graph showing the main contribution to the original data variance for the CHT1 (a) and CHT2 (b) data sets

Table 1. Selection of the correlation matrix showing the correlation between k, EF and Q and the variables defining the CHT1 and CHT2 data sets

		k	EF	U	V	W	T	U'	V'	W'	UV
CHT1	k	1.00	0.26	-0.33	-0.26	-0.02	-0.28	0.93	0.73	0.80	-0.54
	EF	0.26	1.00	-0.20	0.08	0.02	-0.74	0.34	-0.18	0.35	0.17
	Q	0.07	0.21	0.13	0.58	-0.01	-0.19	0.03	-0.14	0.26	0.63
CHT2	k	1	-0.08	0.15	-0.39	0.03	0.21	0.87	0.59	0.64	-0.48
	EF	-0.08	1	-0.09	0.13	0.07	-0.78	0.06	-0.59	0.29	0.47
	Q	-0.14	0.35	0.32	0.80	0.00	-0.30	-0.18	-0.35	0.27	0.83

		UW	VW	T'	UT	VT	WT	Ω	S	Q	q
CHT1	k	-0.10	0.06	0.19	0.20	0.27	-0.08	0.06	0.03	0.03	0.25
	EF	-0.05	0.08	0.72	0.35	0.73	-0.05	0.19	0.20	0.21	0.91
	Q	0.01	-0.10	0.42	-0.27	0.45	-0.02	0.84	0.74	1.00	0.17
CHT2	k	0.00	0.09	-0.19	0.02	-0.18	0.00	-0.23	-0.25	-0.14	-0.05
	EF	-0.03	0.18	0.76	0.18	0.80	-0.01	0.32	0.30	0.35	0.91
	Q	0.05	-0.04	0.62	-0.50	0.69	0.00	0.94	0.88	1	0.31

an exception concerning turbulent kinetic energy. In fact, the values listed in Table 1 for CHT2 indicate almost zero correlations between k and the heat transfer related variables, EF and q. Such result can be explained considering that the CHT2 data set is built by assembling all those observations corresponding to positive values of Q. Therefore, the values listed in Table 1 only indicate that in the range of k compatible with the existence of CS, the heat flux span an almost uniform range of values, leading to correlations close to zero.

Fig. 6 shows histograms of the structure of the first four rotated eigenvectors of S, for the CHT1 and CHT2 data sets, respectively. The analysis of Fig. 6 clearly shows that the main eigenvector structure of the covariance matrix does not significantly change switching from the CHT1 to the CHT2 data sets. It can be observed that the first component is characterized by very large weights on V and UV, and on Ω, S and Q, for both data sets. Such component can be then interpreted as an indicator of organized coherent structures; moreover, it confirms the relation between the latter and the vertical flow motion from the bulk to the wall and vice versa (V, UV). Only a few differences can be emphasized when comparing the two data sets. The CHT2 data set shows, in fact, non-negligible weights for the correlations VW and WT, suggesting a non-negligible influence of three-dimensional effects on the heat transfer process. The second rotated PC is dominated by heat transfer related variables, EF and q (showing a correlation of about 91%, Table 1), T and T, and the correlation VT, confirming the role of the vertical velocity component on the heat transfer mechanism. Likewise the first component, a major relevance of correlations related to three-dimensional effects is observed when analyzing the CHT2 data set. As far as the third and

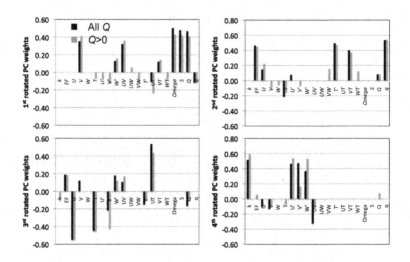

Fig. 6. Scree graph showing the main contribution to the original data variance

fourth PCs are concerned, these can be regarded as representative of the correlation (UT) between mean flow (U) and the heat transfer from the wall to the bulk (T), and of turbulent kinetic energy (k). It is interesting to have a closer look at the third component and particularly at the relative weights of the original variables. In fact, the sign of the coefficients within a PC specifies its orientation in the original multidimensional space, thus clarifying the relation between the original variables of the sample. Given this remark, it becomes clear how the variance on this component is directed towards concordant (positive or negative) values EF, UT, but discordant values of EF and Q, or EF and T. This is in agreement with the observation that the coherent structures in the bulk flow (large values of Q) dissipate at the wall (small values of Q), leading to an increase of EF and heat exchange (UT) and a local decrease of temperature.

4.2   Principal Component Rotation and Principal variables

Fig. 7 shows the distribution of the first four unrotated PCs over the fourth rib (a-d) and the corresponding distribution of the state variables showing highest

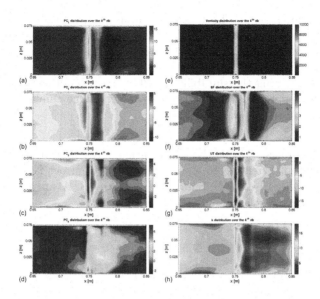

Fig. 7. Distribution of the principal component modes (a-d) and of the variables showing the largest weights on the rotated components (Fig. 6) over the fourth rib. From up to down: vorticity, EF, UT, k.

contribution with the PCs (e-h). It can be clearly observed that the selected variables very well mimic the PC behaviour, indicating that VARIMAX rotation is an effective method to disclose the physical meaning of the PC.

The analysis presented in the previous sections is completed with PV analysis, carried out on the CHT2 data sets, applying PV methods M2 and MC2-MC3. All the employed methods indicate the following variables as PVs: k, T, W, Q and EF. Therefore, PV analysis confirms the results provided by PCA and, in particular, by the rotation of the PCs using the VARIMAX criterion, being the selected variables coincident or highly correlated to the ones dominating the rotated principal components.

5   Conclusions

PCA and PV analysis have been applied to the flow and thermal fields of a ribbed square duct, heated through the ribbed wall. The analysis has shown that most of the information contained in the original thermo-fluid-dynamic field, defined by twenty state variables, can be represented by the first five eigenvectors of the sample covariance matrix, defining the problem principal components. The application of the VARIMAX rotation criterion to the set of principal components allowed isolating the different groups of physical parameters controlling the overall process and to identify the most relevant variables within the initial set of twenty state variables. Importantly, the role of the rotation-dominated regions (coherent structures) in the heat transfer mechanism from the heated wall to the fluid bulk has been highlighted. The PCA analysis has been completed with the application of PV algorithms for the extraction of the most relevant physical quantities driving the process. The outcome of PV analysis fully supported PCA findings, demonstrating the suitability of PCA and PV for the investigation of complex physical processes, characterized by a very large number of potentially relevant physical parameters.

Acknowledgements. The authors acknowledge the financial support of the Air Force Office of Scientific Research (AFOSR), Contract Number: FA8655-08-1-3048, supervised by Dr. S. Surampudi and Dr. G. Abate of the European Office of Aerospace Research and Development.

References

1. Han, J.C., Dutta, S., Ekkad, S.V.: Gas turbine heat transfer and cooling technology. Taylor & Frances (2000)
2. Hunt, J.C.R., Wray, A.A., Moin, P.: Eddies, stream, and convergence zones in turbulent flows. Center for Turbulence Research Report, p. 193 (1988)
3. Jolliffe, I.T.: Principal component analysis. Springer, New York (1986)
4. Kaiser, H.F.: The varimax criterion for analytic rotation in factor analysis. Psychometrika 23, 187–200 (1958)
5. Krzanowski, W.: Selection of variables to preserve multivariate structure, using principal components. Applied Statistics 36, 22–33 (1987)
6. Lohász, M.M., Rambaud, P., Benocci, C.: Flow features in a fully developed ribbed duct flow as a result of les. Journal of Flow, Turbulence and Combustion 77, 59–76 (2007)
7. McCabe, G.P.: Principal variables. Technometrics 26, 137–144 (1984)
8. Nakhle, D., Rambaud, P., Benocci, C., Arts, T.: Numerical investigation of flow and heat transfer in ribbed square duct applying les. In: 10th International Symposium on Experimental and Computational Aerothermodynamics of Internal Flows - Proceeding of ISAIF10, Bruxelles, Belgium, July 4-7 (2011)
9. Parente, A.: Experimental and numerical investigation of advanced systems for hydrogen-based fuel combustion. PhD Thesis, University of Pisa (2008)

Modeling Structural Elements Subjected to Buckling Using Data Mining and the Finite Element Method

Roberto Fernández-Martinez[1], Rubén Lostado-Lorza[2,*],
Marcos Illera-Cueva[2], and Bryan J. Mac Donald[3]

[1] Department of Material Science, University of Basque Country UPV/EHU, Bilbao, Spain
[2] Department of Mechanical Engineering, University of La Rioja, Logroño, Spain
[3] School of Mechanical & Manufacturing Engineering,
Dublin City University, Dublin 9, Ireland
ruben.lostado@unirioja.es

Abstract. Buckling of thin walled welded structures is one of the most common failure modes experienced by these structures in-service. The study of such buckling, to date, has been concentrated on experimental tests, empirical models and the use of numerical methods such as the Finite Element Method (FEM). Some researchers have combined the FEM with Artificial Neural Networks (ANN) to study both open and closed section structures but these studies have not considered imperfections such as holes, weld seams and residual stresses. In this paper, we have used a combination of FEM and ANN to obtain predictive models for the critical buckling load and lateral displacement of the center of the profile under compressive loading. The study was focused on ordinary Rectangular Hollow Sections (RHS) and on the influence of geometric imperfections while taking residual stresses into consideration.

Keywords: Finite Element Method, ANN, Buckling, Geometric Imperfections.

1 Introduction

The collapse of structures or structural elements subjected to buckling normally has a common characteristic: the sudden onset of large displacement and/or deformation of the structure that is achieved once the critical load has been exceeded. Traditionally, the design of structural elements was made following the criterion of mechanical strength and considering that the element was safe if the maximum predicted stress was lower than the value of the characteristic yield strength of the material [1, 2]. In this way, it was certain that the material would not experience plastic, permanent deformation. Failure due to buckling, however, is in most cases, a design flaw by not properly assessed the critical buckling load. This critical load is greatly affected by geometric imperfections (holes, welding seams, and folds in the sheets) or the existence of residual stresses. By Eurocode 3 [3] and through the use of stress concentration factors for the various imperfections, theoretical critical loads are over-predicted, meaning that

---

* Corresponding author.

Á. Herrero et al. (eds.), *International Joint Conference SOCO'13-CISIS'13-ICEUTE'13*,
Advances in Intelligent Systems and Computing 239,
DOI: 10.1007/978-3-319-01854-6_28, © Springer International Publishing Switzerland 2014

structural members under compression are always overdesigned. Many researchers have made various mathematical approaches to the description of the phenomenon for generic scenarios, but today for specific design cases we generally employ numerical techniques such as the FEM in order to obtain a prediction more realistic of the failure. The FEM is a numerical method classified as a hard computing method and is widely used to design and optimize industrial products and processes. This method also has known disadvantages, including the large amount of time and computing power needed in each simulation, particularly if the Finite Element model (FE model) is nonlinear. FE models present some problems where buckling is worked since it tends to be very complex due to the large number of nonlinearities and furthermore requires more computational time. The use of soft computing techniques (such as Neural Networks, Genetic Algorithms and Fuzzy Logic) has also proven to be useful to solve complex and non-lineal problems [4, 5] in a more efficient and faster way.

ANN can be used to help finding complicated dependencies among inputs (shape of the element: number of holes, buckling length, hole diameter) and outputs (critical buckling load, vertical displacement and horizontal displacement), since they have the ability to obtain nonlinear relationships among variables, without prior information about the process [6].

This paper shows how predictive models based on ANN are capable of predicting the value of the critical buckling load for Rectangular Hollow Sections (RHS) under compression by combining soft computing techniques and FEM. These predictive models based on Neural Networks are much simpler and more dependable, mainly due to the number and diameter of holes, as well as irregularities of the weld bead are taken into consideration unlike the current methodology (Eurocode 3, etc.). Furthermore, the rapidity with which the results obtained by ANN models against new parameters (diameter, number of holes, length) compared to FE models makes them much more efficient. The paper is organized as follows. Section 2 presents the proposed FE model. Section 3 presents the ANN models for critical load, vertical displacement and horizontal displacement. Finally, the conclusions are discussed in Section 4.

2   Finite Element Modeling

Usually, the calculation of the critical load of columns subjected to bend by FEM can be performed by a non-linear model or by a linear model (depending on the case studied). In the linear model, the calculation of the critical load is based on the theory of Euler failure [7], while in the non-linear model can be based on the theory of Crisfield [8]. The Nonlinearities that require consideration are geometric nonlinearities, material nonlinearities and nonlinearities due to residual stresses. This paper is focused on Rectangular Hollow Section (RHS) columns embedded in both ends (fixed-end column) subjected to a compressive loading, in order to study the lateral buckling and critical buckling load. The beam studied had a weld cord along its entire lateral length, which introduced a geometric imperfection. On the opposite side to the weld bead, the beam had a number of holes of varying diameter to simulate geometric imperfections. Beams of this type and with similar geometrical imperfections are very commonly used in pallet rack structures, commercial construction and transport industries.

2.1   Parameterized and Details of the Proposed FE Model

The FE model was parameterized with the aim of making a study of the different variables. In this case, the studied lengths (L) were 600, 900 and 1200 mm and the holes, equidistant from each other, were of diameter 10, 15 and 20 mm respectively. The number of holes (N) ranged from 0 to 5, and the distance between holes (l) was calculated according to equation 1.

$$l = \frac{L}{N+1} \tag{1}$$

In all cases, the FE models were formulated with three-dimensional shell elements in order to model correctly the bending. As shown in Table 1, results of 47 FE models simulated were part of the dataset used by ANNs. In this table are shown the number of holes (N Holes), the length of the sections (L), the diameter of the holes (Hole D), the critical load (Crit. L.), the horizontal displacement (Hor. D.) and vertical displacement (Vert. D.).

Table 1. Data from some of the 47 FE models simulated

N Holes	L [mm]	Hole D. [mm]	Crit. L. [N]	Hor. D. [mm]	Vert. D. [mm]
0	600	0	45678.12	0.412	1,024
0	900	0	20840.50	0.902	0,758
0	1200	0	11825.16	1.443	0,534
-	-	-	-	-	-
5	600	10	44209.81	6.832	1.210
5	900	15	19955.33	13.853	1.231
5	1200	20	11156.80	20.864	1.437

Moreover, Figure 1 show, from left to right, 1) a schematic of the parameterized FE model, 2) the longitudinal weld cord in the FE model, and 3) the FE model with 3 holes of 20 mm diameter.

2.2   Experimental Data and FE Validation

The left side of Figure 2 shows the experimental test used to validate the FE model performed in a compression testing machine, which consists of a load cell with vertical and horizontal gauges. The test was performed in order to obtain the buckling force, the displacement of the upper point and the horizontal displacement of the beam at its center point in order to validate the FE models proposed. In this case, only a profile of 600 mm length without holes was validated. The test results showed that buckling of the FE model was proposed quite accurately adjusted to reality.

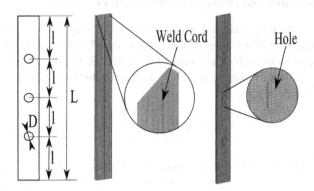

Fig. 1. Parameterized FE model. Detail of the longitudinal weld cord and the FE model with three 20mm diameter holes.

Likewise, the center of Figure 2 shows an image of the deformed beam once the buckling test has concluded. The corresponding deformed FE model is shown on the right of Figure 2.

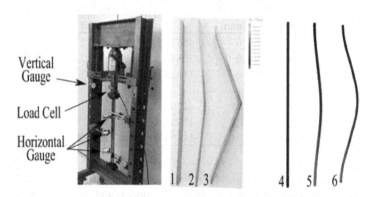

Fig. 2. Compression testing machine, specimen deformation sequence in the testing machine (1, 2, and 3) and deformation sequence in the FE model (4, 5, and 6)

3   Artificial Neural Network Model

ANNs are a powerful mathematical computationally intensive tool for modeling and finding patterns in data sets of material properties, based on the properties of biological neural systems and nervous systems.

Neural network models are formed by a hidden layer, where each node that belongs to the hidden layer receives information from each of inputs, sums the inputs modified by a weight, and adds a constant, called bias, to later transform the result using an activation function like a sigmoid function or a tangential function that allows more differentiation and least squares fitting to tune the weights more finely using the back propagation algorithm (Figure 3). The weights, that modify the input to each neuron, are trained by passing sets of input-output pairs through the model and adjusting the weights to minimize the error between the output of the model and the real value.

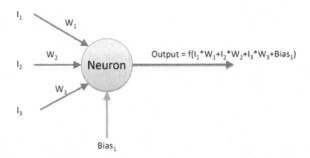

Fig. 3. General structure of the artificial neural network

Originally, the neuron was activated with a step function, however, sigmoid function (Equation 2) allows a more flexible differentiation and least squares fitting, leading to the back propagation algorithm and making it possible to tune the weights more finely.

$$f(x) = \frac{1}{\left(1 + e^x\right)} \qquad (2)$$

This equation is used because the calculation of derivatives has a high computational cost and this function produces almost the same output as an ordinary step function but is mathematically simpler.

The combined use of ANN and FE models has been widely used by many researchers for the design and optimization of industrial processes and products. In the case of products subjected to structural buckling, this combination of techniques has been mainly used for open structural shapes, which have a greater geometric imperfection and therefore less critical buckling load [9, 10, 11]. In this case, a study of closed (RHS) profiles, welded by a lateral longitudinal cord on one side, and with circular holes on the opposing side, was conducted.

3.1   Results of Built and Trained Models

In the study case, ANNs are analyzed to predict critical buckling load, vertical displacement and horizontal displacement. In order to improve the quality of the predicted models, initial data is normalized between 0 and 1. Once data is normalized, and using the 47 instances available from the experiments, 33 are chosen randomly to build and train the models using cross validation of 10 folds. And when the best models are selected, 14 previously separated instances are used to test them.

A single hidden-layer, feed-forward neural network was constructed using the nnet function [12] in the R statistical software environment v2.15.2 [13] for each case.

The number of neurons selected in a hidden layer can be variable. It has to be chosen the right number because otherwise some problems can arise. If the network has too few neurons, it cannot be flexible enough to make an accurate regression. And in the case the network has too many nodes, some problems like over-parameterization, overtraining, and a high computational cost can appear, thus losing the functionality of the models. In this case, a study of the behavior of different errors was made by varying the number of neurons in the hidden layer and varying the initial values of the elements of the network during training.

Trying to avoid finding a local minimum error, the models were built varying the randomness of the parameters that the network uses as initial weight during the training process. A total of 1000 ANNs per configuration were trained, with 9 different configurations of neurons (from 2 to 10 neurons) on the hidden layer, and using a weight decay of 0.00001 and maximum number of iterations of 10,000 to converge.

Figure 4 shows how the Root Mean Squared Error (RMSE) changes depending on these initial random weights and the number of neurons that the hidden layer has, in this case on the critical buckling load attribute.

In order to choose the most accurate models, several criteria were selected [14]: Mean Absolute Error (MAE), RMSE, Relative Absolute Error (RAE), Root Relative Squared Error (RRSE), and correlation coefficient (CORR). From all the models trained using cross validation, it was selected the configuration with the minimum mean RMSE, though taking into account the other criteria.

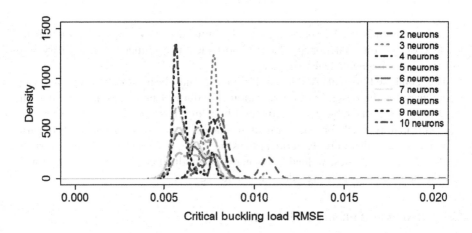

Fig. 4. Root Mean Squared Error (RMSE) obtained from the total number of trained ANNs

3.2   Artificial Neural Network Model Proposed for the Critical Load

Once the whole process of training was made, it was selected an ANN with the most suitable number of neurons in the hidden layer in each case. In the case of the critical buckling load, an ANN with 3 neurons in the hidden layer resulted to be the most accurate network (Figure 5).

This figure is a standard illustration of the generated neural network model. The black lines are positive weights and the grey lines are negative weights. Line thickness is in proportion to magnitude of the weight relative to all others. The hidden layer is labeled as H1 through H3, and B1 and B2 are bias layers that apply constant values to the nodes, similar to intercept terms in a regression model. All the weights of all these elements are shown in Table 2.

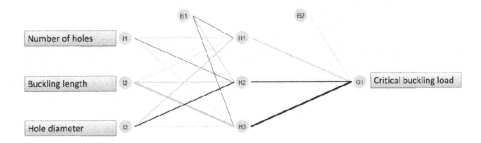

Fig. 5. ANN model proposed for the critical buckling load

Table 2. Weights from the proposed model for the critical buckling load

	Hidden layer				Output layer
	Neuron 1	Neuron 2	Neuron 3		Output 1
B1	0.23	-0.95	0.48	B2	-0.06
I1	-0.89	0.75	-0.26	H1	-1.12
I2	-0.31	-1.48	-2.97	H2	1.03
I3	-1.21	1.07	-0.51	H3	2.25
				I1	-0.27
				I2	0.36
				I3	-0.37

3.3   Artificial Neural Network Model Proposed for Vertical Displacement

In the same way that in the case of the model proposed for the critical load, it was selected the best model to predict the vertical displacement. The ANN that results to be the most accurate network was the one with a 5 neurons hidden layer (Figure 6). The hidden layer is labeled as H1 through H5, and B1 and B2 are bias layers that apply constant values to the nodes. And the weights of all the elements of this network are shown in Table 3.

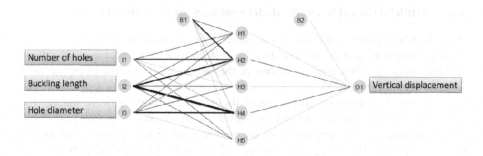

Fig. 6. ANN model proposed for the vertical displacement

Table 3. Weights from the proposed model for the vertical displacement

	Hidden layer					Output layer	
	Neuron 1	Neuron 2	Neuron 3	Neuron 4	Neuron 5		Output 1
B1	4.17	17.84	-6.59	-12.92	-2.79	B2	-1.50
I1	3.30	16.13	-0.51	-10.47	0.02	H1	-1.59
I2	0.84	25.99	2.49	35.51	8.71	H2	3.98
I3	2.42	-29.15	8.07	20.89	-0.27	H3	-0.59
						H4	4.32
						H5	-7.38
						I1	0.00
						I2	3.16
						I3	-0.39

3.4 Artificial Neural Network Model Proposed for Horizontal Displacement

And, as in the previous cases, it was selected the best model to predict the horizontal displacement. The ANN that results to be the most accurate network was the one with a 4 neurons hidden layer (Figure 7), with the weights that can be seen in Table 4. The hidden layer is labeled as H1 through H4, and B1 and B2 are bias layers that apply constant values to the nodes.

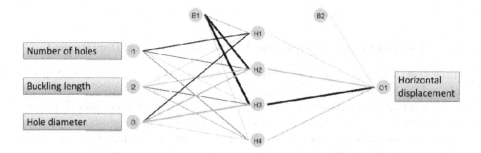

Fig. 7. ANN model proposed for the horizontal displacement

Table 4. Weights from the proposed model for the horizontal displacement

	Hidden layer					Output layer
	Neuron 1	Neuron 2	Neuron 3	Neuron 4		Output 1
B1	-0.93	3.01	3.61	-0.55	B2	-0.08
I1	1.43	0.09	-0.09	0.29	H1	-0.88
I2	-0.49	-1.16	0.03	-1.51	H2	-2.12
I3	0.99	-2.56	-2.85	-0.06	H3	2.79
					H4	-1.01
					I1	0.41
					I2	-0.47
					I3	0.79

3.5   Testing Results

And finally, the selected models were tested to check the accuracy of the models with data not used on the training dataset. Table 5 shows the indices of the selected models in each case using the testing dataset.

Table 5. Results obtained in selected models

	Critical buckling load	Vertical displacement	Horizontal displacement
MAE (%)	0.60	5.65	0.35
RMSE (%)	0.99	6.65	0.43
RAE (%)	1.68	9.81	2.35
RRSE (%)	2.65	11.46	2.39
CORR (%)	99.97	99.42	99.97

4   Conclusions

It is apparent that a neural network approach in combination with FEM to make a regression to predict some mechanical features is a powerful tool for problems such as this one. It is a very useful method for dealing with mechanical engineering problems because of their lack of restrictive assumptions and their ability to cope with combinations of different types of data. In this paper it is shown how the combination of FEM with ANN models could be based a methodology for modeling structural elements subjected to buckling. Furthermore, this ANN models obtained deals properly with geometric nonlinearities such as holes, and welding seams.

Acknowledgements. The authors thank the Autonomous Government of La Rioja for its support through the 3rd Plan Riojano de I+D+I for project MODUVA.

References

1. Akesson, B.: Plate Buckling in Bridges and Other Structures. Taylor & Francis (2007)
2. Johnston, B.: Column Buckling Theory: Historic Highlights. Journal of Structural Engineering 109(9), 2086–2096 (1983)
3. EN 1993 - Eurocode 3: Design of steel structures (1993)
4. Corchado, E., Herrero, Á.: Neural visualization of network traffic data for intrusion detection. Applied Soft Computing 11(2), 2042–2056 (2011)
5. Sedano, J., Curiel, L., Corchado, E., de la Cal, E., Villar, J.: A soft computing method for detecting lifetime building thermal insulation failures. Integrated Computer-Aided Engineering 17(2), 103–115 (2010)
6. Ripley, B.D.: Pattern Recognition and Neural Networks. Cambridge University Press, Cambridge (1996)
7. Timoshenko, S.P., Gere, J.M.: Theory of Elastic Stability, 2nd edn. McGraw-Hill (1961)
8. Crisfield, M.A.: Large-deflection elasto-plastic buckling analysis of eccentrically stiffened plates using finite elements (1976)
9. El-Sawy, K.M., Elshafei, A.L.: Neural network for the estimation of the inelastic buckling pressure of loosely fitted liners used for rigid pipe rehabilitation. Thin-walled Structures 41(8), 785–800 (2003)
10. Waszczyszyn, Z., Bartczak, M.: Neural prediction of buckling loads of cylindrical shells with geometrical imperfections. International Journal of Non-linear Mechanics 37(4), 763–775 (2002)
11. Sadovský, Z., Guedes Soares, C.: Artificial neural network model of the strength of thin rectangular plates with weld induced initial imperfections. Reliability Engineering & System Safety 96(6), 713–717 (2011)
12. Venables, W.N., Ripley, B.D.: Modern Applied Statistics with S, 4th edn. Springer (2002)
13. Team RC: R: A language and environment for statistical computing. R Foundation for Statistical Computing, Vienna, Austria (2012) ISBN 3-900051-07-0, http://www.R-project.org/
14. Fernández, R., Lostado, R., Fernandez, J., Martinez-de-Pison, F.J.: Comparative analysis of learning and meta-learning algorithms for creating models for predicting the probable alcohol level during the ripening of grape berries. Computers and Electronics in Agriculture 80, 54–62 (2012)

Design and Optimization of Welded Products Using Genetic Algorithms, Model Trees and the Finite Element Method

Rubén Lostado-Lorza[1,*], Roberto Fernández-Martínez[2],
Bryan J. Mac Donald[3], and Abdul Ghani-Olabi[3]

[1] Department of Mechanical Engineering, University of La Rioja, Logroño, Spain
[2] Department of Material Science, University of Basque Country UPV/EHU, Bilbao, Spain
[3] School of Mechanical & Manufacturing Engineering,
Dublin City University, Dublin 9, Ireland
ruben.lostado@unirioja.es

Abstract. One of the fundamental requirements in the phases of design and manufacture of any welded product is the reduction of residual stresses and strains. These stresses and strains can cause substantial changes in the geometry of the finished products which often require subsequent machining in order to fit to the dimensions specified by the customer, and are usually caused by the contribution of an external heat flux in a small area. All welded joints contain welding seams with more or less regular geometry. This geometry gives the welded product the strength and quality required to support the mechanical demands of the design, and is affected by the parameters controlling the welding process (speed, voltage and current). Some researchers have developed mathematical models for predicting geometry based on the height, width and cord penetration, but is a difficult task as many of the parameters affecting the quality and geometry of the cord are unknown. As the welded product becomes more and more complex, residual stresses and strains are more difficult to obtain and predict as they depend greatly on the sequence followed to manufacture the product. Over several decades, the Finite Element Method (FEM) has been used as a tool for the design and optimization of mechanical components despite requiring validation with experimental data and high computational cost, and for this reason, the models based on FEM are currently not efficient. One of the potential methodologies used for adjusting the Finite Element models (FE models) is Genetic Algorithms (GA). Likewise, Data Mining techniques have the potential to provide more accurate and more efficient models than those obtained by FEM alone. One of the more common Data Mining techniques is Model Trees (MT). This paper shows the combination of FEM, GA and MT for the design and optimization of complex welded products.

Keywords: Genetic Algorithms, Optimization, Finite Element Method, Model Trees, Welding Process.

---

* Corresponding author.

Á. Herrero et al. (eds.), *International Joint Conference SOCO'13-CISIS'13-ICEUTE'13*,
Advances in Intelligent Systems and Computing 239,
DOI: 10.1007/978-3-319-01854-6_29, © Springer International Publishing Switzerland 2014

1   Introduction

The welding process is a technique widely used in the manufacture of many industrial products. This process requires an external supply of heat flux concentrated in a small area at very high temperature. These high temperatures generate localized residual stresses and strains, which cause harmful defects in manufactured products. Likewise, poor selection of welding process parameters (speed, voltage and current) and a suboptimal manufacturing sequence can further amplify these residual stresses and strains. The design and optimization of any welded product based solely on experimental analysis or trial-and-error results in unacceptable high costs. The advantages of using the Finite Element Method (FEM) to reduce such costs are well known [1]. FEM is classified as a hard computing method and can be used to design and optimize any product or process. This method has also known disadvantages, including the large amount of time and computational cost needed in each simulation especially if the FE model is nonlinear. Generally, FE models of welded product have to take into account the plasticity of materials and must be made in 3D in order to capture the distribution of temperature and residual stresses more accurately. On the other hand, the use of models based on soft computing methods has been proved to be useful to solve complex problems [2, 3]. Soft Computing includes a set of techniques based on handling of imprecise and uncertain information. Artificial Neural Networks (ANN) and Genetic Algorithms (GA) are known branches of soft computing, which can be combined with each other. Some researchers have used ANN models for predicting the geometry of the weld bead based on the parameters of the welding process. In this paper, the prediction of weld bead geometry (width and height) was modeled with models trees (MT), some based on heuristic methods and some based on evolutionary algorithms, because these models had a higher accuracy and effectiveness than ANN in this case. This paper shows the optimized design process based on the combination of Genetic Algorithms (GA), Model Trees (MT) and Finite Element Method (FEM) for industrial products soldered with the Gas metal arc welding process (GMAW process).

2   Proposed Methodology

The proposed methodology for the optimized design process based on the combination of GA, MT and FEM for industrial products welded with GMAW is applied in the following steps (Figure 1).

2.1   Experimental Data

This first stage consists on the manufacture by GMAW of a number of specimens in order to generate data to make the models of the weld cord (with MT). Input Parameters (voltage, current and speed) and Output Parameters (height and width of the cord) are taken into consideration to set up the experiments and in this way to obtain the models of the Weld Geometry. Moreover, during fabrication of the specimens, the temperatures of the cord and welded parts were recorded with a Thermographic camera in order to use the temperature data to validate the FE model.

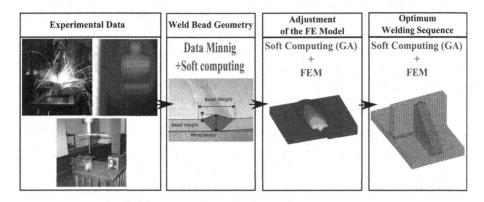

Fig. 1. Proposed methodology for design and optimization of welded products

2.2   Weld Bead Geometry

All welded joints have welding seams with a more or less regular shape. These cords are attached to the parts forming the product manufactured and its geometry is affected by the parameters controlling the welding process (speed, voltage and current) [4]. The geometry of the weld cord gives the welded product the strength and quality required to support mechanical demands of its design.

- **Modeling the Geometry of the Weld Bead**

Research papers discussing the modeling of welding processes are still relatively scarce probably because many of the parameters affecting the quality and geometry of the cord are still unknown. In this sense, some researchers [5] have developed mathematical models for predicting linear cord geometry based on their height, width and cord penetration. Through the widespread use of artificial intelligence, the prediction of the cord geometry has been performed using techniques based on ANN. For instance Srikanthan and Chandel [6] developed one of the first ANN models to predict with sufficiently accurate the bead geometry. More recently better ANN models have been developed to predict and optimize the width of the weld bead using two training algorithms [7].

Other techniques like Regression Trees can be used for numeric prediction. In this methodology, it is going to be used three kinds of regression trees, two built by Heuristic methods and one built by Soft Computing methods in order to determinate which of these machine learning techniques is more suitable to solve the study problem. The three regression trees are:

- Based on a CART algorithm [8, 9] (TC1).
- Based on Quinlan's M5 algorithm [10, 11] for inducing trees of regression models (TC2).
- Based on search over the parameter space of trees using global optimization methods like evolutionary algorithms [12] (TC3).

The first method is based on a CART algorithm, which uses recursive partitioning methods to build the model in a forward stepwise search. This approach is known to be an efficient heuristic where splits are chosen to maximize homogeneity at the next step. Splitting is made in accordance with squared residuals minimization algorithm which implies that expected sum variances for two resulting nodes should be minimized.

The second method improves the idea of a decision-tree induction algorithm using linear regression as a way of making quantitative predictions where a real-valued dependent variable y is modeled as a linear function of several real-valued independent variables $x_1, x_2, ..., x_n$, plus another variable that reflects the noise, ε (Equation 1).

$$y = \varepsilon + \beta_1 x_1 + \beta_2 x_2 + ... + \beta_n x_n \tag{1}$$

In the regression tree M5' each leaf contains a linear regression model based on some of the initial attribute values. In that way, it is combined a conventional decision tree with the possibility of linear regression functions at the nodes.

The third kind of tree uses an alternative way to search over the parameter space of trees using global optimization methods, like evolutionary algorithms. In the first two methods the split rule at each internal node is selected to maximize the homogeneity of its child nodes, without consideration of nodes further down the tree, but using a computationally efficient greedy heuristic that often yields reasonably good results. Therefore in some cases it is interesting to use stochastic optimization methods like evolutionary algorithms, like in the case of evtree which implements an evolutionary algorithm for learning globally optimal classification. These algorithms are inspired by natural Darwinian and are used to optimize a fitness function, such as error rate, varying operators that are modifying the tree structure. In this case, accuracy is measured by Bayesian Information Criterion (BIC) (Equation 2) [13].

$$BIC = -2 \log\left(L\left(\hat{\vartheta}; y\right)\right) + K \log(n) \tag{2}$$

These trees offer the ability to analyze which of the independent variables possess the strongest degree of influence on the tested data, ability that other techniques like artificial neural networks or support vector machines cannot offer.

2.3   Adjustment of the FE Model

One of the first FE models in which the welding process was simulated was in 1995 [14]. The model was formulated in 3D and considered the plasticity of the material and generated the residual stresses and strains in the manufactured product. More recently and based on validation by deformation and strain gauges, a 3D FE model was created to calculate the temperatures, deformations and residual stresses in welded joints [15]. Other researchers [16] used the FEM to model a welding process, validating the model by angular deformations and temperatures measured with thermocouples. The FE model presented in this paper, was a model formulated in 3D, which considered the plasticity of the materials and was simulated as transient nonlinear thermo-mechanical problem.

- **Validating the Proposed FE Model**

In this case the FE model was validated with the temperature values that were collected by a Thermographic camera. The Weld Flux in the FE model was modeled according to the theory of double ellipsoidal shaped [17]. The FE model had 11 different parameters to adjust to allow for validation, so GA was used as the adjustment technique [18]. This adjustment process was conducted as follows: Firstly, a range for the 11 parameters was established. Subsequently, a number of individuals (i.e. FE models) from the initial generation or generation 0 were randomly generated. Once all the individuals were simulated, the objective function J_T was applied (Equation 4). This objective function was defined as the average difference between the temperature obtained from the key nodes of the FE model and the temperature obtained from the thermal camera at each instant of time. The best individuals were those with the lowest value in the objective function and became the first generation or generation 0.

$$J_T = \frac{1}{n}\sum_{i=1}^{n}\left|Y_{TFE_i} - Y_{TTH_i}\right| \tag{3}$$

The next generation (first generation and subsequent generations) were generated using crossing and mutation. The new generation was made up as follows:

- 25% comprised the best individuals from the previous generation (parents of the new generation).
- 60% comprised individuals obtained by crossovers from selected parents.
- The remaining 15% was obtained by random mutation, through a random number used to modify the chromosomes within the pre-determinated ranges.

The aim was to find new solutions in areas not previously explored.

2.4   Optimum Welding Sequence

Since the majority of the mechanical component is complex to manufacturing, the final stresses, and temperature are more difficult to obtain and predict, and depend mainly by the manufacturing path used. Some researchers ([19, 20, 21]) have used experimental data and GA to optimize the welding sequence of complex welded products. This combination requires a significant amount of actual specimens to be welded with different sequences and welding parameters, so that the final cost is very high. In this paper, the combination of the FEM with GA was used to optimize the welding sequence in order that the welded products present the lowest state of stress and deformation possible.

3   Results

- **Results of Weld Bead Geometry**

In the case studied in this work, classic trees, lineal regression trees and evolutionary algorithms trees to predict width and height properties in weld beads are used. In order to improve the quality of the predicted models, the data is normalized between 0 and 1. Once the data is normalized, and using the 33 instances available from the experiments, 23 are chosen randomly to train the model and 10 to test it. The trees are built using different splitting index but with a minimum number of 4 observations that must exist in a node in order for a split to be attempted. The complexity parameter must be 0.01 where any split that does not decrease the overall lack of fit by this factor is not attempted. And the maximum depth of any node of the final tree must be less than 5.

Using CART methodology, where every class value is represented by the average value of instances that reach the leaf, it is got in the test process a MAE = 21.56% and a RMSE = 32.97% for width weld beads, and a MAE = 33.08% and a RMSE = 45.73% for height weld beads (Figure 2).

To improve these poor results, it is used the second kind of trees, where a linear regression model predicts the class value of instances that reach the leaf, it is got in the test process a MAE = 11.66% and a RMSE = 13.51% for width weld beads, and a MAE = 8.43% and a RMSE = 17.43% for height weld beads (Figure 3). In this study case, the obtained model to predict width weld beads contains the 10 linear models that belong at the 10 leaves, labeled LM1 through LM10. Where, for example, the Equation 3 defines the lineal regression that all the cases in leaf LM1.

$$Width = 0.405 \cdot Current - 0.4087 \cdot Speed + 0.3881 \cdot Voltage + 0.3112 \qquad (4)$$

And in order to see what is the different between using classic methods, like case 1 and 2, and methods based on evolutionary algorithms, like case 3, it is used the third kind of trees that provides evolutionary methods for learning globally optimal regression trees. It is got in the test process a MAE = 13.85% and a RMSE = 17.94% for width weld beads, and a MAE = 10.25% and a RMSE = 19.04% for height weld beads (Figure 4).

To compare these three models the following parameters are used: Mean Absolute Error (MAE) and Root Mean Squared Error (RMSE) [22]. In Table 1, it is showed that the use of evolutionary algorithms, when the tree is built, improves a lot the accuracy of the model, but improve more if instead of fix the value of the instances in each leaf with an average value, it is used a lineal regression model in each one.

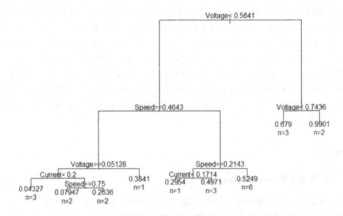

Fig. 2. Tree obtained in the case 1, where every class value is represented by the average value of instances that reach the leaf using CART methodology. In this case tree to predict width beads.

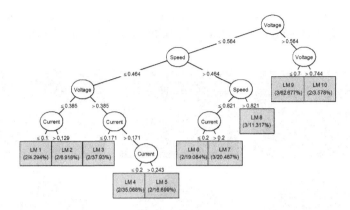

Fig. 3. Tree obtained in the case 2, where a linear regression model predicts the class value of instances that reach the leaf using Quinlan's M5 algorithm. In this case tree to predict width beads.

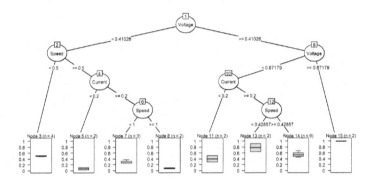

Fig. 4. Tree obtained in the case 3, where every class value is represented by the average value of instances that reach the leaf using evolutionary algorithms in the splitting. Constructed by evtree algorithm. In this case tree to predict width beads.

Table 1. Results obtained in the three models analyzed in the study

	Width weld beads			Height weld beads		
	TC1	TC2	TC3	TC1	TC2	TC3
MAE (%)	21.56	11.66	13.85	33.08	8.43	10.25
RMSE (%)	32.97	13.51	17.94	45.73	17.43	19.04

- **Results of Finite Element Model for the Weld Bead**

In Figure 5 is shown the temperature field obtained by a Thermographic camera and by the FE model in 6 seconds. The FE model shown in these images corresponds to the best individuals obtained from the 3rd generation. In this case, the objective function J_T not vary significantly with respect to the objective function from the 2nd generation and for this reason, the 11 parameters which define the FE model were fixed and used for optimizing the welding sequences to construct then the complex welded product.

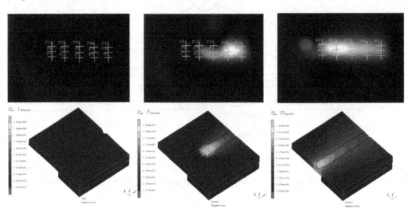

Fig. 5. Comparison between the temperature distribution obtained with thermographic camera and with the FE model for a single weld bead

- **Optimizing the Welding Sequence**

In the left side of Figure 6 is shown two different welding sequences, which involve different points of start and end of the weld, as well as different configuration parameters (speed, current and voltage). The numbers represent the different start points of each of the cords that form the welded product. Each of these welding sequences has been optimized by GA in order that the temperature distribution was as uniform as possible while the deformation was between established ranges. The right side of Figure 6 shows the temperature field induced in the welded product with these two different sequences.

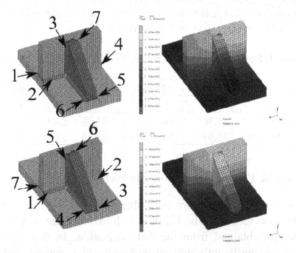

Fig. 6. Different welding sequences used to produce the same complex welded product

4   Conclusions

This paper demonstrates an efficient methodology that may be used to optimize the design of welded joints. An initial FE model is validated against experimental results with the assistance of genetic algorithms. The validated FE model may then be used to optimize more complex welding processes. By combining FEM, GA and MT techniques, it has been shown that it is possible to optimize complex welding processes and to, potentially, automate this process.

Acknowledgements. The authors thank the Autonomous Government of La Rioja for its support through the 3rd Plan Riojano de I+D+I for project MODUVA.

References

1. Shigley, J.E., Mischke, C.R., Budynas, R.G.: Mechanical Engineering Design. McGraw-Hill (2003)
2. Corchado, E., Herrero, A.: Neural visualization of network traffic data for intrusion detection. Applied Soft Computing 11(2), 2042–2056 (2011)
3. Sedano, J., Curiel, L., Corchado, E., de la Cal, E., Villar, J.: A soft computing method for detecting lifetime building thermal insulation failures. Integrated Computer-Aided Engineering 17(2), 103–115 (2010)
4. Reina, M.: Soldadura de los Aceros. Aplicaciones. Weld-Work S.L., Madrid (2003)
5. Kim, I.S., Son, K.J., Yang, Y.S., Yaragada, P.K.: Sensitivity analysis for process parameters in GMA welding processes using a factorial design method. International Journal of Machine Tools & Manufacture 43, 763–769 (2003)
6. Srikanthan, L.T., Chandel, R.S.: Neural network based modelling of GMA welding process using small data sets. In: Proceedings of the Fifth International Conference on Control, Automation, Robotics and Vision, Singapore, pp. 474–478 (1998)
7. Kim, I.S., Son, J.S., Lee, S.H., Yarlagadda, P.K.: Optimal design of neural networks for control in robotic arc welding. Robotics and Computer-Integrated Manufacturing 20(1), 57–63 (2004)
8. Breiman, L., Friedman, J.H., Olshen, R.A., Stone, C.J.: Classification and Regression Trees. Wadsworth International Group (1984)
9. Therneau, T.M., Atkinson, B., Ripley, B.: rpart: Recursive partitioning. R Package Version 3, 1-46 (2010)
10. Quinlan, R.J.: Learning with Continuous Classes. In: 5th Australian Joint Conference on Artificial Intelligence, Singapore, pp. 343–348 (1992)
11. Wang, Y., Witten, I.H.: Induction of model trees for predicting continuous classes. Poster papers of the 9th European Conference on Machine Learning (1997)
12. Grubinger, T., Zeileis, A., Pfeiffer, K.P.: evtree: Evolutionary Learning of Globally Optimal Classification and Regression Trees in R. Research Platform Empirical and Experimental Economics, Universitt Innsbruck (2011)
13. Schwarz, G.: Estimating the dimension of a model. The Annals of Statistics 6(2), 461–464 (1978)
14. McDill, J.M.J., Oddy, A.S.: A nonconforming eight to 26-node hexahedron for three-dimensional thermal-elasto-plastic finite element analysis. Computers & Structures 54(2), 183–189 (1995)

15. Romaní, G., Portolés, A.: Modelo tridimensional de simulación por MEF para estudiar la influencia de variables esenciales de soldadura robotizada GMAW en uniones a tope planas. Soldadura y Tecnologías de Unión 19(109), 22–26 (2008)
16. Chiumenti, M., Cervera, M., Salmi, A., Agelet de Saracibar, C., Dialami, N., Matsui, K.: Finite element modeling of multi-pass welding and shaped metal deposition processes. Computer Methods in Applied Mechanics and Engineering 199(37), 2343–2359 (2010)
17. Goldak, J., Chakravarti, A., Bibby, M.: A new finite element model for welding heat sources. Metallurgical Transactions B 15(2), 299–305 (1984)
18. Lostado, R., Martínez-de-Pisón, F.J., Fernández, R., Fernández, J.: Using genetic algorithms to optimize the material behaviour model in finite element models of processes with cyclic loads. The Journal of Strain Analysis for Engineering Design 46(2), 143–159 (2011)
19. Voutchkov, I., Keane, A.J., Bhaskar, A., Olsen, T.M.: Weld sequence optimization: the use of surrogate models for solving sequential combinatorial problems. Computer Methods in Applied Mechanics and Engineering 194(30), 3535–3551 (2005)
20. Xie, L.S., Hsieh, C.: Clamping and welding sequence optimisation for minimising cycle time and assembly deformation. International Journal of Materials and Product Technology 17(5), 389–399 (2002)
21. Kadivar, M.H., Jafarpur, K., Baradaran, G.H.: Optimizing welding sequence with genetic algorithm. Computational Mechanics 26(6), 514–519 (2000)
22. Fernandez, R., Lostado, R., Fernandez, J., Martinez-de-Pison, F.J.: Comparative analysis of learning and meta-learning algorithms for creating models for predicting the probable alcohol level during the ripening of grape berries. Computers and Electronics in Agriculture 80, 54–62 (2012)

Gene Clustering in Time Series Microarray Analysis

Camelia Chira[1,*], Javier Sedano[1], José R. Villar[2],
Carlos Prieto[3], and Emilio Corchado[4]

[1] Instituto Tecnológico de Castilla y León, Burgos, Spain
{camelia.chira,javier.sedano}@itcl.es
[2] University of Oviedo, Gijón, Spain
villarjose@uniovi.es
[3] Instituto de Biotecnología de León, León, Spain
carlos.prieto@inbiotec.es
[4] Universidad de Salamanca, Salamanca, Spain
escorchado@usal.es

Abstract. A challenging task in time series microarray data analysis is to identify co-expressed groups of genes from a large input space. The overall objective of this study is to obtain knowledge about the most important genes and clusters related to production and growth rate in a real-world microarray data analysis task. Various measures are engaged to evaluate the importance of each gene and to group genes based on their correlation with the output and each other. Some strategies for grouping and selecting genes are integrated resulting in several models tested for real biological data. All proposed models are tested on a real microarray data analysis problem and the results obtained are throughtly presented as well as interpreted from a biological perspective.

1 Introduction

Microarray data analysis (MDA) deals with a large number of features (genes) and needs efficient tools and techniques for the identification and classification of information [1–3]. The number of samples usually available is very low mainly due to the cost associated. This issue combined with the high dimensionality of the feature space make the task of extracting significant knowledge from microarray data an extremely difficult one. Time course (TC) microarray analysis [4–7] aims to find the best gene subset that promotes a certain variable or event when subsequent samples are taken from the same biological data at a certain time rate.

In TC MDA, the overall objective is to provide groups of genes meaningfully correlated and a ranking for each group in some well specified conditions. This paper focuses on a particular TC MDA problem with some specific requirements received from the biological experts. The input data consists of time series

---

* Corresponding author.

Á. Herrero et al. (eds.), *International Joint Conference SOCO'13-CISIS'13-ICEUTE'13*, 289
Advances in Intelligent Systems and Computing 239,
DOI: 10.1007/978-3-319-01854-6_30, © Springer International Publishing Switzerland 2014

samples which contain the expression levels of 8848 genes of a certain bacteria measured at 12 time points, each with 3 replicates. Three output values are available for each sample as follows: (i) the *production* - a real value indicating a production level in the studied bacteria, (ii) the *production growth* - a boolean value indicating if production is produced or not in the current sample, and (iii) the *growth rate* - a real value representing the level of growth in the bacteria. The objective of the problem is to select and group those genes which are the most relevant and related with the changes in the production and growth.

A related problem is that of gene expression classification where each sample has a corresponding output class and the aim is to find the most relevant subset of genes able to correctly classify new samples. A typical approach is to apply a gene selection method in order to reduce dimensionality and then engage a classifier system to evaluate the accuracy of the classification based on the selected genes [8–10]. In the MDA problem considered in this paper, the output does not represent the class label for a sample and the aim is not to classify samples but rather to group them in a meaningful way.

This paper presents framework for MDA that can be used in the case of time course (TC) analysis [4–7] with certain restrictions: groups of genes have to be identified so that genes in the same group are related with each other and with some production or growth output (corresponding to each sample). The proposed generic model to address this task includes three main steps as follows: (i) gene sorting according to some information measures and correlation with the output, (ii) formation of groups using a Markov Blanket (MB) approach, and (iii) validation of groups based on rate of change from one time point to another. Several algorithms result from this model according to the measures (information based or statistic) used in grouping the genes and the strategy selected for the validation step. All resulted methods are applied for a real MDA problem and the experiments performed are discussed.

The paper is structured as follows: information and statistical measures commonly used in microarray analysis are briefly reviewed, a model for gene clustering and selection based on infomation theory measures and new proposed similarity measures are presented, and experiments and results obtained for several model variants of the proposed approach are discussed.

2   Relevant Information Measures for Gene Ranking and Selection

Measures coming from information theory are useful in several fields and often engaged in feature selection [11]. Let X be a random variable and $p(x)$ the probability distribution of X. The entropy $H(X) = -\int p(x) \cdot \log(p(x))dx$ is a measure of the information the feature supports. Similarly, $H(Y|X)$ denots the entropy of a feature y provided the feature x.

The *mutual information* between two features x and y (denoted by $I(X,Y)$) is defined by means of their probability distribution, as stated in Eq. 1. Higher values of the mutual information between two features correspond to higher degrees of relevance between the two features. For our MDA problem, a naive way to select genes would be to calculate the mututal information between each gene and the output and then sort them in descending order. However, this approach would only consider the individual gene contribution and correlation with output.

$$I(X,Y) = \int \int p(x,y) \cdot \log(\frac{p(x,y)}{p(x) \cdot p(y)}) dx dy \tag{1}$$

For feature selection, these measures lack the ability of choosing independent features, particularly in high dimensional datasets. Let us consider two dependent features: if one of them has a high information measure then the second one does too. This results in a disadvantage of the above information metrics that lead to the proposal of other information measures described below.

The *Information Correlation Coefficient* (ICC) measures how independent two features are from each other (see Eq. 2). The higher the value the more relevant the relationship is. This measure is reflexive, symmetric and monotonic.

$$ICC(X,Y) = \frac{I(X,Y)}{H(Y|X)} \tag{2}$$

If $ICC(X,Y) = 1$ then the two variables X and Y are strictly dependent whereas a value of 0 indicates that they are completely irrelevant to each other.

The *Pearson's Correlation Coefficient* (PCC) measures the correlation between two features using statistics [12]. Let Y be the output feature and X is a feature from the input space. Let (x,y) be a pair of values of features X and Y, respectively. The PCC is calculated using Eq. 3.

$$PCC(X,Y) = \frac{\sum (x - \hat{x}) \cdot (y - \hat{y})}{\sqrt{\sum (x - \hat{x})^2 \sum (y - \hat{y})^2}} \tag{3}$$

3 Proposed Methods for Gene Clustering and Validation

The model proposed in this paper to approach the given MDA problem is based on information theory measures engaged to facilitate clustering and a new proposed measure mainly used in cluster validation. Genes are first grouped using different information and statistical measures in connection with the *Markov blanket* concept. In a second phase, groups are validated using a new measure for evaluating the rate of change in time series.

3.1 Gene Clustering

The formation of gene groups has to take into account the degree of relevance between each gene and the output as well as similar changes in gene expression levels in the time series. The phase of gene clustering in the proposed model addresses this problem with an emphasis on the first objective. The gene-output and gene-gene relevant degrees are computed using information correlation measures. Two such measures are considered in this study as follows: *Information Correlation Coefficient (ICC)* and *Pearson Correlation Coefficient (PCC)*.

As described in the previous section, ICC (see Eq. 2) measures the relevance between two variables based on mutual information and joint entropy. ICC takes values between 0 and 1. The higher the ICC value the more relevant the relationship between the two variables is. For instance, if ICC(X,Y) = 1 then X and Y are strictly dependant and relevant. A correlation degree can be expressed by stating that X is relevant to Y with degree ICC(X,Y).

On the other hand, PCC (see Eq. 3) is a statistical measure of the strength of the association between the two variables. PCC values range from -1 to +1. Positive correlation indicates that both variables increase or decrease together, whereas negative correlation indicates that as one variable increases the other decreases (and viceversa).

The basic procedure for gene clustering follows some ideas described in [8]. An ensemble gene selection by grouping (EGSG) method has been proposed in [8] for classification tasks in MDA. In the EGSG method, genes are first clustered by approximate MB and then ensemble classifiers applied. In this study, we adapt the first step from EGSG in order to group similar genes based on the correlation with the production and growth output. Furthermore, clusters are validated continuously during their formation using a newly introduced rate of change measure (detailed in the next subsection).

In the clustering phase, genes are first ranked according to the *Correlation Measure (CM)* with the output. Both ICC and PCC are considered as CM in different combinations for experiments. Groups of genes are formed starting from the highest-ranked gene so that genes in each group are correlated with each other and with the output based on the MB (Markov Blanket) strategy. The CM is used in determining if one gene is the approximate MB of another gene. The first gene added to a group is called the center of that group. A new gene g is accepted in an existing group if the center of the group is the approximate MB of g (otherwise, gene g forms a new group becoming the center of that group). The number of groups emerges from this schema and does not have to be a-priori known.

The main steps of the gene clustering phase are as follows:

A. For each gene $g_i, i = 1 \ldots M$ calculate $CM(g_i, y)$ based on the formula of ICC (Eq. 2) / PCC (Eq. 3).
B. Sort the gene set according to the calculated CM value (starting with the highest value, meaning the most relevant genes to the output y are first in the list). Let S be the sorted gene set.
C. Initialize the number of groups $k = 1$. Initialize the first group G_k with the top ranked gene from S: $G_k = \{S[1]\}$.
D. For each gene $g_i, i = 2 \ldots M$ do
   - *D.1. Grouping phase:*
   -- *D.1.1* If none of the centers of any group is the approximate MB of gene g_i then create a new group for g_i: $k = k + 1$; $G_k = \{g_i\}$.
   -- *D.1.2* If there is a group G_h such that $G_h[1]$ is the approximate MB of gene g_i then add g_i to group G_h: $G_h = G_h \cup \{g_i\}$.

- *D.2. Validation phase:* Check if the membership of gene g_i to the chosen group is validated from a rate of change similarity perspective. The strategy used for validation is detailed in the next subsection.
E. The final number of groups is k and the resulting groups are $G_1 \ldots G_k$.

The first phase results in k groups of genes clustered based on the correlation with the production output and each other.

3.2   Validation of Clusters

After a gene is added to an existing cluster (*grouping phase* - step D.1 described in the previous subsection), the updated cluster is validated by checking if the new gene has the same dynamics with the genes already present in the group (step D.2 - *validation phase*). If the new gene does not actually fit with the cluster then it is moved to a special group of *'unclustered'* genes (denoted by G_0).

The *Rate of Change Similarity (RCS)* measure is proposed to evaluate the similarity of the dynamics between two genes. RCS is defined as the number of significant changes that co-occur in two gene expression profiles. A percentage of the gene span is considered as a parameter to assess if a change is significant or not.

Let the span of a gene or of the output (production or growth) be the full extend of the variable, that is, the difference between its maximum and minimum values. Let τ be the parameter indicating a predefined percentage of the span. A significant change of a variable a (denoted by $\phi_i(a)$ at sample i in the dataset) is considered to occur when the difference of two consecutive values of that variable is higher than the product $\tau \cdot span$ (see Eq. 4). Then, given two variables a and b, the RCS is calculated as stated in Eq. 5.

$$\phi_i(a) = \begin{cases} 1, \ |a_i - a_{i-1}| > \tau \cdot span(a) \\ 0, \ otherwise \end{cases} \tag{4}$$

$$RCS(a,b) = \frac{\sum_{i=2}^{N} \phi_i(a) \cdot max_{j \in \{i-1,i,i+1\}} \phi_j(b)}{\sum_{i=2}^{N} \phi_i(a)} \tag{5}$$

It should be noted that $i = 2$ or $i = N$ represent special situations for Eq. 5. In these extreme situations, the strategy for finding a maximum ϕ value for the second variable b has to be changed from considering three possible rates of change to the only two actually available. In this way, when $i = 2$ the second term in the sum is $max_{j \in \{i,i+1\}} \phi_j(b)$ while for $i = N$ the $max_{j \in \{i-1,i\}} \phi_j(b)$ is considered.

The online validation phase (carried out once a gene is selected to be added to an existing cluster) determines if the new gene has a similar RCS to the output as the most representative gene in the cluster and further between each other. A parameter called δ is used to decide if the RCS for two different pairs of genes (x_1, y_1) and (x_2, y_2) is similar. If $|RCS(x_1, y_1) - RCS(x_2, y_2)| < \delta$ then they are considered similar. The validation phase checks the difference between the

RCS of the center of group and the output with both RCS of the new gene and the center of group and the RCS of the new gene and the output. As already mentioned, if the new gene does not pass the validation step, it is added to a special group G_0 of unclustered genes.

The clustering and validation phase results in k meaningful groups of genes as well as a special group G_0 which contains those genes that have no similarity to any cluster.

3.3 Summary of Proposed Methods

Several variants of the proposed model can be specified according to different measures chosen for clustering and validation phases. In order to allow an extensive analysis, we have selected three different model variants with or without validation and based on ICC, PCC or RCS in different combinations.

The following variants of the model have been selected for the study presented here: (i) *ICC_MB* - genes are sorted and MB clustered based on ICC, (ii) *ICC_MB_PCC* - genes are sorted and MB clustered based on ICC; validation of clusters is based on PCC, and (iii) *ICC_MB_RCS* - genes are sorted and MB clustered based on ICC; validation of clusters is based on RCS.

4 Computational Experiments

The dataset consists of 36 (3x12) samples and 8848 genes. A normalization and an optional discretization step was applied to the dataset. Some experiments use a discretization phase for the data which is applied after normalization. This phase means that the gene expression values are discretized so that insignificant changes are ignored. For this discretization phase, a parameter called *d_step* is used to decide a significant change.

4.1 Normalization

The normalization process was performed with the limma package [13]. Median and none background correction methods were applied for all results reported in this paper. Method none computes M and A values without normalization so the corrected intensities are equal to foreground intensities. On the other hand, method median substracts the weighted median of background intensities from the M-values for each microarray.

4.2 Experiments Setup

Experiments consider the input dataset as follows: the mean value of the 3 samples at each time point providing a dataset of 12 samples with 12 set of outputs (called *Mean12*). With each resulted dataset, the correlation with one of the three available outputs i.e. the production growth output (*Bool*), the production rate (*RealProd*) and the growth output (*RealGrowth*) can be considered in the experiments.

Therefore, experiments and results are grouped in the following categories: *Mean12_Bool*, *Mean12_RealProd* and *Mean12_RealGrowth*. For each experiment category, all three model variants described in the previous section (i.e. *ICC_MB*, *ICC_MB_PCC* and *ICC_MB_RCS*) have been applied and the obtained results are discussed in the following subsections.

The possible parameters in each method include *d_step* (used in the discretization phase), τ and δ (both used in the validation phase). Based on many experiments performed and the results obtained, we selected the following values for each parameter to discuss the results in this paper: $d\_step \in \{0, 0.001, 0.01, 0.1\}$ ($d\_step = 0$ corresponds to no discretization), $\tau \in \{0.01, 0.1, 0.5\}$ and $\delta \in \{0.005, 0.05, 0.1, 0.25, 0.5, 0.75, 0.9\}$.

4.3   Results

Considering the mean value over 3 replicas and the boolean production output, ICC_MB groups all genes in the same cluster except when discretization with step 0.1 is used in which case two clusters are obtained: one with 8830 genes and the second one with 18 genes. ICC_MB_PCC and ICC_MB_RCS further produce a group G_0 for which the size depends on the δ value.

The results obtained considering the real value of production as output are overall better compared to those obtained for the production growth boolean value.

Without discretization, ICC_MB puts all the genes in the same group (similarly with ICC_MB for Mean12_Bool). However, when $d\_step = 0.1$, ICC_MB reports 7 clusters where majority of genes is in the first cluster and the other 6 groups are formed by fewer genes. The result is still poor as the size of one group is too large compared to the rest. This is emphasized by the validation phase (particularly of ICC_MB_RCS) which results in a group G_0 containing many uncorrectly clustered genes from the big size cluster.

To be more specific, ICC_MB_PCC produces a G_0 group which contains from 0 to 8604 genes depending on the value of δ. Again, for $d\_step = 0.1$ best results are obtained: 7 clusters and G_0 with 1590 genes.

ICC_MB_RCS also gives up to 7 clusters depending on parameters τ and δ used in the RCS measure and the cluster validation. Furthemore, the discretization step highly influences the results. When no discretization is used, all genes go in one cluster and the size of G_0 increases with lower values of δ. For discretization step lower than 0.1, two clusters are formed: one with very high number of genes (from 8837 to 8754) and the other with very low number of genes (from 11 to 94). For $d\_step = 0.1$ genes are grouped in 7 clusters and a G_0 group for which the size depends on δ (see Table 1).

From Table 1, it can be seen that for $\delta = 0.005$ the size of G0 is rather large at over 7500 genes whereas at δ 0.5 and 0.9 all clusters are validated by RCS regardless the value of τ. The most balanced results are obtained for $\tau = 0.01$ and $\delta = 0.1$.

Table 1. ICC_MB_RCS ($d\_step = 0.1$) results for Mean12_RealProd with different τ and δ values

τ	δ	Clusters							G_0
		G_1	G_2	G_3	G_4	G_5	G_6	G_7	
0.01	0.005	941	14	110	5	1	17	8	7752
0.01	**0.1**	**4361**	**16**	**357**	**5**	**1**	**17**	**8**	**4083**
0.01	0.5	8207	215	395	5	1	17	8	0
0.1	0.005	1130	73	97	1	1	5	8	7533
0.1	0.1	6832	195	273	5	1	7	8	1527
0.1	0.9	8207	215	395	5	1	17	8	0

Considering the relation of gene values with the growth value output results in more clusters of genes in all methods compared to the Mean12_RealProd where the real production value was considered.

Without discretization, ICC_MB puts all the genes in the same group (same as for Mean12_RealProd). When $d\_step = 0.1$, ICC_MB reports 31 clusters (as opposed to 7 clusters for Mean12_RealProd). The majority of genes go in the first cluster and the other groups are formed by fewer genes (similar behavior with ICC_MB for Mean12_RealProd).

ICC_MB_PCC obtains similar results with ICC_MB except that it also produces the G_0 group which contains from 0 to 8584 genes depending on the value of δ. Again, for $d\_step = 0.1$ best results seem to be obtained: 31 clusters and G_0 with 631 genes at $\delta = 0.25$.

ICC_MB_RCS reports 1 to 31 clusters depending more on the discretization step rather than on the τ parameter (used in the RCS measure) and δ (used in the cluster validation). When no discretization is used, all genes go in one cluster. For discretization step of 0.0001, three clusters are formed: one with very high number of genes (8846) and the other two groups with one gene each. For discretization step of 0.01, 12 clusters are formed: one with very high number of genes (8818) and the other 11 groups having 1 to 8 genes each. For discretization step of 0.1, 31 clusters are formed: one with high number of genes (7731) and the other 30 groups having among 1 and 317 genes each.

4.4   Discussion and Biological Perspective on the Results

Experiments have shown that the model variant ICC_MB is not able to provide any clustering in most scenarios considered. The inclusion of a validation phase (based on either RCS or PCC) is crucial in obtaining a more reliable clustering result starting from ICC_MB. Figure 1 emphasizes the different results obtained by ICC_MB compared to the ICC_MB_PCC where a validation phase is included and also the difference in results between a validation based on PCC and the other based on RCS. A triangular matrix is created as follows: for each pair of genes (g_i, g_j) associate value 0 (corresponding to white color) if none of the two methods grouped genes g_i and g_j together, value 0.5 (corresponding to grey color) if only one of the methods put the two genes in the same group and a

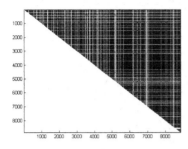 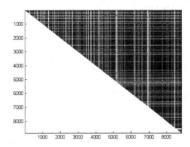

Fig. 1. Comparison of model variants with and without validation: (left) ICC_MB vs. ICC_MB_PCC, and (right) ICC_MB_RCS vs ICC_MB_PCC

value of 1 (corresponding to black color) if both methods produced the same grouping result. Clearly, the RCS provides different kind of groups by checking the rate of change in the gene expression values.

Although the computational results are encouraging, their biological utility is limited due to the big size of resulting groups and the lack of co-expression between the genes of each group. It is known that functionally related genes tend to have similar expression values [14] and hence, the possibility of obtaining groups with a common expression profile is of great interest because it enhances the biological significance. However, it is important to emphasize that gene ranking and selection measures help to identify genes that are involved in the production and growth processes. Therefore, the combination of co-expression and gene ranking approaches could be beneficial because (i) the size of groups is reduced based on a co-expression measure, (ii) genes are ranked based on the growth and production values and (iii) the biological significance is improved based on the assumption in which related biological processes have similar expression patterns.

5   Conclusions and Future Work

The task of gene clustering and selection in connection with a real-world time series microarray problem has been investigated. Several methods based on information theory methods are developed and analysed. Experiments show a poor performance of measures such as ICC in the ability to meaningfully cluster the genes in the considered dataset. However, the importance of validation by similarity measures is clearly emphasized through the comparisons performed.

Future work focuses on development and investigation of methods able to provide gene groups based on the distance between gene expression levels and the correlation with the output.

Acknowledgments. This research has been partially supported through the projects of the Junta de Castilla y Leon CCTT/10/BU/0002 and the projects from Spanish Ministry of Science and Innovation PID 560300-2009-11 and TIN2011-24302.

References

1. Saeys, Y., Inza, I., Larrañaga, P.: A review of feature selection techniques in bioinformatics. Bioinformatics 23(19), 2507–2517 (2007)
2. Larrañaga, P., Calvo, B., Santana, R., Bielza, C., Galdiano, J., Inza, I., Lozano, J.A., Armañanzas, R., Santafé, G., Pérez, A., Robles, V.: Machine learning in bioinformatics. Briefings in Bioinformatics 7(1), 86–112 (2006)
3. Lee, C.-P., Leu, Y.: A novel hybrid feature selection method for microarray data analysis. Applied Soft Computing 11, 208–213 (2011)
4. Peddada, S.D., Lobenhofer, E.K., Li, L., Afshari, C.A., Weinberg, C.R., Umbach, D.M.: Gene selection and clustering for time-course and doseresponse microarray experiments using order-restricted inference. Bioinformatics 19(7), 834–841 (2003)
5. Ernst, J., Bar-Joseph, Z.: Stem: a tool for the analysis of short time series gene expression data. BMC Bioinformatics 7(1), 191 (2006)
6. Storey, J.D., Xiao, W., Leek, J.T., Tompkins, R.G., Davis, R.W.: Significance analysis of time course microarray experiments. Proceedings of the National Academy of Sciences of the United States of America 102(36), 12837–12842 (2005)
7. Liu, T., Lin, N., Shi, N., Zhang, B.: Information criterion-based clustering with order-restricted candidate profiles in short time-course microarray experiments. BMC Bioinformatics 10(1), 146 (2009)
8. Liu, H., Liu, L., Zhang, H.: Ensemble gene selection by grouping for microarray data classification. Journal of Biomedical Informatics 43, 81–87 (2010)
9. Lu, Y., Han, J.: Cancer classification using gene expression data. Information Systems 28(4), 243–268 (2003)
10. Wang, Y., Tetko, I.V., Hall, M.A., Frank, E., Facius, A., Mayer, K.F.X., Mewes, H.W.: Gene selection from microarray data for cancer classification—a machine learning approach. Computational Biology and Chemistry 29, 37–46 (2005)
11. Peng, H., Long, F., Ding, C.: Feature selection based on mutual information: Criteria of max-dependency, max-relevance, and min-redundancy. IEEE Transactions on Pattern Analysis and Machine Learning 27(8), 1226–1238 (2005)
12. Bolboaca, S.-D., Jantschi, L.: Pearson versus spearman, kendall's tau correlation analysis on structure-activity relationships of biologic active compounds. Leonardo Journal of Sciences (9), 179–200 (2006)
13. Smyth, G.K., Speed, T.: Normalization of cdna microarray data. Methods 31(4), 265–273 (2003)
14. Prieto, C., Risueno, A., Fontanillo, C., De Las Rivas, J.: Human gene coexpression landscape: Confident network derived from tissue transcriptomic profiles. PLoS One 3(12), e3911 (2008)

Quality of Microcalcification Segmentation in Mammograms by Clustering Algorithms

Ramón O. Guardado-Medina[1], Benjamín Ojeda-Magaña[1],
Joel Quintanilla-Domínguez[2], Rubén Ruelas[1], and Diego Andina[2]

[1] Departamento de Sistemas de Información CUCEA, Universidad de Guadalajara, C.P. 45100,
Zapopan, Jalisco, México
osvaldo.guardado@cucea.udg.mx, benojed@hotmail.com
[2] E.T.S.I. de Telecomunicación, Universidad Politécnica de Madrid, Avda. Complutense 30,
Madrid 28040, Spain

Abstract. Breast cancer remains a leading cause of death among women worldwide. Mammography is one of the non-invasive methods to find breast tumors, which is very useful in the detection of cancer. Microcalcifications are one of the anomalies of this disease, and these appear as small white spots on the images. Several computer-aided systems (CAD) have been developed for the detection of anomalies related to the disease. However, one of the critical parts is the segmentation process, as the rate of detection of anomalies in the breast by mammography largely depends on this process. In addition, a low detection endangers women's lives, while a high detection of suspicious elements have excessive cost. Hence, in this work we do a comparative study of segmentation algorithms, specifically three of them derived from the family of c-Means, and we use the NU (Non-Uniformity) measure as a quality indicator of segmentation results. For the study we use 10 images of the MIAS database, and the algorithms are applied to the regions of interest (ROI). Results are interesting, the novel method of sub-segmentation allows continuous and gradual adjustment, which is better adapted to the regions of micro calcification, and this results in smaller NU values. The NU measure can be used as an indication of quality, which depends on the number of pixels and the homogeneity of the segmented regions, although it should be put in the context of the application to avoid making misinterpretations.

1 Introduction

Microcalcifications (MCs) are anomalies in the breast of women, and are one of the main indicators of cancer which can be identified through digital images. The digital imaging is an invaluable tool in modern medicine among which we find the Magnetic Resonance Imaging (MRI), the Computed Tomography (CT) and Digital Mammography (DM) among others. This method is very effective as well as being non-invasive and allows analysis of internal anatomy of a subject to identify any anomalies [1]. These technologies have made a significant contribution in the prevention and cure of diseases when detected at an early stage of development. Hence the interest in the development of an automatic detection system, that can serve as a tool for medical support, or as tool for a second opinion. In any case, it is the doctor who has the last word.

Á. Herrero et al. (eds.), *International Joint Conference SOCO'13-CISIS'13-ICEUTE'13*,
Advances in Intelligent Systems and Computing 239,
DOI: 10.1007/978-3-319-01854-6_31, © Springer International Publishing Switzerland 2014

However, the detection of anomalies through images is not so simple, as they often have a very poor contrast and they can show different characteristics. The latter because of the anatomy of the breast tissue density and hormonal conditions change, all of them directly related to the age of patients. The risk that radiologists ignore some subtle abnormalities exist, and in the case of a questionable diagnosis tend to be fairly cautious to avoid any failure. So, in these cases the patient is subjected to invasive diagnosis, as biopsy. The drawback of this method is the high number of unproductive biopsy examinations (no cancer) and the high economic costs they have [2].

Then, images help doctors to do a better detection and localization of anomalies, contributing to have more accurate medical diagnoses. The sensitivity in the diagnosis, that is, the recognition accuracy of all malignancies and specificity, i.e. the possibility of differentiating benign and malignant diseases, can be improved if each mammogram is examined by two radiologists, with the consequence that the process can be inefficient and reduce the productivity of individual specialist. An alternative is to replace one of the radiologists for automatic detection system programmed into a computer. Hence, the objective in the development of CAD systems is helping radiologists to identify all lesions in a mammogram and improve the early diagnosis of breast cancer. In these cases, CAD systems operate normally as a second opinion (automated) or as a dual reading [3–5].

The MCs are small calcium deposits accumulated in breast tissue. They usually appear on the mammogram as small bright embedded within an inhomogeneous background [6]. From certain characteristics such as: size, shape, density, distribution pattern and number of MCs are associated with benign or malignant [18]. Malignant MCs have generally a diameter less than 0.5 mm, fine structure, branching linear, star-shaped, in addition to varying in shape and size. It is very common that the distribution pattern of microcalcification covers more than three microcalcification [19].

Since image segmentation is an essential process in the identification of MCs, in this work we make a comparative study of three segmentation algorithms, all based on the family of c-Means. To do this we use images of MIAS database, and algorithms applied to the ROIs. However, in order to improve the contrast in such images, before segmentation, is applied a pre-processing Top Hat transform, which facilitates the identification of anomalies within the image studied. Furthermore, as also applies NU (Non-Uniformity) thus provides an indicator of the quality of the segmentation obtained in each case. As can be seen in the figures of MCs found, the results are interesting and indicators like this can help set the detection threshold to improve the automatic identification of anomalies, or the sensitivity and specificity of detection.

The paper is structured as follows: Section 2 presents the proposed methodology, the pre-processing method used, and the criteria for assessing the quality of segmentation. In Section 3 shows a comparison of partitional clustering algorithms (k-means, FCM and PFCM), while Section 4 presents some experimental results to identify MCs in the ROIs of 10 images taken from the base MIAS data. Finally, Section 5 provides the main conclusions of this work.

2 Methodology

The problem of MCs identification is very complex and subjective; below there is a description of some characteristics of the MCs. So, we realize a comparative analysis using three partitional clustering algorithms applied to a set of 10 images ROI's. Besides, we evaluate the quality of the segmentation through the NU measure. MCs characteristics: [16]:

- The objects of interest are very small. These are found as small and bright groups, very difficult to identify on mammographies, and sometimes hidden in dense breast tissue.
- These objects, besides being small, are inhomogeneous. These limits applying a simple threshold method for segmentation.
- Sometimes these objects have low contrast and can be interpreted as noise, especially when the image background is not very homogeneous.

Therefore, in this paper evaluates the partitional clustering algorithms in the segmentation process to determine the benefits and deficiencies in the detection of MCs. The proposed methodology is as follows (see Fig. 1):

Quality assessment of image segmentation for ROI's.

1. Get ROI's images of mammograms.
2. Filter the image using Top-Hat transform (1) to enhance contrast.
3. Calculate the intensity distribution, or histogram, for the I_T filtered image.
4. Segment the I_T image with the clustering algorithms: k-Means, Fuzzy c-Means, and the sub-segmentation method based on the PFCM.
5. Calculate the NU value for each image and each algorithm according to (2).

2.1 ROI's Images

Mammograms used for this study were obtained from the database Mammographic Image Analysis Society (MIAS), which contains 322 images of different cases of study, and available as Portable Gray Map (PGM). For each image there are 1024x1024 pixels and a resolution of 200μm/pixel. Furthermore, the database also contains information regarding the location of anomalies, and this was used for the selection of the regions of interest (ROI's) with a size of 256x256 pixels. The 10 images selected from the MIAS database, which have very different characteristics, are: *mdb*058, *mdb*170, *mdb*188, *mdb*204, *mdb*209, *mdb*219, *mdb*223, *mdb*227, *mdb*248 and *mdb*249.

2.2 Enhancemente of Microcalcifications by Top-Hat Transform

In order to improve the results of the segmentation, a pre-processing has been applied which consists in filtering the image using the Top Hat transformed; the result is a contrast enhancement between the objects of the image. This mathematical morphology has been proposed by Matheron and Serra and later extended to the analysis of images

[17]. All mathematical morphology operations work with two sets, one corresponding to the original image to be analyzed, and the other corresponding to a structured element. There exists two mathematical morphologies, the Top-Hat transform and the Bottom-Hat transform. The first is used for bright objects on a background with unequal intensity [19], and this is the one used in this work. This transform is defined by (1).

$$I_T = I_{in} - [(I \odot SE) \oplus SE].$$

(1)

Where, $\mathbf{I}_T$ is the input image, I_T is the transformed image, $\mathbf{SE}$ is the structuring element, $\odot$ represents morphological erosion operation, $\oplus$ represents morphological dilation operation and - image subtraction operation. The term $[(I_{in} \odot SE) \oplus SE]$ is also called morphological opening operation. Fig (2b) shows the results of the Top Hat transform used in this work.

2.3   Region Non-Uniformity (NU)

The NU measure [8] allows to set a reference about the quality of image segmentation, with values within range (0, 1). The measure is defined as:

$$NU = \frac{P \cdot \sigma_P^2}{R \cdot \sigma^2}; \qquad\qquad NU : 0 < NU < 1.$$

(2)

Where $\mathbf{P}$ and $\mathbf{R}$ are the numbers of microcalcifications and total number pixels in the segmented images, σ_P^2 and σ^2 are the variance of grey-scale values in the space micocrocalcification, and the total variance in the image.

3   Segmentation by Clustering Algorithms

Clustering algorithms are classification methods with unsupervised learning [10] [11], used when available data are not labeled. So, they are based on the identification of groups, although no information is available on the optimal number of groups. Hence the goal of clustering algorithms is to partition a database, where data belonging to a group must be as similar or close as possible, whereas these must be as dissimilar as possible of data belonging to the other groups. Clustering techniques are widely used for segmentation of digital images. The main difference to the proper segmentation methods, is that these work in the spatial domain of the image, while the first work in the space of features [18].

Since clustering algorithms work in the space of features, in this particular paper with the gray level intensity of the pixels, they finish identifying prototypes near to the salient points in the histogram, which represents the number of pixels according to gray levels. For this reason, and due to the fact that generally no spatial information is incorporated although there exist some proposals to do it, the segmentation by this method is affected by noise and non-homogeneity of the image. However, as the objects of interest are very small and the image has been enhanced using the Top-Hat transform, spatial information is not considered so necessary for image segmentation.

In this paper the segmentation of mammograms is only based on gray levels. Thus, an image can be represented as $\mathbf{I}(x,y)$, where $x \in [1, N_x]$ and $y \in [1, N_y]$ correspond

Table 1. Comparison between partitional algorithms

Clustering algorithms	Parameters	Partition type	Advantages	Disadvantages
K-means [9]	– c= number of centers. – v_i= random centers.	Hard Partition	– Simple. – Fast. – Works well for disjoint classes	– Very sensitive to fall into local minimum [12]. – Noise sensitive – Starting sensitive prototype
Fuzzy c-means (FCM) [14]	– c= = number of centers. – v_i= random centers. – $m(m > 1)$ = fuzzifier exponent	Fuzzy partition	– Less sensitive to local minimum. – Best partition into disjoint groups. – More information on ownership of the datapoints	– Noise sensitive – The sum of the degrees of membership is equal to 1
Possibilistic Fuzzy c-means (PFCM) [15]	– c= number of centers. – v_i= random centers. – $m(m > 1)$ = fuzzifier exponent (FCM) – $\eta(\eta > 1)$ – $a(a > 1)$ = constant (influences the values of membership and tipicality) – $b(b > 1)$ = constant (influences the values of membership and typicality)	– Fuzzy partition – Possibilistic partition	– Generates ownership matrix U and the matrix of typicality. – Prevents cluster coincide as in the PCM. – You get more and better information about the membership of the datapoints.	– It has several adjustment parameters (a, b, m, η). – It is necessary to know the algorithm to make a good choice of their parameters.

to spatial indices, whereas **i**(x,y) quantized the gray level of a particular pixel. Since digital image segmentation is to partition the total set of pixels S into subsets or regions (**S**$_1$, **S**$_2$,...**S**$_n$), it must meet the following properties:

$$\cup_{k=1,n} \quad \mathbf{S}_k = \mathbf{I}; \tag{3}$$

$$\mathbf{S}_k \cap \mathbf{S}_j = \phi, \quad k \neq j \in [1,n]. \tag{4}$$

Equation (3) indicates that all pixels of the image are associated to the subsets or regions identified by clustering algorithms, while (4) indicates that each pixel is not associated with more than one region.

In this paper we use the most popular clustering algorithms in the literature (k-Means, FCM, and PFCM) for automatic detection of microcalcifications. Table 1 presents the main characteristics of such algorithms where, as you can see, each one has its own peculiarities that make it interesting. The next section contains the results of the application of these algorithms to the segmentation of ROI images.

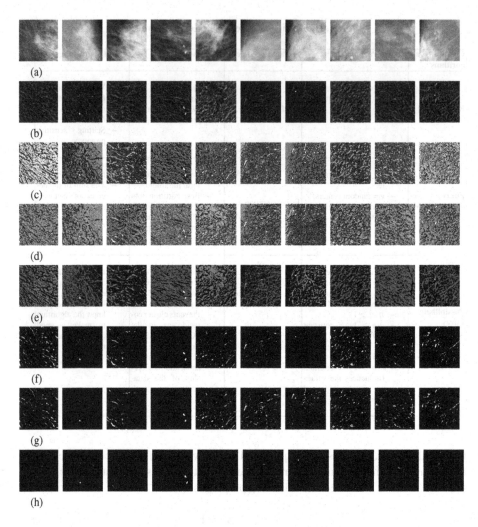

Fig. 1. (a) ROI's of the original images (*mdb*058, *mdb*170, *mdb*188, *mdb*204, *mdb*219, *mdb*223, *mdb*227, *mdb*248 and *mdb*249). (b) Contrast enhancement with Top-Hat transform. Segmentation results: (c) k-means, (d) FCM, (e) Sub-segmentation (PFCM). Identification of MCs by: (c) k-means, (d) FCM, (e) Sub-segmentation based on (PFCM). Micro calcifications identified by (f) k-means, (g) FCM, (h) Sub-segmentation (PFCM).

4   Results and Discussion

Initially 10 images from the MIAS database were selected (*mdb*058, *mdb*170, *mdb*188, *mdb*204, *mdb*209, *mdb*219, *mdb*223, *mdb*227 *mdb*248 and *mdb*249), which have very different characteristics as shown in Figure (1a). Once the ROI's of the original 10 images were available, the next step was to apply the Top Hat transform, or a filtering, in order to improve the contrast between the objects and increase the difference in gray levels between them. Equation (1) was used for this purpose with a structuring element of 15x15 flat disk type. The results can be seen in Fig (1b).

As the Top Hat transform promotes compact groups according to the size of the structuring element, the histogram of the original image is modified with more pronounced peaks around the gray levels of the objects, where more pronounced peak usually occurs very close to black color.

The algorithms selected for this comparative study are the k-Means, the FCM, and the PFCM through the sub-segmentation method. Furthermore, we only have used the gray level feature from the filtered image I_T. Hence, in this case the feature space is equal to the histogram of the image. The optimum number of regions $(S_1, S_2, \ldots S_n)$ in which an image should be segmented is unknown, due to the different objects present in the image, and the capacity of the clustering algorithms. However, for the mammograms this value may be equal to three corresponding to: the background, the tissue, and the MCs.

The partitional clustering algorithms work from prototypes and they fit to the spaces where there is a greater distribution of pixels. Therefore, once the image has been segmented, the prototypes are to be found close to the peaks in the histogram of the image, where these points are concentrated in a larger number of pixels with respect to its environment. However, only a small group of pixels represents the MCs, as they are very small objects, and the number of regions has to be increased significantly until the group of MCs remains more stable. From the experimental results, in this paper we set to 5 the number of regions for image segmentation with the k-Means and the FCM. The sub-segmentation, meanwhile, segment an image into two groups which divides into two subgroups each one using a typicality threshold, where a subgroup corresponds to MCs [20].

In Fig. (1c) contains the segmented images with the k-Means. The drawback of the algorithm is its sensitive to initial values as it is easily trapped in local minima. For this reason it was necessary to run it repeatedly until you get and verify good results. The *mdb*223 image segmentation is an interesting case, as the preprocessing separates very well the MCs from the rest of the image. This shows that the algorithm discriminates very well the objects when these belong to disjoint groups.

In Fig. (1d) sshows the results of the image segmentation with FCM algorithm. As can be seen in Fig. (2f) and Fig. (2g), the results are similar to those of the k-Means. However, FCM algorithm is more stable and is less likely to fall into local minima. Also, the FCM provides more information as it provides the degree of membership of each pixel to each of the identified groups, while the k-Means only provides discrete values on the membership or not to each of these groups.

The latter results in Fig. 2 correspond to those obtained by the sub-segmentation method based on the PFCM. This algorithm first identifies two regions corresponding to the background and the object of interest. Then, these two regions are subdivided according to their degree of typicality with respect to the prototype of the group, and it is precisely the atypical subregion of the object of interest that corresponds to the MCs [21]. The values assigned to PFCM parameters are:$(a = 1, b = 2, m = 2, \eta = 2)$. Typically, a value of 2 is used for m and η, while b must be greater than a in order to reduce the influence of noise [15], while the threshold typicality, which must be defined in the interval $(0, 1)$, was $\alpha = 0.01$ for all the images. As can be seen, in some cases the results are very different and better than those obtained with the k-Means and the FCM.

With the sub-segmentation method we can continuously vary the threshold typicality, which provides a gradual approach to the interest group. Moreover, the k-Means and the FCM identify an integer number of groups, resulting in a discrete approximation, that can lead to the identification of a large number of regions for the correct identification of MCs. This, undoubtedly, contributes to the achievement of better results through sub-segmentation. This is of great importance since an underestimate threatens the lives of women, while an overestimation is very costly as it involves using invasive methods, such as taking a biopsy, and other studies.

4.1 Comparative Analysis

The aim of this study is to conduct a comparative analysis of the segmentation results of 10 mammograms images using three clustering algorithms, in addition to establish a more quantitative criterion directly related to the quality of segmentation results. The NU measure has been proposed for such purpose and it takes values in the interval $(0, 1)$. This measure is related to the homogeneity and size of the overall image and the segmented regions in particular; in this case with microcalcifications. Table 2 contains the values of NU corresponding to each of the segmented images and algorithm used, as well as the number of pixels in the region of the MCs.

Looking at the images in Fig. 2, particularly the filtered images or Fig. (2b), we see that the number of pixels of the MCs in the total of pixels of the image is very low. Therefore, in this type of applications the value of NU, although vary within the range $(0, 1)$, it will be found very close to zero. This can be noted looking at data of Table 2, where the maximum value is less than 0.2. This value is also dependent on the homogeneity of the segmented regions. Therefore, if one wants to further reduce NU the number of segmented regions should be increased in the case of k-Means and FCM, or the typicality threshold must be adjusted in the case of the sub-segmentation. For example, for the *mdb*209 image we have a NU value of 0.0123 and 2266 pixels associated with the MCs for the k-Means, and 0.0181 and 3173 pixels for the FCM, which is a lot of pixels in each case. In order to reduce the NU value the number of segmented regions should be significantly increased. In the case of the sub-segmentation the NU value was 0.00025 and only 90 pixels. In this case, if we increase the threshold value α, also increases the number of pixels corresponding to the group of MCs. In the images where MCs are clearly differentiated from the rest of the objects is very easy to identify them with any segmentation method. As an example of such cases we have the *mdb*170 and *mdb*204 images, where the results are very similar between the different methods used. See Table 2. However, the results of the sub-segmentation have been better in all cases.

Table 2. Evaluation of segmentation results with NU

ROI	Numbers of pixels (MCs)			Value of NU		
	K-means	FCM	Sub-segmentation	K-means	FCM	Sub-segmentation
058	4947	3397	19	0.0432	0.0326	0.0022
170	174	137	137	0.0208	0.0193	0.0193
188	1774	1774	88	0.0286	0.0286	0.0029
204	261	247	247	0.0165	0.0134	0.0134
209	2266	3173	90	0.0123	0.0181	0.00025
219	1326	3303	123	0.1092	0.1389	0.0335
223	42	1694	71	0.0106	0.1902	0.054
227	4597	3047	27	0.0387	0.0276	0.00032
248	1195	2955	69	0.0289	0.0434	0.0058
249	2251	2251	81	0.0569	0.0569	0.0044

5 Conclusions

The segmentation can be substantially improved with the preprocessing of the images, as contrast enhancing filter as this achieves a greater separation between the objects of interest and the background of the image. As has been noted, the identification of MCs poses no particular problem when there is a high contrast among the objects and the algorithms obtain good results. However, if a filtered image results with a high homogeneity, the number of regions to be identified with the k-Means and FCM should be increased significantly, and the changes of the MCs region should be carefully observed as the value of NU decreases. In the case of the sub-segmentation, the identification of the MCs were easier as the MCs are the atypical pixels, and the method has been developed to find this type of data. Furthermore, a proper adjustment of the typicality threshold allows most appropriate adjustment of the algorithm, which offers the possibility to reduce the under or over-estimation of MCs with their respective consequences. Since the number of pixels corresponding to MCs is very small compared with the total pixels of the image, the value of NU is a low value and very close to zero in all cases. However, within the range of values obtained for NU it can be seen that the better identification of MCs is comparatively a value closer to zero. In forthcoming papers we will use this measure to adjust the parameters of the algorithms and establish a searching criterion for the best identification of objects in an image.

Acknowledgement.   The authors wish to thank the The National Council for Science and Technology (CONACyT) in México.

References

1. Pham, D.L., Xu, C., Prince, J.L.: A Survey of Current Methods in Medical Image Segmentation. Annual Review of Biomedical Engineering 2(4), 315–338 (2000)
2. Sampat, P.M., Markey, M.K., Bovik, A.C.: Computer-aided detection and diagnosis in mammography. In: Bovik, A.C. (ed.) Handbook of Image and Video Processing, 2nd edn., pp. 1195–1217. Academic Press, New York (2005)

3. Lee, N., Laine, A.F., Marquez, G., Levsky, J.M., Gohagan, J.K.: Potential of computer-aided diagnosis to improve CT lung cancer screening. IEEE Rev. Biomed. Eng. 2, 136–146 (2009)
4. Hernandez-Cisnero, R.R., Terashima-Marn, H.: Evolutionary neural networks applied to the classification of microcalcification clusters in digital mammograms. Proc. IEEE Congr. Evol. Comput., 2459–2466 (2006)
5. Tang, J., Rangayyan, R.M., Xu, J., El Naga, I., Yang, Y.: Computeraided detection and diagnosis of breast cancer with mammography: Recent advances. IEEE Trans. Inf. Technol. Biomed. 13(2), 236–251 (2009)
6. Bocciglione, G., Chainese, A., Picariello, A.: Computer aided detection of microcalcifications in digital mammograms. Comput. Biol. Med. 30(5), 267–286 (2009)
7. Gonzalez, R.C., Woods, R.E.: Digital Image Processing, 2nd edn. Publishing House of Electronics Industry, Beijing (2007)
8. Zhang, Y.J.: A survey on evaluation methods for image segmentation. Pattern Recogn. 29(8), 1335–1346 (1996)
9. MacQueen, J.B.: Some methods for classification and analysis of multivariate observations. In: Proc. 5th Berkeley Symposium on Mathematical Statistics and Probability, pp. 281–297. University of California Press, Berkeley (1967)
10. Bezdek, J.C., Keller, J., Krishnapuram, R., Pal, N.R.: Fuzzy models and algorithms for pattern recognition and image processing, Boston, London (1999)
11. Jain, A.K., Dubes, R.C.: Algorithms for Clustering Data. Prentice-Hall, Englewood Cliffs (1998)
12. Runkler, T.A.: Ant colony optimization of clustering models. Int. J. Intell. Syst. 20(12), 1233–1251 (2005)
13. Dunn, J.C.: A fuzzy relative of the ISODATA process and its use in detecting compact well-separated clusters. J. Cybernetics 3(3), 32–57 (1973)
14. Bezdek, J.C.: Pattern Recognition with Fuzzy Objective Function Algorithms. Kluwer Academic Publishers, Norwell (1981)
15. Pal, N.R., Pal, K., Keller, J.M., Bezdek, J.C.: A possibilitic fuzzy c-means clustering algorithm. IEEE T. Fuzzy Syst. 13(4), 517–530 (2005)
16. Dengler, J., Behrens, S., Desega, J.F.: Segmentation of Microcalcifications in Mammograms. IEEE T. Med. Imaging 12(4), 634–642 (1993)
17. Serra, J.: Images analysis and mathematical morphological. Academic Press, New York (1982)
18. Pal, N., Pal, S.: A review on image segmentation techniques. IEEE T. Fuzzy Syst. 13(4), 517–530 (1993)
19. Fu, J., Lee, S., Wong, S., Yeh, J., Wang, A., Wu, H.: Image segmentation feature selection and pattern classification for mammographic microcalcifications. Med. Imag. Grap. 29(6), 419–429 (2005)
20. Ojeda-Magaña, B., Quintanilla-Domínguez, J., Ruelas, R., Andina, D.: Images sub-segmentation with the PFCM clustering algorithm. In: Proc. 7th IEEE Int. Conf. Industrial Informatics, pp. 499–503 (2009)

Local Iterative DLT for Interval-Valued Stereo Calibration and Triangulation Uncertainty Bounding in 3D Biological Form Reconstruction*

José Otero and Luciano Sánchez

Computer Science Department
Oviedo University
33204 Gijon
{jotero,luciano}@uniovi.es

Abstract. The use of stereo vision for 3D biological data gathering in the field is affected by constrains in the position of the cameras, the quality of the optical elements and the numerical algorithms for calibration and matching. A procedure for bounding the 3D errors within an uncertainty volume is also lacking.

In this work, this is solved by implementing the whole set of computations, including calibration and triangulation, with interval data. This is in contrast with previous works that rely on Direct Linear Transform (DLT) as a camera model. To keep better with real lens aberrations, a local iterative modification is proposed that provides an on-demand set of calibration parameters for each 3D point, comprising the nearest ones in 3D space. In this way, the estimated camera parameters are closely related with camera aberrations at the lens area through which that 3D point is imaged.

We use real data from previous works in related research areas to judge whether our approach improves the accuracy of other crisp and interval-valued estimations without degrading the precision, and conclude that the new technique is able to improve the uncertainty volumes in a wide range of practical cases.

1   Introduction

The major obstacles to 3D data gathering for biological study include cost, availability and the lack of suitable instrumentation for field studies [1]. According to [2], only optical scanners can recover curved surfaces, and these devices have a high cost. Alternative devices for 3D data capture may involve constraints on positioning of the necessary instruments that are not suitable in the field.

Because of this, techniques from photogrammetry and computer vision are seldom applied to biological systems. Notwithstanding, in [3] the applicability

* This work was supported by the Spanish Ministerio de Economía y Competitividad under Project TIN2011-24302, including funding from the European Regional Development Fund.

Á. Herrero et al. (eds.), *International Joint Conference SOCO'13-CISIS'13-ICEUTE'13*,   309
Advances in Intelligent Systems and Computing 239,
DOI: 10.1007/978-3-319-01854-6_32, © Springer International Publishing Switzerland 2014

of these methods to reconstruct three-dimensional biological forms is discussed. The authors conclude that the precision of stereo vision methods improve photogrammetry by a large margin. Moreover, they assert that the lack of precision in some of their experiments was mainly related to two causes: a) a less precise camera calibration and b) the quality of the camera. To this it may be added that a rigorous procedure for estimating the accuracy of stereo vision-based measurements has not been established yet.

1.1   Sources of Uncertainty in Stereo Vision

The problems found by these authors are not exclusive from bioinformatics. Generally speaking, computer Vision Systems deal with measurement devices that, to a certain extent, are subjected to different kinds of errors. There are many works in this area that propose different models to understand how light refracts through camera lenses. The most simple one is a plain projective model called pin hole [4]. The estimation of the parameters of this model or "calibration" consists in capturing images of an object with several landmarks at known positions. Let the image coordinates of a 3D point be $w = (x, y, z)$, and let the projection of this point at the image be $m = (u, v)$; if a model f with parameters $p_1...p_n$ holds for this data, then $f(w, (p_1...p_n)) = m$. If the model parameters are unknown at least n equations are needed thus n correspondences between 3D points and 2D points are needed. The standard procedure uses more than n point correspondences, in order to partially cancel measurement errors, from which an overconstrained system of equations is obtained.

Real lens depart from the ideal pin-hole camera model and thus, more elaborated models are needed in order to compute where a 3D point is projected in a image when optical aberrations are taken into account. Some good sources of information about available general purpose models are books like [5], [6]. Classical models are Tsai [7], Zhang [8] and Direct Linear Transform (with several additional parameters that account for different kinds of lens imperfections) [9]. More recently, new models have been developed, intended for lens that are far apart ideal ones, like fish-eye lenses [10] [11] [12]. Finally, some papers are devoted to the so called "generalized camera calibration" framework [13], [14] and [11].

When a Computer Vision System comprises several cameras, each one calibrated using a suitable method, 3D points coordinates can be obtained from at least 2D projections on each camera, as in stereo or trinocular system [15]. This process is known as resection or triangulation. In some sense, the model of each camera must be inverted and the subset of 3D space that projects on each 2D point must be identified. Of course, from the geometry of the system, some additional constraints can be obtained that help to solve the problem of matching points between images (also called correspondence) [16], that is, finding pairs of points (one from each camera) that are 2D projection from the same 3D point.

An important issue when dealing with a Computer Vision System is the assessment of the accuracy of the provided measurements. There are some sources of imprecision that lead to measurements with unknown error bounds. This work proposes an improvement of [17] in order to handle real optics. Prior work and open problems are detailed in Section 2. In Section 3 the proposed approach is shown. Some numerical results are provided in 4 and in 5 conclusions and future work are mentioned.

2   Problem Statement and Related Work

Digital cameras have several sources of error and uncertainty:

- Lens imperfections lead to image distortions, blurriness, chromatic aberration and other defects. Usual camera models account for some of these issues, but others are intrinsic to optics.
- Digital images are discrete by definition. Intensity values at each pixel and colour channel and also sampled, usually along a grid evenly distributed in x and y axis. It is commonly assumed that the center of each pixel is the 2D projection of a given 3D point, however this is not entirely true.
- Camera sensor noise is noticed mostly when illumination conditions are not optimal. Due to its stochastic nature, it is not possible to fully remove noise from images.

There are several works in the literature that try to minimize the impact of these sources of error in computer vision systems. In recent years, a plethora of models that try to fit even the most extreme kinds of optics (like fish eye lenses) have emerged [10] [11] [12]. A totally different approach is to provide a generic framework for camera calibration decoupled from lens nature like in [13], [14], [18] and [11]. The effect of digital images quantization has been covered usually in connection with stereo, in works like [19] [20] [21] and [22] or (in lesser extent) [23]. The third source of error, image acquisition noise, has received some attention from Computer Vision comunity in works like [24] or [25].

None of the previously mentioned works tackle the problem of *bounding* the errors that propagate from all the reported sources to the final goal of a Stereo Computer Vision System: the 3D position estimation of a visible point in the scene. A few examples that try to give some information about the tolerance of 3D measurements with stereoscopic systems are [26], [27] and [17].

Based in the latter references, an approach to the problem will be proposed in Section 3 that builds a *local* interval valued estimation that encloses the 3D volume where the 3D triangulation of two 2D interval valued image coordinates are.

2.1   Interval-Valued Estimation of the Tolerance

In [17] both camera calibration and stereo triangulation (3D coordinates recovery from at least two pairs of matched 2D coordinates) are addressed as an interval valued problem. The authors propose the use of pinhole model in equation 1, where $w = (x, y, z)^T$ are 3D coordinates, $m = (u, v)^T$ 2D or image coordinates and $p_i, i = 1 \dots 3$ a 3×4 full-rank matrix shown in equation 2.

$$\begin{cases} u = \frac{p_1^T w + p_{1,4}}{p_3^T w + p_{3,4}} \\ v = \frac{p_2^T w + p_{2,4}}{p_3^T w + p_{3,4}} \end{cases} \quad (1) \qquad P = \begin{pmatrix} p_1^T & p_{1,4} \\ p_2^T & p_{2,4} \\ p_3^T & p_{3,4} \end{pmatrix} \quad (2)$$

The values of this matrix are obtained using equation 3 where there are eleven parameters ($p_{3,4} = 1$). In this equation one pair (u_i, v_i) (2D coordinates) is related with a triplet $w_i = (x, y, z)$ (3D coordinates), that is, for each 2D to 3D correspondence, two equations are obtained. From this follows that at least six 2D to 3D correspondences are needed in order to obtain a overconstrained system of equations. Usually more than six correspondences will be used, in order to overcome measurement errors.

$$\begin{pmatrix} w_i^T & 1 & 0 & 0 & -u_i w_i^T \\ 0 & 0 & w_i^T & 1 & -v_i w_i^T \end{pmatrix} \begin{pmatrix} p_1 \\ p_{1,4} \\ p_2 \\ p_{2,4} \\ p_3 \end{pmatrix} = \begin{pmatrix} u_i \\ v_i \end{pmatrix} \quad (3)$$

A stereoscopic system comprises two cameras, then the previous procedure applied to both cameras yields to different parameter matrices, P and P'. If two 2D coordinates (u, v) and (u', v'), one from each image, are known to be the projection of the same (but with unknown coordinates) 3D point w, then replacing the 2D coordinates and camera parameters for both images in equation 1 leads to the system of equations in 4 with only three unknowns, the 3D coordinates w, related with the known camera parameters and 2D coordinates from both images. This system can be solved using any suitable overconstrained system solution method.

$$\begin{pmatrix} (p_1 - u p_3)^T \\ (p_2 - v p_3)^T \\ (p_1' - u' p_3')^T \\ (p_2' - v' p_3')^T \end{pmatrix} w = \begin{pmatrix} -p_{1,4} + u p_{3,4} \\ -p_{2,4} + v p_{3,4} \\ -p_{1,4}' + u' p_{3,4}' \\ -p_{1,4}' + v' p_{3,4}' \end{pmatrix} \quad (4)$$

In [17], the authors consider that 2D pixel coordinates are discrete and thus that a rounding error of ± 0.5 pixels may occur, that is, pixels are rectangles instead of crisp points. If this is translated to equation 3, is obvious that a interval valued system of linear equations is obtained and thus, P values will be also intervals. The authors state that the numerical solution estimation of interval valued systems of equations using the approach in [28] is the tightest one for this problem. The estimated solution is the smallest hyperrectangle that contains the actual solution. Thus, replacing $p_{i,j}$, (u, v) and (u', v') crisp values with intervals in equation 4 will lead to a interval valued w, in this case a volume.

3   Local Iterative Calibration

A stated before, there are approaches in the literature that tackle optics non-linearities and other issues with an approach called generalized calibration [13], [14] and [11]. Basically, the most extreme case of such approaches handle the problem using a different set of parameters for each image pixel. Given that (usually) the measurement process does not involve all images pixels, we believe that a clever approach could be using a different *local* camera model for each pixel involved in the measurement process. In this way, the additional computational cost is devoted solely to the points required to compute the measurements of interest. In other words, the calibration process is no longer an off-line process but instead, inseparable of the triangulation process.

The procedure is iterative, the number of steps is configurable for each camera. In the following we restrict ourselves to two cameras, but the method is trivially extended to any camera number. For the sake of simplicity, we will first explain the *crisp* local iterative calibration procedure and then we will explain how to extend the method using the interval valued approach in [17].

Suppose that we want to obtain the 3D position (x, y, z) of a point which 2D projection for the right camera is (u^1, v^1) and (u^2, v^2) for the left camera. The first step consists on the usual DLT calibration procedure [9], using all the available calibration points. This leads to an initial, coarse, guess for (x, y, z) as $(\tilde{x}_0, \tilde{y}_0, \tilde{z}_0)$, not valid for usual optics.

With the obtained global model for each camera, the distance from the 3D point estimation $(\tilde{x}_0, \tilde{y}_0, \tilde{z}_0)$, obtained from (u^1, v^1), (u^2, v^2) to each one of the 3D calibration points can be computed. Now, we build a new calibration dataset that comprises the N (a parameter of the local calibration process) closest calibration points to $(\tilde{x}_0, \tilde{y}_0, \tilde{z}_0)$. Using this calibration dataset, a new model is obtained for each camera. Because the model is built using points that are closer to the actual position of the 3D point the new 3D triangulation of (u^1, v^1) and (u^2, v^2), $(\tilde{x}_1, \tilde{y}_1, \tilde{z}_1)$ will be more precise. The procedure can be refined and a new subset of calibration points can be computed again. This procedure can be repeated until the Euclidean distance between 3D triangulation results for (u^1, v^1) and (u^2, v^2) obtained in different successive steps $(\tilde{x}_i, \tilde{y}_i, \tilde{z}_i)$ and $(\tilde{x}_{i+1}, \tilde{y}_{i+1}, \tilde{z}_{i+1})$ falls bellow a threshold or after a pre-set number of iterations is completed.

The whole procedure is illustrated in figure 1, where the full calibration points set comprises the points marked with a cross. From that set, a DLT model can be obtained for each camera and for a point such as the one shown as a black dot in left and right images, a 3D position like 'A' can be obtained. After this, the N closest calibration points to the computed 3D position can be selected, in the figure those are surrounded by a circle. Using that subset of selected points, a new DLT model can be computed, tightly adjusted to the properties of the optics of each camera for the area of the image where the projection of the point to be triangulated lies. In this way, a new triangulated position for that point could be 'B' in the same figure and the same process repeated again.

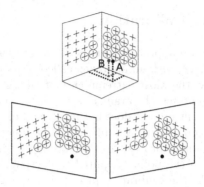

Fig. 1. Calibration points subset selection iterative procedure

Until this point, the *crisp* local calibration procedure has been explained. It is pretty straightforward to extend this method to interval valued data using the method in [17]. We only need to define a distance from interval valued 3D points to crisp 3D points (the calibration points). We will use the distance from the centroid of the interval valued 3D point to each crisp 3D calibration point. In this way the N closest 3D calibration points can be chosen from the whole set of available points. During each iteration, the interval valued procedure explained in section 2.1 will be applied.

4   Results

In this section, the proposed soft method for bounding the uncertainty of the triangulation is tested using real data, using "off the shelf" non-expensive cameras that clearly exhibit barrel distortion, thus the effect of the camera quality in the 3D data gathering is properly accounted for.

4.1   Experimental Setup

We choose to build a portable, detachable calibration object made of pipes and suitable junctions. With this set of elements, a sort of two wire-frame cubes

Fig. 2. Two views of the calibration object used in to calibrate several cameras. As can be seen, barrel distortion is noticeable.

composed structure was constructed. The dimensions of the cubes where 100 × 100 × 100 cm and the height of the structure was 200 cm. Placing that structure at different points on the floor, equally spaced in a grid with 100 cm. gaps, allowed us to cover the full 3D space of a room. In Figure 2 two pictures of the wireframe in the same position, taken from two cameras, can be seen. Barrel distortion can be noticed clearly at the left bottom part of the rightmost picture. Note also that the cameras are much more apart than recommended for this kind of application, in a worst-case scenario. In the left part of Figure 4 the obtained 3D points are shown. From the whole set of available 3D and 2D points that are visible from the two analyzed cameras, we use use 50% of them for calibration purposes and 50% of them for error measurement. Our error measurement for real data is uncertainty volume and gravity center of this volume to control points distance.

4.2   Numerical and Graphical Results

We will compare the size of the uncertainty volume for 3D test points, both for the interval valued standard approach and for the localized approach. This is intended to ascertain which of the approaches is more *precise* when dealing with real imagery where optical aberration is noticeably better than the original approach in [17].

In the left part of Figure 3 boxplots for uncertanty volumes obtained with both approaches are shown. There are significant differences. Using the closest calibration points to the actual 3D point being triangulated is more precise and hence the bounds are tighter.

The next experiment is related with *accuracy*. We will measure this property using the distance from each interval valued point gravity center to actual 3D point position.

In the right part of Figure 3, boxplots of this distances are shown for both approaches. In this case the differences are smaller but clearly there is an accuracy improvement with localized procedure.

In Figure 4 the interval valued estimation for all the test points is shown, using both approaches. In right part of that figure, an example of the interval valued results is shown. In that figure, the box enclosing the sphere that represents the true location of the triangulated point is obtained using the proposed localized approach, the other box is obtained using the standard interval valued approach. A statistical test (Wilcoxon signed rank test) was performed with both center of gravity to true location distance and volume data. Along with mean and standard deviation values the obtained p-values are shown in table 1. The obtained p-values are lower than 0.01 and thus the null hypothesis (there are no differences between results) can be rejected.

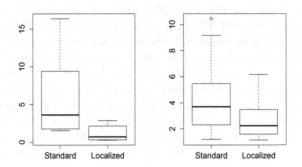

Fig. 3. Left:Uncertainty volume size boxplots for standard and localized approaches. Right:Distances from Interval Valued points gravity center to actual 3D positions. Our approach outperforms the standard approach using both metrics.

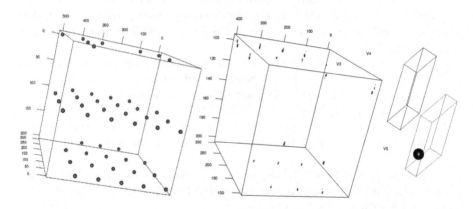

Fig. 4. Left:3D points obtained from several calibration object placements. Center: Interval valued estimation for all the test points using both approaches. Right: detail from a single point interval valued 3D triangulation.

Table 1. Numerical results from both approaches. First two rows: each cell contains the corresponding statistic for the column title measurement for both approaches, first standard interval valued approach, then localized approach. Last row: p-values for Wilcoxon signed rank test.

	CoG Distance	Uncert. Volume
Mean	4.289675, 2.36775	5.633381, 3.729344
SD	2.501068, 1.564342	4.742809, 5.18761
p-value	0.001158	0.008798

5   Concluding Remarks and Future Work

The use of stereo vision techniques for 3D biological data gathering is affected by restrictions in the positioning of the cameras, the distortion of the optical elements, the quantization of the sensors and the numerical algorithms of calibration and triangulation. In this work we have shown an improvement of the method proposed in [17], where an interval-valued procedure for camera calibration and point triangulation is shown. We had found that the mentioned work relies in the DLT calibration method and thus assimilates any camera to a simple pin-hole. Real optics are more complex than a pin-hole and, to some extent, exhibit some degree of optical aberrations, typically barrel type distortion. We have replaced the DLT method with a procedure that interleaves calibration and triangulation using an iterative algorithm. This algorithm selects the closest N calibration points to the actual position of the 3D point being triangulated. In this way, a localized DLT is computed for each point in each camera *on-demand*, that is, just when needed to perform a given measurement. This is also a key aspect of our approach when compared with the so called generalized calibration techniques, that use a different model for *each pixel* in the images. We have demonstrated in our experiments that this approach outperforms the standard Interval Valued approach in two important areas: size of the uncertainty volume (related with precision) and distance of the triangulation to the actual position of the 3D point (related with accuracy).

References

1. Zelditch, M.L., Swiderski, D.L., Sheets, H.D.: Geometric morphometrics for biologists: a primer (2012)
2. Claude, J., Paradis, E., Tong, H., Auffray, J.C.: A geometric morphometric assessment of the effects of environment and cladogenesis on the evolution of the turtle shell. Biological Journal of the Linnean Society (79), 485–501 (2003)
3. Chiari, Y., Wang, B., Rushmeier, H.: Using digital images to reconstruct threedimensional biological forms: a new tool for morphological studies. Biological Journal of the Linnean Society (95), 425–436 (2008)
4. Sonka, M., Hlavac, V., Boyle, R.: Image Processing, Analysis, and Machine Vision. Thomson-Engineering (2007)
5. Hanning, T.: High Precision Camera Calibration (2010)
6. Cyganek, B.: An Introduction to 3D Computer Vision Techniques and Algorithms. John Wiley & Sons (2007)
7. Tsai, R.Y.: A versatile camera calibration technique for high-accuracy 3D machine vision metrology using off-the-shelf TV cameras and lenses. IEEE Journal of Robotics and Automation, 221–244 (1992)
8. Zhang, Z.: A Flexible New Technique for Camera Calibration. IEEE Trans. Pattern Anal. Mach. Intell., 1330–1334 (November 2000)
9. Aziz, A.Y.I., Karara, H.M.: Direct linear transformation into object space coordinates in close-range photogrammetry. In: Proc. of the Symposium on Close-Range Photogrammetry, Urbana, Illinois, pp. 1–18 (1971)
10. Abrahama, S., Förstner, W.: Fish-eye-stereo calibration and epipolar rectification. ISPRS Journal of Photogrammetry and Remote Sensing 59(5), 278–288 (2005)

11. Gennery, D.B.: Generalized Camera Calibration Including Fish-Eye Lenses. Int. J. Comput. Vision 68(3), 239–266 (2006)

12. Schwalbe, E.: Geometric modelling and calibration of fisheye lens camera systems. In: Proceedings 2nd Panoramic Photogrammetry Workshop, Int. Archives of Photogrammetry and Remote Sensing, pp. 5–8 (2005)

13. Dunne, A.K., Mallon, J., Whelan, P.F.: Efficient generic calibration method for general cameras with single centre of projection. Computer Vision and Image Understanding 114(2), 220–233 (2010); Special issue on Omnidirectional Vision, Camera Networks and Non-conventional Cameras

14. Kannala, J., Brandt, S.S.: A generic camera model and calibration method for conventional, wide-angle, and fish-eye lenses. IEEE Transactions on Pattern Analysis and Machine Intelligence 28(8), 1335–1340 (2006)

15. Faugeras, O., Luong, Q.T., Papadopoulou, T.: The Geometry of Multiple Images: The Laws That Govern The Formation of Images of A Scene and Some of Their Applications. MIT Press, Cambridge (2001)

16. Scharstein, D., Szeliski, R.: A Taxonomy and Evaluation of Dense Two-Frame Stereo Correspondence Algorithms. Int. J. Comput. Vision 47(1-3), 7–42 (2002)

17. Fusiello, A., Farenzena, M., Busti, A., Benedetti, A.: Computing rigorous bounds to the accuracy of calibrated stereo reconstruction (computer vision applications). In: Image and Signal Processing, IEE Proceedings Vision, vol. 152(6), pp. 695–701 (December 2005)

18. Ramalingam, S., Sturm, P.: Minimal Solutions for Generic Imaging Models. In: IEEE Conference on Computer Vision and Pattern Recognition, CVPR 2008, Anchorage, Alaska, Etats-Unis. IEEE (June 2008)

19. Blostein, S.D., Huang, T.S.: Error Analysis in Stereo Determination of 3-D Point Positions. IEEE Transactions on Pattern Analysis and Machine Intelligence PAMI-9(6), 752–765 (1987)

20. Rodriguez, J.J., Aggarwal, J.K.: Stochastic analysis of stereo quantization error. IEEE Transactions on Pattern Analysis and Machine Intelligence 12(5), 467–470 (1990)

21. Kim, D.H., Park, R.H.: Analysis of quantization error in line-based stereo matching. Pattern Recognition 27(7), 913–924 (1994)

22. Balasubramanian, R., Das, S., Udayabaskaran, S., Swaminathan, K.: Quantization Error in Stereo Imaging systems. Int. J. Comput. Math. 79(6), 67–691 (2002)

23. Otero, J., Sánchez, L., Alcalá-Fdez, J.: Fuzzy-genetic optimization of the parameters of a low cost system for the optical measurement of several dimensions of vehicles. Soft Comput. 12(8), 751–764 (2008)

24. Kamberova, G., Bajcsy, R.: Sensor Errors and the Uncertainties in Stereo Reconstruction. In: Empirical Evaluation Techniques in Computer Vision, pp. 96–116. IEEE Computer Society Press (1998)

25. Ji, H., Fermüller, C.: Noise causes slant underestimation in stereo and motion. Vision Research 46, 3105–3120 (2006)

26. Mandelbaum, R., Kamberova, G., Mintz, M.: Stereo depth estimation: a confidence interval approach. In: Sixth International Conference on Computer Vision, 1998, pp. 503–509 (January 1998)

27. Egnal, G.: A stereo confidence metric using single view imagery with comparison to five alternative approaches. Image and Vision Computing 22(12), 943–957 (2004)

28. Shary, S.P.: Algebraic Approach in the "Outer Problem" for Interval Linear Equations. Reliable Computing 3(2), 103–135 (1997)

Classification of Chest Lesions with Using Fuzzy C-Means Algorithm and Support Vector Machines

Donia Ben Hassen[1], Hassen Taleb[1], Ismahen Ben Yaacoub[2], and Najla Mnif[2]

[1] LARODEC Laboratory, Higher Institute of Management, University of Tunis, Tunisia
donia_ben_hassen@yahoo.fr
[2] Medical Imaging Department, University Hospital Charles Nicolle, Tunisia

Abstract. The specification of the nature of the lesion detected is a hard task for chest radiologists. While there are several studies reported in developing a Computer Aided Diagnostic system (CAD), they are limited to the distinction between the cancerous lesions from the non-cancerous. However, physicians need a system which is significantly analogous to a human judgment in the process of analysis and decision making. They need a classifier which can give an idea about the nature of the lesion. This paper presents a comparative analysis between the classification results of the Fuzzy C Means (FCM) and the Support Vector Machines (SVM) algorithms. It discusses also the possibility to increase the interpretability of SVM classifier by its hybridization with the Fuzzy C method.

Keywords: Chest lesions, Clustering, Features, FCM, SVM.

1   Introduction

Radiography continues to be the most widely used imaging technique of the initial detection of chest diseases because of its low cost, simplicity, and low radiation dose. Though, uncertainty is widely present in data in this modality. Computer-assisted approaches may be helpful for handling this vagueness and as a support to diagnosis in this field. Therefore, the development of a reliable computer aided diagnosis (CAD) system for lung diseases is one of the most important research topics. Despite, lesion classification systems provide the fundation for lesions diagnosis and patient cure, the studies reported in developing a CAD application was limited to the distinction between the cancerous lesions from the non-cancerous. Physicians need a system which is significantly analogous to a human judgment in the process of analysis and decision making. The design of a classifier which can give an idea about the nature of the lesion, for example the lesion is of 50% an infection, 10% a cancer and 30% a tuberculosis etc... can help the radiologist to be suitable for handling a decision making process concerning.

To reach this goal, we propose a comparison study for chest lesions classification based on Fuzzy C Means (FCM) and Support Vector Machines (SVM) methods.

Á. Herrero et al. (eds.), *International Joint Conference SOCO'13-CISIS'13-ICEUTE'13*,
Advances in Intelligent Systems and Computing 239,
DOI: 10.1007/978-3-319-01854-6_33, © Springer International Publishing Switzerland 2014

The paper is organized as follows: after considering related works in section 1, section 2 describes the different classification systems and section 3 computerized schemes of our CAD system. Section 4 presents the results obtained on a real datasets.

2   Related Works

Many methods have been proposed in the literature for chest lesions classification and diagnosis utilizing a wide variety of algorithms. The majority of researches include the classification process under a description of whole Computer Aided Diagnosis systems. We can discern two main classes of studies concerning the classification of lesions in chest radiographs. The first class considers the classification process as distinction between true lesions and normal tissues in order improve radiologists' accuracy in detecting lung nodules. The work of [1] is an example. The second class adds to the first type of classification another one which distinguishes between the benign lesions and the malign ones. [2] proposed a system which automatically detects lung lesions from chest radiographs. The system extracts a set of candidate regions by applying to the radiograph three different multi-scale schemes. Support Vector Machines (SVMs), using as input different sets of features, has been successfully applied for the classification of chest lesions to benign and malign. [3] has used image processing algorithms for nodule classification to cancerous and non-cancerous tumors. We found that the majority of work in computer aided diagnosis systems in chest radiography has focused on lung cancerous nodule detection. Considering the load of lung diseases and the position of chest radiography in the diagnostic workflow of these diseases, we could argue that the classification of other type of lesions such as tuberculosis should receive more attention. We think also that an adapted framework for extraction of the adopted features describing the different kinds of lesions is the missing part of the methods presented in the literature.

3   Description of the Classification System

The classification problem has been addressed in many computing applications such as medical diagnosis. In the literature we can find the term grouping or clustering instead of classification. Although these two terms are similar in terms of language, they are technically different. In fact, classification is a discriminant analysis that uses a supervised approach where we are provided with a collection of labeled data; the problem is to label a newly unlabeled data. Usually, the given training data is used to learn the descriptions of classes which in turn are used to label a new data. Among these methods are bayes classifier, artificial neural networks, deformable models and support vector machines. Normally, clustering always refers to unsupervised framework which its problem is to group a given collection of unlabeled patterns into meaningful clusters. There is a whole family of unsupervised methods, including probabilistic ones, fuzzy ones, evidential ones. Especially, there are two variants of unsupervised classifiers: Classifiers are known as hard methods and fuzzy methods. The commonly used fuzzy clustering methods are: FCM Fuzzy C- Means and its variants.

3.1 Fuzzy C-Means Algorithm

In this context, the fuzzy diagnosis concept is widely applied [4]. Fuzzy classifiers have been proposed to deal with classification tasks in presence of uncertainty. Fuzzy c-means (FCM) is one of these methods. It is an algorithm of clustering which allows one piece of data to belong to two or more clusters. This method (developed by Dunn [5] and improved by Bezdek[6]) is frequently used in pattern recognition.

The fuzzy c-means algorithm is an extension of the classic k-means algorithm [7] based on the minimization of the following objective function:

$$J_m = \sum_{i=1}^{N} \sum_{j=1}^{C} u_{ij}^m \left\| x_i - c_j \right\|^2 \ , \ 1 \leq m \leq \infty \tag{1}$$

Where N is the number of data points, C represents the number of cluster center, m is any real number greater than 1, u_{ij} is the degree of membership of x_i in the cluster j, x_i is the ith of d-dimensional measured data, c_j is the d-dimension center of the cluster, and $\|*\|$ is any norm expressing the similarity between any measured data and the center.

Fuzzy partitioning is carried out through an iterative optimization of the objective function shown above, with the update of membership u_{ij} and the cluster centers c_j by:

$$u_{ij} = \frac{1}{\sum_{K=1}^{C} \left(\frac{\|x_i - c_j\|}{\|x_i c_k\|} \right)^{\frac{2}{m-1}}} \quad , \ C_j = \frac{\sum_{i=1}^{N} u_{ij}^m \cdot x_i}{\sum_{i=1}^{N} u_{ij}^m} \tag{2}$$

This iteration will stop when $max_{ij}\left\{ \left| u_{ij}^{(k+1)} - u_{ij}^{(k)} \right| \right\} < \varepsilon$, where ε is a termination criterion between 0 and 1, whereas k is the iteration number. This procedure converges to a local minimum or a saddle point of J_m.

With the work of Bezdek, fuzzy clustering methods attain a certain maturity [8]. Other variants of these algorithms were then developed in order to increase performance. These improved versions are often dedicated to a particular application, and still the FCM are generally useful in many situations. Some variants of FCM have been proposed to reduce the influence of data points which do not belong to a cluster.

Recently, many researchers have brought forward new methods to improve the FCM algorithm [9]. The most popular is: Possibilistic C-Means algorithm.

The Fuzzy C means algorithm can provide faster approximate solutions that are suitable for the treatment of issues related to understandability of models and incomplete and noisy data. This makes the technique effective in classification and very close to reasoning of physicians.

3.2   SVM Classification

Support vector machines (SVM) represent a classifier that has been successfully used for chest lesions classification. Moreover, we possess labeled data and it is expected that a supervised classifier achieves a good accuracy.

Let a set of data $(x_1, y_1), \ldots, (x_m, y_m) \in \mathfrak{R}^d \times \{\pm 1\}$ where X={ x_1, x_m} a dataset in $\mathfrak{R}^d$ where each x_i is the feature vector of an image. In the nonlinear case, the idea is to use a kernel function $k(x_i, x_j)$, where $k(x_i, x_j)$ satisfies the Mercer conditions [10]. Here, we used a Gaussian RBF kernel whose formula is:

$$k(x, x') = \exp\left[-\left\| x - x' \right\|^2 \Big/ 2\gamma^2 \right] \tag{3}$$

Where $\|.\|$ indicates the Euclidean norm in $\mathfrak{R}^d$.

Let Ω be a nonlinear function which transforms the space of entry $\mathfrak{R}^d$ to an intern H called a feature space. Ω allows to perform a mapping to a large space in which the linear separation of data is possible [11].

$$\Omega: \mathfrak{R}^d \longrightarrow H$$
$$(x_i, x_{ji}) \longmapsto \Omega(x_i)\Omega(x_j) = k(x_i, x_j) \tag{4}$$

The H space is reproducing Kernel Hilbert space (RKHS). Thus, the dual problem is presented by a Lagrangian formulation as follows:

$$\max W(\alpha) = \sum_{i=0}^{m} \alpha_i - \frac{1}{2} \sum_{i,j=1}^{m} y_i y_j \alpha_i \alpha_j k(x_i, x_j), i = 1, \ldots, m \tag{5}$$

Under the following constraints:

$$\sum_{i=1}^{m} \alpha_i y_i = 0, 0 \leq \alpha_i \leq C \tag{6}$$

They α_i are called Lagrange multipliers and C is a regularization parameter which is used to allow classification errors. The decision function will be formulated as follows:

$$f(x) = \text{sgn}(\sum_{i=1}^{m} \alpha_i y_i k(x, x_i) + b) \tag{7}$$

We hence adopted one approach of multiclass classification: One-against-One. This method consists of creating a binary classification of each possible combination of classes, the result for K classes K (K -1) / 2 .

4   Computerized Scheme for Classification of Lung Lesions

Here, we only describe the final stage of our CAD system: The segmentation and the feature selection are discussed in our previous works [12] and [13] that are briefly presented in the next sections.

4.1   Preprocessing

Preprocessing may lead to better results because it allows more flexibility. For example, the contrast feature is almost meaningless in discriminating normal and lesion pixels, since lesions can occur in regions of both high and low contrast. However, combining this feature with image enhanced intensity can be used for much more effective discrimination.

4.2   Segmentation

We closely follow our previous work described in [12] which is an automatic chest radiography segmentation framework with integration of spatial relations. The algorithm developed as the initial step of our system works under no assumption. The results obtained proving that this is an excellent initialization step for a CAD system aimed at lung lesions detection and recognition of their sites. The segmented lung area includes even those parts of the lungs which are usually excluded from the methods presented in the literature. This decision is motivated by the recognition of the site of the lesion which can help in deducing its nature for example, "lesion is in the right apical" that means that the lesion is localized in the right upper lobe it's an infection. The segmented area is processed with a method that enhances the visibility of the lesions, and an extraction scheme is then applied to select potential features.

4.3   Feature Extraction and Selection

The purpose of features extraction and selection is to reduce the original data set by measuring certain properties that distinguish one input pattern from another pattern. The extracted feature should provide the characteristics of the input type to the classifier by considering the description of relevant properties of the image into a feature space. We believe that the calculation of features is a primordial step to well perform the task of classification. In fact, each pixel should have characteristics used for the differentiation between the lesion and the normal pixels nevertheless for the discrimination between malign and benign lesions. A great variety of features can be computed for each image such that intensities and textures depending on the nature of problem. The description of features in works cited above is not very detailed. Almost features cited in the literature are classical. Between them those of Haralick, were used without really giving details about their meaning. However, it is difficult to interpret what these features are. Based on characteristics given by service of medical imaging of CHU Charles Nicolle (described in table 1), we selected 8 features (size, circularity, x-fraction, y-fraction, skewness, kurtosis, homogeneity, correlation)   that we believe are able to specify the nature of the lesion.

Table 1. Characteristics of principals lesions for radiologists

LESION	METASTASIS	BENIGN TUMORS	MALIGN TUMORS	TUBERCULOSIS	INFECTION
UNIQUE	Exceptional	indifferent	Very common		
MULTIPLES	Very common	indifferent	rare	Common	
LOCALISATION	Basal			Apical	
CONTOURS	sharp	sharp	sharp	blurred	blurred
FORME	Rounded, ovalaire	Rounded, ovalaire polylobed	spiculated	Ill-defined	Ill-defined or fissural limit
OPACITY	homogeneous	homogeneous, dense	heterogeneous +/- excavated	heterogeneous +/- excavated	homogeneous
CALCIFICATIONS	rare except metastasis of sarcoma	frequent central lobulated	rare	Frequent in sequelae stage	Very rare
SIZE		< 1cm	> 1 cm		

5   Experimental Study

The chest radiographs are taken from the JSRT database [14]. This is a publicly available database with 247 Posterior Anterior chest radiographs. 154 images contain exactly one pulmonary lung lesion. The other 93 images contain no lung lesions.

First of all, we defined the number of clusters. We discussed with our collaborators in the service of medical imaging of the university hospital Charles Nicolle and we concluded the five important clusters which are: Lung cancer, Metastasis, Tuberculosis, Infection, Benign tumors.

Table 2. Classes of diseases and number of samples in the database used for performance Evaluation

Classes	Train	Test	Total number
Cancer	61	30	91
Metastasis	6	3	9
Infection	16	8	24
Tuberculosis	12	6	18
Benign tumor	5	3	8

The obtained feature vectors passed for the classification phase by using FCM and SVM's.

In the experiments of FCM, we have used a fuzziness coefficient m = 2.

Step 1. Our dataset contains samples of features belonging to five diseases.

Step 2. The data to be clustered is 8-dimensional data and represents features cited above. From each of the Five groups (Lung cancer Metastasis Tuberculosis Infection Benign tumors), two characteristics (for example, X-fraction vs. Y-fraction as shown in fig.1) of the lesion are plotted in a 2-dimensional plot.

Step 3. Next, the parameters required for Fuzzy C-Means clustering such as number of clusters, exponent for the partition matrix, maximum number of iterations and minimum improvement are defined and set.

Step 4. Fuzzy C-Means clustering is an iterative process. First, the initial fuzzy partition matrix is generated and the initial fuzzy cluster centers are calculated (show the centers in magenta in Fig. 1). In each step of the iteration, the cluster centers and the membership grade point are updated and the objective function is minimized to find the best location for the clusters. The process stops when the maximum number of iterations is reached, or when the objective function improvement between two consecutive iterations is less than the minimum amount of improvement specified.

Table 3 presents the results obtained with fuzzy C Means algorithm.

In table 3 also, we present the results obtained with SVM classifier with parameters C, γ ($2^{(1)}$, $2^{(-7)}$) settings of Gaussian RBF kernel. We have used the grid search which searches the optimal parameters values using cross validation. After learning phase, we test the test data.

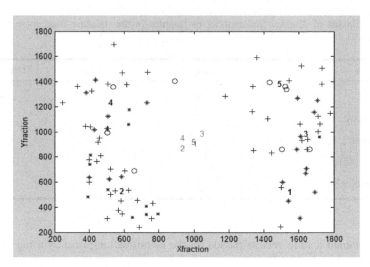

Fig. 1. The two characteristics (X-fraction vs. Y-fraction) of the lesion are plotted in a 2-dimensional plot

Table 3. Performances of FCM and SVM classifier

Classes	Acccuracy (%)		SE(%)		SP(%)	
	FCM	SVM	FCM	SVM	FCM	SVM
Cancer	70.33	74.19	73.00	80.00	71.33	60.00
Metastasis	60.25	50.94	63.88	66.66	61.56	50.00
Infection	55.66	51.72	52.76	50.00	53.00	52.00
Tuberculosis	60.23	56.56	55.85	53.33	57.00	51.88
Benign tumors	53.33	52.00	54.67	51.45	53.89	50.00

We judge the performance of our classification approach using several evaluation criteria often used in the literature. For a five class clustering problem, one can distinguish true positive (TP) (sample correctly classified), false positive (FP) (false sample classified as true sample), false negative (FN) (false sample classifier as false sample), and true negative (TN) (false sample classified as true sample). From these values, measures such as accuracy, sensitivity (SE) and specificity (SP) can be computed given by the following equations.

$$SE = \frac{TP}{TP+FN} \tag{8}$$

$$SP = \frac{TN}{TN + FP} \tag{9}$$

$$Accuracy = \frac{TP + TN}{TP + TN + FP + FN} \tag{10}$$

The results in table 3 show that the clusters are identified well by the FCM algorithm.

The clusters that contain more errors are those that contain fewer samples. The lung cancer is the cluster which is more clearly identified by the FCM algorithm (achieved an averaged accuracy rate of the order 70.33 % for cancer).

The obtained results by SVM are satisfactory. Indeed, we reached a recognition rate of the order 74.19% for the cancer. We remark that the classification rate is less in the other classes because the fewer number of the train and the test data.

We can conclude that the SVM can achieve better accuracy in our classification problem. However, the two methods cannot represent clusters of small size.

The Fuzzy C Means Algorithm uses a fuzzy clustering, in which the input vector x is pre-classified with different membership values. The outputs of the FCM algorithm may present the input vector to the SVM classifier. This last will be used for the automatic recognition of disease.

6   Conclusion and Future Works

The intelligibility is the motive force behind the use of FCM algorithm for this problem. However, a compromise between interpretability and accuracy is met. On the other hand, we focused on a more accurate solution by using SVM. Then we risk losing the linguistic sense defining the fuzzy models. Indeed, we have experiment also the possibility to increase the interpretability of SVM classifier by the hybridization with the clustering method Fuzzy C means.

References

1. Hardie, R., Rogers, S., Wilson, T., Rogers, A.: Performance analysis of a new computer aided detection system for identifying lung nodules on chest radiographs. Medical Image Analysis 12(3), 240–258 (2008)
2. Campadelli, P., Casiraghi, E., Valentini, G.: Lung nodules detection and classification. In: ICIP (1), pp. 1117–1120 (2005)
3. Nehemiah, H., Kannan, A.: An intelligent system for lung cancer diagnosis from chest radiographs. International Journal of Soft Computing, 133–136 (2006)
4. Masulli, F., Schenone, A.: A fuzzy clustering based segmentation system as support to diagnosis in medical imaging. Artificial Intelligence in Medicine 16, 129–147 (1999)
5. Dunn, J.C.: A Fuzzy Relative of the ISODATA Process and Its Use in Detecting Compact Well-Separated Clusters. Journal of Cybernetics 3, 32–57 (1973)
6. Bezdek, J.C.: Pattern Recognition with Fuzzy Objective Function Algorithms. Plenum Press, New York (1981)
7. Lesot, M.J., Bouchon-Meunier, B.: Descriptive concept extraction with exceptions by hybrid clustering. In: Proc. of Fuzz-IEEE 2004, pp. 389–394. IEEE Comp. Intell. Society, Budapest (2004)
8. Khodja, L.: Contribution à la classification floue non supervisée. Thesis. Savoie University, France (1997)

9. Gomathi, M., Thangaraj, P.A.: New Approach to Lung Image Segmentation using Fuzzy Possibilistic C-Means Algorithm. International Journal of Computer Science and Information Security 7 (2010)

10. Vapnik, V., Chapelle, O.: Bounds on error expectation for support vector machines. Neural Computation 12 (2000)

11. Scholkopf, B., Smola, A.: Learning with Kernels. MIT Press (2001)

12. Ben Hassen, D., Taleb, H.: A fuzzy approach to chest radiography segmentation involving spatial relations. IJCA Special Issue on "Novel Aspects of Digital Imaging Applications", 40–47 (2011)

13. Ben Hassen, D., Taleb, H.: Automatic detection of lesions in lung regions that are segmented using spatial relations. Clinical Imaging (2012) (in press)

14. Ginneken, B.V., Stegmann, M.B., Loog, M.: Segmentation of anatomical structures in chest radiographs using supervised methods: a comparative study on a public database. Medical Image Analysis 10, 19–40 (2006)

RBF Neural Network for Identification and Control Using PAC

Ladislav Körösi, Vojtech Németh, Jana Paulusová, and Štefan Kozák

Institute of Control and Industrial Informatics, Faculty of Electrical Engineering
and Information Technology, Slovak University of Technology,
Ilkovičova 3, 812 19 Bratislava, Slovak Republic
{ladislav.korosi,xnemethv,jana.paulusova,stefan.kozak}@stuba.sk

Abstract. In this paper the implementation of RBF online learning algorithm on the Schneider Electric Quantum programmable automation controllers is proposed. Online recursive mean square algorithm with different modifications is proposed for neural network parameter identification. All matrix operations, functions, algorithms and neural network general structure are programmed in the Structured Text programming language in Unity Pro XL software. The proposed method and software implementation is verified on virtual hydraulic system with parameter identification and level control.

Keywords: Neural network, RBF, online learning, programmable automation controllers.

1 Introduction

The most commonly used control systems in industry are programmable logic controller (PLC) or programmable automation controllers (PAC). PAC is a compact controller that has the characteristics of PLC and control systems running on PC. PAC is used generally for continuous control, data collection, remote monitoring, etc. They are able to provide data for process-level for application software and databases. PLC and PAC have the same objective, but it's possible to program them differently. PLC is programmed mostly in the ladder diagram (ladder logic) and PAC can be programmed in more languages (also in higher languages). [10] Neural networks are currently used in various approaches such as signal processing, image recognition, natural speech recognition, identification and others but they are missing from PLC libraries (instruction sets). [4], [6] The proposed paper has the main objective to demonstrate the real deployment of Radial basis function (RBF) neural network for online learning and control in industrial practice. Online learning algorithm is demonstrated on a virtual laboratory hydraulic system implemented directly in PAC.

Á. Herrero et al. (eds.), *International Joint Conference SOCO'13-CISIS'13-ICEUTE'13*,
Advances in Intelligent Systems and Computing 239,
DOI: 10.1007/978-3-319-01854-6_34, © Springer International Publishing Switzerland 2014

2   RBF Neural Network

RBF neural network is an artificial neural network that uses radial basis functions as activation functions in hidden layer. The output of the network is a linear combination of radial basis functions of the hidden layer and neuron parameters. [1], [2], [9] The output of the network is a scalar function of the input vector:

$$\varphi(x) = \sum_{i=1}^{N} a_i \left(\|x - c_i\| \right), \tag{1}$$

Where N is the number of neurons in the hidden layer, c_i is the center vector for neuron i, and a_i is the weight of neuron i in the linear output neuron. The norm is typically the Euclidean distance and the radial basis function is commonly Gaussian functions.

$$\varphi_i(x) = \exp\left(-\frac{\|x - c_i\|^2}{2\sigma^2} \right), \tag{2}$$

Where c_i and σ_i is the center and spread of the i-th neuron.

3   Recursive Mean Square Training Algorithm

The basic recursive mean square algorithm (RMSA) with different modifications (exponential forgetting factor, etc.) is used to train the RBF NN in PAC system. It's an online learning method suitable for real-time applications. The basic RMSA can be summarized as follows:

$$\varepsilon\left(k, \hat{W}(k-1)\right) = y_p(k) - \phi^T(k)\hat{W}(k-1) \tag{3}$$

$$K(k) = \frac{P(k-1)\phi(k)}{1 + \phi^T(k)P(k-1)\phi(k)} \tag{4}$$

$$P(k) = P(k-1) - K(k)\phi^T(k)P(k-1) \tag{5}$$

$$\hat{W}(k) = \hat{W}(k-1) + K(k)\varepsilon\left(k, \hat{W}(k-1)\right) \tag{6}$$

The RMSA minimizes the criteria function (3). For the identification of time-invariant systems, however, older measured values do not correspond to the current system; therefore this criterion is modified so that the older state values weights will have less weight in a criteria function. [5] The criteria function (3) and the update of weights (6) is the same as in basic RMSA.

$$J_N' = \frac{1}{2} \sum_{i=1}^{k} \lambda^{k-1} \varepsilon^2(i,W) \tag{7}$$

$$K(k) = \frac{P(k-1)\phi(k)}{\lambda + \phi^T(k)P(k-1)\phi(k)} \tag{8}$$

$$P(k) = \frac{P(k-1) - K(k)\phi^T(k)P(k-1)}{\lambda} \tag{9}$$

Forgetting is the faster when the value of λ is lower. If $\lambda = 1$, we will get the standard RMSA.

4   Fuzzy c-Means Algorithm

We proposed the Fuzzy c-means clustering (FCM) algorithm to find the optimal centers and spreads for the RBF neural network. This approach reduces the computation complexity for the online training algorithm in PAC system. In fuzzy clustering data elements can belong to more than one cluster, and associated with each element is a set of membership levels. These indicate the strength of the association between that data element and a particular cluster. Fuzzy clustering is a process of assigning these membership levels, and then using them to assign data elements to one or more clusters. [8] Disadvantage of this method is that the results depend on the initial choice of weights. The algorithm was used for offline RBF parameter setup. The algorithm was designed in Matlab.

5   Case Study

Process Description

The proposed methods are illustrated on a virtual hydraulic system for online modeling and control. Virtual model is based on real laboratory model (Fig. 1). The laboratory model is described in more detail in [7].

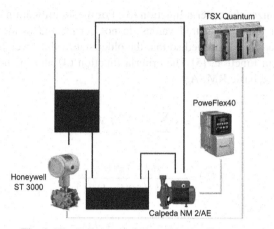

Fig. 1. Block diagram of the real hydraulic system

Consider an open tank with the inlet flow rate M_1 and output flow rate M_2, where the atmospheric pressure acts on water level (Fig. 2)

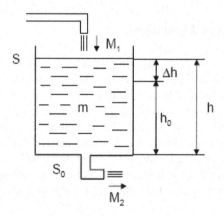

Fig. 2. Hydraulic system with free outlet (S – water surface area [m²], S_0 – cross sectional area of the hole [m²], m – mass of the liquid in the tank [kg], M_i -mass flow of fluid at the inlet and outlet of the tank [kg.s⁻¹], h - water level [m])

The hydraulic model is based on the law of conservation of matter:

$$\frac{dm}{dt} = \rho \frac{dV}{dt} = M_1 - M_2, \tag{8}$$

$$\frac{dV}{dt} = Q_1 - Q_2, \tag{9}$$

Where V is the volume of liquid in the tank [m³], ρ is the density of the liquid [kg.m³], Q_i is the volumetric liquid flow rate at the inlet and outlet of the tank [m³.s⁻¹].

We consider $S = const.$

$$S\frac{dh}{dt} = Q_1 - Q_2 \tag{10}$$

For water flow rate:

$$v_2 = \sqrt{2gh} , \tag{11}$$

Where g is the gravitational acceleration $= 9.81$ ms^{-2}.
   Then

$$Q_2 = S_2 v_2 = \mu S_0 \sqrt{2gh} , \tag{12}$$

Where μ is the coefficient of outflow $\in (0, 1)$, for water $\mu = 0.63$, S_2 is the cross beam of the outflowing fluid [m^2].

Virtual Model Description

The following relation describes the difference of the water level in the tank.

$$\frac{dh}{dt} = \frac{1}{S}\left(Q_1 - \mu S_0 \sqrt{2gh}\right) \tag{13}$$

Idealized virtual model (no dynamic of the pump was considered) was implemented in Unity Pro using *Structured text* programming language in discrete form. The parameters of the virtual model were:

$S = 1$m^2, $s_0 = 0.01$m^2, $T_s = 1$s, where T_s is the sampling time.

Virtual Process Modeling Using RBF NN

To verify the proposed solutions we used the NNARX (neural network auto-regressive model with external input) model structure with a regression vector:

$$y_m(k) = [y(k-1)...y(k-n_a) \quad u(k-n_k)...u(k-n_b-n_k+1)]^T \tag{14}$$

In the first example the RBF NN had 5 input neurons (n_a - 4 past level measurements and n_k - 1 past control action), 25 hidden neurons and one output neuron. The comparison of the time responses of the virtual model and the online trained RBF NN in PAC system are show in Fig. 3 and Fig. 4.
   Increasing the number of input neurons of the RBF neural network improved the quality of the modeling for training samples. RBF NN with 6 input neurons (5 past level measurements and 1 past control action), 25 hidden neurons and one output neuron had the MSE for testing data 0.0000077851 and training data 0.0000075575.

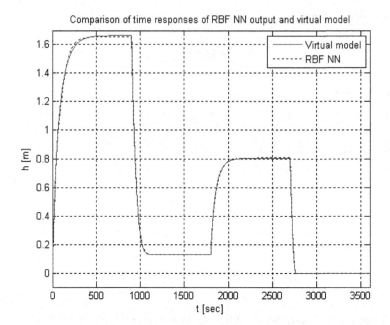

Fig. 3. Comparison of time responses of RBF NN output and virtual model output – training data (Mean square error – MSE = 0.000084498)

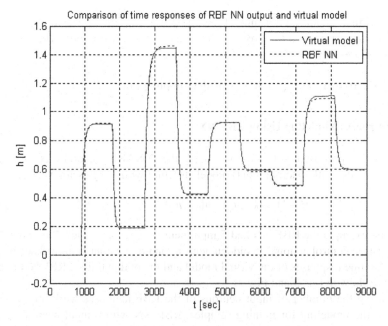

Fig. 4. Comparison of time responses of RBF NN output and virtual model output – testing data (MSE = 0.000070663)

Virtual Process Control Using RBF NN

In this article the RBF NN is used as process model to predict the behavior of the water level according to the control signal.

Standard criteria function includes square of the deviation and control increment:

$$J_r = \frac{1}{2}\alpha(u(k)-u(k-1))^2 + \frac{1}{2}[r(k+1)-y_m(k+1)]^2, \qquad (15)$$

where y_m is the output of the ANN.

The optimization block computes the control signal so that the predicted output of the RBF NN matches the process output. [3] This is an iteration process which has the form:

$$u(k)_{new} = u(k)_{old} - \beta\frac{\partial J_r'}{\partial u(k)_{old}} \qquad (18)$$

The results for different prediction horizons (affecting different control quality) are shown in the following figures. The parameters for all simulation were:

- sampling time: 1sec,
- reference signal: 0.5m,
- RBF NN structure: 5 - 25 – 1,
- number of control iterations: 1,
- $\alpha = 0.01$,
- $\beta = 1$.

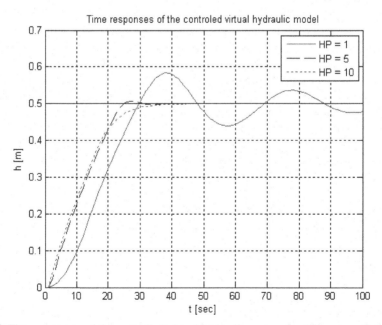

Fig. 5. Time responses of the controlled virtual hydraulic system for different prediction horizons 1, 5 and 10 (r = 0.5m)

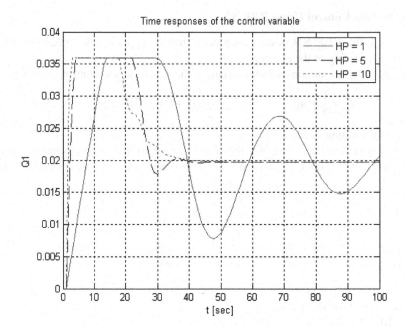

Fig. 6. Time responses of the control variable for different prediction horizons 1, 5 and 10

6   Conclusion

In the presented paper we deal with implementation of PAC based online RBF neural network learning and control. Based on theoretical assumptions about RBF artificial neural networks we proposed and implemented a general RBF neural network structure. Using different online learning methods of RBF NN, we verified the implemented optimal structure with the standard online learning algorithm (RMNS) on various input parameters of neural networks. It can be concluded, that in all cases the RBF NN proved sufficiently accurate approximation of the output signal of the hydraulic process. Proposed application is very convent for real time application of soft computing methods based on RBF NN structures. Based on the verification results is evident that we can achieve high performance for many industrial processes.

Acknowledgement. This paper was supported by the Slovak Scientific Grant Agency VEGA, Grant no.1/1105/11.

References

1. Oravec, M., Polec, J., Marchevský, S.: Neural networks for digital signal processing. FABER, Bratislava (1998) (in Czech)
2. Haykin, S.: Neural networks – A Comprehensive Foundation. Macmillan College Publishing Company, New York (1994)

3. Körösi, L., Kozák, Š.: Optimal Self Tuning Neural Network Controller Design. In: 16th IFAC World Congress, Praha, Czech repubic (2005)
4. Jadlovská, A.: Modeling and control of dynamic processes using neural networks, Košice (2003) ISBN 80-88941 -22 -9 (in Slovak)
5. Körösi, L.: Contribution to the problem of optimizing the structures of artificial neural networks. Institute of Control and Industrial Informatics, FEI STU, Bratislava, 126 p. (2010) (in Slovak)
6. Körösi, L.: Neural Network Modeling and Control Using Programmable Logic Controller. Posterus, 4(12) (2011), ISSN 1338-0087, http://www.posterus.sk/?p=12304
7. Körösi, L., Kelemen, J.: Implementation of Orthogonal Neural Network Learning on PLC. In: Kybernetika a Informatika, 20th International Conference SSKI and FEI STU, Skalka pri Kremnici, Bratislava, Januay 31-February 4, pp. 87–88 (2012)
8. Paulusová, J., Dúbravská, M.: Neuro Fuzzy Predictive Control. International Review of Automatic Control 5(5), s. 667–s. 672 (2012) ISSN 1974-6059 - ISSN 1974-6067
9. Jadlovská, A., Jajčišin, Š.: Using Neural Networks for Physical Systems Behavior Prediction. In: AEI 2012 Conference: Applied Electrical Engineering and Informatics: Proc. FEI TU, Košice (2012) ISBN 978-80-553-1030-5
10. Körösi, L., Mrafko, L., Mrosko, M.: PLC and their Programming - 2. PLC, PAC, DCS – Who Whom? Posterus 4(10) (2011), ISSN 1338-0087, http://www.posterus.sk/?p=11925

Analysis of the Effect of Clustering
on an Asynchronous Motor Fuzzy Model

Pavol Fedor, Daniela Perdukova, and Mišél Batmend

Technical University of Košice, Faculty of Electrical Engineering and Informatics,
Department of Electrical Drives and Mechatronics, Letná 9, 042 00 Košice
{pavol.fedor,daniela.perdukova,misel.batmend}@tuke.sk

Abstract. The paper deals with the analysis of the effect of fuzzy clustering parameters on the setup of a good quality fuzzy model of a complex nonlinear system, such as, for example, an asynchronous motor. On basis of data measured from an analytical model of an asynchronous motor we developed its fuzzy model and then verified the quality of the courses of its individual state quantities in response to the selected input signal. The proposed procedure was applied in the development of a high quality dynamic fuzzy model of an asynchronous motor. Its properties were verified by simulation in the MATLAB programme package.

Keywords: Fuzzy modelling, fuzzy clustering, asynchronous motor.

1   Introduction

Fuzzy models of technological equipment and processes are becoming more widely employed in the application of intelligent process control[1]-[4]. The development of a good quality model of a nonlinear dynamic system only on basis of measured input-output relations is in principle possible [5]-[6], however in some states such models may be very inaccurate and even unstable. This can be due to an insufficient and inconsistent database of acquired data, or its unsuitable processing into qualitative information about the system being modeled.

An asynchronous motor presents a typical example of a nonlinear system in which fuzzy model quality is to a significant extent dependent on the method of processing of the database of measured data, as this is a considerably oscillating system that also includes positive feedbacks[7]-[12].

The paper describes the process of development of an asynchronous motor fuzzy model; it specifies information that is important for the proposal of its fuzzy structures and deals with the verification of its quality on the courses of its internal modelled quantities.

Á. Herrero et al. (eds.), *International Joint Conference SOCO'13-CISIS'13-ICEUTE'13*,
Advances in Intelligent Systems and Computing 239,
DOI: 10.1007/978-3-319-01854-6_35, © Springer International Publishing Switzerland 2014

2   Method of Asynchronous Motor Fuzzy Model Development

A squirrel cage asynchronous motor can be described in a rotating x,y system for example by the following equations [13]-[14]:

$$i_{1x} = K_{11} \cdot \psi_{1x} - K_{12} \cdot \psi_{2x} i_{1y} = K_{11} \cdot \psi_{1y} - K_{12} \cdot \psi_{2y} \tag{1}$$

$$i_{2x} = K_{13} \cdot \psi_{2x} - K_{12} \cdot \psi_{1x} i_{2y} = K_{13} \cdot \psi_{2y} - K_{12} \cdot \psi_{1y} \tag{2}$$

$$\frac{d\psi_{1x}}{dt} = u_{1x} - R_1 . i_{1x} + \omega_k \cdot \psi_{1y} \tag{3}$$

$$\frac{d\psi_{1y}}{dt} = u_{1y} - R_1 . i_{1y} - \omega_k \cdot \psi_{1x} \tag{4}$$

$$\frac{d\psi_{2x}}{dt} = -R_2' . i_{2x} + (\omega_k - \omega) \cdot \psi_{2y} \tag{5}$$

$$\frac{d\psi_{2y}}{dt} = -R_2' . i_{2y} - (\omega_k - \omega) \cdot \psi_{2x} \tag{6}$$

$$m_k = \frac{3.p}{2} \cdot \left(\psi_{1x} . i_{1y} - \psi_{1y} . i_{1x} \right) \frac{J}{p} \cdot \frac{d\omega}{dt} = m_k - m_z \tag{7}$$

where, for the sake of simplification, the following constants have been introduced:

$$\sigma = 1 - \frac{L_h^2}{L_1 . L_2'} \tag{8}$$

$$K_{11} = \frac{1}{\sigma . L_1} \ , \ K_{12} = \frac{L_h}{\sigma . L_1 . L_2'} \ , \ K_{13} = \frac{1}{\sigma . L_2'} \tag{9}$$

List of the used symbols:

i_{1x}, i_{1y}	componentsof stator currentvectori_1
Ψ_{1x}, Ψ_{1y}	componentsof stator fluxvector Ψ_1
Ψ_{2x}, Ψ_{2y}	componentsof rotor fluxvector Ψ_2
u_{1x}, u_{1y}	componentsof stator voltagevectoru_1
R_1	stator resistance
R_2'	rotor resistance converted to stator
ω_k	angular speed of stator voltage
ω	rotor angular speed
J	total moment of inertia
m_k	electric motor torque
m_z	load motor torque
L_1	stator inductance
L_2'	rotor inductance converted to stator
L_h	magnetizing inductance
p	number of poles pairs

The meaning and values of individual parameters are shown in Table 1.

Table 1. Asynchronousmotorparametersformodelling

Parameter	Value	Parameter	Value
P_N	3 kW	R_1	1,8 Ω
U_{1N}	220 V	R_2'	1,85 Ω
I_{1N}	6,9 A	L_h	0,202 H
M_N	20 N.m	$L_{1\sigma} = L_{2\sigma}$'	0,0086 H
n_N	1430 ot.min^{-1}	$K_{11} = K_{13}$	59,3514 H^{-1}
p	2	K_{12}	56,9277 H^{-1}
J	0,1 kg.m^2	k_{tr}	0,1338 N.m.s

For the following analysis we chose the state space representation of an asynchronous motor supplied with normalized constant input signal securing a constant U_1/f_1 ratio, as common in scalar control according to Fig. 1.

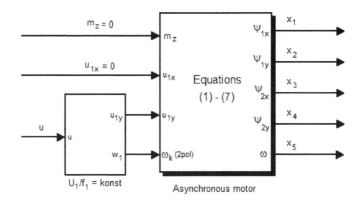

Fig. 1. Selection of AM quantities for state space representation

It applies for discrete dynamic systems that their state in a particular step depends on the state in the preceding step and the change of state according to equation

$$x(k + 1) = x(k) + dx(k) \tag{6}$$

where the change of the state vector $x(t)$ is generated by the effect of the current state and input $u(t)$

$$dx(k) = f(u(k), x(k)) \tag{7}$$

The structural diagram of the dynamic system fuzzy model after division of the state vector into individual components is shown in Fig. 2.

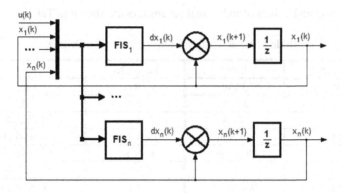

Fig. 2. Structural diagram of dynamic system fuzzy model

The structure of the dynamic system fuzzy model was chosen in accordance with Fig. 2. It is a nonlinear fifth-order ($n = 5$) dynamic system which contains 5 standard Takagi-Sugeno type FIS structures. The database for its construction was sampled with T=1ms. The 5 standardly proposed FIS structures each had 2 rules and statistical error from $0.5*10^{-4}$ up to $1*10^{-3}$.

The comparison of the time courses of state quantity x_5 (rotor speed) of the analytical and the fuzzy model of the asynchronous motor when directly connected to the mains is illustrated in Fig.3.

Note: For the sake of clarity, the state quantity x_5 in Fig. 3 has been normalized.

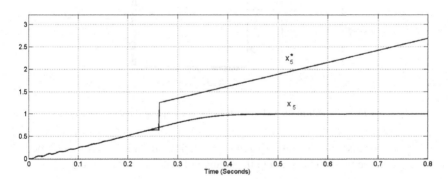

Fig. 3. Comparison of dynamics of state quantities of the analytical (x_5) and the fuzzy model (x_5^*) of the asynchronous motor directly connected to the mains

It is clear from Fig. 3 that the standardly proposed fuzzy model of this nonlinear system approximates the AM analytical model very well until approximately the time 0.23s and then starts to show strong signs of instability.

3   Analysis of AM Fuzzy Model Directly Connected to the Mains

As the structure of the fuzzy model in Fig.2 is universal, it can be assumed that the inconvenient quality of the fuzzy model of AM in steady state is caused by unsuitable processing of the model´s input data database. Processing of the database of input data is carried out in the MATLAB programme environment by partitioning the database into qualitatively significant clusters, i.e. by so called clustering. The individual parameters in clustering have the following meaning:

- *Range of influence*: determines the range of influence of the cluster centre in each of the data dimensions (standardly 0.5)
- *Squash factor*: defines the acceptance of clusters according to mutual position, searching for also near or only distant clusters (standardly 1.25)
- *Accept ratio*: determines the acceptance of points based on their potential of being centres of clusters (standardly 0.5)
- *Reject ratio*: defines the rate of rejection of points without the potential to be centres of clusters (standardly 0.15)

Fig.3 shows that in the development of the model´s FIS structure the data close to the steady state were probably rejected due to their weak potential of being cluster centers. It was therefore necessary to develop modified FIS structures of the model with Reject ratio clustering parameter equal to zero. These structures already have more rules (10 – 11), but their statistical deviation from the referential database is in the order of value 10-5 and the relevant fuzzy model of the AM shows very good dynamic properties within the whole assumed interval (in time 0 – 0.6s), as illustrated in Fig.4.

4   Proposal of AM Fuzzy Model for the Entire Operating Space

In general, the model of a nonlinear dynamic system cannot be built only on data of one single transition characteristic (one single response to a step change of input), as its performance is in principle also dependent on its preceding state. This is clear from Fig. 5, where after disconnection of the motor from the mains at time 0.6s the model completely deviated from the source dataset of the AM, because it entered an area of unidentified inputs space.

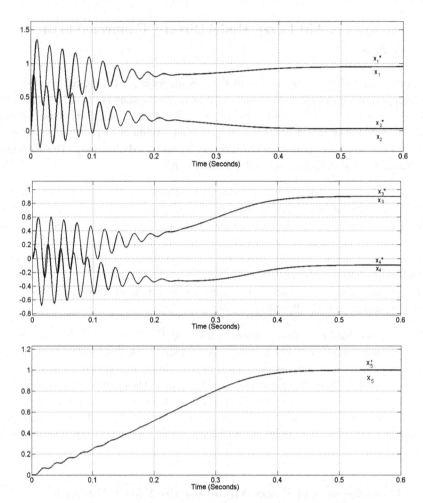

Fig. 4. Comparison of the dynamics of the modified AM fuzzy model with exemplary courses of individual state quantities: a) state quantities x_1 and x_2 b) state quantities x_3 and x_4 c) state quantity x_5

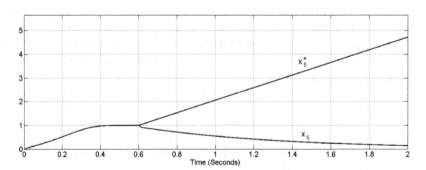

Fig. 5. Dynamics of AM fuzzy model in unidentified input space

The measured database for this pair of input steps will be rather extensive, as a motor not under load has a run-on of as many as 6s. For determining each of the five fuzzy model FIS structures the relevant database matrix has 6600 lines and 7 columns, which considerably increases computational demands for its clustering as well as data training. This is caused by the constant sampling time T=1 ms, which is required in cases of fast changes of motor quantities, but, on the other hand, is unnecessarily short at slow rate of changes. It is therefore advisable to reduce this database prior to the FIS structure proposal, again by means of *clustering*, which will reduce the parameter*Range of influence* (in our case to the value of 0.005) and only the generated clusters will be selected into the database. In this way the number of data for each FIS matrix will be reduced to approx. 450, as shown in Fig. 6.

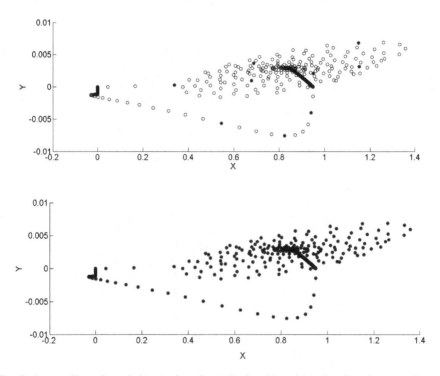

Fig. 6. a) – small number of clusters from large databaseb) – clusters replace large number of data from database

This reduction of referential data database will generate AM fuzzy model FIS structures for each state quantity, containing approx. 15 rules. A detailed analysis of the individual FIS structures can determine whether further modifications of the database are necessary. In our case it was the FIS structure for modelling state quantity x_4 that still showed problematic, and for which new clusters were generated with *Sub clustering*parameters [0.45,1.1,0.5,0]. The comparison of the course of state quantity x_5^* of the AM fuzzy model and state quantity x_5 of the AM analytical model is illustrated in Fig. 7.

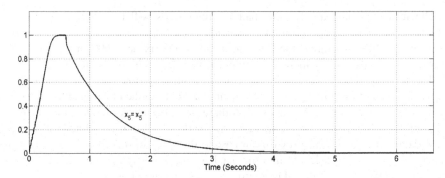

Fig. 7. Dynamics of the course of state quantity x_5^* of AM fuzzy model and state quantityx_5 of AM analytical model for two input steps

For the design of a good quality fuzzy model of a nonlinear dynamic system it is necessary to obtain information on its behaviour in the whole operating range of its inputs. This can be achieved, for example, by regular division of the operating space into a chosen number of levels and generating a referential data database from the transitions between them. In the division of the operating space of input u into 6 levels [0,0.2,0.4,0.6,0.8,1] the number of transitions between them is given by second class variations from 6 elements, which is 30. The time course $u(t)$ of the input signal generator for motor identification was in accordance with Fig. 8.

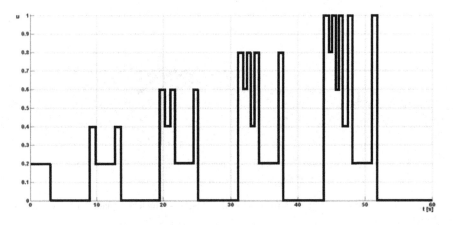

Fig. 8. Generator of steps for obtaining referential data database for AM fuzzy model

With assumed sampling time T=1 ms this database would have approx. 58000 samples and would be too extensive for further processing. It was therefore reduced in the first step by means of *Sub clustering* with parameters [0.01,1.15,0.5,0] to the range of approx. 4000 samples, from which it is possible to generate FIS structures for individual state quantities of the AM fuzzy model with number of rules from 42 to 45. Testing of the model generated in this way showed that structures FIS 3 and FIS 4 that model coupled magnetic fluxes of the rotor still show large inaccuracy and for this reason they were described more precisely by the *Sub clustering* method with *Squash factor* parameter reduced to 0.7 and *Accept ratio* reduced to 0.1. This resulted in the increase of fuzzy rules to 126. Simulation results of the AM fuzzy model generated in this way for state quantity x_5 (motor speed) are illustrated in Fig. 9.

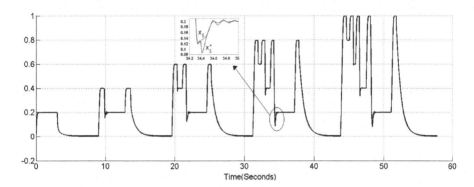

Fig. 9. Comparison of state quantity x_5 from AM fuzzy model with its referential database

The courses of state quantity x_5, which represents the angle velocity of the AM (Fig. 9), prove the high quality of the proposed AM fuzzy model, which is the same also for other state quantities.

5   Discussion and Conclusion

The paper briefly describes the process of designing a good quality fuzzy model of an AM based on the analysis and modification of its FIS structures by means of the Sub clustering method. The AM was modeled on basis of the referential database obtained from its known analytical model and can in principle be obtained by experimental measuring of a real motor.

The process of fuzzy model construction can be summed up into the following steps:

1. Obtaining a suitable referential database, e.g. by measuring carried out on a motor with input signal that covers the whole operating range of inputs (Fig. 8).
2. Reduction of the database by means of the Sub clustering method, where mainly *Range of influence* and *Reject ratio* parameters are adjusted.
3. Analysis of quality and accuracy of fuzzy structures for individual state quantities.
4. Modification (more specific description) of inaccurate FIS structures by repeated application of the Sub clustering method, where mainly the *Squash factor* and *Accept ratio* parameters are adjusted.
5. Testing of the model for randomly generated input signal.

The proposed procedure has been verified on a concrete example of a drive with asynchronous motor. Simulation trial results show that a dynamic system fuzzy model designed in this manner, in spite of its simplicity and simple computer oriented design procedure, can significantly improve the dynamic properties of also strongly nonlinear higher order dynamic systems.

Acknowledgments. The work has been supported by project KEGA 011TUKE-4/2013.

References

1. Guillemin, P.: Fuzzylogicapplied to motor control. IEEE Trans. Ind. Appl. 32(1), 51–56 (1996)
2. Shafei, S.E., Sepasi, S.: Incorporating Sliding Mode and Fuzzy Controller with Bounded Torques for Set-Point Tracking of Robot Manipulators. Elektronika ir Elektrotechnika 8, 3–8 (2010)
3. Li, W., Chang, X.G., Farrell, J., Wahl, F.M.: Designofanenhanced hybrid fuzzy P+ID controller for a mechanical manipulator. IEEE Trans. Syst., Man, Cybern., B, Cybern. 31(6), 938–945 (2001)
4. Tsourdos, A., Economou, J.T., White, A.B., Luk, P.C.K.: Control design for a mobile robot: A fuzzy LPV approach. In: Proc. IEEE Conf. Control Applications, Istanbul, Turkey, pp. 552–557 (2003)
5. Takagi, T., Sugeno, M.: Fuzzy identification of systems and its application to modeling and control. IEEE Trans. Systems, Man and Cybernetics 15, 116–132 (1985)
6. Babuska, R.: Fuzzy Modeling for Control. Kluwer Academic Publishers, Boston (1998)
7. Brandštetter, P., Chlebiš, P., Palacký, P., Škuta, O.: Application of RBF network in rotor time constant adaptation. Elektronika ir Elektrotechnika 7, 21–26 (2011)
8. Nathenas, T.G., Adamidis, G.A.: A Generalized Space Vector Modulation Algorithm for Multilevel Inverters - Induction Motor Fed. International Review of Electrical Engineering – IREE 7(1), 3218–3229 (2012)
9. Mojallali, H., Ahmadi, M.: Hammerstein Based Modeling of Traveling Wave Ultrasonic Motors. International Review of Electrical Engineering – IREE 6(2), 685–697 (2011)
10. Barrero, F., Gonzalez, A., Torralba, A., Galvan, E., Franquelo, L.G.: Speed control of asynchronous motors using a novel fuzzy sliding-modestructure. IEEE Trans. Fuzzy Syst. 10(3), 375–383 (2002)
11. Žilková, J., Timko, J., Kover, S.: DTC on Induction motor Drive. Acta Technica CSAV 56(4), 419–431 (2011)
12. Vittek, J., Bris, P., Makys, P., Stulrajter, M.: Forced dynamics control of PMSM drives with torsion oscillations. COMPEL - The International Journal for Computation and Mathematics in Electrical and Electronic Engineering 29(1), 187–204 (2010)
13. Brandštetter, P.: Sensorless Control of Induction Motor Using Modified MRAS. International Review of Electrical Engineering-IREE 7(3), 4404–4411 (2012)
14. Vittek, J., Pospíšil, M., Minárech, P., Fáber, J.: Forced Dynamics Position Control Algorithm for Drives with Flexible Coupling Including Damping. In: Proc. of the Int. Conf. on Optimisation of Electrical and Electronic Equipment – OPTIM, Brašov, Art. No. 6231800, pp. 403–410 (2012)

Modelling the Technological Part of a Line by Use of Neural Networks

Jaroslava Žilková, Peter Girovský, and Mišél Batmend

Technical University of Košice, Department of Electrical Engineering and Mechatronics,
Letná 9, 042 00 Košice, Slovak Republic
jaroslava.zilkova@tuke.sk

Abstract. The paper deals with the applications of artificial neural networks in modelling and control of a continuous tinning line's technological part. In the conclusion part of the paper, description of the whole model of the tinning line technological section together with the neural speed estimators is presented, along with an evaluation of the achieved simulation results. Training of individual neural networks was performed off-line and adaptation of the network parameters was done by Levenberg-Marquardt's modification of the back-propagation algorithm. The DC drives were simulated in program Matlab with Simulink toolbox and neural networks were proposed in the Matlab environment by use of Neural Networks Toolbox.

Keywords: Mathematical model, technological line, control, DC motor, neural network.

1 Introduction

Modelling is a method enabling the identification of the characteristics of the system under investigation. Using simulation we can then experiment with the model of the line similarly to the real system, without to need intervene in the real system, eliminating the risk of emergency states. An advantage of neural networks is their ability to learn from examples, the ability of abstraction, so their applications are very useful for complex dynamical system modelling [2-5] and control [2,3,5-12], also.

The continuous process line, the model of which we are going to deal with, is a steel strip tinning line producing tin-plated material used in the packaging industry. It is made up of three autonomous sections [1]:

- the entry section, determined for accumulating a stock of material for the technological section and for reduction of traction in the strip,
- the technological section, where electrolytic tinning of steel sheet takes place, according to the technological formula for particular material options in the Ferrostan process technology,
- the exit section, where coiling of the sheet material takes place.

Á. Herrero et al. (eds.), *International Joint Conference SOCO'13-CISIS'13-ICEUTE'13*,
Advances in Intelligent Systems and Computing 239,
DOI: 10.1007/978-3-319-01854-6_36, © Springer International Publishing Switzerland 2014

The entry section of the process line includes the drive of the decoiler device that provides for the uncoiling of the strip at a required tension, independent from the coil diameter and the strip speed. The material is welded into an endless strip that proceeds into the tower accumulator. The subsequent drives of the entry section traction rolls ensure the required strip speed. The remaining drives are provided with traction control and secure the required strip tension over its entire length. Galvanic tinning of the steel strip is carried out in the middle section of the process line. The treated strip is then, at the exit section of the line, recoiled into coils of the specified diameter and width.

The objective of this paper is the development of the neural model of the line technological section. Designing of the neural model of technological tinning lines is based on the mathematical description (of the line without considering a reel decoiler) because on the running production line was not possible to take measurements, but the description of technology and each devices of the line   has been known [1]. Since created mathematical model corresponds to the real process [1], the mathematical model was used for neural networks training sets obtaining and for verification neural model of the line under various conditions without limitation the production.

2   The Line Technological Description and Mathematical Model

Technology part of the continuous line consists of six controlled drives separately excited DC motors that allow independently controlling individual electrical, mechanical and technological quantities so that the production technology would be strictly adhered to. Our intention was to use neural model of the line to observe strip speed at any particular section of the technological line.

Present in the tinning line technology part are six drives:

- drives of traction rolls No.2 (input, middle and output) situated past the tower accumulator and are stretching the strip that enters the middle part of the line,
- drives traction rolls No.3 (input and output) maintain the desired output speed and the desired tension,
- drives of φ rolls, which are used to capture the strip when transiting to the coiler.

The line control allows introducing the strip into the line and supports independent run of each section at desired speed.

During introducing of a new coil, middle part of the line runs at a constant working speed, input part of the line is stopped and the strip is filled from the tower storage. After the input part of the line re-restart-up this part of the line accelerates to a speed about 25% higher than the operating speed of the middle part of the line. Once the tower storage is filled, speeds of them are synchronized.

When designing a neural model of technological tinning lines we used mathematical description of this part of the line.

2.1   Mathematical Model of Middle Part of the Line

Mathematical models of parts of the line along with controls were designed in Simulink, which is part of Matlab, and the line neuron model line was set up using Neural network toolbox of Matlab.

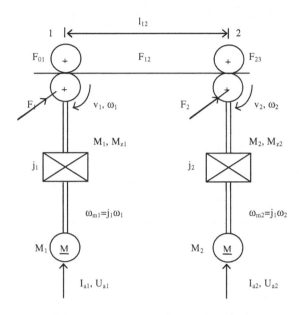

Fig. 1. Part of the continuous line with two motors

To set up an accurate model binding system considered have to be two areas of the model utilization:

- at zero traction, when developing between machines can be material loops,
- at positive tensile stress, when the transportation rollers are mechanically bound together.

In the next section, for more transparent derivation of mathematical expression of the middle part of the line, we are considering only two motors of the line, as illustrated in Fig.1.The part of the continuous line consisting of only two DC separately excited motors 1 and 2, according to Fig.1, can be described by a system of differential equations, in the order for the first drive (equation dynamics as to the driven machine shaft), for the section of the strip between, and the second drive:

$$M_1 + (F_{12} - F_{01})r_1 - M_{z1} = J_1 s.\omega_1 \tag{1}$$

$$\frac{SE}{l_{12}}(r_2\omega_2 - r_1\omega_1) = sF_{12} \tag{2}$$

$$M_2 + (F_{23} - F_{12})r_2 - M_{z2} = J_2 s \omega_2 \tag{3}$$

After supplementing these equations by the equation for the motor armature circuit for the two drives:

$$M_i = j_i C_i \phi_i I_{ai} \tag{4}$$

where $i=1,2$

$$R_{ai} I_{ai} + L_{ai} \frac{dI_{ai}}{dt} + j_i C_i \phi_i \omega_i = U_{ai} \tag{5}$$

We are obtaining the mathematical model of the simplest continuous line, block diagram of which is illustrated in Fig.2. The constant before the parenthesis in equation (2) presents the strip elasticity constant of the strip for the considered section:

$$K_{p12} = \frac{SE}{l_{12}} \tag{6}$$

Whereas the strip material features besides elasticity also dampening effects that are reflected in K_T dampening constant, expressed in % of constant K_p, equation (2) will take on the following form:

$$K_p (r_2 \omega_2 - r_1 \omega_1) + s K_T (r_2 \omega_2 - r_1 \omega_1) = s F_{12} \tag{7}$$

Designed for all of the above-mentioned drives were controllers of the motor armature current and superior speed controllers. Basically, tension control equals controlling the armature current, which determines the motor torque. Direct traction controlling would require the quantity to be precision measured. Yet, the sensors are too expensive and once overloaded they come permanently deformed. Due to the fact, used these days is indirect traction control.

Control is based on equality of mechanical and electrical outputs:

$$P_{mech} = F.v \tag{8}$$

$$P_{el} = U_i.I_a \tag{9}$$

Assuming that the speed of the strip can be observed quantity and value of the induced voltage U_i, and the armature current I_a will be measured, we can determine tension strength from these equations:

$$F = \frac{U_i}{v} I_a \tag{10}$$

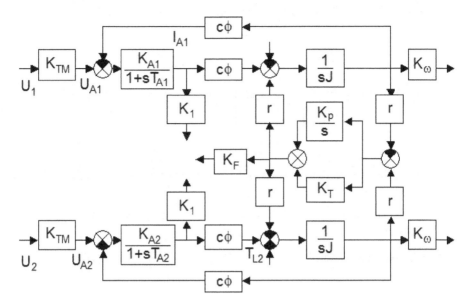

Fig. 2. Block diagram of a part of the continuous line – for two identical drives

The DC motor armature controller maintains the current value I_a proportional to the tension force F and the excitation controller maintains a constant ratio U_i/v. Mathematical models of individual parts of the line along with the control were designed in Simulink.

3   Neural Model of Middle Part of the Line

Neural model of individual parts of the line has been designed as a series-parallel model of drives, shown in Fig. 3.

Whereas to detect speed of the strip, e.g. for determining the tension force, it is necessary to know the speed of the strip in each part of the line, we divided technological line as described above into six parts and for each part of the line a separate neural model was created. Inputs of individual neural models $u(k)$ present the desired values of the strip speed, armature current in the k-th, (k-1) and (k-2) steps for the drive and the observed value of the speed in the (k-1) step.

Basic structure of the neural model of one part of the line approximates speed of the strip based on the following equation:

$$\widehat{v}_s(k) = f\left[v_{\check{z}}(k), i_a(k), i_a(k-1), i_a(k-2), \widehat{v}_s(k-1), w\right] \qquad (11)$$

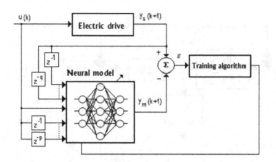

Fig. 3. Block diagram of the series-parallel drive model

Suggested to approximate the relation were two types of neural networks (Figs.4 and 5):

- Multi-layer feed-forward network
- Cascade feed-forward network

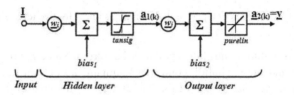

Fig. 4. Neural feed-forward network

Outputs of the first and second layers of the feed-forward network are determined based on relations:

$$\underline{a}_{1(k)} = tansig\left(\underline{I}\,\underline{w}_i + bias_1\right)$$
$$\underline{a}_{2(k)} = purelin\left(\underline{a}_{1(k)}\underline{w}_j + bias_2\right) \tag{12}$$

where I is the input vector and w is vector of synaptic weights of the network.

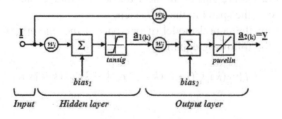

Fig. 5. Cascade feed-forward network

Outputs of the first and second layers of the cascade network are determined based on relations:

$$\underline{a}_1(k) = tansig\left(\underline{I}\underline{w}_i + bias_1\right)$$

$$\underline{a}_2(k) = purelin\left(\underline{I}\underline{w}_k + \underline{a}_1(k)\underline{w}_j + bias_2\right) \tag{13}$$

For both types of networks, used for the output layer neurons were linear activation functions, and for neurons in the hidden layer tansigmoid activation functions:

$$f(n) = \frac{1}{1 + \exp(-an)} \tag{14}$$

where n is the value of the neuron inside activity and a is the slope parameter of the sigmoid. Training of individual networks was carried out off-line, and chosen for adapting the network parameters was the Levenberg-Marquardt modification of back-propagation algorithm.

The optimization objective was seen in minimizing the function (15):

$$e = \frac{1}{Q}\sum\left(v_q - v_q'\right)^2 \tag{15}$$

where: Q - the number of trained samples

v_q - is the q-value of the strip speed based on the given drive model

v_q' - is the q-observed value of the network output.

Synaptic weights adaptation was based on the relation:

$$\Delta\mathbf{w} = \left(\mathbf{J}^T\mathbf{J} + \mu\mathbf{I}\right)^{-1}\mathbf{J}^T\mathbf{e} \tag{16}$$

where:

$\mathbf{J}$	-	is the Jacobian matrix,
$\mathbf{J}^T\mathbf{J}$	-	is the approximated Hessian matrix,
$\mathbf{J}^T\mathbf{e}$	-	is the gradient vector,
$\mathbf{I}$	-	is the unit matrix
μ	-	is the coefficient influencing the rate of adaptation of weights

Neural model of the line middle part was set up using Neural Network Toolbox of Matlab. After comparing the results of testing both types of networks attained by feed-forward and cascade networks, chosen for all six parts was a three-layer cascade feed-forward network with ten neurons in the hidden layer.

4   Results of Simulation and Comparison

Neural models have been tested for a variety of changes in the value of the desired speed and for different load torque changes before and after the tower accumulator.

All tests were conducted under different conditions than they were in the setting up of the training sets of the following desired rates: 1.5 m/ s, 5.1 m/ s, 1.9 m/ s, 2.8 m/ s, 3.6 m/ s; these changes occurred at times: 0s, 10s, 25s, 40s, 45s and at the time 0s, 20s and 35s the load torque changed to the values of 332Nm; 232Nm; 382Nm. Since the results obtained are similar in nature not presented are the testing results of all six neural models but only those of selected models. The following figures show the selected simulation results obtained when testing neural models of the line.

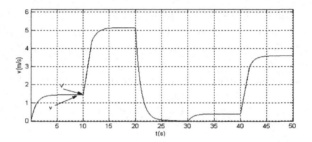

Fig. 6. Traction roll No.1- input, observed and real waveforms of the strip speed

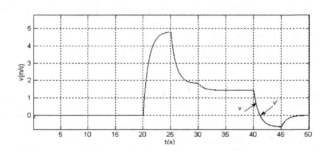

Fig. 7. Observed and real waveform of the strip speed within the tower accumulator

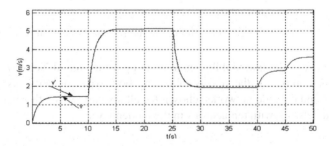

Fig. 8. Traction roll No.2-input: observed and real waveforms of the strip speed

Illustrated in Fig.6 are observed and modeled changes in strip speed at changes in the desired speed in No.1 traction roll.

Waveforms at the identical speed changes in traction rolls No.2, No.3 and φ rolls have the same character (Fig. 8, Fig10). Illustrated in Fig.8 are observed and modeled changes in strip speed within traction roll No. 2.1, corresponding with the waveform shown in Fig. 6.

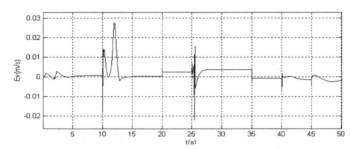

Fig. 9. Detail of the difference in observed and modelled speeds of the strip within traction roll No.2.1

The maximum errors evidenced at dynamic changes fluctuated between 0.5% at the strip speed at φ rolls up to max. 2% at No. 3 traction roll. Max errors at other traction rolls were fluctuating around 1%.Shown in Fig. 9 is the difference between observed and modeled speed zone in the traction rolls No.2.1 corresponding with that shown in Fig.8.

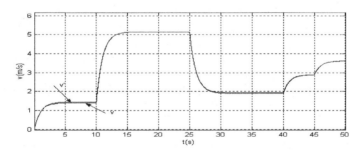

Fig. 10. Traction roll No.3- input: observed and real waveforms of the strip speed

5   Conclusions

Technological lines for treatment of rolled steel or another strip present a significant part of the rolling mill drives. Treatment of rolled strip is based on the longitudinal or transverse divisions or on the surface finish, respectively. The tinning line is intended for producing sheets for packaging. Our objective was to develop a neural model of the line middle part and demonstrate the possibilities of neural networks application.

Mathematical models of parts of the line along with controls were designed in Simulink, and the line neuron model was set up using Neural network toolbox.

Based on easily measurable signals (11) it is possible, using neural model of the line, to observe the strip speed at any particular section of the technological line. Maximum errors that occurred at swift dynamic changes ranged from 0.5% at the strip speed at rolls φ up to maximum of 2% at traction roll No. 3 – the output one. The maximum errors in the other traction rolls and the tower accumulator were around 1%.

As the simulations show, the characteristics of line neural model are similar to standard modeling results. The advantages of neural networks application are mainly in modeling of complex systems, where they allow, the approximation of any continuous function without the precise knowledge of the structure of the modeled system. In our case, the mathematical model of the line was used to obtain the training sets and verification of the proposed neural model without limitation of production process.

Acknowledgements. The authors wish to thank for the support to the R&D operational program Centre of excellence of power electronics systems and materials for their components No. OPVaV-2008/2.1/01-SORO, ITMS 26220120003 funded by European regional development fund (ERDF).The authors wish to thank for the support project APVV-0185-10.

References

1. Company documentation ARTEP VSŽ, k. p. Košice (1999)
2. Vas, P.: Artificial-Intelligence-based electrical machines and drives. Oxford University Press, Oxford (1999)
3. Timko, J., Žilková, J., Balara, D.: Artificial neural networks applications in electrical drives, p. 239. TU of Košice (2002) (in Slovak)
4. Nitu, E., Iordache, M., Marincei, L., Charpentier, I., Le Coz, G., Ferron, G., Ungureanu, I.: FE-modeling of cold rolling by in-feed method of circular grooves. Strojniski Vestnik – Journal of Mechanical Engineering 57(9), 667–673 (2011)
5. Timko, J., Žilková, J., Girovský, P.: Modelling and control of electrical drives using neural networks, p. 202. C-Press, Košice (2009) (in Slovak)
6. Brandštetter, P.: AC drives - Modern control methods. VŠB-TU Ostrava (1999)
7. Levin, A.U., Narendra, K.S.: Control of Nonlinear Dynamical Systems Using Neural Networks: Controllability and Stabilization. IEEE Transactions on Neural Networks 4, 192–206 (1993)
8. Levin, A.U., Narendra, K.S.: Control of Nonlinear Dynamical Systems Using Neural Networks- Part II: Observability, Identification and Control. IEEE Transactions on Neural Networks 7, 30–42 (1996)
9. Hagan, M.T., Demuth, H.B., De Jesús, O.: An introduction to the use of neural networks in control systems. International Journal of Robust and Nonlinear Control 12, 959–985 (2002)
10. Perduková, D., Fedor, P., Timko, J.: Modern methods of complex drives control. Acta Technica CSAV 49, 31–45 (2004)
11. Vittek, J., Dodds, S.J.: Forced dynamics control of electric drives. ZU, Žilina (2003)
12. Žilková, J.: Artificial neural networks in process control, p. 50. TU of Košice, Košice (2001)

Distribution of the Density of Current in the Base
of Railway Traction Line Posts

Lubomir Ivanek[*], Stanislav Zajaczek, Václav Kolář,
Jiri Ciganek, and Stefan Hamacek

VŠB-Technical University of Ostrava, 17. listopadu 15/2172, 708 33 Ostrava - Poruba,
Czech Republic
{lubomir.ivanek,stanislav.zajaczek,
vaclav.kolar,jiri.ciganek.fei,stefan.hamacek}@vsb.cz

Abstract. This paper deals with modelling stray currents in the concrete base of traction line post in one of the most frequent defects - short-circuit of the surge arrester. If these defects occur more frequently, the concrete may be degraded and the post base firmness may deteriorate. The results of the modelling will be subjected to further processing by Construction Material Laboratories at VSB - Technical University in Ostrava.

Keywords: Numerical methods, FEM, COMSOL, railway traction.

1   Introduction

Most of the Czech Republic's DC traction lines feature positive potential connected to the trolley line while the negative pole is connected to the rail. The electric circuit is closed via the trolley, bus, engine, rails and back to the traction transformer. The rails are not perfectly insulated from the earth since they rest on frets in gravel bed. A part of the current flows into the earth and is limited by the cleanness of the gravel bed and the distance between the rail and the gravel bed. The volume of the current flowing into the earth ranges between 15 - 60 %. The Czech traction transformers have nominal voltage of 3.3 kV and nominal rated capacity of 5 - 10 MW. In peak values of current flowing along the traction line are 1,500 A to 3,000 A; therefore, the volume of the leak may reach 500 - 2,000 A. [2, 3]

The posts carrying the traction line are connected to the rails with surge arresters. The surge arrester is a protective equipment which has a conductive connection with the traction line post and which is connected to the rail with a conductor of 10 square mm in diameter. In order to insulate the rail (jamming security equipment and other signals sent through the rails) we use only equipment whose insert will only release voltage higher than:

1. 250 V - publicly accessible places
2. 500 V - places not accessible for the public

---

[*] Corresponding author.

Á. Herrero et al. (eds.), *International Joint Conference SOCO'13-CISIS'13-ICEUTE'13*,
Advances in Intelligent Systems and Computing 239,
DOI: 10.1007/978-3-319-01854-6_37, © Springer International Publishing Switzerland 2014

It is used in areas endangered by the traction line where touch voltage may occur. In such an area it is essential that all conductive places be insulated: railway traction line posts, bridges, fences, drains and everything else which might lead electric current should a conductor touch it. The surge arrester is often defected in such a way that the potential of the rail is transferred to the post. Then, the stray current leaks in the earth not only from the rail, but also from the base of the post. This results in electrolytic processes in the post base concrete and a thermal load of the base.

We have modelled this condition intending to identify places with the greatest current density, i.e. places where the concrete may be damaged. In reality, the crumbling of the concrete does occur.

Based on the manufacturer's documentation of masts, the following parameters were assigned:

Bolts – iron material - $\mu_r = 4000$, $\sigma = 1,12*10^7$ S.m^{-1}

Base plate – soft iron - μ_r– special characteristics of BH from the manufacturer's documentation, $\sigma = 1*10^7$ S.m^{-1}

Concrete - $\mu_r = 1$, $\sigma = 1,5$ S·m^{-1}

Typical soil characteristics of Various terrains:

TYPE OF TERRAIN	RESISTIVITY (Ωm)	PERMITIVITY
Agricultural plains, streams, richest loam soil	30	20
Pastoral, low hills, fertile soil	80	15
Flat, marshy, densely wooded in places	130	13
Pastoral, medium hills with forestation	270	12
Pastoral, heavy clay soils hills	250	12
Rocky soil, steep forested hills, streams	500	10
Rocky, sandy, some rainfall, some vegetation	500	8
Low-rise city suburbs, built-up areas, parks	1000	6
High-rise city centers, industrial areas	3000	4
Sand deserts, arid, no vegetation	>3000	3

1.1   Description of the Post Used

Our modelling took place in 3 kV DC railway traction system. We used a lattice post of steel profiles. The bottom of the post is welded to a steel board with openings for studs; the board is fastened to a stud basket embedded in the post base. The platform of the base is illustrated in Fig. 1. Picture 2 then presents the A-A section of the post base.

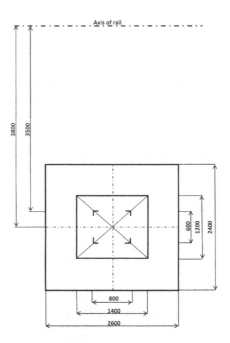

Fig. 1. Ground plan of the post base [1]

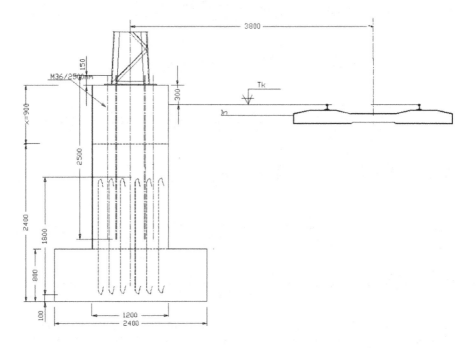

Fig. 2. Section of the post base [1]

1.2   Modelling in COMSOL

The post base was modelled in COMSOL Multiphysic in the AC/DC Module physics. The Electric Currents interface was used to model DC electric current flow in conductive media. The interface solves a current conservation equation for the electric potential. The 3D model computed in approximately 720,000 knots. The 3D model of the area in question is presented in Fig. 3 in the most illustrative view.

The conductivities of the individual sub-areas in question:

- soil $\sigma_e = 0,01\ S \cdot m^{-1}$
- concrete $\sigma_c = 1,5\ S \cdot m^{-1}$
- steel $\sigma_s = 1,12.10^7\ S \cdot m^{-1}$

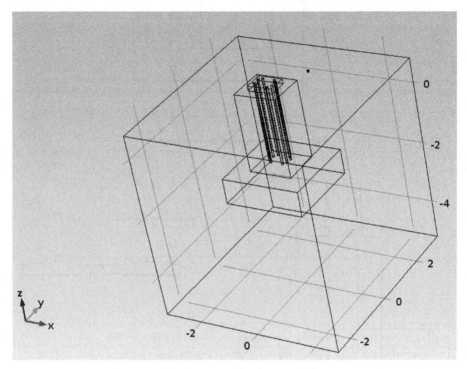

Fig. 3. Model of the area in question

The current density at any point can be expressed on the grounds of knowledge of the electric field at this point.

$$J = \sigma E \tag{1}$$

And there is a simple relation between the intensity of the electric field E and the electric potential φ.

$$E = -\nabla \varphi \tag{2}$$

We selected an artificial boundary around the area of the pole base and the perimeter conditions of this boundary are presented in Fig. 4. The post, its flange and the upper surface of the studs were connected to DC potential against the earth of $\varphi = 50$ - 200V. If, in the course of the approximation of the artificial boundary, the shape of the field did not change, we considered this boundary to be an artificial earth with a zero electric potential. This means that we considered the bottom and sides of the cubes of the boundary delimiting the selected area to be the earth with the zero electric potential. The calculations were made for different distances between the artificial boundary and the post.

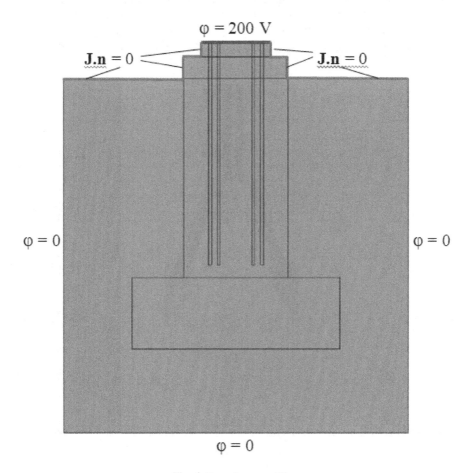

Fig. 4. Boundary conditions

No current flows into the area of the air beyond the earth; therefore, the following condition was selected for the upper side of the cube

$$J.n = 0 \qquad (3)$$

i.e. the derivation of the vector of the electric intensity in the direction of the normal to these sides is equal to zero

$$\sigma . \boldsymbol{E} . \mathbf{n} = 0 \tag{4}$$

$$\left| \sigma \frac{\partial \varphi}{\partial x}_{\mathbf{n}=0} \text{ or} \sigma \frac{\partial \varphi}{\partial y}_{\mathbf{n}=0} \right| \tag{5}$$

The vectors of the current density in the base present: the artificial boundary being located 10 m from the axis of the post in Fig. 5 and the artificial boundary being located 5 m from the axis of the post (a number of vectors selected). Artificial boundaries were chosen from 5m to 20m. Up to 10m, the shape of current field did not change. This distance is therefore decisive. 5m distance was inserted for reference only.

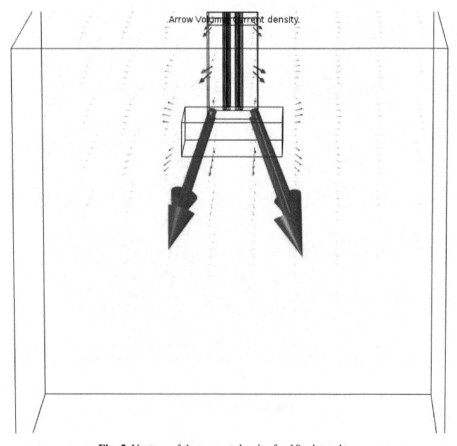

Fig. 5. Vectors of the current density for 10m boundary

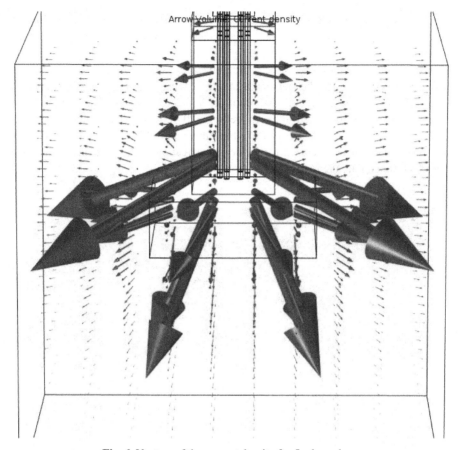

Fig. 6. Vectors of the current density for 5m boundary

2   Summary

As is entailed in reports from the operators of railway systems, the bases of the posts are often damaged, especially whenever the surge arrester has been short-circuited for a long time. In this case, the voltage of the rails goes directly to the post and hence also to the studs at the base. We did several calculations for different input parameters; e.g. the input voltage ranged between 50 and 500 V.

Peak value of the current density:

Table 1.

Boundary 10 m	$\varphi = 10$ (V)	$\varphi = 20$ (V)	$\varphi = 50$ (V)	$\varphi = 100$ (V)	$\varphi = 150$ (V)	$\varphi = 200$ (V)	$\varphi = 500$ (V)
J_{max} (Am^{-2})	1	1,9	4,8	9,7	14,6	19,5	49
J_{axis} (Am^{-2})	0,65	1,3	3,4	6,5	9	13	34

We also altered the conductivity of the selected materials etc. Through modeling we discovered that bases which have sharp edges are really exposed to heavy current loading at the edges - see the circled area in fig. 7. Values of current density loading at the edges of concrete are the same as the maximum current values in Table 1.However, the places of contact between the steel studs and the concrete are exposed to excess load. These places may cause the degradation of the concrete structure. The distribution of the current densities identified in the modelling will be forwarded to the Construction Material Laboratory of the Faculty of Civil Engineering of VSB - Technical University in Ostrava, where it will be used for further investigation of the influence of degradation of concrete under the load effect of direct current.

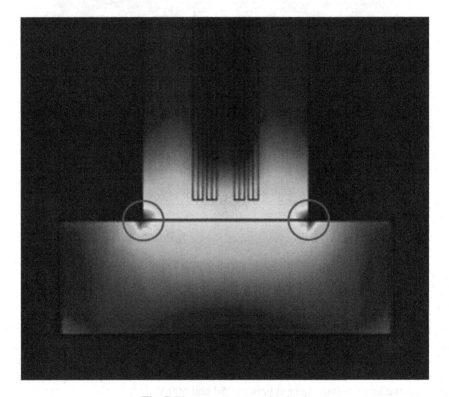

Fig. 7. Places most exposed to load

Rounded edges with a radius of 0.7 m, the maximum value of the current density at the edges is reduced by46%.

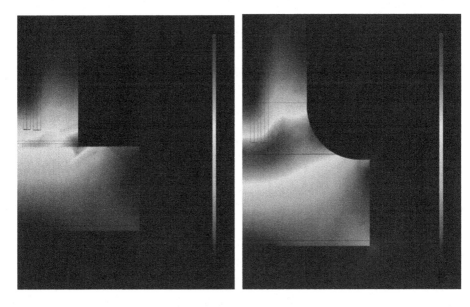

a) J_{max} = 19,576 (Am⁻²) b) J_{max} = 11,418 (Am⁻²)

Fig. 8. Comparison of current density for sharp and rounded edges

Acknowledgements. This paper has been elaborated in the framework of the project Opportunity for young researchers, reg. no. CZ.1.07/2.3.00/30.0016, supported by Operational Programme Education for Competitiveness and co-financed by the European Social Fund and the state budget of the Czech Republic, project SP2013/47 and project MSMT KONTAKT II: LH11125 - Investigation of the ground current fields around the electrified lines.

References

1. SUDOP Praha: Řez základu a stožáru BP 32, EŽ Praha, Praha (2004)
2. Paleček, J.: Vybrané kapitoly z problematiky elektrotechniky v dopravním inženýrství, Skripta, VŠB-TU Ostrava (1996)
3. Ivánek, L.: Modely a přenosové parametry pro šíření zpětných proudů v elektrické trakci, Habilitační práce, Ostrava (1998)
4. ČSN EN 50122-2 "Drážní zařízení – Pevná trakční zařízení – Část2: Ochranná opatření proti účinkům bludných proudů, způsobených DC trakčními proudovými soustavami," platná od (2001)
5. Edwards, R.J.: Typical soil characteristics of Various terrains, http://www.smeter.net/grounds/soil-electrical-resistance.php

Smart Home Modeling with Real Appliances

Miroslav Prýmek[1], Aleš Horák[1], Lukas Prokop[2], and Stanislav Misak[2]

[1] Faculty of Informatics, Masaryk University
Botanicka 68a, 602 00 Brno, Czech Republic
{xprymek,hales}@fi.muni.cz
[2] Faculty of Electrical Power Engineering and Computer Science
VSB – Technical University of Ostrava
17. listopadu 15, 708 33 Ostrava – Poruba, Czech Republic
{lukas.prokop,stanislav.misak}@vsb.cz

Abstract. Renewable energy sources bring new challenges to power consumption planning on all levels (household, city or region). In this paper, we present a smart household appliance-level scheduling system, called PAX. We describe the details of the PAX scheduling core and its usage for optimizing an appliance power consumption profile according to actual power source regime.

The PAX system is tested on data sets from a one year measurements of household appliances and renewable power sources, which are described in detail in the text. These data form a basis for the PAX exploitation in both simulated power system modeling and real-world power consumption scheduling.

1 Introduction

With the world-wide growing demand for ecological sources of electrical energy, there is also a growing number of new problems in the field of the electrical energy distribution and consumption planning. The behaviour of new renewable energy sources (RES) is in many aspects different from classical energy sources [1]. Conventional power plants (coal, gas, nuclear) produce electrical energy in a process which can be well controlled by the grid operators. On the other hand RES use mostly unpredictable and non-controllable sources like wind or sunshine. These sources are of cyclical nature and their power supply profile does not fit well with the profile of (cumulated) power demand [2].

The negative influence of stochastic power sources can be diminished with several approaches. The standard approach uses a backup of the RES with conventional energy sources by usual power system auxiliary services to overcome unpredictable outages of the renewable energy sources. An economic approach uses an energy storage of the surplus energy for future use. The research in the field of efficient energy storage is very intensive, but current technology is still not mature enough for cost-effective usage in most of the real-world scenarios [3].

The approach presented in this paper sees the problem from the opposite perspective: we accept the unpredictability and non-controllable nature of the

Á. Herrero et al. (eds.), *International Joint Conference SOCO'13-CISIS'13-ICEUTE'13*, 369
Advances in Intelligent Systems and Computing 239,
DOI: 10.1007/978-3-319-01854-6_38, © Springer International Publishing Switzerland 2014

renewable sources and search for the needed flexibility on the demand side of the power usage equation. Instead of trying to change the profile of the energy production we try to change the demand/consumption profile. The current *Smart Grid* concept had emerged from this approach. It is based on the idea of a *communication* between power sources and power consumers with intelligent control of their interaction using multi-agent approach. This can be achieved on several levels – a household, municipality or the whole country region. A realistic combined approach is the concept of a hierarchical structure where on each level the aim of the control is to build a predictable entity for interaction with the levels above it.

Besides the multi-agent approach, the RES energy consumption planning on the demand side was solved by computational centralised techniques. For example, Mohsenian-Rad in [4] employed the game theory to solve and formulate an energy consumption scheduling game, where the players are the users and their strategies are the daily schedules of their household appliances and loads. Pedrasa [5] applied particle swarm optimization to schedule a significant number of varied interruptible loads over 16 hours. Unlike these approaches, the system presented in the following text aims at solving the problem with lightweight scheduling rules using very low computational power and straightforward setup.

2   Smart Household Architecture

In the presented research, we have focused on the lowest level of the smart grid structure – a household. The main aims of the control logic on this level are:

- control the operation of particular appliances according to the actual power supply
- minimize demand peaks, make the overall consumption profile as fluent as possible (by coordination of particular appliance operation)
- flexibly react to external effects (outage, brown-out, etc.) to minimize possible losses
- reach the given goals without a significant impact on user experience

In the following text, we describe the *Priority-driven Appliances Control System* (*PAX*). The PAX system is designed for small and cheap microcontrollers for particular appliance control, yet flexible enough to fulfill the above-stated criteria well. The controlling core of the system can mix real and virtual appliances, so the system can be used as a smart home simulator as well as a real, physical appliance controller. Later in the paper we describe how real world data are used as a basis for smart home modeling.

2.1   Data Sources for Initial Testing

Several measurements were performed to collect real household data to test and demonstrate PAX functionalities. Each household appliance in a selected real household was monitored using power networks analyzer MDS-U [6] during a long time period. MDS-U is a power analyzer which is able to measure voltage, current and power factor and calculate electrical quantities like active, reactive and apparent power for selected time interval. In this case, one minute time interval was used for data collection. Averaged power consumption curves were defined based on long time power consumption data for 20 most common household appliances (refrigerator, cook top, wall oven, personal computer, washer, dishwasher, vacuum cleaner, etc.). During the long time measurement, the switching scheme was evaluated for all monitored household appliances. Usual power consumption for each time can be defined based on averaged power consumption curves and the switching scheme for most common household appliances. The power consumption curve of each household appliance are the fundamental data for PAX testing.

Together with household appliances monitoring there were selected renewable power sources monitored during 1 year time period. According to the actual trends in renewable energy sources utilization, photovoltaic (PV) and wind power plants (WPP) were chosen. PV consists of mono crystalline panels Aide Solar (P_MAX 180 W, I_MP 5 A, V_MP 36 V, I_SC 5.2 A, V_OS 45 V) and PV has 2 kWp rated power. WWP uses synchronous generator with permanent magnets of 12 kVA, voltage 560 V, current 13,6 A, torque 780 Nm and 180 rpm. A detailed description of the small off-grid power system test bed can be found in [7] or [8].

In the current PAX testing experiment, two data sources were used – a data set of power consumption and a data set of power production. Both the mentioned data sets are based on one year measurements of real data.

2.2 The PAX Implementation

The PAX system is developed using the Erlang programming language [9]. Erlang is excellent in dealing with large amounts of discrete data. It is a functional language with no shared memory. The program consists of many autonomous units which communicate with each other only by message passing. That is why applications written in Erlang can be parallelized in natural way and run easily on multi-core processors or even separated machines. The parallelization scalability is almost linear. Erlang also has very good support for runtime code compilation and loading which is used extensively for particular virtual appliance control in the PAX system and for graphical user interface customizations.

As we have mentioned above, the PAX system is designed to support research of the smart home automation from the simulation phase up to device control in real-time. For the later usage scenario, the scheduler is driven by real-time clock, the events are induced by external sources or occur at predefined times.

When used as a simulator, without connections to real devices, there is no need to restrain the process with real-time clock. The (simulated) events should occur and be processed as fast as possible to make mass data processing possible. For this purpose, the PAX core implements an *event queue*. The queue is filled with simulated or real-world measured data and the simulation consists of sequential processing of the queue events. Each simulation step has several phases:

1. fetch subsequent event from the event queue
2. deliver it to the appropriate agent
3. the agent receives the event and/or
   (a) changes its inner state
   (b) inserts a new event into the event queue
4. the simulation ends when the event queue is empty

Since the PAX core heavily depends on (asynchronous) message passing, it is possible to convert almost any code sequence into a sequence of events in the event queue without any special changes in the code. The same code can thus be used for offline data processing (simulation) and online device control. With this pure event-driven design, it is possible to implement arbitrary time precision simulation and simulate every possible aspect of the control system.

2.3   From Real Devices to Virtual and Back

A development of a successful energy consumption control scheme consists in several steps:

1. measure power consumption profiles of typical household appliances
2. pre-process the data to remove any non-significant fluctuation (at a desired precision level)
3. convert the data into power consumption change events
4. test different consumption planning algorithms on the simulated appliances
5. test the best algorithm back on the real appliances

The main advantage of the PAX system is that the steps 2 to 5 can be done using the same software system thus keeping the simulation results as close to the real deployment as possible.

3   Model Construction – Accumulator Type Appliance

In this section, we will illustrate the above-described process of transferring real-appliance measured data into a virtual appliance model suitable for the PAX simulation.

Figure 1 shows parts of real power consumption profiles of a typical refrigerator and a personal computer, measured with one minute resolution. The black line along the measured data (represented by a grey line) depicts the cleaned data used for the model development.

The main objective of the data cleaning is to identify the consumption level *breaking points* which can then be transformed to the consumption-change events with specific time values and either converted into a repeating pattern or statistically analyzed and reproduced on a random basis.

refrigerator   computer

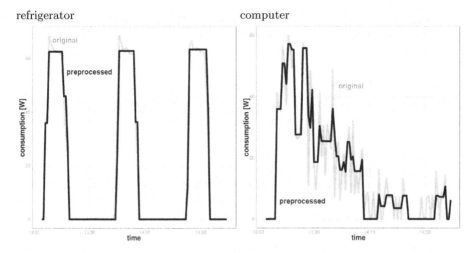

Fig. 1. Measured power consumptions and preprocessed profiles of a refrigerator and a computer

The consumption data preprocessing algorithm is as follows:

1. Set a relative threshold value (typically 1/5 of the maximal consumption).
2. Find the next time point where the difference between the last mean value and an actual consumption exceeds the threshold (the next breaking point).
3. Compute the mean value between the last two breaking points.
4. Repeat until end of data.

The number of breaking points (and thus the similarity of original and preprocessed curves) is determined by the chosen threshold, which is can be adapted to achieve the desired accuracy of the model.

The power consumption profile of a refrigerator consists of periodic repeating peaks of (nearly) the same length, height and distance. This behaviour is caused by the thermostat which keeps the inner temperature of the refrigerator in given limits. Generally, we consider the refrigerator to be an instance of a *feedback-controlled device* which acts like a perpetual (small amount) energy accumulator (see [10] for the PAX appliance types). Whenever the accumulated energy falls below a given value, the device consumes the electrical power from the grid and accumulates it up to the upper limit value.

A conventional refrigerator does not communicate with its environment and simply repeats the accumulation cycle forever in the safe range. Such behaviour brings two main disadvantages:

– If there are several appliances of this type, they produce unwanted random consumption peaks when their accumulation phases overlap.

– In case of an emergency state of the grid (blackout, brownout), such device can only be switch on or off. There is no intermediate state which would not substantially impact the device's function (to keep the food cooled) while lowering the load on the grid.
– the appliance cannot precede an anticipated/planned power loss.

To harness the potential of the smart grid concept, a PAX model of a smart refrigerator derived from real data is presented further. The smart refrigerator changes its behaviour according to the state of the grid and cooperates with other power consuming devices.

In a virtual intelligent refrigerator model based on real data, we must first identify the breaking points on the measured power profile of a real refrigerator and fit the profile to the fluctuation of the controlled value (the inner temperature or the accumulated energy). The measured refrigerator temperature oscillates between 1 and 4°C. At 4°C the cooling system motor is switched on and works for 10 minutes until the temperature of 1°C is reached (see Figure 1). We can approximate the cooling speed to be 0.005°C per second. The motor-idle phase then lasts 30 minutes. The heating speed can be approximated as 0.0017°C per second. Of course, the real speed of cooling and heating is not linear but linearization is a sufficient approximation in this case.

As mentioned above, the PAX system consumption scheduling is based on priorities of particular power consumption requests and sources (for details see [10]). The conventional non-intelligent refrigerator operates steadily in the given temperature range and therefore its consumption requests have only one given priority P. The appliance agent then periodically asks for power consumption of 64 W with the priority P for 10 minutes. In the model input data, the measured values from Figure 1 correspond to the following simulator events:

```
{{{2013,03,04},{0,3,0}},{set_consumption,fridge1,4,63.9043}}
{{{2013-03-04},{0,13,0}},{set_consumption,fridge1,0,0}}
{{{2013-03-04},{0,42,0}},{set_consumption,fridge1,4,63.4513}}
{{{2013-03-04},{0,53,0}},{set_consumption,fridge1,0,0}}
{{{2013-03-04},{1,21,0}},{set_consumption,fridge1,4,63.7587}}
{{{2013-03-04},{1,32,0}},{set_consumption,fridge1,0,0}}
```

For each time moment, the actual consumption of the appliance *fridge1* is set to the value of ≈ 64 watts with constant priority 4 or to the value of 0 watts (with 0 priority).

Representation of such a behaviour in a general device model (in pseudocode) looks like:

1. start → insert_event(Now,set_consumption(4,64))
2. event set_consumption(_,64) →
   insert_event(Now+10m,set_consumption(0,0))
3. event set_consumption(_,0) →
   insert_event(Now+30m,set_consumption(0,64))

Now we have obtained a model of the standard refrigerator. With the inter-appliance communication ability brought by the PAX smart grid, we turn it into

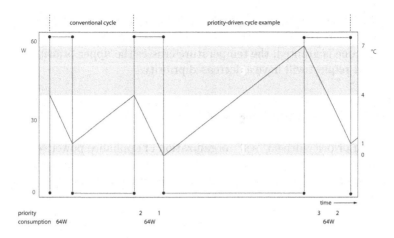

Fig. 2. Priority ranges

a more flexible model of the refrigerator behaviour. The new model is based on the idea that with decrease of the accumulated-energy value (in the refrigerator case it equals the decrease of the inner temperature), the urgency of the power consumption increases because of a risk of a malfunction. In the PAX model this means that the priority of the consumption requests is proportional to the temperature.

The temperature range is divided into priority zones according to the risk that the device could miss its primary purpose (i.e. the risk of the food to rotten). If we approximate the temperature change function as stated above, the device agent can in each time interval estimate the time when the temperature crosses the priority border (if the actual consumption does not change). The priority ranges and the consumption prediction is illustrated in Figure 2.

The behaviour model of the smart refrigerator controlling agent is as follows:

1. start → make_request
2. make_request →
   ConsumptionEnd=time_to_upper_limit(Now,PrioNow),
   request(Now,ConsumptionEnd,PrioNow,64)
3. request_allowed(From,To,Prio,Amount) →
   insert_event(From,set_consumption(PrioNow,64)),
   insert_event(To,set_consumption(0,0))
4. request_denied →
   {Prio,ConsumptionStart}=time_to_lower_limit(Now),
   ConsumptionEnd=time_to_upper_limit(ConsumptionStart,Prio),
   request(ConsumptiontStart,ConsumptionEnd,Prio,64)
5. event set_consumption(_,0) → make_request

At start, the agent requests the standard consumption time frame from the scheduler. If this request is denied, the agent computes the time when (with no power supplied) the temperature will cross the next lower priority limit and requests the consumption from that time with a new (increased) priority. If the requested consumption frame is granted, the temperature crosses the upper priority boundary and the next request will have a decreased priority.

3.1   The Effect of Appliance Optimization

The effect of priority-driven "self" organization of appliance power consumption can be demonstrated on a concurrent run of several accumulator-like devices. We have used three refrigerators with slightly different power consumptions and cooling effectiveness.

The three appliances have constant consumption of 70, 60 and 50 watts and constant cooling speed of 450, 500 and 300 seconds to lower the temperature down one degree Celsius. The consumption priority ranges are $> 7°C$: priority 5, $4–7°C$: prio.4, $1–4°C$: prio.3, $0–1°C$: prio.2 and $< 0°C$: priority 0.

When there is no consumption limit, the appliances operate in the narrow temperature range. Due to the different cooling effectiveness, each refrigerator has different cycle length and therefore there are consumption peeks whenever the operation of two or more refrigerators overlaps. The resulting summary consumption profile can be very fuzzy even for constant consumption and only three appliances (see Figure 3a).

In a limited capacity regime of the power source, the appliances cannot operate simultaneously and only the highest priority consumption requests are approved by the dispatcher. In the case of two or more concurrent requests with the same priority, the lowest-power-amount requests are satisfied first. As stated above, the request priority is a function of the refrigerator inner temperature hence the refrigerators with the most urgent cooling need are served first. In the meantime, the temperature in the other refrigerators increases, which automatically causes higher priority requests from them in subsequents scheduling cycles.

This mechanism automatically leads to a flat consumption profile under the given power source limit with regularly alternating consumption of particular appliances. The result is illustrated in Figure 3b.

There is no need for an explicit plan or system state negotiation, or complicated and hard-to-maintain rules. The fluency of the power consumption profile, peak elimination and postponement of the consumption according to the amount of available power, is the function of private priority-changing rules of particular agents on the one hand and priority-driven scheduling of the consumption on the other.

The process of "flattening" the consumption profile is automatically imposed by decreased power source capacity and can therefore change dynamically as a reaction to the actual power supply without any switching of consumption profiles, rules etc. We have demonstrated this effect on the appliance whose standard operation is mostly regular. The priority-driven control process makes it more or less irregular to fit better to the actual power supply. The control process works in the same way for appliances with a naturally irregular power profile.

a)

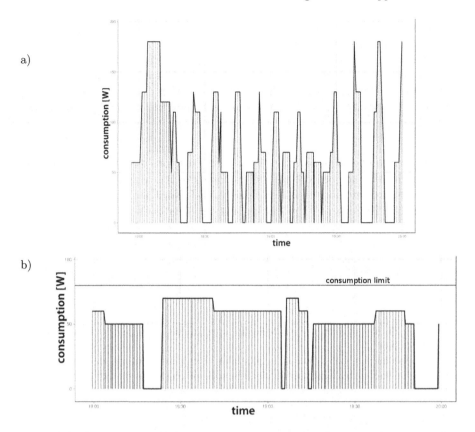

b)

Fig. 3. a) Original power consumption profile of three different refrigerators (each appliance with different bar color), and b) the optimized form of this profile

The regular behaviour appliance was chosen for the purpose of a comprehensible demonstration.

This intelligent self-organizing behaviour emerges from the interaction between the appliance controllers and the scheduler without the need to develop complicated centralized rules for all the devices in the grid.

4   Conclusions

We have demonstrated the method of intelligent appliance-controlling rules in the Priority-Driven Appliance Control System (PAX). The PAX system is based on the multi-agent paradigm of autonomous "lightweight" agents keeping the amount of knowledge about the surrounding world as low as possible. Every appliance controller works with the information about its own state and the power consumption urgency in the given time interval and in the near future. The communication with other agents is based mainly on the request-answer schema. Thanks to these two properties, particular appliance controllers can be implemented using really cheap and low computation power microcontrollers (for details about the prototype hardware see [11]).

Another advantage of the presented system is that the system is usable for real appliance control as well as pure simulation, or in a mixed-mode combining online data from real appliance with the virtual ones. Thus the development cycle from a simulation model to the real controller is as smooth and fast as possible.

Acknowledgments. This work has been partly supported by the Czech Science Foundation under the project 102/09/1842 and by the Ministry of Education, Youth and Sports of the Czech Republic (ENET No. CZ.1.05/2.1.00/03.0069) and project SP2013/68 and by IT4Innovations Centre of Excellence project, reg. no. CZ.1.05/1.1.00/02.0070 supported by Operational Programme 'Research and Development for Innovations' funded by Structural Funds of the European Union and state budget of the Czech Republic.

References

1. Lund, H.: Renewable energy strategies for sustainable development. Energy 32(6), 912–919 (2007)
2. Georgopoulou, E., Lalas, D., Papagiannakis, L.: A multicriteria decision aid approach for energy planning problems: The case of renewable energy option. European Journal of Operational Research 103(1), 38–54 (1997)
3. Liu, C., Li, F., Ma, L.P., Cheng, H.M.: Advanced materials for energy storage. Advanced Materials 22(8), E28–E62 (2010)
4. Mohsenian-Rad, A., Wong, V.W.S., Jatskevich, J., Schober, R., Leon-Garcia, A.: Autonomous demand-side management based on game-theoretic energy consumption scheduling for the future smart grid. IEEE Transactions on Smart Grid 1(3), 320–331 (2010)
5. Pedrasa, M.A.A., Spooner, T.D., MacGill, I.F.: Scheduling of demand side resources using binary particle swarm optimization. IEEE Transactions on Power Systems 24(3), 1173–1181 (2009)
6. Egu Brno, S.R.O.: Power Network Analyzer,
   http://www.egubrno.cz/sekce/s005/pristroje/mds/mds_ostatni_3_5_u.html
7. Prokop, L., Misak, S.: Energy Concept of Dwelling. In: 13rd International Scientific Conference Electric Power Engineering, pp. 753–758 (2012)
8. Misak, S., Prokop, L.: Technical-Economical Analysis of Hybrid Off-grid Power System. In: 11th International Scientific Conference Electric Power Engineering, pp. 295–300 (2010)
9. Cesarini, F., Thompson, S.: Erlang programming. O'Reilly Media (2009)
10. Prýmek, M., Horák, A.: Priority-based smart household power control model. In: Electrical Power and Energy Conference 2012, pp. 405–411. IEEE Computer Society, London (2012)
11. Prýmek, M., Horák, A.: Modelling Optimal Household Power Consumption. In: Proceedings of ElNet 2012 Workshop, Ostrava, Czech Republic, VSB Technical University of Ostrava (2012)

Implementation of Fuzzy Speed Controller in Control Structure of A.C. Drive with Induction Motor

Pavel Brandstetter, Jiri Friedrich, and Libor Stepanec

VSB - Technical University of Ostrava, Department of Electronics, 17. listopadu 15/2172, 70833 Ostrava-Poruba, Czech Republic
pavel.brandstetter@vsb.cz

Abstract. The paper deals with application of the fuzzy logic in a control structure of A.C. drive with an induction motor. The paper shows the own authors´ approach in implementing fuzzy speed controller into a specific control system with digital signal processor. It was realized important simulations and experiments which confirm the rightness of proposed structure and good behavior of developed fuzzy speed controller. Explanation of obtained results is described and main features of used implementation methods are also summarized.

Keywords: Fuzzy logic, speed control, vector control, induction motor, A.C. drive, digital signal processor.

1   Introduction

Modern A.C. variable-speed drives (VSDs) have come a long way as an alternative solution to D.C. drives. Their development is characterized by process made in various areas including electrical machines, power electronics, control systems etc.

Initially, D.C. motors were used in the VSDs because they could easily achieve the required speed and torque without the need for sophisticated control methods and complicated electronics. However, the evolution of A.C. variable-speed drive technology was driven partly by the desire to emulate the excellent performance of the D.C. motor, such as fast torque response and speed accuracy [1].

Vector-controlled drives providing high-dynamic performance are finding increased number of industrial applications. The induction motors are often preferred choice in variable-speed drive applications. Nowadays low cost digital signal processors (DSPs) enable the development of cost effective VSDs and the widespread availability of DSPs enable the development of new control methods. The applications of soft computing methods in electrical drives, which include fuzzy logic (FL) applications, discuss FL based speed or position controller applications. The FL controllers can lead to improved performance, enhanced tuning and adaptive capabilities [2]-[4].

In the first part of the paper there is described a basic issue of the speed control of the vector-controlled induction motor and its control structure. Next there is described used fuzzy controller and its implementation into the control system with a digital signal processor.

Á. Herrero et al. (eds.), *International Joint Conference SOCO'13-CISIS'13-ICEUTE'13*,
Advances in Intelligent Systems and Computing 239,
DOI: 10.1007/978-3-319-01854-6_39, © Springer International Publishing Switzerland 2014

2   List of Symbols

i_{Sx}, i_{Sy}	components of stator current space vector i_S
v_{Sx}, v_{Sy}	components of stator voltage space vector v_S
i_m	magnetizing current
i_{Sxref}	reference value of the magnetizing current component
i_{Syref}	reference value of the torque current component
F_{Ci}	transfer function of the closed current control loop
$F_{o\omega}$	transfer function of current and speed control loop
J	total moment of inertia
K_{AD}	transfer constant of the A/D converter
K_I, K_{IS}	transfer constant of the current sensor and incremental sensor IS
K_T	torque constant
K_p, K_i	proportional and integral gain of the PI controller
$K_{R\omega}, T_{R\omega}$	parameters of the speed controller
L_h	magnetizing inductance
$T_{S\omega}$	sampling period of the speed control loop
T_R	rotor time constant
T_L	load torque
ω_{mref}, ω_m	reference and real rotor angular speed
$e_\omega, \Delta e_\omega$	speed error and speed error change
ε	rotor angle
γ	angle between real axes of coordinate system $[\alpha, \beta]$ and $[x, y]$
σ_R	total and rotor leakage constant

3   Speed Control Structure of Induction Motor

For variable-speed drive with the induction motor (IM), a cascade structure is often used. The cascade control structure consists of several control loops, whereas the current control loops for the magnetizing and torque current components i_{Sx}, i_{Sy} are subordinate and the flux and speed control loops are superior loops (see Fig. 1). Using subordinate current control loop we ensure a quick response to its reference value, which comes from the superior flux and speed control loop.

The speed controller determines the reference quantity i_{Syref} for the current controller. Its design is based on the motion equation:

$$J\frac{d\omega_m}{dt} = \frac{3}{2}\frac{L_h}{(1+\sigma_R)}i_m i_{Sy} - T_L = K_T i_m i_{Sy} - T_L \tag{1}$$

The open speed control loop is described by transfer function:

$$F_{o\omega}(s) = F_{R\omega}(s)F_{Ci}(s)\frac{K_{IS}K_T i_m}{sK_I K_{AD}J}\frac{1}{(1+s\frac{T_{S\omega}}{2})} \tag{2}$$

For the speed control we can design PI controller with the parameters $K_{R\omega}$ and $T_{R\omega}$:

$$F_{R\omega}(s) = K_{R\omega} \frac{(1+sT_{R\omega})}{sT_{R\omega}}$$ (3)

The parameters of the classical PI controllers have been set by various mathematical and experimental methods. Modern design approaches include soft computing methods, such as application of neural networks, fuzzy logic, genetic algorithms, etc.

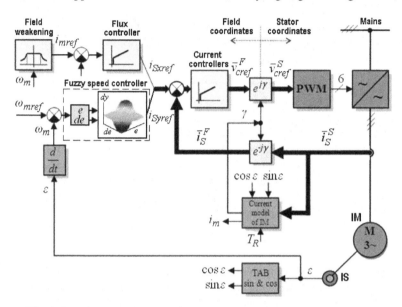

Fig. 1. Speed control structure of induction motor with the fuzzy controller

4   Fuzzy Speed Controller

The fuzzy logic control is receiving great interest world-wide. There are various types of fuzzy logic systems. In the most frequently used fuzzy logic systems (FLS) there are three parts. There is a first transformation which converts crisp input values into fuzzy values, thus it is also called the fuzzification. This is followed by an inference engine which is simply a rule interpreter, and infers the output values using the inputs and linguistic expert rules. Finally there is a second transformation which converts fuzzy values into crisp output values, thus is also called the defuzzification. This part is most time-consuming part for implementation to DSP control system. There are three main types of fuzzy systems: Zadeh-Mamdani fuzzy system, Sugeno fuzzy system and Tsukamoto fuzzy system [5], [6].

The main purpose of application fuzzy controllers is to reduce the tuning efforts associated with the controllers and also to obtain improved responses. It should be noted that, in general, fuzzy logic controller contains four main parts: fuzzifier, knowledge base (rule base and data base), inference engine and defuzzifier. Real fuzzy logic controller (FLC) consists of preprocessing (matching input value) and post processing which is very important for example for tuning FLC.

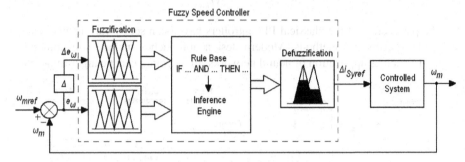

Fig. 2. Block structure of the fuzzy speed controller

The basic structure of the fuzzy speed controller in closed control loop is shown in Fig. 2. In our case, the controlled variable is the rotor angular speed ω_m. The input variables of the fuzzy speed controller are: speed error e_ω and speed error change Δe_ω. The output variable is a change of torque current component Δi_{Syref}:

$$e_{\omega(k)} = \omega_{mref(k)} - \omega_{m(k)} \tag{4}$$

$$\Delta e_{\omega(k)} = e_{\omega(k)} - e_{\omega(k-1)} \tag{5}$$

The classical PI speed controller for the induction motor drive can be expressed using proportional and integral gain K_p, K_i. The change of the torque current component in discrete form using sampling period $T_{S\omega}$ is described as follows:

$$\Delta i_{Sy(k)} = i_{Syref(k)} - i_{Syref(k-1)} = K_p \Delta e_{\omega(k)} + K_i T_{S\omega} e_{\omega(k)} \tag{6}$$

$$i_{Syref(k)} = i_{Syref(k-1)} + \Delta i_{Sy(k)} \tag{7}$$

We can obtain a simplified description of the fuzzy speed controller using fuzzification process **F** and defuzzification process **D**:

$$i_{Sy(k)} = \mathbf{D}\left\{\mathbf{F}\left[K_p \Delta e_{\omega(k)} + K_i T_{S\omega} e_{\omega(k)}\right]\right\} + i_{Syref(k-1)} \tag{8}$$

The shared disadvantage of such derived fuzzy controllers is the method of their setup. The primary crude setup is carried out by changing the scope of the universe for the speed error e_ω, speed error change Δe_ω and for the action quantity Δi_{Syref}. Further debugging is then carried out by editing the table or shifting the membership functions. Setup is mostly intuitive, and there is no generally applicable method.

To simplify design, it is a good idea to normalize the scopes of the universe for input and output variables, for instance in the interval <-1; 1>. The input or output variable is then multiplied by a constant expressing the actual scope of the universe.

We introduce the scale m for the scope of the universe ($m > 0$). This scale will set the scope of the universe for the speed error e_ω and speed error change Δe_ω. By modifying equation (8) we obtain the following equation:

$$i_{Sy(k)} = m\mathbf{D}\left\{\mathbf{F}\left[\frac{1}{m}K_p\Delta e_{\omega(k)} + \frac{1}{m}K_i T_{S\omega}e_{\omega(k)}\right]\right\} + i_{Syref(k-1)} \tag{9}$$

The physical significance of the parameters in fuzzy PI controller is the same as in the classical PI controller, both for the controller gain and for the integration time constant. When setting up a fuzzy PI controller, one can follow the same steps used to set the parameters of the classical PI controller. If we use non-linear or asymmetrical distributions of membership functions, it is clear that the controller will become non-linear. A suitable distribution of the membership functions in a specific case may lead to a control process which has more advantageous properties than when using a fuzzy regulator with an approximately linear distribution of the membership functions. Changing the membership functions or adjusting the rule basis usually also requires a change of the controller parameters [7].

5   Simulation Results

Design of the fuzzy controller was carried out in Matlab - Simulink with an extension Fuzzy Logic Toolbox. One possible initial rule base which consists of 49 linguistic rules is shown in Tab. 1 and gives the change of torque current component Δi_{Syref} in terms of two inputs: the speed error e_ω and speed error change Δe_ω.

Table 1. Fuzzy rule base with 49 rules

$\Delta e_\omega \backslash e_\omega$	NB	NM	NS	ZO	PS	PM	PB
PB	ZO	PS	PM	PB	PB	PB	PB
PM	NS	ZO	PS	PM	PB	PB	PB
PS	NM	NS	ZO	PS	PM	PB	PB
ZO	NB	NM	NS	ZO	PS	PM	PB
NS	NB	NB	NM	NS	ZO	PS	PM
NM	NB	NB	NB	NM	NS	ZO	PS
NB	NB	NB	NB	NB	NM	NS	ZO

In Tab.1, the following fuzzy sets are used: NB negative big, NM negative medium, NS negative small, ZO zero, PS positive small, PM positive medium and PB positive big. For example, it follows from Tab. 1 that the following rules are:

$$\text{IF } e_\omega \text{ is NS and } \Delta e_\omega \text{ is PB then } \Delta i_{Syref} \text{ is PM} \tag{10}$$

A number of simulations of the speed control structure with the fuzzy speed controller (see Fig. 1.) were carried to verify functionality after setup and debugging.

The result of the designed fuzzy controller is a control surface which specifies the response of the output to a change of the input variables, which is then easier to implement into a microcomputer control system, e.g. in the form of a table [7].

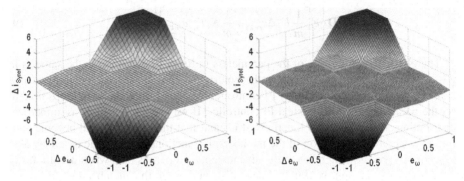

Fig. 3. Control surface of the fuzzy speed controller - 41x41 (left) and 77x77 elements (right)

6   Implementation of Fuzzy Controller into Control System

For implementation into DSP control system, three ways were used: implementation using a table, an approximated table and implementation using Sugeno FLS. The Sugeno FLS avoids the use of defuzzification, which is most time-consuming part for implementation to DSP [7].

6.1   Implementation Using the Table

A basic prerequisite for implementation is the use of certain pre-obtained data for the creation of the table. A suitable example of this is the characteristic surface obtained by simulations, or as the result of practical realizations of the classical structure of the fuzzy controller containing the appropriate fuzzification, inference and defuzzification blocks and its debugging on a real controlled system. The latter method is more suitable in cases where emphasis is placed on computation speed.

Table 2. Part of the table resulting by the transfer of the control surface

$e_\omega \backslash \Delta e_\omega$	-0.3	-0.2	-0.1	0	0.1	0.2	0.3
-0.2	-0.4667	-0.4	-0.3429	-0.24	-0.0857	0	0.0667
-0.1	-0.4286	-0.3429	-0.2571	-0.12	0	0.0857	0.1714
0	-0.36	-0.24	-0.12	0	0.12	0.24	0.36
0.1	-0.1714	-0.0857	0	0.12	0.2571	0.3429	0.4286
0.2	-0.0667	0	0.0857	0.24	0.3429	0.4	0.4667
0.3	0	0.0667	0.1714	0.36	0.4286	0.4667	0.4667

For the application, a combination of Sugeno type FLS with modified output membership functions to singletons is selected. This method then enables to create a table from the control surface. A further criterion of the table is its scale. By transfer of the control surface (see Fig. 3), the table of 41 x 41 elements is created, whose part is shown in Tab. 2.

6.2   Implementation Using the Approximated Table

This implementation method differs from the previous one by having a greater scope of the table due to searching the values for approximation. Linear 4-point approximation is used. A simplified diagram of the approximation is provided in Tab. 2 (gray marked part). First, we choose a coordinate which corresponds to the output values adjusted to the scope of the table, e.g. 0.24. Then the calculation of two values is carried out (basically the values between 0.12 and 0.2571, 0.24 and 0.3429). These values are then used to carry out the resulting approximation with adjustment corresponding to the necessary modification of the input deviation value.

6.3   Implementation Using Sugeno FLS

This implementation is based on the classical fuzzy logic scheme and its parts, such as fuzzification, inference and defuzzification. Fuzzification itself is carried out via trapezoidal membership functions or any membership functions comprising straight-line segments. In this case, one may realize a membership function comprised of 3 straight lines entered by 4 points. Of course, it is also possible to use triangular membership functions. During fuzzification, no restrictions are placed here as long as the membership functions are correctly entered - contrary to the restrictions e.g. in tables used to minimize the number of elements. The entered functions are only restricted by the scope of the data processed by the microcomputer.

The MAX-MIN method modified for Sugeno FLS is used for inference, and the rules are written down in a table whose size copies the number of appropriate terms, or membership functions for individual input variables. The size of the action interference is then selected with respect to controller limits, which corresponds to the selection of a universe of appropriate singletons. For the input variables, we use a normalized scope of universes of input variables. Defuzzification is then carried out by computing the weighted average from the values obtained by the rules and individual correspondence scales. The method requires a significantly greater amount of computational time. Its advantage is that it does not need to know e.g. the characteristic control surface in advance.

7   Experimental Results

To verify the simulation models and principles as well as vector control, experimental laboratory workplace with the IM drive was created. The basic parts are machine set, frequency converter, control system with DSP, personal computer and the necessary measuring instruments. The IM is mechanically connected to the frame and together with D.C. motor make up machine set. Both machines are mechanically connected by a flexible coupling. The D.C. motor is used as the load machine. In the control system, a Texas Instruments DSP TMS320C40 is used. The DSP works with 32-bit data in floating-point arithmetic. The instruction cycle interval is 50 ns.

The speed control of the vector controlled IM was tested in the different conditions of the rotor angular speed and load torque. The following figures show the time responses of important quantities of the IM drive with the vector control which were obtained by measurement on the laboratory stand with the active load unit.

In the Figs. 4-6 there are time responses of quantities for the implementation using the table of 77x77 elements (left) and using the approximated table of 77x77 elements (right). The Fig. 7 and 8 show time responses of quantities for the implementation using Sugeno FLS [7].

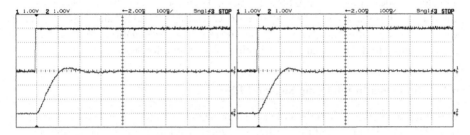

Fig. 4. Reference and real speed, speed change $0 \rightarrow 300$ rpm without load, (CH1: $\omega_{mref} = f(t)$, CH2: $\omega_m = f(t)$, $m_\omega = 100$ rpm/d, $m_t = 100$ ms/d)

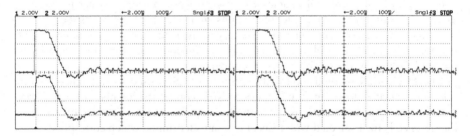

Fig. 5. Reference and real torque current component, speed change $0 \rightarrow 300$ rpm without load, (CH1: $i_{Syref} = f(t)$, CH2: $i_{Sy} = f(t)$, $m_i = 2$ A/d, $m_t = 100$ ms/d)

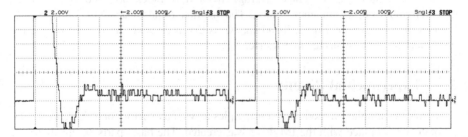

Fig. 6. Speed error, speed change $0 \rightarrow 300$ rpm without load, (CH1: $e_\omega = f(t)$, $m_e = 10$ rpm/d, $m_t = 100$ ms/d)

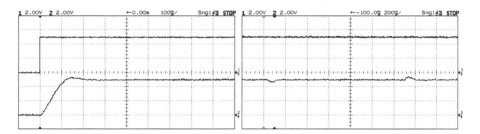

Fig. 7. Reference and real speed, speed change $0 \rightarrow 300$ rpm without load (left), jump of the load at 300 rpm (right),
(CH1: $\omega_{mref} = f(t)$, CH2: $\omega_m = f(t)$, $m_\omega = 120$ rpm/d, $m_t = 100$ ms/d)

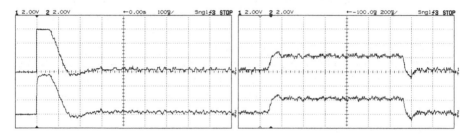

Fig. 8. Reference and real torque current component, speed change $0 \rightarrow 300$ rpm without load (left), jump of the load at 300 rpm (right)
(CH1: $i_{Syref} = f(t)$, CH2: $i_{Sy} = f(t)$, $m_i = 2$ A/d, $m_t = 100$ ms/d)

The following Tab. 3 shows computational complexity of implemented fuzzy controllers expressed using the calculation period of the used DSP. The computational complexity of the algorithm influences the sampling period of the control loop [7].

Table 3. Computational complexity of implemented fuzzy controllers

Implementation of the fuzzy controller	Calculation period for DSP TMS320C40 [μs]
Implementation using the table	6.0
Implementation using the approximated table	7.8
Implementation using Sugeno FLS	32.5 - 42.5

8 Conclusion

The experimental measurements carried out for three implementation variants allow us to evaluate individual variants. The first two variants are both relatively simple, and rely on transferring the characteristic control surface for the given controller into a table. The scope of the table should depend primarily on the complexity of the control surface. The number of elements in the table affects especially the first tested method, which does not approximate the table leading to a permanent control error in the stable state and, in some cases, a worse transition process than in the case of the other two methods. However, this method is the easiest to apply and requires the least amount of computational time.

The second implementation method uses linear approximation and differs from the first method by a zero deviation in the stable state, but this comes at a cost of more computational time. These two application methods could be used to design a composite method, which compromises between the required amount of computational time and of microcomputer memory.

The last applied fuzzy logic method is classic fuzzification, inference and defuzzification. This method requires the largest amount of time, but the amount of time required varies - the method does not have a fixed computation time. This is an inherent property of the process and of determining how many rules there are at a given point of time. In case of the activation of one term from each of the two input variables, the lower of the specified times is used. In case of the activation of both terms from each variable, the greater of the specified times is used. With respect to setup and experimenting, this method is more suitable for practical realization, since it is relatively simple to setup the placement of individual terms in the universe and to change their shape. In the table version, it is always necessary to rewrite elements in the table, making this method useless if there are too many. It is thus recommended to use the classical fuzzy controller structure for experiments, and then carry out the final application e.g. via a table.

Acknowledgments. The article was elaborated in the framework of IT4Innovations Centre of Excellence project, reg. no. CZ.1.05/1.1.00/02.0070 funded by Structural Funds of the EU and state budget of the CR and in the framework of the project SP2013/118 which was supported by VSB-Technical University of Ostrava.

References

1. Vas, P.: Artificial-Intelligence-Based Electrical Machines and Drives. Oxford science publication (1999)
2. Kromer, P., Platos, J., Snasel, V., Abraham, A.: Fuzzy Classification by Evolutionary Algorithms. In: IEEE International Conf. on Systems, Man and Cybernetics, pp. 313–318 (2011)
3. Birou, I., Maier, V., Pavel, S., Rusu, C.: Indirect Vector Control of an Induction Motor with Fuzzy-Logic based Speed Controller. Advances in Electrical and Computer Engineering 10(1), 116–120 (2010)
4. Palacky, P., Hudecek, P., Havel, A.: Real-Time Estimation of Induction Motor Parameters Based on the Genetic Algorithm. In: Herrero, Á., et al. (eds.) Int. Joint Conf. CISIS'12-ICEUTE'12-SOCO'12. AISC, vol. 189, pp. 401–409. Springer, Heidelberg (2013)
5. Fedor, P., Perdukova, D., Ferkova, Z.: Optimal Input Vector Based Fuzzy Controller Rules Design. In: Herrero, Á., Snášel, V., Abraham, A., Zelinka, I., Baruque, B., Quintián, H., Calvo, J.L., Sedano, J., Corchado, E. (eds.) Int. Joint Conf. CISIS'12-ICEUTE'12-SOCO'12. AISC, vol. 189, pp. 371–380. Springer, Heidelberg (2013)
6. Perdukova, D., Fedor, P.: Simple Method of Fuzzy Linearization of Non-Linear Dynamic System. Acta Technica CSAV 55(1), 97–111 (2010)
7. Stepanec, L.: Applications of the Fuzzy Logic in the Control of Electrical Drives. PhD. Thesis, VSB-Technical University of Ostrava (2003)

Artificial Neural Network Modeling of a Photovoltaic Module

Jose Manuel López-Guede, Jose Antonio Ramos-Hernanz, and Manuel Graña

Grupo de Inteligencia Computacional, Universidad del Pais Vasco (UPV/EHU)

Abstract. This paper deals with the problem of designing an accurate and computationally fast model of a particular real photovoltaic module. There are a number of well known theoretical models, but they need the fine tuning of several parameters, whose values are often is difficult or impossible to estimate. The difficulty of these calibration processes has driven the research into approximation models that can be trained from data observed during the working operation of the plant, i.e. Artificial Neural Network (ANN) models. In this paper we derive an accurate ANN model of a real ATERSA A55 photovoltaic module, showing all the steps and electrical devices needed to reach that objective.

Keywords: Photovoltaic module, Photovoltaic cell, Neural Network Model, Atersa A55.

1   Introduction

The increasing world's energy demands, and the need of revising the energy policies in order to fight against the emissions of CO_2 and environmental pollution are some reasons for the increasing interest in the development of renewable energy sources. This interest has motivated research and technological investments devoted to improve energy efficiency and generation. Photovoltaic energy is a clean energy, with long service life and high reliability. Thus, it has been considered as one of the most sustainable renewable energie sources. Photovoltaic systems may be located close to the points of consumption, avoiding transmission losses and contributing to the reduction of CO_2 emissions in urban centers. However, due to the high cost and the low efficiency of commercial modules (about 15%), it is essential to ensure that they work at their peak production regime. To reach this objective it is neccesary to develop appropriate control algorithms, and to have an accurate model of the real (not ideal) photovoltaic elements behavior is mandatory. There are a number of theoretical photovoltaic cell models [1,2,3,4,5], however they are not easily fit to a given real particular cell, because parameter values are either unknown or difficult to estimate. Lack of calibration thus render these models useless. The main objective of this paper is to describe the process that to obtain a model based on artificial neural networks (ANN) training over acquired data of a real ATERSA A-55 photovoltaic module.

Á. Herrero et al. (eds.), *International Joint Conference SOCO'13-CISIS'13-ICEUTE'13*,   389
Advances in Intelligent Systems and Computing 239,
DOI: 10.1007/978-3-319-01854-6_40, © Springer International Publishing Switzerland 2014

The paper is structured as follows. Section 2 gives a background on the most used theoretical models of photovoltaic modules and on ANNs. Section 3 describes the experimental design that we have followed, detailing the photovoltaic module characteristics, the electrical installation and the devices used to the data capture and the ANNs training process. Section 4 discusses the experimental results and finally, section 5 summarizes our conclusions and future work.

2   Background

2.1   Ideal Photovoltaic Cell

The ideal photovoltaic cell can be modeled as an electric current source with an anti-parallel diode, as shown in the sub-circuit enclosed by a dotted line on left part of the Fig. 1. The direct electric current generated when the cell is exposed to light varies linearly with the solar radiation. The whole circuit of the the Fig. 1 shows an improved model that includes the effects of a shunt resistor and the other one in series, where the involved physical magnitudes are: I_{ph} is the photogenerated current or photocurrent, I_d is the current of the diode, I_P is the shunt current, R_S is the series resistance (Ω) and R_p is the shunt resistance (Ω). This equivalent circuit can be used either for an individual photovoltaic cell, for a photovoltaic module that consists of several cells or for a photovoltaic matrix that is composed of several modules.

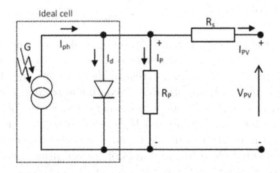

Fig. 1. Basic and improved equivalent model of an ideal photovoltaic cell

2.2   Theoretical Photovoltaic Models

Typical Equation Model. Based on the ideal equivalent circuit of Fig. 1, we can describe the relationship between the voltage (V_{PV}) and the current (I_{PV}) supplied by the photovoltaic cell by means of eq. (1)[6], so that expanding each current we obtain eq. (2)[7]:

$$I_{PV} = I_{ph} - I_d - I_P, \tag{1}$$

$$I_{PV} = I_{ph} - I_0 \left(e^{\frac{q(V_{PV} + I_{PV} R_S)}{aKT}} - 1 \right) - \frac{V_{PV} + I_{PV} R_S}{R_P}, \tag{2}$$

where:

I_0 is the saturation current of the diode (A),

q is the charge of the electron, 1.6×10^{-19} (C),

a is the diode ideality constant,

K is the Boltzmann's constant, 1.38×10^{-23} (j/K),

T is the cell temperature (ïżœC).

Since this is a theoretical model, we have to use an estimation of all the involved parameters and this circumstance leads necessarily only to approximate values when the study of a particular photovoltaic module is being carried out.

Characteristic Curves. Another theoretical model are the I-V curves provided by manufacturers. These curves give the manufacturer's specification of the relation between the current (I_{PV}) and the voltage (V_{PV}) supplied by a particular photovoltaic module. In Fig. 2 the I-V curves of the ATERSA A-55 photovoltaic module are shown, at a specific temperature for a few irradiance values. Temperature and irradiance are relevant magnitudes in the relation between I_{PV} and V_{PV}, but the construction a detailed 4-D graphical representation is a task not undertaken by the manufacturing companies, therefore users and researchers lack some detailed information that may be key to obtain efficient control of the photovoltaic regime.

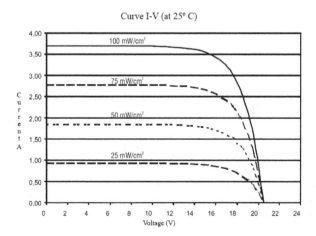

Fig. 2. I-V curve of the Atersa A-55 photovoltaic module

2.3   Artificial Neural Networks

The use of ANNs is motivated by its ability to model systems [8]. These bio-inspired computational devices have several advantages, and among others, these are the most outstanding to our problem:

- Learning capabilities: If they are properly trained, they can learn complex mathematical models. There are several well known training algorithms and good and tested implementations of them. The main challenge concerning this issue is to choose appropriate inputs and outputs to the black box model and the internal structure.
- Generalization capabilities: Again, if they are properly trained and the training examples cover a variety of different situations, the response of a neural network in unseen situations (i.e., with unseen inputs) will probably be acceptable and quite similar to the correct response. So it is said that they have the *generalization property*.
- Real time capabilities: Once they are trained, and due to their parallel internal structure, their response is always very fast. Their internal structure could be more or less complex, but in any case, all the internal operations that must be done are several multiplications and additions if it is a linear neural network. This fast response is independent of the complexity of the learned models.

3   Experimental Design

In this section we describe the experimental design that we have followed, giving the specifications of the photovoltaic module that has been used, the description of the devices used to obtain measurements of all the magnitudes that characterize its I-V curve, and the procedure that has been followed to train and test an accurate neural network based model.

3.1   Photovoltaic Module

In this subsection we introduce the characteristics of the photovoltaic module of which we are going to obtain an empirical ANN model. The module has been manufactured by Atersa, a pioneer company in Spain within the photovoltaic solar power sector, with more than 35 years of experience. This company offers a 10 years mechanical and linear power warranty, because the long time performance of the photovoltaic modules is a big issue. In this experiment we have chosen the ATERSA A55 model, and several of these modules can be seen in Fig. 3. The specifications of this module are shown in Table 1.

3.2   Electrical Installation for Data Capture

In this subsection we describe the electrical installation needed to obtain the value of the magnitudes involved and the data logging system. On one hand,

Fig. 3. ATERSA A55 photovoltaic modules

Table 1. ATERSA A55 photovoltaic module characteristics

Attribute	Value
Model	Atersa A-55
Cell type	Monocrystalline
Maximum Power [W]	55
Open Circuit Voltage Voc [V]	20,5
Short circuit Current Isc [A]	3,7
Voltage, max power Vmpp [V]	16,2
Current, max power Impp [A]	3,4
Number of cells in series	36
Temp. Coeff. of Isc [mA/ï¿œC]	1,66
Temp. Coeff. of Voc [mV/ï¿œC]	-84,08
Nominal Operation Cell Temp. [ï¿œC]	47,5

Fig. 4(a) shows the conceptual disposition of the measuring devices: the voltmeter is placed in parallel with the module and the amperemeter in series. Besides, there is a variable resistance to act as a variable load and obtain different pairs of voltage and current with the same irradiance and temperature. The variable resistance value is controlled according to our convenience, but the temperature and the irradiance depends on the climatological conditions. On the other hand, Fig. 4(b) shows the devices that have been used to capture the data. The first device is the data logger Sineax CAM, and it was configured to generate records with the irradiance and temperature of the environment and the voltage and current supplied by the photovoltaic module under those environmental conditions. The second element is the multimeter TV809, which helps to isolate the data logging device from the photovoltaic module and converts voltage and current magnitudes to a predefined range. The third element is the irradiance sensor Si-420TC-T-K. It is placed outside close to the photovoltaic module and it is used to provide the irradiance and temperature conditions under which the module is working outside. Finally, the fourth element of the figure are current clamps Chauvin Arnous PAC12 used to measure direct currents provided by the module.

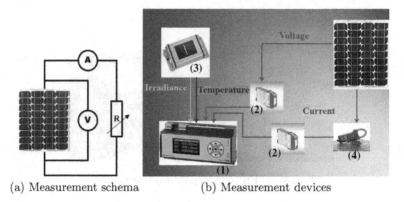

(a) Measurement schema   (b) Measurement devices

Fig. 4. Real photovoltaic module data measurement

3.3 ANN Model Generation

In this subsection we provide a detailed specification of the process followed for the training of the accurate approximation of the I-V curve by an ANN. The trained ANNs are feedforward models, with one or two hidden layers and have been trained with the classical backpropagation. The size of the hidden layers was varying from 1 to 500 for the case of single layer, and from 1 to 100 for each layer in the case of two hidden layers. For each network structure, two activation functions of the neural units has been tested: linear and tan-sigmoid. Five independent training/test processes have been performed for each network, to assess its generalization. Training is performed applying Levenberg-Marquardt algorithm for speed reasons. The model to learn by the ANNs is specified by three single inputs, i.e., the environmental temperature, the environmental irradiance and the voltage supplied by the module, while the output is the current supplied. As summary, the input patterns are 3 dimensional real vectors, while output patterns are 1 dimensional real vectors. This specification of the problem leads to a single variable regression problem, and it involves strong no-linearities. Finally, we have to determine which data and how they are used for the training process. All the ANN input vectors are presented once per iteration in a batch. We have used the raw data, i.e., without normalization. The input vectors and target vectors have been divided into three sets using random indices as follows: 60% are used for training (116 vectors), 20% are used for validation (39 vectors), and finally, the last 20% (39 vectors) are used for testing.

4   Experimental Results

In this section we discuss the results that have been obtained once we have carried out the experiments designed in section 3. In Fig. 5 we show a summary of the learning of all trained ANNs. On one hand, Fig. 5(a) and Fig. 5(b) show the mean squared error (MSE) for the training, validation and test sets of the best ANN model as a function of the number of hidden units using the linear and the tan-sigmoid activation function, respectively. In these figures an overfitting effect can be observed when the number of hidden nodes is larger than 10^1. On the other hand, Fig. 5(c) and Fig. 5(d) show the mean squared error on the test set for ANN with two hidden layers as a function of the number of hidden units in each layer using the linear and the tan-sigmoid activation function, respectively. Again, the number of hidden units per layer most appropriate is close to 10^1. Anyway, from Fig. 5 it is clear that the ANNs with two hidden layers does not have a better performance than the ANNs with one hidden layer. Moreover, the training of these double layered networks is more complex and slower than that of one layer, and the response time when they are in the exploitation phase is also larger. Therefore, we will be using single hidden layer ANN with three hidden units in the following, which gives good generalization and quick response in the exploitation phase. Fig. 6 shows the performance of the best three hidden neurons ANN. Fig. 6(a) and Fig. 6(b) show the original values to learn and the output learned by the network overlapped with all patters and only with the test patters respectively, while Fig. 6(c) and Fig. 6(d) show absolute error between the original and learned values for the same previous cases.

Despite of the very good learning showed by the chosen network, we have represented in Fig. 7 the correlation coefficient (R-value). Its value is over 0.996 for the total response and over 0.997 for the test dataset, and taking into account that $R = 1$ means perfect correlation between the network response and the values to learn, we can conclude that the model is very accurate.

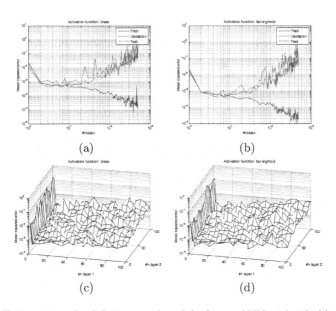

Fig. 5. (a) Train, test and validation results of the linear ANN with 1 hidden layer, (b) idem with tan-sigmoid activation, (c) Test results of the linear with 2 hidden layers, (d) idem with tan-sigmoid activation

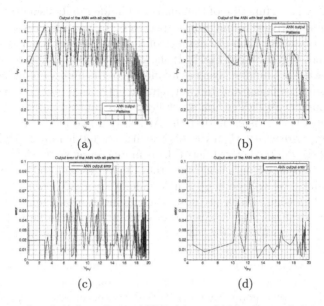

Fig. 6. Test results of the best linear ANN with one hidden layer of three neurons. (a) All patterns output, (b) Test patterns output, (c) All patterns error, (d) Test patterns error.

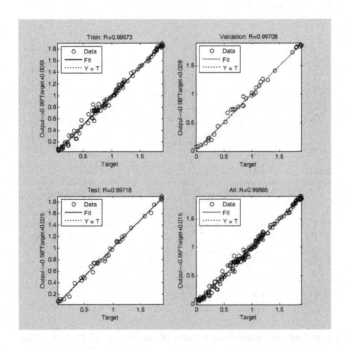

Fig. 7. Accuracy of the selected network (3 hidden nodes)

5   Conclusions and Future Work

In this paper we have discussed the problem that arises when we want to work (in a wide sense, i.e., designing control algorithms) with a real photovoltaic module but without having to use physically that module. In the background section we have reviewed the two most used models to reach this objective, but they have the problem that they are theoretical, lacking the accuracy needed in some cases regarding to a particular module. To overcome this problem, our approach is to train ANN models as approximations to the characteristic I-V curve of a given module. This approach has shown several advantages. The first of them is the correspondence with a particular module, not with an ideal and probably non-existent module. Besides, after following the experimental design described in this paper, the obtained models reached a MSE of 3.10^{-4} on independent dataset, so the accuracy and generalization capability of the model has been verified. The third advantage is that this model is built using a very simple network structure, which gives its response very fast allowing extensive simulations in a reasonable time.

For future work, we plan to gather more experimental data under a variety of environmental conditions different from the winter conditions of the data reported in this paper. A broader range of temperature and irradiance values will improve generalization of the ANN model.

Acknowledgments.   The research was supported by Grant UFI11-07 of the Research Vicerectorship, Basque Country University (UPV/EHU).

References

1. Khezzar, R., Zereg, M., Khezzar, A.: Comparative study of mathematical methods for parameters calculation of current-voltage characteristic of photovoltaic module. In: International Conference on Electrical and Electronics Engineering, ELECO 2009, pp. I-24–I-28 (2009)
2. Ramos, J., Zamora, I., Campayo, J.: Modelling of photovoltaic module. In: International Conference on Renewable Energies and Power Quality, ICREPQ 2010 (2010)
3. Tsai, H.: Development of generalized photovoltaic model using matlab/simulink. In: Proceedings of the World Congress on Engineering and Computer Science, WCECS 2008 (2008)
4. Villalva, M., Gazoli, J., Filho, E.: Modeling and circuit-based simulation of photovoltaic arrays. In: Power Electronics Conference, COBEP 2009, Brazilian, pp. 1244–1254 (2009)
5. Walker, G.R.: Evaluating MPPT converter topologies using a MATLAB PV model. Journal of Electrical & Electronics Engineering, Australia 21(1), 49–56 (2001)
6. Gow, J., Manning, C.: Development of a photovoltaic array model for use in power-electronics simulation studies. IEE Proceedings on Electric Power Applications 146(2), 193–200 (1999)
7. Luque, A., Hegedus, S. (eds.): Handbook of Photovoltaic Science and Engineering. John Wiley & Sons Ltd. (2003)
8. Widrow, B., Lehr, M.: 30 years of adaptive neural networks: perceptron, madaline, and backpropagation. Proceedings of the IEEE 78(9), 1415–1442 (1990)

A Toolbox for DPA Attacks to Smart Cards

Alberto Fuentes Rodríguez,[1,*], Luis Hernández Encinas[1],
Agustín Martín Muñoz[1], and Bernardo Alarcos Alcázar[2]

[1] Instituto de Seguridad de la Información (ISI)
Consejo Superior de Investigaciones Científicas (CSIC)
C/ Serrano 144, 28006-Madrid, Spain
{alberto.fuentes,luis,agustin}@iec.csic.es

[2] Departamento de Automática, Escuela Politécnica Superior
Universidad de Alcalá (UAH)
Carretera A2, km 32, 28871–Alcalá de Henares, Spain
bernardo.alarcos@uah.es

Abstract. Theoretical security of cryptographic systems does not guarantee its security in practice when those systems are implemented in physical devices. In this work we present initial stages of a software tool under development to carry out differential power analysis attacks against smart cards. Once properly configured, the toolbox controls a digital oscilloscope which acquires the power traces during the operation of the device and automatically performs the necessary traces alignment.

1 Introduction

During the last years the use of different electronic devices which implement cryptographic features to perform different operations (personal identification, payment cards, etc.) has increased worldwide. Many of those uses are relevant enough to require important security guarantees.

Until the publication in 1996 of the paper by Kocher ([1]), the cryptographic community considered that the security of a cryptosystem lied in the strength of the mathematical problem in which it was based on. For instance, the security of the RSA algorithm is based on the difficulty of factorizing the two prime numbers which are used, and the security can be enhanced by using larger prime numbers, for example. With the quick and widespread development of portable cryptographic tokens, typically smart cards, which usually have limited memory and computational capabilities, several cryptosystems, as those based in elliptic curves, have been developed; its main virtue is allowing the use of much shorter keys to achieve a similar level of security to that of RSA ([2], [3, Ch.8]).

1.1 General Concepts about Attacks to Physical Devices

Kocher's work demonstrated that it was possible to break the security of embedded cryptographic systems, even easily, by means of an attack which, instead

---

* Corresponding author.

Á. Herrero et al. (eds.), *International Joint Conference SOCO'13-CISIS'13-ICEUTE'13*, 399
Advances in Intelligent Systems and Computing 239,
DOI: 10.1007/978-3-319-01854-6_41, © Springer International Publishing Switzerland 2014

of trying to solve the underlying mathematical problem, took advantage of the information which could be obtained by the fact that the cryptosystem was physically implemented in a device.

In other cases, the information could be leaked by several *side channels* as could be the power consumption —thus enabling Simple Power Analysis (SPA) or Differential Power Analysis (DPA) attacks ([4], [5], [6])—, and the electromagnetic field radiated by the device during its operation —enabling Simple Electro-Magnetic Analysis (SEMA) or Differential ElectroMagnetic Analysis (DEMA) ([7], [8], [9])—. The hypothesis on which side channel attacks are based is that the magnitudes of these characteristics directly depend on the instructions, mathematical operations, and data used by the processor during the cryptographic operations. In this way, the cryptographic key used can be deduced by an adequate analysis of the information obtained by measuring the leakage by these side channels. The procedures to develop both a power analysis and a electromagnetic analysis attack are quite the same; the only difference is the measured magnitude because, as it is well known, the electromagnetic field radiated by the chip circuits are caused, according to Maxwell's equations, by the displacement of charges inside the circuits (the current), which are specially significant when transistors switch their state ([8]).

The above mentioned attacks, which only capture the information leaked from the device without altering it, are known as passive or non-invasive attacks. Moreover, active attacks, invasive or semi-invasive, have also been developed to tamper with the correct behaviour of the device in order to obtain secret information (fault attacks); some of these attacks can alter or even destroy the device ([10], [11]).

In physical attacks it is assumed, as in the case of classical cryptanalysis, that Kerckhoffs' ([12]) principle is verified, that is, the potential eavesdropper has access to the device and knows the cryptographic algorithm that is running in the chip, so as the details of the implementation; the only thing she doesn't know is the key. Furthermore, it is supposed that she can operate the device the number of times she needs, choosing the input values, and obviously, that she can interact with the device or measure certain parameters in its surroundings.

1.2   Example of a SPA Attack

One of the most widely used technique among all side channel attacks is power analysis which employs power consumption traces measured during the operation of the cryptographic device. These traces are captured by using a digital oscilloscope that measures the voltage across the resistor connected in serial to the power source terminal of the device which communicate with the chip, or by using specially designed measurement boards as shown in Figure 1.

By means of Simple Power Analysis (SPA) attacks, the attacker succeeds in obtaining a cryptographic key after capturing a single trace, or a set of a few traces, if he has a detailed knowledge of the algorithm which is being executed. As an example, let's consider the case of a RSA algorithm which executes a modular exponentiation, $y = x^k \mod n$, where the key k is represented in binary

Fig. 1. Picture of a board to measure the power consumption of a smart card

as $k = (k_r k_{r-1} \ldots k_0)_2$. The modular exponentiation, computed with the square and multiply algorithm, processing the bits from left to right, is:

1. $y \leftarrow x$.
2. For i from $(r - 1)$ to 0 do:
   (a) $y \leftarrow y^2 \mod n$.
   (b) If $(k_i = 1)$ then $y \leftarrow (y \cdot x) \mod n$.
3. Return (y).

This way, if for example $k = 23 = (10111)_2$, then $x^k = x^{23} = (((x^2)^2 x)^2 x)^2 x$. As can be observed, when the algorithm processes a '0' bit, the multiplication included in step 2(b) is not executed and, thus, the power consumption will be lower than what would be needed for a '1' bit. If an attacker knows that a smart card is executing a RSA which implements the above algorithm and, measuring the consumption, he captures the SPA trace represented in Figure 2, he would easily identify the key $k = 653642 = (10011111100101001010)_2$.

Without the adequate countermeasures, this kind of attack is very effective and requires a small amount of resources. If the relationship between the consumed power and the cryptographic key is not clear, the captured signal uses to have a very low level as compared to noise, and SPA attacks are not feasible. In those cases, statistical techniques are used in order to perform a DPA attack, requiring a huge amount of data to be captured and processed.

In this work a software tool which is being developed to carry out DPA attacks is presented. It is an open source project intended to provide a baseline from where new techniques or approaches could be shared and reviewed. It is oriented to be clear, user-friendly, and easy to understand, in order to easily choose between different known techniques and options without the need of knowing its internal characteristics. Another key feature is that it is modular because, being DPA attacks a very active research field with different aspects (i.e., alignment techniques, statistical analysis, etc.), the toolbox should provide the capability of integrating new findings in an easy way. It should also be as efficient as possible so as to minimize the resources needed to carry out the attack (main memory,

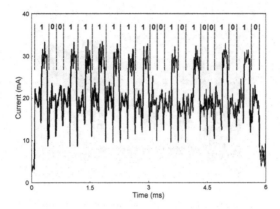

Fig. 2. Power consumption during the execution of a RSA with a key $k = 653642$

cache usage, multithreading programming, etc.), being open to new optimization techniques.

In the following section a description of the general procedure to develop a DPA attack is outlined. In section 3, the classes currently included in the toolbox, which allow the storage and alignment of traces for a DPA attack, are explained. Finally, preliminary results and conclusions are presented.

2 Description of a DPA Attack

In order to obtain a secret key an eavesdropper must have the targeting device and know or guess the model of power consumption (the better the guess, the better the result); it is also assumed that she knows the type of cryptographic algorithms which is being executed, the plaintexts or the ciphertexts, and can measure all the power traces she needs.

To carry out a DPA attack, a large amount of plaintext are ciphered and the corresponding power consumption traces are measured, stored and syncronized. In other words, an alignment procedure is made with the traces captured by means of the digital oscilloscope. The goal is to guarantee that a correct comparison between the traces is performed (the values of all traces must be compared in their correct time instant, to ensure that they are caused by the same operation) all along the execution of the algorithm ([13]). The procedure for a DPA attack is:

1. Choose an intermediate result of the executed algorithm from a function that uses as inputs part of the cryptographic key and known data, usually the plaintext or ciphertext.
2. Create a power profile, measuring the power consumption for D encryption/decription operations with different plaintexts/ciphertexts. For each run the power trace is created with T samples. Thus, a $D \times T$ matrix, P, is obtained.

3. Calculate a hypothetical intermediate value for every possible choice of the key k. If K is the total number of possible choices for k, the result of this step is a $D \times K$ matrix, V.
4. Map the hypothetical intermediate values V to a matrix H of hypothetical power consumption values which are simulated according to a certain power model, commonly the Hamming-distance or the Hamming-weight model ([13, §3.3]). These calculations result in a matrix H of size $D \times K$ which contains these hypothetical power consumption values.
5. Compare the hypothetical power consumption values with the power traces at each position. This results a $K \times T$ matrix R containing the results of the comparison. Different algorithms are used for the comparison (difference or distance of means, generalized maximum-likelihood testing, etc.).

If the statistical method uses a correlation coefficient, the attack is known as Correlation Power Analysis (CPA). This kind of attack was first introduced in [14]. Compared with DPA, CPA requires less number of power traces to launch a successful attack because in DPA all unpredicted data bits contribute to generate a worse Signal to Noise Ratio (SNR). The SNR of DPA could be improved if multiple bits are used in prediction ([15]).

Figure 3 shows a power trace measured with our PicoScope 5204 digital oscilloscope with a Pintek DP-30HS high-sensitivity differential probe during the execution of a DES in a NXP JCOP41/72k smart card. Thousands of similar traces are captured with our software tool and then processed to carry out a DPA attack. DPA attacks can be generalized to Higher-Order Differential Power Analysis (HODPA), where several points in a power trace are used instead of a single one. Usually, a DPA attack of nth order uses n samples simultaneously which correspond to n different intermediate values in the same captured trace. HODPA are specially useful to break implementations which include countermeasures ([16], [17], [18]).

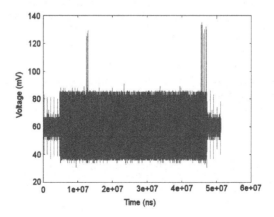

Fig. 3. Power consumption captured during the execution of DES

3   Toolbox Storage and Alignment Capabilities

Each trace is captured by the digital oscilloscope and is stored as a vector of values for further processing. The main parameters that have to be properly adjusted when storing the traces are vertical sensitivity, sampling rate, resolution, memory size and DC offset. The size of the elements of the vector depends on the resolution of the voltage values provided by the oscilloscope. Once the capture of the power values has finished, data are transferred from the oscilloscope memory to the PC memory. If streaming mode is used, the data are transferred during the capture process, although this is only possible with low sampling rates.

Being the data corresponding to the traces the biggest amount of information that the toolbox has to deal with, all the rank of resolutions that the architecture can manage (8 bits, 16 bits, 32 bits) must be provided. In this way, the toolbox would be as versatile as possible and it would minimize the memory consumption and cache usage (the internal representation of traces is a key factor determining the performance).

As an example, for a sampling rate of 4 ns, a computation process that requires 25 ms by the crypto device would require a trace with $25 \cdot 10^{-3}/(4 \cdot 10^5) = 62.5 \cdot 10^5$ sampled values. The memory needed to store 1000 traces in an 8-bit architecture is 5.96 MB. The noisiest the signal is, the higher the number of traces that must be measured. Thus, the internal trace representation can determine if a computer can be used to carry out an attack or not.

In order to represent traces as objects, three possible options have been analyzed, either using *C++ templates*, *objet oriented inheritance*, or *float values*.

1. **C++ templates:** A class *trace* has been created that operates with a generic type which specifies the resolution (i.e., *uint8_t* for 8 bits). As the unique abstraction required for trace representation is the resolution of the elements stored, it fits with the C++ template feature.
2. **Objet oriented inheritance:** Inheritance in object oriented programming is one of the main building blocks together with encapsulation, polymorphism and abstraction. Inheritance provides an *Is-a* relationship between objects. From this point of view, an "abstract" class *CTrace* has been created. This class has a *derived* class for each resolution option. Using this abstraction mechanism, other classes can use *CTrace* class without the need of knowing which resolution is being used.
3. **Float representation of values:** Keeping the values as float does not require abstraction between the different resolutions. Thus, values are represented as 32-bit floats and the internal values represent the voltage, that is, conversion from raw values to voltage values is done only once, while traces are being captured.

Currently the toolbox includes storage and alignment capabilities. The main trace-related classes created so far are :

***CTrace*:** This class is used as an abstraction to caller objects of the resolution of the trace. It is implemented as *CTrace8*, *CTrace16*, or *CTrace32*, depending on resolution of the values (8, 16, or 32 bits, respectively).

CStatTrace: This class represents statistical data obtained from several traces (i.e., mean, variance).

CTimeSlice: A time slice contains the values taken at a time point from several traces. This class is used as an abstraction to caller objects of the resolution of the time slice. It is implemented as *CTimeSlice8*, *CTimeSlice16*, or *CTimeSlice32*, depending on resolution of the values (8, 16, or 32 bits, respectively).

CTraceSet: This class contains sets of traces of the same device taken with the same timing, in order to do statistical analysis. Trace Sets can be in two states Statistical Mode and Alignment Mode. Method availability or performance may depend on the mode. It has a subclass, *CPreProcTraceSet*, with the added property that the traces can be preprocessed: aligned, compressed, etc. Traces alignment through least square matching pattern technique is made by the subclass *CAlignMatchSqrTraceSet* —the displacement between the pattern (first trace of the set) and the power trace is calculated—. For traces alignment through integration, subclass *CAlignSumTraceSet* is used —the displacement between the pattern (first trace of the set) and the power trace is calculated—.

A schematic diagram of the above described classes is shown in Figure 4, where two interface classes, *CFileStorage* (which specifies that the subclasses can be stored in a file) and *CGraphicable* (which specifies that its subclasses can be printed out if gnuplot is installed), are also included.

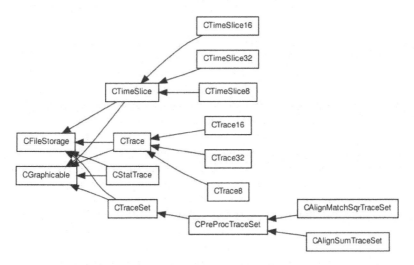

Fig. 4. Hierarchy of classes to manage trace acquisition and alignment

4   Results and Conclusions

This section discusses different implementation issues to optimize the toolbox. Using C++ templates, type abstraction is resolved at compile time. Thus, the compilation time is longer but there is not need of solving polymorphism at execution time that could impact the performance, but as it is one of the toolbox objectives this option must be taken into account. Also templates are checked at compile time for type consistency, so it avoids type errors at execution time.

Template definitions cannot be separated into a header (.h) and source file (.cpp) because templates are instantiated at compile time, not at link time. Thus, declaration and implementation must be located in the same file. The solution to this problem is the explicit instantiation. When templates are explicitly instantiated the programmer defines which possible values the type can acquire. With this approach the compiler will create object code for each specified value and neither compiler nor linker errors will appear. Although currently C++ allows implicit template instantiation to be defined in a separated source file by using the *export* keyword, this feature is not allowed by most C++ compilers.

Another important aspect which must be considered about using C++ templates is that the trace class is used to form other classes in the toolbox (i.e., *CTraceSet*). These must be able to deal with all possible type values that the traces may acquire, but the only way to do that is to define such classes as templates as well. This imply that the templates will extend to many parts of the toolbox source code, which would become less clear. As the toolbox is expected to be used and modified by other developers, clearness of source code is essential.

Using inheritance, resolution operations are done at *CTrace* class level, so other classes can be build without taking resolution into account. In this case, the code is clear and easy to understand, an important advantage with respect to templates. However, abstract methods are solved at execution time. Thus, part of the processing time will be spent in the *dynamic dispatch*.

The option of creating an abstract class representing an unique data element (voltage value) has also been evaluated. In case of using object inheritance, an operation that affects the data elements of a trace requires to call an abstract method for each data element instead of just calling the abstract method of the *CTrace* class. As each call to an abstract method may be solved dynamically, it requires much more computation than in the case of using templates.

As mentioned above, keeping the values as float does not require abstraction between the different resolutions, simplifying source code. In addition, as values stored are the voltage values themselves, conversion from raw to voltage values is done once, while traces are being captured, requiring less computational power than the previously analyzed options. However, the size of internal values is 32 bits. Usually, oscilloscopes offer a maximum resolution of 12 or 16 bits. In those cases the memory and disk spent to store trace information is at least twice the required. Also, storing values with a 32-bit size has a computational disadvantage. Most toolbox operations access trace values sequentially. So, less elements will be found at cache memory and the probability of finding the next value in cache is low. Each time that a value is not found in cache (cache miss), it has to be fetched from its original storage location (RAM memory) which is comparatively slower. So, cache miss delays may counteract the benefits of computing only once the conversion from raw to voltage values.

Table 1. Dyn/Stat benchmark output

| | Time (s) | | | |
| | -O0 optimization | | -O3 optimization | |
Length (elements)	Static	Dynamic	Static	Dynamic
10 000 000	0.177548	0.189265	0.135924	0.120644
100 000 000	1.776202	1.891002	1.333785	1.201456

Performance has been analyzed by means of a benchmark, *Dyn/Stat*, which has been designed to compare the computational time spent in executing the same operations with the different approaches used to internally represent trace values. The aim of *Dyn/Stat* is to check the performance differences between dynamic binding (binding is solved at execution time) and static binding (binding is solved at compile time). Dynamic binding is done with object oriented inheritance with virtual classes (*Object oriented inheritance* approach), while static binding is done when virtual classes are not used (*C++ templates* approach).

Traces are vectors, so the benchmark checks the performance of storing values in vectors and executing vectorial operations. It has been run in a Intel(R) Core(TM)2 Quad CPU Q9550 @ 2.83GHz with GCC compiler version 4.4.3. *Dyn/Stat* benchmark outputs the time of computation for assigning values to two vectors for both approaches with different vector lengths and compilation parameters. Results presented in Table 1 show that, when -O0 minimum optimization is used, compilation time is reduced. In this case, the time penalty for dynamic binding represents a 6.60% in case of a vector with 10^7 elements and 6.46% for 10^9 elements. Thus, it can be assumed that the number of elements does not impact in the dynamic binding time penalty. For -O3 maximum optimization, all compiler optimizations are used. One of these optimizations is -fdevirtualize that allows the compiler to convert the dynamic binding into static binding. Devirtualization can be done when the subclass that implements the virtual function can be defined at compilation time [19]. Thus, the virtual (Dynamic in Table 1) solution is 12.67% faster in case of a vector with 10^7 elements and 11.01% for 10^9 elements.

It can be pointed out that the system works properly if the number of data to process is increased, i.e., processing time grows linearly. Moreover, when the minimum optimization is used (-O0), the virtual solution has a time penalty of about 6.5%, whereas if maximum optimization (-O3) is considered, the time spent in computations is about 11–12.7% shorter with respect to the template-based solution, for vectors of length 10^7–10^9.

Acknowledgements. This research was partly supported by Ministerio de Ciencia e Innovación (Spain) under the grant TIN2011-22668.

References

1. Kocher, P.C.: Timing attacks on implementations of Diffie-Hellman, RSA, DSS, and other systems. In: Koblitz, N. (ed.) CRYPTO 1996. LNCS, vol. 1109, pp. 104–113. Springer, Heidelberg (1996)

2. NIST: Digital Signature Standard (DSS). National Institute of Standard and Technology, Federal Information Processing Standard Publication, FIPS 186-3 (2009)
3. Fúster Sabater, A., Hernández Encinas, L., Martín Muñoz, A., Montoya Vitini, F., Muñoz Masqué, J.: Criptografía, protección de datos y aplicaciones. Una guía para estudiantes y profesionales, RA-MA, Madrid, Spain (2012)
4. Kocher, P., Jaffe, J., Jun, B.: Introduction to differential power analysis and related attacks. Technical report, Cryptography Research Inc. (1998), http://www.cryptography.com/resources/whitepapers/DPATechInfo.pdf
5. Kocher, P.C., Jaffe, J., Jun, B.: Differential power analysis. In: Wiener, M. (ed.) CRYPTO 1999. LNCS, vol. 1666, pp. 388–397. Springer, Heidelberg (1999)
6. Kocher, P.C., Jaffe, J., Jun, B., Rohatgi, P.: Introduction to differential power analysis. J. Cryptograp. Eng. 1, 5–27 (2011)
7. Quisquater, J.J., Samyde, D.: A new tool for non-intrusive analysis of smart cards based on electromagnetic emissions, the SEMA and DEMA methods. In: EURO-CRYPT 2000 Rump Session (2000)
8. Quisquater, J.J., Samyde, D.: ElectroMagnetic Analysis (EMA): Measures and counter-measures for smart cards. In: Attali, S., Jensen, T. (eds.) E-smart 2001. LNCS, vol. 2140, pp. 200–210. Springer, Heidelberg (2001)
9. Quisquater, J.J., Samyde, D.: Eddy current for magnetic analysis with active sensor. In: Proc. of 3rd Conference on Research in SmartCards, E-Smart 2002, Nice, France, pp. 185–194 (2002)
10. Boneh, D., DeMillo, R.A., Lipton, R.J.: On the importance of checking cryptographic protocols for faults. In: Fumy, W. (ed.) EUROCRYPT 1997. LNCS, vol. 1233, pp. 37–51. Springer, Heidelberg (1997)
11. Skorobogatov, S.: Semi-invasive attacks-A new approach to hardware security analysis. PhD thesis, University of Cambridge, Darwin College. UK (2005), http://www.cl.cam.ac.uk/techreports/UCAM-CL-TR-630.pdf
12. Kerckhoffs, A.: La cryptographie militaire. Journal des Sciences Militaires IX, 1–2, 5–38, 161–191 (1883)
13. Mangard, S., Oswald, E., Popp, T.: Power analysis attacks: Revealing the secrets of smart cards. Advances in Information Security. Springer Science+Business Media, NY (2007)
14. Brier, E., Clavier, C., Olivier, F.: Correlation power analysis with a leakage model. In: Joye, M., Quisquater, J.-J. (eds.) CHES 2004. LNCS, vol. 3156, pp. 16–29. Springer, Heidelberg (2004)
15. Messerges, T., Dabbish, E., Sloan, R.: Examining smart-card security under the threat of power analysis attacks. IEEE Trans. Comput. 51(4), 541–552 (2002)
16. Peeters, E., Standaert, F.-X., Donckers, N., Quisquater, J.-J.: Improved higher-order side-channel attacks with FPGA experiments. In: Rao, J.R., Sunar, B. (eds.) CHES 2005. LNCS, vol. 3659, pp. 309–323. Springer, Heidelberg (2005)
17. Muller, F., Valette, F.: High-order attacks against the exponent splitting protection. In: Yung, M., Dodis, Y., Kiayias, A., Malkin, T. (eds.) PKC 2006. LNCS, vol. 3958, pp. 315–329. Springer, Heidelberg (2006)
18. Standaert, F.-X., Veyrat-Charvillon, N., Oswald, E., Gierlichs, B., Medwed, M., Kasper, M., Mangard, S.: The world is not enough: Another look on second-order DPA. In: Abe, M. (ed.) ASIACRYPT 2010. LNCS, vol. 6477, pp. 112–129. Springer, Heidelberg (2010)
19. Namolaru, M.: Devirtualization in GCC. In: Proceedings of the GCC Developers' Summit (2006), http://ols.fedoraproject.org/GCC/Reprints-2006/namolaru-reprint.pdf

Sensitive Ants for Denial Jamming Attack on Wireless Sensor Network

Camelia-M. Pintea and Petrica C. Pop

Technical University Cluj-Napoca North University Center
76 Victoriei, 430122 Baia-Mare, Romania
cmpintea@yahoo.com, petrica.pop@cunbm.utcluj.ro

Abstract. The *Quality of Service* in security networks systems is constantly improved. The current paper propose a newly defense mechanism, based on ant system, for different jamming attack on *Wireless Sensor Network (WSN)*. The sensitive ants react on attacks based on several reliable parameters including their level of sensitivity. The information in network is re-directed from the attacked node to its appropriate destination node. The paper analyzes how are detected and isolated the jamming attacks using mobile agents, ant systems in general and with the newly ant-based sensitive approach in particular.

1 Introduction

Denial of Service (DoS) is preventing or is inhibiting the normal use or management of communications in a network through flooding it with useless information. Jamming attacks on wireless networks, special cases of *DoS*, are the attacks that disturb the transceivers operations on wireless networks [1]. A *Radio Frequency (RF)* signal emitted by a jammer corresponds to the useless information received by the sensor nodes of a network.

Nowadays are several techniques used to reduce the effect of jamming attacks in wireless networks. In [2] is proposed a traffic rerouting scheme for *Wireless Mesh Network (WMN)*. There are determined multiple candidates of a detour path which are physically disjoint. In a stochastic way, is selected just one candidate path as a detour path to distribute traffic flows on different detour paths. The mechanism on packet delivery ratio and end-to-end delay is improved when compared with a conventional scheme.

In [3] is introduced *Mitigating Colluding Collision Attacks (MCC)* in *Wireless Sensor Networks*. It is provided an analysis for the probability of isolating malicious nodes from *WSN* indulging in packet dropping through colluding collision. The *MCC* protocol redefines the notion of guards from *Basic Local Monitoring* (BLM) [4]. At first extends the number of guards from only the common neighbors of the relaying node and the next hop including all the neighbors of the relaying node. Further, use a counter, at each node for each neighbor, for counting the number of forwards by that neighbor. When compared with *BLM*, *MCC* proved to be more efficient on detecting packet dropping through colluding collision attacks.

Á. Herrero et al. (eds.), *International Joint Conference SOCO'13-CISIS'13-ICEUTE'13*,
Advances in Intelligent Systems and Computing 239,
DOI: 10.1007/978-3-319-01854-6_42, © Springer International Publishing Switzerland 2014

Securing *WSN*s against jamming attacks is a very important issue. Several security schemes proposed in the *WSN* literature are categorized in: *detection techniques, proactive countermeasures, reactive countermeasures* and *mobile agent-based countermeasures*. The advantages and disadvantages of each method are described in [5]. Our current interest is in *mobile agent-based solutions*.

Artificial intelligence has today a great impact in all computational models. Multi-agent systems are used for solving difficult Artificial Intelligence problems [6, 7]. Multi-agents characteristics includes organization, communication, negotiation, coordination, learning, dependability, learning and cooperation [8, 6, 9]. Other features are their knowledge [10] and their actual/future relation between self awareness and intelligence. There are also specific multi-agents with their particular properties, as robots-agents [11, 12]. One of the commonly paradigm used with *MAS* systems is the pheromone, in particular cases artificial ants pheromone. Ant-based models are some of the most successfully nowadays techniques used to solve complex problems [13–15, 11, 16, 17].

Our goal is to improve the already existing ant systems on solving jamming attacks on *WSN* using the ants sensitivity feature. The structure of the current paper follows. The second section describes the already known mechanisms of Jamming Attack on Wireless Sensor Network and the particular unjamming techniques based on Artificial Intelligence. The next section is about ants sensitivity. The section includes also the newly introduced sensitive ant-model for detecting and isolated jamming in a *WSN*. Several discussions about the new methods and the future works concludes the paper.

2   About Jamming Attack on Wireless Sensor Network Using Artificial Intelligence

The current section describes the main concepts and the software already implemented on Jamming Attack on Wireless Sensor Network including the *Artificial Intelligence* models.

2.1   Jamming Attack on Wireless Sensor Network

Jamming attacks are particular cases of Denial of Service (DoS) attacks [18]. The main concepts related to this domain follows. At first several considerations on Wireless Sensor Network (WSN).

Definition 1. *A Wireless Sensor Network consists of hundreds/thousands of sensor nodes randomly deployed in the field forming an infrastructure-less network.*

Definition 2. *A sensor node of a WSN collects data and routes it back to the Processing Element (PE) via ad-hoc connections with neighbor sensor nodes.*

Definition 3. *A Denial of Service attack is any event that diminishes or eliminates a networks capacity to perform its expected function.*

Definition 4. *Jamming is defined as the emission of radio signals aiming at disturbing the transceivers operation.*

There are differences between jamming and radio frequency interference.

- the jamming attack is *intentional* and *against a specific target*;
- the radio frequency interference is unintentional, as a result of nearby transmitters, transmitting in the same or very close frequencies levels.

An example of radio frequency interference is the coexistence of multiple wireless networks on the same area with the same frequency channel [5].

Definition 5. *Noise is considered the undesirable accidental fluctuation of electromagnetic spectrum, collected by an antenna.*

Definition 6. *The Signal-to-Noise Ratio is*

$$SNR = \frac{P_{signal}}{P_{noise}}$$

where P is the average power.

A jamming attack can be considered *effective* if the *Signal-to-Noise Ratio(SNR)* is less than one $(SNR < 1)$.

There are several jamming techniques [5]: Spot Jamming, Sweep Jamming, Barrage Jamming and Deceptive Jamming shortly described in Table 1.

Definition 7. *Jammer refers to the equipment and its capabilities that are exploited by the adversaries to achieve their goal.*

There are several types of jammers, from simple transmitter or jamming stations with special equipment, used against wireless networks.[19]

- the constant jammer - emitting totally random continuous radio signals; *target*: keeping the *WSN*s channel busy; disrupting nodes communication; causing interference to nodes that have already commenced data transfers and corrupt their packets.
- the deceptive jammer;
- the random jammer - sleeps for a random time and jams for a random time;
- the reactive jammer - in case of activity in a *WSN* immediately sends out a random signal to collide with the existing signal on the channel.

As a result of jamming attacks the transmitted packets of data from/to *WSN* will be corrupted. Figure 1 illustrates a Jamming Attack on Wireless Sensor Network and how are excluded the jammed nodes from the wireless network.

2.2 Artificial Intelligence in Jamming Attack on Wireless Sensor Network

A short overview of the main artificial intelligent techniques already used and their benefits for *DoS* and Jamming Attack on *WSN* follows.

Table 1. A list of several jamming techniques [5] with their attack scheme, their main characteristics and if there are / or not efficient ways to avoid or limit the attacks

Technique	The attack	Characteristics	Avoiding possibility
Spot Jamming	override the original signal on the same frequency and modulation that the target;	very powerful;	avoided: by changing to another frequency;
Sweep Jamming	a jammers power changes very fast from one frequency to another;	is able to jam multiple frequencies in quick succession;	limited: does not affect all the frequencies at the same time;
Barrage Jamming	a range of frequencies is jammed at the same time;	jam many frequencies at once to decrease SNR of receivers;	limited: as the range of jammed frequencies grows, the power of jamming is reduced proportionally;
Deceptive Jamming	applied in a single frequency or in a set of frequencies;	floods WSN with useless or fake data without leaving traces;	very dangerous, cannot be easily detected;

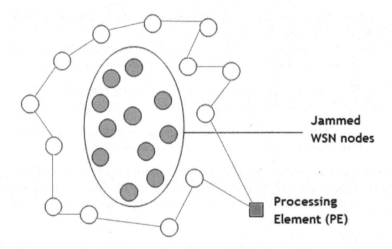

Fig. 1. Graphical symbolic illustration of avoiding Jamming Attack on Wireless Sensor Network

A general overview of several research papers about critical issue of jamming in *WSN*s is shown in [5]. The authors observed that there are still many open research issues including *mobile agents(MA)* for defending against jamming attacks is partially unexplored and some promising method are in [20, 21]. The mobile agent (MA) term refers to "an autonomous program with the ability to move from host to host and act on behalf of users towards the completion of an assigned task" [22]. An example of using *MA*s under jamming attacks is that mobile agents can temporary remain on their current position and return to the Processing Element with the collected data when they sense clear terrain.

In [20] is proposed Jam Avoidance Itinerary Design (JAID). The algorithm first calculate near-optimal routes for agents as they visit the nodes and when facing attacks, modify the routes of agents in order to avoid the jammed nodes. The mobile agents will no longer visit the jammed areas. A disadvantage of the model is that it cannot defend the *WSN* when the jamming attack is against all nodes simultaneously. It has also several advantages as: the mobile agents exclude the jammed nodes from the *WSN* itineraries and they will continue to deliver in good condition the data from the working nodes to the PE; the JAID technique minimize the energy cost needed for transmission when computes the best routes for mobile agents.

In [21] is introduced the *Jamming Attack Detection and Countermeasures* in *WNS* using Ant System. Ants are a particular class of mobile agents. At first, a group of ants deposit an initial quantity of pheromone when traverse the *WSN* nodes in a random manner. That is how they indirectly communicate with the others ants, in a stigmergic manner [15].

The amount of pheromone left by the previous ant agents increases the probability that the same route is taken during the current iteration. Beside the several parameters from ant-based systems [15] specific parameters of wireless networks are: hops, energy, distance, packet loss, SNR, Bit Error Rate (BER) and packet delivery. These parameters will affect the way is selected a specific route of ants in *WSN*. As in all ant-based system, pheromone evaporation is preventing the suboptimal solutions.

Some still "unknown" of ant-based system when unjamming a *WSN* are: the the energy cost required by ants and how quickly the pheromone trails are able to react to nimble attackers. It is possible that as in JAID algorithm, when a wide area of the wireless sensor network is jammed then ants will not be able to keep uninterrupted the rest of networks operations.

Several advantages [5] of ant-based mechanism against Jamming Attack on Wireless Sensor Network are following.

- The ants are spread all over the *WSN* and have the control of the WSN;
- The ants try to find the best jamming-free routes for data transferring;
- The ants take into account crucial node;
- The ants take into account the network parameters, as nodes remaining energy, packet loss, SNR.
- The ants can adapt more easily to a large-scale *WSN* jammed environment, based on the pheromone indirect communication.
- An ant will not move from an attacked node until it finds a neighbor node with a clear communication channel.

Both considered mobile agent techniques have a medium defense effectiveness, a medium value of expected implementation deployment cost and have compatibility with existing software and a low impact to energy efficiency.

3   Sensitive Ant-Based Technique for Jamming Attack on Wireless Sensor Network

The current section starts with several concepts on sensitive ants and include the newly introduce technique based on sensitive artificial ants for solving the Jamming Attack on Wireless Sensor Network (WSN).

3.1   About Sensitive Ants

The current section describes the concept of sensitive ants proposed in [23, 11]. Effective metaheuristics for complex problems, as large scale routing problems (e.g. the *Generalized Traveling Salesman Problem*) based on sensitive ants are illustrated in [8, 11, 23–25]. Several concepts are defined and described further.

Definition 8. *Sensitive ants refers to artificial ants with a Pheromone Sensitivity Level (PSL) expressed by a real number in the unit interval* $0 \leq PSL \leq 1$.

Definition 9. *A pheromone blind ant is an ant completely ignoring stigmergic information, with the smallest PSL value, zero.*

Definition 10. *A maximum pheromone sensitivity ant has the highest possible PSL value, one.*

Definition 11. *The sensitive-explorer ants have small Pheromone Sensitivity Level values indicating that they normally choose very high pheromone level moves.*

Definition 12. *The sensitive-exploiter ants have high Pheromone Sensitivity Level values indicating that they normally choose any pheromone marked move.*

The sensitive-explorer ants are also called small PSL-ants, hPSL and sensitive-exploiter ants are called high PSL-ants, hPSL. They intensively exploit the promising search regions already identified by the sensitive-explorer ants. For some particular problems the sensitivity level of hPSL ants have been considered to be distributed in the interval (0.5, 1) and for the sPSL ants the sensitivity level in the interval (0, 0.5).

3.2   Newly Sensitive Ant Technique for Jamming Attack on WSN

Based on the already described notion in previous section 3.1, it is introduced a new ant-based concept with sensitivity feature. The *Sensitive Ant Algorithm for Denial Jamming Attack on Wireless Sensor Network* is further called *Sensitive Ant Denial Jamming* on *WSN* algorithm.

As we know, not all ants react in the same way to pheromone trails and their sensitivity levels are different there are used several groups of ants with different levels of sensitivity. For an easy implementation are used just two groups-colonies. The algorithm should be easily modified if there are used several ant-colonies with specific pheromone levels.

In [21] is firstly introduced an ant system for jamming attack detection on *WSN*. The performance of the ant system is given by the node spacing and several parameters: Q an arbitrary parameter, ρ trail memory, α power applied to the pheromones in probability function and β power of the distance in probability function.

It is considered a *WSN* in a two dimensional Euclidean space. There are several key elements of AS for keeping the network robust and de-centralized. One is the information on the resource availability on every node used to predict the link for the ants next visit. Other key elements are the pheromones intensity and dissipate energy of ants as they traverse the nodes based on path probabilities. The key factor for making decisions in [21] is the transition probability (1). In the newly *Sensitive Ant Denial Jamming* on *WSN* only the ants with small pheromone level are using this probability.

$$P_{ij} = \frac{(\varphi_{ij} \cdot \eta_{ij})^\alpha \cdot (\frac{1}{D_{ij}})^\beta}{\sum_k (\varphi_{ik} \cdot \eta_{ik})^\alpha \cdot (\frac{1}{D_{ik}})^\beta} \tag{1}$$

where η_{ij} is the normalized value (2) of Hop, H_{ij}, Energy, E_{ij}, Bit Error Rate, B_{ij}, Signal to Noise ratio, SNR_{ij}, Packet Delivery, Pd_{ij} and Packet Loss, Pl_{ij} [26].

$$\eta_{ij} = H_{ij} \cdot E_{ij} \cdot B_{ij} \cdot SNR_{ij} \cdot Pd_{ij} \cdot Pl_{ij} \tag{2}$$

In the sensitive ant model the ants with high pheromone level are choosing the next node based on (3) from the neighborhood J of node j.

$$j = argmax_{u \in J_{ik}} \{ (\varphi_{iu} \cdot \eta_{iu})^\alpha \cdot (\frac{1}{D_{iu}})^\beta \}. \tag{3}$$

φ_{ij} is the pheromone intensity between the source node i and destination node j; The normalized value is the difference between total and actual value of the performance parameters. The performance value is used to compute the transition probabilities in a route. The link being active or dead in a tour taken by an ant is incorporated in the pheromone. The pheromone is globally updated [27] following each complete tour by ant system with the update rule (4) is following [21].

$$\varphi_{ij}(t) = \rho(\varphi_{ij}(t-1)) + \frac{Q}{D_t \cdot \eta_t} \tag{4}$$

where D_t is the total distance traveled by ants during the current tour. The trails formed by the ant is dependent on the link factor. The tabu list includes now updated values of the energy available in the nodes for a particular sub-optimal route with high reachability. The pseudo-code description of the *Sensitive Ant Denial Jamming* on *WSN* follows.

```
Sensitive Ant Denial Jamming on WSN
Begin
 Set parameters and place the ants randomly on network
 For i=1 to Niter do
    if (not (j<=hops)or(reached destination node))
       if (jamming attack)
         Compute probability transition (1) for each sPSL-ant
         Compute the transition rule (2) for each hPSL-ant
       else move ant to next node
       endif
       if (reached destination node)
         node is penalized for t seconds
         Compute probability transition (1) for each sPSL-ant
         Compute the transition rule (3) for each hPSL-ant
       endif
    endif
    Update pheromone (4)
    Update tabu list: the distance, energy, SNR, number of hops
    packet delivery, packet loss and bit error rate
    Validate the best path
    Communicate the jammed region to the Processing Element (PE)
    Endfor
End
```

A run of the algorithm returns the valid path of the wireless network. That is how the information in *WSN* is re-routed. Termination criteria is given by a given number of iterations, N_{iter}

3.3   Discussions about the Newly Concept

The performance of *Sensitive Ant Denial Jamming* on *WSN* is discussed in the following. Sensitive ants with lower pheromone level are able to explore the wireless network and the ants with high pheromone level intensively exploit the promising search regions already identified in the network. The ants behavior emphasizes search intensification. The ants "learn" during their lifetime and are capable to improve their performances. That is how they modifies their level of sensitivity: the PSL value increase or decrease based on the search space topology encoded in the ants experience.

The s-PSL ants are more independent than h-PSL ants, with the potential to autonomously discover new promising regions of the solution space. If the current node of the sensitive ant is jammed, the ant will change its sensitivity level, based on its experience and will move to another node from neighborhood. The Processing Element (PE) considered, in addition to the initial ant system for jamming attacks [21], is a special node of the wireless network. PE is aware about jamming attack based on the indirect information of the ants from its neighborhood. This lead to a re-routed information on *WSN* only on the valid nodes.

In the future numerical experiments to assess the performance of the new algorithm will be provided. The introduced model *Sensitive Ant Algorithm for Denial Jamming Attack on WSN* seems to have more chances to improve the jamming attack detection and re-routing in wireless sensor network.

4 Conclusions

Jamming attacks are particular *Denial of Service* attacks. The paper shows the main jamming attacks and several countermeasure. It is introduced a new *Sensitive Ant Denial Jamming* on *Wireless Sensor Network* based on ant system. Ant systems are a mobile-agents class, one of the security schemes proposed in the literature. In general mobile agent techniques including ant models proved to have a medium defense effectiveness, a medium cost but a good compatibility with existing software. The introduced sensitive model brings a new feature that improves the reactions of agents in the network in case of jamming attacks and redirect the information also to the processing element in order to re-routing information.

Acknowledgment. This work was supported by a grant of the Romanian National Authority for Scientic Research, CNCS-UEFISCDI, project number PN-II-RU-TE-2011-3-0113.

References

1. Adamy, D.: EW 102: A second course in electronic warfare. Artech House Publishers (2004)
2. Lim, Y., Kim, H.M., Kinoshita, T.: Traffic rerouting strategy against jamming attacks in wsns for microgrid. International Journal of Distributed Sensor Networks (2012)
3. Khalil, I.: Mpc: mitigating stealthy power control attacks in wireless ad hoc networks. In: Global Telecommunications Conference, GLOBECOM 2009, pp. 1–6. IEEE (2009)
4. Khalil, I., Bagchi, S., Nina-Rotaru, C.: Dicas: detection, diagnosis and isolation of control attacks in sensor networks. In: Security and Privacy for Emerging Areas in Communications Networks, SecureComm 2005, pp. 89–100. IEEE (2005)
5. Mpitziopoulos, A., Gavalas, D., Konstantopoulos, C., Pantziou, G.: A survey on jamming attacks and countermeasures in wsns. IEEE Communications Surveys & Tutorials 11(4), 42–56 (2009)
6. Wooldridge, M.: An introduction to multiagent systems. Wiley (2008)
7. Iantovics, B., Enăchescu, C.: Intelligent complex evolutionary agent-based systems. In: AIP Conference Proceedings, vol. 1117, pp. 116–124 (2009)
8. Chira, C., Pintea, C.M., Dumitrescu, D.: Sensitive stigmergic agent systems a hybrid approach to combinatorial optimization. In: Advances in Soft Computing, vol. 44, pp. 33–39 (2008)
9. Stoean, C., Stoean, R.: Evolution of cooperating classification rules with an archiving strategy to underpin collaboration. In: Teodorescu, H.-N., Watada, J., Jain, L.C. (eds.) Intelligent Systems and Technologies. SCI, vol. 217, pp. 47–65. Springer, Heidelberg (2009)

10. Popescu-Bodorin, N., Balas, V.E.: From cognitive binary logic to cognitive intelligent agents. In: 2010 14th International Conference on Intelligent Engineering Systems (INES), pp. 337–340. IEEE (2010)
11. Pintea, C.M.: Combinatorial optimization with bio-inspired computing. Ed.Edusoft (2010)
12. Pintea, C.-M., Pop, P.C.: Sensor networks security based on sensitive robots agents: A conceptual model. In: Herrero, Á., et al. (eds.) Int. Joint Conf. CISIS'12-ICEUTE'12-SOCO'12. AISC, vol. 189, pp. 47–56. Springer, Heidelberg (2013)
13. Crisan, G.C.: Ant Algorithms in Artificial Intelligence. PhD thesis, "Al. I. Cuza" University of Iasi (2007)
14. Crisan, G.C., Nechita, E.: Solving fuzzy tsp with ant algorithms. International Journal of Computers Communications & Control 3(S), 228–231 (2008)
15. Dorigo, M., Gambardella, L.M.: Ant colony system: A cooperative learning approach to the traveling salesman problem. IEEE Transactions on Evolutionary Computation 1(1), 53–66 (1997)
16. Karapetyan, D., Reihaneh, M.: An efficient hybrid ant colony system for the generalized traveling salesman problem. Algorithmic Operations Research 7(1), 21–28 (2012)
17. Stoean, R., Stoean, C.: Evolution and artificial intelligence. Modern paradigms and applications, Ed. Albastra (2010)
18. Wood, A.D., Stankovic, J.A.: Denial of service in sensor networks. Computer 35(10), 54–62 (2002)
19. Xu, W., Trappe, W., Zhang, Y., Wood, T.: The feasibility of launching and detecting jamming attacks in wireless networks. In: Proceedings of the 6th ACM International Symposium on Mobile Ad Hoc Networking and Computing, pp. 46–57. ACM (2005)
20. Mpitziopoulos, A., Gavalas, D., Konstantopoulos, C., Pantziou, G.: Jaid: An algorithm for data fusion and jamming avoidance on distributed sensor networks. Pervasive and Mobile Computing 5(2), 135–147 (2009)
21. Muraleedharan, R., Osadciw, L.A.: Jamming attack detection and countermeasures in wireless sensor network using ant system. In: Defense and Security Symposium, International Society for Optics and Photonics (2006)
22. Pham, V.A., Karmouch, A.: Mobile software agents: An overview. IEEE Communications Magazine 36(7), 26–37 (1998)
23. Chira, C., Pintea, C.M., Dumitrescu, D.: Sensitive ant systems in combinatorial optimization. Studia Universitatis Babes-Bolyai. Series Informatica, pp. 185–192 (2007)
24. Pintea, C.-M., Chira, C., Dumitrescu, D., Pop, P.C.: A sensitive metaheuristic for solving a large optimization problem. In: Geffert, V., Karhumäki, J., Bertoni, A., Preneel, B., Návrat, P., Bieliková, M. (eds.) SOFSEM 2008. LNCS, vol. 4910, pp. 551–559. Springer, Heidelberg (2008)
25. Pintea, C.-M., Chira, C., Dumitrescu, D.: Sensitive ants: Inducing diversity in the colony. In: Krasnogor, N., Melián-Batista, M.B., Pérez, J.A.M., Moreno-Vega, J.M., Pelta, D.A. (eds.) NICSO 2008. SCI, vol. 236, pp. 15–24. Springer, Heidelberg (2009)
26. Liang, Y.C., Smith, A.E.: An ant system approach to redundancy allocation. In: Proceedings of the 1999 Congress on Evolutionary Computation, CEC 1999, vol. 2. IEEE (1999)
27. Dorigo, M., Maniezzo, V., Colorni, A.: Ant system: optimization by a colony of cooperating agents. IEEE Transactions on Systems, Man, and Cybernetics, Part B: Cybernetics 26(1), 29–41 (1996)

Supervised Machine Learning for the Detection of Troll Profiles in Twitter Social Network: Application to a Real Case of Cyberbullying

Patxi Galán-García, José Gaviria de la Puerta, Carlos Laorden Gómez,
Igor Santos, and Pablo García Bringas

DeustoTech Computing, University of Deusto
{patxigg,jgaviria,claorden,isantos,pablo.garcia.bringas}@deusto.es

Abstract. The use of new technologies along with the popularity of so-cial networks has given the power of anonymity to the users. The ability to create an alter-ego with no relation to the actual user, creates a situation in which no one can certify the match between a profile and a real person. This problem generates situations, repeated daily, in which users with fake accounts, or at least not related to their real identity, publish news, reviews or multimedia material trying to discredit or attack other people who may or may not be aware of the attack. These acts can have great impact on the affected victims' environment generating situations in which virtual attacks escalate into fatal consequences in real life. In this paper, we present a methodology to detect and associate fake profiles on Twitter social network which are employed for defamatory activities to a real profile within the same network by analysing the content of comments generated by both profiles. Accompanying this approach we also present a successful real life use case in which this methodology was applied to detect and stop a cyberbullying situation in a real elementary school.

Keywords: On-line Social Networks, Trolling, Information Retrieval, Identity Theft, Cyberbullying.

1 Introduction

On-line Social Networks (OSNs) are some of the most frequently used Internet services. There is not a generic definition of these platforms, although Boyd et al [1] defined them as web services that allow an individual to do three things: i) generate a public or semi-public profile in a specific system, ii) create a list of users to interact with and browse through the list of contacts and iii) see what was done by others within the system.

The massive presence that users have in this platforms and the relative easiness to hide a user's real identity implies that false profiles and "troll users"[1] are spreading, becoming a nuisance to legitimate users of these services.

---

[1] Users posting inflammatory, extraneous, or off-topic messages in an on-line community.

Á. Herrero et al. (eds.), *International Joint Conference SOCO'13-CISIS'13-ICEUTE'13*, 419
Advances in Intelligent Systems and Computing 239,
DOI: 10.1007/978-3-319-01854-6_43, © Springer International Publishing Switzerland 2014

In some cases, these "malicious" users, use OSN platforms to commit crimes like identity impersonation, defamation, opinion polarisation or cyberbullying. Despite all of them present real and worrying dangers [2,3] the later, cyberbullying, has actually become one of the most hideous problems in our society, generating even more side effects than "real life bullying" [4] due to the impact and the permanent nature of the comments flooding OSN platforms. As a hard to forget, but not isolated, example recently one teenager, Amanda Todd, committed suicide due to harassment in the form of blackmailing, bullying, and physical assaults[2] using OSNs as the channel of abuse.

This hurtful event, which raised the discussion on criminalizing cyberbullying in several areas, was kind of aided by the aforementioned possibility of easily creating false profiles in social network platforms. Thanks to these anonymous and not-linked-to-real-life profiles some twisted minded individuals are able to torment their victims without even being brought to justice. In a similar vein, it was recently brought to our attention an incident in an Spanish elementary school where an alleged student was writing defamatory comments in Twitter using a fake profile, causing anxiety attacks and depression episodes among the affected students.

Some social platforms are trying to manually identify the real person behind their profiles but, until the job is done, being able to correlate or link a false profile to a real person within the network is the only option to fight the problem.

In light of this background, we present a methodology to associate a false profile's tweets with one real individual, provided he/she has another profile created with real information. The assumption that the trolling user will have another "real" profile is not fortuitous, it relies on the fact that these kind of users like to interact with the fake identity and stay updated and participate in parallel conversations. Moreover, we apply the presented methodology to a real life cyberbullying situation inside an elementary school.

2 Background

2.1 On-Line Social Networks

On-line Social Networks (OSNs) are platforms that provide users with some useful tools to interact with other users through connections. The connections start by creating a profile in an OSN, which consists of a user's representation and can be private, semi-public or public.

The importance of the connections users make within these social platforms resides on the amount of information, usually private, that is published in these usually public profiles. Therefore, privacy in OSNs is a serious issue. Despite the importance of this fact it is commonly ignored by the users of social platforms, resulting in the publication of too much personal information, information which can be (and in some cases is) used by sexual predators, criminals, large corporations and governmental bodies to generate personal and behavioural profiles of the users.

---

[2] http://en.wikipedia.org/wiki/Suicide_of_Amanda_Todd

In fact, one of the currently most worrying problems in OSNs is cyberbullying. This problem is a relatively new and widespread situation and most of the times implies an emotional trauma for the victim. Teenagers use these platforms to express their feelings and life experiences because they are not able to do it in their real lives, while other users abuse and torment others with impunity [1]. This common misuse of OSNs, the emotional abuse, is commonly referred to as trolling [5] and can happen in many different ways like defacement of deceased person pages, name calling, controversial comments with the intention to cause anger and cause arguments. To solve this type of problems, some OSNs have chosen the approach of age verification systems with real ID cards, but not always with successful results [6].

2.2 Cyberbullying

Cyberbullying, according to [7], refers to any harassment that occurs via the internet, cell phones or other devices. This type of bullying uses communication technologies to intentionally harm others through hostile behaviour such as sending text messages and posting ugly comments on the Internet. But, usually, the definition of this phenomenon starts using the traditional definition of bullying.

In the literature of cyberbullying detection, the main focus has been directed towards the content of the conversations. For example using text mining paradigms to identify on-line sexual predators [8] [9], vandalism detection [10], spam detection [11] and detection of internet abuse and cyberterrorism [12]. This type of approaches are very promising, but not always applicable to every aspect of cyberbullying detection because some attacks can only be detected by analysing user's contexts.

In a recent study on cyberbullying detection, Dinakar et al. [13] applied a range of binary and multiclass classifiers on a manually labelled corpus of YouTube comments. The results showed that a binary individual topic-sensitive classifiers approach can outperform the detection of textual cyberbullying compared to multiclass classifiers. They showed the application of common sense knowledge in the design of social network software for detecting cyberbullying. The authors treated each comment on its own and did not consider other aspects to the problem as such the pragmatics of dialogue and conversation and the social networking graph. They concluded that taking into account such features will be more useful on social networking websites and which could to a better modelling of the problem.

3   Proposed Approach

The main idea underlying our approach is that *every trolling profile is followed by the real profile of the user behind the trolling one.* This assumption is based on the fact that this kind of users want to stay updated on the activity that surrounds the fake profile. Besides, each individual writes in a characteristic way. Studying different features of the written text is possible to determine the authorship of, for example, e-mails [14]. In this case, despite users behind fake profiles may try to write in a different way to avoid detection, Twitter provides other characteristics that, in conjunction with the text analysis, may be used to link a trolling account to a real user's profile.

Therefore, with these ideas in mind, we postulated the following hypothesis: *It is possible to link a trolling account to the corresponding real profile of the user behind the fake account, analysing different features present in the profile, connections' data and tweets' characteristics, including text, using machine learning algorithms.*

To prove the hypothesis, we first prepared the method to determine the authorship of twitter profiles based on their published tweets and then applied these techniques to a real cyberbullying situation in one elementary school.

The authorship identification step was performed studying a group of profiles which have some kind of relation among them (to replicate to the best possible extent the conditions of a classroom cyberbullying event). The methodology would follow the next steps: i) select the profiles under study, ii) collect all the information of the profile and its tweets, iii) select the features to be extracted from the retrieved tweets and iv) apply machine learning methods to build the models that will determine the authorship of the gathered tweets.

3.1   Selecting Different Profiles

The number of selected profiles for this study was 19. The idea was to gather several profiles with social relations both inside the social network and in real life. We wanted to avoid retrieving very different Twitter accounts, with respect to their content, which would ease the task of determining the authorship. In this way, the conditions found in cyberbullying situations, with respect to the participants and conditions, are sufficiently replicated.

In order to select the profiles we checked that it was not private, the number of own tweets (less than 50-100 samples would imply almost no activity), number of followers and following users (no connections would show no interaction with other users) and the relation between other selected profiles (we wanted the selected accounts to be connected).

Therefore, the first selected profile was of one of the authors, and then we continued analysing their followers and followings, keeping the desired ones. The final dataset was varied, we had men and women, different ages, different studies and different behaviours of their Twitter account.

3.2   Collecting Profiles Data and Tweets

It is important to notice that there is one important limitation imposed by the Twitter API. The number of requests could not exceed 350 per hour, which limits considerably the possibility to retrieve a large amount of samples, so we had to use several accounts to gather them.

Our Java-based collecting method obtained, from the selected profiles, the users' ID and the timeline tweets, until having at least 100 *genuine tweets*.

A genuine tweet is the tweet that is generated by the user itself (i.e., written by itself) and is not one retweet of another user's tweet.

3.3 Features

Our dataset contains the following features extracted from each of the profiles the tweets, time of publication, language, geoposition and Twitter client. The first feature, the tweet, is the text published by the user, which gives us the possibility of determine a writing style, very characteristic of each individual. The time of publication helps determining the moments of the day in which the users interact in the social network. The language and geoposition also help filtering and determining the authorship because users have certain behaviours which can be extrapolated analysing these features. Finally, despite being possible that users have several devices from where they tweet (e.g., PC, smartphone or tablet), they usually choose to do it using their favourite Twitter client, which gives us another filtering mechanism.

3.4 Supervised Learning

Once we have the profile data, user's tweets and the chosen features, the next step is to generate an ARFF [15] file (i.e., Attribute Relation File Format) to classify the profiles according to the writing style of the tweets using WEKA (i.e., the Waikato Enviroment for Knowledge Analysis) [16].

In this experiment, we have chosen to compare the performance of different classification algorithms included in WEKA: i) Random Forest, ii) J48 (WEKA's C4.5 implementation), iii) K-Nearest Neighbor (KNN), iv) Sequential Minimal Optimization (SMO) and v) Bayes Theorem-based algorithms.

To optimize the results, before training the classifiers, we filtered the *tweets* text with stopwords [17] in Spanish[3].

To validate the suitability of the results, we employed K-fold cross-validation [18], a technique which consists on dividing the dataset into K folds, using the instances corresponding to $K-1$ folds for training the model, and the instances in the remaining fold for testing. K training rounds are performed using a different fold for testing each time, and thus, training and testing the model with every possible instance in the dataset.

At last, to evaluate the results, we used *True Positive Ratio (TPR)*, *False Positive Ratio (FPR)* and *Area Under ROC Curve (AUC)*.

4 Experiments

To evaluate the capabilities of the method to assign the correct authorship to the Twitter profiles, we used a dataset comprising 1,900 tweets corresponding to 19 different twitter accounts (100 tweets per profile).

---

[3] http://paginaspersonales.deusto.es/claorden/resources/
SpanishStopWords.txt

For the experiments, we modelled the tweets using the Vector Space Model (VSM) [19]. VSM is an algebraic approach for Information Filtering (IF), Information Retrieval (IR), indexing and ranking. This model represents natural language documents mathematically by vectors in a multidimensional space where the axes are terms within messages. We used the *Term Frequency – Inverse Document Frequency* (TF–IDF) [19] weighting schema, where the weight of the ith term in the jth document, denoted by $weight(i, j)$, is defined by $weight(i, j) = tf_{i,j} \cdot idf_i$ where *term frequency* $tf_{i,j}$ is defined as $tf_{i,j} = n_{i,j} / \sum_k n_{k,j}$ where $n_{i,j}$ is the number of times the term $t_{i,j}$ appears in a document d, and $\sum_k n_{k,j}$ is the total number of terms in the document d. The inverse term frequency idf_i is defined as $idf_i = |\mathcal{D}||\mathcal{D} : t_i \in d|$ where $|\mathcal{D}|$ is the total number of documents and $|\mathcal{D} : t_i \in d|$ is the number of documents containing the term t_i.

Moreover, VSM requires a pre-processing step in which messages are divided into tokens by separator characters (e.g., space, tab, colon, semicolon, or comma). The *tokenisation*, the process of breaking the stream of text into the minimal units of features (i.e., the tokens) [19], was based in an n-gram selection with sizes of $n = 1$, $n = 2$ and $n = 3$. This process is performed to construct the VSM representation of the messages and it is required for the learning and testing of classifiers [20].

Table 1 shows the results obtained after applying the selected algorithms to our dataset, measured in Accuracy, *FPR*, *TPR* and *AUC*.

Table 1. Obtained results for the selected machine learning algorithms. It must be noted that we applied the default configurations under WEKA for each of the algorithms.

Algorithm	Accuracy	FPR	TPR	AUC
SMO-PolyKernel	68.47	0.02	0.68	0.96
J48	65.81	0.02	0.66	0.94
SMO-NormalizedPolyKernel	65.29	0.02	0.65	0.94
RandomForest	66.48	0.02	0.66	0.93
KNN $k = 10$	59.79	0.02	0.60	0.92
KNN $k = 3$	59.7	0.02	0.60	0.90
KNN $k = 5$	59.39	0.02	0.59	0.90
NaiveBayes	33.91	0.04	0.34	0.90
KNN $k = 2$	61.06	0.02	0.61	0.89

The results show that SMO and Decision Trees are the most appropriate algorithms. More precisely, the best results are obtained using a PolyKernel with 68.47% accuracy and 0.96 of *AUC*. In second and third position, very close, we have J48 with 65.81% accuracy and 0.94 and NormalizedPolyKernel with 65.29% accuracy and 0.94 of *AUC*. Random Forest, in fourth position, obtains 66.48% accuracy and 0.93 *AUC*. Finally, *KNN* and Naive Bayes algorithms do not have remarkable results, with values from 59.39% to 61.06% of accuracy for *KNN* and of 33.91 for Naive Bayes in terms of accuracy and from 0.89 to 0.92 and 0.90 respectively in terms of *AUC*.

5   Real Case Study

The proposed methodology, has been tested in a real situation. In one school in the city of Bilbao (Spain), some students were implicated in a cyberbullying situation. The staff of this school proposed to us whether it was possible to find which of the students had been the author/s behind the trolling profile or not.

In this case the profile was named "Gossip", in a clear reference to the popular TV Show *Gossip Girl*, and, for two weeks, the student using this profile commented personal indiscretions about his/her classmates. Initially, it was only the publication of not hurtful tweets. These comments included events or facts such as one student not doing the assigned homework.

Two weeks later, the profile started publishing things a little more private but not very important. At that moment, all the classmates were following that profile and in the school hallways the students theorised about who could be the responsible but never had the certainty to prove it.

Then, the teachers at the school started to fear the relevance of what at first seemed as a childish game, but that had evolved into a serious problem. They did not know what they could do with it and decided it was time to ask for help.

Once we were introduced in the situation, we first analysed the trolling profile, the published comments and the interaction with other profiles. We noticed a repeated behaviour, most of the contents were referred to a particular girl and had a lot of personal and school related information. Those facts revealed that the author behind the fake profile had to be a member of the same school or even the same class. Moreover, we took the assumption that the real person behind the "Gossip Girl" was following the fake profile, which is consistent with the theory that most of these users want to keep track of the activities and parallel conversations surrounding the trolling profile.

With these considerations in mind, we retrieved all the tweets from the "gossip profile" and their followers and followings profiles. The idea was to train our classifiers with the tweets published by all the users interacting with the trolling profile and then try to identify the authorship of Gossip's tweets using the acquired knowledge.

As a result, we obtained 17,536 tweets corresponding to the 92 users who were followers and/or followings of Gossip Girl, and 43 tweets from the trolling profile, Gossip Girl.

Table 2 offers the results of the authorship identification carried out by the best four classifiers analysed in Section 4: SMO-PolyKernel, J48, SMO-NormalizedPolyKernel and RandomForest.

The table shows the level of authorship attributed to each subject from the whole collection of messages (43 tweets) published by Gossip Girl. It can be appreciated that three subjects appear among the top 4 in the fourth classifiers (highlighted cells). These results made us realise that those three subjects had a great probability of being the responsible ones behind the trolling profile. It is interesting to add, that what seemed to be a one-person misbehaviour had turned, apparently, into a group abuse.

Table 2. Results of the authorship identification for the 92 users followers and/or followings of Gossip Girl's trolling profile. The profile name of the account has been replaced with a subject number due to anonymity issues. The authorship percentage corresponds to the number of tweets published as Gossip Girl, out of the total 43 tweets from the trolling profile, that have been related to the subject. Note that only 20 of the subjects are presented in this table, as most of them have absolutely no indication of being behind the trolling profile.

SMO PolyKernel PolyKernel		J48		SMO NormalizePolykernel NormalizePolykernel		Random Forest	
Sub.#	Authorship	Sub.#	Authorship	Sub.#	Authorship	Sub.#	Authorship
34	23%	34	21%	34	21%	42	21%
66	19%	42	16%	42	14%	34	16%
42	14%	87	14%	66	14%	87	14%
87	14%	46	12%	87	14%	46	12%
8	12%	8	7%	30	12%	31	7%
31	5%	50	7%	31	7%	50	7%
63	2%	31	5%	33	5%	8	2%
20	2%	83	5%	50	2%	14	2%
29	2%	14	2%	8	2%	39	2%
53	2%	39	2%	20	2%	53	2%
83	2%	53	2%	29	2%	83	2%
91	2%	29	2%	53	2%	20	2%
33	0%	30	2%	63	2%	29	2%
49	0%	91	2%	24	0%	44	2%
52	0%	33	0%	28	0%	48	2%
1	0%	48	0%	49	0%	63	2%
2	0%	66	0%	52	0%	6	0%
3	0%	6	0%	1	0%	49	0%
4	0%	57	0%	2	0%	56	0%
5	0%	63	0%	3	0%	66	0%

After the analysis, we reported our findings to the school's staff. With the knowledge of the three names of the alleged abusers the managing office in the school summoned all the students, warning them about the consequences of this misconduct, trying to reduce the impact of their acts, should they confess before it was too late. With fear in the body, the perpetrators revealed their identity, which matched with the three names we identified.

Finally the staff of the school did not reported the event but required them to publicly apologize.

6   Conclusions

More and more children are nowadays connected to the Internet. Although this communication channel provides a lot of important advantages, in many cases, because of the anonymity, different kind of abuses may arise, being one of them the cyberbullying. A rapid identification of this type of users on the Internet is crucial, giving a lot of importance to the systems and/or tools able to identify these threats, in order to protect this population segment on the Internet.

Therefore, we consider that the hereby proposed methodology offers a safe way to identify the real user or users behind a trolling account given some previous conditions: i) the real user/s behind the fake profile has/have a "real" and active account in the social network, ii) the real account of the user/s behind the fake profile is/are somehow connected to the fake profile. These conditions are in theory easy to fulfil due to the assumption that a real person behind a trolling profile wants to keep track of the activities and parallel conversations surrounding the trolling profile.

However, the proposed mechanisms have several limitations. First, despite we assume the previous conditions will be fulfilled, there could be the case in which a user behind a trolling account has no relation with the fake profile to avoid rising suspicions. In this case, it would be necessary to enlarge the circle of users to be analysed or even find a more specific circle based on specific characteristics of the trolling profile. Second, expert abusive users can intentionally change their writing style and/or behaviour to avoid detection. Being the behaviour (e.g., device, time, location) the most difficult to change due to the unconsciousness nature of most human acts, it would also be a really effective way of avoiding detection with no clear solution. On the other hand, the change on writing style could be tackled by analysing the language in more depth, finding for example the use of synonyms or word alterations.

Therefore, as future work, it would be interesting to expand this work in three main directions. Firstly, we would like to analyse different language characteristics and semantics present in the tweets. That analysis could/should include more NLP techniques such as language phenomena study (e.g., synonymity, metonymy or homography), Word Sense Disambiguation (WSD) or Opinion Mining, among others. Besides, given the nature of social networks, it is "easy" to hide among the bast number of users populating these platforms. A possible approach would be to create a kind of *writing style/behaviour signature* able to identify twitter users by the published content. In case of detecting an abuse, that information could be used to reduce the number of users to be further analysed. Finally, we would like to adopt this work to other social networks, chat rooms, and similar environments.

References

1. Boyd, D., Ellison, N.: Social network sites: Definition, history, and scholarship. Journal of Computer-Mediated Communication 13(1), 210–230 (2007)
2. Sanz, B., Laorden, C., Alvarez, G., Bringas, P.G.: A threat model approach to attacks and countermeasures in on-line social networks. In: Proceedings of the 11th Reunion Española de Criptografía y Seguridad de la Información (RECSI), Tarragona, Spain, September 7-10, pp. 343–348 (2010)
3. Laorden, C., Sanz, B., Alvarez, G., Bringas, P.G.: A threat model approach to threats and vulnerabilities in on-line social networks. In: Herrero, Á., Corchado, E., Redondo, C., Alonso, Á. (eds.) Computational Intelligence in Security for Information Systems 2010. AISC, vol. 85, pp. 135–142. Springer, Heidelberg (2010)
4. Smith, P.K., Mahdavi, J., Carvalho, M., Fisher, S., Russell, S., Tippett, N.: Cyberbullying: Its nature and impact in secondary school pupils. Journal of Child Psychology and Psychiatry 49(4), 376–385 (2008)
5. Bishop, J.: Scope and limitations in the government of wales act 2006 for tackling internet abuses in the form of 'flame trolling'. Statute Law Review 33(2), 207–216 (2012)

6. Palfrey, J., Sacco, D., Boyd, D., DeBonis, L., Tatlock, J.: Enhancing child safety & online technologies (2008), cyber.law.harvard.edu/sites/cyber.law.harvard.edu/files/ ISTTF_Final_Report.pdf (accessed)

7. Vandebosch, H., Van Cleemput, K.: Defining cyberbullying: A qualitative research into the perceptions of youngsters. CyberPsychology & Behavior 11(4), 499–503 (2008)

8. Laorden, C., Galán-García, P., Santos, I., Sanz, B., Hidalgo, J.M.G., Bringas, P.G.: Negobot: A conversational agent based on game theory for the detection of paedophile behaviour. In: Herrero, Á., Snášel, V., Abraham, A., Zelinka, I., Baruque, B., Quintián, H., Calvo, J.L., Sedano, J., Corchado, E. (eds.) Int. Joint Conf. CISIS'12-ICEUTE'12-SOCO'12. AISC, vol. 189, pp. 261–270. Springer, Heidelberg (2013)

9. Kontostathis, A.: Chatcoder: Toward the tracking and categorization of internet predators. In: Text Mining Workshop 2009 Held in Conjunction with the 9th Siam International Conference on Data Mining (SDM 2009), Sparks, NV (May 2009)

10. Smets, K., Goethals, B., Verdonk, B.: Automatic vandalism detection in wikipedia: Towards a machine learning approach. In: AAAI Workshop on Wikipedia and Artificial Intelligence: An Evolving Synergy, pp. 43–48 (2008)

11. Tan, P.N., Chen, F., Jain, A.: Information assurance: Detection of web spam attacks in social media. In: Proceedings of Army Science Conference, Orland, Florida (2010)

12. Simanjuntak, D.A., Ipung, H.P., Lim, C., Nugroho, A.S.: Text classification techniques used to faciliate cyber terrorism investigation. In: 2010 Second International Conference on Advances in Computing, Control and Telecommunication Technologies (ACT), pp. 198–200. IEEE (2010)

13. Dinakar, K., Reichart, R., Lieberman, H.: Modeling the detection of textual cyberbullying. In: International Conference on Weblog and Social Media-Social Mobile Web Workshop (2011)

14. De Vel, O., Anderson, A., Corney, M., Mohay, G.: Mining e-mail content for author identification forensics. ACM Sigmod Record 30(4), 55–64 (2001)

15. Holmes, G., Donkin, A., Witten, I.H.: Weka: A machine learning workbench. In: Proceedings of the 1994 Second Australian and New Zealand Conference on Intelligent Information Systems, pp. 357–361. IEEE (1994)

16. Garner, S.R., et al.: Weka: The waikato environment for knowledge analysis. In: Proceedings of the New Zealand Computer Science Research Students Conference, pp. 57–64. Citeseer (1995)

17. Wilbur, W.J., Sirotkin, K.: The automatic identification of stop words. Journal of Information Science 18(1), 45–55 (1992)

18. Kohavi, R., et al.: A study of cross-validation and bootstrap for accuracy estimation and model selection. In: International Joint Conference on Artificial Intelligence, vol. 14, pp. 1137–1145. Lawrence Erlbaum Associates Ltd. (1995)

19. Salton, G., McGill, M.: Introduction to modern information retrieval. McGraw-Hill, New York (1983)

20. Baeza-Yates, R.A., Ribeiro-Neto, B.: Modern Information Retrieval. Addison-Wesley Longman Publishing Co., Inc., Boston (1999)

Deterministic Tableau-Decision Procedure via Reductions for Modal Logic K

Joanna Golińska-Pilarek[1], E. Muñoz-Velasco[2], and Angel Mora[2]

[1] Institute of Philosophy, University of Warsaw, Poland
j.golinska@uw.edu.pl
[2] Dept. Matemtica Aplicada. Universidad de Mlaga, Spain
{emilio,amora}@ctima.uma.es

Abstract. A deterministic tableau decision procedure via reductions, TK, for verification of validity of modal logic K is presented. The system TK is a deterministic tableau decision procedure defined in the original methodology of tableau systems which does not use any additional kind of branching (apart from the required branching for disjunctions) nor any external techniques such as backtracking, backjumping, loop-checking, etc. A nice feature of system TK is its uniqueness; given a formula it generates in a deterministic way only one tableau tree for it.

Keywords: modal logics, tableau methods, decision procedures, prefixed tableau systems.

1 Introduction

Modal logics are extensions of classical propositional logic with modal operators such as *it is possible that*, *it is necessary that*, *it will be always the case that*, *an agent knows (believes) that*, etc. In the last fifty years modal logics have become very popular and extremely useful in many areas of computer science. These areas include security for information systems, knowledge and belief representation, database theory and distributed systems, program verification, cryptography and agent based systems, computational linguistics, and nonmonotonic formalisms. Thus, many efforts have been made to design deduction systems for modal logics, and – in the case of decidable logics – to design effective and simple-to-use decision procedures (cf. [1–3]).

There are several approaches for logics in security for information systems. For instance, in [4], a logic for specifying and reasoning about secure distributed systems is described; this logic combines modalities for knowledge and time with modalities for permission and obligation. In [5], some of the concepts, protocols, and algorithms for access control in distributed systems from a logical perspective are studied. In [6], the influence of trust on the assimilation of acquired information into an agent's belief is considered by using modal logic. More recently, in [7], a formal language with modal logic to prevent unauthorized and malicious access to information systems is proposed, where knowledge related modal operators are employed to represent agents' knowledge in reasoning. Furthermore, in [8], a version of distributed temporal logic is introduced, that is well-suited both for verifying security protocols and as a metalogic for reasoning

Á. Herrero et al. (eds.), *International Joint Conference SOCO'13-CISIS'13-ICEUTE'13*,
Advances in Intelligent Systems and Computing 239,
DOI: 10.1007/978-3-319-01854-6_44, © Springer International Publishing Switzerland 2014

about different security protocol models. Recent advances in the area of security for information systems can be found, for instance, in [9, 10].

Let us focus on modal logic K. It is the minimal decidable modal logic and its tableau system is the core of all tableau systems for normal modal logics. As a consequence, it can be considered as a basis for many logic based approaches in security for information systems. A comprehensive survey on tableau methods for modal logics can be found in [1, 2, 11], among others. The famous tableau system for K is Fitting's tableau (see e.g., [12, 13]). It is the extension of the tableau for the classical propositional calculus with two rules (ϑ) and (κ) which are nondeterministic. As an example, consider a tableau tree with the formula $\Box p \land \Diamond p \land \Diamond \neg p$ at its root. First, we apply the rule $(\land)$ and we obtain the node with the formulas $\Box p, \Diamond p, \Diamond \neg p$. Then, if we apply the rule (ϑ) and (κ) to $\Diamond p$, we will obtain an unclosed one-branching tree with a formula p. However, if we apply (ϑ) and (κ) to $\Diamond \neg p$, we will obtain a closed tree with formulas p and $\neg p$. Thus, valid formula $\neg(\Box p \land \Diamond p \land \Diamond \neg p)$ has at least two different tableau trees, and one of them is closed, while the other is not closed. Therefore, in order to verify satisfiability of that formula, it does not suffice to construct any tableau tree as in classical calculus. We must check all possible tableau trees, until we find a closed one, but the tableau does not include any rules for finding the proper tree. Within the framework of Fitting's tableau calculus it is impossible to avoid this nondeterminism. To solve this problem, the prover needs to backtrack over the choices of $\Diamond$-formulas. Whenever there is more than one $\Diamond$-formula on a branch one systematically has to consider the application of (κ) rule to each of them. However, the Fitting's tableau has also another kind of nondeterminism. We must care not only to which formula we apply the rules, but also at which state we do so, since the preceding application of the rules might delete information which is necessary for finding a closed tree. Hence, one of the main challenges in designing a generic tableau prover is to identify and to remove the nondeterminism implicit in the pen-and-paper formulation of tableau systems.

The aim of this paper is to present a new tableau system TK for modal logic K. The motivation for considering the system TK is rather theoretical than practical. We are interested in a tableau which is not only a base for an algorithm verifying validity of a formula, but is *itself* a deterministic decision procedure. To be precise, by a tableau we mean a system determined by rules and axioms, where comma in the rules represents conjunction and branching represents disjunction. Having this notion of a tableau system, any extension of a tableau with another kind of branching, for example with *and-branching* as it is in some decision procedures for K, is not a tableau in our sense. The same holds for any extension of tableaux with external techniques, such as backtracking, backjumping or simplifications. Second, by a decision procedure we mean an algorithm which in finite number of steps decides whether a formula is valid or not. Thus, we are interested in the 'traditional' tableau (with one kind of branching and without external techniques) which is itself an algorithm for deciding validity of formulas. That is, the tableau algorithm has to be of the following type: construct *any* proof tree in the tableau and then you will know the answer whether a formula is valid or not. With these assumptions, any extension of tableaux with other kind of branching or external techniques is in our terminology a decision procedure *based* on tableaux. Of course, we are aware that the elimination of additional tools in the construction of a decision procedure may increase the space complexity. It is the cost of limitations on the construction of an algorithm to the pure tableau methodology. That is, if additional tools are allowed (i.e., and-branching or external techniques), a possible algorithm may be better than that constructed with the use of weaker tools. However, our aim is not to construct any *tableau-based decision procedure*, possibly with the best space complexity. Our aim is to construct the best possible *tableau decision procedure* in a sense explained above. The problem of existence of such a decision procedure with good space complexity is interesting enough from theoretical point of view. However, it may be also noteworthy for practical reasons, since in practice deterministic decision procedures may be more effective than non-deterministic ones.

The paper is organized as follows. In Section 2 we present the syntax and semantics of our approach. Section 3 is the main part of the paper where the system TK which is a deterministic tableau decision procedure for validity of K-formulas is presented. First, we present with details the system TK. We show that every finite set of clauses has a finite unique tableau tree. Then, we show the soundness and completeness of the system and explain how our system works on the basis of an example. Final remarks and prospects of future work are described in Section 4.

2   Syntax and Semantics

The language of logic K consists of the symbols from the following pairwise disjoint sets: $\mathbb{V} = \{p_1, p_2, p_3, \ldots\}$ – an ordered countable infinite set of propositional variables indexed with natural numbers; $\{\neg, \vee\}$ – the set of classical propositional operations of negation ($\neg$) and disjunction ($\vee$); $\{\Box\}$ – the set consisting of modal propositional operation called the *necessity* operation. The set of K-formulas is the smallest set including the set of propositional variables and closed with respect to all the propositional operations.

A K-*model* is a structure $\mathcal{M} = (U, R, m)$ such that U is a non-empty set (of states), R is a binary relation on U, and m is a meaning function such that $m(p) \subseteq U$, for any propositional variable $p \in \mathbb{V}$. The relation R is referred to as the *accessibility relation*. The *satisfaction relation* is defined as usual in modal logics. Recall that for $\Box$-formulas it is defined as:

$$\mathcal{M}, w \models \Box\varphi \text{ if and only if for all } w' \in U, (w, w') \in R \text{ implies } \mathcal{M}, w' \models \varphi.$$

A K-formula φ is said to be *true* in a K-model $\mathcal{M} = (U, R, m)$, $\mathcal{M} \models \varphi$, whenever for every $w \in U$, $\mathcal{M}, w \models \varphi$, and it is K-*valid* whenever it is true in all K-models. A formula φ is said to be K-*satisfiable* whenever there exist a K-model $\mathcal{M} = (U, R, m)$ and $w \in U$ such that $\mathcal{M}, w \models \varphi$. A finite set Φ of K-formulas is said to be K-*satisfiable* whenever there exist a K-model $\mathcal{M} = (U, R, m)$ and $w \in U$ such that for every $\varphi \in \Phi$, $\mathcal{M}, w \models \varphi$. A finite set Φ of K-formulas is true in a K-model $\mathcal{M}$ whenever each formula from Φ is true in $\mathcal{M}$.

From now on, we will call formulas of the form p or $\neg p$ *classical literals* and we use letters like a, b, c (possible with indices) as their meta-representations. By $\neg a$ we denote the formula $\neg p$ if $a = p$, and p if $a = \neg p$. A *simple modal clause*, or simply a *simple clause*, is a K-formula either of the form: $\delta_m = a_1 \vee \cdots \vee a_m, \Box a \vee b, \neg\Box a \vee b,$

or $\neg\Box a$, where $a, b, a_1, \ldots, a_m$ are classical literals and $m \geq 1$. Simple clauses of the form $a_1 \vee \ldots \vee a_m$, $m \geq 2$, $\Box a \vee b$, $\neg\Box a \vee b$ are referred to as *disjunctive clauses*. A *modal clause* or simply a *clause* is a K-formula of the form $\Box^s \phi$, where ϕ is a simple clause and $s \geq 0$. In what follows $\varphi \vee \psi$ denotes any disjunctive clause, where clauses of the form $a_1 \vee \cdots \vee a_m$, for $m \geq 2$, are meant as $a_1 \vee (a_2 \vee (\ldots \vee (a_{m-1} \vee a_m) \ldots))$.

The following result states equisatisfiablity between a K-formula and its corresponding set of clauses. In fact, it is true for any normal modal logic (cf. [14–16]).

Proposition 1. *For every K-formula φ there exists a finite set of clauses $\{\phi_1, \ldots, \phi_r\}$ such that φ is K-satisfiable if and only if $\{\phi_1, \ldots, \phi_r\}$ is K-satisfiable. Moreover, the set $\{\phi_1, \ldots, \phi_r\}$ can be obtained in quadratic time.*

Now, we extend the language of K to *prefixed* K-formulas for which the system TK is defined. A *prefixed K-formula* is a formula of the form $X : \phi$, where ϕ is a K-clause and X is a finite non-empty subset of natural numbers. Given a prefixed K-formula $X : \phi$, the set X is referred to as its *prefix set*. The informal idea of this notation is that a formula prefixed with a set of natural numbers asserts that it is satisfied at all the worlds named by the natural numbers of its prefix set. A prefixed K-formula of the form $X : a$, where X is a prefix set and a is a classical literal, is referred to as a *prefixed K-literal*. Similarly, a prefixed K-formula the form $X : \phi$, where X is a prefix set and ϕ is a simple clause (resp. disjunctive clause) is referred to as a *prefixed simple K-clause* (resp. *prefixed disjunctive K-clause*).

Let $\mathcal{M} = (U, R, m)$ be a K-model. A valuation in $\mathcal{M}$ is a function $v: \mathbb{N} \to U$. The satisfaction relation between prefixed K-formulas and K-models and valuations is defined as:

$$\mathcal{M}, v \models X : \phi \text{ if and only if for every } i \in X, \mathcal{M}, v(i) \models \phi.$$

A prefixed K-formula $X : \phi$ is said to be *true* in a K-model $\mathcal{M} = (U, R, m)$ whenever for every valuation v in $\mathcal{M}$, $\mathcal{M}, v \models X : \phi$, and it is K-*valid* whenever it is true in all K-models. A formula $X : \phi$ is said to be K-*satisfiable* whenever there exist a K-model $\mathcal{M} = (U, R, m)$ and a valuation v in $\mathcal{M}$ such that $\mathcal{M}, v \models X : \phi$. A finite set Φ of K-formulas is said to be K-*satisfiable* whenever there exist a K-model $\mathcal{M} = (U, R, m)$ and a valuation v in $\mathcal{M}$ such that $\mathcal{M}, v \models X : \phi$, for every $X : \phi \in \Phi$. Clearly, we have the following:

Proposition 2. *For every K-formula φ there exists a finite set of clauses $\{\phi_1, \ldots, \phi_r\}$ such that formula φ is K-satisfiable if and only if the set $\{\{1\} : \phi_1, \ldots, \{1\} : \phi_r\}$ is K-satisfiable.*

Now, we define a kind of lexicographical *order* on the set of all clauses. The order will be used in the rules to guarantee uniqueness of their conclusions. Although any order on clauses will suffice, the order defined below seems to be one of natural possibilities to ensure good effectiveness for future implementations of the system TK. The main motivation for this kind of order is an observation that decompositions of simpler formulas first gives in many cases shorter proof trees.

Let a and b be classical literals. Then $a < b$ whenever either of the following holds:

- $a = p_i$ and $b = p_j$ and $i < j$
- $a = p_i$ and $b = \neg p_j$
- $a = \neg p_i$ and $b = \neg p_j$ and $i < j$,

where p_i, p_j are propositional variables and $i, j \in \mathbb{N}$.

Let $m, k > 1$ and let $\delta_m = a_1 \vee \ldots \vee a_m$ and $\delta_k = b_1 \vee \ldots \vee b_k$ be disjunctive clauses. Then, $\delta_m < \delta_k$ whenever either $m < k$ or both $m = k$ and there exists $i \in \{1, \ldots, m\}$ such that $a_i < b_i$ and $a_j = b_j$ for all $j < i$.

For all classical literals a, b, c, and d, we define:

$\Box a \vee b < \Box c \vee d$ (resp. $\neg \Box a \vee b < \neg \Box c \vee d$) if either $a < c$ or both $a = c$ and $b < d$.

Let $m \geq 1$. For all classical literals $a_0, a_1, \ldots, a_m, a, b, c, d$, and e:

$$a_0 < a_1 \vee \ldots \vee a_m < \neg \Box a < \Box b \vee c < \neg \Box d \vee e.$$

Finally, let $s_i, s_j \geq 0$ and let ϕ_i, ϕ_j be simple clauses. Then:

$$\Box^{s_i} \phi_i < \Box^{s_j} \phi_j \text{ iff either } s_i < s_j \text{ or both } s_i = s_j \text{ and } \phi_i < \phi_j.$$

Let X and Y be finite non-empty subsets of $\mathbb{N}$. Then, $X < Y$ whenever either $\mathrm{card}(X) < \mathrm{card}(Y)$ or both $\mathrm{card}(X) = \mathrm{card}(Y)$ and there exists $x \in X$ such that for every $y \in Y$, $x < y$. Now, we extend the order $<$ to all prefixed K-clauses. We define:

$X : \phi < Y : \psi$ if and only if either $X < Y$ or both $X = Y$ and $\phi < \psi$.

The following proposition is a direct consequence of definition of $<$.

Proposition 3. *The set of all prefixed K-clauses is linearly ordered by $<$.*

If Φ is a finite set of prefixed K-formulas, then a prefixed disjunctive K-clause $X : \phi$ is said to be *minimal with respect to Φ* whenever $X : \phi \leq Y : \psi$, for all prefixed disjunctive K-clauses $Y : \psi \in \Phi$.

Now, we introduce notions used in the construction of the system TK presented in the next section. Let $\Phi = \{X_i : \Box^{s_i} \phi_i\}_{i \in I} \cup \{Y_j : \neg \Box a_j\}_{j \in J}$ be a set of prefixed K-clauses such that J is a non-empty set of indexes and for every $i \in I$, $s_i \geq 1$. Let $N = \bigcup_{i \in I} X_i \cup \bigcup_{j \in J} Y_j$ and let $k, k' \in N$ be such that $k \neq k'$. Then, $k \in N$ is said to be a *subcopy of $k' \in N$ with respect to $j \in J$* whenever $k, k' \in Y_j$, for every $i \in I$, if $k \in X_i$, then $k' \in X_i$, and in addition, if for all $i \in I$, $k \in X_i$ iff $k' \in X_i$, then $k > k'$. An element $k \in N$ is *essential with respect to $j \in J$* whenever $k \in Y_j$ and it is not a subcopy with respect to j of any element in N. Thus, for every $k \in N$ and for every $j \in J$, either $k \notin Y_j$ or k is a subcopy of some element in Y_j or k is essential with respect to j.

For $i \in I$ and $j \in J$, we define:

- $red(X_i \cap Y_j) \stackrel{\mathrm{df}}{=} \{k \in X_i \cap Y_j \mid k \text{ is essential with respect to } j\}$
- $red(Y_j) \stackrel{\mathrm{df}}{=} \{k \in Y_j \mid k \text{ is essential with respect to } j\}$.

Let $\Phi = \{X_i : \Box^{s_i} \phi_i\}_{i \in I} \cup \{Y_j : \neg \Box a_j\}_{j \in J}$ be a set of prefixed K-clauses such that J is a non-empty set of indexes and for every $i \in I$, $s_i \geq 1$. Let $N = \bigcup_{i \in I} X_i \cup \bigcup_{j \in J} Y_j$. Let us define:

$$red(\Phi) \stackrel{\mathrm{df}}{=} \{\bigcup_{j \in J} red(X_i \cap Y_j) : \square^{s_i}\phi_i\}_{i \in I} \cup \{red(Y_j) : \neg\square a_j\}_{j \in J}.$$

In the above, we assume that if $\bigcup_{j \in J} red(X_i \cap Y_j) = \emptyset$, for some $i \in I$, then $\square^{s_i}\phi_i$ does not belong to $red(\Phi)$. Furthermore, for every $j \in J$, there is a function σ_j : $Y_j \setminus red(Y_j) \to red(Y_j)$ such that for every $k \in Y_j \setminus red(Y_j)$ it satisfies the following: $\sigma_j(k)$ is the smallest number in $red(Y_j)$ such that k is its subcopy with respect to j.

3   Tableau System TK

In this section, we present tableau system TK for the modal logic K. The tableau system TK belongs to the family of prefixed tableaux. Prefixed tableaux provide additional machinery to retain information about the worlds during the construction of the proof. Namely, each prefix occurring at some stage of the proof contains some information about part of the current proof. Prefixed tableau systems were introduced in [17]. Their present modular form can be found in [13, 18].

The system TK is defined for formulas prefixed with sets of natural numbers. It is determined by axiomatic sets of K-clauses prefixed with sets and rules which apply to finite sets of K-clauses prefixed with sets.

A finite set of prefixed K-formulas is said to be TK-*axiomatic* whenever it is a super-set of $\{X : \phi, Y : \neg\phi\}$, for a K-formula ϕ and for any non-empty sets $X, Y \subset \mathbb{N}$ such that $X \cap Y \neq \emptyset$. Clearly, every TK-axiomatic set is K-unsatisfiable. Sets of formulas which are not axiomatic are referred to as *non-axiomatic*. Clearly, every non-axiomatic set of prefixed K-literals is K-satisfiable.

The rules of TK-system are of the form $(*)$ $\frac{\Gamma}{\Gamma_1 | \dots | \Gamma_n}$, where $\Gamma, \Gamma_1, \dots, \Gamma_n, n \geq 1$, are finite non-empty sets of prefixed K-formulas. A rule of the form $(*)$ is *applicable* to a finite set Y if and only if $Y = \Gamma$. If $n > 1$, then a rule of the form $(*)$ is an n-fold branching rule. In a rule, the set above the line is referred to as its *premise* and the set(s) below the line is(are) its *conclusion(s)*.

Below we list the rules of the system TK, which are the rules $(\vee)$ and (K). In what follows, if ϕ is a K-formula and X is empty, then $\{X : \phi\}$ denotes the empty set.

The rule $(\vee)$ for disjunctive clauses:

$$(\vee) \quad \frac{F \cup \{X : (\phi \vee \psi)\} \cup \{Y : \phi, Y' : \psi\}}{F \cup D_{Z_1} | \dots | F \cup D_{Z_n}}, \text{ where}$$

- F is a finite (possibly empty) set of prefixed K-formulas.
- $X : (\phi \vee \psi)$ is a prefixed K-disjunctive clause that is minimal with respect to F.
- $\{Y : \phi, Y' : \psi\}$ is a (possibly empty) set of prefixed K-formulas.
- $n = 2^{\mathrm{card}(X)}$.
- For every $1 \leq i \leq n$, Z_i is the i-th subset of X with respect to the lexicographical order on the set of all subsets of X with the usual ordering on natural numbers.
- $D_{Z_i} \stackrel{\mathrm{df}}{=} \{Y \cup Z_i : \phi\} \cup \{Y' \cup (X \setminus Z_i) : \psi\}$, for every $1 \leq i \leq n$.

The rule $(\vee)$ behaves as the classical rule for disjunction. Indeed, we may treat the set D_{Z_i} defined above as the result of maximal decomposition of the set $\{i : (\phi \vee \psi) \mid i \in X\} \cup \{i : \phi \mid i \in Y\} \cup \{i : \psi \mid i \in Y'\}$. Moreover, observe that any choice of a set $Z_i \subseteq X$ is uniquely determined. Thus, conclusions of the rule $(\vee)$ are uniquely ordered.

The rule (K):

$$(K) \quad \frac{L \cup \{X_i : \Box^{s_i} \phi_i\}_{i \in I} \cup \{Y_j : \neg \Box a_j\}_{j \in J}}{\bigcup_{i \in I, j \in J} (G_i \cup H_j)}, \text{ where}$$

- L is a finite (possibly empty) non-axiomatic set of prefixed K-literals.
- $\{X_i : \Box^{s_i} \phi_i\}_{i \in I}$ is a finite (possibly empty) set of prefixed K-clauses, where $s_i \geq 1$ and I is an ordered set of indexes such that $i < i'$ iff $(X_i : \Box^{s_i} \phi_i) < (X_{i'} : \Box^{s_{i'}} \phi_{i'})$, for all $i, i' \in I$.
- $\{Y_j : \neg \Box a_j\}_{j \in J}$ is a finite non-empty set of prefixed simple K-clauses, where J is an ordered set of indexes such that $j < j'$ iff $(Y_j : \neg \Box a_j) < (Y_{j'} : \neg \Box a_{j'})$, for all $j, j' \in J$.
- We define the set $\mathbb{N} \stackrel{\text{df}}{=} \{n_{kj} \mid j \in J, k \in red(Y_j)\}$ as a set of natural numbers such that:
  - Its smallest element n is the smallest natural number that does not occur in any prefix set of the premise of the rule (K),
  - For each $x \in \mathbb{N}$, if $n \neq x$, then there exists $y \in \mathbb{N}$, such that $x = y + 1$,
  - For all $n_{kj}, n_{k'j'} \in \mathbb{N}$, $n_{kj} < n_{k'j'}$ iff either $j < j'$ or both $j = j'$ and $k < k'$.
- For every $i \in I$ such that for all $j \in J$, $\Box^{s_i - 1} \phi_i \neq \neg a_j$, a set G_i has the form:

$$G_i \stackrel{\text{df}}{=} \{Z_i : \Box^{s_i - 1} \phi_i\}, \text{ where } Z_i \stackrel{\text{df}}{=} \{n_{kj} \mid j \in J \text{ and } k \in red(X_i \cap Y_j)\} \subseteq \mathbb{N}.$$

- For every $j \in J$ such that for all $i \in I$, $\neg a_j \neq \Box^{s_i - 1} \phi_i$, a set H_j has the form:

$$H_j \stackrel{\text{df}}{=} \{Z_j : \neg a_j\}, \text{ where } Z_j \stackrel{\text{df}}{=} \{n_{kj} \mid k \in red(Y_j)\} \subseteq \mathbb{N}.$$

- For all $i \in I, j \in J$ such that $\Box^{s_i - 1} \phi_i = \neg a_j$:

$$G_i = H_j = \{Z_i \cup Z_j : \neg a_j\}, \text{ where } Z_i \text{ and } Z_j \text{ are defined as above.}$$

Note that each prefix set of the conclusion rule (K) is included in the set $\mathbb{N}$ which is the set of new natural numbers, i.e., numbers that do not occur in any prefix set of the premise. The order on $\mathbb{N}$ guarantees uniqueness of the conclusion of the rule (K).

As usual in tableau systems, given a finite set of prefixed K-formulas, successive applications of the rules of TK-system result in a tree whose nodes consist of finite sets of prefixed K-formulas. Let φ be a K-formula and let $\{\phi_1, \ldots, \phi_r\}$ be a set of clauses equisatisfiable with $\neg \varphi$ as ensured in Proposition 1. A TK-*proof tree* of φ is a tree with the following properties: (1) The formulas $\{1\} : \phi_1, \{1\} : \phi_2, \ldots, \{1\} : \phi_r$ are at the root of the tree; (2) Each node except the root is obtained by an application of a TK-rule to its predecessor node; (3) A node does not have successors whenever its set of prefixed formulas is TK-axiomatic or none of the rules applies to it.

A branch of a TK-proof tree is *closed* if it contains a node with a TK-axiomatic set of prefixed formulas. A TK-proof tree is closed if and only if all of its branches are closed. A K-formula φ is TK-*provable* whenever there is a closed TK-proof tree of it, which is then referred to as its TK-*proof*. The rules of TK-tableau guarantee that:

Theorem 1. *For every finite set* $\Phi = \{\{1\} : \phi_1, \ldots, \{1\} : \phi_r\}$ *of* K-*clauses there exists exactly one finite* TK-*proof tree with* Φ *at the root.*

Example 1. Consider the following set Φ of prefixed K-clauses:

$$\Phi = \{\{1\} : \neg\Box p_2, \{1\} : \neg\Box\neg p_2, \{1\} : \Box(\Box p_1 \vee p_2), \{1\} : \Box(\neg\Box p_1 \vee \neg p_2)\}$$

To fix the notation, let $X_1 = X_2 = Y_1 = Y_2 = \{1\}$ and let:

- $X_1 : \Box(\Box p_1 \vee p_2)$ denotes the formula $\{1\} : \Box(\Box p_1 \vee p_2)$
- $X_2 : \Box(\neg\Box p_1 \vee \neg p_2)$ denotes the formula $\{1\} : \Box(\neg\Box p_1 \vee \neg p_2)$
- $Y_1 : \neg\Box p_2$ denotes the formula $\{1\} : \neg\Box p_2$
- $Y_2 : \neg\Box\neg p_2$ denotes the formula $\{1\} : \neg\Box\neg p_2$

Observe that $red(Y_j) = red(X_i \cap Y_j) = \{1\}$, for all $i, j \in \{1, 2\}$. Thus, after the application of the rule (K) to Φ we obtain the node with the set $\bigcup_{i,j \in \{1,2\}} G_i \cup H_j$, where:

- $G_1 = \{\{n_{11}, n_{12}\} : (\Box p_1 \vee p_2)\}$ and $G_2 = \{\{n_{11}, n_{12}\} : (\neg\Box p_1 \vee \neg p_2)\}$
- $H_1 = \{\{n_{11}\} : \neg p_2\}$ and $H_2 = \{\{n_{12}\} : p_2\}$
- $2 = n_{11} < n_{12} = 3$.

Thus, the conclusion of Φ is:

$$\Phi_1 = \{\{2\} : \neg p_2, \{3\} : p_2, \{2, 3\} : (\Box p_1 \vee p_2), \{2, 3\} : (\neg\Box p_1 \vee \neg p_2)\}.$$

Now, we apply to Φ_1 the rule $(\vee)$. Note that the minimal disjunctive clause in Φ_1 is $\{2, 3\} : (\Box p_1 \vee p_2)$. Let $D = \Phi_1 \setminus \{\{2, 3\} : (\Box p_1 \vee p_2)\}$. The application of the rule $(\vee)$ to Φ_1 results with the following four conclusions:

- $D_1 = (D \setminus \{3 : p_2\}) \cup \{\{2, 3\} : p_2\}$ – axiomatic, since it contains $\{2, 3\} : p_2$ and $\{2\} : \neg p_2$
- $D_2 = (D \setminus \{3 : p_2\}) \cup \{\{2\} : \Box p_1, \{3\} : p_2\}$
- $D_3 = D \cup \{\{3\} : \Box p_1, \{2\} : p_2\}$ – axiomatic, since it contains $\{2\} : p_2$ and $\{2\} : \neg p_2$
- $D_4 = D \cup \{\{2, 3\} : \Box p_1\}$

Thus, two of four conclusions of Φ_1, namely D_2 and D_4, are still not closed. The only rule that applies to D_2 and D_4 is the rule $(\vee)$. Let $S = D_2 \setminus \{\{2, 3\} : (\Box p_1 \vee p_2)\}$ and $S' = D_4 \setminus \{\{2, 3\} : (\Box p_1 \vee p_2)\}$. First, we decompose D_4. It has four conclusions of the following forms:

- $S_1' = (S' \setminus \{2 : \neg p_2\}) \cup \{\{2, 3\} : \neg p_2\}$ – axiomatic, since it contains $\{2, 3\} : \neg p_2$ and $\{3\} : p_2$
- $S_2' = S' \cup \{\{2\} : \neg\Box p_1, \{3\} : \neg p_2\}$ – axiomatic, since it contains $\{3\} : \neg p_2$ and $\{3\} : p_2$
- $S_3' = (S' \setminus \{2 : \neg p_2\}) \cup \{\{3\} : \neg\Box p_1, \{2\} : \neg p_2\}$ – axiomatic, since it contains $\{3\} : \neg\Box p_1$ and $\{2, 3\} : \Box p_1$
- $S_4' = S' \cup \{\{2, 3\} : \neg\Box p_1\}$ – axiomatic, since it contains $\{2, 3\} : \neg\Box p_1$ and $\{2, 3\} : \Box p_1$

Hence, all the conclusions of D_4 are closed. Now, we decompose D_2 and we get the following conclusions:

- $S_1 = (S \setminus \{2 : \neg p_2\}) \cup \{\{2,3\} : \neg p_2\}$ – axiomatic, since it contains $\{2,3\} : \neg p_2$ and $\{3\} : p_2$
- $S_2 = (S \setminus \{2 : \neg p_2\}) \cup \{\{2\} : \neg \Box p_1, \{2,3\} : \neg p_2\}$ – axiomatic, since it contains $\{3\} : \neg p_2$ and $\{3\} : p_2$
- $S_3 = (S \setminus \{2 : \neg p_2\}) \cup \{\{3\} : \neg \Box p_1, \{2\} : \neg p_2\}$
- $S_4 = S \cup \{\{2,3\} : \neg \Box p_1\}$ – axiomatic, since it contains $\{2,3\} : \neg \Box p_1$ and $\{2\} : \Box p_1$

Three of four conclusions of D_2 are closed. Still we have one node with a non-axiomatic set $T = \{\{2\} : \neg p_2, \{3\} : p_2, \{2\} : \Box p_1, \{3\} : \neg \Box p_1\}$. The only rule that applies to T is the rule (K). Since $\{2\} \cap \{3\} = \emptyset$, we have $\mathbb{N} = \{n_1 = 4\}$. Thus, the conclusion of T is the set $\{\{4\} : \neg p_1\}$, which is not axiomatic and to which none of the rules applies. Finally, we may conclude that the tree for Φ is not closed.

In order to obtain the soundness and completeness of our system, we show that the rules of TK-tableau preserve satisfiability of sets of their premises. Formally, a rule is said to be TK-correct whenever K-satisfiability of its premise implies K-satisfiability of some of its conclusions.

Proposition 4
1. The TK-axiomatic sets of prefixed K-formulas are K-unsatisfiable.
2. The TK-rules are TK-correct.

TK-rules have a stronger property. In fact, they are K-invertible, that is they preserve and reflect K-unsatisfiability of their premises and conclusions. For space reasons, we omit the proof. As a consequence, we have the following result.

Theorem 2. *The system* TK *is a sound and complete deterministic decision procedure for the logic* K.

4   Conclusions and Future Work

We have presented a sound and complete tableau decision procedures TK for modal logic K. The system TK is deterministic. It does not use any external technique such as backtracking, backjumping, etc., and what is more important, it is defined in the original methodology of tableau systems with one kind of branching representing a decomposition of disjunctions.

We are already working on the implementation of TK (see [19]) and its comparison with similar provers (e.g., [20–22]), together with a study about how the prover scales as the problem size increases. The other natural problem for future work is whether TK-system could be extended to decision procedures of other normal modal logics and its applications to Description Logics.

Acknowledgements. The first author of the paper is supported by the Polish National Science Centre research project DEC-2011/02/A/HS1/00395. This work is partially supported by the Spanish research projects TIN2009-14562-C05-01, TIN2012-39353-C04-01 and Junta de Andalucía project P09-FQM-5233.

References

1. Fitting, M.: Modal proof theory. In: Patrick Blackburn, J.V.B., Wolter, F. (eds.) Studies in Logic and Practical Reasoning, vol. 3, pp. 85–138. Elsevier (2007)
2. Massacci, F.: Single step tableaux for modal logics. Journal of Automated Reasoning 24(3), 319–364 (2000)
3. Negri, S.: Proof analysis in modal logic. Journal of Philosophical Logic 34(5-6), 507–544 (2005)
4. Glasgow, J., Macewen, G., Panangaden, P.: A logic for reasoning about security. ACM Trans. Comput. Syst. 10(3), 226–264 (1992)
5. Abadi, M., Burrows, M., Lampson, B., Plotkin, G.: A calculus for access control in distributed systems. ACM Trans. Program. Lang. Syst. 15(4), 706–734 (1993)
6. Liau, C.J.: Belief, information acquisition, and trust in multi-agent systems—A modal logic formulation. Artificial Intelligence 149(1), 31–60 (2003)
7. Bai, Y., Khan, K.M.: Ell secure information system using modal logic technique. International Journal of Secure Software Engineering 2(2), 65–76 (2011)
8. Caleiro, C., Viganò, L., Basin, D.: Relating strand spaces and distributed temporal logic for security protocol analysis. Logic Journal of IGPL 13(6), 637–663 (2005)
9. De Paz, J.F., Navarro, M., Pinzón, C.I., Julián, V., Tapia, D.I., Bajo, J.: Mathematical model for a temporal-bounded classifier in security environments. Logic Journal of IGPL 20(4), 712–721 (2012)
10. Amato, F., Casola, V., Mazzocca, N., Romano, S.: A semantic approach for fine-grain access control of e-health documents. Logic Journal of IGPL (2012)
11. Horrocks, I., Hustadt, U., Sattler, U., Schmidt, R.: Computational modal logic. In: Patrick Blackburn, J.V.B., Wolter, F. (eds.) Studies in Logic and Practical Reasoning, vol. 3, pp. 181–245. Elsevier (2007)
12. Fitting, M.: Proof Methods for Modal and Intuitionistic Logics. Springer (1983)
13. Goré, R.: Tableau methods for modal and temporal logics. In: D'Agostino, M., Gabbay, D.M., Hähnle, R., Posegga, J. (eds.) Handbook of Tableau Methods, pp. 297–396. Springer, Netherlands (1999)
14. Goré, R., Nguyen, L.A.: Clausal tableaux for multimodal logics of belief. Fundamenta Informaticae 94(1), 21–40 (2009)
15. Mints, G.: Gentzen-type systems and resolution rules part i propositional logic. In: Martin-Löf, P., Mints, G. (eds.) COLOG 1988. LNCS, vol. 417, pp. 198–231. Springer, Heidelberg (1990)
16. Nguyen, L.A.: A new space bound for the modal logics K4, KD4 and S4. In: Kutyłowski, M., Wierzbicki, T., Pacholski, L. (eds.) MFCS 1999. LNCS, vol. 1672, pp. 321–331. Springer, Heidelberg (1999)
17. Kanger, S.: Provability in logic. PhD thesis, Stockholm (1957)
18. Massacci, F.: Strongly analytic tableaux for normal modal logics. In: Bundy, A. (ed.) CADE 1994. LNCS, vol. 814, pp. 723–737. Springer, Heidelberg (1994)
19. Golińska-Pilarek, J., Muñoz-Velasco, E., Mora-Bonilla, A.: Relational dual tableau decision procedure for modal logic K. Logic Journal of IGPL 20(4), 747–756 (2012)
20. Gasquet, O., Herzig, A., Longin, D., Sahade, M.: LoTREC: Logical Tableaux Research Engineering Companion. In: Beckert, B. (ed.) TABLEAUX 2005. LNCS (LNAI), vol. 3702, pp. 318–322. Springer, Heidelberg (2005)
21. Abate, P., Goré, R.: The tableau workbench. Electronic Notes in Theoretical Computer Science 231, 55–67 (2009)
22. Horrocks, I.: The FaCT system. In: de Swart, H. (ed.) TABLEAUX 1998. LNCS (LNAI), vol. 1397, pp. 307–312. Springer, Heidelberg (1998)

Real-Time Polymorphic Aho-Corasick Automata for Heterogeneous Malicious Code Detection

Ciprian Pungila and Viorel Negru

West University of Timisoara
Blvd. V. Parvan 4, Timisoara 300223, Timis, Romania
{cpungila,vnegru}@info.uvt.ro
http://info.uvt.ro/

Abstract. We are proposing a new, heterogeneous approach to performing malicious code detection in intrusion detection systems using an innovative hybrid implementation of the Aho-Corasick automaton, commonly used in pattern-matching applications. We are introducing and defining the Aho-Corasick polymorphic automaton, a new type of automaton which can change its nodes and transitions in real-time on adequate hardware, using an approach we designed for heterogeneous hardware and which easily scales to hybrid heterogeneous systems with multiple CPUs and GPUs. Using as a test-bed a set of the latest virus signatures from the ClamAV database, we analyze the performance impact of several different types of heuristics on the new type of automata and discuss its feasibility and potential applications in real-time intelligent malicious code detection.

Keywords: aho-corasick, intelligent malicious code detection, heterogeneous hardware, parallel algorithm.

1 Introduction

Intelligent malicious code detection nowadays is primarily focused on two important aspects in modern intrusion detection systems (IDS): static approaches (usually equivalent to matching virus signatures in the static analysis process, or to finding malicious data packets sent through the network based on different aspects) and dynamic approaches (usually referring to intelligently deducting whether a running program is behaving maliciously, or whether incoming data packets through a network may cause potential damage to the software or to the hardware of the system involved).

While multiple pattern-matching algorithms have existed for a long time now, and various types of algorithms are used for different types of applications, the Aho-Corasick algorithm [1] is still preferred due to its simplicity and speed in various fields of research: intrusion detection systems (including regular expression matching), bioinformatics (DNA matching), language processing (approximate string matching), etc.

Á. Herrero et al. (eds.), *International Joint Conference SOCO'13-CISIS'13-ICEUTE'13*, 439
Advances in Intelligent Systems and Computing 239,
DOI: 10.1007/978-3-319-01854-6_45, © Springer International Publishing Switzerland 2014

We are proposing a new approach to implementing custom heuristics for malicious code detection on top of the Aho-Corasick automaton, by efficiently constructing, storing and parallelizing it on heterogeneous hardware and by employing different types of heuristics to the pattern-matching stage throughout the process. We discuss related work in section 2 of this paper, describing the architecture and methodology involved in applying it to the problem of malicious code detection in section 3. Finally, we present experimental results of our architecture and implementation process in section 4.

2   Related Work

The problem of pattern-matching is widely spread in most IDS systems nowadays, both in the static, signature-based approaches (for antivirus engines in [2] and [3], as well as network IDS in [4] and [5]) as well as in heuristic-based approaches to supervised and unsupervised learning stages for dynamic analysis (using system-call analysis in [6], supervised-learning approaches in the cloud in [7], pattern recognition approaches in [8], along with decision trees and stochastic models in [9]). It is also worth mentioning that the approach is widely spread in literature for solving several other related problems, such as approximate string searching [10] or data recovery in digital forensics [11].

One important limitation of all approaches we know of to date using the Aho-Corasick automaton in pattern-matching applications is the inability to rapidly rebuild the automaton, a necessity for adaptive heuristics (heuristics which change or alter the nodes/transitions of the automaton during processing, e.g. when performing an update to an antivirus engine which requires some signatures - considered to be false positives - to be removed, or others to be added; when removing certain keywords from the automaton during dynamic analysis because of user confirmation that they are false positives; when inserting new keywords in the automaton as a result of the analysis of the behavior of a program known to be malicious, etc.). So far, this has been a significant challenge, especially in real-time systems where such automata need to be rebuilt as fast as possible to continue the processing.

Our main paper's contributions are: a) implementing a parallel version of the Aho-Corasick automaton for pattern matching which accepts custom heuristics on different layers of the processing stages involved (both when constructing and parsing the automaton); b) proposing a new software architecture for heterogeneous systems for implementing custom heuristics aimed at detecting malicious code, based on our own parallel implementation of the Aho-Corasick automaton, along with an an efficient storage model for it in both CPU (central processing unit) and GPU (graphics processing unit) memory; c) implementing a real-time approach to rebuilding the automaton when using adaptive heuristics; d) proposing efficient methods (performance-wise) of running different heuristics on hybrid, heterogeneous hardware systems, on top of the pattern-matching process implemented in the previous steps.

2.1   The Aho-Corasick Algorithm

The Aho-Corasick algorithm is based on the trie tree concept and represents an automaton which adds, in addition to the usual transitions present in the trie tree, an additional failure transition which gets called whenever a mismatch on a node occurs. This ensures that mismatches are spread throughout the automaton properly, minimizing the parsing of the automaton as much as possible and therefore increasing processing speed. The failure transition is computed as being the longest suffix of the word at the current node, which also exists in the automaton; if there is no suffix with such a property, the failure transition simply points to the root of the automaton (Figure 1). It is easy to observe that failures for all level 0 and 1 nodes are always pointing back to the root.

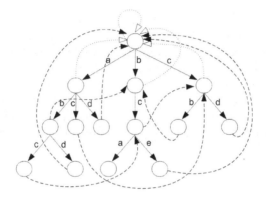

Fig. 1. An Aho-Corasick automaton for the input set {abc, abd, ac, ad, bca, bce, cb, cd}. Dashes transitions are failure pointers. [12]

We define a polymorphic automaton as being an Aho-Corasick automaton which can be created/reconstructed/adapted in real-time, either as a result of an insertion/removal of one or more patterns, or as a result of a modification of these. Adaptive heuristics are the ones usually altering the automaton to ensure that the supervised/unsupervised learning processes are properly carried out.

2.2   The CUDA Architecture

The CUDA (Compute Unified Device Architecture) was introduced by NVIDIA back in 2007 as a parallel computing platform, as well as a programming framework, which allows executing custom-built code easily on the GPU. GPUs are comprised usually of one or more stream processors, which are operating on SIMD (Single Instruction Multiple Data) programs. Programs issued by the host computer to the GPU are called kernels and are executed on the device as one or more threads, organized in one or more blocks. Each multiprocessor executes one or more thread blocks, those active containing the same number of threads and being scheduled internally.

2.3   Automaton Compression

Automaton compression is a common approach to reducing the storage space required for storing the automaton. While some techniques used in the process, such as path compression, significantly increase the processing time and also decrease performance when executed on GPUs by increasing memory divergence, others, such as limiting the automaton in depth, introduce false positives (which may be still desired in the case of approximate string searching) or increase the processing required to accurately find a match (for exact string matching).

A very efficient automaton compression technique through serialization was proposed in [13] and was designed for heterogeneous systems. It uses a stack of nodes (Figure 3a) in order to construct a compact storage model which could be easily transferred between the host (CPU) and the device (GPU) at maximum bandwidth. In this approach, all child nodes of a parent node are stored in consecutive offsets in the stack, with the stack being transferred between the host and the device in a single burst. We are using the same approach in this paper, where pointers to other nodes in the automaton are therefore replaced by a simple offset in the stack, with each node storing a bitmapped field specifying the total number of transitions possible for that node (computing popcount for each bit returns the position in the offset stack), and a failure offset which points to the node we are jumping to in case of a mismatch. The automaton can be compressed this way using either a pre-order or post-order parsing of the Aho-Corasick tree.

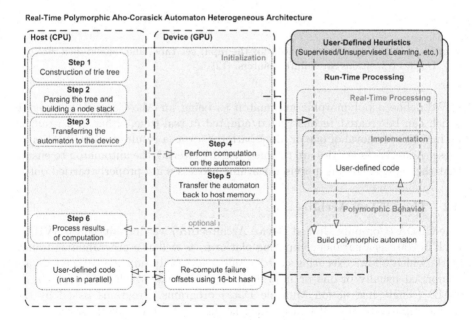

Fig. 2. The proposed architecture

3 Implementation

Heterogeneous hardware systems use one or more CPUs and GPUs to work and perform multiple tasks concurrently, as much as possible. A CPU-GPU combination is mostly preferred nowadays, since it is found in most computers and furthermore, GPU-executed code can be ran in parallel with the CPU code. Additionally, on systems with interchangeable graphics (e.g. with both an integrated and a dedicated graphics solution), such approaches are practically ideal for huge computational tasks. There are however specific constraints here, such as those related to GPU processing, or to the scalability of the work performed on multiple CPU cores. In essence, we are discussing as follows two types of heterogeneity for the malicious code detection problem: the first on a software level (for implementing both the static and dynamic analysis processes discussed earlier), and the second on a hardware level (for efficiently transferring information between the CPU and the GPU, ensuring optimal load balancing between the same two elements, and efficiently implementing our architecture to solve the problem at hand). We will discuss both aspects as follows.

3.1 Proposed Architecture

Our proposed architecture is based on the simple observation that the primary bottleneck when constructing the Aho-Corasick automaton for performing the pattern-matching process lies in the computation of the failure function. The more nodes the automaton has, the longer it takes to compute the failure function for all of them, especially when the automaton is comprised of tens or hundreds of millions of nodes. As they stand, such approaches are not feasible for implementation in real-time systems simply because of their long response time when having to reconfigure the automaton, either by removing, adding or modifying nodes and/or arcs.

Our proposed architecture is the first we know of to explore the full parallel potential of GPU architectures for implementing polymorphic automata. The overall layout of the architecture is present in Figure 2. The CPU is used to construct the trie tree for the automaton, after which a parsing of the automaton creates the stack of nodes and copies them to the GPU memory in one single burst (using the full bandwidth of the PCI-Express architecture). We define a *unique suffix* in the automaton as being the longest suffix of a pattern for which the corresponding nodes in the automaton have at most one direct child node. Any patterns being added or removed from the automaton at this stage may result in the stack of sequential nodes being modified, as follows: in case of an addition, a new unique suffix will be added to the automaton, which requires a relocation of the nodes in the automaton (inserting at least one node somewhere inside the stack, and relocating the following nodes to the right of this one and rebuilding the offsets of the nodes; to avoid the offset computation stage, we add an additional 2 empty nodes for all lists of direct children of nodes having a depth of at most 10 in the automaton); deletion requires a unique suffix to be deleted, which does not necessarily require a relocation (similar to a file-system,

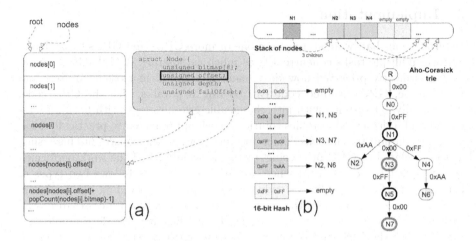

Fig. 3. a) The stack-based representation of very large Aho-Corasick automata as presented in [12]. b) An example of the stack of nodes, the 16-bit hashing technique and the Aho-Corasick trie.

we can leave the removed node areas unoccupied, until a new pattern, which fits in those spots, is added).

We have implemented two different approaches to the failure re-computation stage of the Aho-Corasick automaton: the first, which we call the "no-hash" technique, is taking advantage of the linear disposal of nodes in the stack in GPU memory, and assigns a thread in the GPU to each node. This way, a thread runs a kernel which handles, for a particular node, the re-computation of the failure offset. This has the advantage that it does not require any additional memory, at the expense of processing all nodes in the automaton. The second approach (which we call "16-bit hashing", see Figure 3b) uses a $2 \times log_2 N$-bit hash, where N is the size of the alphabet (e.g. $N = 256$ for our experiments using the ASCII charset) and the key represents a two-character suffix in the pattern, pointing to the list of nodes of depth 2 or more, which have this suffix (we call this a "look-ahead" hash). Since the failure offset re-computation stage involves finding the longest suffix of a word which also exists in the tree, whenever we add/remove a pattern we actually insert or delete a unique suffix, and since the only nodes affected are those who end in the suffixes of the affected nodes, all we have to do is recompute these nodes' failure offsets, leaving the others alone). This significantly reduces the computation stage since it reduces the amount of nodes required to have the offset recomputed, but additionally requires memory for our hash and also depends on the pattern configuration. For example, if removing the unique suffix *cd* from the pattern *abcd*, all nodes in the stacks corresponding to the hash values of *bc, cd* will be affected and will need to have the failure offset recomputed. Larger hashes may of course be used to reduce the processing time even further, at the higher expense of more memory required in the GPU.

3.2   Experimental Testbed

We have performed our testing using a mobile quad-core Ivy-Bridge i7 3610QM CPU (built in the 22nm process), clocked at 2.3 GHz, and an average GPU card based on the new Kepler architecture that NVIDIA released not long ago, the GT650M, backed up by 2 GB of RAM GDDR5. The GDDR5-based GT650M is built in the 28nm process and has a peak memory bandwidth of 64 GB/s, with a peak of 564.5 GLFOPS, while the i7 CPU has a peak memory bandwidth of 25.6 GB/s and 77.74 GFLOPS. These numbers, as also pointed out by Lee at al [14], reflect the actual performance of the GPU vs. that of the CPU in real-world situations. Our GPU implementation used a number of 128 threads per block and always assigned a single thread to each node whose failure offset needed to be computed.

We used different sets of virus signatures (of 10^4, 2×10^4, 5×10^4, 9×10^4 and 11×10^4 patterns, average signature length was 250 bytes in each) from the latest up-to-date ClamAV [15] database in order to create a very large automaton (the actual number of patterns and the resulting number of nodes in the automaton are shown in Figure 4a). We then began to evaluate the no-hash implementation on both the CPU (single-core) and the GPU, after which we moved to the 16-bit hash implementation applied for adding and removing 1, 5 and 10 patterns, consecutively. Each experiment was reproduced three times and the average times were computed, to ensure a better evaluation of the outcome. For real-time activity, we are mostly interested in the performance applied to a single pattern (for instance, it is common to remove a false positive from a list of patterns, or to add a new pattern when a new potential threat is found), but we also evaluated the same approach for 5 and then 10 patterns.

In order to efficiently assess the performance of our proposed architecture, we have used several different types of heuristics applied to the patterns in order to evaluate the reconstruction time needed for the automaton. Our experiments included both additions to the automaton, as well as removals. The heuristics were chosen in order to select the best candidate(s) for removal, while additions

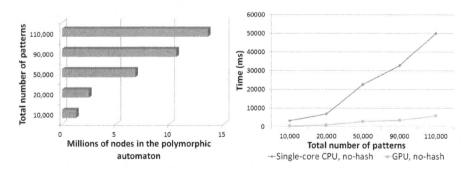

Fig. 4. a) Number of nodes resulting from the different set of patterns used in the automaton during testing. b) The no-hash running times of the algorithm on both the CPU and the GPU.

were performed through direct insertion in the automaton. Since the heuristics
depend on the type of application and are therefore not the subject of the paper
(e.g. for system-call analysis as pointed out in [6], required to assess the threat
potential of an application in real-time, a heuristic may be used to compare
different sequences of system-calls and remove those considered to be false positives, or to add new threats to the database if a sequence call is very similar with
another existing one which is known to be malicious), for removals, we picked a
random signature from the database, we introduced additional noise to a random
percentage of up to 50%, and then compared the similarity of this new obtained
signature to the others existing, using the Hamming[16], Needleman-Wunsch[17]
and Bray-Curtis[18] distances. Finally, we picked the best candidate(s) from the
resulting set, meaning the best candidates with the closest resemblance to the
modified signature and removed them. For additions, we followed the same approach of randomly generating up to 10 signatures, adding them after the process
to the automaton.

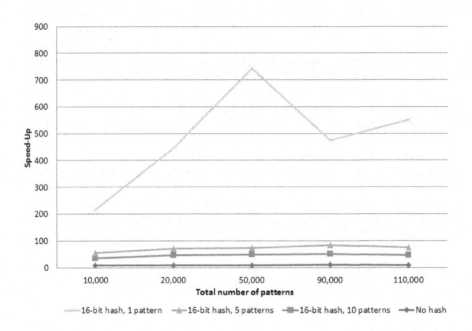

Fig. 5. Speed-ups achieved for the 16-bit hash implementation on the GPU, relative
to the no-hash single-core CPU implementation

3.3   Results

The first results are related to no-hash the GPU vs. CPU implementation and
are listed in Figure 4b. The speed-ups observed here are consistent with the
bandwidth capabilities of the hardware involved, and ranges from a 8.1x for
10,000 patterns, to a peak of 9.5x for 90,000 patterns.

After implementing the 16-bit hash support in the architecture, we were able to dramatically reduce the processing times to just a few milliseconds, obtaining, for a single-pattern operation for instance, a speed-up of the GPU 16-bit hash implementation of over 700 times (Figure 5) vs. the CPU no-hash variant (for 50,000 patterns). The average running times obtained on the GPU were 15, 16, 31, 73 and 94 milliseconds for the five sets of patterns used, making the approach highly feasible for real-time implementations. As expected, the 16-bit hash significantly reduced the number of nodes per stack which need to be processed and for which the failure computation needs to take place, by a factor varying from 25x to 88x. It is however important to note that the actual numbers depend on several decisive factors, as pointed out earlier: the length of the unique suffixes of the patterns being added/removed and the density of the characters in the alphabet from these unique suffixes, as computed by the 16-bit hash.

4   Conclusion

We have proposed an innovative and highly-efficient architecture for implementing real-time polymorphic Aho-Corasick pattern-matching automata on hybrid hardware, which makes full use of the massive parallelism offered by GPUs nowadays, in order to significantly reduce the processing time required to reconstruct the automaton when adding or removing patterns, as a result of applying one or more heuristics. Our experimental results have shown that the speed-ups achieved using the 16-bit hashing technique we proposed are reducing processing time dramatically (by a few hundred times on average) even on moderate GPU hardware, making our implementation highly feasible for real-time processing applications on heterogeneous hardware. Future research we are undertaking looks into even better hashing techniques, potentially using textured memory which is cached and much faster, and efficient ways to parallelize different heuristics in GPU implementations which reduce memory divergence as much as possible.

Acknowledgements. This work was partially supported by the grant of the European Commission FP7-REGPOT-CT-2011-284595 (HOST), and Romanian national grant PN-II-ID-PCE-2011-3-0260 (AMICAS).

References

1. Aho, A., Corasick, M.: Efficient string matching: An Aid to blbiographic search. CACM 18(6), 333–340 (1975)
2. Cha, S.K., Moraru, I., Jang, J., Truelove, J., Brumley, D., Andersen, D.G.: Split Screen: Enabling Efficient, Distributed Malware Detection. In: Proc. 7th USENIX NSDI (2010)
3. Lee, T.H.: Generalized Aho-Corasick Algorithm for Signature Based Anti-Virus Applications. In: Proceedings of 16th International Conference on Computer Communications and Networks, ICCN (2007)
4. Snort, http://www.snort.org/

5. Paxson, V.: Bro: A system for detecting network intruders in real-time. Computer Networks 31, 2435–2463 (1999)
6. Pungila, C.: A Bray-Curtis Weighted Automaton for Detecting Malicious Code Through System-Call Analysis. In: 11th International Symposium on Symbolic and Numeric Algorithms for Scientific Computing (SYNASC), pp. 392–400 (2009)
7. Arshad, J., Townend, P., Xu, J.: A novel intrusion severity analysis approach for Clouds. Future Generation Computer Systems (2011), doi:10.1016/j.future.2011.08.009
8. Corchado, E., Herrero, A.: Neural visualization of network traffic data for intrusion detection. Applied Soft Computing 11(2), 2042–2056 (2011)
9. Panda, M., Abraham, A., Patra, M.R.: Hybrid Intelligent Approach for Network Intrusion Detection. Procedia Engineering 30, 1–9 (2012), doi:10.1016/j.proeng.2012.01.827
10. Wang, Z., Xu, G., Li, H., Zhang, M.: A Fast and Accurate Method for Approximate String Searc. In: Proceedings of the 49th Annual Meeting of the Association for Computational Linguistics: Human Language Technologies, vol. 1, pp. 52–61 (2011)
11. Pungila, C.: Improved file-carving through data-parallel pattern matching for data forensics. In: 7th IEEE International Symposium on Applied Computational Intelligence and Informatics (SACI), pp. 197–202 (2012)
12. Pungila, C.: Hybrid Compression of the Aho-Corasick Automaton for Static Analysis in Intrusion Detection Systems. In: Herrero, Á., Snášel, V., Abraham, A., Zelinka, I., Baruque, B., Quintián, H., Calvo, J.L., Sedano, J., Corchado, E. (eds.) Int. Joint Conf. CISIS'12-ICEUTE'12-SOCO'12. AISC, vol. 189, pp. 77–86. Springer, Heidelberg (2013)
13. Pungila, C., Negru, V.: A Highly-Efficient Memory-Compression Approach for GPU-Accelerated Virus Signature Matching. In: Gollmann, D., Freiling, F.C. (eds.) ISC 2012. LNCS, vol. 7483, pp. 354–369. Springer, Heidelberg (2012)
14. Lee, V.W., Kim, C., Chhugani, J., Deisher, M., Kim, D. Nguyen, A.D., Satish, N., Smelyanskiy, M., Chennupaty, S., Hammarlund, P., Singhal, R., Dubey, P.: Debunking the 100X GPU vs. CPU myth: an evaluation of throughput computing on CPU and GPU. In: Proceedings of the 37th Annual International Symposium on Computer Architecture (ISCA 2010), pp. 451–460. ACM, New York, http://doi.acm.org/10.1145/1815961.1816021, doi:10.1145/1815961.1816021
15. Clam AntiVirus, http://www.clamav.net
16. Hamming, R.W.: Error detecting and error correcting codes. Bell System Technical Journal 29(2), 147–160 (1950)
17. Needleman, S.B., Wunsch, C.D.: A general method applicable to the search for similarities in the amino acid sequence of two proteins. J. Mol. Biol. 47, 443–453 (1970)
18. Bray-Curtis dissimilarity, http://www.code10.info/index.php?view=article&id=46

Twitter Content-Based Spam Filtering

Igor Santos, Igor Miñambres-Marcos, Carlos Laorden, Patxi Galán-García,
Aitor Santamaría-Ibirika, and Pablo García Bringas

DeustoTech-Computing, Deusto Institute of Technology (DeustoTech)
Avenida de las Universidades 24, 48007 Bilbao, Spain
{isantos,claorden,patxigg,a.santamaria,pablo.garcia.bringas}@deusto.es,
igor.m@opendeusto.es

Abstract. Twitter has become one of the most used social networks.
And, as happens with every popular media, it is prone to misuse. In
this context, spam in Twitter has emerged in the last years, becoming
an important problem for the users. In the last years, several approaches
have appeared that are able to determine whether an user is a spammer or
not. However, these blacklisting systems cannot filter every spam message
and a spammer may create another account and restart sending spam. In
this paper, we propose a content-based approach to filter spam tweets.
We have used the text in the tweet and machine learning and compression
algorithms to filter those undesired tweets.

Keywords: spam filtering, Twitter, social networks, machine learning,
text classification.

1 Introduction

Similarly as happened with e-mail, online social networks have emerged and become a powerful communication media, where users can share links, discuss and
connect with each other. In particular, Twitter is one of the most used social networks, a very popular social system that serves also as a powerful communication
channel, a popular news source; and as happens with every powerful channel, it
is prone to misuse and, therefore, a spam marketplace has been created.

This new type of spam has several particularities (such as the limitation of 140
characters) and, despite the Twitter's effort to combat these spam operations,
there is still no a proper solution to this problem [1]. And it is known that
spam is not only very annoying to every-day users, but also constitutes a major
computer security problem that costs billions of dollars in productivity losses [2].
It can also be used as a medium for phishing (i.e., attacks that seek to acquire
sensitive information from end-users) [3] and the spread of malicious software
(e.g., computer viruses, Trojan horses, spyware and Internet worms) [2].

Regarding Twitter spam filtering, there has been an increment in the research
activity in the last years. In particular, several methods have appeared to detect
spammer accounts in Twitter. Benevenuto et al. [4] developed a system to detect
spammers in Twitter by using the properties of the text of the users' tweets and

user behavioural attributes, being able to detect about a 70% of the spammer accounts. Grier et al. [5] proposed a schema based on URL blacklisting. Wang [6] proposed a new approach that employed features from the social graph from Twitter users to detect and filter spam. Gao et al. [7] presented a generic Online Social Network spam filtering system that they tested using Facebook and Twitter data. Their approach used several features using clustering to detect spam campaigns. Similarly, Ahmed and Abulaish [8] presented a new approach to identify spammer accounts using classic statistical methods.

Although these approaches are able to deal with the problem, a new spammer account may emerge to substitute the filtered accounts. Hence, these blacklisting systems, similarly as happens with e-mails, should be complemented with content-based approaches commonly used in e-mail spam filtering. In this context, a recent work by Martinez-Romo and Araujo proposed a method for detecting spam tweets in real time using different language models and measuring the divergences [9].

Against this background, we present here a study of how classic e-mail content-based techniques can filter spam in Twitter. To this end, we have comprised a dataset (publicly available) of spam and ham (not spam) tweets. Using these data, we have tested several content-based spam filtering methods. In particular, we have tested statistical methods based on the bag of word models, and also compression-based text classification algorithms.

In summary, we advance the state of the art through the following contributions: (i) a new and public dataset of Twitter spam, to serve as evaluation of Twitter spam filtering systems; (ii) we adapt content-based spam filtering to Twitter; (iii) we present a new compression-based text filtering library for the well-known machine-learning tool WEKA; and (iv) we show that the proposed method achieves high filtering rates, even on completely new, previously unseen spam, discussing the weakness of the proposed model and explain possible enhancements.

The remainder of this paper is organised as follows. Section 2 explains the methods used in this study. Section 3 details the process to build the dataset, the experiments performed and presents the results. Section 4 discusses the main shortcomings of the proposed method and proposes possible improvements, outlining avenues for future work.

2   Content-Based Spam Filtering Methods

In order to find a solution to the problem of spam, the research community has undertaken a huge amount of work. Because *machine learning* approaches have succeeded in text categorisation problems [10], these techniques have been adopted in spam filtering systems. Consequently, substantial work has been dedicated to the *Naïve Bayes* classifier [11], with studies in anti-spam filtering confirming its effectiveness [12–14]. Another broadly embraced machine-learning-based technique is *Support Vector Machines* (SVM) [15]. The advantage of SVM is that its accuracy does not degrade even with many features [16]. Therefore, such approaches have been applied to spam filtering [17, 18]. Likewise, *Decision Trees* that classify by means of automatically learned rule-sets [19], have also been used for spam filtering [20]. All of these machine-learning-based spam filtering approaches are termed *statistical approaches* [21].

Machine learning approaches model e-mail messages using the *Vector Space Model* (VSM) [22], an algebraic approach for *Information Filtering* (IF), *Information Retrieval* (IR), indexing and ranking. This model represents natural language documents in a mathematical manner through vectors in a multidimensional space.

However, this method has its shortcomings. For instance, in spam filtering, *Good Words Attack* is a method that modifies the term statistics by appending a set of words that are characteristic of legitimate e-mails. In a similar vein, *tokenisation attacks* work against the feature selection of the message by splitting or modifying key message features rendering the term-representation no longer feasible [23]. Compression-based text classification methods have been applied for spam filtering [2], with good results solving this issue.

In the remainder of this section, we detail the methods we have employed.

2.1   Machine-Learning Classification

Machine-learning is an active research area within *Artificial Intelligence* (AI) that focuses on the design and development of new algorithms that allow computers to reason and decide based on data (i.e. computer learning). We use supervised machine-learning; however, in the future, we would also like to test unsupervised methods for spam filtering. In the remainder of this section, we review several supervised machine-learning approaches that have succeeded in similar domains.

- **Bayesian Networks:** Bayesian Networks [24] are based on *Bayes' Theorem* [25]. They are defined as graphical probabilistic models for multivariate analysis. Specifically, they are directed acyclic graphs that have an associated probability distribution function [26]. Nodes within the directed graph represent problem variables (they can be either a premise or a conclusion) and the edges represent conditional dependencies between such variables. Moreover, the probability function illustrates the strength of these relationships in the graph [26].
- **Decision Trees:** These models are a type of machine-learning classifiers that are graphically represented as trees. Internal nodes represent conditions regarding the variables of a problem, whereas final nodes or leaves represent the ultimate decision of the algorithm [19]. Different training methods are typically used for learning the graph structure of these models from a labelled dataset. We used *Random Forest*, an ensemble (i.e., combination of weak classifiers) of different randomly-built decision trees [27], and *J48*, the WEKA [28] implementation of the *C4.5* algorithm [29].
- **K-Nearest Neighbour:** The *K-Nearest Neighbour* (KNN) [30] classifier is one of the simplest supervised machine learning models. This method

classifies an unknown specimen based on the class of the instances closest to it in the training space by measuring the distance between the training instances and the unknown instance. Even though several methods to choose the class of the unknown sample exist, the most common technique is to simply classify the unknown instance as the most common class amongst the K-nearest neighbours.

- **Support Vector Machines (SVM):** SVM algorithm divide the data space into two regions using a *hyperplane*. This hyperplane always maximises the *margin* between those two regions or classes. The margin is defined by the farthest distance between the examples of the two classes and computed based on the distance between the closest instances of both classes, which are called *supporting vectors* [15]. Instead of using linear hyperplanes, it is common to use the so-called *kernel functions*. These kernel functions lead to non-linear classification surfaces, such as polynomial, radial or sigmoid surfaces [31].

2.2   Compression-Based Text Classifier

In this section, we describe how we have performed the text classification using the compression models. In particular, given a set of training documents $\mathcal{D}$, we can generate a compression model $\mathcal{M}$ using these documents.

The training documents are previously labelled in n different classes, dividing our training documents $\mathcal{D}$ in n different document sets $\mathcal{D} = (\mathcal{D}_1, \mathcal{D}_2, ..., \mathcal{D}_{n-1}, \mathcal{D}_n)$ depending on their labelled class (in our task 2: spam or ham, that is $\mathcal{D}_{spam}$ and $\mathcal{D}_{ham}$)). In this way, we generate n different compression models $\mathcal{M}$, one for each class: $\mathcal{M} = (\mathcal{M}_1, \mathcal{M}_2, ..., \mathcal{M}_{n-1}, \mathcal{M}_n)$.

When a training document $d_{i,j}$, where i denotes the class and j denotes the document number, is analysed, we add it to the compression model of its class $\mathcal{M}_i$, training and updating the compression model. In this way, if we proceed with the training of every document in $\mathcal{D}$, each model will be trained with documents of that class, therefore being adapted and prepared to compress only documents of that particular class.

In particular, for the compression models used, each model is given a training sequence in order to learn the model $\mathcal{M}$ that provides each future outcome a probability value for each future symbol. The prediction performance is usually measured by the average log-loss [32] $\ell(\hat{M}, x_1^T)$ that given a test sequence $d_1^T = (d_1, d_2, ..., d_{n-1}, d_n)$:

$$\ell(\hat{M}, x_1^T) = \frac{1}{T} \log \hat{\mathcal{M}}(d_i | d_1, d_2, ..., d_{i-1}) \tag{1}$$

A small average log-loss over a test sequence means that the sequence is going to be well compressed. In this way, the mean of the log-loss of each model $\mathcal{M}$ achieves the best possible entropy:

$$LogEval(d, \mathcal{M}) = \frac{1}{T} \sum_{i=1}^{T} \log \mathcal{M}(d_i | d_1^{i-1}) \tag{2}$$

In this case, it is also possible to adapt the model with the symbols d_i of the test sequence, generating another way of measuring considering the sample of the type of data generated by the model:

$$LogEval(d, \mathcal{M}) = \frac{1}{T} \sum_{i=1}^{T} \log \mathcal{M}'(d_i|d_1^{i-1})) \qquad (3)$$

where M' is the model adapted with the d_{i-1} symbol.

In both cases, the class of tested document is selected as the minimal value of each compression model tested. When a testing document d arrives, our system can also evaluate the probability of a document belonging to a class using each compression model $\mathcal{M}_i$ and then we generate a probability with values between 0 and 1, using the next formula:

$$P(d \in \mathcal{M}_i) = \frac{\frac{1}{LogEval(d,\mathcal{M}_i)}}{\sum_{j=1}^{n} \frac{1}{LogEval(d,\mathcal{M}_i)}} \qquad (4)$$

Similarly, we may use the minimum value of $LogEval$:

$$\mathcal{M}_i(d) = \underset{\mathcal{M}_i \in \mathcal{M}}{\operatorname{argmin}} \, LogEval(d, \mathcal{M}_i) \qquad (5)$$

Using the implementation of Prediction by Partial Matching (PPM) [33], Lempel-Ziv 78 (LZ78) [34], an improved LZ78 algorithm (LZ-MS) [35], Binary Context Tree Weighting Method (BI-CTW) [36], Decomposed Context Tree Weighting (DE-CTW) [37], Probabilistic Suffix Trees (PST) [38], presented in [32][1] and adapting the original version of DMC [39][2], we have adapted these models and provided a WEKA[28] 3.7 package[3] named `CompressionTextClassifier` used in this paper.

We briefly describe the 2 compression models used for Twitter spam filtering. We used DMC and PPM because they have been the most used ones in e-mail spam filtering:

- **Dynamic Markov Chain (DMC):** DMC [39] models information with a finite state machine. Associations are built between every possible symbol in the source alphabet and the probability distribution over those symbols. This probability distribution is used to predict the next binary digit. The DMC method starts in an already defined state, changing the state when new bits are read from the entry. The frequency of the transitions to either 0 or 1 are summed when a new symbol arrives. The structure can also be updated using a state cloning method. DMC has been previously used in e-mail spam filtering tasks [2] and in SMS spam filtering [40, 41] obtaining high filtering rates.

---

[1] Available at: `http://www.cs.technion.ac.il/~ronbeg/vmm/code_index.html`
[2] Available at: `http://www.jjj.de/crs4/dmc.c`
[3] Available at: `http://paginaspersonales.deusto.es/isantos/Resources/`
`CompressionTextClassifier-0.4.3.zip`

- **Prediction by Partial Match (PPM):** Prediction by partial matching (PPM) algorithm [33] is one of the best lossless compression algorithms. The implementation is based on a prefix tree [32]. In this way, using the training character string, the algorithm constructs the tree. In a similar vein as the LZ78 prefix tree, each node within the tree represents a symbol and has a counter of occurrences. The tree starts with a root symbol node for an empty sequence. It parses each symbol of the training sequence and the parsed symbol and its context, allowing the model to define a potential path in the tree. Once the tree is generated, the resultant data structure can be used to predict each symbol and context, transversing the tree according to the longest suffix of the testing sequence.

3 Evaluation

To evaluate the proposed method, we constructed a dataset comprising 31,457 tweets retrieved from a total of 223 user accounts, from which 143 were spam accounts verified in the TweetSpike site[4] while 80 were randomly selected among legitimate users. Besides, the accounts were manually checked to determine if they were correctly classified. Regarding the tweets, after removing all the duplicated instances, the final dataset used for evaluation of the proposed algorithms had 25,846, from which 11,617 correspond to spam tweets and 14,229 to legitimate tweets[5].

- **Cross Validation:** In order to evaluate the performance of machine-learning classifiers, *k-fold cross validation* is commonly used in machine-learning experiments.

  For each classifier tested, we performed a k-fold cross validation with $k = 10$. In this way, our dataset was split 10 times into 10 different sets of learning sets (90% of the total dataset) and testing sets (10% of the total data).
- **Learning the Model:** For each fold, we performed the learning phase for each of the algorithms presented in Section 2.1 with each training dataset, applying different parameters or learning algorithms depending on the concrete classifier. If not specified, the default ones in WEKA were used.

  To evaluate each classifier's capability we measured *accuracy*, which is the total number of the classifier's hits divided by the number of messages in the whole dataset (shown in equation 6).

$$Accuracy(\%) = \frac{TP + TN}{TP + FP + FN + TN} \cdot 100 \tag{6}$$

where TP is the amount of correctly classified spam (i.e., true positives), FN is the amount of spam misclassified as legitimate mails (i.e., false negatives),

---

[4] Available at: http://www.tweetspike.org

[5] Available online at: http://paginaspersonales.deusto.es/isantos/resources/twitterspamdataset.csv

FP is the amount of legitimate mail incorrectly detected as spam, and TN is the number of legitimate mail correctly classified.

Furthermore, we measured the precision of the spam identification as the number of correctly classified spam e-mails divided by the number of correctly classified spam e-mails and the number of legitimate e-mails misclassified as spam:

$$S_P = \frac{N_{s \to s}}{N_{s \to s} + N_{l \to s}} \tag{7}$$

where $N_{s \to s}$ is the number of correctly classified spam messages and $N_{l \to s}$ is the number of legitimate e-mails misclassified as spam.

Additionally, we measured the recall of the spam e-mail messages, which is the number of correctly classified spam e-mails divided by the number of correctly classified spam e-mails and the number of spam e-mails misclassified as legitimate:

$$S_R = \frac{N_{s \to s}}{N_{s \to s} + N_{s \to l}} \tag{8}$$

We also computed the F-measure, which is the harmonic mean of both the precision and recall, as follows:

$$F\text{-}measure = \frac{2N_{s \to s}}{2N_{s \to s} + N_{s \to l} + N_{l \to s}} \tag{9}$$

Finally, we measured the *Area Under the ROC Curve* (AUC), which establishes the relation between false negatives and false positives. The ROC curve is represented by plotting the rate of true positives (TPR) against the rate of false positives (FPR).

Table 1 shows the results obtained with both the common machine-learning algorithms and the compression models proposed. On the one hand, we can appreciate how one of the most used classifiers in e-mail spam filtering, Naive Bayes, performs poorly when compared to the rest of the models, obtaining a 0.76 of AUC. In a similar vein, PPM with adaptation obtains a 0.86 of AUC against the 0.92-0.99 of the rest of the classifiers. Later SVM, with different kernels, and KNN algorithms, obtain a lowest value of AUC of 0.92 for SVM with Radial Basis Function and a highest value of AUC of 0.97 for KNN with a k of 3. The decision tree C4.5 also obtains a 0.97 of AUC but improving every other measure when compared to the previous algorithms. On the next scale, there are algorithms with 0.98-0.99 of AUC. Amongst them, Random Forest are the ones with the best behaviour, obtaining not only a 0.99 of AUC but also significant results in the other measures. The only algorithms with the same performance are DMC and PPM, both without adaptation, that also obtain a 0.99 of AUC. Finally, with 0.98 of AUC we can find the rest of the Bayes-based algorithms and DMC with adaptation.

Table 1. Results of the evaluation of common machine learning classifiers and the compression models

Classifier	Acc.	S_P	S_R	F − Measure	AUC
Random Forest N=50	96.42	0.98	0.94	0.96	0.99
Random Forest N=30	96.41	0.98	0.94	0.96	0.99
DMC without Adaptation	95.99	0.96	0.95	0.96	0.99
Random Forest N=10	95.96	0.97	0.94	0.95	0.99
PPM without Adaptation	94.80	0.97	0.91	0.94	0.99
Naive Bayes Multinomial Word Frequency	94.94	0.95	0.93	0.94	0.98
Naive Bayes Multinomial Boolean	94.75	0.95	0.93	0.94	0.98
Bayes K2	94.12	0.99	0.88	0.93	0.98
DMC with Adaptation	93.11	0.94	0.90	0.92	0.98
C4.5	95.79	0.98	0.92	0.95	0.97
KNN K=3	93.71	0.97	0.89	0.93	0.97
KNN K=5	92.61	0.97	0.86	0.91	0.97
SVM PVK	95.81	0.97	0.93	0.95	0.96
KNN K=1	94.22	0.95	0.92	0.93	0.96
SVM Lineal	94.38	0.92	0.96	0.94	0.95
SVM RBF	93.20	0.99	0.85	0.92	0.92
PPM with Adaptation	76.50	0.78	0.69	0.72	0.86
Naive Bayes	72.72	0.64	0.89	0.75	0.76

4   Conclusions

In this paper, we presented a study of how classic e-mail content-based techniques can filter Twitter spam, adapting the algorithms to Twitter's peculiarities, but, while analysing the text of the messages has proven as a great approach to identifying the type of the communications. In particular: (i) we have compiled a new and public dataset of spam in Twitter, (ii) we evaluated the classic content-based spam filtering techniques to this type of spam, and (iii) as a technical contribution, we have presented a new and free software compression-based text filtering library for the well-known machine-learning tool WEKA.

Future versions of this spam filtering system will move in two main directions. First, we will enhance this approach using social network features. Second, we plan to enhance the semantic capabilities, as already studied in e-mail spam filtering [42, 43], by studying the linguistic relationships in tweets.

References

1. Thomas, K., Grier, C., Song, D., Paxson, V.: Suspended accounts in retrospect: an analysis of twitter spam. In: Proceedings of the 2011 ACM SIGCOMM Conference on Internet Measurement Conference, pp. 243–258. ACM (2011)
2. Bratko, A., Filipič, B., Cormack, G., Lynam, T., Zupan, B.: Spam filtering using statistical data compression models. The Journal of Machine Learning Research 7, 2673–2698 (2006)

3. Jagatic, T., Johnson, N., Jakobsson, M., Menczer, F.: Social phishing. Communications of the ACM 50(10), 94–100 (2007)
4. Benevenuto, F., Magno, G., Rodrigues, T., Almeida, V.: Detecting spammers on twitter. In: Annual Collaboration, Electronic messaging, Anti-Abuse and Spam Conference, CEAS (2010)
5. Grier, C., Thomas, K., Paxson, V., Zhang, M.: @spam: The underground on 140 characters or less. In: Proceedings of the 17th ACM Conference on Computer and Communications Security, pp. 27–37. ACM (2010)
6. Wang, A.H.: Don't follow me: Spam detection in twitter. In: Proceedings of the 2010 International Conference on Security and Cryptography (SECRYPT), pp. 1–10. IEEE (2010)
7. Gao, H., Chen, Y., Lee, K., Palsetia, D., Choudhary, A.: Towards online spam filtering in social networks. In: Symposium on Network and Distributed System Security, NDSS (2012)
8. Ahmed, F., Abulaish, M.: A generic statistical approach for spam detection in online social networks. Computer Communications (in press, 2013)
9. Martinez-Romo, J., Araujo, L.: Detecting malicious tweets in trending topics using a statistical analysis of language. Expert Systems with Applications (2012)
10. Sebastiani, F.: Machine learning in automated text categorization. ACM Computing Surveys (CSUR) 34(1), 1–47 (2002)
11. Lewis, D.: Naive (Bayes) at forty: The independence assumption in information retrieval. In: Nédellec, C., Rouveirol, C. (eds.) ECML 1998. LNCS, vol. 1398, pp. 4–18. Springer, Heidelberg (1998)
12. Schneider, K.: A comparison of event models for Naive Bayes anti-spam e-mail filtering. In: Proceedings of the 10th Conference of the European Chapter of the Association for Computational Linguistics, pp. 307–314 (2003)
13. Androutsopoulos, I., Koutsias, J., Chandrinos, K., Spyropoulos, C.: An experimental comparison of naive Bayesian and keyword-based anti-spam filtering with personal e-mail messages. In: Proceedings of the 23rd Annual International ACM SIGIR Conference on Research and Development in Information Retrieval, pp. 160–167 (2000)
14. Seewald, A.: An evaluation of naive Bayes variants in content-based learning for spam filtering. Intelligent Data Analysis 11(5), 497–524 (2007)
15. Vapnik, V.: The nature of statistical learning theory. Springer (2000)
16. Drucker, H., Wu, D., Vapnik, V.: Support vector machines for spam categorization. IEEE Transactions on Neural Networks 10(5), 1048–1054 (1999)
17. Blanzieri, E., Bryl, A.: Instance-based spam filtering using SVM nearest neighbor classifier. Proceedings of FLAIRS 20, 441–442 (2007)
18. Sculley, D., Wachman, G.: Relaxed online SVMs for spam filtering. In: Proceedings of the 30th Annual International ACM SIGIR Conference on Research and Development in Information Retrieval, pp. 415–422 (2007)
19. Quinlan, J.: Induction of decision trees. Machine Learning 1(1), 81–106 (1986)
20. Carreras, X., Márquez, L.: Boosting trees for anti-spam email filtering. In: Proceedings of RANLP 2001, 4th International Conference on Recent Advances in Natural Language Processing, pp. 58–64. Citeseer (2001)
21. Zhang, L., Zhu, J., Yao, T.: An evaluation of statistical spam filtering techniques. ACM Transactions on Asian Language Information Processing (TALIP) 3(4), 243–269 (2004)
22. Salton, G., Wong, A., Yang, C.: A vector space model for automatic indexing. Communications of the ACM 18(11), 613–620 (1975)

23. Wittel, G., Wu, S.: On attacking statistical spam filters. In: Proceedings of the 1st Conference on Email and Anti-Spam, CEAS (2004)
24. Pearl, J.: Reverend bayes on inference engines: a distributed hierarchical approach. In: Proceedings of the National Conference on Artificial Intelligence, pp. 133–136 (1982)
25. Bayes, T.: An essay towards solving a problem in the doctrine of chances. Philosophical Transactions of the Royal Society 53, 370–418 (1763)
26. Castillo, E., Gutiérrez, J.M., Hadi, A.S.: Expert Systems and Probabilistic Network Models, Erste edn., New York, NY, USA (1996)
27. Breiman, L.: Random forests. Machine Learning 45(1), 5–32 (2001)
28. Garner, S.: Weka: The Waikato environment for knowledge analysis. In: Proceedings of the 1995 New Zealand Computer Science Research Students Conference, pp. 57–64 (1995)
29. Quinlan, J.: C4. 5 programs for machine learning. Morgan Kaufmann Publishers (1993)
30. Fix, E., Hodges, J.L.: Discriminatory analysis: Nonparametric discrimination: Small sample performance. technical report project 21-49-004, report number 11. Technical report, USAF School of Aviation Medicine, Randolf Field, Texas (1952)
31. Amari, S., Wu, S.: Improving support vector machine classifiers by modifying kernel functions. Neural Networks 12(6), 783–789 (1999)
32. Begleiter, R., El-Yaniv, R., Yona, G.: On prediction using variable order markov models. J. Artif. Intell. Res. (JAIR) 22, 385–421 (2004)
33. Cleary, J., Witten, I.: Data compression using adaptive coding and partial string matching. IEEE Transactions on Communications 32(4), 396–402 (1984)
34. Ziv, J., Lempel, A.: Compression of individual sequences via variable-rate coding. IEEE Transactions on Information Theory 24(5), 530–536 (1978)
35. Nisenson, M., Yariv, I., El-Yaniv, R., Meir, R.: Towards behaviometric security systems: Learning to identify a typist. In: Lavrač, N., Gamberger, D., Todorovski, L., Blockeel, H. (eds.) PKDD 2003. LNCS (LNAI), vol. 2838, pp. 363–374. Springer, Heidelberg (2003)
36. Willems, F.: The context-tree weighting method: Extensions. IEEE Transactions on Information Theory 44(2), 792–798 (1998)
37. Volf, P.A.J.: Weighting techniques in data compression: Theory and algorithms. Citeseer (2002)
38. Ron, D., Singer, Y., Tishby, N.: The power of amnesia: Learning probabilistic automata with variable memory length. Machine Learning 25(2), 117–149 (1996)
39. Cormack, G., Horspool, R.: Data compression using dynamic markov modelling. The Computer Journal 30(6), 541–550 (1987)
40. Cormack, G., Gómez Hidalgo, J., Sánz, E.: Spam filtering for short messages. In: Proceedings of the 16th ACM Conference on Conference on Information and Knowledge Management, pp. 313–320. ACM (2007)
41. Cormack, G., Hidalgo, J., Sánz, E.: Feature engineering for mobile(sms) spam filtering. In: Proceedings of the 30th Annual International ACM SIGIR Conference on Research and Development in Information Retrieval, vol. 23, pp. 871–872 (2007)
42. Santos, I., Laorden, C., Sanz, B., Bringas, P.G.: Enhanced topic-based vector space model for semantics-aware spam filtering. Expert Systems With Applications 39(1), 437–444, doi:10.1016/j.eswa.2011.07.034
43. Laorden, C., Santos, I., Sanz, B., Alvarez, G., Bringas, P.G.: Word sense disambiguation for spam filtering. Electron. Commer. Rec. Appl. 11(3), 290–298 (2012)

Content-Based Image Authentication Using Local Features and SOM Trajectories

Alberto Peinado, Andrés Ortiz, and Guillermo Cotrina

Departamento de Ingeniería de Comunicaciones
E.T.S. Ingeniería de Telecomunicación, Universidad de Málaga
Campus de Teatinos s/n, 29071 Málaga, Spain

Abstract. In this paper we propose a content-based image authentication mechanism using SOM trajectories and local features extracted from the image. The features computed are used as input to the SOM which computes a number of prototypes. Then, the prototypes corresponding to each image block define a trajectory in the SOM space which is used to define the hash. Moreover, modifications in the texture of intensity distribution can be identified as they impose changes in hash.

1 Introduction

Digital communications allow the distribution and interchange of digital images. Currently, there is an increasing utilization of digital images obtained (legally or illegally) from repositories or by searching in Internet. Since many of those images are subject to copyright licenses it is very important to provide authentication tools in order to verify the integrity and the authorship of the images.

The traditional cryptographic techniques used to provide authentication are mainly based on conventional hash functions, such as MD5 or SHA-1, in which the modification of one bit in the input message leads to a different output. In this way, the modification of only one pixel of the image produces a completely different hash value, and consequently, the original image and the 1-pixel modified version will be considered distinct although perceptually identical. In this paper, we propose an image authentication mechanism that allows the owner to recognize the authorship of an image when it is applied to a modified or altered version of the original. The algorithm follows the general scheme of content-based authentication mechanisms [1], which contains an image feature extraction phase and a secret key-dependent hash computation. Considering H an image hash function, the value $h = H_K(I)$ is the hash value of the image I computed by means of H using the secret key K.

The image authentication algorithm proposed in this paper produces similar or identical hash values for perceptually similar images, but different hash values for perceptually dissimilar images, as it is shown in Section 7. Moreover, only the holder of the secret key K employed to produce the hash value h will be able to verify the ownership of the images. Sections 2 deals with the pre-computation phase. Sections 3 gives the details about SOM trajectories computation and

Á. Herrero et al. (eds.), *International Joint Conference SOCO'13-CISIS'13-ICEUTE'13*, 459
Advances in Intelligent Systems and Computing 239,
DOI: 10.1007/978-3-319-01854-6_47, © Springer International Publishing Switzerland 2014

how they are used for hash calculation. Section 4 presents the security aspects included in the algorithm proposed. Section 5 describes the experimental results and section 6 shows the conclusions.

2   Preprocessing and Feature Extraction

In this section, preprocessing and feature extraction stages are described. Feature space composed as presented here are used in the next section for hash calculation.

2.1   Preprocessing

Preprocessing stage aims to normalize the images and homogeneize the intensity, minimizing the probability of abrupt changes in the image histogram. This way, the image is resized to 150x150 pixels, ensuring the same number of blocks are extracted regardless of the image resolution. Moreover, a 3x3 gaussian smoothing filter is applied in order to avoid noise effects, avoiding abrupt changes in the histogram. The small size of the gaussian kernel (3x3) removes noise while yields only a slight blurring effect on the image.

2.2   Feature Extraction

In this work we does not use the intensity value as features but local features extracted from image blocks. Thus, the image is divided into blocks of 25x25 pixels. This block size is enough to capture textural properties locally as well as moment invariants which provide rotation and scaling invariance. Features extracted can be grouped into three categories:

Local Gradients. This feature consist in computing the two principal directions in which the variation of the intensity is maximum. This is computed by means of the local intensity gradients, calculating the partial derivative of the image in x and y directions. Let $I(x,y)$ be the intensity of the image at coordinates (x,y). Thus, the direction of maximum variation can be computed as

$$\nabla(I) = \left(\frac{\partial I(x,y)}{\partial x}, \frac{\partial I(x,y)}{\partial y} \right) \tag{1}$$

However, as $I(x,y)$ is a discrete function, partial derivatives can be only calculated at some coordinates (x,y). Thus, the equation 1 can be rewritten using the difference between contiguous pixels:

$$\frac{\partial I(x,y)}{\partial x} = \frac{I(x+1,y) - I(x-1,y)}{\partial x} \tag{2}$$

$$\frac{\partial I(x,y)}{\partial y} = \frac{I(x+1,y) - I(x-1,y)}{\partial y} \tag{3}$$

In addition, Equation 1 can be expressed in polar coordinated to compute the magnitude and direction (angle) of maximum variation. In this work we computed the two principal directions for maximum intensity variation (i.e. maximum gradient magnitude).

Moment Invariant. The second group of features extracted from the image aims to provide translation, scaling and rotation invariance. It is achieved through the HU moments [2,3] computed over each image block. Hu invariants are derived from the central moments [3]. Mathematical details on the moment calculation can be found in [2].

Histogram Features. Histogram features extracted from each image block aim to describe the distribution of the gray levels on that block. This is addressed by computing mean, variance, kurtosis and entropy of the local histogram. These features are described in equations 4, 5, 6 and 7, respectively.

$$\mu = \frac{1}{N} \sum_{i=1}^{N} l_i \tag{4}$$

$$\sigma^2 = \frac{1}{N} \sum_{i=1}^{N} (l_i - \mu)^2 \tag{5}$$

$$Kurtosis = \frac{\sum_{i=1}^{N} (l_i - \mu)^4}{(N-1)\sigma^4} \tag{6}$$

$$Entropy = -\sum_{i=1}^{N} l_i log(l_i) \tag{7}$$

3   Clustering the Feature Space with SOM

Self-Organizing Maps (SOM) [4] is one of the most used artificial neural network models for unsupervised learning and inspired in the animal brain. The main purpose of SOM is to group the similar data instances close in into a low dimensional lattice (output map), while the topology is preserved. Thus, data instances which are far away in the feature space will be apart in the output space.

This is addressed in SOM during the training phase, by measuring the Euclidean distance between an input vector and the weights associated to the units on the output map. Thus, the unit $U_\omega(t)$, closer to the input vector $x(t)$ is referred as winning unit and the associated weight ω_i is updated. Moreover, the weights of the units in the neighbour of the winning unit $(h_{U_{\omega_i}})$ are also updated as in Equation 9. The neighbour function defines the shape of the neighbourhood and usually, a Gaussian function which shrinks in each iteration is used as shown in Equation 10. This deals with a competitive process in which the winning neuron on each iteration is called Best Matching Unit (BMU). At map convergence, the weights of the map units will not change significantly with iterations.

$$U_\omega(t) = argmin_i \|x(t) - \omega_i(t)\| \tag{8}$$

$$\omega_i(t+1) = \omega_i(t) + \alpha_i(t) h_{U_i}(t) \Big(x(t) - \omega_i(t) \Big) \tag{9}$$

$$h_{U_i}(t) = e^{-\frac{\|r_U - r_i\|}{2\sigma(t)^2}} \tag{10}$$

3.1   Clustering the SOM

SOM can be seen as a clustering method as it quantizes the feature space by a number of prototypes, and each prototype can be considered as the most representative vector of a class. On the other hand, the prototypes are projected onto a two or three dimensional space while topology (i.e. distribution of SOM units in the projection space) is preserved and each unit acts as a single cluster. However, this simple model that considers each unit as a different cluster does not exploit valuable information contained in SOMs. Indeed, SOM provides extra advantages over classical clustering algorithms if more than one unit represents the same class, and a range of SOM units act as BMU for a subset of the data manifold. This add flexibility to the clustering algorithm and allows to compute a set of prototype vectors for the same class. Nevertheless, since each cluster can be prototyped by a set of model vectors, grouping SOM units is necessary to define cluster borders [5]. Although specific algorithms such as [6] have been developed for clustering the SOM, in this work we use the k-means algorithm [7] over the prototypes ω_i for simplicity.

3.2   Trajectory Calculation

During the training process, a sequence of inputs are applied to SOM. This yields a BMU-trajectory, which is a projection of the feature sequence onto the low dimensional map. However, as dimensionality reduction is performed, some information is lost. This problem can be mitigated considering groups of winner units or even the whole activation map instead a single BMU for computing the trajectory [8]. In this work we address this problem by replacing single BMUs by groups of winner units in order to preserve more information. Groups of BMUs are computed by clustering the SOM as commented before. On the other hand, clustering the SOM has additional advantages as described previously.

Thus, the original BMU sequence

$$f_s = \{\omega_1, \omega_2, ..., \omega_k\} \tag{11}$$

is replaced by

$$f_g = \{\zeta_1, \zeta_2, ..., \zeta_l\} \tag{12}$$

where ζ_i denotes the i-cluster computed in the SOM layer.

As an example, Figure 1 shows the trajectory of the first 15 BMUs in the map, with 3 clusters.

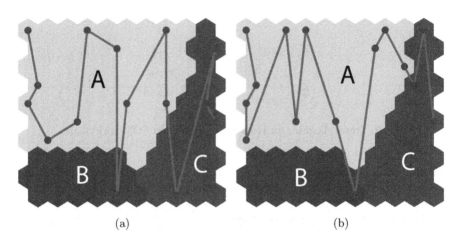

(a)   (b)

Fig. 1. Trajectories on the SOM map for (a) original image and (b) image cropped at 30%

3.3   Distance between Sequences

The process described in previous sections can be summarized as a projection method which maps a set of features computed from each image block to a low dimensional map (i.e. SOM). Moreover, these features are treated as a sequence in the feature space and therefore, a sequence in the low dimensional projection space is obtained for each image. Thereby, image similarity can be addressed by measuring similarity between sequences, and deviations from the original trajectory can be assessed by measuring distances between BMUs sequences. In this work we used the *Levenshtein* or *edit* distance which can be considered as a generalization of the *Hamming distance*, and defined as

$$lev_{ab}(|a|, |b|) = \begin{cases} max(|a|, |b|) & min(|a|, |b|) = 0 \\ min \begin{cases} lev_{a,b}(|a| - 1, |b|) + 1 \\ lev_{a,b}(|a|, |b| - 1) + 1 & else \\ lev_{a,b}(|a| - 1, |b| - 1) + [a_i \neq b_j] \end{cases} \end{cases} \tag{13}$$

for two sequences a and b. This measure deals with differences between sequences regarding deletion, insertion or substitution of a single element at different positions in the sequence.

3.4   Hash Calculation Using SOM Trajectory

Once the SOM has been trained, the units are clustered into three groups. Thus, SOM trajectory can be seen as transitions among three states.

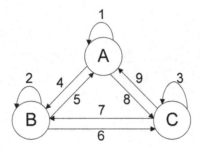

Fig. 2. States diagram for hash calculation from SOM trajectories

Hence, SOM trajectories can be turned into state transitions codified by 4 bits each. As an example, the cluster sequence $ACCACAAABAACAAAA$ is converted to 839891145189111, that is, 1000 0011 1001 1000 1001 0001 0001 0100 0101 0001 1000 1001 0001 0001 0001.

4   Providing Security

As it is defined in previous sections, the content-based image authentication produces a user key-dependent hash value, in such a way that the same image produces different hash values for different values of the key K. Since K is a secret key, known only by the image owner, the effect of modifications on the original image cannot be quantified by the potential attackers from the hash values. Security is provided in the present image authentication method once the pre-computation phase has been performed. Figure 3 shows the pre-computation phase, in which the input image is normalized to a 150 x 150 pixels image, and then divided into 36 blocks of 25 x 25 pixels.

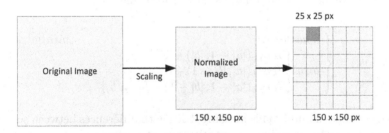

Fig. 3. Pre-computation phase

The image features are extracted sequentially from the 36 blocks. Each block produces a feature vector of 24 components, corresponding to the features described in subsection 2.2. In this step, the image owner uses his key K to shuffle the features by means of a permutation applied on the feature vector. This permutation affects also to the SOM, which is trained taking the permutation into account, determining different winning units (BMU) and finally, different trajectories, as it is presented in section 3.2.

Since features are represented as 24-dimensional elements, there exists 24! distinct permutations to be selected by the image owner. Thus, the length of the key K_f used to select the permutation is $log_2(24!) \approx 79$ bits.

The permutated feature vectors corresponding to each block constitute the sequence of inputs applied to the SOM. The output is the sequence of clusters to which belongs the BMU produced by the 36 image blocks. Since three clusters have been considered, the output is a ternary sequence represented by the symbols A, B and C (see Figure 1).

The last operation is the binary codification of the ternary sequence. This sequence can also be represented as the sequence of state transitions, following the diagram of Figure 2. The image owner selects at this step the identification of state transition; that is, the numbers in Figure 2 assigned to transitions are not static, but depends on the user key. The 9 state transitions can be identified by 9 symbols (1 to 9) in 9! different forms. So, the length of the key K_t used to select the identification is $log_2(9!) \approx 18$ bits. Finally, the sequence of 35 transitions is codified to binary form spreading each symbol to 4 bits. This yields to a binary hash value of 140 bits.

The user key K in Figure 4 corresponds to the concatenation of K_f and K_t. The total length is 97 bits. This length could be increased by selecting more features than those described in subsection 2.2. In this way, the number of possible permutation increases and consequently, the length of K_f.

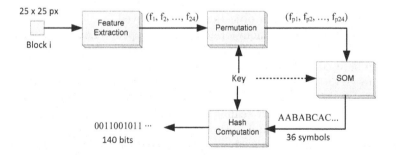

Fig. 4. Key-dependent hash generation

5   Experimental Results

Experiments using three representative images from the *USC-SIPI* [9] database were conducted to test stability and robustness of the proposed method. In the experiments performed, SOM prototypes have been initialized along the two directions indicated by the two principal eigenvectors of the training data. This avoids differences in the results due to random effects in the initialization process. As it is described earlier, SOM used in the experiments is composed of 10 x 10 units connected in an hexagonal lattice. All the experiments were performed using *matlab* [10] and *som toolbox* [11].

These three images are shown in Figure 5.

(a)   (b)   (c)

Fig. 5. Test images from the USC-SIPI database, (a) Baboon, (b) Lena, (c), Pepper

Hash robustness has been assessed using two different image transformations. The first one consists in scaling the image by 2%, 5%, 20% and 30%. Thus, image hash has been computed for the original image and the rest of images, and distance between different hash results allows identifying the image. As shown in Figure 6 distance between scaled versions of the image is 0 in all the cases. This indicates that hash computed from the original image matches the hash of each scaled version. The second transformation applied to the original image consist in cropping the images. Figure 7. This causes a noticeable effect in the image as some relevant objects may be removed.

However, hash distance among original and cropped versions is high enough to separate them as shown in Figure 7. Moreover, the distance between hash results can be used as a similarity measurement between images.

In Figures 6 and 7, distance of 0 indicates no difference between the original image and the scaled or cropped version, respectively. These results show that the proposed method support scaling as it yields a distance of 0 for the tested scale rates. On the other hand, although distance between the original image and its cropped versions is not always 0, it is possible to differentiate between images as the distance is always smaller for the cropped versions of the original image than for cropped versions of other images.

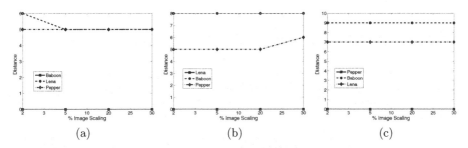

Fig. 6. Distances between hash sequences for scaled images(a) Baboon, (b) Lena, (c), Pepper

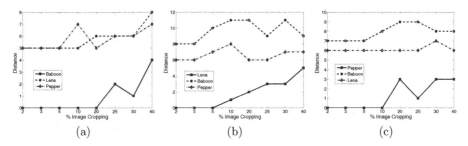

Fig. 7. Distances between hash sequences for cropped images(a) Baboon, (b) Lena, (c), Pepper

6   Conclusions and Future Work

In this paper we present an image hashing method based on local features extracted from the image and the BMU trajectory on SOM. The static problem can be turned into temporal if each block is treated as a new image. Thus, training the self organizing map with all blocks composing an image provide a sequence of BMUs defining a trajectory which can be used to identify differences between different versions of the same image. In addition, trajectory identifying an image is used to derive a 140-bit hash. As a future work we plan to include more SOM levels in order to reduce distance between transformed versions of the same image (specially for morphological transformations) and increase the distance between transformed versions of different images. This will also allow avoiding wining units at cluster borders increasing system robustness.

Acknowledgements. This work has been supported by the MICINN under project "TUERI: Technologies for secure and efficient wireless networks within the Internet of Things with applications to transport and logistics", TIN2011-25452 and by Universidad de Málaga. Campus de Excelencia Andalucía Tech.

References

1. Han, S., Chu, H.: Content-based image authentication: current status, issues and challenges. International Journal of Information Security 9, 19–32 (2010)
2. Hu, M.K.: Visual pattern recognition by moment invariants. IRE Transactions on Information Theory 8(2), 179–187 (1962)
3. Huang, Z., Leng, J.: Analysis of hu's moment invariants on image scaling and rotation. In: Proc. 2nd Int. Computer Engineering and Technology (ICCET) Conf., vol. 7 (2010)
4. Kohonen, T.: Self-Organizing Maps. Springer (2001)
5. Ortiz, A., Górriz, J., Ramírez, J., Salas-González, D., Llamas-Elvira, J.: Two fully-unsupervised methods for mr brain segmentation using som-based strategies. Applied Soft Computing (2012)
6. Tasdemir, K., Milenov, P., Tapsall, B.: Topology-based hierarchical clustering of self-organizing maps. IEEE Transactions on Neural Networks 22(3), 474–485 (2011)
7. Theodoridis, S., Koutroumbas, K.: Pattern Recognition. Academic Press (2009)
8. Somervuo, P., Kohonen, T.: Self-organizing maps and learning vector quantization for feature sequences. Neural Processing Letters 10, 151–159 (1999)
9. USC-SIPI: Signal & image processing institute: The usc-sipi image database. University of southern california (2007)
10. Matlab software: Mathworks, 2013 (2012)
11. Vesanto, J., Himberg, J., Alhoniemi, E., Parhankangas, J.: Som toolbox. Helsinki University of Technology (2000)

Anomaly Detection Using String Analysis for Android Malware Detection

Borja Sanz, Igor Santos, Xabier Ugarte-Pedrero, Carlos Laorden,
Javier Nieves, and Pablo García Bringas

S³Lab, University of Deusto
Avenida de las Universidades 24, 48007 Bilbao, Spain
{borja.sanz,isantos,xabier.ugarte,claorden,
jnieves,pablo.garcia.bringas}@deusto.es

Abstract. The usage of mobile phones has increased in our lives because they offer nearly the same functionality as a personal computer. Specifically, Android is one of the most widespread mobile operating systems. Indeed, its app store is one of the most visited and the number of applications available for this platform has also increased. However, as it happens with any popular service, it is prone to misuse, and the number of malware samples has increased dramatically in the last months. Thus, we propose a new method based on anomaly detection that extracts the strings contained in application files in order to detect malware.

Keywords: malware detection, anomaly detection, Android, mobile malware.

1 Introduction

Smartphones have become extremely useful gadgets, very rich in functionality. Their operating systems have evolved, becoming closer to the desktop ones. We can read our email, browse the Internet or play games with our friends, wherever we are. In addition, the smartphone functionality can be enhanced, similarly to desktop computers, through the installation of software applications.

In recent years, a new approach to distribute applications has gained popularity: application stores. These stores, which distribute software and manage the payments, have become very successful. Apple's AppStore was the first online store to bring this new paradigm to users and now offers more than 800,000 applications[1]. In a similar way, Google's Play Store, Android's official application store, hosts 675,000 apps[2].

Nevertheless, this success has also drawn attention to criminals and many malware samples have emerged. According to Kaspersky, more than 35,000 new samples were identified in 2012, which represents six times the number detected in 2011.

---

[1] http://www.apple.com/pr/library/2013/01/28Apple-Updates-iOS-to-6-1.html
[2] http://officialandroid.blogspot.com.es/search?q=675000

Á. Herrero et al. (eds.), *International Joint Conference SOCO'13-CISIS'13-ICEUTE'13*, 469
Advances in Intelligent Systems and Computing 239,
DOI: 10.1007/978-3-319-01854-6_48, © Springer International Publishing Switzerland 2014

Several approaches have been proposed to deal with this issue. Crowdroid [1] is an approach that analyses the behaviour of the applications through device usage features. Blasing et al. created AASandbox [2], which is a hybrid (i.e., dynamic and static) approximation. The approach is based on the analysis of the logs for the low-level interactions obtained during execution. Shabtai and Elovici [3] also proposed a Host-Based Intrusion Detection System (HIDS) which used machine learning methods to determine whether the application is malware or not. Google itself has also deployed a framework for the supervision of applications called Bouncer. Oberheide and Miller 2012[3] revealed how the system works: it is based in QEMU and performs both static and dynamic analysis. In previous work, we used permissions to train machine-learning algorithms in order to detect malware [4], obtaining more than 86% accuracy and 0.92 of Area Under ROC curve.

However, machine-learning classifiers (or supervised-learning methods) require a high number of labelled applications for each of the classes (i.e., malware or benign applications) to train the different models. Unfortunately, it is quite difficult to acquire this amount of labelled data for a real-world problem such as malware detection. In order to compose such data-set, a time-consuming process of analysis is mandatory that renders in a cost increment.

In light of this background, we present an anomaly detection method for the detection of malware in Android. This method employs the strings contained in the disassembled Android applications, constructing a bag of words model in order to generate an anomaly detection model that measures deviations from normality (i.e., legitimate applications).

In summary, our main contributions are: (i) we present a new technique for the representation of Android applications, based on the bag of words model formed by the strings contained in the disassembled applications; (ii) we propose a new anomaly-based malware detection method for Android; and (iii) we show that this approach can provide detection of malicious applications in Android using the strings contained in the disassembled application as features.

The remainder of this paper is organised as follows. Section 2 presents and details our approach to represent applications in order to detect malware in Android. Section 3 describes the anomaly detection techniques that our approach is using. Section 4 describes the empirical evaluation of our method. Finally, section 5 concludes and outlines the avenues of further work in this area.

2   Representation of Applications Using String Analysis

One of the most widely-used techniques for classic malware detection is the usage of strings contained in the files [5,6]. This technique extracts every printable string within an executable file. The information that may be found in these strings can be, for example, options in the menus of the application or malicious URLs to connect to. In this way, by means of an analysis of these data, it is possible to extract valuable information in order to determine whether an application is malicious or not.

---

[3] http://jon.oberheide.org/files/summercon12-bouncer.pdf

The process adopted in our approach is the following. First, we disassemble the application using the open-source Android disassembler `smali`[4]. Afterwards, we search for the `const-string` operational code within the disassembled code.

Using this disassembler, the representation of Android binaries is semantically richer than common desktop binaries. For example, the strings extraction in desktop binaries is complex and, usually, malware writers use obfuscation techniques to hide relevant information. Instead, the obfuscation of strings in Android binaries is more difficult, given the internal structure of binaries in this platform.

In order to conform the strings, we tokenise the symbols found using the classic separators (e.g., dot, comma, colon, semi-colon, blank space, tab, etc.). In this way, we construct a text representation of an executable $\mathcal{E}$, which is formed by strings s_i, such that $\mathcal{E} = (s_1, s_2, ..., s_{n-1}, s_n)$ where n is the number of strings within a file.

$\mathcal{C}$ is the set of Android executables $\mathcal{E}$, $\{\mathcal{E} : \{s_1, s_2, ...s_n\}\}$, each comprising n strings $s_1, s_2, \ldots, s_n$. We define the weight $w_{i,j}$ as the number of times the string s_i appears in the executable $\mathcal{E}_j$; if s_i is not present in $\mathcal{E}$, $w_{i,j} = 0$. Therefore, an application $\mathcal{E}_j$ can be represented as the vector of weights $\mathcal{E}_j = (w_{1,j}, w_{2,j}, ...w_{n,j})$.

In order to represent a string collection, a common approach in text classification is to use the Vector Space Model (VSM) [7], which represents documents algebraically, as vectors in a multidimensional space.

This space consists only of positive axis intercepts. Executables are represented by a string-by-executable matrix, where the $(i, j)^{th}$ element illustrates the association between the ith string and the jth executable. This association reflects the occurrence of the ith string in the executable j. Strings can be individually weighted, allowing the strings to become more or less important within a given executable or the executable collection $\mathcal{C}$ as a whole.

We used the *Term Frequency – Inverse Document Frequency* (TF–IDF) [8] weighting schema, where the weight of the ith string in the jth executable, denoted by $weight(i, j)$, is defined by:

$$weight(i, j) = tf_{i,j} \cdot idf_i \tag{1}$$

where *term frequency* $tf_{i,j}$ is defined as:

$$tf_{i,j} = \frac{n_{i,j}}{\sum_k n_{k,j}} \tag{2}$$

where $n_{i,j}$ is the number of times the string s_i appears in a executable $\mathcal{E}_j$, and $\sum_k n_{k,j}$ is the total number of strings in the executable $\mathcal{E}_j$. The inverse term frequency idf_i is defined as:

$$idf_i = \log\left(\frac{|\mathcal{C}|}{|\mathcal{C} : t_i \in \mathcal{E}|}\right) \tag{3}$$

where $|\mathcal{C}|$ is the total number of executables and $|\mathcal{C} : s_i \in \mathcal{E}|$ is the number of executables containing the string s_i.

---

[4] http://code.google.com/p/smali/

3 Anomaly Detection Techniques

Anomaly detection models what it is a normal application and every deviation to this model is considered anomalous. Our method represents Android applications as points in the feature space, using the method described in the previous section. When an application is being inspected, our method starts by computing the values of the points in the feature space. This point is then compared with the previously calculated points of the legitimate applications. To this end, distance measures are required. In this study, we have used the following distance measures:

- *Manhattan Distance*: This distance between two points x and y is the sum of the lengths of the projections of the line segment between the points onto the coordinate axes:

$$d(x,y) = \sum_{i=0}^{n} |x_i - y_i| \qquad (4)$$

  where x is the first point; y is the second point and x_i and y_i are the ith component of the first and second point, respectively.
- *Euclidean Distance*: This distance is the length of the line segment connecting two points. It is calculated as:

$$d(x,y) = \sum_{i=0}^{n} \sqrt{v_i^2 - u_i^2} \qquad (5)$$

  where x is the first point; y is the second point and x_i and y_i are the ith component of the first and second point, respectively.
- *Cosine Similarity*: It consists of measuring the similarity between two vectors by finding the cosine of the angle between them [9]. Since we are measuring distance and not similarity, we have used $1 - CosineSimilarity$ as a distance measure:

$$d(x,y) = 1 - cos(\theta) = 1 - \frac{\boldsymbol{v} \cdot \boldsymbol{u}}{||\boldsymbol{v}|| \cdot ||\boldsymbol{u}||} \qquad (6)$$

  where $\boldsymbol{v}$ is the vector from the origin of the feature space to the first point x, $\boldsymbol{u}$ is the vector from the origin of the feature space to the second point y, $\boldsymbol{v} \cdot \boldsymbol{u}$ is the inner product of $\boldsymbol{v}$ and $\boldsymbol{u}$. $||\boldsymbol{v}|| \cdot ||\boldsymbol{v}||$ is the cross product of $\boldsymbol{v}$ and $\boldsymbol{u}$. This distance ranges from 0 to 1, where 1 means that the two evidences are completely different and 0 means that the evidences are the same (i.e., the vectors are orthogonal between them).

By means of these measures, we can compute the deviation of an application with respect to a set of benign ones.

Since we have to compute this measure with the points representing valid apps, a combination metric is required in order to obtain a final distance value which considers every measure performed. To this end, our system employs 3 simplistic rules: (i) select the mean value, (ii) select the lowest distance value and (iii) select the highest value of the computed distances.

In this way, when our method inspects an application, a final distance value is acquired, which will depend on both the chosen distance measure and a combination rule.

4 Empirical Validation

To evaluate our method, we used a dataset composed of 666 samples: 333 malicious applications and 333 legitimate apps. Malicious applications were gathered from the company VirusTotal[5]. VirusTotal offers a series of services called Virus-Total Malware Intelligence Services, which allow researchers to obtain samples from their databases.

To evaluate our anomaly-based approach, we followed the next configuration for the empirical validation:

1. *Cross validation*: We performed a 5-fold cross-validation over benign samples to divide them into 5 different divisions of the data into training and test sets.
2. *Computation of distances and combination rules*: We extracted the strings from the applications and used the 3 different measures and the 3 different combination rules described in Section 3 to obtain a final measure of deviation for each testing evidence. More accurately, we applied the following distances: (i) the Manhattan distance, (ii) the Euclidean distance and (iii) the Cosine distance. For the combination rules, we tested the following ones: (i) the mean value, (ii) the lowest distance and (iii) the highest value.
3. *Defining thresholds*: For each measure and combination rule, we established 10 different thresholds to determine whether a sample is valid or not.
4. *Testing the method*: We evaluated the method by measuring these parameters:
   - *True Positive Ratio* (TPR), also known as sensitivity.

$$TPR = \frac{TP}{(TP + FN)} \tag{7}$$

   where TP is the number of applications correctly classified as malware and FN is the number of applications misclassified as benign software.
   - *False Positive Ratio* (FPR), which is the number of legitimate applications misclassified as malware.

$$FPR = \frac{FP}{(FP + TN)} \tag{8}$$

   where FP is the number of valid applications incorrectly detected as malicious and TN is the number of valid applications correctly classified.

---

[5] http://www.virustotal.com

Table 1. Results for the different combination measures using Manhattan Distance. The results in bold are the best for each combination rule and distance measure.

Comb.	Thres.	TPR	FPR	AUC	Acc.
Average	1042516.95554	1.00000	1.00000		50.00%
	1297301.02116	0.56456	0.89189		33.63%
	1552085.08679	0.47988	0.82282		32.85%
	1806869.15241	0.37117	0.73273		31.92%
	2061653.21803	0.23964	0.55856	0.29121	34.05%
	2316437.28365	0.15856	0.34835		40.51%
	2571221.34927	0.11051	0.17117		46.97%
	2826005.41490	**0.07568**	**0.03303**		**52.13%**
	3080789.48052	0.00240	0.02102		49.07%
	3335573.54614	0.00000	0.00300		49.85%
Max.	2115684.24601	1.00000	1.00000		49.85%
	2386946.41796	0.90871	0.98198		50.00%
	2658208.58992	0.54294	0.87387		46.34%
	2929470.76187	0.43724	0.79880		33.45%
	3200732.93383	0.30931	0.67868	0.30224	31.92%
	3471995.10578	0.21201	0.43243		31.53%
	3743257.27774	0.14114	0.19520		38.98%
	4014519.44969	0.08468	0.07207		47.30%
	4285781.62165	0.06006	0.00601		50.63%
	4557043.79360	**0.00000**	**0.00000**		**52.70%**
Min.	**0.00000**	**1.00000**	**1.00000**		**50.00%**
	274472.12573	**0.54655**	**0.85285**		**50.00%**
	548944.25146	0.44805	0.75976		34.68%
	823416.37718	0.33333	0.64565		34.41%
	1097888.50291	0.20721	0.50751	0.29987	34.38%
	1372360.62864	0.14114	0.32733		34.98%
	1646832.75437	0.09610	0.15315		40.69%
	1921304.88009	0.00300	0.03604		47.15%
	2195777.00582	0.00000	0.02102		48.35%
	2470249.13155	0.00000	0.00000		48.95%

- *Accuracy*, which is the total number of hits divided by the number of instances in the dataset.

$$Accuracy = \frac{TP + TN}{(TP + FP + TN + FN)} \qquad (9)$$

- *Area Under ROC Curve* [10], establishes the relation between FNR and FPR for the different thresholds stablished.

Table 1 shows the results obtained using the Manhattan distance, Table 2 shows the results for the Euclidean distance and Table 3 shows the results for

Table 2. Results for the different combination measures using Euclidean Distance. The results in bold are the best for each combination rule and distance measure.

Comb.	Thres.	TPR	FPR	AUC	Acc.
Average	1292555.44	0.99880	1.00000		49.94%
	1494979.97	0.53814	0.86787		33.51%
	1697404.49	0.39880	0.76577		31.65%
	1899829.02	0.27327	0.56757		35.29%
	2102253.55	0.19700	0.25526	0.321240159	47.09%
	2304678.08	0.12613	0.13213		49.70%
	2507102.61	**0.10511**	**0.04805**		**52.85%**
	2709527.14	0.02402	0.01802		50.30%
	2911951.67	0.01502	0.00601		50.45%
	3114376.19	0.00000	0.00000		50.00%
Max.	2486071.46	1.00000	1.00000		50.00%
	2643030.09	0.88829	0.95796		46.52%
	2799988.72	0.49189	0.82282		33.45%
	2956947.35	0.28889	0.63664		32.61%
	3113905.98	0.19339	0.30030	0.326824121	44.65%
	3270864.60	**0.13874**	**0.10811**		**51.53%**
	3427823.23	0.08889	0.06306		51.29%
	3584781.86	0.02222	0.01201		50.51%
	3741740.49	0.01201	0.00000		50.60%
	3898699.12	0.00000	0.00000		50.00%
Min.	0.00	1.00000	1.00000		50.00%
	308138.33	0.59760	0.91892		33.93%
	616276.67	0.53754	0.84384		34.68%
	924415.01	0.43664	0.75676		33.99%
	1232553.35	0.29730	0.60961	0.320127335	34.38%
	1540691.69	0.21021	0.35135		42.94%
	1848830.03	0.12613	0.12312		50.15%
	2156968.37	**0.10511**	**0.02703**		**53.90%**
	2465106.71	0.01502	0.00601		50.45%
	2773245.05	0.00000	0.00000		50.00%

the Cosine distance. In this way, both the Manhattan and Euclidean distances achieved very low accuracy results. However, the results of our anomaly-based system using the Cosine similarity are considerably sounder. In particular, it obtained a best result of 83.51% of accuracy, with an FPR of 27% and a TPR of 94%, using the average combination rule.

Table 3. Results for the different combination measures using Cosine Distance. The results in bold are the best for each combination rule and distance measure.

Comb.	Thres.	TPR	FPR	AUC	Acc.
Average	0.83693246	1.00	1.00		50.00%
	0.85505219	0.99	1.00		49.70%
	0.87317191	0.99	1.00		49.61%
	0.89129164	0.98	1.00		49.13%
	0.90941137	0.98	0.96	**0.884413242**	50.90%
	0.92753109	0.98	0.80		58.86%
	0.94565082	0.97	0.52		72.40%
	0.96377055	**0.94**	**0.27**		**83.51%**
	0.98189027	0.77	0.11		82.79%
	1.00001000	0.00	0.00		50.00%
Max.	**1.00000000**	**1.00**	**1.00**		**50.00%**
	1.00000111	0.00	0.00		50.00%
	1.00000222	0.00	0.00		50.00%
	1.00000333	0.00	0.00		50.00%
	1.00000444	0.00	0.00	0.5	50.00%
	1.00000556	0.00	0.00		50.00%
	1.00000667	0.00	0.00		50.00%
	1.00000778	0.00	0.00		50.00%
	1.00000889	0.00	0.00		50.00%
	1.00001000	0.00	0.00		50.00%
Min.	0.00000000	1.00	1.00		50.00%
	0.11111222	1.00	0.98		51.20%
	0.22222444	1.00	0.95		52.70%
	0.33333667	1.00	0.92		53.75%
	0.44444889	1.00	0.84	0.854307461	57.69%
	0.55556111	0.94	0.70		61.98%
	0.66667333	0.87	0.33		76.94%
	0.77778556	**0.58**	**0.04**		**77.30%**
	0.88889778	0.41	0.02		69.64%
	1.00001000	0.00	0.00		50.00%

5   Conclusions and Future Work

Smartphones and tablets are flooding both consumer and business markets and, therefore, manage a large amount of information. For this reason, malware writers have found in these devices a new source of income and therefore the number of malware samples has grown exponentially in these platforms.

In this paper, we present a new malicious software detection approach that is inspired on anomaly detection systems. In contrast to other approaches, this method only needs to previously label goodware and measures the deviation of a new sample with respect to normality (applications without malicious intentions). Although anomaly detection systems tend to produce high error rates (specially, false positives), our experimental results show low FPR values. The number of samples that exist today is assumable by existing systems, but is growing very rapidly. This approach reduces the necessity to collect malware samples and is trained using benign ones. In addition, our method is based on features that are extracted from string analysis of the application, making possible to prevent the installation of malicious software.

However, this approach also has several limitations. Through an internet connection, a benign application can download a malicious payload and change its behaviour. To detect these kind of changes, a dynamic approach is necessary to monitor the behaviour of the applications. Nevertheless, these approaches require considerable computational effort.

Future work is oriented in three main directions. First, there are other features that could be used to improve the detection ratio. These features could be obtained from the `AndroidManifest.xml` file. Second, other distance measurements and combination rules could be tested. Finally, the effectiveness of the method relies on the appropriate choice of the thresholds, making necessary to improve these metrics.

References

1. Burguera, I., Zurutuza, U., Nadjm-Tehrani, S.: Crowdroid: behavior-based malware detection system for android. In: Proceedings of the 1st ACM Workshop on Security and Privacy in Smartphones and Mobile Devices, pp. 15–26. ACM (2011)
2. Blasing, T., Batyuk, L., Schmidt, A.D., Camtepe, S.A., Albayrak, S.: An android application sandbox system for suspicious software detection. In: 2010 5th International Conference on Malicious and Unwanted Software (MALWARE), pp. 55–62. IEEE (2010)
3. Shabtai, A., Elovici, Y.: Applying behavioral detection on android-based devices. In: Mobile Wireless Middleware, Operating Systems, and Applications, pp. 235–249 (2010)
4. Sanz, B., Santos, I., Laorden, C., Ugarte-Pedrero, X., Bringas, P., Alvarez, G.: Puma: Permission usage to detect malware in android. In: Proceedings of the 5th International Conference on Computational Intelligence in Security for Information Systems, CISIS (2012)
5. Santos, I., Penya, Y., Devesa, J., Bringas, P.: N-Grams-based file signatures for malware detection. In: Proceedings of the 11th International Conference on Enterprise Information Systems (ICEIS), vol. AIDSS, pp. 317–320 (2009)
6. Santos, I., Devesa, J., Brezo, F., Nieves, J., Bringas, P.G.: Opem: A static-dynamic approach for machine-learning-based malware detection. In: Herrero, Á., et al. (eds.) Int. Joint Conf. CISIS'12-ICEUTE'12-SOCO'12. AISC, vol. 189, pp. 271–280. Springer, Heidelberg (2013)
7. Baeza-Yates, R.A., Ribeiro-Neto, B.: Modern Information Retrieval. Addison-Wesley Longman Publishing Co., Inc., Boston (1999)
8. Salton, G., McGill, M.: Introduction to modern information retrieval. McGraw-Hill, New York (1983)

9. Tata, S., Patel, J.M.: Estimating the selectivity of tf-idf based cosine similarity predicates. ACM SIGMOD Record 36(2), 7–12 (2007)
10. Singh, Y., Kaur, A., Malhotra, R.: Comparative analysis of regression and machine learning methods for predicting fault proneness models. International Journal of Computer Applications in Technology 35(2), 183–193 (2009)

Classification of SSH Anomalous Connections

Silvia González[1], Javier Sedano[1], Urko Zurutuza[2], Enaitz Ezpeleta[2],
Diego Martínez[3], Álvaro Herrero[3], and Emilio Corchado[4]

[1] Instituto Tecnológico de Castilla y León
C/ López Bravo 70, Pol. Ind. Villalonquejar, 09001 Burgos, Spain
`javier.sedano@itcl.es`
[2] Electronics and Computing Department, Mondragon University
Goiru Kalea, 2, 20500 Arrasate-Mondragon, Spain
`{uzurutuza,eezpeleta}@mondragon.edu`
[3] Department of Civil Engineering, University of Burgos, Spain
C/ Francisco de Vitoria s/n, 09006 Burgos, Spain
`ahcosio@ubu.es`
[4] Departamento de Informática y Automática, Universidad de Salamanca
Plaza de la Merced, s/n, 37008 Salamanca, Spain
`escorchado@usal.es`

Abstract. The Secure Shell Protocol (SSH) is a well-known standard protocol for remote login and used as well for other secure network services over an insecure network. It is mainly used for remotely accessing shell accounts on Unix-liked operating systems to perform administrative tasks. For this reason, the SSH service has been for years an attractive target for attackers, aiming to guess root passwords performing dictionary attacks, or to directly exploit the service itself. To test the classification performance of different classifiers and combinations of them, this study gathers and analyze SSH data coming from a honeynet and then it is analysed by means of a wide range of classifiers. The high-rate classification results lead to positive conclusions about the identification of malicious SSH connections.

Keywords: Secure Shell Protocol, SSH, Honeynet, Honeypot, Intrusion Detection, Classifier, Ensemble.

1 Introduction

A network attack or intrusion will inevitably violate one of the three computer security principles -availability, integrity and confidentiality- by exploiting certain vulnerabilities such as Denial of Service, Modification and Destruction [1]. One of the most harmful issues of attacks and intrusions, which increases the difficulty of protecting computer systems, is precisely the ever-changing nature of attack technologies and strategies.

Intrusion Detection Systems (IDSs) [2-4] have become an essential asset in addition to the computer security infrastructure of most organizations. In the context of computer networks, an IDS can roughly be defined as a tool designed to detect

Á. Herrero et al. (eds.), *International Joint Conference SOCO'13-CISIS'13-ICEUTE'13*,
Advances in Intelligent Systems and Computing 239,
DOI: 10.1007/978-3-319-01854-6_49, © Springer International Publishing Switzerland 2014

suspicious patterns that may be related to a network or system attack. Intrusion Detection (ID) is therefore a field that focuses on the identification of attempted or ongoing attacks on a computer system (Host IDS - HIDS) or network (Network IDS - NIDS).

ID has been approached from several different points of view up to now; many different Computational Intelligence techniques - such as Genetic Programming [5], Data Mining [6-8], Expert Systems [9], Fuzzy Logic [10], or Neural Networks [11-13] among others - together with statistical [14] and signature verification [15] techniques have been applied mainly to perform a 2-class classification (normal/anomalous or intrusive/non-intrusive).

The Secure Shell Protocol (SSH) is a standard protocol for remote login and used as well for other secure network services over an insecure network. It is an Application Layer protocol under the TCP/IP stack. The SSH protocol consists of three major components: The Transport Layer Protocol that provides server authentication, confidentiality, and integrity with perfect forward secrecy. TheUser Authentication Protocol which authenticates the client to the server. And the Connection Protocol that multiplexes the encrypted tunnel into several logical channels.

The main usage of SSH protocol is for remotely accessing shell accounts on Unix-liked operating systems with administrative purposes. For this reason, the SSH service has been for years an attractive service for attackers, aiming to guess root passwords performing dictionary attacks, or to directly exploit the service itself. The SANS Institute's Internet Storm Center [16] keeps monitoring an average of 100,000 targets being attacked every day in Internet. Being able of distinguishing among malicious SSH packets and benign SSH traffic for server administration may play an indispensable role in defending system administrators against malicious adversaries.

The aim of the present study is to assess classifiers and ensembles in the useful task of identifying bad-intentioned SSH connections. To do so, real data, coming from the Euskalerthoneynet is analysed as described in the remaining sections of the paper. In this contribution, section 2 presents the proposed models that are applied to SSH data as described in section 3, together wit the obtained results. Some conclusions and lines of future work are introduced in section 4.

1.1   SSH and Honeynets

A honeypot has no authorised function or productive value within the corporate network other than to be explored, attacked or compromised [17]. Thus, a honeypot should not receive any traffic at all. Any connection attempt with a honeypot is then an attack or attempt to compromise the device or services that it is offering- is by default illegitimate traffic. From the security point of view, there is a great deal of information that may be learnt from a honeypot about a hacker's tools and methods in order to improve the protection of information systems.

In a honeynet, all the traffic received by the sensors is suspicious by default. Thus every packet should be considered as an attack or at least as a piece of a multi-step attack. Numerous studies propose the use of honeypots to detect automatic large scale attacks; honeyd [18] and nepenthes [19] among others. The first Internet traffic monitors known as Network Telescopes, Black Holes or Internet Sinks were presented by Moore *et al.* [20].

The Euskalerthoneynet [21] has been monitoring attacks against well-known services, including SSH. Furthermore, the sensors have recorded the SSH sessions used to administer and maintain the different devices of the infrastructure.

Having both malicious and real administrative SSH traffic recorded, we perform a classification of such traffic to detect attacks against the SSH service.

1.2   Previous Work

Attacks to SSH service have attracted researchers' attention for a long time. Song et al. [22] analysed timing and keystroke attacks. Researchers have also used honeypots to study and analyse attacks to this protocol, focusing on login attempts and dictionary attacks [23], [24]. In [24] authors analyse SSH attacks on honeypots focusing on visualisation of the data gathered. The honeypots collect real attacks, making experiments and analysis results applicable to real deployments.

Considering the data capture, as previously introduced, the present study takes advantage of the Euskalert project [21]. It has deployed a network of honeypots in the Basque Country (northern Spain) where eight companies and institutions have installed one of the project's sensors behind the firewalls of their corporate networks. The honeypot sensor transmits all the traffic received to a database via a secure communication channel. These partners can consult information relative to their sensor (after a login process) as well as general statistics in the project's website. Once the system is fully established, the information available can be used to analyse attacks suffered by the honeynet at network and application level. Euskalert is a distributed honeypot network based on a Honeynet GenIII architecture [25].

2   Proposal

One of the most interesting features of IDSs would be their capability to automatically detect whether a portion of the traffic circulating the network is an attack or normal traffic. This task is more challenging when confronting brand-new bad intentioned activities with no previous examples. Automated learning models(classifiers) [26] are well-known algorithms designed specifically for the purpose of deciding about previously-unseen data. This issue makes them suitable for the IDS task. Going one step further, ensemble methods [27] combine multiple algorithms into one usually more accurate than the best of its components. So, the main idea behind ensemble learning is taking advantage of classification algorithms diversity to face more complex data. For this reason, present study proposes the combination of classifiers to get more accurate results when detecting anomalous and intrusive events.

A wide variety of automated learning techniques have been applied in this study to classify SSH connections. Several base classifiers as well as different ways of combining them have been considered for the analysis of Euskalert data. 35 base classifiers have been applied in present study, comprising neural models such as the MultiLayer Perceptron and Voted-Perceptron [28], decision trees such as CART [31] or REP-Tree [32], and traditionalclustering algorithms such as the k-Nearest Neighbours (K-NN) [30].

These base classifiers have been combined according to the ensemble paradigm by 19 different strategies. The applied ensemble schemes range from basic ones such as Bagging [33] orboosting-based [34] (Adaboost) to some other, more modern algorithms such as the LogitBoost [35] or the StackingC [36]. As results prove, ensemble learning adds an important value to the analysis, as almost all variants consistently improve results obtained by the single classifier.

3   Experimental Validation on Real Data

As previously mentioned, the performance of automated learning techniques have been assessed using real datasets, coming from the Euskalert project. The detailed information about the data and the run experiments is provided in this section.

3.1   Datasets

We have performed the experimental study by extracting SSH data related to 34 months of real attacks and administration tasks that reached the 8 sensors of the Euskalert project [25]. Data from a so long time period guarantees that a broad variety of situations are considered.

This honeynet system receives 4,000 packets a month on average. The complete dataset contains a total of 2,647,074 packets, including TCP, UDP and ICMP traffic received by the distributed honeypot sensors. For this experiment, we have analysed SSH connections happened between May 2008 and March 2011. First, we have filtered out traffic containing real attacks to the SSH port (22), and SSH connections to the system management port (2399).

Then, the traffic has been processed in order to obtain the Secure Shell sessions out of the packets. Two different approaches have been used in order to identify the sessions:

The approach for defining an SSH session was based on the TCP logic, using packets with the same source IP, same destination IP and a common source port. This last value is a non-privileged port number that remains the same during any TCP session.Out of the 2,647,074 packets, the TCP-based dataset was summarized as 8,478 attack sessions and 82 administration sections.

The features that were extracted from each one of the sessions in the dataset are described in table 1.

Table 1. Features for SSH sessions

Feature	Description
Src	IP address of the source host
Time	duration of the session
Numpac	number of packets that the source host sent
Minlen	minimum size of the packets
Maxlen	maximum size of the packets
Avglen	average size of the packets
Numflags	amount of different flags used

Table 2 shows the range of each feature, depending on the nature of the session (administrator or attack).

Table 2. Range of features for SSH sessions

Feature	Type	Attack	Administrator
Src	inet	---	---
Time	interval	00:00:00 – 352 days 09:48:19.891	00:00:00.004 – 519 days 18:24:05.446
Numpac	integer	1 - 95	1 - 23
Minlen	integer	40 - 64	40 - 380
Maxlen	integer	40 - 220	40 - 380
Avglen	numeric(8,2)	40 - 96	40 - 380
Numflags	integer	1 - 6	1 - 4

3.2   Practical Settings

The experimentation has been based on the performance of 1,534 tests, carried out through 35 different classifiers (such as "NaiveBayes", "Ibk", "LinearRegression", "JRip", "RBFNetwork", "SMO", etc.) combined by means of the following ensembles: Base classifier, Bagging, Adaboost, MultiBoostAB, RandomSubSpace, Dagging, Decorate, MultiClassClassifier, CVParameterSelection, AttributeSelected-Classifier, ThresholdSelector, Vote, FilteredClassifier, Grading, MultiScheme, Ordi-nalClassClassifier, RotationForest, Stacking, and StackingC.

For testing purposes, each ensemble processes a combination of 10 same type base classifiers. The data sets were trained and classified with ensembles and classifiers by means of WEKA software [37].

3.3   Results

To summarize the results data, only the classification rate from the base classifier and the highest rate from the different ensembles are shown below in Table 3 (comprising training results) and Table 4 (comprising classification results).

From Table 4, it can be seen that the best classification result (1) is obtained by the DecisionTable classifier combined by the Adaboost ensemble.

Table 3. Training results on SSH sessions

#	Classifier	Base Classifier	Max
1	MultilayerPerceptron	0,99325	0,998053
2	NaiveBayes	0,992861	0,99325
3	K-nn IBK	0,997793	0,997923
4	DecisiontreeSImpleCart	0,993899	0,998053
5	Rule InductionJrip	0,995846	0,998183
6	RBF network	0,994159	0,996495
7	REPTree	0,995717	0,997274
8	NaiveBayesMultinomial	0,870457	0,990395
9	IB1	0,997923	0,997923
10	PART	0,996885	0,997923
11	ZeroR	0,990395	0,990395
12	BayesianLogisticRegression	0,990395	0,991044
13	ComplementNaiveBayes	0,754154	0,990395
14	DMNBtext	0,990395	0,990395
15	NaiveBayesMultinomialUpdateable	0,872144	0,990395
16	NaiveBayesUpdateable	0,992861	0,99325
17	Logistic	0,992082	0,997534
18	SMO	0,990524	0,998183
19	SPegasos	0,990395	0,990395
20	VotedPerceptron	0,990395	0,997664
21	DTNB	0,997534	0,998053
22	DecisionTable	0,998053	0,998183
23	NNge	0,995976	0,997274
24	OneR	0,997534	0,997793
25	Ridor	0,996625	0,997534
26	ADTree	0,996366	0,998183
27	BFTree	0,99338	0,998183
28	DecisionStump	0,99351	0,995457
29	FT	0,994548	0,997793
30	J48	0,995846	0,998183
31	LADTree	0,995846	0,997923
32	LMT	0,995067	0,997923
33	NBTree	0,994029	0,997923
34	RandomForest	0,996885	0,997793
35	RandomTree	0,996495	0,997404

Table 4. Classification results on SSH sessions

#	Classifier	Base Classifier	Max
1	MultilayerPerceptron	0,992982	0,997661
2	NaiveBayes	0,992982	0,992982
3	K-nn IBK	0,996491	0,997661
4	DecisiontreeSImpleCart	0,994152	0,997661
5	Rule InductionJrip	0,995322	0,997661
6	RBF network	0,992982	0,996491
7	REPTree	0,994152	0,997661
8	NaiveBayesMultinomial	0,854971	0,990643
9	IB1	0,996491	0,997661
10	PART	0,997661	0,99883
11	ZeroR	0,990643	0,990643
12	BayesianLogisticRegression	0,990643	0,991813
13	ComplementNaiveBayes	0,753216	0,990643
14	DMNBtext	0,990643	0,990643
15	NaiveBayesMultinomialUpdateable	0,85614	0,990643
16	NaiveBayesUpdateable	0,992982	0,992982
17	Logistic	0,991813	0,997661
18	SMO	0,990643	0,997661
19	SPegasos	0,990643	0,997661
20	VotedPerceptron	0,990643	0,997661
21	DTNB	0,997661	0,997661
22	DecisionTable	0,997661	**1**
23	NNge	0,996491	0,997661
24	OneR	0,997661	0,99883
25	Ridor	0,997661	0,997661
26	ADTree	0,995322	0,997661
27	BFTree	0,994152	0,997661
28	DecisionStump	0,992982	0,995322
29	FT	0,992982	0,997661
30	J48	0,995322	0,99883
31	LADTree	0,994152	0,997661
32	LMT	0,991813	0,997661
33	NBTree	0,992982	0,997661
34	RandomForest	0,997661	0,997661
35	RandomTree	0,997661	0,99883

4   Conclusions and Future Work

Classification of benign and malicious SSH sessions is extremely valuable for preventing unauthorized users to access production networks. The successful classification results obtained in this study can efficiently discover a malicious connection attempt and make possible to discard the session before a dictionary attack becomes a major problem to the network assets.

It has been shown how base classifiers provide good results in differentiating real administering SSH sessions from attacks, but the use of ensemble classifiers even improve the effectiveness up to a 100% in at least one case.

This may be due to the fact that a real SSH session can be comprised by an increasing number of different bash commands, generated by few different IP addresses (administrators). This would derive in a more specific behaviour than the rest of SSH attacks gathered by the honeynet.

Those exceptional classification results obtained by ensemble classifiers can be applied to other protocols and services of the attacks received by the honeynets, such as HTTP, SNMTP, or even FTP, learning from the honeypots classification models that will later prevent detected attacks surpass the organization networks causing any damage.

Acknowledgments. This research is partially supported through projects of the Spanish Ministry of Economy and Competitiveness with ref: TIN2010-21272-C02-01 (funded by the European Regional Development Fund), and SA405A12-2 from Junta de Castilla y León.

References

1. Myerson, J.M.: Identifying Enterprise Network Vulnerabilities. International Journal of Network Management 12, 135–144 (2002)
2. Computer Security Threat Monitoring and Surveillance. Technical Report. James P. Anderson Co. (1980)
3. Denning, D.E.: An Intrusion-Detection Model. IEEE Transactions on Software Engineering 13, 222–232 (1987)
4. Chih-Fong, T., Yu-Feng, H., Chia-Ying, L., Wei-Yang, L.: Intrusion Detection by Machine Learning: A Review. Expert Systems with Applications 36, 11994–12000 (2009)
5. Abraham, A., Grosan, C., Martin-Vide, C.: Evolutionary Design of Intrusion Detection Programs. International Journal of Network Security 4, 328–339 (2007)
6. Julisch, K.: Data Mining for Intrusion Detection: A Critical Review. In: Barbará, D., Jajodia, S. (eds.) Applications of Data Mining in Computer Security, pp. 33–62. Kluwer Academic Publishers (2002)
7. Giacinto, G., Roli, F., Didaci, L.: Fusion of Multiple Classifiers for Intrusion Detection in Computer Networks. Pattern Recognition Letters 24, 1795–1803 (2003)
8. Chebrolu, S., Abraham, A., Thomas, J.P.: Feature Deduction and Ensemble Design of Intrusion Detection Systems. Computers & Security 24, 295–307 (2005)
9. Kim, H.K., Im, K.H., Park, S.C.: DSS for Computer Security Incident Response Applying CBR and Collaborative Response. Expert Systems with Applications 37, 852–870 (2010)

10. Tajbakhsh, A., Rahmati, M., Mirzaei, A.: Intrusion Detection using Fuzzy Association Rules. Applied Soft Computing 9, 462–469 (2009)

11. Sarasamma, S.T., Zhu, Q.M.A., Huff, J.: Hierarchical Kohonen Net for Anomaly Detection in Network Security. IEEE Transactions on Systems Man and Cybernetics, Part B 35, 302–312 (2005)

12. Herrero, Á., Corchado, E., Gastaldo, P., Zunino, R.: Neural Projection Techniques for the Visual Inspection of Network Traffic. Neurocomputing 72, 3649–3658 (2009)

13. Zhang, C., Jiang, J., Kamel, M.: Intrusion Detection using Hierarchical Neural Networks. Pattern Recognition Letters 26, 779–791 (2005)

14. Marchette, D.J.: Computer Intrusion Detection and Network Monitoring: A Statistical Viewpoint. Springer-Verlag New York, Inc. (2001)

15. Roesch, M.: Snort–Lightweight Intrusion Detection for Networks. In: 13th Systems Administration Conference (LISA 1999), pp. 229–238 (1999)

16. SANS Institute's Internet Storm Center,
    https://isc.sans.edu/port.html?port=22

17. Charles, K.A.: Decoy Systems: A New Player in Network Security and Computer Incident Response. International Journal of Digital Evidence 2 (2004)

18. Provos, N.: A Virtual Honeypot Framework. In: 13th USENIX Security Symposium, vol. 132 (2004)

19. Baecher, P., Koetter, M., Holz, T., Dornseif, M., Freiling, F.: The Nepenthes Platform: An Efficient Approach to Collect Malware. In: Zamboni, D., Kruegel, C. (eds.) RAID 2006. LNCS, vol. 4219, pp. 165–184. Springer, Heidelberg (2006)

20. Moore, D., Shannon, C., Brown, D.J., Voelker, G.M., Savage, S.: Inferring Internet Denial-of-service Activity. ACM Transactions on Computer Systems 24, 115–139 (2006)

21. Herrero, Á., Zurutuza, U., Corchado, E.: A Neural-Visualization IDS for Honeynet Data. International Journal of Neural Systems 22, 1–18 (2012)

22. Song, D.X., Wagner, D., Tian, X.: Timing Analysis of Keystrokes and Timing Attacks on SSH. In: Proceedings of the 10th Conference on USENIX Security Symposium, vol. 10, p. 25. USENIX Association, Washington, D.C. (2001)

23. Coster, D.D., Woutersen, D.: Beyond the SSH Brute Force Attacks. In: 10th GOVCERT.NL Symposium (2011)

24. Koniaris, I., Papadimitriou, G., Nicopolitidis, P.: Analysis and Visualization of SSH Attacks Using Honeypots. In: IEEE European Conference on Computer as a Tool (IEEE EUROCON 2013) (2013)

25. Friedman, J.H., Tukey, J.W.: A Projection Pursuit Algorithm for Exploratory Data-Analysis. IEEE Transactions on Computers 23, 881–890 (1974)

26. Bishop, C.M.: Pattern Recognition and Machine Learning. Springer (2007)

27. Seni, G., Elder, J.: Ensemble Methods in Data Mining: Improving Accuracy Through Combining Predictions. Morgan and Claypool Publishers (2010)

28. Freund, Y., Schapire, R.E.: Large Margin Classification Using the Perceptron Algorithm. Mach. Learn. 37, 277–296 (1999)

29. Moody, J., Darken, C.J.: Fast Learning in Networks of Locally-tuned Processing Units. Neural Computation 1, 281–294 (1989)

30. Bailey, T., Jain, A.: A Note on Distance-Weighted k-Nearest Neighbor Rules. IEEE Transactions on Systems, Man and Cybernetics 8, 311–313 (1978)

31. Breiman, L., Friedman, J.H., Olshen, R.A., Stone, C.J.: Classification and Regression Trees, p. 358. Wadsworth Inc., Belmont (1984)

32. Zhao, Y., Zhang, Y.: Comparison of Decision Tree Methods for Finding Active Objects. Advances in Space Research 41, 1955–1959 (2008)

33. Breiman, L.: Bagging Predictors. Machine Learning 24, 123–140 (1996)
34. Freund, Y., Schapire, R.E.: Experiments with a New Boosting Algorithm. In: International Conference on Machine Learning, pp. 148–156 (1996)
35. Friedman, J., Hastie, T., Tibshirani, R.: Additive Logistic Regression: a Statistical View of Boosting. The Annals of Statistics 28, 337–407 (2000)
36. Seewald, A.K.: How to Make Stacking Better and Faster While Also Taking Care of an Unknown Weakness. In: Nineteenth International Conference on Machine Learning. Morgan Kaufmann Publishers Inc. (2002)
37. Hall, M., Frank, E., Holmes, G., Pfahringer, B., Reutemann, P., Witten, I.H.: The WEKA Data Mining Software: An Update. ACM SIGKDD Explorations Newsletter 11, 10–18 (2009)

Provable Secure Constant-Round Group Key Agreement Protocol Based on Secret Sharing

Ruxandra F. Olimid

Department of Computer Science, University of Bucharest, Romania
ruxandra.olimid@fmi.unibuc.ro

Abstract. Group Key Agreement (GKA) allows multiple users to collaboratively compute a common secret key. Motivated by the very few existing GKA protocols based on secret sharing with formal security proofs, we propose a new method to build such protocols. We base our construction on *secret n-sharing*, an untraditional perspective of secret sharing that brings several advantages. Our proposal achieves better security than the existing work while it maintains a constant number of communication rounds regardless the group size.

Keywords: group key agreement, secret sharing, provable security.

1 Introduction

Besides popular examples like chat, digital conferences or file sharing, group applications have rapidly grown as (computational intelligent) distributed systems: grids, distributed artificial intelligence, collaborative problem solving, multi-agent systems, peer-to-peer networks. Reliable communication (as secure conversation between agents) represents a critical aspect of group applications, which may rely on a private common key obtained by all qualified participants to a Group Key Establishment (GKE) protocol. GKE divides into *Group Key Transfer* (GKT) - a privileged party selects the key and securely distributes it to the other members - and *Group Key Agreement* (GKA) - all participants collaborate to compute the key. The current work restricts to GKE based on secret sharing.

1.1 Related Work

Secret Sharing Schemes. Secret sharing was introduced by Blakley [1] and Shamir [18] as a solution to backup cryptographic keys. From the multitude of existing work we recall a single secret sharing scheme, which we will later use in this paper: Karnin et al.'s scheme that permits secret reconstruction by performing a sum modulo a prime [11]. Traditionally in the literature, each participant receives a share and the secret can be recovered only when an authorized set of shares belonging to distinct users are combined together. Recently, Sun et al. proposed a different approach: to split the same secret multiple times and give each user a qualified set of shares (under the appropriate circumstances) [19]. Although their construction is insecure [14], we show that their idea can be successfully used to build strong GKE protocols.

Á. Herrero et al. (eds.), *International Joint Conference SOCO'13-CISIS'13-ICEUTE'13*, 489
Advances in Intelligent Systems and Computing 239,
DOI: 10.1007/978-3-319-01854-6_50, © Springer International Publishing Switzerland 2014

GKE Protocols Based on Secret Sharing. Pieprzyk and Li showed the benefits of using secret sharing in GKE protocols and gave a couple of examples based on Shamir's scheme [16]. Recently, Harn and Lin [9] and Yuan et. al. [20] also used Shamir's scheme for GKT. Some other examples from the current literature include: Sáez's protocol [17] (based on a family of vector space secret sharing schemes), Hsu et al.'s protocol [10] (based on linear secret sharing schemes) and Sun et al.'s protocol [19] (based on the different approach on secret sharing we have previously mentioned). We remind Bresson and Catalano [2] and Cao et al.'s [7] as protocols based on secret sharing with formal security proofs.

Provable Secure GKE. The first security model for GKE (BCPQ) [5] represents a generalization of the models designed for two or three party protocols. It was further improved to allow dynamic groups (BCP) [3] and strong corruption (BCP+) [4]. Katz and Shin were the first to consider the existence of malicious participants within the Universally Composability (UC) framework [12]. At PKC'09, Gorantla, Boyd and Nieto described a stronger model (GBG[1]) that stands against *key compromise impersonation* (KCI) attacks [8]. Two years later, Zhao et al. extended their model (eGBG) and considered *ephemeral key leakage* (EKL) to derive the session key [21].

Unfortunately, recent GKE protocols based on secret sharing lack formal security proofs and hence become susceptible to vulnerabilities: Nam et al. revealed a replay attack on Harn et Lin's protocol [13], Olimid exposed an insider attack and a known key attack on Sun et al.'s construction [14] and Olimid reported an insider attack on Yuan et al.'s scheme [15].

1.2 Our Contribution

Although secret sharing brings several advantages as a building block of GKE, not much research has been done on the formal security of such protocols. This motivates our work: we give a method to build secure GKA protocols based on secret sharing in the GBG model [8]. To the best of our knowledge, no other construction was proved secure in a similar or stronger security model. In addition, our protocol maintains a constant number of rounds regardless the number of participants and its performance is comparable with the existing work.

We base our protocol on the untraditional perspective on secret sharing inspired by the work of Sun et al. [19], which we call *secret n-sharing*. We define the notion of *perfect secret n-sharing* as a natural generalization of perfect secret sharing.

Secret *n*-sharing brings several advantages to GKE protocols: users communicate through broadcast channels only, key computation is efficient, key confirmation is achieved by default, the number of rounds remains constant regardless of the group size, each participant restores the key from his own shares.

---

[1] We adopt the notation from [21].

2 Preliminaries

2.1 Background

We assume that the reader is familiar with the following notions, but (informally) remind them to introduce the notations that we will use for the rest of the paper.

Let $\{U_1, \ldots, U_n\}$ be the set of n users that may take part in the GKE protocol. We denote by (pk_i, sk_i) the public-private key pair an user U_i, $i = 1 \ldots n$ uses for signing under a signature scheme $\Sigma = (\mathsf{Gen}, \mathsf{Sign}, \mathsf{Verify})$, where: $\mathsf{G}(1^k)$ is a randomized algorithm that on input a security parameter k outputs a public-private key pair (pk_i, sk_i); $\mathsf{Sign}(sk_i, \cdot)$ is a randomized signing algorithm under the private key pair sk_i; $\mathsf{Verify}(pk_i, \cdot, \cdot)$ is a deterministic algorithm that outputs 1 for a valid message-signature pair and 0 otherwise. We require that Σ is unforgeable under chosen message attack (UF-CMA) and denote by $\mathsf{Adv}_{\mathcal{A}, \Sigma}^{\mathsf{UF-CMA}}$ the advantage of an adversary $\mathcal{A}$ to win the UF-CMA game.

Let $\mathcal{F} = (\mathsf{G}, \mathsf{F}, \mathsf{F}^{-1})$ be a trapdoor function, where: $\mathsf{G}(1^k)$ is a randomized algorithm that on input a security parameter k outputs a public-private key pair (pk, sk); $\mathsf{F}(pk, \cdot)$ is a deterministic function depending on pk; $\mathsf{F}^{-1}(sk, \cdot)$ inverts $\mathsf{F}(pk, \cdot)$. We require that $\mathcal{F}$ is secure and denote by $\mathsf{Adv}_{\mathcal{A}, \mathcal{F}}$ the advantage of an adversary $\mathcal{A}$ to invert $\mathsf{F}(pk, \cdot)$ without the knowledge of sk.

Let $\mathsf{H} : \{0, 1\}^l \to \{0, 1\}^k$ be a collision resistant hash function modeled as a random oracle and q_r the maximum number of possible queries to H.

2.2 GBG Model

In the GBG model [8], each user U has multiple instances (*oracles*) denoted by Π_U^s, where U participates in the sth run of the protocol (*session*) that is unique identified by the *session id* sid_U^s. Let q_s be the upper bound for the number of sessions. Every instance Π_U^s computes a session key K_U^s and enters an *accepted* state or terminates without computing a session key. The *partner id* pid_U^s of an oracle Π_U^s is the set of parties to whom U wishes to establish a common session key, including himself. Each party U_i, $i = 1, \ldots, n$ owns a public-private *long-term key* pair (pk_i, sk_i) known to all instances $\Pi_{U_i}^{s_i}$ and maintains an internal state that contains all private ephemeral information used during the session.

An adversary $\mathcal{A}$ is a PPT (Probabilistic Polynomial Time) algorithm with full control over the communication channel (he can modify, delete or insert messages) that interacts with the group members by asking queries (Execute, Send, RevealKey, RevealState, Corrupt, Test). The model introduces revised notions of AKE (*Authenticated Key Exchange*) security, MA (*Mutual Authentication*) security and *contributiveness* (key unpredictability by equal contribution of the parties to the key establishment). We denote by $\mathsf{Adv}_{\mathcal{A}}^{\mathsf{AKE}}$, $\mathsf{Adv}_{\mathcal{A}}^{\mathsf{MA}}$, $\mathsf{Adv}_{\mathcal{A}}^{\mathsf{Con}}$ the advantage of an adversary $\mathcal{A}$ to win the AKE security game, MA security game and respectively the contributiveness game. We skip the definitions[2], but strongly invite the reader to address the original paper [8].

---

[2] Because lack of space.

$$
\begin{array}{|ll|}
\hline
\multicolumn{2}{|c|}{S \in \mathbb{Z}_q,\ q > 2 \text{ prime},\ n = 2,\ m = 2.} \\
\textbf{Share 1 - The dealer:} & \textbf{Share 2 -The dealer:} \\
\quad 1.\ \text{chooses } s_1^1 \leftarrow^R \mathbb{Z}_q; & \quad 1.\ \text{chooses } s_1^2 \leftarrow^R \mathbb{Z}_q; \\
\quad 2.\ \text{computes } s_2^1 = S - s_1^1 \pmod q; & \quad 2.\ \text{computes } s_2^2 = S - s_1^2 \pmod q; \\
\textbf{Rec 1} & \textbf{Rec 2} \\
\quad S = s_1^1 + s_2^1 \pmod q; & \quad S = s_1^2 + s_2^2 \pmod q; \\
\quad \mathcal{AS}_1 = \{\{s_1^1, s_2^1\}\} & \quad \mathcal{AS}_2 = \{\{s_1^2, s_2^2\}\} \\
\multicolumn{2}{|c|}{\mathcal{AS} = \{\{s_1^1, s_2^1\}, \{s_1^2, s_2^2\}\} = \mathcal{AS}_1 \cup \mathcal{AS}_2} \\
\hline
\end{array}
$$

Fig. 1. Perfect Secret 2-Sharing Scheme based on Karnin et al.'s scheme

2.3   Secret n-Sharing Schemes

Let $\mathcal{SS} = (\mathsf{Share}, \mathsf{Rec})$ be a secret sharing scheme, where: $\mathsf{Share}(S, m)$ is a randomized sharing algorithm that splits a secret S into m shares $s_1, \ldots, s_m$; $\mathsf{Rec}(s_{i_1}, \ldots, s_{i_t})$, $t \leq m$ is a deterministic reconstruction algorithm from shares that outputs S if $\{s_{i_1}, \ldots, s_{i_t}\}$ is an *authorized* subset and halts otherwise.

The set of all authorized subsets is called *access structure*. The access structure of an (m, m) *all-or-nothing* secret sharing scheme consists of the single set with cardinality m. A secret sharing scheme is *perfect* if it provides no information about the secret to unauthorized subsets.

We follow the idea of Sun et al. [19] and extend secret sharing schemes: a *secret n-sharing scheme* is a scheme that splits a secret n times into the same number of shares m using the same Share algorithm. In other words, it runs a secret sharing scheme n times on the same input. Let $\mathcal{AS}$ be the access structure of a secret n-sharing scheme and $\mathcal{AS}_i$, $i = 1, \ldots, n$, be the access structure of the i-th run of the scheme which is based on. It is immediate that an authorized subset in any of the n splits remains authorized in the extended construction:

$$\mathcal{AS}_1 \cup \ldots \cup \mathcal{AS}_n \subseteq \mathcal{AS}. \tag{1}$$

Definition 1. *(Perfect Secret n-Sharing) A secret n-sharing scheme is called perfect if the following conditions hold: (1) its access structure is $\mathcal{AS} = \mathcal{AS}_1 \cup \ldots \cup \mathcal{AS}_n$; (2) it provides no information about the secret to unauthorized subsets.*

Definition 1 states that no authorized subsets exist except the ones already authorized within the n perfect sharing instances and that combining shares originating from distinct runs give no additional information about the secret.

We affirm that perfect secret n-sharing exists. In order to support our claim we introduce in Fig.1 an example based on Karnin et al.'s scheme [11].

3   Our Proposal

We introduce a new GKA protocol based on perfect secret n-sharing in Fig.2.

The main idea is that each user U_i uniformly selects a random r_i (later a share in a $(2, 2)$ all-or-nothing scheme) and broadcasts its secured value through a trapdoor function F (Rounds 1 and 2). The initiator (U_1, without loss of

Round 1 - User U_1:
  1.1. runs $\mathcal{F}.\mathsf{G}(1^k)$ to obtain a public-private key pair (pk, sk);
  1.2. chooses $r_1 \leftarrow^R \{0,1\}^k$;
  1.3. broadcatsts $U_1 \rightarrow^*: (\mathcal{U}, pk, \mathsf{F}(pk, r_1), \sigma = \Sigma.\mathsf{Sign}(sk_1, \mathcal{U}||pk||\mathsf{F}(pk, r_1)))$;
Round 2 - Each user U_i, $i = 2, \ldots, n$:
  2.1. checks if $\Sigma.\mathsf{Verify}(pk_1, \mathcal{U}||pk||\mathsf{F}(pk, r_1), \sigma) = 1$.
     If the equality does not hold, he quits;
  2.2. chooses $r_i \leftarrow^R \{0,1\}^k$;
  2.3. broadcasts $U_i \rightarrow^*: (\mathsf{F}(pk, r_i), \sigma_i = \Sigma.\mathsf{Sign}(sk_i, \mathcal{U}||pk||\mathsf{F}(pk, r_i)))$;
Round 3 - User U_1:
  3.1. checks if $\Sigma.\mathsf{Verify}(pk_i, \mathcal{U}||pk||\mathsf{F}(pk, r_i), \sigma_i) = 1$, $i = 2, \ldots, n$.
     If at least one equality does not hold, he restarts the protocol;
  3.2. computes $\mathsf{sid}_{U_1} = \mathsf{F}(pk, r_1)||\ldots||\mathsf{F}(pk, r_n)$, $r_i = \mathsf{F}^{-1}(sk, \mathsf{F}(pk, r_i))$, the session key
     $K = \mathsf{H}(\mathsf{sid}_{U_1}||r_1||r_2||\ldots||r_n)$ and r_i' such that $\mathcal{SS}.\mathsf{Share}(K, 2) = \{r_i, r_i'\}$, $i = 2, \ldots, n$;
  3.3. broadcasts $U_1 \rightarrow^*: (r_2', \ldots r_n', \sigma' = \Sigma.\mathsf{Sign}(sk_1, \mathcal{U}||pk||r_2'||\ldots||r_n'||\mathsf{sid}_{U_1}))$;
Key Computation - Each user U_i, $i = 2, \ldots, n$:
  4.1. checks if $\Sigma.\mathsf{Verify}(pk_j, \mathcal{U}||pk||\mathsf{F}(pk, r_j), \sigma_j) = 1$, $j = 2, \ldots n$, $j \neq i$.
     If at least one equality does not hold, he quits;
  4.2. computes $\mathsf{sid}_{U_i} = \mathsf{F}(pk, r_1)||\ldots||\mathsf{F}(pk, r_n)$ and $K = \mathcal{SS}.\mathsf{Rec}(r_i, r_i')$;
  4.3. checks if $\Sigma.\mathsf{Verify}(pk_1, \mathcal{U}||pk||r_2'||\ldots||r_n'||\mathsf{sid}_{U_i}, \sigma') = 1$.
     If the equality holds, he accepts the key K; otherwise he quits;
Key Confirmation - Each user U_i, $i = 2, \ldots, n$:
  5.1. computes r_j such that $\mathcal{SS}.\mathsf{Share}(K, 2) = \{r_j, r_j'\}$, $j = 2, \ldots n$, $j \neq i$;
  5.2. checks if $\mathsf{F}(pk, r_j)$ equals the one sent in step 2.3.
     If at least one equality does not hold, he quits.

Fig. 2. GKA Protocol based on Secret n-Sharing

generality) computes the session key K based on the received values and the session id, invokes secret n-sharing on inputs K and $r_2, \ldots, r_n$ and extracts the second shares $r_2', \ldots, r_n'$, which he then broadcasts (Round 3). Each user U_i can recover the agreed session key K from only his own shares r_i and r_i' (Key Computation).

Key confirmation is achieved by default due to secret n-sharing: U_i eavesdrops r_j' (step 3.3), computes the corresponding r_j such that $\{r_j, r_j'\}$ represents a valid set of shares for K (step 5.1) and verifies that r_j is the genuine value chosen by U_j (step 5.2). Hence, U_i is sure that U_j uses the correct values $\{r_j, r_j'\}$ as inputs for the reconstruction algorithm and obtains the same key. However, an adversary may prevent the last broadcast message (step 3.3) to arrive to one or more users, who become unable to recover the key. This represents a DoS (Denial of Service) attack, which is always detected, but leads to the futility of the protocol. We do not consider this as a weakness of our proposal, since GKE security models ignore DoS scenarios [6] and therefore all existing provable secure GKA protocols are susceptible to such attacks.

The construction requires a secret n-sharing scheme that given as input a secret K and the shares $r_2, \ldots, r_n$ permits to compute the second shares $r_2', \ldots, r_n'$ such that $\mathsf{Share}(K, 2) = \{r_i, r_i'\}$ (or, equivalent $\mathsf{Rec}(r_i, r_i') = K$), $i = 2, \ldots, n$. We emphasize that this does not restrict the applicability, since efficient schemes with such property exist - for example the secret n-sharing scheme in Fig.1.

4 Security Proofs

Theorem 1. *(AKE Security) If the signature scheme Σ is UF-CMA, the trapdoor function $\mathcal{F}$ is secure, the hash function H is a random oracle and the secret n-sharing scheme $\mathcal{SS}$ is perfect then our protocol is AKE secure and*

$$\mathsf{Adv}_{\mathcal{A}}^{\mathsf{AKE}} \leq 2n^2 \mathsf{Adv}_{\mathcal{A},\Sigma}^{\mathsf{UF-CMA}} + \frac{(q_s + q_r)^2}{2^{k-1}} + \frac{(n+1)q_s{}^2}{2^{k-1}} + 2\mathsf{Adv}_{\mathcal{A},\mathcal{F}}.$$

Proof. We prove by a sequence of games. Let $\mathsf{Win}_i^{\mathsf{AKE}}$ be the event that the adversary $\mathcal{A}$ wins Game $i, i = 0, \ldots, 4$.

Let **Game 0** be the original AKE security game. By definition, we have:

$$\mathsf{Adv}_{\mathcal{A}}^{\mathsf{AKE}} = |2Pr[\mathsf{Win}_0^{\mathsf{AKE}}] - 1|. \tag{2}$$

Let **Game 1** be the same as Game 0, except that the simulation fails if an event Forge occurs, where Forge simulates a successful forgery on an honest user signature. We follow the idea from [8] to estimate $Pr[\mathsf{Forge}]$ and obtain:

$$|Pr[\mathsf{Win}_1^{\mathsf{AKE}}] - Pr[\mathsf{Win}_0^{\mathsf{AKE}}]| \leq Pr[\mathsf{Forge}] \leq n^2 \mathsf{Adv}_{\mathcal{A},\Sigma}^{\mathsf{UF-CMA}}. \tag{3}$$

Let **Game 2** be the same as Game 1, except that the simulation fails if an event Collision occurs, meaning that the random oracle H produces a collision for any of its inputs. Since the total number of random oracle queries is bounded by $q_s + q_r$:

$$|Pr[\mathsf{Win}_2^{\mathsf{AKE}}] - Pr[\mathsf{Win}_1^{\mathsf{AKE}}]| \leq Pr[\mathsf{Collision}] \leq \frac{(q_s + q_r)^2}{2^k}. \tag{4}$$

Let **Game 3** be the same as Game 2, except that the simulation fails if an event Repeat occurs, where Repeat simulates a replay attack: it appears when the same public-private key pair (pk, sk) is used in different sessions (event bounded by $q_s^2/2^k$) or when a party uses the same value r_i in different sessions (event bounded by $nq_s^2/2^k$). Hence, we get:

$$|Pr[\mathsf{Win}_3^{\mathsf{AKE}}] - Pr[\mathsf{Win}_2^{\mathsf{AKE}}]| \leq Pr[\mathsf{Repeat}] \leq \frac{(n+1)q_s{}^2}{2^k}. \tag{5}$$

Note that this last game eliminates forgeries and replay attacks.

Let **Game 4** be the same as Game 3, except that the simulation fails if $\mathcal{A}$ is able to compute at least one value $r_i, i = 1, \ldots, n$. Hence:

$$|Pr[\mathsf{Win}_4^{\mathsf{AKE}}] - Pr[\mathsf{Win}_3^{\mathsf{AKE}}]| \leq \mathsf{Adv}_{\mathcal{A},\mathcal{F}}. \tag{6}$$

Since only $r_i', i = 2, \ldots, n$ are available for the adversary in this last game and $\mathcal{SS}$ is perfect, $\mathcal{A}$ has no advantage in finding the secret and $|Pr[\mathsf{Win}_4^{\mathsf{AKE}}]| = 0$. We conclude by combining (2) - (6).

Theorem 2. *(MA Security) If the signature scheme Σ is UF-CMA and the hash function H is a random oracle then our protocol is MA secure and*

$$\mathsf{Adv}_{\mathcal{A}}^{\mathsf{MA}} \leq n^2 \mathsf{Adv}_{\mathcal{A},\Sigma}^{\mathsf{UF-CMA}} + \frac{(q_s + q_r)^2}{2^k} + \frac{(n+1)q_s{}^2}{2^k}.$$

Proof. We prove by a sequence of games. Let $\mathsf{Win}_i^{\mathsf{MA}}$ be the event that the adversary $\mathcal{A}$ wins Game i, $i = 0, \ldots, 3$.

Let **Game 0** be the original MA security game. By definition, we have:

$$\mathsf{Adv}_{\mathcal{A}}^{\mathsf{MA}} = Pr[\mathsf{Win}_0^{\mathsf{MA}}]. \tag{7}$$

Let **Game 1** be the same as Game 0, except that the simulation fails if the event Forge defined in Game 1 of Theorem 1 occurs:

$$|Pr[\mathsf{Win}_1^{\mathsf{MA}}] - Pr[\mathsf{Win}_0^{\mathsf{MA}}]| \leq Pr[\mathsf{Forge}] \leq n^2 \mathsf{Adv}_{\mathcal{A},\Sigma}^{\mathsf{UF-CMA}}. \tag{8}$$

Let **Game 2** be the same as Game 1, except that the simulation fails if the event Collision defined in Game 2 of Theorem 1 occurs:

$$|Pr[\mathsf{Win}_2^{\mathsf{MA}}] - Pr[\mathsf{Win}_1^{\mathsf{MA}}]| \leq Pr[\mathsf{Collision}] \leq \frac{(q_s + q_r)^2}{2^k}. \tag{9}$$

Let **Game 3** be the same as Game 2, except that the simulation fails if the event Repeat defined in Game 3 of Theorem 1 occurs:

$$|Pr[\mathsf{Win}_3^{\mathsf{MA}}] - Pr[\mathsf{Win}_2^{\mathsf{MA}}]| \leq Pr[\mathsf{Repeat}] \leq \frac{(n+1)q_s{}^2}{2^k}. \tag{10}$$

This last game excludes both forgeries and replay attacks. If Game 3 does not aboard, it is impossible for honest partnered parties to accept with different keys. Hence $Pr[\mathsf{Win}_3^{\mathsf{MA}}] = 0$.

We conclude by combing (7) - (10).

Theorem 3. *(Contributiveness) If the trapdoor function $\mathcal{F}$ is secure and the hash function H is a random oracle then our protocol is contributive and*

$$\mathsf{Adv}_{\mathcal{A}}^{\mathsf{Con}} \leq \frac{(n+1)q_s{}^2}{2^k} + \frac{q_r}{2^k}.$$

Proof. We prove by a sequence of games. Let $\mathsf{Win}_i^{\mathsf{Con}}$ be the event that the adversary $\mathcal{A}$ wins Game $i, i = 0 \ldots 2$.

Let **Game 0** be the original contributiveness game. By definition, we have:

$$\mathsf{Adv}_{\mathcal{A}}^{\mathsf{Con}} = Pr[\mathsf{Win}_0^{\mathsf{Con}}]. \tag{11}$$

Let **Game 1** be the same as Game 0, except that the simulation fails if the event Repeat defined in Game 3 of Theorem 1 occurs:

$$|Pr[\mathsf{Win}_1^{\mathsf{Con}}] - Pr[\mathsf{Win}_0^{\mathsf{Con}}]| \leq Pr[\mathsf{Repeat}] \leq \frac{(n+1)q_s{}^2}{2^k}. \tag{12}$$

Let **Game 2** be the same as Game 1, except that the simulation fails if $\mathcal{A}$ can find a collision for $K = H(\text{sid}_{U_1}||r_1||r_2||\ldots||r_n)$ on an input r_i, $i = 1,\ldots,n$:

$$|Pr[\text{Win}_2^{\text{Con}}] - Pr[\text{Win}_1^{\text{Con}}]| \leq \frac{q_r}{2^k}. \tag{13}$$

If Game 2 does not abort, the output of the random oracle is uniformly distributed. Hence $Pr[\text{Win}_2^{\text{Con}}] = 0$.

We conclude by combining (11) - (13).

5   Protocol Analysis and Future Work

Table 1 analysis the complexity of the proposed protocol from three different perspectives: storage, computational cost and overall transmission cost. Let l_x be the length (in bits) of x and c_y be the cost to execute y. First, our proposal is storage efficient: each user maintains his long-lived secret key and a session ephemeral value r_i; in addition, the initiator keeps secret a session trapdoor key. Second, the computational cost is acceptable for a regular party. In case Key Confirmation phase is performed, extra cost is required. However, we stress that $c_{\mathcal{SS}.\text{Share}}$ and $c_{\mathcal{SS}.\text{Rec}}$ can be neglected as they reduce to a sum modulo a prime when the scheme in Fig.1 is used. We also remark that the untraditional perspective on secret n-sharing scheme permits the parties to efficiently recover the key by themselves (with no need to interact with the other parties in Key Computation Phase). Third, the overall dimension of the exchanged messages during one session of the protocol is cost-efficient.

Our proposal introduces different storage and computational costs for the initiator and the rest of the users - a property of GKT rather than GKA protocols. This suggests to introduce an online high performance entity that runs the computation instead of the initiator, while the initiator plays the role of a regular party (after he requests the key establishment). Therefore, the costs of the initiator become lower, while key contributiveness is maintained.

Table 2 compares our proposal with the existing work (we restrict the comparison to GKA protocols based on secret sharing with formal security proofs).

Our protocol maintains a constant number of rounds regardless the group size, while it achieves better security: Bresson and Catalano [2] miss the strong corruption (the adversary is not allowed to reveal private internal state information of participants) and Cao et al. [7] miss a contributiveness proof. Our proposal is secure in the GBG model and hence it stands against KCI attacks.[3]

We emphasize two more advantages of our work: the session id is computed at runtime (the environment of the protocol does not generate it in advance) and participants communicate through broadcast channels only (highly appreciated in distributed and multi-agent systems as well as ad-hoc mobile networks due to concurrency increase and transmission overhead reduction).

We consider as topics for further research the improvements of the protocol such that it admits dynamic groups, anonymity and robustness.

---

[3] We are confident that it can be improved to become secure in the eGBG model; because of space limitations we consider this for an extended version of the paper.

Table 1. Complexity Analysis of the Proposed Protocol

	Storage	Computation	Transmission
U_1	$l_{sk} + l_{sk_1} + k$	$c_{\mathcal{F}.\mathsf{G}} + c_\mathsf{F} + (n-1)c_{\mathsf{F}-1} + c_\mathsf{H}$ $2c_{\Sigma.\mathsf{Sign}} + (n-1)c_{\Sigma.\mathsf{Verify}} + (n-1)c_{\mathcal{SS}.\mathsf{Share}}$	$l_\mathcal{U} + l_{pk} + nl_\mathsf{F} +$
U_i $(i \neq 1)$	$l_{sk_i} + k$	$c_\mathsf{F} + c_{\Sigma.\mathsf{Sign}} + nc_{\Sigma.\mathsf{Verify}} + c_{\mathcal{SS}.\mathsf{Rec}}$ $(+(n-2)c_{\mathcal{SS}.\mathsf{Share}})$	$+(n+1)l_{\Sigma.\mathsf{Sign}} + (n-1)k$

Table 2. Comparison to the Existing Work

	No. of Rounds	Transmission Type	Group Type	sid Generation	Security Model
Our Protocol	3	broadcast	static	at runtime	GBG / ROM
Bresson and Catalano [2]	3	unicast, broadcast	static	in advance	BCP / ROM
Cao et al. [7]	3	broadcast	static	in advance	UC / ROM

6 Conclusions

We proposed a new method to build GKA protocols based on secret sharing. We relied our construction on a slightly different perspective of secret sharing inspired by the idea of Sun et al. [19], which we call *secret n-sharing*. We introduced the notion of *perfect n-sharing* as a natural generalization of perfect sharing. Secret *n*-sharing brings several advantages as a building block of GKA protocols: broadcast communication is sufficient, key computation is efficient, key confirmation is achieved by default, users recover the key from their own shares. Our protocol achieves better security than the existing work, while it maintains the same number of communication rounds regardless the number of participants.

Acknowledgments. This paper is supported by the Sectorial Operational Program Human Resources Development (SOP HRD), financed from the European Social Fund and by the Romanian Government under the contract number SOP HDR/107/1.5/S/82514.

References

1. Blakley, G.: Safeguarding Cryptographic Keys. In: Proceedings of the 1979 AFIPS National Computer Conference, pp. 313–317 (1979)
2. Bresson, E., Catalano, D.: Constant Round Authenticated Group Key Agreement via Distributed Computation. In: Bao, F., Deng, R., Zhou, J. (eds.) PKC 2004. LNCS, vol. 2947, pp. 115–129. Springer, Heidelberg (2004)
3. Bresson, E., Chevassut, O., Pointcheval, D.: Provably Authenticated Group Diffie-Hellman Key Exchange - The Dynamic Case. In: Boyd, C. (ed.) ASIACRYPT 2001. LNCS, vol. 2248, pp. 290–309. Springer, Heidelberg (2001)

4. Bresson, E., Chevassut, O., Pointcheval, D.: Dynamic Group Diffie-Hellman Key Exchange under Standard Assumptions. In: Knudsen, L.R. (ed.) EUROCRYPT 2002. LNCS, vol. 2332, pp. 321–336. Springer, Heidelberg (2002)
5. Bresson, E., Chevassut, O., Pointcheval, D., Quisquater, J.J.: Provably Authenticated Group Diffie-Hellman Key Exchange. In: Proceedings of the 8th ACM Conference on Computer and Communications Security (CCS 2001), pp. 255–264 (2001)
6. Bresson, E., Manulis, M.: Securing group key exchange against strong corruptions. In: Proceedings of ASIA CSS 2008, pp. 249–260 (2008)
7. Cao, C., Yang, C., Ma, J., Moon, S.J.: Constructing UC Secure and Constant-Round Group Key Exchange Protocols via Secret Sharing. EURASIP J. Wireless Comm. and Networking (2008)
8. Gorantla, M.C., Boyd, C., González Nieto, J.M.: Modeling Key Compromise Impersonation Attacks on Group Key Exchange Protocols. In: Jarecki, S., Tsudik, G. (eds.) PKC 2009. LNCS, vol. 5443, pp. 105–123. Springer, Heidelberg (2009)
9. Harn, L., Lin, C.: Authenticated Group Key Transfer Protocol based on Secret Sharing. IEEE Trans. Comput. 59(6), 842–846 (2010)
10. Hsu, C., Zeng, B., Cheng, Q., Cui, G.: A Novel Group Key Transfer Protocol. Cryptology ePrint Archive, Report 2012/043 (2012)
11. Karnin, E.D., Greene, J.W., Hellman, M.E.: On Secret Sharing Systems. IEEE Transactions on Information Theory 29(1), 35–41 (1983)
12. Katz, J., Shin, J.S.: Modeling Insider Attacks on Group Key-Exchange Protocols. In: Proceedings of the 12th ACM Conference on Computer and Communications Security (CCS 2005), pp. 180–189 (2005)
13. Nam, J., Kim, M., Paik, J., Jeon, W., Lee, B., Won, D.: Cryptanalysis of a Group Key Transfer Protocol based on Secret Sharing. In: Kim, T.-h., Adeli, H., Slezak, D., Sandnes, F.E., Song, X., Chung, K.-i., Arnett, K.P. (eds.) FGIT 2011. LNCS, vol. 7105, pp. 309–315. Springer, Heidelberg (2011)
14. Olimid, R.F.: On the Security of an Authenticated Group Key Transfer Protocol Based on Secret Sharing. In: Mustofa, K., Neuhold, E.J., Tjoa, A.M., Weippl, E., You, I. (eds.) ICT-EurAsia 2013. LNCS, vol. 7804, pp. 399–408. Springer, Heidelberg (2013)
15. Olimid, R.F.: Cryptanalysis of a Password-based Group Key Exchange Protocol Using Secret Sharing. Appl. Math. Inf. Sci. 7(4), 1585–1590 (2013)
16. Pieprzyk, J., Li, C.H.: Multiparty Key Agreement Protocols. In: IEEE Proceedings - Computers and Digital Techniques, pp. 229–236 (2000)
17. Sáez, G.: Generation of Key Predistribution Schemes using Secret Sharing Schemes. Discrete Applied Mathematics 128(1), 239–249 (2003)
18. Shamir, A.: How to Share a Secret. Commun. ACM 22(11), 612–613 (1979)
19. Sun, Y., Wen, Q., Sun, H., Li, W., Jin, Z., Zhang, H.: An Authenticated Group Key Transfer Protocol based on Secret Sharing. Procedia Engineering 29, 403–408 (2012)
20. Yuan, W., Hu, L., Li, H., Chu, J.: An Efficient Password-based Group Key Exchange Protocol Using Secret Sharing. Appl. Math. Inf. Sci. 7(1), 145–150 (2013)
21. Zhao, J., Gu, D., Gorantla, M.C.: Stronger Security Model of Group Key Agreement. In: Proceedings of the 6th ACM Symposium on Information, Computer and Communications Security (ASIACCS 2011), pp. 435–440 (2011)

Analysis and Implementation of the SNOW 3G Generator Used in 4G/LTE Systems

J. Molina-Gil[1], P. Caballero-Gil[1], C. Caballero-Gil[1], and Amparo Fúster-Sabater[2]

[1] Department of Statistics, O.R. and Computing,
University of La Laguna, Spain
{jmmolina,pcaballe,ccabgil}@ull.es
[2] Institute of Applied Physics,
Spanish National Research Council, Madrid, Spain
amparo@iec.csic.es

Abstract. The fourth generation of cell phones, marketed as 4G/LTE (Long-Term Evolution) is being quickly adopted worldwide. Given the mobile and wireless nature of the involved communications, security is crucial. This paper includes both a theoretical study and a practical analysis of the SNOW 3G generator, included in such a standard for protecting confidentiality and integrity. From its implementation and performance evaluation in mobile devices, several conclusions about how to improve its efficiency are obtained.

Keywords: 4G/LTE encryption, stream cipher, SNOW 3G.

1 Introduction

The large increase of mobile data use and the emergence of broadband demanding applications and services are the main motivations for the proposal of progressive substitution of 3G/UMTS by 4G/LTE technology. Nowadays, commercial LTE networks have been launched in many countries. In particular they include: four countries of Africa, between 2012 and 2013; eleven countries of America, including USA from 2010; nineteen countries in Asia, where Japan was the main technology promoter; twenty-nine countries in Europe, excluding Spain even though being one of the largest countries; and two countries in Oceania.

In general, each evolution of telecommunications systems has involved the improvement of security features thanks to the learning from weaknesses and attacks suffered by their predecessors. Regarding the encryption systems used to protect confidentiality in mobile phone conversations, the evolution has been the following. First, the stream cipher A5/1 and its A5/2 version were developed for the 2G/GSM cell phone standard. Serious weaknesses in both ciphers were identified so the encryption system listed in the 3G/UMTS standard substituted them by a completely different scheme, the Kasumi block cipher. In 2010, Kasumi was broken with very modest computational resources. Consequently, again the encryption system was changed in the new standard 4G/LTE, where the stream cipher SNOW 3G is used for protecting confidentiality and integrity.

Á. Herrero et al. (eds.), *International Joint Conference SOCO'13-CISIS'13-ICEUTE'13*, 499
Advances in Intelligent Systems and Computing 239,
DOI: 10.1007/978-3-319-01854-6_51, © Springer International Publishing Switzerland 2014

The main issue of this work is the practical analysis of the SNOW 3G generator, which is the core of both the confidentiality algorithm UEA2 and the integrity algorithm UIA2, published in 2006 by the 3GPP Task Force [1]. Its choice allows higher speed data rates in cell phones thanks to its efficiency when implemented in devices with limited resources. The theoretical analysis of the security level of the SNOW 3G is out of the scope of this paper.

This work is organized as follows. A brief discussion on related work is included in Section 2. Then, Section 3 introduces the main concepts and notations used throughout this work, together with a theoretical description of the SNOW 3G generator. Section 4 gives some details of the implementation carried out in the iPhone Operating System (iOS), and its performance evaluation. Finally, Section 5 closes this paper with some conclusions and future work.

2   Related Work

The predecessors of SNOW 3G are SNOW 1.0 [2] and SNOW 2.0 [3].

The original version, SNOW 1.0, was submitted to the NESSIE project, but soon a few attacks were reported. One of the first published attacks was a key recovery requiring a known output sequence of length 2^{95}, with expected complexity 2^{224} [4]. Another cryptanalysis was a distinguishing attack [5], also requiring a known output sequence of length 2^{95} and about the same complexity.

Those and other attacks demonstrated some weaknesses in the design of SNOW 1.0, so a more secure version called SNOW 2.0, was proposed. SNOW 2.0 is nowadays one of two stream ciphers chosen for the ISO/IEC standard IS 18033-4 [6]. Also, SNOW 2.0 uses similar design principles to the stream cipher called SOSEMANUK, which is one of the final four Profile 1 (software) ciphers selected for the eSTREAM Portfolio [7].

Afterwards, during its evaluation by the European Telecommunications Standards Institute (ETSI), the design of SNOW 2.0 was further modified to increase its resistance against algebraic attacks [8] with the result named SNOW 3G. Full evaluation of the design of SNOW 3G has not been made public, but a survey of it is given by ETSI in [9].

The designers and external reviewers show that SNOW 3G has remarkable resistance against linear distinguishing attacks [10, 11], but SNOW 3G have suffered other types of attacks. One of the first and simplest cryptanalytic attempts was the fault attack proposed in [12]. An approach to face that problem includes employing nonlinear error detecting codes. A cache-timing attack [13] on SNOW 3G, based on empirical timing data, allows recovering the full cipher state in seconds without the need of any known keystream. Such an attack is based on the fact that operations like the permutations and multiplications by the constant α and its inverse are actually implemented using lookup tables. The work [14] describes a study of the resynchronization mechanism of SNOW 3G using multiset collision attacks, showing a simple 13-round multiset distinguisher with complexity of 2^8 steps.

The SNOW 3G generator has been subject of a few review works [15, 16]. The present paper provides a new study, more focused on a practical view.

3   Theoretical Description of the SNOW 3G Generator

Stream ciphers are based on generators of pseudo-random keystream sequence whose bits are bitwise XORed with the plaintext in order to generate the ciphertext. The main advantage of stream ciphers is that they are lightweight and can operate at a high speed, making them extremely suitable for power-constrained devices such as mobile phones. The stream generator analysed in this work has a typical nonlinear structure based on a Linear Feedback Shift Register (LFSR).

The following terms and notation are used within this paper to describe the stream cipher SNOW 3G and its implementation:

$GF(2)=\{0,1\}$	Galois Field with two elements 0 and 1.
$GF(2)[x]$	Ring of polynomials in the variable x with coefficients in $GF(2)$.
d	Degree of a polynomial.
$p(x)$	Primitive polynomial of degree d in $GF(2)[x]$.
$GF(2^d)$	Extension field of $GF(2)$ defined by $p(x)$, with 2^d elements.
$GF(2^d)[x]$	Ring of polynomials in the variable x with coefficients in $GF(2^d)$.
$\beta \in GF(2^8)$	Root of the $GF(2)[x]$ polynomial $x^8 + x^7 + x^5 + x^3 + 1$.
$\alpha \in GF(2^{32})$	Root of the $GF(2^8)[x]$ polynomial $x^4 + \beta^{23}x^3 + \beta^{245}x^2 + \beta^{48}x + \beta^{239}$.
s_t	32-bit stage of an LFSR.
=	Assignment operator.
$\oplus$	Bitwise XOR operation.
$\boxplus$	Integer addition modulo 2^{32}.
$\parallel$	Concatenation of two operands.

As shown in Fig. 1, the SNOW 3G generator consists of two main components: an LFSR and a Finite State Machine (FSM).

The LFSR component has 16 stages s_0, s_1, s_2,..., s_{15}, each holding 32 bits. Its feedback is defined by a primitive polynomial over the finite field $GF(2^{32})$, and involves two multiplications, one by a constant $\alpha \in GF(2^{32})$ and another by its inverse, as described by the following relation:

$$s_{t+16} = \alpha\, s_t \oplus s_{t+2} \oplus \alpha^{-1}\, s_{t+11}, \text{ for } t \geq 0. \tag{1}$$

The FSM component constitutes the nonlinear part of the generator. The FSM involves two input data from the LFSR, which are the s_5 and s_{15} stages contents. The FSM is based on three 32-bit registers R1, R2 and R3, and two substitution boxes S1 and S2 that are used to update the registers R2 and R3. Both S-Boxes S1 and S2 map each 32-bit input to a 32-bit output by applying several combinations of a basic S-box on each one of the 4 bytes of the input. However, while box S1 is based on the AES (Advanced Encryption Standard) S-box, the basic S-box of S2 was specially designed for SNOW 3G. The mixing operations in the FSM are bitwise XOR operations and integer additions modulo 2^{32}.

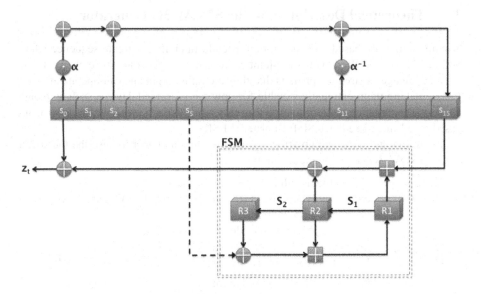

Fig. 1. SNOW 3G Generator

The clocking of the LFSR component of SNOW 3G has two different modes of operation, the initialisation mode and the keystream mode. On the one hand, when the initialisation is performed, the generator is clocked without producing any output. On the other hand, in the keystream mode, with every clock tick the generator produces a 32-bit word. Thus, SNOW 3G is a word-oriented generator that outputs a sequence of 32-bit words under the control of a 128-bit key and a 128-bit Initialization Vector IV.

Regarding the implementation of SNOW 3G, which is the main object of the following section, several observations can be done. First, the two multiplications involved in the LFSR can be implemented as a byte shift together with an unconditional XOR with one of 2^8 possible patterns, as shown below.

Since β is a root of the primitive polynomial $x^8 + x^7 + x^5 + x^3 + 1$, the extension field $GF(2^8)$ can be generated through successive powers of β so that $\{0, 1, \beta, \beta^2, \beta^3, ..., \beta^{2^8-2}\}$ is the entire field $GF(2^8)$. Thus, any element of $GF(2^8)$ can be represented either with a polynomial in $GF(2)[x]$ of degree less than 8, or with a byte whose bits correspond to the coefficients in such a polynomial. Operations in $GF(2^8)$ correspond to operations with polynomials modulo $x^8 + x^7 + x^5 + x^3 + 1$. This means that, in particular, the multiplication of two elements in $GF(2^8)$ results from the multiplication of the two corresponding polynomials, which is then divided by the polynomial $x^8 + x^7 + x^5 + x^3 + 1$, so that the remainder is the resulting output. The implementation of this operation as a binary multiplication is as follows. Considering both multiplier bytes, for each 1 bit in one of the multipliers, a number of left shifts are run on the other multiplier byte followed, every time the leftmost bit of the original byte before the shift is 1, by a conditional bitwise XOR with $A9_{16}=10101001_2$, which is the byte corresponding to the polynomial $x^8 + x^7 + x^5 + x^3 + 1$. The number of left shifts is given by the position of the 1 bit in the first multiplier.

Since α is a root of the primitive $GF(2^8)[x]$ polynomial $x^4 + \beta^{23}x^3 + \beta^{245}x^2 + \beta^{48}x + \beta^{239}$, the finite extension field $GF(2^{32})$ can be generated through successive powers of α so that $\{0, 1, \alpha, \alpha^2, \alpha^3, ..., \alpha^{2^{32}-2}\}$ is the entire field $GF(2^{32})$. Thus, we can represent any element of $GF(2^{32})$ either with a polynomial in $GF(2^8)[x]$ of degree less than 4, or with a word of 4 bytes corresponding to the 4 coefficients in such a polynomial. Operations in $GF(2^{32})$ correspond to operations with polynomials modulo $x^4 + \beta^{23}x^3 + \beta^{245}x^2 + \beta^{48}x + \beta^{239}$. This means that, in particular, the multiplication of α and any 4-byte word (c_3, c_2, c_1, c_0) in $GF(2^{32})$ results from the multiplication of x and the polynomial $c_3x^3 + c_2x^2 + c_1x + c_0$, which is then divided by the polynomial $x^4 + \beta^{23}x^3 + \beta^{245}x^2 + \beta^{48}x + \beta^{239}$, so that the resulting output is the remainder $(c_2+c_3\beta^{23}) x^3 + (c_1+c_3\beta^{245}) x^2 + (c_0+c_3\beta^{48}) x + c_3\beta^{239}$, or equivalently, the 4-byte word $(c_2+c_3\beta^{23}, c_1+c_3\beta^{245}, c_0+c_3\beta^{48}, c_3\beta^{239})$. Thus, a fast binary implementation of this operation can be based on precomputed tables $(c\beta^{23}, c\beta^{245}, c\beta^{48}, c\beta^{239})$, $\forall c \in GF(2^8)$. Similarly, the multiplication of α^{-1} and any 4-byte word (c_3, c_2, c_1, c_0) in $GF(2^{32})$ results from the multiplication of x^{-1} and the polynomial $c_3x^3 + c_2x^2 + c_1x + c_0$, which is $c_3x^2 + c_2x + c_1 + c_0 x^{-1}$. Since $xx^{-1}=1$ and $\beta^{255}=1$, x^{-1} can be expressed as $\beta^{255-239}x^3 + \beta^{255-239+23}x^2 + \beta^{255-239+245}x + \beta^{255-239+48} = \beta^{16}x^3 + \beta^{39}x^2 + \beta^6x + \beta^{64}$. Thus, the resulting output of the product is the remainder $(c_0\beta^{16})x^3+(c_3+c_0\beta^{39})x^2+(c_2+c_0\beta^6)x+(c_1+c_0\beta^{64})$, or equivalently, the 4-byte word $(c_0\beta^{16}, c_3+c_0\beta^{39}, c_2+c_0\beta^6, c_1+c_0\beta^{64})$. Thus, a fast binary implementation of this operation can be based on precomputed tables $(c\beta^{16}, c\beta^{39}, c\beta^6, c\beta^{64})$, $\forall c \in GF(2^8)$.

4 iOS Implementation and Evaluation

This work analyses a software implementation of SNOW 3G in cell phone platform. In particular, we have implemented it for iOS platform and the used programming language has been Objective C.

The first aspect we have taken into account is that LFSRs have been traditionally designed to operate over the binary Galois field $GF(2)$. This approach is appropriate for hardware implementations but its software efficiency is quite low. Since microprocessors of most cell phones have a word length of 32 bits, the LFSR implementation is expected to be more efficient for extended fields $GF(2^{32})$. Thus, since the implementation of SNOW 3G is over the finite field $GF(2^{32})$, it is more suitable for the architecture that supports current cell phones. The second aspect is related to arithmetic operations and specifically, the multiplication on extension fields of $GF(2)$ because the feedback function in SNOW 3G involves several additions and multiplications, and the multiplication is the most computationally expensive operation.

In this section, we study and compare different software implementation in order to find the optimal one for devices with limited resources, such as smartphones. We have performed several studies on an iPhone 3GS whose main characteristics are described in Table 1.

Table 1. Device used for the evaluation

iPhone 3GS			
Architecture	*CPU Frequency*	*Cache L1I/L1D/L2*	*RAM*
Armv7-A	600 MHz	16 Kb/16 Kb/256 Kb	256 MB

All the results shown in this work have been obtained using Instruments, which is a tool for analysis and testing of performance of OS X and iOS code. It is a flexible and powerful tool that lets track one or more processes and examine the collected data. The tests correspond to the average of 10 runs in which 10^7 bytes of keystream sequence are generated using the platform described above. Table 2 shows the total time (in milliseconds) for each SNOW 3G function explained below. The evidences indicate that the multiplication is the most expensive function. The second most expensive function is the shift register, which is performed in each clock pulse.

Below we study two different techniques to perform multiplications and several techniques for LFSR software implementation proposed in [11].

Table 2. Function Performance in Recursive Mode

Summary		
Function	*Time(ms)*	*%*
MULxPow	29054,9	92,88
ClockLFSRKeyStreamMode	572	1,77
DIValpha	356,6	1,1
main	264,7	0,8
MULalpha	326,8	0,99
GenerateKeystream	243,8	0,73
ClockFSM	180,3	0,54
S2	128,1	0,34
S1	129,9	0,37
Generator	1,3	0
Total Time	30258,5	

4.1   Multiplication

As shown in Table 2, according to the implementation proposed in [1] the most consuming time function in SNOW 3G is the MULxPow used in both the multiplication by α and by α^{-1}. Each multiplication can be implemented either as a series of recursive byte shifts plus additional XORs, or as a lookup table with precomputed results. In each clocking of the LFSR, the feedback polynomial uses two functions MUL_α and DIV_α which are defined as:

$$MUL_\alpha = MUL_X POW(c, 23, 0xA9) \,||\, MUL_X POW(c, 245, 0xA9)$$
$$MUL_X POW(c, 48, 0xA9) \,||\, MUL_X POW(c, 239, 0xA9)$$

$$DIV_\alpha = MUL_X POW(c, 16, 0xA9) \;||\; MUL_X POW(c, 39, 0xA9)$$
$$MUL_X POW(c, 6, 0xA9) \;||\; MUL_X POW(c, 64, 0xA9)$$

The first method might be more appropriate for systems with limited memory resources, as it does not require a large storage. However, as we can see in Table 2, it has a significant computational cost.

The second method involving precomputed tables provides optimal time results, as can be seen in Table 3. Indeed, it can be considered the fastest procedure for multiplication because it results in an improvement of 96% in time consumption with respect to the first recursive method. However, one of the biggest problems with this proposal could be the needed storage in devices with limited resources. In particular, for SNOW 3G, the table has 256 elements, each of 32 bits, what results in a total of 32*256 bits. Furthermore, the implementation uses the two functions MULα and DIVα, so it involves two tables, what means a total of 2048 bytes. Consequently, this method seems quite adequate for the characteristics of the chosen device.

Table 3. Function Performance With precomputed tables

Computational Cost		
Function	*Time(ms)*	*%*
ClockLFSRKeyStreamMode	347,3	28,69
main	277,2	22,35
ClockFSM	182,2	14,95
S1	146	12,01
S2	138	11,3
GenerateKeystream	107,3	8,84
Generator	1,3	0,04
Total Time	1199,4	

4.2   LFSR

The LFSR structures are difficult to implement efficiently in software. The main reason is the shift of each position during each clock pulse. This shift in hardware implementation occurs simultaneously, so the whole process can be performed in a single clock pulse. However, in software implementation, the process is iterative and costly.

As we saw in Table 3, once optimized the multiplication, it is the ClockLFSRKeyStreamMode function the most time consuming. Thus, we have used different software optimization techniques, proposed in [17] together with the hardcode technique presented in the specifications in order to improve the LFSR's final performance.

The hardcode method consists in embedding the data directly into the source code, instead of using loops or indices as the rest of techniques do. The cost of this proposal corresponds to 15 assignments. This technique, despite being longer, seems to require less time. Below is the implementation of this method.

```
void ClockLFSRKeyStreamMode()
{
  u32 v = ( ( (LFSR_S0 << 8) & 0xffffff00 ) ^
  ( MULalpha( (u8)((LFSR_S0>>24) & 0xff) ) ) ^
  ( LFSR_S2 ) ^
  ( (LFSR_S11 >> 8) & 0x00ffffff ) ^
  ( DIValpha( (u8)( ( LFSR_S11) & 0xff ) ) )
  );
  LFSR_S0 = LFSR_S1;
  LFSR_S1 = LFSR_S2;
  LFSR_S2 = LFSR_S3;
  LFSR_S3 = LFSR_S4;
  LFSR_S4 = LFSR_S5;
  LFSR_S5 = LFSR_S6;
  LFSR_S6 = LFSR_S7;
  LFSR_S7 = LFSR_S8;
  LFSR_S8 = LFSR_S9;
  LFSR_S9 = LFSR_S10;
  LFSR_S10 = LFSR_S11;
  LFSR_S11 = LFSR_S12;
  LFSR_S12 = LFSR_S13;
  LFSR_S13 = LFSR_S14;
  LFSR_S14 = LFSR_S15;
  LFSR_S15 = v;
}
```

The analysis carried out with the precomputed multiplication method involves an experiment to assess 10^7 bytes of the keystream generated by the LFSR proposed for SNOW 3G. The result values are summarized in Table 4, which shows the functions' time and the total implementation time.

The results show that the hardcode method is not the best implementation. Although it represents an 11% of improvement over the traditional method, the sliding windows method presents an improvement of 29% with respect to the traditional, and 20% compared to the hardcode method.

Table 4. Performance of Different LFSR Implementation Methods

Function	Traditional Time (ms)	HardCode Time (ms)	Circular Buffers Time (ms)	Sliding Windows Time (ms)	Loop Unrolling Time (ms)
ClockLFSRKeyStreamMode	491,1	342,5	834	184,1	291,3
Generator	1,4	1,3	1,3	1,8	1,1
GenerateKeystream	65,8	65,8	198,3	88,2	68,4
main	246,6	306,8	296,8	297	294,4
Total Time	804,9	716,4	1330,4	**571,1**	655,2

From the obtained results, we conclude that the circular buffer method is not applicable because the update of different indices involves modular arithmetic, which is not very efficient.

The new LFSR proposal can affect other SNOW 3G parts like the FSM. For this reason our main aim is to determine whether improving LFSR shift times negatively affects other code parts, and in that case to state the improvement that can be achieved. In order to do it, we have implemented in SNOW 3G with precomputed tables using sliding windows for shift register, Table 5 shows the summary of the results. If we compare them with the results of Table 3, it is clear that this implementation improves the time for the *ClockLFSRKeyStreamMode*, *S1* and *GenerateKey*. However, other functions like *ClockFSM, S2, GenerateKeystream* have increased slightly their values. The function with the worst time result is *S2*, as its value has increase 26% related to the previous proposal. Moreover, the greatest improvement has been in *ClockLFSRKeyStreamMode* function, with a 47%. All this results in an overall improvement of 10% compared to the implementation proposed in the specifications.

Table 5. Function Performance in optimized Mode

Computational Cost		
Function	*Time(ms)*	*%*
ClockLFSRKeyStreamMode	184,7	15,28
main	282,3	22,23
ClockFSM	195,4	17,68
S1	135,9	12,65
S2	163,4	16,33
GenerateKeystream	118,2	10,95
Generator	1,2	1,07
Total Time	1081,1	

5 Conclusions and Future Work

This paper has provided an analysis both from a theoretical and practical point of view, of the generator used for the protection of confidentiality and integrity in the 4G/LTE generation of mobile phones. In particular, after an introduction to the theoretical basis of the SNOW 3G generator, the implementation of the generator on the iOS mobile platform and several experiments have been carried out, obtaining from a comparison with similar software, several interesting conclusions on how to improve efficiency through the optimization of the software. Since this is an on-going work, there are still many open problems such as the analysis of other parameters not yet analyzed in this work, the implementation using different architectures and a comparative study. Also other future works are the proposal of a lightweight version of the SNOW 3G generator for devices with limited resources, and the analysis of several theoretical properties of the generator.

Acknowledgements. Research supported by Spanish MINECO and European FEDER Funds under projects TIN2011-25452 and IPT-2012-0585-370000, and the FPI scholarship BES-2009-016774.

References

1. ETSI/SAGE. Specification of the 3GPP Confidentiality and Integrity Algorithms UEA2 & UIA2. Document 2: SNOW 3G Specification, version 1.1 (September 2006), http://www.3gpp.org/ftp/
2. Ekdahl, P., Johansson, T.: SNOW - a new stream cipher. In: Proceedings of NESSIE Workshop (2000)
3. Ekdahl, P., Johansson, T.: A New Version of the Stream Cipher SNOW. In: Nyberg, K., Heys, H.M. (eds.) SAC 2002. LNCS, vol. 2595, pp. 47–61. Springer, Heidelberg (2003)
4. Hawkes, P., Rose, G.G.: Guess-and-determine attacks on SNOW. In: Nyberg, K., Heys, H.M. (eds.) SAC 2002. LNCS, vol. 2595, pp. 37–46. Springer, Heidelberg (2003)
5. Coppersmith, D., Halevi, S., Jutla, C.S.: Cryptanalysis of stream ciphers with linear masking. In: Yung, M. (ed.) CRYPTO 2002. LNCS, vol. 2442, pp. 515–532. Springer, Heidelberg (2002)
6. ISO/IEC 18033-4:2005. Information technology - Security techniques - Encryption algorithms - Part 4: Stream ciphers, http://www.iso.org/iso/home/store/catalogue_ics/
7. Berbain, C., Billet, O., Canteaut, A., Courtois, N., Gilbert, H., Gouget, A., Sibert, H.: Sosemanuk, a fast software-oriented stream cipher. In: eSTREAM, ECRYPT Stream Cipher. ECRYPT-Network of Excellence in Cryptology, Call for stream Cipher Primitives-Phase 2 (2005), http://www.ecrypt.eu.org/stream
8. Billet, O., Gilbert, H.: Resistance of SNOW 2.0 Against Algebraic Attacks. In: Menezes, A. (ed.) CT-RSA 2005. LNCS, vol. 3376, pp. 19–28. Springer, Heidelberg (2005)
9. ETSI/SAGE Technical report: Specification of the 3GPP Confidentiality and Integrity Algorithms UEA2 & UIA2. Document 5: Design and Evaluation Report, Version 1.1 (September 2006)
10. Nyberg, K., Wallén, J.: Improved Linear Distinguishers for SNOW 2.0. In: Robshaw, M. (ed.) FSE 2006. LNCS, vol. 4047, pp. 144–162. Springer, Heidelberg (2006)
11. Watanabe, D., Biryukov, A., De Canniere, C.: A Distinguishing Attack of SNOW 2.0 with Linear Masking Method. In: Matsui, M., Zuccherato, R.J. (eds.) SAC 2003. LNCS, vol. 3006, pp. 222–233. Springer, Heidelberg (2004)
12. Debraize, B., Corbella, I.M.: Fault Analysis of the Stream Cipher Snow 3G. In: Proceedings of Workshop on Fault Diagnosis and Tolerance in Cryptography, pp. 103–110 (2009)
13. Brumley, B.B., Hakala, R.M., Nyberg, K., Sovio, S.: Consecutive S-box Lookups: A Timing Attack on SNOW 3G. In: Soriano, M., Qing, S., López, J. (eds.) ICICS 2010. LNCS, vol. 6476, pp. 171–185. Springer, Heidelberg (2010)
14. Biryukov, A., Priemuth-Schmid, D., Zhang, B.: Multiset collision attacks on reduced-round SNOW 3G and SNOW 3G⊕. In: Zhou, J., Yung, M. (eds.) ACNS 2010. LNCS, vol. 6123, pp. 139–153. Springer, Heidelberg (2010)
15. Orhanou, G., El Hajji, S., Bentaleb, Y.: SNOW 3G stream cipher operation and complexity study. Contemporary Engineering Sciences 3(3), 97–111 (2010)
16. Kitsos, P., Selimis, G., Koufopavlou, O.: High performance ASIC implementation of the SNOW 3G stream cipher. In: IFIP/IEEE VLSI-SOC (2008)
17. Delgado-Mohatar, O., Fúster-Sabater, A.: Software Implementation of Linear Feedback Shift Registers over Extended Fields. In: Proceedings of CISIS/ICEUTE/SOCO Special Sessions, pp. 117–126 (2012)

Cryptanalysis of the Improvement of an Authentication Scheme Based on the Chinese Remainder Theorem for Multicast Communications

Alberto Peinado, Andrés Ortiz, and Guillermo Cotrina

Dept. Ingeniería de Comunicaciones,
E.T.S.I. Telecomunicación, Universidad de Málaga,
Campus de Teatinos, 29071 Málaga, Spain
{apeinado,aortiz}@ic.uma.es

Abstract. Recently, Antequera and López-Ramos have proposed an improvement on the secure multicast protocol based on the extended Euclidean algorithm in order to overcome the weaknesses and breaches detected by Peinado and Ortiz. The improvement defines a new authentication scheme based on the Chinese Remainder Theorem. However, we show in this work that the protocol is still vulnerable to impersonation attack due to the relationships between the authentication message of different multicast keys.

Keywords: Cryptanalysis, Key refreshment, Key Distribution, Multicast.

1   Introduction

In 2010, Naranjo et al proposed a key refreshment scheme [1] for multicast networks composed by several protocols. The security of that scheme relied mainly on the Extended Euclidean (EE) algorithm. The scheme was composed by the key refreshment procedure itself, an authentication protocol to verify the keys and a zero-knowledge protocol to authenticate nodes trying to detect illegal peers.

In 2011, Peinado and Ortiz presented a cryptanalysis [5] showing several weaknesses and vulnerabilities of the scheme [1] and its later applications [2], [3]. That cryptanalysis reveals three main breaches: a) the legal members can impersonate the Key server against the rest of users; b) the authentication protocol fails because forged refreshments are not detected; c) the secret keys (tickets) of the members can be recovered by other members using the zero-knowledge protocol originally proposed to detect illegal peers.

Furthermore, Peinado and Ortiz presented in [6] the practical cryptanalysis of the examples proposed in [1], using genetic algorithm, in which the secret keys are recovered by factorization of integers composed by 64 bits prime factors.

Antequera and López-Ramos, co-authors of the original scheme [1], have presented several improvements to overcome the vulnerabilities [7]. The main improvement resides on the modification of the authentication protocol. The new scheme employs one more key per user to apply the Chinese Remainder Theorem. However,

Á. Herrero et al. (eds.), *International Joint Conference SOCO'13-CISIS'13-ICEUTE'13*,
Advances in Intelligent Systems and Computing 239,
DOI: 10.1007/978-3-319-01854-6_52, © Springer International Publishing Switzerland 2014

we show in this work that the new authentication protocol does not avoid impersonation attacks, allowing dishonest users to cheat the rest of users.

The next section describes the key refreshment defined in [1]. Section 3 deals with the cryptanalysis in [5,6] showing the main weakness of the scheme. Section 4 describes the improved authentication protocol defined in [7] to detect forged refreshments. In section 5, the cryptanalysis of the new authentication mechanism is presented. Finally, conclusions appear in section 6.

2   Key Refreshment Scheme

The original key refreshment scheme is presented in this section. Let r be the symmetric encryption key to be multicast, and let n be the number of members at a given time. The process can be divided in several phases.

Phase 1. Member Ticket Assignment. When a member i enters the system, he joins the group and a member ticket x_i is assigned by the Key server. Every ticket x_i is a large prime and is transmitted to the corresponding member under a secure channel. It is important to note that this communication is performed once per member only, in such a way that the ticket x_i is used by the member i for the whole time he is in the group. Furthermore, all tickets must be different from each other.

Therefore, x_i is only known by its owner and the Key server, while r will be shared by all group members and the Key server.

Phase 2. Distribution. This phase is performed by several steps.

Step 1. The Key server selects the parameters of the system to generate the encryption key r. It selects:

- Two large prime numbers, m and p, such that p divides $m - 1$.
- $\delta < x_i$ for every $i = 1, ..., n$
- k such that $\delta = k + p$.
- g such that $g^p = 1 \bmod m$.

The encryption key r is computed as $r = g^k \bmod m$.

Step 2. The Key server calculates

$$L = \prod_{i=1}^{n} x_i \tag{1}$$

The parameter L is kept private in the Key server.

Step 3. The Key server computes u and v by means of the Extended Euclidean algorithm [4], such that

$$u \cdot \delta + v \cdot L = 1 \tag{2}$$

Step 4. The Key server multicasts g, m and u in plain text.

Step 5. Each member i calculates $\delta = u^{-1}$ mod x_i to obtain the encryption key r by means of the equation

$$g^\delta \bmod m = g^k \bmod m = r \tag{3}$$

Each refreshment of the encryption key r, implies the generation of new values for m, g, p and/or k. As a consequence, δ, u and v will be also refreshed.

Phase 3. Arrival of a Member j. In this case, the ticket x_j is assigned to member j by the Key server. Then, the ticket is included in the encryption key generation by means of equation 1, since the parameter L is the product of all the tickets. In this way, the encryption key r does not change, and thus the rest of the members do not need to refresh the key. The only operation the Key server must perform is a multiplication to obtain the new value for L.

Phase 4. Departure of a Member j. When a member leaves the group, he should not be able to decrypt contents anymore. Hence, a key refreshment is mandatory. This is achieved by dividing L by the ticket x_j, and generating a new encryption key afterwards. In this case, the new key must be distributed to every member using the procedure of phase 2.

Phase 5. Other Refreshments. For security reasons, the Key server might decide to refresh the encryption key r after a given period of time with no arrivals or departures. This refreshment is performed by the procedure of phase 2.

3   Vulnerability of the Key Refreshment Scheme

We recall here the main vulnerability of the original key refreshment, since the improved version presented in [7] does not modify substantially the scheme. Let us consider a legal member h which receives the key refreshment parameters (g, m and u) in a regular way. He performs the distribution phase, as described in section 2. Since the member h computes $\delta = u^{-1}$ mod x_h, he can obtain a multiple of L by the following equation

$$v \cdot L = 1 - u \cdot \delta \tag{4}$$

The knowledge of that multiple allows the member h to impersonate the server. The only thing he has to do is the following

Step 1. Member h generates a new value $\delta' < \delta$, and computes u' and v' applying the extended Euclidean algorithm, such that

$$u' \cdot \delta' + v' \cdot (v \cdot L) = 1 \tag{5}$$

Step 2. Member h sends g, m and u' to the other members. Those members will obtain the new value $\delta' = u^{r1} \bmod x_i$, and compute the refreshed key by equation 3. The effect of this fake refreshment is a malfunctioning of the system (DoS attack), since the legal members cannot identify legal refreshments.

Although the knowledge of a multiple of L is enough to impersonate the server, the member h could obtain the parameter L when a new refreshment arrives, as it is explained in detail in [5,6]. Hence, anyone who knows a multiple of L could generate arbitrary values m', p', g', k' and δ for the parameters and computes u'. This way the legal users agree on a fake key $r' = g'^{k'} \bmod m$.

It is important to note that the knowledge of a multiple of L does not allow the recovering of the tickets x_i and the exponent k.

4   Authentication Protocol

The authentication protocol proposed in [1] to verify the multicast key was broken in [5,6]. As a countermeasure, Antequera and López-Ramos have proposed in [7] a new scheme based on the Chinese Remainder Theorem. It is as follows.

Every legal user holds a ticket x_i as before and a new secret non-negative value $b_i < x_i$. The session key to be distributed is $r = g^k \bmod m$, since the key generation and distribution has not been modified. The authentication protocol is divided into two steps.

Step 1. The Key server computes $s = (g^k)^{-1} \bmod L$. Then the server chooses a random number a, such that $a < x_i$, for every x_i, and computes $h(a)$, for h a hash function, as in [1]. Finally, the server solves the system of congruences $x = a \cdot s + b_i \bmod x_i$, $i=1,\ldots,n$, getting a solution S.

The authenticating message corresponding to the key r is $(S,h(a))$.

Step 2. Every member i receives the authentication message and computes the value

$$S_i = S - b_i \bmod x_i \tag{6}$$

Then she verifies the equation

$$h(S_i \cdot g^k \bmod x_i) = h(a). \tag{7}$$

If the equation holds true, then the r is considered authentic.

5   Breaking the Authentication Protocol

As it is described in section 3, the main weakness of the key refreshment scheme proposed in [1] resides on the fact that any user could impersonate the server, generating forged refreshment messages, from the knowledge of a multiple of L. The protocol presented in [7] still suffers from the same weakness because the key generation and distribution phases have not been modified (see Steps 1-3 in section 2). So, every legal member knows u, δ and is able to compute a multiple $v{\cdot}L$. With these parameters any user could impersonate the server and generate a fake key, following the steps described in section 3.

The new authentication protocol proposed in [7] has been designed to avoid the server impersonation attack. This is one of the most relevant modifications presented in [7] of the original scheme in [1]. However, we show that the new authentication protocol is also insecure. To do so, we consider that the attack will be performed by a legal member of the system based on the following facts.

Fact 1. The verification equation defined in (7) to check the validity of the authentication message received by the users presents an inconsistency. The users are not able to compute g^k, since k is kept secret by the Key Server, and consequently the verification process could not be performed. Instead, we rewrite equation (7) as

$$h(S_i{\cdot}r \bmod x_i) = h(a) \tag{8}$$

where $r = g^k \bmod m$ is the multicast key computed by the users. Otherwise, the authentication protocol in [7] could not be performed.

Fact 2. Each legal member knows u, δ and is able to compute a multiple $v{\cdot}L$ of L.

Fact 3. Parameters S and $h(a)$ are public. Those parameters constitute the authentication message of a given secret key r (see Step 1 in section 4).

Fact 4. Each legal member knows a, since he has to compute it to verify the authentication message as it is stated in equations (7) and (8) . Hence,

$$a = S_i{\cdot}r \bmod x_i = S_i{\cdot}(g^k \bmod m) \bmod x_i \tag{9}$$

Fact 5. Each legal member knows $g^k \bmod m$, but not the parameter k.

Taking into account those facts, the impersonation attack will be performed in two steps: the generation of the false key and the falsification of the authentication message corresponding to the false key.

5.1   Generation of the False Key

The attacker (a legal user j) computes a false key $r' = g^{k'} \bmod m$ related to a previous valid multicast key $r = g^k \bmod m$ in the following way.

Step 1. The attacker chooses a new value δ' such that

$$\delta' = \delta + \Delta \tag{10}$$

Note that from step 1 in section 2 this implies that $k' = k + \Delta$, although k is not known, since m and p are not modified.

Step 2. The attacker applies the extended Euclidean algorithm to compute u' and v' such that equation (5) is satisfied.

Step 3. The attacker impersonates the server and multicasts the message (u', g, m) as in the original protocol.

Step 4. The members of the system compute the key r' applying the regular equations. Let us consider user i. Then, $\delta' = u'^{-1} \bmod x_i$ and

$$r' = g^{\delta'} \bmod m = g^{\delta + \Delta} \bmod m = g^{k + \Delta} \bmod m = g^{k'} \bmod m \qquad (11)$$

5.2   Falsification of the Authentication Message

The attacker (a legal user j) generates the parameters S' and a' as follows

$$S' = S + \beta r^{r^1} \bmod v \cdot L \qquad (12)$$

$$a' = a\alpha + \beta, \qquad (13)$$

where β is an integer and α is also integer such that

$$r' = r\alpha \quad \bmod v \cdot L \qquad (14)$$

It is important to note that a' must be less than x_i for $i = 1, \dots, n$. The attacker does not know the tickets x_i. However, he knows the values of a, sent in previous authentications. Since a is always less than x_i for $i = 1, \dots, n$, the attacker uses the highest value of a (named a_{max}) instead of the unknown tickets as the upper bound of equation (13). In order to facilitate that condition, the attacker selects a small value of a to compute a'. Note that when the attacker selects a value of a, he is really selecting the tuple (a, S, r, g, m).

On the other hand, α always exist for any value of r and r'. However, the attacker has to check if α satisfied the upper bound imposed to equation (13).

Next, the attacker broadcasts S' and $h(a')$ impersonating the server.

Finally, the users verify the key r' using the authentication message $(S', h(a'))$ applying the equation (8)

$$h((S' - b_i)r' \bmod x_i) = h((S + \beta r^{r^1} - b_i)r' \bmod x_i)$$

$$= h((S - b_i)r' + \beta r^{r^1}r' \bmod x_i)$$

$$= h(asr\alpha + \beta \bmod x_i)$$

$$= h(a\alpha + \beta \bmod x_i)$$

$$= h(a') \qquad (15)$$

As one can observe from equations (12) and (13), there exists a relationship between all authentication messages corresponding to every distinct key. The reason resides upon the fact that any two distinct keys $r \neq r'$ are related by $\alpha \bmod vL$.

A very important result can be derived from the case $\beta = 0$. If the attacker selects $\beta = 0$, then the authentication message corresponding to $r' \neq r$ can be computed from

$$S' = S \tag{16}$$

$$a' = a\alpha \tag{17}$$

This implies that the same value of S is used to authenticate every possible key such that $a\alpha < x_i$ for $i = 1, \ldots, n$, where $r' = r\alpha \bmod vL$.

6   Conclusions

We prove that the authentication protocol proposed in [7] to overcome the weaknesses detected in [1] is not secure. On one hand, the definition of the protocol presents an inconsistency turning it into non-implementable. Considering that the inconsistency could be derived from a typographical mistake, the protocol produces, for every multicast key, an authentication message which can be easily forged. That is, the attacker could select new false keys and compute the corresponding authentication message. The users would accept those keys as valid.

This attack is a server impersonation attack performed by a legal user. However, the information known by the users allow them to perform the attack later, when they have left the system. The effect of this attack is a desynchronization or denial of service.

On the other hand, we prove an important weakness derived from the fact that the authentication messages are related to each other. Hence, if the attacker knows the authentication message of a given key, he can compute the authentication message of any other key.

Acknowledgments. This work has been supported by the MICINN under project "TUERI: Technologies for secure and efficient wireless networks within the Internet of Things with applications to transport and logistics", TIN2011-25452.

References

1. Naranjo, J.A.M., López-Ramos, J.A., Casado, L.G.: Applications of the Extended Euclidean Algorithm to Privacy and Secure Communications. In: Proc. of 10th International Conference on Computational and Mathematical Methods in Science and Engineering (2010)
2. Naranjo, J.A.M., López-Ramos, J.A., Casado, L.G.: Key Refreshment in overlay networks: a centralized secure multicast scheme proposal. In: XXI Jornadas de Paralelismo, Valencia, Spain, pp. 931–938 (2010)
3. Naranjo, J.Á.M., Ramos, J.A.L., Casado, L.G.: A Key Distribution scheme for Live Streaming Multi-tree Overlays. In: Herrero, Á., Corchado, E., Redondo, C., Alonso, Á. (eds.) Computational Intelligence in Security for Information Systems 2010. AISC, vol. 85, pp. 223–230. Springer, Heidelberg (2010)
4. Menezes, A., Oorschot, P., Vanstone, S.: Handbook of applied cryptography. CRC Press (1996)

5. Peinado, A., Ortiz, A.: Cryptanalysis of Multicast protocols with Key Refreshment based on the Extended Euclidean Algorithm. In: Herrero, Á., Corchado, E. (eds.) CISIS 2011. LNCS, vol. 6694, pp. 177–182. Springer, Heidelberg (2011)
6. Peinado, A., Ortiz, A.: Cryptanalysis of a Key Refreshment Scheme for Multicast protocols by means of Genetic Algorithm. Logic Journal of the IGPL (August 2012), doi:10.1093/jigpal/jzs031
7. Antequera, N., López-Ramos, J.A.: Remarks and countermeasures on a cryptanalysis of a secure multicast protocol. In: Proc. 7th International Conference on Next Generation Web Services Practices, pp. 210–214 (2011)

Disclosure of Sensitive Information in the Virtual Learning Environment Moodle

Víctor Gayoso Martínez[1], Luis Hernández Encinas[1],
Ascensión Hernández Encinas[2], and Araceli Queiruga Dios[2]

[1] Information Security Institute (ISI)
Spanish National Research Council (CSIC), Madrid, Spain
{victor.gayoso, luis}@iec.csic.es
[2] Department of Applied Mathematics, E.T.S.I.I. of Béjar
University of Salamanca, Spain
{ascen, queirugadios}@usal.es

Abstract. In recent years, the use of Virtual Learning Environments (VLEs) has greatly increased. Due to the requirements stated by the Bologna process, many European universities are changing their education systems to new ones based on information and communication technologies. The use of web environments makes their security an important issue, which must be taken into full consideration. Services or assets of the e-learning systems must be protected from any threats to guarantee the confidentiality of users' data. In this contribution, we provide an initial overview of the most important attacks and countermeasures in Moodle, one of the most widely used VLEs, and then we focus on a type of attack that allows illegitimate users to obtain the username and password of other users when making a course backup in specific versions of Moodle. In order to illustrate this information we provide the details of a real attack in a Moodle 1.9.2 installation.

Keywords: Cryptography, Learning Environment, Moodle, Secure Communications, Web Security

1 Introduction

The educational model in many European universities is changing to a new qualification system in accordance with the proposals of the Bologna Process. This Process aims to create since 2010 a European Higher Education Area (EHEA), and it is based on cooperation between ministries, higher education institutions, and students from 47 European countries [1]. The Bologna agreement has supposed a lot of changes in higher education, more specifically in the quality assurance, the contents of the subjects, the duration of the bachelor's degree, and the education model itself.

Universities have started the implementation of the new teaching-learning techniques moving from a blackboard-based education to a computer-based one, and from a teaching system, where teachers were basically lecturers, to a new

Á. Herrero et al. (eds.), *International Joint Conference SOCO'13-CISIS'13-ICEUTE'13*,   517
Advances in Intelligent Systems and Computing 239,
DOI: 10.1007/978-3-319-01854-6_53, © Springer International Publishing Switzerland 2014

teaching-learning system, where students must plan their education and training carefully. In this sense, the use of computers and online platforms becomes an essential tool for the student [2].

There are several platforms in the market that offer the features needed for on-line education. Sometimes they are used as the sole instrument for teaching, but in other cases they are used to support courses combining online and traditional classes [3]. These type of products are usually known as Course Management System (CMS), Learning Management System (LMS) or Virtual Learning Environment (VLE). In particular, a report about the VLEs used by the Spanish universities, published in 2009, includes a comparative analysis of the situation regarding online platforms in each university [4]. In that report, it is stated that 55.55% of the universities used Moodle as their corporate platform for classes in 2008, 30% of the universities used no platform, and the rest used Dokeos, Sakai, LRN, and Illias in different small percentages.

Moodle (Modular Object-Oriented Dynamic Learning Environment) is an e-learning software platform whose first version was released in August 2002 [5]. Moodle was originally developed by Martin Dougiamas to help educators create online courses with a focus on interaction and collaborative construction of contents, and is provided freely as open source software under the GNU General Public License. As of April 2013, it had a user base of almost 80,000 registered sites across 232 countries [6], serving more than 7 million courses. In addition to the standard features of Moodle, developers can extend its functionality by creating specific plugins [7]. Due to all its features, Moodle is probably the most popular platform for online education.

Moodle can be run on Windows, Mac, and Linux operating systems. The most widely used version of Moodle is 1.9.x, and the latest stable version as of April 2013 is 2.4.3. Moodle is one of the most used platforms in the world, specially in USA, Spain, and Brazil [6]. For this reason, its security is a crucial aspect which must be analyzed in depth.

The information stored at the Moodle platform includes elements such as user profiles, students' results and grades, examination and assessment questions, discussion forums contents, assignments submitted by the students, news and announcements, wikis, etc. This information is critical, and so it must be protected accordingly to its importance.

After presenting several security aspects related to the use of Moodle, this contribution focuses in one of the most important threats, the information disclosure. This disclosure is even more serious in education institutions where users access several services and information (e-mail, payroll data, student's grades) with the same password that they use in Moodle.

In particular, we have analyzed the possibility of obtaining the usernames and passwords of users in specific versions of Moodle which are currently installed in many educational centres and institutions. In fact, we prove that is very easy to obtain such information and get access, impersonating other users, to the system where those specific versions of Moodle are being used, unless the site administrators have taken the appropriate measures to avoid this risk.

The rest of the paper is organized as follows: In Section 2, we provide some information about the security of the VLEs in general and, more specifically, we describe some potential vulnerabilities of Moodle. A practical attack against some versions of Moodle is presented in Section 3. Finally, we summarize our conclusions in Section 4.

2   Moodle Security

The proliferation of VLEs and the use of computers as part of daily classes in most levels of the educational system implied the necessity of providing communications between users and protecting the information delivered through those communication channels. VLE systems allow to share information and data among all the users, and they often manage one-to-many and many-to-many communications, providing powerful capabilities for learning [8].

E-learning can be considered as a special form of e-business, where its digital content has to be distributed, maintained, and updated. Moreover, these contents have to be adequately protected from unauthorized use and modification. At the same time, they must be available to the students in a flexible way [9].

Discussions about how to protect web applications and the e-learning content from being used without permission, including the authentication system, and different types of availability, integrity, and confidentiality attacks, are presented in [10], [11], and [12].

Moodle, as any other VLE, is open to attacks if vulnerabilities are found and they are not revised by the developers [13, 14]. Fortunately, most of the vulnerabilities of Moodle have already been satisfactorily corrected with the latest versions. At the Moodle website, the history of each version can be found, as well as the new features included in them [15].

A particularization to Moodle security vulnerabilities using the AICA model (whose name refers to the concepts of Availability, Integrity, Confidentiality, and Authentication) is analyzed in [16]. After this analysis, the authors recommend some security and privacy protection mechanisms in order to minimize the vulnerabilities found. The suggestions about the setting configuration are intended for Moodle administrators, but there are no recommendations in that article for end-users. Given that Moodle is managed as a corporative tool, teachers are not typically allowed to change settings or to decide upon any other change that could make the VLE secure against external or internal threats.

From a broader point of view, some of the attacks against Moodle and the measures that must be taken into account in order to secure the virtual environment installation are presented in the following paragraphs [17].

An organization's servers provide a wide variety of services to internal and external users, and they typically store and process sensitive information for the organization. Servers are frequently targeted by attackers because of the value of their data and services [18]. Some of the practical countermeasures to the most common attacks on servers are: using firewalls, updating the operating system and software, and removing unnecessary software packages.

The term authentication refers to the means and processes used to corroborate the identity of a person, a computer terminal, a credit card, etc. Some of the threats related to the authentication phase in Moodle are the following: using weak passwords, maintaining the same password for a long time, undistinguishing authentication roles, and session hijacking.

Internet bots are software applications that run automated tasks applied to any content publicly available on the Internet. They usually perform tasks that are repetitive and trivial at a much higher speed than any human being can do. So, it is important to protect Moodle from unwanted search bots, against spam bots, and against brute force attacks.

In confidentiality attacks, the main purpose of an attacker is not data modification, but data access and dissemination. This flaw is based on the fact that sensitive information does not have an appropriate encryption.

Regarding this threat, a new solution, named VAST, was proposed in 2005. VAST uses large, structured files to improve the secure storage of secret data, and preserves both the confidentiality and integrity of the data [19]. The VAST storage system uses extremely large (e.g., terabyte-sized) files to store secret information. A secret is broken into shares distributed over a large file, so that no single portion of the file holds recoverable information.

3   A Practical Attack

Data backup and recovery are essential parts of any professional service operation, as they permit to recover information from a loss originated by either human error or infrastructure failure.

The backup procedure implemented in Moodle allows to save in a compressed file all the relevant information pertaining to a certain course. This is an important feature as teachers might need to move their courses to another server, backup the courses before an upgrade, or just protect the course contents against potential errors. This procedure is typically managed by administrators or by the teachers themselves, who can select the specific elements that will be included in the backup bundle by using a web menu.

The output of the backup process is a compressed file that, among other elements, includes the file `moodle_backup.xml` (or `moodle.xml`, depending on the version used). Though there is no official Moodle documentation about `moodle_backup.xml`, we will explain the structure and contents of this XML file.

The first level of information of the XML file is formed by the elements INFO, ROLES, and COURSE, whose meaning is described next:

- INFO includes the name of the backup file (NAME), the Moodle version and release (MOODLE_VERSION and MOODLE_RELEASE, respectively), the date of the backup (DATE), the backup tool version (BACKUP_VERSION) and release (BACKUP_RELEASE), the storage method for the backup (ZIP_METHOD), the URL of the site (ORIGINAL_WWWROOT), and general information about a set of elements included in the course such as assignments, quizzes, forums, lessons, resources, surveys, wikis, etc. (DETAILS), each one presented as a MOD element.

- ROLES is composed of several ROLE items. Each ROLE element gathers the information about the capabilities possessed by users of a certain type (e.g. teachers, students, etc.).
- COURSE is where the actual information about the course is located, and it is comprised of several elements: HEADER includes general information about the course (e.g. official name, starting date, etc.), BLOCKS informs about the different modules that can be accessed in the course web page, SECTION presents the information about several elements of the course (questionnaires, etc.), USERS includes important data about the users of the course (e-mail address, etc.), QUESTION_CATEGORIES includes information about the questions developed by the teacher that must be answered by the students, LOGS records the details of every user access (user ID, IP of the computer from which the user connected, etc.), GRADEBOOK stores the students' grades, and finally, MODULES includes information such as the resources (files uploaded by the teacher, etc.), and the posts published at the forums.

In order to illustrate the previous information, Figure 1 shows the folded tree with the XML structure of the moodle_backup.xml file of one of our courses.

Next, we will analyze in more detail the USERS module, which contains critical information that a potential hacker would like to obtain. Within USERS, information about each user of the course is stored inside a USER element.

Figure 2 presents the content related to one of the USER elements of the example file moodle_backup.xml that we have used as the target of the attack, and that has been obtained from a real Moodle 1.9.2 installation.

Among this information, we can find the following items:

- USERNAME: The name that uniquely identifies any user in a particular Moodle deployment (the actual value of the username used in our attack is not displayed for security reasons).
- PASSWORD: The hash of the user's access code computed with the MD5 algorithm.
- IDNUMBER: An identification number provided to the system for each user, for example, the national identity number, a corporate identification number, etc. (the actual value of this field is not displayed in this contribution for security reasons).
- FIRSTNAME: Given name of the user as registered in Moodle.
- LASTNAME: Family name of the user as registered in Moodle.
- EMAIL: User's e-mail address.

It is very important to point out that passwords are stored in Moodle as hashes processed by the MD5 algorithm, a 128-bit hash function designed by Ron Rivest in 1992 and published as a RFC (Request for Comments) document by the IETF (Internet Engineering Task Force) [20].

```
<?xml version="1.0" encoding="UTF-8" ?>
- <MOODLE_BACKUP>
  - <INFO>
      <NAME>copia_de_seguridad-teoria_de_codigos-20120131-1657.zip</NAME>
      <MOODLE_VERSION>2007101522</MOODLE_VERSION>
      <MOODLE_RELEASE>1.9.2+ (Build: 20080903)</MOODLE_RELEASE>
      <BACKUP_VERSION>2008030300</BACKUP_VERSION>
      <BACKUP_RELEASE>1.9</BACKUP_RELEASE>
      <DATE>1328025465</DATE>
      <ORIGINAL_WWWROOT>https://moodle.          </ORIGINAL_WWWROOT>
      <ZIP_METHOD>external</ZIP_METHOD>
    + <DETAILS>
    </INFO>
  - <ROLES>
    + <ROLE>
    </ROLES>
  - <COURSE>
    + <HEADER>
    - <BLOCKS>
    - <SECTIONS>
      + <SECTION>
      </SECTIONS>
    - <USERS>
      + <USER>
      </USERS>
    - <QUESTION_CATEGORIES>
      + <QUESTION_CATEGORY>
      </QUESTION_CATEGORIES>
    - <LOGS>
      + <LOG>
      </LOGS>
    - <GRADEBOOK>
      - <GRADE_CATEGORIES>
        + <GRADE_CATEGORY>
        </GRADE_CATEGORIES>
      - <GRADE_ITEMS>
        + <GRADE_ITEM>
        </GRADE_ITEMS>
      </GRADEBOOK>
    - <MODULES>
      + <MOD>
      </MODULES>
      <FORMATDATA />
    </COURSE>
</MOODLE_BACKUP>
```

Fig. 1. XML structure of the `moodle_backup.xml` file

We recall that hash functions do not encrypt data as they use no keys; in other words, they only determine a summary (digest) of the data [21].

Before Moodle v1.9.7 (November 2009), hashed passwords were automatically stored in the backup files of the courses (nevertheless, in the configuration it is possible to select items not to be included in the backup). Starting in version 1.9.7, this feature was disabled, so in the latest versions hashed passwords are not stored as part of the backup process. In fact, if a course is restored to a new site, users will need to reset their password the first time they connect. We have checked this problem in a Moodle 1.9.8 backup from another education centre and, as it was expected, the hashed passwords are not included in the backup file.

```
<USER>
    <ID>751</ID>
    <AUTH>ldap</AUTH>
    <CONFIRMED>1</CONFIRMED>
    <POLICYAGREED>0</POLICYAGREED>
    <DELETED>0</DELETED>
    <USERNAME>            </USERNAME>
    <PASSWORD>ef73781effc5774100f87fe2f437a435</PASSWORD>
    <IDNUMBER>            </IDNUMBER>
    <FIRSTNAME>            </FIRSTNAME>
    <LASTNAME>            </LASTNAME>
    <EMAIL>                  </EMAIL>
    <EMAILSTOP>0</EMAILSTOP>
    <PHONE1>2223</PHONE1>
    <PHONE2>C1S12</PHONE2>
    <INSTITUTION>NIF</INSTITUTION>
    <DEPARTMENT>Matemática Aplicada</DEPARTMENT>
    <ADDRESS>                      </ADDRESS>
    <CITY>        </CITY>
    <COUNTRY>ES</COUNTRY>
    <LANG>es_utf8</LANG>
    <TIMEZONE>99</TIMEZONE>
    <FIRSTACCESS>0</FIRSTACCESS>
    <LASTACCESS>1328025540</LASTACCESS>
    <LASTLOGIN>1328006894</LASTLOGIN>
    <CURRENTLOGIN>1328023518</CURRENTLOGIN>
    <LASTIP>85.59.198.105</LASTIP>
</USER>
```

Fig. 2. User information in the `moodle_backup.xml` file used in the attack

The problem with MD5 is that it has been proved that it is a weak algorithm. Several attacks have shown that nowadays it is not a secure choice for a hashing algorithm [22–24]. On the internet there exist several websites that can retrieve the original text or message related to a given MD5 hash, either directly as an online service (e.g. [25]), or as a downloadable binary application (e.g. hashcat [26], a free tool to recover plain text strings for a variety of hashing methods).

Almost 25% of Moodle versions deployed in education institutions are 1.8.x or previous [6]. Though no statistics are available about the exact version of Moodle that has been installed at the different registered sites, taking into account that the latest 1.9.x version was 1.9.18, and that the Moodle versions vulnerable to this problem are versions previous to 1.9.7, it is reasonable to expect that a proportion of 1.9.x sites are vulnerable, though with the available data it is not possible to provide an exact figure. In summary, with the data publicly available at [6], it can be stated that the percentage of affected sites could range from 25% to almost 80%. Figure 3 shows the distribution of all Moodle registrations by version as of April 2013.

Going back to the presented attack, Table 1 and Table 2 show the information about two different users whose details are stored in the `moodle_backup.xml` target file.

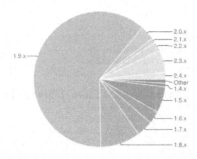

Fig. 3. Moodle registrations by version (source: Moodle website)

Table 1. Data from user 1 retrieved from the `moodle_backup.xml` target file

Backup data	User 1 information
USERNAME	X12345 (fictitious)
PASSWORD	`ef73781effc5774100f87fe2f437a435`
FIRSTNAME	Alice (fictitious)
LASTNAME	Smith (fictitious)
EMAIL	alice@edu.com (fictitious)

Table 2. Data from user 2 retrieved from the `moodle_backup.xml` target file

Backup data	User 2 information
USERNAME	X54321 (fictitious)
PASSWORD	`8fe87226333c05d4996c46a0d4165cb7`
FIRSTNAME	Bob (fictitious)
LASTNAME	Smith (fictitious)
EMAIL	bob@edu.com (fictitious)

If we take the passwords and feed one of the MD5 cracking tools with that information, we will obtain almost immediately the actual passwords. Table 3 present the real passwords recovered using this method.

Table 3. Actual passwords retrieved with a MD5 cracking tool

Password	Decrypted
`ef73781effc5774100f87fe2f437a435`	1234abcd
`8fe87226333c05d4996c46a0d4165cb7`	maria08

In other words, in general, each teacher in a Moodle course is able to create a backup file that will include all the previously mentioned information from other teachers involved in the course and from the students attending the course.

As a result of this research, we have proved that the possibility of accessing the backup files of some versions of Moodle permits an attacker to obtain the MD5 hashed passwords, which is a very important weakness. In fact, such information allows the attacker to disclose the password of any user in a very simple way.

This risk is even greater in education institutions where users have the same password not only to login into Moodle, but also to login into the e-mail service, to get the payroll data, or to grade the students' work, to name just a few examples.

4   Conclusions

One of the most widely used e-learning platforms, particularly in North American, Spanish, and Brazilian educational centres, is the Virtual Learning Environment Moodle. This virtual campus allows online interactions between teachers and students, either as a complement of traditional education or as the teaching tool for distance learning.

This type of services through the Internet must be completely secure, since the information exchanged between users is usually confidential, and the data stored at the platform can be used to validate the students' grades.

In this paper we have analyzed the security of Moodle regarding a specific type of data storage attack which can lead to valid authentications from illegitimate users. The example that illustrates this attack shows that it is very easy for a teacher to create a backup file from a course and obtain the username and password of other users.

As a result of that research, we have proved that a significant proportion of Moodle installations are vulnerable to this attack, so it is of paramount importance that Moodle administrators update their installations at least to version 1.9.7, or that they configure the backup file in order to avoid sensible information be included in it.

Acknowledgment. This work has been partially supported by Ministerio de Ciencia e Innovación (Spain) under the grant TIN2011-22668, and by Fundación Memoria D. Samuel Solórzano Barruso under the project FS/19-2011.

References

1. EHEA: European Higher Education Area website 2010–2020 (2010), http://www.ehea.info.
2. González, J., Jover, L., Cobo, E., Muño, P.: A web-based learning tool improves student performance in statistics: A randomized masked trial. Computers & Education 55(2), 704–713 (2010)
3. McCray, G.: The hybrid course, merging on-line instruction and the traditional classroom. Inform. Tech. Managem. 1(4), 307–327 (2000)
4. Prendes Espinosa, M.: Plataformas de campus virtual de software libre. Análisis comparativo de la situación actual en las universidades españolas (2009)

5. Moodle: Moodle.org, About (2012), http://moodle.org/about/
6. Moodle: Moodle.org, Moodle Statistics (2012), http://moodle.org/stats/
7. Gutiérrez, E., Trenas, M., Ramos, J., Corbera, F., Romero, S.: A new Moodle module supporting automatic verification of VHDL-based assignments. Computers & Education 54(2), 562–577 (2010)
8. Luminita, D.: Information security in e-learning platforms. Procedia-Social and Behavioral Sciences 15(15), 2689–2693 (2011)
9. Zamzuri, Z.F., Manaf, M., Ahmad, A., Yunus, Y.: Computer security threats towards the e-learning system assets. In: Zain, J.M., Wan Mohd, W.M.B., El-Qawasmeh, E. (eds.) ICSECS 2011, Part II. CCIS, vol. 180, pp. 335–345. Springer, Heidelberg (2011)
10. Nickolova, M., Nickolov, E.: Threat model for user security in e-learning systems. Int. J. Inform. Tech. Knowledge 1, 341–347 (2007)
11. Bradbury, D.: The dangers of badly formed websites. Computer Fraud & Security, 12–14 (January 2012)
12. Scholte, T., Balzarotti, D., Kirda, E.: Have things changed now? An empirical study on input validation vulnerabilities in web applications. Computers & Security 31(3), 344–356 (2012)
13. Diaz, J., Arroyo, D., Rodriguez, F.B.: An approach for adapting Moodle into a secure infrastructure. In: Herrero, Á., Corchado, E. (eds.) CISIS 2011. LNCS, vol. 6694, pp. 214–221. Springer, Heidelberg (2011)
14. Kumar, S., Dutta, K.: Investigation on security in LMS Moodle. Int. J. Inform. Tech. Knowledge Managem. 4(1), 233–238 (2011)
15. Moodle: Moodle.org, Open-source community-based tools for learning (2012), http://moodle.org
16. Stapic, Z., Orehovacki, T., Danic, M.: Determination of optimal security settings for LMS Moodle. In: 31st MIPRO International Convention on Information Systems Security, pp. 84–89 (2008)
17. Miletić, D.: Moodle Security. Packt Publishing, Birmingham (2011)
18. NIST: Guide to General Server Security. National Institute of Standard and Technology, SP 800-123 (2008)
19. Dagon, D., Lee, W., Lipton, R.: Protecting secret data from insider attacks. In: S. Patrick, A., Yung, M. (eds.) FC 2005. LNCS, vol. 3570, pp. 16–30. Springer, Heidelberg (2005)
20. Rivest, R.: The MD5 message-digest algorithm. Technical Report RFC 1321, Internet Activities Board (1992)
21. Menezes, A.J., van Oorschot, P.C., Vanstone, S.A.: Handbook of Applied Cryptography. CRC Press, Inc., Boca Raton (1996)
22. Wang, X., Yu, H.: How to break MD5 and other hash functions. In: Cramer, R. (ed.) EUROCRYPT 2005. LNCS, vol. 3494, pp. 19–35. Springer, Heidelberg (2005)
23. Sotirov, A., Stevens, M., Appelbaum, J., Lenstra, A., Molnar, D., Osvik, D., de Weger, B.: MD5 considered harmful today. In: Announced at the 25th Chaos Communication Congress (2008)
24. Sasaki, Y., Aoki, K.: Finding preimages in full MD5 faster than exhaustive search. In: Joux, A. (ed.) EUROCRYPT 2009. LNCS, vol. 5479, pp. 134–152. Springer, Heidelberg (2009)
25. Forchino, L.: MD5 Decrypt online (2012), http://www.md5decrypt.org
26. Domains By Proxy: Hashcat-advanced password recovery (2012), http://hashcat.net

Using Identity-Based Cryptography
in Mobile Applications

V. Mora-Afonso and P. Caballero-Gil

Department of Statistics, O.R. and Computing, University of La Laguna, Spain
`alu3966@etsii.ull.es`, `pcaballe@ull.es`

Abstract. This work includes a review of two cases study of mobile applications that use Identity-Based Cryptography (IBC) to protect communications. It also describes a proposal of a new mobile application that combines the use of IBC for Wi-Fi or Bluetooth communication between smartphones, with the promising Near Field Communication (NFC) technology for secure authentication. The proposed scheme involves NFC pairing to establish as public key a piece of information linked to the device, such as the phone number, so that this information is then used in an IBC scheme for peer-to-peer communication. This is a work in progress, so the implementation of a prototype based on smartphones is still being improved.

Keywords: Identity-Based Cryptography, Mobile Application, Near Field Communication.

1   Introduction

The use of smartphones has been increasing rapidly worldwide for the last years so that they have overtaken computers. Moreover, the figures say that this tendency is to be accelerated in the next future [1]. Smartphones are nowadays used for everything: checking email, social networking, paying tickets, communication between peers, data sharing, etc. However, in general they have lower computing capabilities than a standard computer, as well as power and battery limitations. Thus, every operation or computation in a smartphone must be implemented taking into account these constraints, maximizing efficiency and memory usage in order to avoid possible problems. In spite of these difficulties, still secure communication mechanisms and protocols need to be provided in different layers and applications in order to offer security to end users in a transparent way so that they can find functional mobile applications both user-friendly and robust.

The so-called Identity-Based Cryptography (IBC) may be seen as an evolution of Public-Key Cryptography (PKC) because IBC gets rid of most of the problems of traditional Public-Key Infrastructures (PKIs), as they are usually related to certificates. Furthermore, it provides the same security level, but using shorter encryption keys and more efficient algorithms. In the past decade, IBC has been subject of intensive study, and many different encryption, signature, key agreement and signcryption schemes have been proposed [2, 3, 4].

Á. Herrero et al. (eds.), *International Joint Conference SOCO'13-CISIS'13-ICEUTE'13*,
Advances in Intelligent Systems and Computing 239,
DOI: 10.1007/978-3-319-01854-6_54, © Springer International Publishing Switzerland 2014

Near Field Communication (NFC) is a short-range high frequency wireless communication technology [5] that enables simple and safe interactions between electronic devices at a few centimeters, allowing consumers to perform contactless transactions, access digital content, and connect electronic devices with a single touch.

This paper aims to review some mobile applications based on mechanisms and protocols that use IBC to provide security, and to propose a new NFC-based mobile application for Bluetooth/Wi-Fi pairing for peer-to-peer communication whose security is based on IBC. The characteristics of these schemes perfectly fit with the demand and requirements of not only smartphones but also backend applications and servers in terms of simplicity, security, efficiency and scalability. In particular, two known proposals are here analyzed and discussed. The first one aims to protect information sharing among mobile devices in dynamic groups whereas the second one proposes a secure protocol for communications in social networks, both using Identity Based Cryptography as the main security mechanism.

The remainder of this paper is organized as follows. In Section 2 a brief introduction to the cryptographic primitives used throughout the document is included. Then, Section 3 reviews the two mobile applications chosen for being studied. Section 4 presents the proposed scheme and discusses its design, security and implementation. Finally, Section 5 concludes the paper, including some future work.

2   Background

Identity-Based Cryptography is a type of public-key cryptography where a public piece of information linked to a node is used as its public key. Such information may be an e-mail address, domain name, physical IP address, phone number, or a combination of any of them. Shamir described in [6] the first implementation of IBC. In particular, such a proposal is an identity-based signature that allows verifying digital signatures by using only public information such as the user's identifier. Shamir also proposed identity-based encryption, which appeared particularly attractive since there was no need to acquire an identity's public key prior to encryption. However, he was unable to come up with a concrete solution, and identity-based encryption remained an open problem for many years.

2.1   Identity-Based Encryption

Identity-Based Encryption (IBE) is a public cryptographic scheme where any piece of information linked to an identity can act as a valid public key. A Trusted Third Party (TTP) server, the so-called Private Key Generator (PKG), first stores a secret master key used to generate a set of public parameters and the corresponding users' private keys. After a user's registration, it receives the public parameters and its private key. Then, a secure communication channel can be established without involving the PKG. Boneh and Franklin proposed in [2] the first provable secure IBE scheme. Its security is based on the hardness of the Bilinear Diffie-Hellman Problem [7].

IBE schemes are usually divided into four main steps:

1   Setup: This phase is executed by the PKG just once in order to create the whole IBE environment. The master key is kept secret and used to obtain users' private keys, while the remaining system parameters are made public.
2   Extract: The PKG performs this phase when a user requests a private key. The verification of the authenticity of the requestor and the secure transport of the private key are problems with which IBE protocols do not deal.
3   Encrypt: This step is run by any user who wants to encrypt a message M in order to send the encrypted message C to a user whose public identity is ID.
4   Decrypt: Any user that receives an encrypted message C runs this phase to decrypt it using its private key d and recovering the original message M.

2.2   Hierarchical Identity-Based Encryption

Hierarchical Identity-Based Encryption (HIBE) may be seen as a generalization of IBE that reflects an organizational hierarchy. Thus, it is an organizational-hierarchy oriented generalization of IBE [8][9]. It lets a root PKG distribute the workload by delegating private key generation and identity authentication to lower-level PKGs, improving scalability. In this way, a possible disclosure of a domain PKG's secret does not compromise the secrets of higher-level PKGs.

2.3   Attribute-Based Encryption

Attribute-Based Encryption (ABE) is a type of public-key encryption in which a user's public key or a ciphertext is associated with a set of attributes (e.g. the country the user lives in, the kind of subscription the user has, etc.). This scheme provides complete access control on encrypted data by setting up policies and attributes on ciphertexts or keys. Two different types of ABE can be distinguished: Ciphertext Policies ABE (CP-ABE) and Key Policies ABE (KP-ABE). In CP-ABE the decryption of a ciphertext is only possible if the set of attributes of the user key matches the attributes of the ciphertext [10]. On the other hand, in KP-ABE, ciphertexts are labeled with sets of attributes, and private keys are associated with access structures that control which ciphertexts a user is able to decrypt [11].

3   Two Cases Study

Through the following description of two-cases study, it is demonstrated that the implementation of Identity-Based Cryptosystems, which have many advantages over traditional PKIs, is possible, practical and efficient not only in smartphones but in everyday applications such as social networks, securing users communications transparently.

3.1   Secure Information Sharing in Mobile Groups

The work [12] proposes an application for mobile devices that can be used to share information in groups thanks to a combination of HIBE and ABE, and describes its implementation for the Android operating system. In particular, it describes a protocol called Trusted Group-based Information Sharing (TGIS) to establish group-based trust relationships in order to share information within a group or among groups. TGIS defines a cryptographic access control scheme to provide secure communication among mobile devices of collaborators who belong to the same or different organizations, and/or with different ranks or privileges. It assumes that each device belongs to an organization that has implemented some hierarchical system, what allows deploying credentials that can be used as public keys to distribute group keys. On the other hand, for controlling information shared within a group, secure access control is based on attributes. With this application, groups can be created dynamically and by any user, who acts as group leader in charge of generating the corresponding private key for each group member. Thus, since these members can be from different organizations, the group leader in fact acts as group PKG.

An interesting use case of this application is in emergency situations such as an earthquake or a volcanic eruption because in this kind of environments many groups from very different types need to communicate and collaborate in order to reach a common goal. Police officers, emergency services and firefighters may have to share information, data, locations, etc. without making this knowledge publicly available.

Going into detail, TGIS protocol is divided into six phases.

1. The first stage, named Domain Setup, is when each user registers its device with an identity in its organization's hierarchical domain, in order to receive a HIBE private key from a PKG. The PKG makes public a set of parameters used to generate HIBE public keys from user identities, and keeps secret the master private key used to generate each user's HIBE private key.
2. The second stage, named Group Setup, starts when a group has to be created because peers need to share information. Then, one of these users becomes the group leader in order to generate a public/private key pair and a set of group attributes for the group and signs with its HIBE private key the generated public key and the group attributes so that other users can verify this message using the group leader's HIBE identity.
3. Once a group has been created, the third phase, named User Enrollment, begins. Whenever a user wants to join a group, the leader decides what privileges it will have, assigns privileges to the group set of attributes, creates the group private key for the new user and sends it securely using the new user's HIBE identity.
4. When a group has been created, information can be shared among members, so the fourth phase, called Intra-Group Communication, can be executed. Users can share data and decrypt them complying with the group access policies defined in the group setup phase, so that only group members with the proper permissions can access the shared information. When a user encrypts a message, it defines which attributes can decrypt it using the group public key and an access tree describing these policies.

5. Communication among different groups can be achieved through the fifth phase called Inter-Group Collaboration, where every group leader makes available the group public key and the set of attributes.
6. Finally, in order to achieve message authentication in the sixth defined phase, every message exchanged within a group or among groups must be signed using an identity-based signature scheme.

In order to generate trust in dynamic collaborative groups for exchanging information between mobile devices, the described work proposed a distributed security protocol based on HIBE to provide flexible and secure access control. Such a proposal was implemented and evaluated in Android phones, what proves that this type of cryptography may be used in different applications.

3.2   Secure Smartphone-Based Social Networking

Given the use of social networks nowadays and the lack of secure mechanisms to allow communication among users of different social providers, if a user's account gets compromised, all their contacts get indirectly compromised and could become victims of different attacks such as phishing. There are a few applications [13, 14, 15] that prevent users from being scammed, provide access control, etc., but usually current social networks security schemes assume third parties are honest-but-curious and so they are not as scalable as current social networks demand. In [16] a security-transparent smartphone-based solution to this problem is presented. This approach uses IBC as mechanism to protect data, which can be applied in any third-party social network provider. An Android implementation of such a scheme is presented as a mobile application using WeChat [17].

In every social network scheme usually many different parts are involved. They are: a social network provider, a PKG to which users connect in order to get their private keys based on their public ID (mostly phone number) and users, who send and/or receive data from other users through their smartphone (in which they can perform reasonable computing operations efficiently). In order to gain access control, before sending data, the user must encrypt them using the receiver's identity. On the other side of the channel, when the user receives data, they must get their private key from the PKG in order to be able to decrypt data.

Given that an adversary can compromise a server or a PKG, in the security model of this scheme, certain assumptions are made. A compromised social network provider might tamper the communication, inserting, deleting or impersonating any user. However, a compromised PKG is assumed to be honest, i.e. it has no intention of starting a Man-In-the-Middle Attack (MIMA).

The steps towards secured social networking in the described scheme are as follows. If two users A and B want to securely share data through the proposed application, when the third party social networking application is started, its Application Programming Interfaces (APIs) are hooked with the described method, allowing users to encrypt data transparently. Whenever the application quits, original APIs are re-established so that user A can start communicating, but before sending any data, a secure channel needs to be set up by generating a session key. In order to do so a protocol integrating IBE and Diffie-Hellman key exchange is used. Both users A and B can then check whether the process was right by encrypting the peers ID and received messages using the session key. If one of the users is offline, a session key cannot be established so the initiator sets the session key by their choice.

As aforementioned, this scheme is not secure against a MIMA attack initiated by a malicious PKG, but this is supposed in the work that it will not to happen. Thanks to the use of IBE, non-PKG adversaries cannot start a MIMA attack. Furthermore, thanks to the use of Diffie-Hellman protocol, they neither can retrieve the session key.

The scheme proposed in that work includes a concrete application in which the described scheme is used. It was implemented for Android smartphones, using WeChat, and based on the Pairing-Based Cryptography library [18].

4   Proposal of NFC Authentication for IBE Communication

Bluetooth is a great technology, but experience so far has shown some of its weaknesses, which should be solved for a better use. One of its main drawbacks is the pairing process [19], because pairing two unknown devices requires over ten seconds, what makes that the user loses time in this operation. Furthermore, such a procedure should be transparent, letting the users focus on what they want to do and not on painful side operations. In conclusion, Bluetooth pairing lacks user-friendliness. In this regard, NFC is a breath of fresh air because it can be used for Bluetooth secure pairing [5]. In the following we propose a scheme using NFC-based Bluetooth pairing for providing a communication channel between smartphones, which is more secure than others described in previous works based on E0 or E1 stream ciphers [20]. In particular, the scheme here described applies IBE using phone numbers as public identities to deploy secure communication between devices.

Let us suppose that two users A and B want to securely share information through the Bluetooth of their smartphones. Assuming the existence of a TTP playing the role of the PKG server, which can register them as users of the system and provide them with corresponding private keys matching their identities, our protocol is divided into two main phases, whose main characteristics are show in Figures 1 and 2.

1. Setup phase. As soon as users tap their mobile phones, the pairing process begins. Following the NFC forum recommendation, smartphones transparently share their Bluetooth addresses, device names, and any other relevant information needed for the pairing process in an NFC Data Exchange Format (NDEF) message according to the Extended Inquiry Response (EIR) format specified in the Bluetooth Core Specification. Another important piece of information, the phone numbers, is also shared within this phase because they will be used as public identities later on in the scheme. Once finished this initial data transmission, the Bluetooth channel is already set up so that any information can be exchanged between the devices.

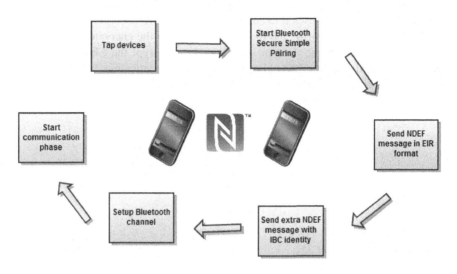

Fig. 1. Setup Phase

2. Communication phase. Once both devices have been paired, this stage begins. It is important to establish a real secure and trustworthy channel due to the known vulnerabilities of the Bluetooth cipher. IBC is a perfect approach to this problem so it is used in the proposal to guarantee confidentiality, integrity and authenticity. Depending on the size of the data to be shared, the initiating user can decide to send information directly by using an IBC or signcryption scheme, and/or other cryptographic schemes. For instance, in order to avoid a drastic increase in processing time, a session key is stated to be agreed for sharing large amounts of data through an efficient symmetric cipher such as AES by using an identity-based key agreement protocol like the ones used in [21, 22]. On the other hand, in order to prevent an increase in implementation complexity, an elegant but powerful Diffie-Hellman protocol is run by using the same IBC algorithm and by generating random keys through the inspection of data with smartphone sensors such as gyroscope, accelerometer, compass or GPS location, what allows obtaining enough randomness to establish a secure session key to be used in this communication phase.

The proposed protocol is simple, secure, transparent and extensible to be used with any other communication technologies such as Wi-Fi, so that the only necessary change is the NFC pairing detail during the setup phase.

In particular, the IBE algorithm used in the implementation of the communication phase basically follows the description of Boneh-Franklin (BF) scheme [2] based on bilinear pairings, with a few variations. Our implementation is based on a family of supersingular elliptic curves over finite fields F_p of large prime characteristic p (between 512-bit and 7680-bit), known as type-1 curves. In the designed IBE, each user calculates its public key from its phone number, and the PKG calculates the corresponding private key from that public key by using its master key, as explained in the aforementioned extract step. In the encrypt step, the sender A chooses a Content En-cryption Key (CEK) and uses it to encrypt the message M, encrypts the CEK with the public key of the recipient B, and then sends to B both the encrypted message C and the encrypted CEK. In the decrypt step, B uses to decrypt the CEK its private key securely obtained from the PKG after a successful authentication, which is then used to decrypt the encrypted message C.

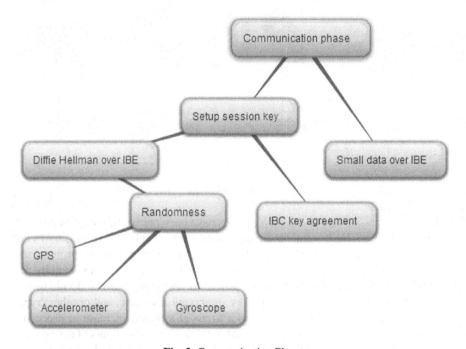

Fig. 2. Communication Phase

Regarding the elliptic curves used in the IBE schemes, the implementation of the application is based on type-1 curves E of the form $y^2=x^3+1$ defined over F_p for primes p congruent to 11 modulo 12, mainly because these curves can be easily gen-erated at random. The algorithm uses both the group $E(F_p)$ of points (x,y) satisfying the curve equation with affine coordinates x,y in F_p, and the group $E(F_p^2)$ of points (x,y) satisfying the curve equation with affine coordinates x,y in F_p^2, and with corres-ponding Jacobian projective coordinates X,Y,Z also in F_p^2.

Under the aforementioned conditions, the Tate pairing e takes as input two points P and Q in $E(F_p)$, and produces as output an integer in F_p^2. Thus, the Tate pairing is the core of the implemented IBE because it is a map that is hard to invert, but efficiently computable using Miller's algorithm and Solinas primes, thanks to its linearity in both arguments. However, when implementing it with smartphones, we found that al-though pairing in the projective system is faster than in the affine system, the cost is still very high in some cases, so pre-computation of some intermediate results, such as lambda for the sum of points, was used as solution to speed-up the whole process.

According to the implemented BF IBE, during the setup stage the PKG computes the master key as an integer s, and the public system parameters such as a prime p, a hash function h, the elliptic curve E, a point P in $E(F_p)$ and another point obtained by multiplying s times the point P. Afterwards, from the set of public parameters each public key is obtained by each user with phone number ID as a point Q_{ID} in $E(F_p)$. During the extract step, the PKG returns to each applicant user with phone number ID, its private key in the form of another point S_{ID} in $E(F_p)$. In this way, the encryption of a message M with the public key of a receiver whose phone number is ID, involves both a multiple rP of the point P (being r a randomly chosen integer) and the XOR of M and the hash of the r-power of the pairing result on both the public key Q_{ID} and the public identity ID. In this way, the decryption of C is possible by adding to its second element, the hash of the pairing result of the private key and its first element.

The implementation of the proposed scheme has taken advantage of many primitives included in the pairing-based cryptographic library [8] because it is fully functional and ready-to-use. A few details concerning the technical features of the preliminary implementation of the proposed application are provided below.

The first implementation has been in the Windows Phone platform [23], and in particular in Windows Phone 8. We have used the Windows Phone 8 SDK, together with Visual Studio Ultimate 2012. All tests have been run either in the WP emulator or in a Nokia Lumia 920 smartphone. The specifications for the used smartphone are: processor is Qualcomm Snapdragon™ S4, processor type is Dual-core 1.5 GHz, 1 GB of RAM, and Bluetooth 3.0. The external libraries used in the Windows implementation have been: 32feet.NET.Phone, which is a shared source library to simplify Bluetooth development; and Bouncy Castle, which is a lightweight cryptography API for Java and C#. The average times obtained with the Nokia device are shown below:

Message size (bytes)	Time to encrypt	Time to decrypt
128	7497,198 ms	7368,289 ms
512	7498,221 ms	6998,858 ms

The implementation of the prototype is still being improved but, as shown above, the preliminary results obtained till now are promising.

5   Conclusions and Future Work

The general objective of this work has been the development of new ways to secure direct communications between smartphones, keeping them simple, efficient, energy-saving, functional and practical. Identity-Based Cryptography is perfect for that goal. Thus, the particular main aim of this paper has been to review known IBC-based protocols for mobile devices, in order to propose a new Bluetooth/Wi-Fi scheme using NFC for pairing and IBE for communications. Current work in progress is aimed at implementing the proposal in different testing environments and platforms, and at doing an in-depth analysis of its security. Our future research will focus on developing extensions such as signcryption and practical applications.

Acknowledgements. Research supported by Spanish MINECO and European FEDER Funds under projects TIN2011-25452 and IPT-2012-0585-370000.

References

1. Anson, A.: Smartphone Usage Statistics (2012), http://ansonalex.com/infographics/smartphone-usage-statistics-2012-infographic
2. Boneh, D., Franklin, M.: Identity-Based Encryption from the Weil Pairing. In: Kilian, J. (ed.) CRYPTO 2001. LNCS, vol. 2139, pp. 213–229. Springer, Heidelberg (2001)
3. Hess, F.: Efficient identity based signature schemes based on pairings. In: ACM SAC (2002)
4. Barreto, P.S.L.M., Libert, B., McCullagh, N., Quisquater, J.-J.: Efficient and provably-secure identity-based signatures and signcryption from bilinear maps. In: Roy, B. (ed.) ASIACRYPT 2005. LNCS, vol. 3788, pp. 515–532. Springer, Heidelberg (2005)
5. NFC forum, http://www.nfc-forum.org
6. Shamir, A.: Identity-Based Cryptosystems and Signature Schemes. In: Blakely, G.R., Chaum, D. (eds.) Advances in Cryptology - CRYPTO 1984. LNCS, vol. 196, pp. 47–53. Springer, Heidelberg (1985)
7. Boneh, D., Boyen, X., Shacham, H.: Short Group Signatures. In: Franklin, M. (ed.) CRYPTO 2004. LNCS, vol. 3152, pp. 41–55. Springer, Heidelberg (2004)
8. Gentry, C., Silverberg, A.: Hierarchical ID-based cryptography. In: Zheng, Y. (ed.) ASIACRYPT 2002. LNCS, vol. 2501, pp. 548–566. Springer, Heidelberg (2002)
9. Boneh, D., Boyen, X., Goh, E.-J.: Hierarchical identity based encryption with constant size ciphertext. In: Cramer, R. (ed.) EUROCRYPT 2005. LNCS, vol. 3494, pp. 440–456. Springer, Heidelberg (2005)
10. Bethencourt, J., Sahai, A., Waters, B.: Ciphertext-policy attribute based encryption. In: IEEE Symposium on S&P (2007)
11. Goyal, V., Pandey, O., Sahai, A., Waters, B.: Attribute-based encryption for fine-grained access control of encrypted data. In: ACM CCS (2006)
12. Chang, K., Zhang, X., Wang, G., Shin, K.G.: TGIS: Booting Trust for Secure Information Sharing in Dynamic Group Collaborations. In: IEEE PASSAT, pp. 1020–1025 (2011)
13. Jahid, S., Nilizadeh, S., Mittal, P., Borisov, N., Kapadia, A.: DECENT: A decentralized architecture for enforcing privacy in online social networks. In: IEEE PerCom, pp. 326–332 (2012)
14. Feldman, A.J., Blankstein, A., Freedman, M.J., Felten, E.W.: Social Networking with Frientegrity: Privacy and Integrity with an Untrusted Provider. USENIX Security, 647–662 (2012)
15. Rahman, M.S., Huang, T.-K., Madhyastha, H.V., Faloutsos, M.: Efficient and Scalable Software Detection in Online Social Networks. USENIX Security, 663–678 (2012)
16. Wu, Y., Zhao, Z., Wen, X.: Transparently Secure Smartphone-based Social Networking. In: IEEE WCNC (2013)
17. Tencent: WeChat, http://www.wechat.com
18. The Pairing Based Cryptography library, http://crypto.stanford.edu/pbc
19. Barnickel, J., Wang, J., Meyer, U.: Implementing an Attack on Bluetooth 2.1+ Secure Simple Pairing in Passkey Entry Mode. In: IEEE TrustCom, pp. 17–24 (2012)
20. Lu, Y., Vaudenay, S.: Cryptanalysis of Bluetooth Keystream Generator Two-Level E0. In: Lee, P.J. (ed.) ASIACRYPT 2004. LNCS, vol. 3329, pp. 483–499. Springer, Heidelberg (2004)
21. Zhang, L., Wu, Q., Qin, B., Domingo-Ferrer, J.: Provably secure one-round identity-based authenticated asymmetric group key agreement protocol. Information Sciences 181(19), 4318–4329 (2011)
22. Guo, H., Li, Z., Mu, Y., Zhang, X.: Provably secure identity based authenticated key agreement protocols with malicious private key generators. Information Sciences 181(3), 628–647 (2011)
23. NFC Secure Notes, http://www.windowsphone.com/s?appid=c72e51ce-fdda-452b-84c8-523cc27c76d5

SlowReq: A Weapon for Cyberwarfare Operations. Characteristics, Limits, Performance, Remediations

Maurizio Aiello, Gianluca Papaleo, and Enrico Cambiaso

National Research Council
CNR-IEIIT
Genoa, Italy
{maurizio.aiello,gianluca.papaleo,
enrico.cambiaso}@ieiit.cnr.it

Abstract. In the last years, with the advent of the Internet, cyberwarfare operations moved from the battlefield to the cyberspace, locally or remotely executing sabotage or espionage operations in order to weaken the enemy. Among the technologies and methods used during cyberwarfare actions, Denial of Service attacks are executed to reduce the availability of a particular service on a network. In this paper we present a Denial of Service tool that belongs to the Slow DoS Attacks category. We describe in detail the attack functioning and we compare the proposed threat with a similar one known as slowloris, showing the enhancements provided by the proposed tool.

Keywords: slow dos attack, denial of service, cyberwarfare.

1 Introduction

The Internet network is today the most important communication medium, since it provides a worldwide communication to billions of computers around the world. In the last years, the spread of devices able to connect to the Internet emphasized communication as a need for users. Because of this, the Internet has to be kept a safe place, protecting it from malicious operations.

In addition, *cyberwarfare* operations are radically changing modus-operandi, bringing the attack operations from the battle field to the cyberspace [1]. This increasing phenomenon executes politically motivated hacking to conduct military operations, such as sabotage or espionage, against an informative system owned by the adversary. It is known that governments are crossing this street, allowing them to become a more relevant threat without having to be a superpower, equipping their attack resources with computer and software weapons [2]. For instance, Stuxnet malware has been designed to affect Iranian nuclear facilities, silently spying and subverting industrial systems, causing real-world physical damage [3].

There are several technologies used for cyberwarfare operations and the choice of the technology is bounded to the operation the attacker has to accomplish. Among the available techniques, Denial of Service (DoS) attacks represent an interesting solution, since they consume the victim's resources with the malicious objective of

Á. Herrero et al. (eds.), *International Joint Conference SOCO'13-CISIS'13-ICEUTE'13*,
Advances in Intelligent Systems and Computing 239,
DOI: 10.1007/978-3-319-01854-6_55, © Springer International Publishing Switzerland 2014

reduce availability for a particular service to legitimate users. For example, during the Iranian presidential election in 2009, *slowloris* emerged as a prominent tool for Denial of Service attacks execution.

In the arena of DoS attacks, a first evolution of such menaces consisted in the amplification of the attack, to the so called Distributed Denial of Service (DDoS) attacks. A DDoS attack involves more attacking machines, which collaborate (voluntarily or not) to the same operation, thus causing a more relevant damage and hardening the attack mitigation.

Although there are several attacks/approaches which could lead to a DoS (i.e. physical, exploit, flooding, etc...), we are interested in analyzing Slow DoS Attacks (SDAs), which use low bandwidth rate to accomplish a Denial of Service. Since they could also affect the application layer of the victim host, SDAs are more difficult to counter, in relation to flooding DoS attacks. Indeed, unlike flooding based DoS, SDAs make use of reduced attack bandwidth, allowing to take down a particular server even by a single attacking host, using minimal resources.

In the last years, after the birth of hacktivist groups such as *Anonymous*, Denial of Service attacks are also considered as a relevant menace for Internet services, enterprises, organizations, and governments. Indeed, this kind of attacks may lead to financial and economical losses or create personal damage to the targeted victim.

For instance, in March 2013, an important DDoS attack has been executed against Spamhaus, an anti-spam company. This attack exploited the open DNS resolvers to amplify the attack, making the victim's website offline but also affecting the performance of the global Internet [4].

In Section 2 we report the related works relatively to the slow DoS field: instead of analyzing the various tools and methods used for cyberwarfare, we choose to focus our work on this category of attacks, including the menace proposed in this paper. In Section 3 we describe the attack and its characteristics, while in Section 4 we report the mitigation techniques for the proposed menace. In Section 5 we describe the test environment we have executed. Section 6 refers to the results of the executed tests. Finally, Section 7 reports the conclusions.

2   Related Work

Slow DoS Attacks emerged in the last few years and consolidated as a severe threat on the Internet. If we consider this category of attacks, the first menace is the Shrew attack, which sends an attack burst to the victim, giving to the transport layer the illusion of a high congestion on the network link [5]. Mácia-Fernández et al. [6] propose instead the low-rate DoS attack against application servers (LoRDAS), able to lead a DoS reducing the attack burst and concentrating it to specific instants.

Although they aren't related to a research work, some attacks got directly published on the web, obtaining popularity due to the high impact offered, requiring relatively low attack bandwidth. Among these attacks, slowloris, implemented by Robert "RSnake" Hansen, could be considered as one of the most known Slow DoS Attack. It exploits the HTTP protocol, sending a large amount of pending requests to the victim [7].

The Apache Range Header attack, no longer considered as a menace, has been also published in the Internet as a script, by a user known as KingCope [4]. This attack exploits the byte-range parameter of the HTTP protocol, forcing the web server to replicate in memory the requested resource.

In 2012, Sergey Shekyan [4] released slowhttptest, a tool to accomplish a set of Slow DoS Attacks. Together with the tool, his team introduces the Slow Read attack, able to slow down the responses of a web server through the sending of legitimate requests.

In this work we enrich the available set of SDA tools, proposing, executing, and analyzing the SlowReq attack.

3   The SlowReq Attack

The SlowReq attack exploits a vulnerability on most server applications implementations, designed to assign a thread to each request, in order to serve it, thus requiring the server a large amount of resources. As a consequence, such implementations lead to limit the number of simultaneous threads on the machine, thus decreasing the number of requests manageable by the server. Differently to flooding DoS attacks, which aim to overwhelm the network of the victim host, Slow DoS Attacks adopt a smarter approach, directly affecting the application layer of the victim, opening more request than the one the server/daemon is able to accept, thus requiring less attack bandwidth.

Particularly, SlowReq is an application layer DoS attack that sends a large amount of slow (and endless) requests to the server, saturating the buffer resources of the server while it is waiting for the completion of the requests. As an example we report the case of the HTTP protocol, where the characters \r\n\r\n represent the end of the request: in the SlowReq attack these characters are never sent, forcing the server to an endless wait. Additionally, the request is sent abnormally slowly. Similar behavior could be adopted for other protocols as well (SMTP, FTP, etc.). As a consequence of the strategy previously described, if the same attack is applied to a large amount of connections with the victim, a Denial of Service could be obtained on the victim machine. A deeper explanation of this kind of attacks can be found in [4].

The client requests to the server are spread over a wide period of time, and must be considered *slow* in terms of "little amount of bytes sent per second by the client": indeed, for such attacks this value approaches an extremely low value.

Analyzing the velocity for the attack to be completely effective, we must stress that in the moments immediately following the beginning of the attack all available slots of the server are occupied, so new clients' connections experience a Denial of Service. Nevertheless, after the launch of a SlowReq attack the server may have already established active connections with other (legitimate) clients, which are perfectly working until they are closed. As the attack aims to obtain all the available connections on the server, the attacker would probably seize those connections as they become available, after the legitimate clients finish their communications with the server. Indeed, our implementation of SlowReq tries to detect a connection close as it happens, re-establishing the connection with the victim as soon as possible. Actually there could be a race condition between the attacker and other legitimate clients, in order to connect to the server through those connections. In practice, sooner or later the attacker would obtain the connections, since it continuously try to connect to the victim.

From the stealth perspective, an important feature of the proposed attack is that it is difficult to detect it while it is active. Indeed, log files on the server are often updated only when a complete request is received: in our case requests are typically endless, and during the attack log files don't contain any trace of such a behavior. Therefore, a log analysis is not sufficient enough to produce an appropriate warning in reasonable times. Of course, as the attack proceeds and connections are closed due to some circumstances (forced reset, custom configurations on the server, timeout period occurrence etc.), the log files are updated.

3.1   How the Attack Works

The SlowReq attack works by sending slow endless requests to the targeted system. In particular, the attack executes three different program flows:

- the first one aims to continuously establish connections with the server, thus opening a large amount of connections, without sending any data to the server;
- the purpose of the second flow is to maintain the connections with the server alive by slowly sending data to the victim through the established connections, preventing a server connection close. The exploited resource, according to taxonomy reported in [4] is the timeout.
- the third flow aims at detecting connections that have been closed by the server.

It follows a brief description of each flow.

As already said, first flow's purpose is to seize all the connections available at the application layer on the victim machine. Assuming that the maximum number of simultaneous requests the attacker wants to leave opened with the server is m, this flow continuously check the currently established connections with the victim, opening the remaining ones, in order to reach the m value. Moreover, a sleep method is used in case the m value is reached, with the purpose of reducing CPU consumption on the attacking machine, making the program sleep for some (a few) instants.

The maintaining flow makes instead use of a *Wait Timeout* parameter to manage the slowness of sending, which is accomplished through the waiting of the Wait Timeout expiration, thus sending a single character (default one is a single space; in general, each character is good) to the victim on the established channel.

Finally, the third flow, control flow, repeatedly checks the status of the established connections. In particular, thanks to this flow the attack is able to re-establish the closed connections, as soon as they are closed. This flow is directly related with the connection flow. Indeed, some instants later the number of established connection is decreased by the check flow (when a connection is closed), a new connection is established from the connection flow. As a consequence, the attack is able to autonomously establish and maintain the m connections with the server along the time.

Assuming the m value is equal to the maximum number of connections accepted by the server and assuming that the server is vulnerable to a Long Request DoS attack, the attack is able to successfully reach a DoS. Moreover, it can successfully and quickly detect when a "full DoS" (the number of connections established by the attacker is equal to m) is not reached anymore, trying to reach it again as soon as possible.

3.2   SlowReq vs. Slowloris

If we consider the taxonomy proposed in [4], we can state that the SlowReq attack is a Slow Request attack belonging to the Long Requests category. As a consequence, since to the best of our knowledge currently there are no Slow Requests threats, the menace proposed on this document represents a first menace working in such way.

The slowloris attack offers some similarities with SlowReq [7]. Nevertheless, if we carefully analyze the two menaces, attacks differ for their methodologies, leading to different results in terms of stealth, flexibility and bandwidth usage. While the slowloris approach leaves requests pending, the SlowReq one sends slow requests, and both the pending and endless characteristics are a consequence of this behavior.

Talking about flexibility, the proposed attack is more general due to the potential protocol independence characteristic. Indeed, slowloris is not a protocol independent attack, since the sent messages are bounded to the HTTP/HTTPS protocols, and the venomous characteristic is due to the lack of \r\n final line, but a complete message (including newlines) is sent to the server, which is potentially able to parse the request: in other words, in order to avoid a connection close. legitimate requests are sent, even if some tricks are used.

4   SlowReq Mitigation

Relatively to flooding DoS ones, Slow DoS Attacks are more difficult to detect and mitigate, since they need less resources to reach the Denial of Service state. Indeed, while flooding DoS attacks can be detected through some well known filters, actually there is not a general and effective algorithm able to detect a Slow DoS Attack.

In order to reduce the impact of a SlowReq attack, servers may need to be extended to support additional features. A software solution could directly involve the service software/process/daemon. Although there are various approaches that can be applied for this purpose, we focus on the following methods.

- Blocking the maximum amount of connections coming from the same IP address, in order to reduce the damage carried out by the attack
- Applying temporal limits to the received requests, avoiding the acceptance of long requests
- Applying bandwidth limits to the received requests, limiting the requests sending to a particular minimum bit-rate

There are many different options for the implementation of these mitigation techniques (i.e. these countermeasures could be applied on firewall, operating system, proper software modules, etc...). However, if we consider the HTTP protocol (which is often exploited by Slow DoS Attacks), there are some modules that could implement the features exposed above. Among all the available ones, we have found two modules which implements some peculiarities described above.

mod-security

This module can be used to manage the security of a web server. From version 2.5.13, a new defense capability has been implemented, helping in the prevention of a pending requests based Slow DoS Attack. In fact, the new directive SecReadStateLimit allows to hook into the connection level filter of Apache, reducing the number of threads in the state SERVER_BUSY_STATE for each IP address (a similar behavior is given by other modules, such as the limitipconn module). Indeed, with HTTP/1.1 pipelining capability there is no legitimate scenario where a single client will be initiating a large amount of simultaneous connections.

Reqtimeout

This module gives the ability to set temporal and bandwidth limits for the received requests. In particular, it's possible to set:

1. the minimum time the server waits for receiving the first byte of the request header/body

2. the maximum time the server waits for receiving the whole request

3. a minimum reception bandwidth rate for each request

If the clients do not respect the configured limits, a 408 Request Timeout error is sent.

4.1   Server Platforms Immunity

A Slow DoS Attack often exploits the HTTP protocol to reach the DoS state: considering this protocol, some SDAs (i.e. slowloris) do not affect particular web server platforms. Therefore, we can't exclude a similar behavior for SlowReq.

In particular, from tests we have executed we noticed that IIS servers are resistant to the SlowReq attack (the same applies for slowloris). Indeed, although IIS isn't able to detect a SlowReq attack by default, a non distributed attack can't tie up the resources of such servers, since IIS is able to deprioritize [8] and to manage an extremely high number of simultaneous connections (thousand of connections, instead of the few hundreds managed by Apache) [9].

Except for some specific daemon versions, we have observed that lighttpd based servers are also immune to a SlowReq attack. Nevertheless, lighttpd version 1.5 appears to be affected by SlowReq, since it always allocates a 16 KB buffer for a read, regardless its length [10]. Indeed, each packet received by SlowReq, which delivers a single byte to the application layer, would require a memory allocation of 16 KB.

5   Test Environment

We executed two different sets of experiments. During the tests we have analyzed parameters related to the number of connections established with the application server during the time (a sample per second). Such approach provides us an accurate way to report the attack status/success during the time.

5.1   Modules Tests

As we have described above, slowloris is the most similar attack to SlowReq. In the first experiment we try to highlight the difference of the two attacks, comparing the victim status during both a SlowReq and a slowloris attack. Trials are executed comparing the two different attacks by changing the configuration on targeted system, sequentially enabling the modules described in Section 4.2.

We have attacked an Apache 2 web server running on a Linux based host. The Apache server has been configured to serve at most 150 simultaneous connections (through the MaxClient directive). This value usually has a default value not greater than 256 [11]. Having a known MaxClient value we are able to retrieve a percentage of attack connections served by the victim. In particular, a percentage equal to 100 would identify a full DoS, while a percentage equal to 0 would represent the attack ineffectiveness. In the other cases, a partial DoS would be reached.

We have changed the web server configuration by sequentially enabling the modules described in Section 4.2. Moreover, we have tested the proposed tool against a pure web server, without any protection module enabled.

In particular, for mod-security we have configured the SecReadStateLimit parameter to 5. As a consequence, any IP addresses will be served for at most 5 maximum simultaneous requests.

The reqtimeout module has been configured to wait at most 20 seconds for the first byte of the request header and to wait no longer than 40 seconds in total. Moreover, we have configured a minimum bandwidth rate in reception of 500 bytes/s.

Relatively to the attack operations, we choose a 600 seconds long attack and a 60 seconds long Wait Timeout. Therefore, during the attack execution multiple slow sending phases are executed. Moreover, we have also limited the maximum number of simultaneous active connections to the maximum value accepted by the victim.

6   Performance Results

We now report the results for both the test sets we have executed. Each graph shows the percentage of connections established with the server by the attacker vs the time variable, in seconds.

In order to retrieve an accurate value for the established connections number, we have analyzed all the established connections from the victim host perspective, selecting and counting the connections that are associated with the targeted daemon. Indeed, since protection filters are applied by application modules, there could be

established connections with the server which are not bounded to the attacked daemon, in case protection modules are blocking them. The daemon bound filter allows
us to select the connections that have effectively reached the application.

6.1   Modules Tests

We now report the attack success for both SlowReq and slowloris tools, analyzing the
differences between the two approaches.

Figure 1 shows the attack results against a web server without any protection
module enabled.

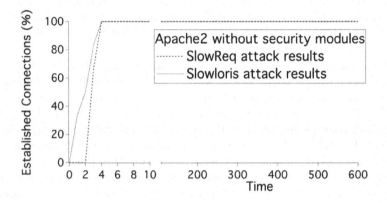

Fig. 1. Apache 2 without any protection module

It's evident that the DoS is reached by both the tools after a few seconds, and it's
maintained for all the duration of the attacks.

Figure 2 shows the results when `mod-security` module is running on the attacked server.

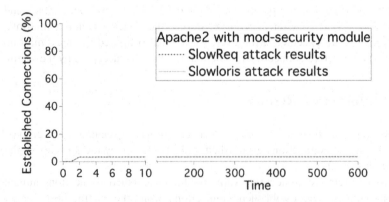

Fig. 2. Apache 2 with mod-security module

This module can successfully mitigate both the attacks. In fact, figure shows that in this case the DoS state is never reached and, as in module configuration, the number of connections coming from the attacker's IP address is limited to 5, a non-distributed version of the attacks can't tie up the resources of the victim.

Finally, Figure 3 reports the results of the attacks against a server equipped with the reqtimeout module.

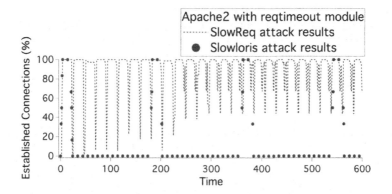

Fig. 3. Apache 2 with reqtimeout module

This graph represents the most important difference between the two analyzed tools. In particular, in the first seconds the DoS is reached for both the attacks. Then, after 20 seconds (as configured on reqtimeout module), the connections established by both the tools are closed. Nevertheless, differently form slowloris, SlowReq is able to detect the connection closes, thus trying to reach the DoS again, with success. Therefore, we can state that the SlowReq attack is more effective than slowloris, thus being more difficult to mitigate.

It's important to notice that in the last case the attacker isn't able to maintain the established connections with the server. This functionality becomes an important achievement, since the SlowReq attack is able to detect the connection closes faster than slowloris, leading to another Denial of Service after a few seconds, thus reducing the efficacy of the installed module.

7   Conclusions and Future Work

In this paper we have proposed a new Denial of Service attack. We have illustrated a possible adoption of the tool for cyberwarfare operations.

Analyzing its functioning, we have categorized this menace as a Slow DoS Attack, placing it into the Slow Requests DoS sub-category. Since, at writing time, this category doesn't include any threat, our work represent an innovation of the attacking tools. Moreover, we have noticed that in some cases the SlowReq attack is able to affect different protocols. Therefore, further study may be needed.

Since the most similar menace to SlowReq is the slowloris attack, we have compared the two tools, showing the advantages of SlowReq and the enhanced efficiency, provided by the ability of the attack to detect connection closes in reasonable times.

Even if we showed how some techniques can successfully mitigate a SlowReq DoS attack, in these cases a distributed version of the attack would be successful. Therefore, a possible extension of our work may concern the development of a distributed SlowReq attack.

In the executed attacks we have pre-configured the maximum number of simultaneous connections opened by the attacker (the m parameter described in Section 3.1). A possible extension of the work may use instead a dynamical value, trying to automatically retrieve it analyzing the connections status during the time.

References

[1] Greengard, S.: The new face of war. Communications of the ACM 53, 20–22 (2010)

[2] Chen, T.M.: Stuxnet, the real start of cyber warfare? IEEE Network 24, 2–3 (2010)

[3] Combs, M.M.: Impact of the Stuxnet Virus on Industrial Control Systems. In: XIII International Forum Modern Information Society Formation Problems, Perspectives, Innovation Approaches, pp. 5–10 (2012)

[4] Cambiaso, E., et al.: Slow DoS Attacks: Definition and Categorization. International Journal of Trust Management in Computing and Communications (in press article, 2013)

[5] Kuzmanovic, A., Knightly, E.W.: Low-rate TCP-targeted denial of service attacks: the shrew vs. the mice and elephants. In: Proceedings of the 2003 conference on Applications, Technologies, Architectures, and Protocols for Computer Communications, pp. 75–86 (2003)

[6] Macia-Fernandez, G., et al.: Evaluation of a low-rate DoS attack against iterative servers. Computer Networks 51, 1013–1030 (2007)

[7] Cambiaso, E., Papaleo, G., Aiello, M.: Taxonomy of Slow DoS Attacks to Web Applications. In: Thampi, S.M., Zomaya, A.Y., Strufe, T., Alcaraz Calero, J.M., Thomas, T. (eds.) SNDS 2012. CCIS, vol. 335, pp. 195–204. Springer, Heidelberg (2012)

[8] Damon, E., et al.: Hands-on denial of service lab exercises using SlowLoris and RUDY. In: Proceedings of the 2012 Information Security Curriculum Development Conference, pp. 21–29 (2012)

[9] MaxConnections - IIS 6.0,
http://msdn.microsoft.com/en-us/library/ms524491(v=vs.90).aspx
(accessed in 2013)

[10] Slow request dos/oom attack,
http://download.lighttpd.net/lighttpd/security/
lighttpd_sa_2010_01.txt
(accessed in 2013)

[11] Apache MPM Common Directives - MaxClients Directive,
http://httpd.apache.org/docs/2.2/mod/
mpm_common.html-maxclients (accessed in 2013)

Identification of Degeneracies in a Class of Cryptographic Sequences

Amparo Fúster-Sabater

Information Security Institute, C.S.I.C.
Serrano 144, 28006 Madrid, Spain
amparo@iec.csic.es

Abstract. In this work, the parameter linear complexity for a class of filtered sequences has been considered and analyzed. The study is based on the handling of bit-strings that permit identify potential degeneracies or linear complexity reductions in the sequences generated from this kind of nonlinear filters. Numerical expressions to determine the linear complexity of such sequences have been developed as well as design rules to generate sequences that preserve maximal linear complexity are also provided. The work complete the analysis of the linear complexity for these sequence generators found in the literature.

Keywords: filter design, pseudorandom sequence, LFSR, cryptography.

1 Introduction

Pseudorandom binary sequences are typically used in a wide variety of applications such as: spread spectrum communication systems, multiterminal system identification, global positioning systems, software testing, error-correcting codes or cryptography. This work deals specifically with this last application.

Secret-key encryption functions are usually divided into two separated classes: stream ciphers and block-ciphers depending on whether the encryption function is applied either to each individual bit or to a block of bits of the original message, respectively. At the present moment, stream ciphers are the fastest among the encryption procedures so they are implemented in many technological applications: the encryption algorithm RC4 [15] used in Wired Equivalent Privacy (WEP) as a part of the IEEE 802.11 standards, the encryption function E0 in Bluetooth specifications [1] or the recent proposals HC-128 or Rabbit coming from the eSTREAM Project [2] and included in the latest release versions of CyaSSL (lightweight open source embedded implementation of the SSL/TLS protocol) [17]. Stream ciphers try to imitate the mythic *one-time pad cipher* or *Vernam cipher* [13] that remains as the only absolutely unbreakable cipher. They are designed to generate from a short key a long sequence (the *keystream sequence*) of pseudorandom bits. Some of the most recent designs in stream ciphers can be found in [2]. This keystream sequence is XORed with the original message (in emission) in order to obtain the ciphertext or with the ciphertext (in reception) in order to recover the original message. References [12,14,16] provide a solid introduction to the study of stream ciphers.

Á. Herrero et al. (eds.), *International Joint Conference SOCO'13-CISIS'13-ICEUTE'13*, 547
Advances in Intelligent Systems and Computing 239,
DOI: 10.1007/978-3-319-01854-6_56, © Springer International Publishing Switzerland 2014

Most keystream generators are based on maximal-length Linear Feedback Shift Registers (LFSR) [4] whose output sequences or m-sequences are combined by means of nonlinear combinators, irregularly decimated generators, typical elements from block ciphers, etc to produce sequences of cryptographic application. One general technique for building a keystream generator is to use a nonlinear filter, i.e. a nonlinear function applied to the stages of a single maximal-length LFSR. That is the output sequence (*filtered sequence*) is generated as the image of a nonlinear Boolean function F in the LFSR stages. Among other properties [3], a large Linear Complexity (LC) is a necessary requirement that every filtered sequence must satisfy in order to be accepted as cryptographic sequence. Linear complexity of a sequence is defined as the length of the shortest LFSR able to generate such a sequence. In cryptographic terms, LC must be as large as possible in order to prevent the application of the Berlekamp-Massey algorithm [10]. The problem of determining the exact value of the linear complexity attained by filtered sequences is still open [7,8]. At any rate, several basic references concerning these features can be quoted: In [5], Groth presented the linear complexity of a sequence as a controllable parameter which increases with the order of the filtering function. The author applied Boolean functions of degree 2 to the stages of a maximal-length LFSR. Nevertheless, in his work there is no explicit mention to the degeneracies that may occur in the linear complexity of the filtered sequences. In [6], Key gave an upper bound on the linear complexity of the sequences obtained from *kth*-order nonlinear filters. Moreover, Key stated that "a *2nd*-order product of two distinct phases of the same m-sequence never degenerates" although, as it was pointed by Rueppel in [16], this result is restricted to phase differences less that the length of the LFSR. Using the DFT (Discrete Fourier Transform) technique, Massey *et al.* [11] proved that Key's upper bound holds for any arbitrary choice of phase difference, provided that L the length of the LFSR is a prime number.

In this paper, a method that completes the computation of LC for *2nd*-order nonlinear filters has been developed. The procedure is based on the handling of bit-strings that permit identify potential degeneracies (reductions) in the linear complexity of the filtered sequence. Both choices of L (prime or composite number) have been separately considered. In the prime case, Massey nondegeneration result is also derived. In the composite case, the phase differences that reduce the value of the linear complexity have been identified. Moreover, the cosets that degenerate and, consequently, the final value of the linear complexity are also computed. Finally and as a direct consequence of the previous results, practical rules to select phase differences that preserve Key's upper bound in this king of nonlinear filters are stated. The generalization to Boolean functions of degree greater than 2 seems to be the natural continuity of this work.

2   Notation and Basic Concepts

Notation and basic concepts that will be used throughout the work are now introduced.

Maximal-sequence (m-sequence). Let $\{s_n\}$ be the binary output sequence of a maximal-length LFSR of L stages, that is a LFSR whose characteristic polynomial $P(x)$ is primitive of degree L, see [16], [4]. In that case, the output sequence is a m-sequence of period $2^L - 1$. Moreover, $\{s_n\}$ is completely determined by the LFSR initial state and the characteristic polynomial $P(x)$.

The roots of $P(x)$ are α^{2^i} ($i = 0, 1, \ldots, L-1$) where α is a primitive element in $GF(2^L)$ that is an extension of the binary field $GF(2)$ with 2^L elements, see [9]. Any generic term of the sequence, s_n, can be written by means of the roots of the polynomial $P(x)$ as:

$$s_n = \sum_{j=0}^{L-1} (C\alpha^n)^{2^j}, \quad n \geq 0 \tag{1}$$

where $C \in GF(2^L)$. Furthermore, the $2^L - 1$ nonzero choices of C result in the $2^L - 1$ distinct shifts of the same m-sequence. If $C = 1$, then $\{s_n\}$ it is said to be in its *characteristic phase*.

Nonlinear filter. It is a Boolean function denoted by $F(x_0, x_1, \ldots, x_{L-1})$ in L variables of degree k. For a subset $A = \{a_0, a_1, \ldots, a_{r-1}\}$ of $\{0, 1, \ldots, L-1\}$ with $r \leq k$, the notation $x_A = x_{a_0} x_{a_1} \ldots x_{a_{r-1}}$ is used. The Boolean function can be written as:

$$F(x_0, x_1, \ldots, x_{L-1}) = \sum_A c_A\, x_A, \tag{2}$$

where $c_A \in \{0, 1\}$ are binary coefficients and the summation is taken over all subsets A of $\{0, 1, \ldots, L-1\}$. In particular, a *2nd*-order nonlinear filter applied to the L stages of a LFSR is the product of 2 distinct phases of $\{s_n\}$, that is $s_{n+t_0} \cdot s_{n+t_1}$, where t_0, t_1 are integers satisfying $0 \leq t_0 < t_1 < 2^L - 1$.

Filtered sequence. It is the sequence $\{z_n\}$ or output sequence of the nonlinear filter F applied to the L stages of the LFSR. It is currently used as keystream sequence in stream cipher. The keystream bit z_n is computed by selecting bits from the m-sequence such that

$$z_n = F(s_n, s_{n+1}, \ldots, s_{n+L-1}). \tag{3}$$

Cyclotomic coset. Let Z_{2^L-1} denote the set of integers $[1, \ldots, 2^L - 1]$. An equivalence relation R is defined on its elements $q_1, q_2 \in Z_{2^L-1}$ such as follows: $q_1 R q_2$ if there exists an integer j, $0 \leq j \leq L - 1$, such that

$$2^j \cdot q_1 = q_2 \bmod 2^L - 1. \tag{4}$$

The resultant equivalence classes into which Z_{2^L-1} is partitioned are called the *cyclotomic cosets* mod $2^L - 1$, see [4]. All the elements q_i of a cyclotomic coset have the same number of ones in their binary representation; this number is called the *coset weight*. The leader element, E, of a coset is the smallest integer in such an equivalence class. Every cyclotomic coset can be univocally represented by its leader element E or by a L-bit string corresponding to the binary representation of the integer E. Moreover, the cardinal of any coset is L or a proper divisor of L.

Uniform coset. A cyclotomic coset mod $(2^L - 1)$ is called uniform coset if its cardinal p is a proper divisor of L. In fact, an uniform coset e is a set of integers of the form $\{e, e \cdot 2, e \cdot 2^2, \dots, e \cdot 2^{(p-1)}\}$ mod $(2^L - 1)$, where the smallest positive integer p satisfying

$$e \cdot 2^p \equiv e \ mod \ (2^L - 1) \tag{5}$$

is a proper divisor of L. This type of coset belongs to the group of improper cosets defined by Golomb in [4] although both groups may not coincide.

As any coset, an uniform coset e of binary weight w can be univocally represented by a L-bit string with w ones. Let p_i $(i = 1, \dots, r)$ be the proper divisors of L, then an uniform coset e associated with p_i is made out of $d_i = L/p_i$ groups of p_i bits with w/d_i ones in each group. The cardinal of coset e is p_i. More than one uniform coset may be associated with each divisor p_i. The name of uniform coset is due to the uniform distribution of ones all along the L-bit string. Clearly if L is a prime number, then there are no uniform cosets. Next, an illustrative example of this type of cosets is presented.

Example 1. For $L = 12$, $w = 6$ the different uniform cosets mod $(2^L - 1)$ in terms of their corresponding L-bit strings can be written such as follows:

$$
\begin{aligned}
coset\ e_1 &= 2^0 + 2^1 + 2^2 + 2^6 + 2^7 + 2^8 &\Leftrightarrow\ 000111\ 000111, \\
coset\ e_2 &= 2^0 + 2^1 + 2^3 + 2^6 + 2^7 + 2^9 &\Leftrightarrow\ 001011\ 001011, \\
coset\ e_3 &= 2^0 + 2^1 + 2^4 + 2^6 + 2^7 + 2^{10} &\Leftrightarrow\ 010011\ 010011, \\
coset\ e_4 &= 2^0 + 2^1 + 2^4 + 2^5 + 2^8 + 2^9 &\Leftrightarrow\ 0011\ 0011\ 0011, \\
coset\ e_5 &= 2^0 + 2^2 + 2^4 + 2^6 + 2^8 + 2^{10} &\Leftrightarrow\ 01\ 01\ 01\ 01\ 01\ 01,
\end{aligned}
$$

where cosets e_1, e_2, e_3 are associated with the proper divisor 6, coset e_4 with 4 and coset e_5 with 2. Thus, their corresponding cardinals are 6, 4 and 2, respectively.

In brief, a uniform coset can be interpreted as a L-bit string such that when it is circularly shifted p_i positions remains unchanged.

3 Linear Complexity for *2nd*-Order Nonlinear Filtering Functions

Key's upper bound on LC [6] for the filtered sequence from the application of a *kth*-order nonlinear filter is

$$LC = \sum_{i=1}^{k} \binom{L}{i}, \tag{6}$$

and the linear complexity of the filtered sequence can be computed as [6]

$$LC = \sum_{i} card\ (coset\ E_i), \tag{7}$$

where *card* means cardinal of the coset E_i and the sum is extended to the non-degenerate cosets, that is those cyclotomic cosets whose associated coefficients are different from zero.

Thus, for *2nd*-order nonlinear filters the LC of the filtered sequence can be obtained from the study of the coefficients associated with the cyclotomic cosets of binary weight $w \leq 2$. Now, we introduce their expressions.

3.1 Cyclotomic Coset of Weight w=1

According to (1), the generic terms s_{n+t_0} and s_{n+t_1} of the m-sequence $\{s_n\}$ can be written as:

$$s_{n+t_0} = \alpha^{t_0} \cdot \alpha^n + \alpha^{2t_0} \cdot \alpha^{2n} + \ldots + \alpha^{2^{(L-1)}t_0} \cdot \alpha^{2^{(L-1)}n}, \tag{8}$$

$$s_{n+t_1} = \alpha^{t_1} \cdot \alpha^n + \alpha^{2t_1} \cdot \alpha^{2n} + \ldots + \alpha^{2^{(L-1)}t_1} \cdot \alpha^{2^{(L-1)}n}. \tag{9}$$

Then, regrouping terms the generic element $z_n = s_{n+t_0} \cdot s_{n+t_1}$ of the filtered sequence is given by

$$\begin{aligned} z_n &= \alpha^{2^{(L-1)}t_0} \cdot \alpha^{2^{(L-1)}t_1} \cdot \alpha^n + \alpha^{t_0} \cdot \alpha^{t_1} \cdot \alpha^{2n} + \ldots \\ &= C_1 \cdot \alpha^n + (C_1)^2 \cdot \alpha^{2n} + \ldots. \end{aligned}$$

C_1 being the coefficient associated with the coset 1. Identifying coefficients in both sides of the equation, we get:

$$C_1 = \alpha^{2^{(L-1)}t_0} \cdot \alpha^{2^{(L-1)}t_1}. \tag{10}$$

Hence, C_1 is different from zero, the coset 1 will always be nondegenerate and will contribute to the LC of the sequence $\{z_n\}$. The value of such a contribution will be its cardinal L that is $\binom{L}{1}$ in equation (6).

3.2 Cyclotomic Cosets of Weight w=2

Let E be the leader of a generic cyclotomic coset of weight 2, that is E is an integer of the form $E = 2^{c_0} + 2^{c_1}$, where c_i are integers satisfying [4]

$$0 = c_0 < c_1 \leq \frac{1}{2} L \ (if \ L \ is \ even), \tag{11}$$

$$0 = c_0 < c_1 \leq \frac{1}{2}(L-1) \ (if \ L \ is \ odd). \tag{12}$$

Then C_E, the coefficient associated with the coset E, can be computed from the root presence test of Rueppel [16] such as follows:

$$C_E = \begin{vmatrix} \alpha^{t_0 \cdot 2^{c_0}} & \alpha^{t_1 \cdot 2^{c_0}} \\ \alpha^{t_0 \cdot 2^{c_1}} & \alpha^{t_1 \cdot 2^{c_1}} \end{vmatrix}. \tag{13}$$

Taking $t_1 = t_0 + d$, equation (13) can be rewritten as

$$C_E = \alpha^{t_0 \cdot 2^{c_0}} \cdot \alpha^{t_0 \cdot 2^{c_1}} \cdot \begin{vmatrix} 1 & \alpha^{d \cdot 2^{c_0}} \\ 1 & \alpha^{d \cdot 2^{c_1}} \end{vmatrix}. \tag{14}$$

We can take without loss of generality $t_0 = 0$ and $t_1 = d$ as the interest of these parameters is focussed on the relative distance between phases rather than in the particular values of t_0, t_1. Thus,

$$C_E = M \cdot \begin{vmatrix} 1 & \alpha^d \\ 1 & \alpha^{d \cdot 2^{c_1}} \end{vmatrix}. \tag{15}$$

Now from equation (15) we see that the coefficient C_E will be zero if and only if

$$\alpha^{d \cdot 2^{c_1}} = \alpha^d, \tag{16}$$

or equivalently

$$d \cdot 2^{c_1} \equiv d \, mod \, (2^L - 1). \tag{17}$$

Hence, equations (5) and (17) have the same solutions and the values of the phase differences d that may produce degeneracies in the cosets E of weight 2 are the elements of the uniform cosets mod $(2^L - 1)$.

4   Linear Complexity in Terms of Bit-Strings

Now two distinct cases can be considered.

4.1   L Is a Prime Number

Next result follows in a very natural way from (17).

Theorem 1. *Let L be the length of a maximal-length LFSR. If L is a prime number, then the linear complexity of the sequence product of two distinct phases, $\{z_n\} = \{s_{n+t_0} \cdot s_{n+t_1}\}$, always meets Key's upper bound.*

Proof. The result follows from the fact that if L is prime, then there are no uniform cosets or equivalently equation (17) has no solution. In this case, all the cyclotomic cosets of binary weight ≤ 2 contribute to the LC of the filtered sequence.   □

The previous theorem confirms Massey's nondegeneracy result obtained from the DFT technique.

4.2   L Is a Composite Number

Recall that in equation (17) d gives information on the phase differences that might produce degeneracies as well as c_1 gives information on the cosets E of weight 2 that might degenerate. Let us consider both variables separately.

Computation of d

Let $p_1 > p_2 > \ldots > p_r$ be proper divisors of L. For each p_i we can obtain different uniform cosets whose elements are univocally defined by L-bit strings made out of $d_i = L/p_i$ repetitions of groups of p_i bits with $w_i \in [1, 2, \ldots p_i - 1]$ ones. For each value of w_i there will be

$$\binom{p_i}{w_i} \tag{18}$$

distinct L-bit strings (i.e., w_i ones placed in p_i different positions) grouped in $\binom{p_i}{w_i}/p_i$ uniform cosets.

So we get in total

$$\sum_{i=1}^{r} \sum_{j=1}^{p_i-1} \binom{p_i}{j} \tag{19}$$

different L-bit strings or values of the phase difference d that correspond to all the elements of all the uniform cosets mod $(2^L - 1)$.

Computation of c_1:

Notice that for any coset E of weight 2 the integer E can be written as $E = 2^{c_0} + 2^{c_1} = 2^0 + 2^{c_1}$. Therefore, c_1 represents the position of the most significant 1 in the binary representation of E. If d is an integer belonging to any uniform coset and $c_1 = p_i$, then equation (17) holds and the corresponding coset E will be degenerate. In a more general way, $c_1 = \dot{p}_i = N \cdot p_i$ (where N is an integer) satisfies equation (17) too as

$$d \cdot 2^{N p_i} = d \cdot 2^{(N-1) p_i} = \ldots = d \cdot 2^{p_i} \equiv d \bmod (2^L - 1). \tag{20}$$

Thus, for any d in the set of uniform cosets, denoted by $[e_j]$, and $E = 2^0 + 2^{c_1} = 2^0 + 2^{\dot{p}_i}$ the corresponding coset E of weight 2 will be degenerate for the product sequence $\{s_n \cdot s_{n+d}\}$. The procedure can be repeated for each one of the proper divisor p_i of L ($i = 1, \ldots, r$) as well as for every $c_1 = \dot{p}_i$ in the intervals given in equations (11) and (12). A simple and illustrative example is presented.

Example 2. For $L = 15$, we have that the proper divisors are $p_1 = 5$, $p_2 = 3$.

- The 15-bit strings corresponding to the uniform cosets e_j associated with p_1 are:

$$coset\ e_1, w_1 = 1 \Leftrightarrow 00001\ 00001\ 00001,$$
$$coset\ e_2, w_1 = 2 \Leftrightarrow 00011\ 00011\ 00011,$$
$$coset\ e_3, w_1 = 2 \Leftrightarrow 00101\ 00101\ 00101,$$
$$coset\ e_4, w_1 = 3 \Leftrightarrow 00111\ 00111\ 00111,$$
$$coset\ e_5, w_1 = 3 \Leftrightarrow 01011\ 01011\ 01011,$$
$$coset\ e_6, w_1 = 4 \Leftrightarrow 01111\ 01111\ 01111.$$

– The 15-bit strings corresponding to the uniform cosets e_j associated with p_2 are:

$$coset\ e_7, w_2 = 1 \Leftrightarrow 001\ 001\ 001\ 001\ 001,$$
$$coset\ e_8, w_2 = 2 \Leftrightarrow 011\ 011\ 011\ 011\ 011.$$

Now we can write the cyclotomic cosets E of weight 2.

$$coset\ E_1 \Leftrightarrow 00000\ 00000\ 00011,$$
$$coset\ E_2 \Leftrightarrow 00000\ 00000\ 00101,$$
$$coset\ E_3 \Leftrightarrow 00000\ 00000\ 01001,$$
$$coset\ E_4 \Leftrightarrow 00000\ 00000\ 10001,$$
$$coset\ E_5 \Leftrightarrow 00000\ 00001\ 00001,$$
$$coset\ E_6 \Leftrightarrow 00000\ 00010\ 00001,$$
$$coset\ E_7 \Leftrightarrow 00000\ 00100\ 00001.$$

According to the previous considerations, we have:

– For $p_1 = 5$, the cyclotomic coset E_5 ($c_1 = 5$) will be degenerate for any phase difference d taking values in the elements of the uniform cosets $e_1, e_2, \ldots e_6$.
– For $p_2 = 3$, the cyclotomic cosets E_3 ($c_1 = 3$) and E_6 ($c_1 = 6$) will be degenerate for any phase difference d taking values in the elements of the uniform cosets e_7 and e_8.

From the previous considerations next results can be stated.

Theorem 2. *Let p_i be a proper divisor of L and let $[e_j^i]$ be the subset of uniform cosets associated with p_i. If $d \in [e_j^i]$, then the number of cyclotomic cosets of binary weight 2 that will be degenerate is given by*

$$N_{p_i} = \lfloor \frac{\lfloor L/2 \rfloor}{p_i} \rfloor. \tag{21}$$

Proof. According to the previous interpretation of the degeneracy of a cyclotomic coset of weight 2, it is easy to see that N_{p_i} coincides with the number of times that a structure of p_i bits is contained in the maximal distance between the two ones in the binary representation of a cyclotomic coset of weight 2. The fact that this maximal distance is $\lfloor L/2 \rfloor$ for the coset $2^0 + 2^{\lfloor L/2 \rfloor}$ completes the proof. □

From the previous theorem the following corollaries can be derived.

Corollary 1. *For a phase difference d under the conditions of Theorem (2), the value of the linear complexity of the filtered sequence is given by*

$$\frac{1}{2}(L^2 + L) - (N_{p_i} - 1)L - \frac{1}{2}L \le LC \le \frac{1}{2}(L^2 + L) - (N_{p_i} \cdot L), \tag{22}$$

$\frac{1}{2}(L^2 + L)$ *being Key's upper bound for a 2nd-order nonlinear filter and N_{p_i} defined as in Theorem (2). In fact, LC meets the lower bound if L is even and $p_i = \frac{1}{2}L$ (or a divisor). Otherwise, LC equals the upper bound.*

Proof. The result is a straight consequence of Theorem (2). Indeed, the linear complexity will take the maximal value given by Key's upper bound except for the contribution to the complexity of the cyclotomic cosets of weight 2 that are degenerate. The quantitative value of this decreasing will be the number of degenerate cosets multiplied by their corresponding cardinals. By construction, all the cyclotomic cosets of weight 2 have cardinal L except for coset $E = 2^0 + 2^{L/2}$, for L even, whose cardinal is $\frac{1}{2}L$. The lower bound is due to the fact that if L is even and $p_i = \frac{1}{2}L$ (or a divisor), then the cyclotomic coset $E = 2^0 + 2^{L/2}$ is degenerate. Otherwise, as all the cyclotomic cosets of weight 2 have cardinal L, the linear complexity takes the value given by the upper bound.
□

Corollary 2. *If $L = p^k$, p being a prime number, then the uniform cosets associated with the smaller proper divisor of L, that is p, reach the highest degeneracies as they include the degeneracies of the uniform cosets associated with all the other proper divisors of L.*

Proof. As $p|p_i \ \forall i$, then a structure of p bits always is contained in a structure of p_i bits. Therefore, the uniform cosets associated with p include the degeneracies of the uniform cosets associated with $p_i \ \forall i$. To the previous degeneracies, we have to add the degeneracies due exclusively to the cosets associated to p.   □

5   Selection of Phase Differences with Guaranteed Maximal Linear Complexity

According to the previous sections, it is known that the phase difference d satisfying $d \in [e_j]$ (where $[e_j]$ as before is the set of uniform cosets mod $(2^L - 1)$) may produce degeneracies in the filtered sequence. Thus, a general rule to avoid such degeneracies consists in selecting $d \notin [e_j]$. More specific rules can be presented.

1. Let p_1 be the greatest proper divisor of L. The smaller leader of the uniform coset will be
$$e_{min} = 2^0 + 2^{p_1} + 2^{2p_1} + \ldots + 2^{(d_1-1)p_1} \tag{23}$$
    with $d_1 = L/p_1$. Therefore, taking d in the interval $0 < d < e_{min}$ the maximal LC of the filtered sequence, that is Key's upper bound, is guaranteed.
2. If the binary expression of d contains k ones but $(L, k) = 1$, then d is not an element of a uniform coset and the maximal LC of the filtered sequence is guaranteed.

In brief, the number and distribution of the ones in the binary representation of d is a parameter to be handled in order to preserve Key's upper bound.

6   Conclusion

A method of analyzing the linear complexity of nonlinearly filtered sequences has been introduced. The procedure is based on the concept of uniform cosets and allows one to identify easily the degenerate cosets as well as to develop numerical expressions for the linear complexity of the generated sequences. Finally, practical rules to design *2nd*-order nonlinear filters whose output sequences preserve the maximal linear complexity have been derived. The same method can be generalized to Boolean function of degree greater than 2 what is left as the natural expansion of this work.

Acknowledgment. This work has been supported by CDTI (Spain) Project Cenit-HESPERIA as well as by Ministry of Science Project TIN2011-25452.

References

1. Bluetooth, Specifications of the Bluetooth system, Version 1.1,
   http://www.bluetooth.com/
2. eSTREAM, the ECRYPT Stream Cipher Project, Call for Primitives,
   http://www.ecrypt.eu.org/stream/
3. Fúster-Sabater, A., Caballero-Gil, P., Delgado-Mohatar, O.: Deterministic Computation of Pseudorandomness in Sequences of Cryptographic Application. In: Allen, G., Nabrzyski, J., Seidel, E., van Albada, G.D., Dongarra, J., Sloot, P.M.A. (eds.) ICCS 2009, Part I. LNCS, vol. 5544, pp. 621–630. Springer, Heidelberg (2009)
4. Golomb, S.W.: Shift Register-Sequences. Aegean Park Press, Laguna Hill (1982)
5. Groth, E.J.: Generation of binary sequences with controllable complexity. IEEE Trans. Informat. Theory 17(3), 288–296 (1971)
6. Key, E.L.: An Analysis of the Structure and Complexity of Nonlinear Binary Sequence Generators. IEEE Trans. Informat. Theory 22(6), 732–736 (1976)
7. Kolokotronis, N., Kalouptsidis, N.: On the linear complexity of nonlinearly filtered PN-sequences. IEEE Trans. Informat. Theory 49(11), 3047–3059 (2003)
8. Limniotis, K., Kolokotronis, N., Kalouptsidis, N.: On the Linear Complexity of Sequences Obtained by State Space Generators. IEEE Trans. Informat. Theory 54(4), 1786–1793 (2008)
9. Lidl, R., Niederreiter, H.: Introduction to Finite Fields and Their Applications. Cambridge University Press, Cambridge (1986)
10. Massey, J.L.: Shift-Register Synthesis and BCH Decoding. IEEE Trans. Informat. Theory 15(1), 122–127 (1969)
11. Massey, J.L., Serconek, S.: A Fourier transform approach to the linear complexity of nonlinearly filtered sequences. In: Desmedt, Y.G. (ed.) CRYPTO 1994. LNCS, vol. 839, pp. 332–340. Springer, Heidelberg (1994)
12. Menezes, A.J., et al.: Handbook of Applied Cryptography. CRC Press, New York (1997)
13. Nagaraj, N.: One-Time Pad as a nonlinear dynamical system. Commun. Nonlinear Sci. Numer. Simulat. 17, 4029–4036 (2012)
14. Paar, C., Pelzl, J.: Understanding Cryptography. Springer, Heildeberg (2010)
15. Paul, G., Maitra, S.: RC4 Stream Cipher and Its Variants. Discrete Mathematics and Its Applications. CRC Press, Taylor & Francis Group, Boca Raton (2012)
16. Rueppel, R.A.: Analysis and Design of Stream Ciphers. Springer, New York (1986)
17. Yet Another SSL (YASSL), http://www.yassl.com

Design of Nonlinear Filters with Guaranteed Lower Bounds on Sequence Complexity*

Amparo Fúster-Sabater

Information Security Institute, C.S.I.C.
Serrano 144, 28006 Madrid, Spain
amparo@iec.csic.es

Abstract. Sequence generators based on LFSRs are currently used to produce pseudorandom sequences in cryptography. In this paper, binary sequences generated by nonlinearly filtering maximal length sequences are studied. Emphasis is on the parameter linear complexity of the filtered sequences. In fact, a method of computing all the nonlinear filters that generate sequences with a guaranteed linear complexity ($LC \geq \binom{L}{k}$), where L is the LFSR length and k the filter's degree) is introduced. The method provides one with a good structural vision on this type of generators as well as a practical criterium to design cryptographic sequence generators for stream ciphers.

Keywords: Nonlinear filter, linear complexity, cyclotomic coset, Boolean function, stream cipher, cryptography.

1 Introduction

A stream cipher is an encryption algorithm that encrypts individual bits of a plaint text or original message with a time-varying transformation. Stream ciphers are very popular encryption procedures due to many attractive features: they are the fastest among encryption procedures, can be efficiently implemented in hardware and are suitable for environments where the bits need to be processed individually. A stream cipher consists of a keystream generator whose pseudorandom output sequence (*keystream sequence*) is added modulo 2 to the plaintext bits. Some of the most recent designs in stream ciphers can be found in [1] and [9]. References [3], [10] provide a solid introduction to the study of stream ciphers. Desirable properties for the sequences obtained from keystream generators can be enumerated as follows: a) Long period, b) Large linear complexity, c) Good statistical properties.

One general technique for building a keystream generator is to use a nonlinear filter, i.e. a nonlinear function applied to the stages of a single Linear Feedback Shift Register (LFSR). Generally speaking, sequences obtained in this way are supposed to accomplish all the previous properties. Period and statistical properties of the filtered sequences are characteristics deeply studied in the

---

* Work was supported by Ministry of Science and Innovation under Project TIN2011-25452/TSI.

Á. Herrero et al. (eds.), *International Joint Conference SOCO'13-CISIS'13-ICEUTE'13*, 557
Advances in Intelligent Systems and Computing 239,
DOI: 10.1007/978-3-319-01854-6_57, © Springer International Publishing Switzerland 2014

literature, see [8] and the references above mentioned. Nevertheless, the problem of determining the exact value of the linear complexity (a measure of its unpredictability) attained by nonlinear filtering is still open [4], [2], [6]. At any rate, several contributions to the linear complexity study can be found and quoted:

1. In [10], the root presence test proved that the output sequence of nonlinear filters including a unique term of equidistant phases has a linear complexity lower bounded by $LC \geq \binom{L}{k}$, where L is the LFSR length and $k \simeq L/2$ the order of the filter. For (L, k) in a cryptographic range, e.g. $(128, 64)$, the lower bound is a very large number.

2. More recently, in [5], the authors provide with a larger lower bound $LC \geq \binom{L}{k} + \binom{L}{k-1}$ on the linear complexity of filtered sequences. At any rate, this lower bound is applicable only for filters of degree $k \in (2, 3, L - 1, L)$, what is not a real cryptographic application.

In this paper, a method of computing all the nonlinear filters with $LC \geq \binom{L}{k}$ has been developed. The procedure is based on the concept of equivalence classes of nonlinear filters and on the handling of filters from different classes.

The paper is organized as follows: specific notation and the concept of sequential decomposition in cosets is introduced in Section 2. In Section 3, three different representations of nonlinear filters as well as two operations applied to such representations are given. The construction of all possible nonlinear filters preserving the cosets of weight k is developed in Section 4 with an illustrative example. Finally, conclusions in Section 5 end the paper.

2   Basic Concepts and Notation

Specific notation and different basic concepts are introduced as follows:

PN-sequence. Let $\{s_n\}$ be the binary output sequence of a maximal-length LFSR of L stages, that is a LFSR whose characteristic polynomial $P(x)$ is primitive of degree L, see [10], [3]. In that case, the output sequence is a *PN*-sequence of period $2^L - 1$. Its generic element, s_n, can be written as $s_n = \alpha^n + \alpha^{2n} + \ldots + \alpha^{2^{(L-1)}n}$, where $\alpha \in GF(2^L)$ is a root of the polynomial $P(x)$ and $GF(2^L)$ denotes an extension field of $GF(2)$ with 2^L elements [7].

Nonlinear Filter. The Boolean function F defined as $F : GF(2)^L \to GF(2)$ denotes a kth-order nonlinear filter $(k \leq L)$ applied to the L stages of the previous LFSR. That is, $F(s_n, s_{n+1}, \ldots, s_{n+L-1})$ includes at least a product of k distinct elements of the sequence $\{s_n\}$.

Filtered Sequence. The sequence $\{z_n\}$ is the output sequence of the nonlinear filter F applied to the L stages of the LFSR in successive time instants.

Cyclotomic Coset. Let Z_{2^L-1} denote the set of integers $[1, \ldots, 2^L - 1]$. An equivalence relation R is defined on its elements $q_1, q_2 \in Z_{2^L-1}$ such as follows: $q_1 R q_2$ if there exists an integer j, $0 \leq j \leq L - 1$, such that $2^j \cdot q_1 = q_2 \bmod 2^L - 1$. The resultant equivalence classes into which Z_{2^L-1} is partitioned are called the *cyclotomic cosets* mod $2^L - 1$, see [3]. All the elements q_i of a cyclotomic coset have the same number of 1's in their binary representation, this number is called the *coset weight*. The leader element, E, of every coset is the smallest integer in such an equivalence class and its cardinal is L or a proper divisor of L.

Characteristic Polynomial of a Cyclotomic Coset. It is a polynomial $P_E(x)$ defined by $P_E(x) = (x+\alpha^E)(x+\alpha^{2E}) \ldots (x+\alpha^{2^{(r-1)}E})$, where the degree r ($r \leq L$) of $P_E(x)$ equals the cardinal of the cyclotomic coset E.

Characteristic Sequence of a Cyclotomic Coset. It is a binary sequence $\{S_n^E\}$ defined by the expression $\{S_n^E\} = \{\alpha^{En} + \alpha^{2En} + \ldots + \alpha^{2^{(r-1)}En}\}$ with $n \geq 0$. Recall that the previous sequence $\{S_n^E\}$ satisfies the linear recurrence relationship given by $P_E(x)$, see [3], [7].

Sequential Decomposition in Cyclotomic Cosets. The generic element z_n of the filtered sequence $\{z_n\}$ can be written as:

$$
\begin{aligned}
z_n = F(s_n, s_{n+1}, \ldots, s_{n+L-1}) = \\
C_1\alpha^{E_1 n} + (C_1\alpha^{E_1 n})^2 + \ldots + (C_1\alpha^{E_1 n})^{2^{(r_1-1)}} + \\
C_2\alpha^{E_2 n} + (C_2\alpha^{E_2 n})^2 + \ldots + (C_2\alpha^{E_2 n})^{2^{(r_2-1)}} +
\end{aligned}
$$

$$
\vdots \tag{1}
$$

$$
C_N\alpha^{E_N n} + (C_N\alpha^{E_N n})^2 + \ldots + (C_N\alpha^{E_N n})^{2^{(r_N-1)}},
$$

where r_i is the cardinal of coset E_i, the subindex i ranges in the interval $1 \leq i \leq N$ and N is the number of cosets of weight $\leq k$. At this point different features can be pointed out. Note that the i-th row of (1) corresponds to the nth-element of the sequence $\{C_i\alpha^{E_i n} + (C_i\alpha^{E_i n})^2 + \ldots + (C_i\alpha^{E_i n})^{2^{(r_i-1)}}\}$. The coefficient $C_i \in GF(2^{r_i})$ determines the starting point of such a sequence. If $C_i = 1$, then $\{S_n^{E_i}\}$ is in its *characteristic phase*. If $C_i = 0$, then $\{S_n^{E_i}\}$ would not contribute to the filtered sequence $\{z_n\}$. In that case, the cyclotomic coset E_i would be degenerate.

3   Representation of Nonlinear Filters

According to the previous section, nonlinear filters can be characterized by means of three different representations:

1. *Algebraic Normal Form (ANF)*: a nonlinear filter $F(s_n, s_{n+1}, \ldots, s_{n+L-1})$ can be uniquely expressed in ANF as the sum of distinct products in the variables $(s_n, s_{n+1}, \ldots, s_{n+L-1})$. The nonlinear order, k, of the filter is the maximum order of the terms appearing in its ANF. This is the representation currently used by the designer as variables and filter's order are easily handled.

2. *Bit-wise sum of the characteristic sequences*: a nonlinear filter $F(s_n, s_{n+1}, \ldots, s_{n+L-1})$ can be represented in terms of the N characteristic sequences $\{S_n^{E_i}\}$ that appear in its sequential decomposition in cosets, see equation (1). This representation enhances how the filtered sequenced is nothing but the addition of elementary sequences.

3. *A N-tuple of coefficients*: this is a representation very close to the previous one. In fact, a nonlinear filter $F(s_n, s_{n+1}, \ldots, s_{n+L-1})$ can be represented in terms of a N-tuple of coefficients $(C_1, C_2, \ldots, C_N)$ with $C_i \in GF(2^{r_i})$ where each coefficient determines the starting point of its sequence $\{S_n^{E_i}\}$. This representation gives information about the linear complexity of the filtered sequence.

3.1   Shifted Sequences

Let A be the set of the kth-order nonlinear filters applied to a LFSR of length L. We are going to group the elements of A producing the filtered sequence $\{z_n\}$ or a shifted version of $\{z_n\}$, notated $\{z_n\}^*$. From equation (1), it is clear that if we substitute C_i for $C_i \cdot \alpha^{E_i}$ $\forall i$, then we will obtain $\{z_{n+1}\}$. In general, $C_i \rightarrow C_i \cdot \alpha^{jE_i}$ $\forall i \Rightarrow \{z_n\} \rightarrow \{z_{n+j}\}$. This fact enables us to define an equivalence relationship $\sim$ on the set A as follows: $F \sim F'$ with $F, F' \in A$ if $\{F(s_n, \ldots, s_{n+L-1})\} = \{z_n\}$ and $\{F'(s_n, \ldots, s_{n+L-1})\} = \{z_n\}^*$. Therefore, two different nonlinear filters F, F' in the same equivalence class will produce shifted versions of the same filtered sequence. In addition, it is easy to see that the relation defined above is an equivalence relationship and that the number of elements in every equivalence class equals the period of the filtered sequence, T, so that the index j satisfies $0 \le j \le T - 1$. As the sequence $\{z_n\}^*$ is nothing but a shifted version of the filtered sequence $\{z_n\}$, then both filters F, F' will include the same characteristic sequences (starting at different points) as well as the same cosets of weight $\le k$ in their sequential decomposition in cosets.

3.2   Basic Operations with Nonlinear Filters

Now two different operations on nonlinear filters can be defined.

1. *Shifting operation $\mathcal{S}$*:

$$\mathcal{S}(F_0(s_n, s_{n+1}, \ldots, s_{n+L-1})) = F_1(s_n, s_{n+1}, \ldots, s_{n+L-1}),$$

that is the shifting operation applied to a nonlinear filter F_0 producing the filtered sequence $\{z_n\}$ gives rise to a new filter F_1 in the same equivalence class producing the sequence $\{z_{n+1}\}$.

2. *Sum operation*:

$$F_0(s_n, \ldots, s_{n+L-1}) + F_1(s_n, \ldots, s_{n+L-1}) = F_{01}(s_n, \ldots, s_{n+L-1}),$$

that is the sum of two consecutive nonlinear filters F_0, F_1 in the same equivalence class gives rise to a new filter F_{01} in a different equivalence class.

For a simple example, we consider the previous operations applied to the three different representations of nonlinear filters.

Example 1. For the pair $(L, k) = (3, 2)$, that is a nonlinear filter of second order applied to the stages of a LFSR of length $L = 3$ and primitive characteristic polynomial $P(x) = x^3 + x^2 + 1$, we have:

- The generic element, s_n, of the PN-sequence that can be written as $s_n = \alpha^n + \alpha^{2n} + \alpha^{4n}$, where α is a root of $P(x)$ so that $\alpha^3 = \alpha^2 + 1$.
- Two cyclotomic cosets: coset $1 = \{1, 2, 4\}$ of binary weight 1 and coset 3 $= \{3, 6, 5\}$ of binary weight 2.
- Two characteristic sequences of period $T = 7$: $\{S_n^1\} = \{1, 1, 1, 0, 1, 0, 0\}$ and $\{S_n^3\} = \{1, 0, 0, 1, 0, 1, 1\}$ both of them in their characteristic phases.
- Two characteristic polynomials: $P_1(x) = P(x)$ and $P_3(x)$ where

$$P_1(x) = (x + \alpha)(x + \alpha^2)(x + \alpha^4) = x^3 + x^2 + 1,$$
$$P_3(x) = (x + \alpha^3)(x + \alpha^6)(x + \alpha^5) = x^3 + x + 1.$$

- The extension field $GF(2^3) = \{0, \alpha, \alpha^2, \alpha^3, \ldots, \alpha^6, \alpha^7 = 1\}$.

Let F_0 be a nonlinear filter in ANF, $F_0(s_n, s_{n+1}, s_{n+2}) = s_n s_{n+1}$, applied to the previous LFSR starting at the initial state $(1, 1, 1)$. Therefore, the filtered sequence is $\{z_n\} = \{1, 1, 0, 0, 0, 0, 0\}$, the sequential decomposition in cosets is

$$\{z_n\} = C_1\{S_n^1\} \oplus C_3\{S_n^3\}$$
$$= \{1, 0, 0, 1, 1, 1, 0\} \oplus \{0, 1, 0, 1, 1, 1, 0\},$$

and the 2-tuple of coefficients is $(C_1, C_3) = (\alpha^4, \alpha^6)$ so that the sequence $\{S_n^1\}$ is shifted 4 positions $(C_1 = \alpha^4)$ regarding its characteristic phase while the sequence $\{S_n^3\}$ is shifted 2 positions $(C_3 = \alpha^6 = (\alpha^3)^2)$ regarding its characteristic phase too. For F_0 the previous operations are analyzed.

Shifting operation $\mathcal{S}$:

1. *ANF*: the shifting operation $\mathcal{S}$ increases by 1 all the indexes of the variables that appear in the filter expression,

$$\mathcal{S}(F_0(s_n, s_{n+1}, s_{n+2})) = F_1(s_n, s_{n+1}, s_{n+2})$$
$$\mathcal{S}(s_n s_{n+1}) \quad = \quad s_{n+1} s_{n+2}.$$

2. *Bit-wise sum of the characteristic sequences*: the operation $\mathcal{S}$ shifts simultaneously one position to all the characteristic sequences included in the nonlinear filter representation,

$$\mathcal{S}(F_0) = \mathcal{S}(\{1, 0, 0, 1, 1, 1, 0\} \oplus \{0, 1, 0, 1, 1, 1, 0\})$$
$$= \{0, 0, 1, 1, 1, 0, 1\} \oplus \{1, 0, 1, 1, 1, 0, 0\} = F_1.$$

3. *A N-tuple of coefficients*: the operation $\mathcal{S}$ multiplies each coefficient C_i of the N-tuple by the corresponding factor α^{E_i},

$$\mathcal{S}(C_1, C_3) = (C_1 \cdot \alpha, C_3 \cdot \alpha^3)$$
$$\mathcal{S}(\alpha^4, \alpha^6) = \quad (\alpha^5, \alpha^2).$$

Sum Operation

1. *ANF*: the sum of two nonlinear filters in ANF gives rise to a new nonlinear filter that is the logic sum of the two previous filters.

$$F_{01}(s_n, \ldots, s_{n+L-1}) = F_0(s_n, \ldots, s_{n+L-1}) + F_1(s_n, \ldots, s_{n+L-1})$$
$$F_{01}(s_n, \ldots, s_{n+L-1}) = s_n s_{n+1} \oplus s_{n+1} s_{n+2} \; .$$

2. *Bit-wise sum of the characteristic sequences*: the bit-wise sum of characteristic sequences included in the nonlinear filter representation gives rise to new characteristic sequences starting at different points,

$$
\begin{aligned}
F_0 \quad &= C_1\{S_n^1\} \oplus C_3\{S_n^3\} \\
F_1 \quad &= (C_1\,\alpha)\{S_n^1\} \oplus (C_3\,\alpha^3)\{S_n^3\} \\
F_0 + F_1 &= C_1(1+\alpha)\{S_n^1\} \oplus C_3(1+\alpha^3)\{S_n^3\} \\
&= (\{1,0,0,1,1,1,0\} \oplus \{0,1,0,1,1,1,0\}) \\
&\quad \oplus (\{0,0,1,1,1,0,1\} \oplus \{1,0,1,1,1,0,0\}) \\
&= \{1,0,1,0,0,1,1\} \oplus \{1,1,1,0,0,1,0\} \\
&= \{0,1,0,0,0,0,1\},
\end{aligned}
$$

thus the resulting sequence is not $\{z_n\}^*$ anymore but a different sequence. Consequently, the filter $F_0 + F_1$ belongs to a different equivalence class.

3. *A N-tuple of coefficients*: the N-tuple of coefficients of the sum of filters is the sum of the N-tuples of each nonlinear filter

$$
\begin{aligned}
F_0 &\to (C_1, C_3) \;, \quad F_1 \to (C_1 \cdot \alpha, C_3 \cdot \alpha^3) \\
F_0 + F_1 &\to (C_1 \cdot (1+\alpha), C_3 \cdot (1+\alpha^3)) = \\
&\quad (\alpha^4 \cdot \alpha^5, \alpha^3 \cdot \alpha^2) = (\alpha^2, \alpha^5) \; .
\end{aligned}
$$

4 Construction of All Possible Nonlinear Filters with Cosets of Weight k

In order to generate all the nonlinear filters with guaranteed cosets of weight k, we start from a filter with a unique term product of k equidistant phases of the form:

$$F_0(s_n, s_{n+1}, \ldots, s_{n+L-1}) = s_n s_{n+\delta} \cdots s_{n+(k-1)\delta} \tag{2}$$

with $1 \le k \le L$ and $gcd(\delta, 2^L - 1) = 1$. According to [10], the sequence obtained from this type of filters includes all the k-weight cosets. In the sequel, we will focus exclusively on the N_k cosets of weight k with N_k-tuple representation. In fact, they are the only cosets whose presence is guaranteed [4].

Given F_0 in ANF, the computation of its N_k-tuple is carried out via the root presence test described in [10]. Next, the N_k-tuple representations for $F_1 = \mathcal{S}(F_0)$ and $F_0 + F_1$ are easily computed too. The key idea in this construction method is shifting the filter $F_0 + F_1$ through its equivalence class and summing it with F_0 in order to cancel the successive components of its N_k-tuple.

The final result is:

1. A set of N_k basic filters of the form $(0, 0, \ldots, d_i, \ldots, 0, 0)$ $(1 \leq i \leq N_k)$ with $d_i \in GF(2^L), d_i \neq 0$.
2. Their corresponding ANF representations.

The combination of all these basic filters with d_i $(1 \leq i \leq N_k)$ ranging in $GF(2^L)$ (with the corresponding ANF representations) gives rise to all the possible terms of order k that preserve the cosets of weight k. Later, the addition of terms of order $< k$ in ANF permits the generation of all the nonlinear filters of order k that guarantee a linear complexity $LC \geq \binom{L}{k}$.

An algorithm for computing the basic nonlinear filters with cosets of weight k is depicted in Fig. 1. The employed notation is now introduced:

- F_0 is the initial filter with guaranteed cosets of weight k. Its N_k-tuple coefficient representation can be written as:

$$F_0 = (C_1^0, C_2^0, \ldots, C_{N_k}^0) = (C_i^0) \ (1 \leq i \leq N_k).$$

- $F_1 = \mathcal{S}(F_0)$ is the consecutive filter in the same equivalence class. Its N_k-tuple coefficient representation can be written as:

$$F_1 = (C_1^1, C_2^1, \ldots, C_{N_k}^1) = (C_i^1) \ (1 \leq i \leq N_k).$$

- $F_{01} = F_0 + F_1$ is a new filter in a different equivalence class whose N_k-tuple coefficient representation is:

$$F_{01} = (C_1^2, C_2^2, \ldots, C_{N_k}^2) = (C_i^2) \ (1 \leq i \leq N_k).$$

- The filter (C_i^2) ranges in its equivalence class until the j-th component $(C_j^2) = (C_j^0)$. The resulting filter is:

$$(C_1^3, C_2^3, \ldots, C_j^0, \ldots, 0) = (C_i^3) \ (1 \leq i \leq N_k),$$

where $C_l^3 = 0$ for $(j + 1 \leq l \leq N_k)$.
- The filter (C_i^4) is the sum of:

$$(C_i^0) + (C_i^3) = (C_i^4) = (C_1^4, C_2^4, \ldots, 0, \ldots, 0) \ (1 \leq i \leq N_k),$$

where $C_l^4 = 0$ for $(j \leq l \leq N_k)$.
- (I_i^j) is an intermediate filter where (C_i^4) is stored for the corresponding value of the index j.

$$(I_1^j, I_2^j, \ldots, I_{N_k}^j) = (I_i^j) \ (1 \leq i \leq N_k).$$

- (B_i^j) is a basic filter whose components are 0 except for the j-th component $d_j \neq 0$.

$$(B_1^j, B_2^j, \ldots, B_{N_k}^j) = (B_i^j) = (0, 0, \ldots, d_j, \ldots, 0) \ (1 \leq i \leq N_k).$$

The symbol $(B_i^j)'$ means that the initial filter (B_i^j) has been shifted through its equivalence class.

Example 2. Let $(L, k) = (5, 3)$ be a nonlinear filter of third order applied to the stages of a LFSR of length $L = 5$ and primitive characteristic polynomial $P(x) = x^5 + x^3 + 1$ where α is a root of $P(x)$ so that $\alpha^5 = \alpha^3 + 1$. We have $N_3 = 2$ cyclotomic cosets of weight 3: coset $7 = \{7, 14, 28, 25, 19\}$ and coset $11 = \{11, 22, 13, 26, 21\}$. The initial filter with guaranteed cosets of weight 3 is $F_0(s_0, s_1, s_2) = s_0 s_1 s_2$. The algorithm described in Fig. 1 is applied.

INPUT: The nonlinear filter $F_0(s_0, s_1, s_2) = s_0 s_1 s_2 \rightarrow (C_i^0) = (\alpha^{20}, \alpha^{13})$
Compute: $F_1(s_0, s_1, s_2) = s_1 s_2 s_3 \rightarrow (C_i^1) = (\alpha^{20} \cdot \alpha^7, \alpha^{13} \cdot \alpha^{11}) = (\alpha^{27}, \alpha^{24})$
Initialize: $(I_i^2) = (C_i^0) = (\alpha^{20}, \alpha^{13})$
for $j = N_3$ to 2
– Step 1: Addition of two filters $F_0 + F_1 = F_{01}$

$$(C_i^0) + (C_i^1) = (C_i^2)$$
$$(\alpha^{20}, \alpha^{13}) + (\alpha^{27}, \alpha^{24}) = (\alpha^5, \alpha^5)$$

– Step 2: Comparison $F_0 : F_1$

$$(C_i^0) : (C_i^2)$$
$$(\alpha^{20}, \alpha^{13}) : (\alpha^5, \alpha^5)$$

– Step 3: Shifting of (C_i^2) until $(C_2^2) = (C_2^0)$

$$(C_1^2, C_2^2) \rightarrow (C_1^3, C_2^0)$$
$$(\alpha^5, \alpha^5) \rightarrow (\alpha^{27}, \alpha^{13}) = (C_i^3)$$

– Step 4: Addition

$$(C_i^0) + (C_i^3) = (C_i^4)$$
$$(\alpha^{20}, \alpha^{13}) + (\alpha^{27}, \alpha^{13}) = (\alpha^5, 0)$$
$$(I_i^1) = (C_i^4)$$

end for
Introduce $(B_i^1) = (I_i^1) = (\alpha^5, 0)$
for $j = 2$ to N_3
– Step 6: Comparison $(B_i^1) : (I_i^2)$

$$(\alpha^5, 0) : (\alpha^{20}, \alpha^{13})$$

– Step 7: Shifting of (B_i^1) until $(B_1^1) = (I_1^2) = \alpha^{20}$

$$(B_i^1) \rightarrow (B_i^1)'$$
$$(\alpha^5, 0) \rightarrow (\alpha^{20}, 0)$$

– Step 8: Addition

$$(B_i^1)' + (I_i^2) = (B_i^2)$$
$$(\alpha^{20}, 0) + (\alpha^{20}, \alpha^{13}) = (0, \alpha^{13})$$
$$(B_1^2, B_2^2) = (0, \alpha^{13})$$

end for

```
Input: One nonlinear filter with guaranteed k-weight cosets,
       F₀(sₙ,...,s_{L-1}) → (C₁⁰,...,Cᵢ⁰,...,C_{N_k}⁰),

Compute F₁ = S(F₀(sₙ,...,s_{L-1})) → (C₁¹,...,Cᵢ¹,...,C_{N_k}¹),
for j = N_k to 2 do
    Step 1: Addition of the two filters: F₀ + F₁ = F₀₁ →
            (Cᵢ⁰) + (Cᵢ¹) = (Cᵢ²)
    Step 2: Comparison F₀ : F₀₁
            (C₁⁰,...,Cᵢ⁰,...,C_{N_k}⁰) : (C₁²,...,Cᵢ²,...,C_{N_k}²)
    Step 3: Shifting of (C₁²,...,Cᵢ²,...,C_{N_k}²) through its equivalence class
            until Cⱼ² = Cⱼ⁰
            (C₁²,...,Cⱼ²,...,C_{N_k}²) → (C₁³,...,Cⱼ³,...,0)
    Step 4: Addition
            (Cᵢ⁰) + (Cᵢ³) = (Cᵢ⁴) = (C₁⁴,...,0,...,0)
            keep (Iᵢ^{j-1}) = (Cᵢ⁴)
    Step 5: Substitution
            (Cᵢ⁰) ← (Cᵢ⁴)
end for

(Bᵢ¹) = (Iᵢ¹); Display the ANF.
for j = 2 to N_k do
    Step 6: Comparison (Bᵢ¹),...,(Bᵢ^{j-1}) : (Iᵢ^j)
        for l = 1 to j-1 do
            Step 7: Shifting of (Bᵢˡ)
                    until Bₗˡ = Iₗʲ
        end for
    Step 8: Addition
            Σ_{l=1}^{j-1} (Bᵢˡ)' + (Iᵢʲ) = (Bᵢʲ)
            Display the ANF.
end for

Output: N_k basic filters (Bᵢʲ) = (0,0,...,dⱼ,...,0) to generate
        all the nonlinear filters preserving the k-weight cosets
        and their ANF representations.
```

Fig. 1. Pseudo-code of the algorithm to generate filters with guaranteed k-weight cosets

OUTPUT: $N_3 = 2$ basic nonlinear filters and their corresponding ANF representations.

1. $(B_i^1) = (\alpha^5, 0)$
   ANF: $s_0 s_1 s_2 \oplus s_0 s_1 s_3 \oplus s_0 s_1 s_4 \oplus s_0 s_2 s_4 \oplus s_0 s_3 s_4 \oplus s_1 s_2 s_3 \oplus s_1 s_3 s_4$
2. $(B_i^2) = (0, \alpha^{13})$
   ANF: $s_0 s_2 s_4 \oplus s_0 s_3 s_4 \oplus s_1 s_2 s_4$.

Ranging (B_i^1) and (B_i^2) in their corresponding equivalence class and summing all the possible combinations in ANF representation, we get the 31×31 possible combinations of terms of order 3 that guarantee the cosets of weight 3 (coset 7 and coset 11). Next, the addition of terms of order < 3 in ANF representation permits us the generation of all the nonlinear filters of order 3.

5   Conclusions

A method of computing all the nonlinear filters applied to a LFSR that guarantee the cosets of weight k has been developed. The final result is a set of basic nonlinear filters whose combination allows one to generate all the nonlinear filters preserving a large linear complexity. That means a practical criterium to design cryptographic sequence generators. The method increases by far the number of nonlinear filters with a guaranteed linear complexity found in the literature. The generalization to other cosets of weight $< k$ is left as the natural continuity of this work.

References

1. eSTREAM, the ECRYPT Stream Cipher Project, The eSTREAM Portfolio (2012), http://www.ecrypt.eu.org/documents/D.SYM.10-v1.pdf
2. Caballero-Gil, P., Fúster-Sabater, A.: A wide family of nonlinear filter functions with large linear span. Inform. Sci. 164, 197–207 (2004)
3. Golomb, S.: Shift-Register Sequences, revised edn. Aegean Park Press (1982)
4. Kolokotronis, N., Kalouptsidis, N.: On the linear complexity of nonlinearly filtered PN-sequences. IEEE Trans. Inform. Theory 49, 3047–3059 (2003)
5. Kolokotronis, N., Limniotis, K., Kalouptsidis, N.: Lower Bounds on Sequence Complexity Via Generalised Vandermonde Determinants. In: Gong, G., Helleseth, T., Song, H.-Y., Yang, K. (eds.) SETA 2006. LNCS, vol. 4086, pp. 271–284. Springer, Heidelberg (2006)
6. Limniotis, K., Kolokotronis, N., Kalouptsidis, N.: On the Linear Complexity of Sequences Obtained by State Space Generators. IEEE Trans. Inform. Theory 54, 1786–1793 (2008)
7. Lidl, R., Niederreiter, H.: Finite Fields. In: Enciclopedia of Mathematics and Its Applications 20, 2nd edn. Cambridge University Press, Cambridge (1997)
8. Peinado, A., Fúster-Sabater, A.: Generation of pseudorandom binary sequences by means of linear feedback shift registers (LFSRs) with dynamic feedback. Mathematical and Computer Modelling 57, 2596–2604 (2013)
9. Robshaw, M., Billet, O. (eds.): New Stream Cipher Designs. LNCS, vol. 4986. Springer, Heidelberg (2008)
10. Rueppel, R.A.: Analysis and Design of Stream Ciphers. Springer, New York (1986)

The Isomorphism of Polynomials Problem Applied to Multivariate Quadratic Cryptography

Marta Conde Pena[1], Raúl Durán Díaz[2],
Luis Hernández Encinas[1], and Jaime Muñoz Masqué[1]

[1] Instituto de Seguridad de la Información, CSIC, E-28006 Madrid, Spain
{marta.conde,luis,jaime}@iec.csic.es
[2] Universidad de Alcalá, E-28871 Alcalá de Henares, Spain
raul.duran@uah.es

Abstract. The threat quantum computing poses to traditional cryptosystems (such as RSA, elliptic-curve cryptosystems) has brought about the appearance of new systems resistant to it: among them, multivariate quadratic public-key ones. The security of the latter kind of cryptosystems is related to the isomorphism of polynomials (IP) problem. In this work, we study some aspects of the equivalence relation the IP problem induces over the set of quadratic polynomial maps and the determination of its equivalence classes. We contribute two results. First, we prove that when determining these classes, it suffices to consider the affine transformation on the left of the central vector of polynomials to be linear. Second, for a particular case, we determine an explicit system of invariants from which systems of equations whose solutions are the elements of an equivalence class can be derived.

Keywords: Equivalence classes, Equivalent keys, Isomorphism of polynomials problem, Multivariate cryptography, System of invariants.

1 Introduction

The public-key cryptosystems that are most widely implemented at present are based on the classic integer factorization problem (RSA cryptosystem) or on the discrete logarithm problem (ElGamal and elliptic curve cryptosystems ECC).

While the above cryptosystems are currently considered to display good security properties for a set of appropriately chosen parameters, Peter Shor published in 1997 ([1]) an algorithm that, assuming the existence of a quantum computer powerful enough, is able to break them.

Since then, the threat of quantum computing has encouraged the cryptographic community to revisit old trapdoor one-way functions or design new ones and derive from them new public-key systems over which quantum computation appears to have no advantage. Among the usually called post-quantum cryptosystems, we can mention hash-based ([2]), code-based ([3]), lattice-based ([4]) or multivariate quadratic-based public-key cryptographic systems ([5]). In this paper we focus in the latter.

Á. Herrero et al. (eds.), *International Joint Conference SOCO'13-CISIS'13-ICEUTE'13*, 567
Advances in Intelligent Systems and Computing 239,
DOI: 10.1007/978-3-319-01854-6_58, © Springer International Publishing Switzerland 2014

While systems of linear equations over finite fields are solvable in polynomial time via the row reduction method, solving a system of quadratic equations over a finite field is known to be an NP-complete problem (see [6], p.251; [7], section 2.5; [8]) and no efficient method to solve it is known as long as the number of unknowns and equations is chosen conveniently. Multivariate quadratic public-key cryptosystems ($\mathcal{MQ}$ cryptosystems for short) make use of multivariate quadratic polynomials over finite fields, and even though some of them are broken as of today, they are considered of interest since they are based on two problems which are believed to be hard, namely the $\mathcal{MQ}$ (Multivariate Quadratic) problem and the IP (Isomorphism of Polynomials) problem (see [7], [9], and [10] for novel algorithms that refine the computational time of a brute force attack).

In this work we study some aspects of the polynomial equivalence relation induced by the IP problem. The determination of these classes would have considerable impact in $\mathcal{MQ}$ cryptography: on the one hand, it provides us with a better understanding about which public keys are good choices for cryptanalysis purposes; for example, it is desirable that the equivalence class of the public key has as few elements as possible. On the other hand, considering that 'equivalent' keys exist for $\mathcal{MQ}$ schemes ([7]) and hence the private key space can be reduced, it lets us precise which exact reduction of the private key space is possible. This is relevant since we must examine if the resulting private key space is still large enough for the $\mathcal{MQ}$ scheme to be secure. In this paper, we identify these polynomial equivalence classes as the orbits of a group action on the set of quadratic functions and then determine, in a particular case, a system of invariants that leads us to a system of equations from which the orbits can be derived.

The rest of the contents are organized as follows: in section 2 we give a quick overview of $\mathcal{MQ}$ cryptography and the two hard problems that underlie the one-way security of this kind of cryptosystems. In section 3 we fix some notation, present the problem and the related work done on the subject. We present our results in section 4, and the conclusions and future work in section 5.

2   Multivariate Cryptography and One-Way Security

Introduced by Matsumoto and Imai in 1988 ([11]), multivariate quadratic cryptography is based on the difficulty of the $\mathcal{MQ}$ problem, defined as follows:

Definition 1. *The $\mathcal{MQ}$ problem over a finite field of q elements $\mathbb{F}_q$ consists in finding a solution $x \in \mathbb{F}_q^n$ to a given system of m quadratic polynomial equations $y = (p_1, \dots, p_m)$ over $\mathbb{F}_q$ in n indeterminates. That is, the goal is to find a solution of the following system of equations*

$$y_1 = p_1(x_1, \dots, x_n)$$
$$y_2 = p_2(x_1, \dots, x_n)$$
$$\vdots$$
$$y_m = p_m(x_1, \dots, x_n)$$

for a given $y = (y_1, y_2, \ldots, y_m) \in \mathbb{F}_q^m$ *and unknown* $x = (x_1, x_2, \ldots, x_n) \in \mathbb{F}_q^n$.

This problem is NP-complete over any finite field for a randomly selected polynomial vector $p = (p_1, \ldots, p_m)$ ([6], p.251; [7], section 2.5; [8]), and even though this sole fact does not guarantee that a potential $\mathcal{MQ}$-based cryptosystem is secure, there is strong evidence pointing out that the $\mathcal{MQ}$ problem is also hard on average (see [12], [13]), thus suggesting that the $\mathcal{MQ}$ problem can indeed be used as a basis for a secure public-key cryptosystem.

To construct a generic multivariate quadratic public-key cryptosystem, choosing a central vector of quadratic polynomials $p' : \mathbb{F}_q^n \mapsto \mathbb{F}_q^m$ easy to invert, and then make it look random at the eyes of an adversary composing it with two affine invertible transformations in the way $p = T \circ p' \circ S$ seems like a good starting point. Indeed, if we make p public, keep T, p', S private and encrypt a message $x \in \mathbb{F}_q^n$ by computing $y = p(x)$, an eavesdropper should fail to decrypt as he would need to solve an instance of the $\mathcal{MQ}$-problem; however, the legitimate user can decrypt in polynomial time using his private key. We now describe a generic $\mathcal{MQ}$ cryptosystem precisely:

- **Key generation:** Using a cryptographically secure pseudorandom number generator over $\mathbb{F}_q$, matrices are created until a random matrix in $\mathbb{F}_q^{m \times m}$ with non-zero determinant is generated, denote it by T_l. Then, a random column vector of length m is created, denote it by T_c. The affine transformation $T : \mathbb{F}_q^m \mapsto \mathbb{F}_q^m$ is set to be $T = T_l x + T_c$. The affine map $S : \mathbb{F}_q^n \mapsto \mathbb{F}_q^n$ is created in an analogous manner. To generate a vector of quadratic polynomials $p' : \mathbb{F}_q^m \mapsto \mathbb{F}_q^n$ so that for a known output y of the function p', the pre-image of p' is easily computable, p' is usually chosen uniformly at random from the set of invertible quadratic functions. However, this is not the only way to do so, as building a small amount of redundancy into messages sent (usually by concatenating part or all of a message's hash to the message itself) makes it possible to use p' that are non-invertible. The *private key* is defined to be the ordered set (T, p', S). The *public key* is formed by $p = T \circ p' \circ S$.
- **Encryption:** Given the public key of a user, p, the *encryption process* consists in transforming a given message or plaintext $x \in \mathbb{F}_q^n$ (possibly with the redundancy required by the choice of the key) into a ciphertext $y = p(x)$.
- **Decryption:** Given the secret key of a user, (T, p', S), the *decryption process* consists in transforming a given ciphertext y into a plaintext x. In general (depending on the form of building p'), the decryption process goes as follows: first compute

$$T^{-1}(y) = (p' \circ S)(x),$$

and the set of pre-images of p':

$$p'^{-1}(T^{-1}(y)).$$

Then, each element $\tau \in p'^{-1}(T^{-1}(y))$ is given as an input to S^{-1} (which exists by the very definition of S). The exact redundancy specified by the encryption scheme will almost certainly occur in only one such pre-image $S^{-1}(\tau)$. Remove the redundancy to retrieve the decrypted message $x \in \mathbb{F}_q^n$.

If p' is invertible, the plaintext, x, is obtained by simply applying the three inverses S^{-1}, p'^{-1}, and T^{-1} in order to the ciphertext y:

$$x = S^{-1} \circ p'^{-1} \circ T^{-1}(y).$$

A public-key cryptosystem is said to be one-way secure if with access to the public information and a given ciphertext it is hard to recover the message, in the sense that there exists no polynomial time algorithm that outputs the message with non-negligible probability. The one-way security of multivariate quadratic cryptosystem rests upon the $\mathcal{MQ}$ problem and, as Patarin pointed out in ([14]), upon what is called the IP problem:

Definition 2. *Given a pair of (not necessarily quadratic) polynomial vectors, p and p', where each vector is an m-tuple of polynomials over the same set of n indeterminates as in Definition 1, the IP problem is to find a pair of affine transformations $T : \mathbb{F}_q^m \mapsto \mathbb{F}_q^m$ and $S : \mathbb{F}_q^n \mapsto \mathbb{F}_q^n$ such that*

$$p = T \circ p' \circ S.$$

Solving the IP problem is equivalent to solving for the unknowns of S and T using the entries of p and p' as constants. This produces a system of total degree one larger than the total degree of p'. In particular, when solving problems in the situation of $\mathcal{MQ}$-based encryption schemes, the total degree of these equations is 3 in the unknown entries of T and S. This problem is NP-hard (see [15]).

3   Mathematical Problem

3.1   Notation

We denote the ground field of q elements $\mathbb{F}_q$ by $\mathbb{F}$, the set of affine transformations in $\mathbb{F}^n$ by $GA(n, \mathbb{F})$, the set of linear transformations in $\mathbb{F}^n$ by $GL(n, \mathbb{F})$, and by $P(n, m, \mathbb{F})$, or simply by $P(n, m)$ if there is no ambiguity, the set of all quadratic polynomial maps from $\mathbb{F}^n$ to $\mathbb{F}^m$. Note that

$$\dim P(n, m) = \tfrac{1}{2}mn(n+1) + mn + m = \tfrac{1}{2}m(n+2)(n+1). \tag{1}$$

In what follows we work under the hypothesis $n \leq m$, as we are mainly interested in injective quadratic maps in $P(n, m)$ for purposes of uniqueness in decryption.

Recall that in the context of multivariate quadratic cryptography, the public key $p = (p_1, \ldots, p_m) \in P(n, m)$ is a vector of length m whose components are quadratic polynomials, this is,

$$p_i(x_1, \ldots, x_n) = \sum_{1 \leq j \leq k \leq n} \gamma_{i,j,k} x_j x_k + \sum_{j=1}^{n} \mu_{i,j} x_j + \nu_i \tag{2}$$

for $1 \leq i \leq m$, $1 \leq j \leq k \leq n$ and $\gamma_{i,j,k}, \mu_{i,j}, \nu_i \in \mathbb{F}$ (they are the quadratic, linear and constant terms respectively). The expression (2) of each component can be rewritten, using the Einstein summation convention, as

$$y^\alpha(p(x)) = \sum_{j \leq k} q_{jk}^\alpha x^j x^k + \ell_j^\alpha x^j + k^\alpha,$$

where $x = (x^1, \ldots, x^n) \in \mathbb{F}^n$, $y = (y^1, \ldots, y^m) \in \mathbb{F}^m$, and $q_{jk}^\alpha, \ell_j^\alpha, k^\alpha \in \mathbb{F}$.

Moreover, it is often convenient to be able to express a system of equations $y = p(x)$ in matrix form. To do this, we order the products $x^j x^k$ lexicographically according to their superindices, we write them in a column vector of length $\frac{1}{2}n(n+1)$ denoted by $\mathcal{C}(x)$, and we define the following matrices

$$Q_p = \left(q_{jk}^\alpha\right)_{1 \le j \le k \le n, 1 \le \alpha \le m}, \quad L_p = \left(l_j^\alpha\right)_{1 \le j \le n, 1 \le \alpha \le m},$$

$$K_p = (k^\alpha)_{1 \le \alpha \le m}^t, \quad X = (x^\alpha)_{1 \le \alpha \le m}^t, \quad Y = (y^\alpha)_{1 \le \alpha \le m}^t. \tag{3}$$

The equations derived from $y = p(x)$ can then be expressed as

$$Y(p(x)) = Q_p \mathcal{C}(x) + L_p X + K_p.$$

3.2   Statement of the Problem and Related Work

From Definition 2, it follows that the IP problem induces the equivalence relation:

$$p \sim p', \; p, p' \in P(n, m) \iff \exists T \in GA(m, \mathbb{F}), \exists S \in GA(n, \mathbb{F}) : p = T \circ p' \circ S.$$

Mathematical Problem: Determine all such polynomial equivalence classes.

Application to MQ Cryptography: Suppose we have an $\mathcal{MQ}$ scheme, a private key (T, p', S) and a public key $p = T \circ p' \circ S$. On the one hand, the equivalence class of $p' \in P(n, m)$ is

$$[p'] = \{p \in P(n, m) : p = T \circ p' \circ S \text{ for some } T \in GA(m, \mathbb{F}), S \in GA(n, \mathbb{F})\},$$

this is, the equivalence class of $p' \in P(n, m)$ is formed by all the different public keys that can be derived from a fixed central polynomial vector p' just varying the affine transformations T and S. From this point of view the problem of determining the equivalence classes is appealing as we can then assess which central polynomial vectors produce public keys with good cryptographic properties.

On the other hand, the equivalence class of $p \in P(n, m)$ is

$$[p] = \{p' \in P(n, m) : p = T \circ p' \circ S \text{ for some } T \in GA(m, \mathbb{F}), S \in GA(n, \mathbb{F})\},$$

this is, the equivalence class of p is formed by all the central polynomial vectors p' that give rise to the same public key p. For each $p' \in [p]$, there exist $T_{p'} \in GA(m, \mathbb{F})$ and $S_{p'} \in GA(n, \mathbb{F})$ such that $p = T_{p'} \circ p' \circ S_{p'}$. This way, for cryptanalysis purposes, finding (T, p', S) is equivalent to finding any of the triples $(T_{p'}, p', S_{p'})$, and in fact, these keys are said to be equivalent. As a consequence, we can store just one representative of a set of equivalent keys, which is advantageous for the implementation of $\mathcal{MQ}$ schemes in devices with restricted memory. From this point of view, obtaining the equivalence classes is interesting as we can determine which exact reduction of the private key space is possible and then assess if the resulting private key space is large enough not to compromise the security. Furthermore, it is clear that the most interesting public keys are those whose equivalence class has as few elements as possible.

An equivalence relation can be defined on the set of private keys under which the private keys (T_1, p_1', S_1) and (T_2, p_2', S_2) are indeed equivalent if they lead to the same public key, this is, if $T_1 \circ p_1' \circ S_1 = T_2 \circ p_2' \circ S_2$. The study of such an equivalence relation is done in [16], [17], [18].

A good mathematical framework to study the polynomial equivalence problem is proposed in [19]. In particular, consider the group action

$$GA(m, \mathbb{F}) \times GA(n, \mathbb{F}) \times P(n, m) \to P(n, m)$$
$$((T, S), p) \mapsto (T, S) \cdot p$$

where

$$((T, S) \cdot p)(x) = (T \circ p \circ S)(x) \, \forall x \in \mathbb{F}^n, \forall p \in P(n, m),$$

Two elements of $P(n, m)$ are defined to be equivalent with respect to the action of the group $GA(m, \mathbb{F}) \times GA(n, \mathbb{F})$ as follows:

$$p \, R \, p', \, p, p' \in P(n, m) \iff \exists (T, S) \in GA(m, \mathbb{F}) \times GA(n, \mathbb{F}) : p = (T, S) \cdot p'$$

The relation R is equivalent to the relation $\sim$, and therefore the problem of determining the polynomial equivalence classes is equivalent to determining the orbit space of the above action of $GA(m, \mathbb{F}) \times GA(n, \mathbb{F})$ on $P(n, m)$.

In [20], work has been done on enumerating or bounding the number of polynomial equivalence classes for homogeneous quadratic polynomials, but their description remains open. It also explains how the determination of the polynomial equivalence classes problem relates to the problem of equivalent keys studied in ([16], [17], [18], [21]).

We focus on the description of the classes: we first give a qualitative result guaranteeing that the orbits with respect to the left the action of $GA(m, \mathbb{F})$ are the same as the orbits with respect to the action of $GL(m, \mathbb{F})$. We next determine a system of invariants from which certain equivalence classes in the case $n = m$ can be derived.

4   Results

We divide the study of the action of $GA(m, \mathbb{F}) \times GA(n, \mathbb{F})$ on $P(n, m)$ into the study of a left and a right action on $P(n, m)$, as follows:

$$GA(m, \mathbb{F}) \times P(n, m) \to P(n, m), \qquad P(n, m) \times GA(n, \mathbb{F}) \to P(n, m),$$
$$(T, p) \mapsto T \cdot p, \qquad (p, S) \mapsto p \cdot S,$$

where:

$$(T \cdot p)(x) = (T \circ p)(x), \qquad (p \cdot S)(x) = (p \circ S)(x),$$
$$\forall x \in \mathbb{F}^n, \forall p \in P(n, m), \forall T \in GA(m, \mathbb{F}), \forall S \in GA(n, \mathbb{F}).$$

The left action induces the following equivalence relation on $P(n, m)$:

$$p \sim_l p', \, p, p' \in P(n, m) \iff \exists T \in GA(m, \mathbb{F}) : p = T \cdot p',$$

and the right action induces the next equivalence relation on $P(n, m)$:

$$p \sim_r p', \, p, p' \in P(n, m) \Longleftrightarrow \exists S \in GA(n, \mathbb{F}) : p = p' \cdot S.$$

It is just a computational matter to obtain the matrices associated to the quadratic polynomial maps $T \cdot p$ and $p \cdot S$ as a function of the matrices associated to p defined in (3), and we state this in a lemma.

Lemma 1. *The matrices associated to the transformed quadratic maps $T \cdot p$ y $p \cdot S$ are, respectively:*

$$Q_{T \cdot p} = A Q_p, \qquad L_{T \cdot p} = A L_p, \qquad K_{T \cdot p} = A K_p + a^t,$$

where $T(x) = A(x) + a$, $\forall x \in \mathbb{F}^m$, $A \in GL(m, \mathbb{F})$, $a = (a^1, \ldots, a^m) \in \mathbb{F}^m$.

$$Q_{p \cdot S} = Q_p \tilde{B}, \quad L_{p \cdot S} = Q_p \dot{S} + L_p B, \quad K_{p \cdot S} = Q_p \mathcal{C}(b) + L_p(b) + K_p,$$

with $S(x) = B(x) + b$, $x \in \mathbb{F}^n$, $B = (b_j^i)_{i,j=1}^n \in GL(n, \mathbb{F})$, $b = (b^1, \ldots, b^n) \in \mathbb{F}^n$, where $\tilde{B}$ is the square matrix of order $\frac{1}{2}n(n+1)$ given by:

$$\tilde{B}_{hl}^{jk} = \frac{1}{1 + \delta_{hl}} \left(b_h^j b_l^k + b_l^j b_h^k \right), \qquad 1 \leq j \leq k \leq n, \quad 1 \leq h \leq l \leq n,$$

and $\dot{S}$ the matrix of order $\frac{1}{2}n(n+1) \times n$ defined as:

$$\dot{S}_h^{jk} = b^j b_h^k + b^k b_h^j, \qquad 1 \leq j \leq k \leq n.$$

4.1   Equivalence with Respect to the Left Action of $GA(m, \mathbb{F})$

In this section, we present our results with respect to the study of the left action of $GA(m, \mathbb{F})$ on $P(m, n)$:

Lemma 2. *Given a map $p \in P(n, m)$, we denote by $\bar{p} \in P(n, m)$ the map defined as $\bar{p}(x) = p(x) - p(0_n)$, $\forall x \in \mathbb{F}^n$, where $0_n = (0, \overset{(n)}{\ldots}, 0)$. Two maps $p, p' \in P(n, m)$ are equivalent with respect to the left action of the affine group $GA(m, \mathbb{F})$ if and only if $\bar{p}, \bar{p}' \in P(n, m)$ are equivalent with respect to the left action of the linear group $GL(m, \mathbb{F})$. Therefore, the study of the left action of the affine group $GA(m, \mathbb{F})$ on $P(n, m)$ reduces to the study of the left action of the linear group $GL(m, \mathbb{F})$ on the subspace*

$$\bar{P}(n, m) = \{\bar{p} \in P(n, m) : \bar{p}(0_n) = 0_m\} \subset P(n, m).$$

In a nutshell, this means that to check if p, p' are equivalent with respect to $\sim_l$ we can just ignore their affine parts and check if their linear parts are equivalent or not.

A consequence of the lemma is that for cryptanalysis purposes of $\mathcal{MQ}$ cryptosystems, we can consider a public key p to be $p = T \circ p' \circ S$, with $T \in \text{Hom}^{-1}(\mathbb{F}^m)$, $S \in GA(n, \mathbb{F})$. A stronger result for cryptanalysis purposes is proved in ([16]): it actually suffices to consider $T \in \text{Hom}^{-1}(\mathbb{F}^m)$, $S \in \text{Hom}^{-1}(\mathbb{F}^n)$, this is, both linear.

Recall now the concept of invariant:

Definition 3. *Let* $G \times \mathbb{X} \to \mathbb{X}$, $(g, x) \mapsto g \cdot x$, $\forall g \in G$, $\forall x \in \mathbb{X}$, *the left action of a group* G *on a space* $\mathbb{X}$. *A function* $I \colon \mathbb{X} \to \mathbb{F}$ *is said to be* G-*invariant if it satisfies the condition* $I(g \cdot x) = I(x)$, $\forall g \in G$, $\forall x \in \mathbb{X}$. *(Analogous definition for a right action).*

A system of G-*invariant functions* $I_i \colon \mathbb{X} \to \mathbb{F}$, $1 \le i \le \nu$ *is said to be complete if the equations* $I_i(x) = I_i(x')$, $1 \le i \le \nu$, *for any* $x, x' \in \mathbb{X}$ *imply the existence of an element* $g \in G$ *such that:* $x' = g \cdot x$. *(Analogous definition for a right action).*

In other words, a complete system of invariants determines the orbits of the action of G on $\mathbb{X}$: determining the orbit of x_0 is reduced to solving the system of equations $I_1(x) = I_1(x_0), \ldots, I_\nu(x) = I_\nu(x_0)$.

Going back to our case, set $\bar{P}^r(n, m) \subset \bar{P}(n, m)$ to be the subset of all maps $\bar{p}$ such that the rank of the matrix $L_{\bar{p}}$ is r, for $0 \le r \le n$. We have a partition of $\bar{P}(n, m)$:

$$\bar{P}(n, m) = \cup_{r=0}^n \bar{P}^r(n, m),$$
$$\emptyset = \bar{P}^r(n, m) \cap \bar{P}^s(n, m), \quad 0 \le r < s \le n.$$

Note that this partition induces a partition in $P(n, m)$. Since $n \le m$ by hypothesis, $\bar{P}^n(n, m)$ is the set of polynomial quadratic maps such that their linear part has maximum rank. This set is of particular interest in $\mathcal{MQ}$ cryptography since we should focus on maps with as little linear dependence as possible. We find a complete system of invariants for the set $\bar{P}^n(n, m)$ in the particular case $n = m$:

Theorem 1. *Consider* $m = n$. *We define a system of* $\frac{1}{2}n^2(n+1)$ *functions* $I_{i,jk} \colon \bar{P}^n(n, n) \to \mathbb{F}$, $j \le k$ *as follows:*

$$(I_{i,jk}(\bar{p}))_{1 \le j \le k \le n}^{1 \le i \le n} = (L_{\bar{p}})^{-1} Q_{\bar{p}}, \quad \forall \bar{p} \in \bar{P}^n(n, n).$$

The above functions form a complete system of invariants for the left action of the linear group $GL(n, \mathbb{F})$ *on* $\bar{P}^n(n, n)$.

In a nutshell, the solutions of the systems of equations $I_{i,jk}(\bar{p}) = I_{i,jk}(p)$ are precisely the elements of the orbit of $\bar{p}$ under $\sim_l$, this is, the elements of the equivalence class of $\bar{p}$ under $\sim_l$.

Corollary 1 (Normal forms). *Every quadratic map of* $\bar{P}^n(n, n)$ *is* $GL(n, \mathbb{F})$-*equivalent to a unique quadratic map* $\bar{p} \in \bar{P}^n(n, n)$ *such that* $L_{\bar{p}}$ *is the identity.*

4.2   Equivalence with Respect to the Right Action of $GA(m, \mathbb{F})$

One does not get so lucky with the right action of $GA(m, \mathbb{F})$ on $P(n, m)$ as Lemma 2 is not true in this case: we see an easy counterexample. Let $m = n = 1$ and $p(x) = qx^2 + \ell x + k$, $p'(x) = q'x^2 + \ell'x + k'$, $q \ne 0$, $q' \ne 0$, two quadratic maps. Set $S(x) = Bx + b$ an affine transformation; here, $B \neq 0$ and b are scalars. We have:

$$p' = p \cdot S \iff \begin{cases} q' = qB^2 \\ \ell' = (2qb + \ell)B \\ k' = qb^2 + \ell b + k \end{cases}, \quad \bar{p}' = \bar{p} \cdot B \iff \begin{cases} q' = qB^2 \\ \ell' = \ell B \end{cases}$$

Therefore:

– The maps p and p' are equivalent with respect to the right action of $GA(1, \mathbb{F})$ if and only if the following two conditions are satisfied:

(i) $\ell^2 - 4q(k - k') \in (\mathbb{F}^*)^2$,
(ii) $q' \{\ell^2 - 4q(k - k')\} = q\ell'^2$.

– The maps $\bar{p}$ and $\bar{p}'$ are equivalent to the right action of $GL(1, \mathbb{F})$ if and only if: $q'\ell^2 = q\ell'^2$, which is the former condition (ii) for $k = k' = 0$. Note that in this case the condition (i) is automatically verified.

On the other hand, according to (1), we have:

$$\dim(P(n, m)/GA(n, \mathbb{F})) = m - n + \tfrac{1}{2}n \left[(m - 2)n + 3m\right],$$
$$\dim(\bar{P}(n, m)/GL(n, \mathbb{F})) = \tfrac{1}{2}n \left[(m - 2)n + 3m\right],$$

from where: $\dim(P(n, m)/GA(n, \mathbb{F})) - \dim(\bar{P}(n, m)/GL(n, \mathbb{F})) = m - n$. Hence, if $m > n$, then the orbit spaces have different dimension.

5 Conclusions and Future Work

We studied some aspects of the equivalence relation induced by the IP problem and saw that the determination of these equivalence classes results in a better understanding of the security of $\mathcal{MQ}$ schemes and an improvement of their efficiency. In particular, we proved that when determining these classes, it suffices to consider the affine transformation on the left of the central vector of polynomials to be linear. Then we obtained an explicit system of invariants for a particular case, from which systems of equations whose solutions are the elements of an equivalence class can be derived.

Concerning the right action the result obtained is negative, so further study is required. The study of the two actions acting simultaneously and hence the determination of the classes remains open.

Acknowledgment. This work has been partially supported by Ministerio de Ciencia e Innovación (Spain) under the grant TIN2011-22668.

References

1. Shor, P.W.: Polynomial-time algorithms for prime factorization and discrete logarithms on a quantum computer. SIAM J. Comput. 26(5), 1484–1509 (1997)

2. Merkle, R.C.: Secrecy, authentication, and public key systems. PhD thesis, Stanford University (1979)
3. McEliece, R.J.: A public-key cryptosystem based on algebraic coding theory. Technical Report 42-44, Jet Propulsion Laboratory (1978)
4. Goldreich, O., Goldwasser, S., Halevi, S.: Public-key cryptosystems from lattice reduction problems. In: Kaliski Jr., B.S. (ed.) CRYPTO 1997. LNCS, vol. 1294, pp. 112–131. Springer, Heidelberg (1997)
5. Ding, J., Gower, J.E., Schmidt, D.: Multivariate Public Key Cryptosystems. Advances in Information Security, vol. 25. Springer (2006)
6. Garey, M.R., Johnson, D.S.: Computer and Intractability: A Guide to the Theory of NP-Completness. W. H. Freeman & Co. (1990)
7. Wolf, C.: Multivariate Quadratic Polynomials in Public Key Criptography. PhD thesis, Katholieke Universiteit Leuven (November 2005)
8. Patarin, J.: Cryptanalysis of the Matsumoto and Imai public key scheme of Eurocryptp'88. In: Coppersmith, D. (ed.) CRYPTO 1995. LNCS, vol. 963, pp. 248–261. Springer, Heidelberg (1995)
9. Feldmann, A.T.: A Survey of Attacks on Multivariate Cryptosystems. PhD thesis, University of Waterloo (2005)
10. Bouillaguet, C., Fouque, P.-A., Véber, A.: Graph-theoretic algorithms for the 'isomorphism of polynomials' problem. In: Johansson, T., Nguyen, P.Q. (eds.) EUROCRYPT 2013. LNCS, vol. 7881, pp. 211–227. Springer, Heidelberg (2013)
11. Matsumoto, T., Imai, H.: Public quadratic polynomial-tuples for efficient signature-verification and message-encryption. In: Günther, C.G. (ed.) EUROCRYPT 1988. LNCS, vol. 330, pp. 419–453. Springer, Heidelberg (1988)
12. Courtois, N., Klimov, A., Patarin, J., Shamir, A.: Efficient algorithms for solving overdefined systems of multivariate polynomial equations. In: Preneel, B. (ed.) EUROCRYPT 2000. LNCS, vol. 1807, pp. 392–407. Springer, Heidelberg (2000)
13. Courtois, N., Goubin, L., Meier, W., Tacier, J.-D.: Solving underdefined systems of multivariate quadratic equations. In: Naccache, D., Paillier, P. (eds.) PKC 2002. LNCS, vol. 2274, pp. 211–227. Springer, Heidelberg (2002)
14. Patarin, J.: Hidden fields equations (HFE) and isomorphisms of polynomials (IP): Two new families of asymmetric algorithms. In: Maurer, U.M. (ed.) EUROCRYPT 1996. LNCS, vol. 1070, pp. 33–48. Springer, Heidelberg (1996)
15. Patarin, J., Goubin, L., Courtois, N.: Improved algorithms for isomorphisms of polynomials. In: Nyberg, K. (ed.) EUROCRYPT 1998. LNCS, vol. 1403, pp. 184–200. Springer, Heidelberg (1998)
16. Wolf, C., Preneel, B.: Equivalent keys in multivariate quadratic public key systems. Journal of Mathematical Cryptology 4(4), 375–415 (2005)
17. Wolf, C., Preneel, B.: Large superfluous keys in Multivariate Quadratic asymmetric systems. In: Vaudenay, S. (ed.) PKC 2005. LNCS, vol. 3386, pp. 275–287. Springer, Heidelberg (2005)
18. Wolf, C., Preneel, B.: Equivalent keys in HFE, C*, and variations. In: Dawson, E., Vaudenay, S. (eds.) Mycrypt 2005. LNCS, vol. 3715, pp. 33–49. Springer, Heidelberg (2005)
19. Faugère, J.-C., Perret, L.: Polynomial equivalence problems: Algorithmic and theoretical aspects. In: Vaudenay, S. (ed.) EUROCRYPT 2006. LNCS, vol. 4004, pp. 30–47. Springer, Heidelberg (2006)
20. Lin, D., Faugère, J.-C., Perret, L., Wang, T.: On enumeration of polynomial equivalence classes and their application to MPKC. Finite Fields and Their Applications 18(2), 283–302 (2012)
21. Mingjie, L., Lidong, H., Xiaoyun, W.: On the equivalent keys in multivariate cryptosystems. Tsinghua Science & Technology 16, 225–232 (2011)

Social News Website Moderation through Semi-supervised Troll User Filtering

Jorge de-la-Peña-Sordo, Igor Santos,
Iker Pastor-López, and Pablo García Bringas

S^{3}Lab, DeustoTech Computing, University of Deusto
Avenida de las Universidades 24, 48007, Bilbao, Spain
{jorge.delapenya,isantos,iker.pastor,pablo.garcia.bringas}@deusto.es

Abstract. Recently, Internet is changing to a more social space in which all users can provide their contributions and opinions to others via websites, social networks or blogs. Accordingly, content generation within social webs has also evolved. Users of social news sites make public links to news stories, so that every user can comment them or other users' comments related to the stories. In these sites, classifying users depending on how they behave, can be useful for web profiling, user moderation, etc. In this paper, we propose a new method for filtering trolling users. To this end, we extract several features from the public users' profiles and from their comments in order to predict whether a user is troll or not. These features are used to train several machine learning techniques. Since the number of users and their comments is very high and the labelling process is laborious, we use a semi-supervised approach known as collective learning to reduce the labelling efforts of supervised approaches. We validate our approach with data from 'Menéame', a popular Spanish social news site, showing that our method can achieve high accuracy rates whilst minimising the labelling task.

Keywords: User Profiling, Content Filtering, Web Mining, User Categorisation, Machine-learning.

1 Introduction

Multimedia content is widely spread in the web, thanks to the technological evolution, specially mobile devices; however, textual content has still a major role in web content. Regardless to the fact that most of the social websites have multimedia files, such video or images, a large section is still occupied by textual content published in ideas, opinions and beliefs. This kind of content can be comments or stories reflecting different users' behaviours. Therefore, in a certain way, the users' dynamic interaction and collaboration has been drastically enhanced.

In particular, social news websites such as Digg[1] or 'Menéame'[2] are very popular among users. These sites work in a very simple and intuitive way: users

---

[1] http://digg.com/
[2] http://meneame.net/

Á. Herrero et al. (eds.), *International Joint Conference SOCO'13-CISIS'13-ICEUTE'13*,   577
Advances in Intelligent Systems and Computing 239,
DOI: 10.1007/978-3-319-01854-6_59, © Springer International Publishing Switzerland 2014

submit their links to stories online, and other users of these systems rate them by voting. The most voted stories appear, finally, in the front-page [1].

In the context of web categorisation, supervised machine-learning is a common approach. Specifically, in a similar domain as social news, such as filtering spam in reviews, supervised machine-learning techniques have also been applied [2]. However, supervised machine-learning classifiers require a high number of labelled data for each of the classes (i.e., troll user or normal user). Labelling this amount of information is quite arduous for a real-world problem such as web mining. To generate these data, a time-consuming process of analysis is mandatory and, in the process, some data may avoid filtering.

One type of machine-learning technique specially useful in this case, is semi-supervised learning. This technique is appropriate when a limited amount of labelled data exists for each class [3]. In particular, the approach of collective classification [4] employs the relational structure of labelled and unlabelled datasets combination, to increase the accuracy of the classification. With these relational models, the predicted label will be influenced by the labels of related samples. The techniques of collective and semi-supervised learning have been applied satisfactorily in several fields of computer science like text classification [4] or spam filtering [5].

In light of this background, we propose a novel user categorisation approach based on collective classification techniques to optimise the classification performance when filtering controversial users of social news website. This method employs a combination of several features from the public users' profiles and from their comments.

In summary, our main contributions are: (i) a new method to represent users in social news websites, (ii) an adaptation of the collective learning approach to filter troll users and (iii) an empirical validation which shows that our method can maintain high accuracy rates, while minimising the efforts of labelling.

The remainder of this paper is structured as follows. Section 2 describes the extracted features of the users. Section 3 describes the collective algorithms we applied. Section 4 describes the experimental procedure and discusses the obtained results. Finally, Section 5 concludes and outlines the avenues of the future work.

2 Method Description

'Menéame' is a Spanish social news website, in which news and stories are promoted. It was developed in later 2005 by Ricardo Galli and Benjamín Villoslada and it is currently licensed as free software. It ranks users using the 'karma' value whose boundaries are 1 and 20. When a new user is registered, a default value of 6 point of 'karma' is assigned. This value of 'karma' is computed and re-computed based on the activity of the user in the previous 2 days. The algorithm for the computation combines 4 different components: (i) received positive votes regarding the sent news, (ii) the number of positive votes the users made, (iii) negative votes made and (iv) the number of votes their comments have received.

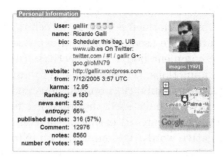

Fig. 1. An example of a user profile in 'Menéame.net'

2.1 Extracted Features

In this sub-section, we describe the features that we extract from the users' profile and their comments. We divide these features into 2 different categories: profile and comment related.

- **Profile Related Features:** Several of the features have been gathered from the public profile of the user. In Figure 1, we can see an example of user profile. The different features used are: (i) from, the registered date; (ii) karma, a number between 1 and 20; (iii) ranking, user position in terms of 'karma'; (iv) the number of news submitted; (v) the number of published news; (vi) entropy, diversity ratio of the posts; (vii) the number of comments; (viii) the number of notes; (ix) number of votes and (x) text avatar.

  Using the new Google Images[3], we have utilised the avatar of the user to translate it into text. The avatar is used as a query of the system and the text of the most similar images is used as representation. For instance, we retrieve the URL of the user's avatar image and paste it as a query in the Google Images service. Hereafter, the service retrieves the most similar images and the most probable query for that image. If Google Images service does not find any possible query for the given image, our system leaves the avatar feature blank. However, there is a high number of users that did not change the default 'Menéame' avatar. To minimise the computing overhead, our system directly employs the string 'Meneame'.

- **Comment Related Features:** We have also analysed the comments using our comment categorisation approach [6] to count the number of comments in each category: (i) focus of the comment (focus on the news story or on other comments); (ii) type of information (contribution, irrelevant or opinion) and (iii) controversy level (normal, controversial, very controversial or joke).

The approach used different syntactic, opinion and statistical features to build a representation of the comments. In particular, the following features were used in order to categorise each comment of a particular user:

---

[3] http://images.google.com

- **Syntactic:** We count the number of words in the different syntactic categories. To this end, we performed a Part-of-Speech tagging using FreeLing[4]. The following features were used, all of them expressed in numerical values extracted from the comment body: (i) adjectives, (ii) numbers, (iii) dates, (iv) adverbs, (v) conjunctions, (vi) pronouns, (vii) punctuation marks, (viii) interjections, (ix) determinants, (x) abbreviations and (xi) verbs.
- **Opinion:** Specifically, we have used the following features: (i) number of positive votes of the comment, (ii) karma of the comment and (iii) number of positive and negative words. We employed an external opinion lexicon[5]. Since the words in that lexicon are in English and 'Menéame' is written in Spanish, we have translated them into Spanish.
- **Statistical:** We used: (i) the information contained in the comment using the Vector Space Model (VSM) [7] approach, which was configured using words or n-grams as tokens; (ii) the number of references to the comment (in-degree); (iii) the number of references from the comment (out-degree); (iv) the number of the comment in the news story; (v) the similarity of the comment with the description of the news story, using the similarity of the VSM of the comment with the model of the description; (vi) the number of coincidences between words in the comment and the tags of the commented news story and (vii) the number of URLs in the comment body.

Therefore, using our approach [6], we have categorised users' comments, generating 9 new features that count the number of comments in these comment categories: (i) referring to news stories, (ii) referring to other comments, (iii) contribution comments, (iv) irrelevant comments, (v) opinion comments, (vi) normal comments, (vii) controversial comments, (viii) very controversial comments and (ix) jokes comments.

3   Collective Classification

Collective classification is a combinatorial optimisation problem, in which we are given a set of users, or nodes, $\mathcal{U} = \{u_1, ..., u_n\}$ and a neighbourhood function N, where $N_i \subseteq \mathcal{U} \setminus \{\mathcal{U}_i\}$, which describes the underlying network structure [8]. Being $\mathcal{U}$ a random collection of users, it is divided into 2 sets $\mathcal{X}$ and $\mathcal{Y}$, where $\mathcal{X}$ corresponds to the users for which we know the correct values and $\mathcal{Y}$ are the users whose values need to be determined. Therefore, the task is to label the nodes $\mathcal{Y}_i \in \mathcal{Y}$ with one of a small number of labels, $\mathcal{L} = \{l_1, ..., l_q\}$. We employ the *Waikato Environment for Knowledge Analysis* (WEKA) [9] and its Semi-Supervised Learning and Collective Classification plugin[6].

- **Collective IBK:** Internally, it applies an implementation of the *K-Nearest Neighbour* (KNN), to determine the best k instances on the training set and

---

[4] Available in: http://nlp.lsi.upc.edu/freeling/
[5] Available in: http://www.cs.uic.edu/~liub/FBS/opinion-lexicon-English.rar
[6] Available at: http://www.cms.waikato.ac.nz/~fracpete/projects/collective-classification

builds then, for all instances from the test set, a neighbourhood consisting of k instances from the training pool and test set (either a naïve search over the complete set of instances or a k-dimensional tree is used to determine neighbours) that are sorted according to their distance to the test instance they belong to. The neighbourhoods are ordered with respect to their 'rank'(the different occurrences of the two classes in the neighbourhood). The class label is determined by majority vote or, in tie case, by the first class for every unlabelled test instance with the highest rank value. This is implemented until no further test instances remain unlabelled. The classification ends by returning the class label of the instance that is about to be classified.

- **CollectiveForest:** The WEKA's implementation of RandomTree is utilised as base classifier to divide the test set into folds containing the same number of elements. The first repetition trains the model using the original training set and generates the distribution for all the instances in the test set. The best instances are then added to the original training set (choosing the same number of instances in a fold). The next repetitions train with the new training set and then produce the distributions for the remaining instances in the test set.

- **CollectiveWoods & CollectiveTree:** CollectiveWoods in association with the algorithm CollectiveTree operates like CollectiveForest by using the RandomTree algorithm. CollectiveTree is similar to WEKA's original version of RandomTree classifier.

- **RandomWoods:** This classifier works like WEKA's classic RandomForest but using CollectiveBagging (classic Bagging, a machine learning ensemble meta-algorithm to improve stability and classification accuracy, extended to make it available to collective classifiers) in combination with CollectiveTree algorithm instead of Bagging and RandomTree algorithms.

4 Empirical Validation

We have retrieved the information of the users' profiles from the users that appear in the downloaded comments [6], generating a combined dataset of 3,359 users' profiles and their comments. Afterwards, we labelled each user into *Normal* or *Controversial*. *Normal* means that the user is not hurtful or hurting, using in its argument a restrained tone. *Controversial* refers to a troll user which seeks to create polemic. To this end, we built a dataset, following the next distribution: 1,997 number of normal users and 1,362 of controversial users.

4.1 Methodology

We developed an application to obtain all the features described in Section 2. We implemented 2 different procedures to construct the VSM of the text avatar body: (i) VSM with words as tokens and (ii) n-grams with different values of

n (n=1, n=2 and n=3) as tokens. Furthermore, we removed every word devoid of meaning in the text, called stop-words, (e.g., 'a','the','is') [10]. To this end, we employed an external stop-word list of Spanish words[7]. Subsequently, we evaluated the precision of our proposed method. To this end, by means of the dataset, we conducted the following methodology:

- **Cross Validation:** This method is generally applied in machine-learning evaluation [11]. In our experiments, we utilised $k = 10$ as a K-fold cross validation refers. As a consequence, our dataset is split 10 times into 10 different sets of learning and testing. We changed the number of labelled instances from 10% to 90% for each fold. Performing this operation, we measured the effect of the number of previously labelled instances on the final performance of collective classification.
- **Information Gain Attribute Selection:** For each training set, we extracted the most important features for each of the classification types using *Information Gain* [12], an algorithm that evaluates the relevance of an attribute by measuring the information gain with respect to the class. Using this measure, we removed any features in the training set that had a IG value of zero: (i) word VSM remove 99.01% of the features and (ii) n-gram VSM the 99.42%.
- **Learning the Model:** We accomplished this step using different learning algorithms depending on the specific model, for each fold. As discussed above, we employed the implementations of the collective classification provided by the *Semi-Supervised Learning and Collective Classification* package for machine-learning tool WEKA. In our experiments, we used the following models: (i) *Collective IBK*, with k=10; (ii) *CollectiveForest*, with N=100; (iii) *CollectiveWoods*, with N=100 and (iv) *RandomWoods*, with N=100.
- **Testing the Model:** We measured the *True Positive Rate* (TPR) to test our procedure; i.e., the number of the controversial users correctly detected divided by the total controversial users: $TPR = TP/(TP + FN)$; where TP is the number of controversial users correctly classified (true positives) and FN is the number of controversial users misclassified as normal users (false negatives). In the other hand, we also took into account the *False Positive Rate* (FPR); i.e., the number of normal users misclassified divided by the total normal users: $FPR = FP/(FP + TN)$; where FP is the number of normal instances incorrectly detected and TN is the number of normal users correctly classified. In addition, we obtained *Accuracy*; i.e., the total number of hits of the classifiers divided by the number of instances in the whole dataset: $Accuracy(\%) = (TP + TN)/(TP + FP + FN + TN)$. Finally, we recovered the *Area Under the ROC Curve* (AUC). This measure establishes the relation between false negatives and false positives [13]. By plotting the TPR against the FPR, we can obtain the ROC curve.

---

[7] Available in: http://paginaspersonales.deusto.es/isantos/resources/
stopwords.txt

Table 1. Results of the Controversy Level for Word VSM.

Dataset	Accuracy (%)	TPR	FPR	AUC
KNN K = 10	89.43 ± 4.93	0.80 ± 0.12	0.04 ± 0.01	0.97 ± 0.02
BN: Bayes K2	82.52 ± 2.00	0.82 ± 0.03	0.17 ± 0.02	0.87 ± 0.02
BN: Bayes TAN	90.47 ± 1.47	0.99 ± 0.02	0.15 ± 0.02	0.96 ± 0.01
Naïve Bayes	67.10 ± 4.33	0.28 ± 0.12	0.06 ± 0.02	0.82 ± 0.03
SVM: Polynomial Kernel	75.50 ± 4.46	0.43 ± 0.12	0.03 ± 0.01	0.70 ± 0.06
SVM: Normalise Polynomial	95.00 ± 1.81	0.93 ± 0.04	0.04 ± 0.01	0.95 ± 0.02
SVM: Pearson VII	95.47 ± 1.31	0.96 ± 0.03	0.05 ± 0.02	0.96 ± 0.01
SVM: Radial Basis Function	59.57 ± 0.31	0.00 ± 0.01	0.00 ± 0.00	0.50 ± 0.00
DT: J48	96.58 ± 0.86	0.97 ± 0.02	0.04 ± 0.01	0.97 ± 0.01
DT: Random Forest N = 100	96.43 ± 0.94	0.97 ± 0.02	0.04 ± 0.01	0.99 ± 0.00

Table 2. Results of the Controversy Level for N-gram VSM

Dataset	Accuracy (%)	TPR	FPR	AUC
KNN K = 10	89.55 ± 4.81	0.80 ± 0.12	0.04 ± 0.01	0.97 ± 0.02
BN: Bayes K2	82.52 ± 2.00	0.82 ± 0.03	0.17 ± 0.02	0.87 ± 0.02
BN: Bayes TAN	90.47 ± 1.47	0.99 ± 0.02	0.15 ± 0.02	0.96 ± 0.01
Naïve Bayes	67.31 ± 4.63	0.27 ± 0.12	0.05 ± 0.02	0.82 ± 0.03
SVM: Polynomial Kernel	75.56 ± 4.43	0.44 ± 0.12	0.03 ± 0.01	0.70 ± 0.06
SVM: Normalise Polynomial	95.09 ± 1.59	0.94 ± 0.04	0.04 ± 0.01	0.95 ± 0.02
SVM: Pearson VII	95.68 ± 1.13	0.96 ± 0.03	0.05 ± 0.02	0.96 ± 0.01
SVM: Radial Basis Function	59.57 ± 0.31	0.00 ± 0.01	0.00 ± 0.00	0.50 ± 0.00
DT: J48	96.58 ± 0.86	0.97 ± 0.02	0.04 ± 0.01	0.97 ± 0.01
DT: Random Forest N = 100	96.49 ± 0.97	0.97 ± 0.02	0.04 ± 0.01	0.99 ± 0.00

4.2   Results

In our experiments, we examined various configurations of the collective algorithms with different sizes of the $\mathcal{X}$ set of known instances; the latter varied from 10% to 90% of the instances utilised for training (i.e., instances known during the test). On the other hand, we compared the filtering capabilities of our method with some of the most used supervised machine-learning algorithms. Specifically, we use the following models:

- *Bayesian Networks (BN):* We used different structural learning algorithms: K2 [14], Tree Augmented Naïve (TAN) [15] and Naïve Bayes Classifier [11].
- *Support Vector Machines (SVM):* We performed experiments with different kernels: (i) polynomial [16], (ii) normalised polynomial [17], (iii) Pearson VII function-based (PVK) [18] and (iv) radial basis function (RBF) [19].
- *K-Nearest Neighbour (KNN):* We launched experiments with $k = 10$.
- *Decision Trees (DT):* We executed experiments with J48 (the *Weka* implementation of the *C4.5* algorithm [20]) and Random Forest [21], an ensemble of randomly constructed decision trees. In particular, we employed $N = 100$.

Table 1 shows the results with words as tokens and supervised learning, Table 2 shows the results with n-grams as tokens and supervised learning. These two tables contain the supervised results about both VSM approaches. Figure 2 shows the results with word-based VSM and collective learning, Figure 3 shows the results with VSM generated with n-grams and collective learning.

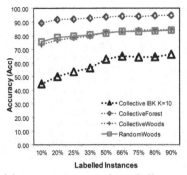

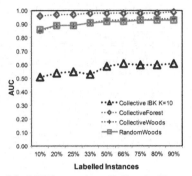

(a) **Accuracy results.** Collective-Forest was the best classifier with 90% of labelling and achieving a value of 94.76%.

(b) **AUC results.** CollectiveForest achieved of 0.98, with a 33% of labelling instances.

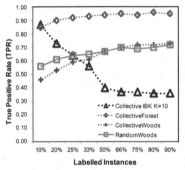

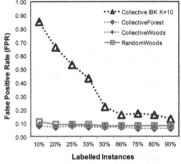

(c) **TPR results.** The best classifier was CollectiveForest, with an accuracy of 96%, labelling the 80%.

(d) **FPR results.** CollectiveWoods, obtained values close to zero and with 20% labelling effort.

Fig. 2. Results of our collective-classification-based for users categorisation using words as tokens. In resume, *CollectiveForest* was the best classifier.

Regarding the results obtained both word and n-gram VSM applying supervised algorithms and collective classifications, Random Forest with 100 trees using n-grams was the best classifier. It obtained similar results in Accuracy (96.49% and 96.58%), TPR (0.97) and FPR (0.04) than J48, but in AUC terms achieved the highest value: 0.99. CollectiveForest using n-grams as tokens with the 66% of known instances, obtained 94.08% of accuracy, 0.99 of AUC, 0.94 of TPR and 0.06 of FPR. With regards to the use of collective classification, comparing with the supervised approaches, it achieved close results. We can maintain the results of the best supervised learning algorithm whilst the labelling efforts are reduced significantly, in this case a 33% of the dataset.

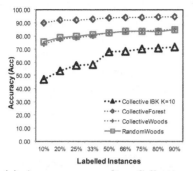

(a) **Accuracy results.** Collective-Forest, using a 90% of labelled instances, achieved a 94.70% of accuracy.

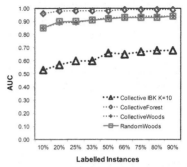

(b) **AUC results.** The best algorithm was CollectiveForest. This classifier achieved 0.99, labelling only 66% of the dataset.

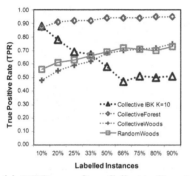

(c) **TPR results.** CollectiveForest was the best classifier obtaining 0.95 labelling 80% of the dataset.

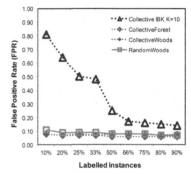

(d) **FPR results.** CollectiveForest achieved results close to zero, being the best classifier.

Fig. 3. Results of our collective-classification-based for users categorisation using n-grams as tokens. Summarising, *CollectiveForest* obtained the best results.

5   Conclusions

One problem about supervised learning algorithms is that a preceding work is required to label the training dataset. When it comes to web mining, this process may originate a high performance overhead because of the number of users that post comments any time and the persons who create new accounts everyday. In this paper, we have proposed the first trolling users filtering method system based on collective learning. This approach is able to determine when a user is controversial or not, employing users' profile features and statistical, syntactic and opinion features from their comments. We have empirically validated our method using a dataset from 'Menéame', showing that our technique obtains the same accuracy than supervised learning, despite having less labelling requirements.

The avenues of future work are oriented in three main ways. Firstly, we project to extend the study of our semi-supervised classification by applying additional algorithms to the problem of filter trolling users in social news websites. Secondly, incorporating new different extracted features from the user dataset to train the models. And finally, we will focus on executing an extended analysis of the effects of the labelled dataset dimension.

References

1. Lerman, K.: User participation in social media: Digg study. In: Proceedings of the 2007 IEEE/WIC/ACM International Conferences on Web Intelligence and Intelligent Agent Technology-Workshops, pp. 255–258. IEEE Computer Society (2007)
2. Jindal, N., Liu, B.: Review spam detection. In: Proceedings of the 16th International Conference on World Wide Web, pp. 1189–1190. ACM (2007)
3. Chapelle, O., Schölkopf, B., Zien, A., et al.: Semi-supervised learning, vol. 2. MIT Press, Cambridge (2006)
4. Neville, J., Jensen, D.: Collective classification with relational dependency networks. In: Proceedings of the Second International Workshop on Multi-Relational Data Mining, pp. 77–91 (2003)
5. Laorden, C., Sanz, B., Santos, I., Galán-García, P., Bringas, P.G.: Collective classification for spam filtering. In: Herrero, Á., Corchado, E. (eds.) CISIS 2011. LNCS, vol. 6694, pp. 1–8. Springer, Heidelberg (2011)
6. Santos, I., de-la Peña-Sordo, J., Pastor-López, I., Galán-García, P., Bringas, P.: Automatic categorisation of comments in social news websites. Expert Systems with Applications (2012)
7. Baeza-Yates, R.A., Ribeiro-Neto, B.: Modern Information Retrieval. Addison-Wesley Longman Publishing Co., Inc., Boston (1999)
8. Namata, G., Sen, P., Bilgic, M., Getoor, L.: Collective classification for text classification. Text Mining, 51–69 (2009)
9. Garner, S.: Weka: The waikato environment for knowledge analysis. In: Proceedings of the 1995 New Zealand Computer Science Research Students Conference, pp. 57–64 (1995)
10. Salton, G., McGill, M.: Introduction to modern information retrieval. McGraw-Hill, New York (1983)
11. Bishop, C.M.: Neural Networks for Pattern Recognition. Oxford University Press (1995)
12. Kent, J.: Information gain and a general measure of correlation. Biometrika 70(1), 163–173 (1983)
13. Singh, Y., Kaur, A., Malhotra, R.: Comparative analysis of regression and machine learning methods for predicting fault proneness models. International Journal of Computer Applications in Technology 35(2), 183–193 (2009)
14. Cooper, G.F., Herskovits, E.: A bayesian method for constructing bayesian belief networks from databases. In: Proceedings of the 1991 Conference on Uncertainty in Artificial Intelligence (1991)
15. Geiger, D., Goldszmidt, M., Provan, G., Langley, P., Smyth, P.: Bayesian network classifiers. Machine Learning, 131–163 (1997)
16. Amari, S., Wu, S.: Improving support vector machine classifiers by modifying kernel functions. Neural Networks 12(6), 783–789 (1999)

17. Maji, S., Berg, A., Malik, J.: Classification using intersection kernel support vector machines is efficient. In: IEEE Conference on Computer Vision and Pattern Recognition (CVPR), pp. 1–8. IEEE (2008)
18. Üstün, B., Melssen, W., Buydens, L.: Visualisation and interpretation of support vector regression models. Analytica Chimica Acta 595(1-2), 299–309 (2007)
19. Park, J., Sandberg, I.W.: Universal approximation using radial-basis-function networks. Neural Computation 3(2), 246–257 (1991)
20. Quinlan, J.: C4. 5 programs for machine learning. Morgan Kaufmann Publishers (1993)
21. Breiman, L.: Random forests. Machine learning 45(1), 5–32 (2001)

Secure Android Application in SOA-Based Mobile Government and Mobile Banking Systems

Milan Marković[1] and Goran Đjorđjević[2]

[1] Banca Intesa ad Beograd, Bulevar Milutina Milankovića 1c, 11070 Novi Beograd, Serbia
milan.z.markovic@bancaintesa.rs
[2] Institute for Manufacturing banknotes and coins NBS, Pionirska 2, 11000 Beograd, Serbia
djg_goran@mail.com

Abstract. In this paper, we consider an overview of a possible secure model for m-government and m-banking systems. The proposed model is based on secure mobile application and SOA-Based central platform. The model additionally consists of external entities, such as: PKI, XKMS, Authentication Server and Time Stamping server. The proposed model could be used in different local and/or cross-border m-government scenarios. Besides, specifics of m-banking systems based on the same or similar secure model are also presented. As a possible example of described secure mobile application we considered and experimentally evaluated a secure Android based Web services application.

Keywords: Android based mobile phone application, m-government, m-banking, Web Service, SOAP protocol, XML-Security, WS-Security, XKMS protocol, SAML, Timestamp.

1 Introduction

This work is related to the consideration of possible secure m-government and m-banking models and applying a secure Android Web services based application in them. An overview of possible secure m-government and m-banking systems realized according to the similar model based on secure JAVA mobile Web service application and the SOA-Based central platform is given in [1]. In this paper, a possibility of using the secure Android based mobile application is considered and experimentally evaluated.

First, we consider a possible model of secure SOA-based m-government online systems, i.e. about secure mobile communication between citizens and companies with the small and medium governmental organizations, such as municipalities. This model will be considered in both local and cross-border case. The latter means either crossing borders of municipalities in the same country or crossing borders between countries (e.g. some municipalities in different countries). The work presented related to the m-government systems and examples described have been partially included in the general framework of the EU IST FP6 SWEB project (Secure, interoperable cross border m-services contributing towards a trustful European cooperation with the non-EU member Western Balkan countries, SWEB) [2], [3], [4].

As a second goal of this paper, we consider a possible usage of similar secure model in the m-banking systems.

Á. Herrero et al. (eds.), *International Joint Conference SOCO'13-CISIS'13-ICEUTE'13*,
Advances in Intelligent Systems and Computing 239,
DOI: 10.1007/978-3-319-01854-6_60, © Springer International Publishing Switzerland 2014

As a third goal of this paper, we consider a possible usage of the Android-based secure mobile application in m-government and m-banking systems previously described. A feasibility of using such Android based secure mobile application is experimentally evaluated in the paper.

The paper is organised as follows. The architecture of the proposed model is given in Section 2. A description and some experimental results obtained by the secure Android-based application is given in Section 3 while conclusions are given in Section 4.

2   Possible Secure m-Government and m-Banking Model Architecture

The proposed m-government and m-banking model consists of:

- **Mobile users** (citizen, companies) who send some Web services requests to m-government or m-banking platform for a purpose of receiving some governmental documents (e.g. residence certificate, birth or marriage certificates, etc.) or doing some banking transactions. These users use secure mobile Web service application on their mobile devices (mobile phones, smart phones, tablets, etc.) for such a purpose.
- **SOA based Web service endpoint implementation** on the platform's side that implements a complete set of server based security features. Well processed requests with all security features positively verified, the Web service platform's application proceeds to other application parts of the proposed SOA-Based platform, including the governmental Legacy system for issuing actual governmental certificates requested or to the back office system in the Bank for processing the banking transactions. In fact, the proposed platform could change completely the application platform of some governmental or banking organization or could serve as the Web service „add-on" to the existing Legacy system implementation. In the latter case, the Legacy system (or back office in the Bank) will not be touched and only a corresponding Web service interface should be developped in order to interconnect the proposed SOA-Based platform and the Legacy governmental of Bank's back office system.
- **External entities**, such as: PKI server with XKMS server as a front end, the Authentication server, and TSA (Time Stamping Authority).

Functions of the proposed external entities are following:

- **STS server** – is responsible for strong user authentication and authorization based on PKI X.509v3 electronic certificate issued to users and other entities in the proposed model. The communication between STS server and the user's mobile Web service application is SOAP-based and secured by using WS-Security features. After the succesful user authentication and authorization, the STS server issues a SAML token to the user which will be subsequently used for the user authentication and authorization to the Web service of the proposed m-government platform. The SAML token is signed by the STS server and could consist of the user role for platform's user authentication and authorization. The alternative is that it could be a general-purpose Authentication server which will authenticate users by using any kind of authentication credentials, such as: user credentials (username/password), OTP, PKI digital certificates, etc.

- **PKI server** - is responsible for issuing PKI X.509v3 electronic certificates for all users/entities in the proposed m-governmental and m-banking model (users, civil servants, administrators, servers, platforms, etc.). Since some certificate processing functions could be too heavy for mobile users, the PKI services are exposed by the XKMS server which could register users, as well as locate or validate certificates on behalf of the mobile user. This is of particular interests in all processes that request signature verification on mobile user side.
- **TSA server** - is responsible for issuing time stamps for user's requests as well as for platform's responses (signed m-documents).

3 Experimental Analysis

This Section is dedicated to the experimental analysis of the cryptographic operations implemented on Android mobile phone as a possible example of the secure mobile client application. We analysed implementation of different PKI functions that could be main elements of the considered secure Android based web service application on three different test platforms: smart mobile phone, tablet and PC. Namely, we compared experimental results obtained on some mobile devices (smart phone, tablet) with the same experiments did on the PC computer in order to test feasibility of the analysed PKI functions implementation on mobile devices

The presented experimental results are generated using:

1. LG E610 mobile phone that has following characteristics (Mobile Phone):
- CPU   Core: ARM Cortex-A5
- CPU Clock: 800 MHz
- RAM capacity: 512 MB
- Embedded Operating System: Android 4.0.3 Ice Cream Sandwich
- NFC Functions.
2. Tablet Ainol Novo 7 Paladin device that has following characteristics (hereafter Tablet):
- XBurst CPU
- CPU Clock: 1 GHz
- RAM capacity: 1 GB
- Embedded Operating System: Android 4.0.1
3. PC Desktop Computer that has following characteristics (PC Desktop):
- Intel Pentium CPU G620
- CPU Clock: 2.60 GHz
- RAM capacity: 2 GB
- Operating System: Windows XP with Service Pack 3
4. PC Laptop computer that has following characteristics (PC Laptop):
- Intel Celeron CPU P4600
- CPU Clock: 2 GHz

- RAM capacity: 3 GB
- Operating System: Windows 8

The Android platform ships with a cut-down version of Bouncy Castle - as well as being crippled. It also makes installing an updated version of the libraries difficult due to class loader conflicts. The different version of Android operating system has implemented different version of Bouncy Castle library releases. In order to avoid lack of interoperability between different devices that have implemented different operating systems and get more flexible code we used Spongy Castle functions (http://rtyley.github.com/spongycastle/). In order to achieve smaller and faster implementation we partly modified Spongy Castle functions. Experimental results that are presented in this Section are based on the modified version of Spongy Castle functions.

We considered possibility to create asymmetric RSA private/public key in real-time using the standard mobile phone device. During generation of private component of RSA key pair it is used Miller-Rabin big number primary test. It is probabilistic algorithm for determining whether a number is prime or composite. The number of iteration of Miller-Rabin primary test that we used during private key pair component generation is 25. The error probability of Miller-Rabin's test for $T = 25$ times is $8.88 * 10^{-14}\%$. Time needed for generation RSA private components and calculation of Chinese Remainder Theorem's parameters is shown in Table 1.

After generation of RSA private/public key pair we stored the private component of the key in PKCS#5 based key store using a mechanism of the Password Based Encryption Scheme 2 (PBE S2). An algorithm used for protection of RSA private

Table 1. RSA private/public key pair generation

RSA public/private key length (bits)	Device	Time (ms)
512	Mobile Phone	286.23
	Tablet	159.97
	PC Laptop	53.39
	PC Desktop	40.18
1024	Mobile Phone	1 282.46
	Tablet	822.29
	PC Laptop	367.23
	PC Desktop	273.29
2048	Mobile Phone	7 437.70
	Tablet	5 727.04
	PC Laptop	3 552.45
	PC Desktop	2 647.51
3072	Mobile Phone	23 167.97
	Tablet	21 612.54
	PC Laptop	15 964.68
	PC Desktop	12 038.61
4096	Mobile Phone	62 274.3
	Tablet	57 545.29
	PC Laptop	47 630.71
	PC Desktop	35 593.62

component in the created key store is AES. The number of iteration count was 1000. Time needed for creation of key store, encrypting private key component using AES algorithm where number of iterations was 1000, is shown in Table 2.

Table 2. RSA private key protecting in AES based keystore

AES key length (bits)	Device	Time (ms)
128	Mobile Phone	167.46
	Tablet	1 116.39
	PC Laptop	7.84
	PC Desktop	5.02
256	Mobile Phone	332.94
	Tablet	2 217.99
	PC Laptop	15.60
	PC Desktop	10.01

We measured the time needed for creation X509 v3 self-signed certificate comprising a creation of PKCS#10 certificate request (Table 3). As a signature algorithm we used SHA-1 hash algorithm and RSA asymmetric cryptographic algorithm.

In order to evaluate the possibility of using the mobile phone in m-Government/m-Banking application based on Web service we measured time need for creation of XML-Signature and Web Service Security (WSS) Signature (Table 4, Table 5), respectively. In all these experiments, we use a file of 1KB and SHA-1 hash function.

Table 3. Create X509 v3 self-signed certificate

RSA public/private key length (bits)	Device	Time (ms)
512	Mobile Phone	18.11
	Tablet	34.49
	PC Laptop	1.78
	PC Desktop	1.33
1024	Mobile Phone	27.41
	Tablet	46.27
	PC Laptop	9.01
	PC Desktop	6.78
2048	Mobile Phone	84.41
	Tablet	116.05
	PC Laptop	58.07
	PC Desktop	43.92
3072	Mobile Phone	216.89
	Tablet	283.22
	PC Laptop	180.97
	PC Desktop	137.36
4096	Mobile Phone	467.95
	Tablet	548.40
	PC Laptop	414.14
	PC Desktop	312.90

Table 4. XML-Signature creation

RSA public/private key length (bits)	Device	Time (ms)
512	Mobile Phone	29.64
	Tablet	59.73
	PC Laptop	2.12
	PC Desktop	1.57
1024	Mobile Phone	38.15
	Tablet	73.78
	PC Laptop	9.38
	PC Desktop	7.05
2048	Mobile Phone	95.87
	Tablet	144.08
	PC Laptop	58.50
	PC Desktop	43.85
3072	Mobile Phone	228.20
	Tablet	319.65
	PC Laptop	181.54
	PC Desktop	137.87
4096	Mobile Phone	479.10
	Tablet	586.54
	PC Laptop	414.74
	PC Desktop	312.79

Table 5. WSS-Signature creation

RSA public/private key length (bits)	Device	Time (ms)
512	Mobile Phone	63.76
	Tablet	126.99
	PC Laptop	2.79
	PC Desktop	2.02
1024	Mobile Phone	74.51
	Tablet	147.68
	PC Laptop	10.07
	PC Desktop	7.50
2048	Mobile Phone	131.00
	Tablet	216.81
	PC Laptop	59.18
	PC Desktop	44.57
3072	Mobile Phone	266.29
	Tablet	384.48
	PC Laptop	182.1
	PC Desktop	138.03
4096	Mobile Phone	507.47
	Tablet	663.93
	PC Laptop	415.19
	PC Desktop	311.66

The time needed for verification of WSS-Signed message is shown in Table 6.

Table 6. WSS-Signature verification

RSA public/private key length (bits)	Device	Time (ms)
512	Mobile Phone	34.67
	Tablet	94.83
	PC Laptop	0.88
	PC Desktop	0.61
1024	Mobile Phone	34.81
	Tablet	95.61
	PC Laptop	1.16
	PC Desktop	0.84
2048	Mobile Phone	34.87
	Tablet	101.27
	PC Laptop	2.42
	PC Desktop	1.74
3072	Mobile Phone	34.94
	Tablet	107.58
	PC Laptop	4.58
	PC Desktop	3.29
4096	Mobile Phone	50.20
	Tablet	111.22
	PC Laptop	7.29
	PC Desktop	5.31

Table 7. WSS-Encryption mechanism

RSA public/private key length (bits)	Device	Time (ms)
512	Mobile Phone	38.03
	Tablet	80.86
	PC Laptop	0.98
	PC Desktop	0.67
1024	Mobile Phone	38.39
	Tablet	85.13
	PC Laptop	1.29
	PC Desktop	0.90
2048	Mobile Phone	38.54
	Tablet	89.96
	PC Laptop	1.78
	PC Desktop	2.63
3072	Mobile Phone	45.36
	Tablet	100.32
	PC Laptop	4.59
	PC Desktop	3.19
4096	Mobile Phone	48.94
	Tablet	109.08
	PC Laptop	6.99
	PC Desktop	5.36

We also analyzed possibility of encryption XML based message using WSS Encryption as well as WSS Decryption mechanisms (Tables 7, 8), respectively. In all these experiments, we use a file of 1KB and SHA-1 hash function, as well as 3DES symmetric cryptographic algorithm with symmetric key length of 168 bits.

Table 8. WSS-Decryption mechanism

RSA public/private key length (bits)	Device	Time (ms)
512	Mobile Phone	34.96
	Tablet	80.91
	PC Laptop	2.41
	PC Desktop	1.77
1024	Mobile Phone	44.20
	Tablet	119.74
	PC Laptop	9.67
	PC Desktop	7.28
2048	Mobile Phone	102.48
	Tablet	169.87
	PC Laptop	58.88
	PC Desktop	44.36
3072	Mobile Phone	232.75
	Tablet	339.54
	PC Laptop	181.79
	PC Desktop	138.01
4096	Mobile Phone	486.20
	Tablet	609.42
	PC Laptop	415.98
	PC Desktop	313.88

In the above experimental analysis we presented experimentally obtained results/ times required for implementation of different kind of cryptographic operations that could be main elements of the considered Secure Android Web services application on mobile phone, tablet, laptop and PC desktop computer. Some of the observations are:

- The operation of digital signing of XML message (XML-Signature mechanism), using 2048-bit private RSA key, takes 95.87 ms on mobile phone. It means that in one second can be implemented 10 operations of generation XML-Signature using 2048-bit private key pair component. The fact led to the conclusion that mobile phone can be used in real time for implementation of digital signature operations.
- The operation of encryption/decryption of RSA keypar as well as appropriate X509 certificate that have stored in PKCS#5 based key store (PBE S2 scheme) using AES algorithm with 1000 iterations takes 167.46 ms on mobile phone. Bearing in mind the above mentioned, we can

conclude that protected PKCS#5 based key store, that contains RSA keypairs and X509 certificates, can be used in mobile phone in real time.

- The operation of encryption of XML message (WSS-Encryption mechanism), using 2048-bit public RSA key, extracted from X509 certificate, takes 38.54 ms on mobile phone. The operation of decryption of WSS-Encrypted message using 2048-bit private RSA key, takes 102.48 ms. It means that in one second can be implemented about 10 operations of decryption WSS-Encrypted message using 2048-bit RSA private key. Regarding to the above mentioned, we can also conclude that mobile phone could be used in real time in process encryption/decryption WSS messages that are used in protected Web Service communications.

4   Conclusions

In this paper, we presented an overview of possible secure model of m-government and m-banking systems as well as an analysis of possibility and feasibility of using secure Android-based web service mobile application in it.

First, this work is related to the consideration of some possible SOA-based m-government online systems, i.e. about secure mobile communication between citizens and companies with the small and medium governmental organizations, such as municipalities.

Second, the paper refers to the application of the same or similar proposed secure model in m-banking systems.

Third, the paper presented a possible example of an Android-based secure mobile client application that could be used in the described m-government and m-banking model.

Presented experimental results justify that security operations related to asymmetric key pair generation, primary tests on the random numbers, X.509v3 digital certificate generation, XML/WSS digital signature/verification and XML/WSS encryption/decryption are feasible for usage on some current smart phones. Thus, we could conclude that this application could serve as a basis for implementing secure m-government/m-banking system based on the model described in this paper.

Main contributions of the paper are:

- A proposal of the possible secure model for m-government and m-banking systems based on secure Android based mobile Web service application and SOA-Based platform.
- Usage of secure mobile Web service application in which all modern security techniques are implemented (XML security, WS-Security, Authentication/ SAML, Time Stamping, PKI, XKMS).
- Usage of SOA-Based request-response m-government and m-banking platform (Web Services) which is more suitable for usage of secure mobile application compared to the session-based Web application platform [5].

- Usage of XKMS service which is more suitable for mobile PKI system since it outsources complex operations such as PKI certificate validation services to the external entity, the XKMS server, compared to other techniques [5].
- Usage of the Android-based mobile client application for which a feasiility of implementing security operations is experimentally presented and verified.

Future researching directions in domain of m-government systems are:

- Full implementation of secure mobile Web service applications for all other mobile platforms (JAVA, iPhone, BlackBerry, Windows Mobile)
- Full implementation of advanced electronic signature formats (e.g. XAdeS, PAdeS)
- Integration of PKI SIM technology in the secure mobile client Web service application
- Application based (Android, JAVA, iPhone, Windows Mobile) digital signature by using the asymmetric private key on the PKI smart cards and usage of the integrated NFC (Near Field Communication) security element as a smart card reader.
- Using the proposed secure model for other PKI based e/m-governmental services (strong user authentication to other e-government web portals, signing documents prepared through some other communication channels, qualified signatures, etc.)

Future researching directions in domain of m-banking systems are:

- PKI SIM cards with asymmetric private key and signature JAVA applet on it
- Application based (Android, JAVA, iPhone, Windows Mobile) digital signature by using the asymmetric private key on the PKI SIM cards
- One single channel for mobile banking – mobile phone with the equivalent security mechanisms as in the case of Internet Banking – Application based Mobile Banking functionalities with PKI SIM card
- Application based (Android, JAVA, iPhone, Windows Mobile) digital signature by using the asymmetric private key on the PKI smart cards and usage of the integrated NFC (Near Field Communication) security element as a smart card reader.

References

1. Marković, M., Đorđević, G.: On Secure m-government and m-banking model. In: Proc. of 6th International Conference on Methodologies, Technologies and Tools Enabling e-Government, Belgrade, Serbia, July 3-5, pp. 100–111 (2012)
2. Marković, M., Đorđević, G.: On Possible Model of Secure e/m-Government System. In: Information Systems Management, vol. 27, pp. 320–333. Taylor & Francis Group, LLC (2010)
3. Marković, M., Đorđević, G.: On Secure SOA-Based e/m-Government Online Services. In: Handbook "Service Delivery Platforms: Developing and Deploying Converged Multimedia Services", ch. 11, pp. 251–278. Taylor & Francis (2008)

4. Marković, M., Đorđević, G.: On Possible Secure Cross-Border M-Government Model. In: Proceedings of the 5th International Conference on Methodologies, Technologies and Tools Enabling e-Government, Camerino, Italy, June 30-July 1, pp. 381–394 (2011)
5. Lee, Y., Lee, J., Song, J.: Design and implementation of wireless PKI technology suitable for mobile phone in mobile-commerce. Computer Communication 30(4), 893–903 (2007)

Extending a User Access Control Proposal for Wireless Network Services with Hierarchical User Credentials

Juan Álvaro Muñoz Naranjo[1], Aitor Gómez-Goiri[2], Pablo Orduña[2], Diego López-de-Ipiña[2], and Leocadio González Casado[1]

[1] Dpt. of Computer Science,
University of Almería,
Agrifood Campus of International Excellence (ceiA3), Spain
{jmn843,leo}@ual.es
[2] Deusto Institute of Technology - DeustoTech,
University of Deusto, Spain
{aitor.gomez,pablo.orduna,dipina}@deusto.es

Abstract. We extend a previous access control solution for wireless network services with group-based authorization and encryption capabilities. Both the basic solution and this novel extension focus on minimizing computation, energy, storage and communications required at sensors so they can be run in very constrained hardware, since the computations involved rely on symmetric cryptography and key derivation functions. Furthermore, no additional messages between users and sensors are needed. Access control is based on user identity, group membership and time intervals.

Keywords: access control, group-based authorization, wireless network services, Internet of Things, ubiquitous computing.

1 Introduction

The Internet of Things (IoT) initiative advocates for providing identities to everyday objects by representing them on the Internet. A way to achieve that is to physically connect them to the Internet so the objects can interact with Internet services and vice-versa. Together with mobile computing, IoT constitutes the clearest sign of the Ubiquitous Computing prominence in our current lives [1]. On the other hand, security has been an ever-present concern in Internet communications, and will keep being in the new scenario: if we want the IoT paradigm to reach all its possibilities then we need to provide reliable routines for information encryption and user authentication and authorization. Furthermore, these routines must be able to run seamlessly in very constrained hardware: small and cheap devices with limited processing capabilities and sometimes energy restrictions. For example, a typical mote in a wireless sensor network is not able to make use of public key cryptography (on a frequent manner at least) given the high computational and energy demands of the latter. Hence, very lightweight

Á. Herrero et al. (eds.), *International Joint Conference SOCO'13-CISIS'13-ICEUTE'13*, 601
Advances in Intelligent Systems and Computing 239,
DOI: 10.1007/978-3-319-01854-6_61, © Springer International Publishing Switzerland 2014

security routines are needed. In [3,4] we presented an access control solution for wireless environments in which users access services offered by constrained devices (e.g. wireless sensors). This solution provides efficient encryption, authentication and authorization on a per-user basis, i.e. a given user can access the services offered by a given sensor based on her identity. Furthermore, it needs no additional messages in the user-sensor communication. In this work, we extend the solution in order to differentiate groups of credentials in the authorization process, i.e. users can access the services offered by a group of sensors when they have the corresponding group credentials. The groups can be either hierarchical or non-hierarchical. In the latter, members in different privilege groups enjoy different non-hierarchical sets of services. In the former, members in higher privilege groups enjoy more services than lower level users. We present the basic protocol, the novel group credentials extension, and a discussion in terms of security and both message overhead and storage requirements. Experimental results in [4] confirm the applicability of our proposal.

The article is organized as follows. Section 2 discusses some proposals from the literature. Sections 3 and 4 present the scenario we are addressing here and recall the basic protocol, respectively. Section 5 introduces the novel groups extension, while Sections 6 and 7 discuss it in terms of security, message overhead, storage and efficiency. Section 8 concludes the article.

2   Related Work

The popular SPINS solution [6] provides lightweight symmetric encryption and authentication in wireless sensor networks where a Base Station is actively involved. It is composed of two sub-protocols: SNEP, which provides encryption, authentication and data freshness evidence between two parties, and μTESLA, used for authenticating broadcast messages to the whole network. LEAP+ [8] proposes an authentication and encryption framework for similar scenarios. Apart from its own protocols, μTESLA is used for authentication of broadcast messages from the Base Station. Ngo et al [5] proposed an access control system for the scenario we address here: wireless networks that provide services to users supported by an Authorization Service. It allows both individual and group-based authentication thanks to the combination of user keys and group keys. The recent MAACE [2] also focuses on the same scenario with individual and per-group authentication. However its storage requirements at every sensor are very large (sensors must store all keys shared with online users at a given time). The authors solve the storage problem by involving the Base Station in frequent communications, which is not a proper solution from our point of view since sending information is by far the most energy-consuming operation for sensors.

3   Scenario

The scenario we address in this work involves three kinds of players: sensors, Base Stations and user devices (e.g. smartphones), interacting together in a given facility (buildings, factories, greenhouses, homes, etc).

Sensors are extremely constrained wireless devices, frequently battery-powered and with reduced computational capabilities, which provide users with services of any kind. Their reduced equipment and power supply prevents them from carrying out the complex arithmetic operations involved in public-key cryptography. However, symmetric cryptography is an option, either in software or hardware since many sensor models include an AES coprocessor. Note that under this category we also consider actuators, which are devices able to perform actions related to physical access control (opening gates to authorized users), ventilation, emergencies, etc.

Base Stations are better equipped devices that handle groups of sensors for message routing purposes, data collection and also for key management in our case. They are assumed to have a more powerful hardware and a permanent (or at least much larger) power supply and large storage space. They are also assumed to handle public-key cryptography routines and certificates.

Finally, users communicate with Base Stations and sensors through their powerful smart devices, such as mobile phones or tablets.

The key point here is that sensors need to perform access control on users, however they have to face several limitations: 1) they are not able to handle complex public-key authentication nor encryption routines and 2) they do not have enough memory space so as to keep large sets of user keys. The goal of our basic protocol is to provide an access control mechanism with symmetric encryption and authentication routines which minimizes storage requirements. On the other hand, the goal of the groups extension introduced in this work

Table 1. Notation

MS_S	Master secret for sensor S
$Kenc_{S,A},\ Kauth_{S,A}$	Encryption and authentication keys for communication between sensor S and user A
$Kenc_{S,A}\{x,\ ctr\}$	x is encrypted in counter mode using key $Kenc_{S,A}$ and counter ctr
$MAC_{Kauth_{S,A}}(x)$	A MAC is done on x using $Kauth_{S,A}$
$KDF(x,\ \{a,\ b\})$	A Key Derivation Function is applied to master secret x using a as public salt and b as user-related information
$H(x)$	A hash function is applied to x
$x\|\|y$	Concatenation of x and y
ID_A	Identifier of user A
a	Random integer salt
$init\_time,\ exp\_time$	Absolute initial and expiration time of a given key
MS_p	Master secret for privilege group p
$Kenc_{p,A},\ Kauth_{p,A}$	Encryption and authentication keys between sensors offering services for group p and user A
ID_p	Identifier of privilege group p
$A \rightarrow *$	User A sends a message to any listening sensor
$S_p \rightarrow A$	One sensor giving services from privilege group p sends a message to A

is to manage users on a per-group basis: each user group has a different set of privileges, meaning that they can access different sets of the services provided by the sensors. Table 1 shows the notation used throughout the article.

4   The Basic Protocol

Here we briefly summarize the initial version of the protocol as showed in [3,4]. It provides encryption and user access control to user $\leftrightarrow$ sensor one-to-one communications. The Base Station, a more powerful device, performs high-level authentication on the user (with authorization certificates based in public key cryptography, for example) and provides her with two symmetric keys (for encryption and authentication, respectively) and parameters for their generation at the sensor. If those parameters are attached to the first message of a conversation then the sensor can input them to a Key Derivation Function in order to obtain an identical pair of symmetric keys that make communication possible. Figure 1 depicts the message exchange in the protocol. Let us explain it with more detail.

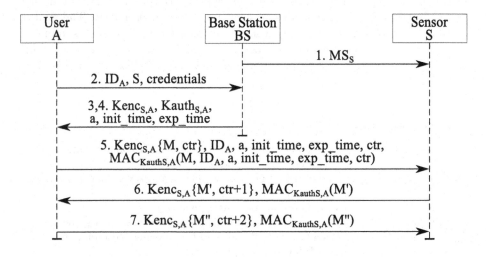

Fig. 1. Messages involved in the original protocol

1. At the time of sensor deployment, the latter receives a master secret MS_S, which is *secretly shared* (see Section 6) by the Base Station BS and the sensor S. This step is run only once in the life of the sensor (unless the master secret needs to be changed).
2. Upon arrival, user A sends her credentials (e.g. an authorization certificate) to BS so high-level access control can be performed, and the list of sensors she wants to communicate with (in Fig. 1 we only consider S). This step is run only at user arrival.

3. BS computes:
   (a) a, random integer salt
   (b) $(init\_time, exp\_time)$, keying material validity interval
   (c) $Kenc_{S,A}$, $Kauth_{S,A} = KDF(MS_S, \{a,\ ID_A\|init\_time\|exp\_time\})$
4. BS sends the information generated in the previous step to A under a secure channel (see Section 6).
5. A encrypts her first message to S with $Kenc_{S,A}$ in counter mode (thus using a fresh counter ctr), attaches parameters ID_A, a, $init\_time$, exp_time, ctr in plain text and a MAC obtained with $Kauth_{S,A}$.
6. Upon reception of the message, S obtains the key pair $Kenc_{S,A}$, $Kauth_{S,A}$ by feeding the Key Derivation Function with the attached parameters; S can now decrypt the message. The reply is encrypted in counter mode with $Kenc_{S,A}$ and $ctr + 1$ and authenticated with a MAC using $Kauth_{S,A}$.
7. Any subsequent message is encrypted and authenticated with the same key pair after increasing the counter by one.

When the message exchange finishes the sensor deletes all information related to the user since it can be easily and quickly recomputed at the beginning of the next exchange, thus saving space at the sensor. The sensor is sure of the authenticity of the user since the only way of knowing ($Kenc_{S,A}$, $Kauth_{S,A}$) is either knowing MS_S (which is kept secret) or obtaining it from the Base Station (which is actually the case). What is more, the MAC at the end of the message provides integrity assurance in addition to authentication. We refer the reader to [3,4] for more considerations on security, efficiency, message overhead and storage.

5 A Groups Extension

In this section we address a scenario with different groups of users, each group giving its members access privilege to a given set of services provided by sensors. Services provided by a sensor may (but not necessarily) belong to more than one group. The associated access control routines should not be intensive in terms of computations or message exchanges.

Let us assume that there are $l > 0$ groups. The main idea is that there exists a different master secret MS_p for every privilege group $p \in [1,\ l]$, hence sensors should only reply to service requests encrypted and/or authenticated with a key pair derived from the corresponding master secret. From here, we propose two different approaches based on how services are arranged into groups. In Approach 1 privilege groups are not hierarchical, like in the case of employees that are allowed to enter different areas of a facility based on their activity (though some services might be in more than one group). In Approach 2 privilege groups are hierarchical, hence a user with privilege level p should enjoy all privileges from groups $[1,\ p]$. An example of this scenario is a smart house with different privilege groups based on age: children would have access to certain services of the house, while parents should have full control of the house.

5.1   Approach 1: Non-hierarchical Privilege Groups

In this case, the Base Station generates l independent random master secrets $MS_1, \ldots, MS_l$ assuming there exist l different privilege groups. Sensors offering services from any privilege group p receive MS_p from the Base Station under a secure channel. In this scenario, users will typically belong to one group only, and sensors will provide services to one group as well. Figure 2(a) shows an example with three users and three sensors. However, if a sensor offers services to different privilege groups (or if a given service is included in more than one group), then the sensor should store each group's master secret. In a similar way, users assigned to more than one group (if that occurred) should receive a different pair of keys per group, and use the appropriate one to the requested service.

When user A arrives at the system the Base Station authenticates her and generates a different pair of symmetric keys $(Kenc_{p,A}, Kauth_{p,A})$ for the privilege group A belongs to (group p in this case). These keys are generated by the BS and sensors assigned to group p in the same way as in the basic protocol: the user identifier, a random salt a and a key validity interval $(init\_time, exp\_time)$ are fed to a Key Derivation Function along with the corresponding master secret as shown in Eq. (1).

$$Kenc_{p,A}, \ Kauth_{p,A} = KDF(MS_p, \{a, \ ID_A \| init\_time \| exp\_time\}) \quad (1)$$

These keys are sent to A by the BS under a secure channel (see Section 6). When user A wants to request a service from privilege group p she needs to encrypt and authenticate her message with that pair of keys like in the basic protocol (note that ID_p has been added).

$$A \rightarrow * : [Kenc_{p,A}\{M, \ ctr\}, \ ID_A, \ ID_p, \ a, \ init\_time, \ exp\_time, \ ctr,$$
$$MAC_{Kauth_{p,A}}(M, \ ID_A, \ ID_p, \ a, \ init\_time, \ exp\_time, \ ctr)] \quad (2)$$

Any nearby sensor providing services from group p (let us name it S_p) can now reply to A after deriving the appropriate pair of keys from the received information and MS_p. The counter is explicitly stated on plain text so synchronization is not lost due to an arbitrary sequence of messages if more than one sensor is involved in the conversation.

$$S_p \rightarrow A : [Kenc_{p,A}\{M', \ ctr+1\}, \ ctr+1, \ MAC_{Kauth_{p,A}}(M', \ ctr+1)] \quad (3)$$

5.2   Approach 2: Hierarchical Privilege Groups

In this case, services are arranged in hierarchical groups: users assigned to privilege group p should be granted access to all services in groups $[1, p]$. Here every sensor in the system receives the lowest group's level master secret MS_1 from the BS. The rest are obtained by hashing the immediately lower master secret, i.e. $MS_p = H(MS_{p-1})$. This requires lower permanent storage requirements at the cost of a slightly higher computational demand and more security risks as

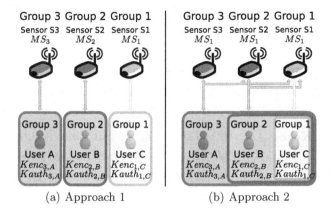

(a) Approach 1   (b) Approach 2

Fig. 2. Examples of the two approaches with three groups

we will see later. Note that every sensor can obtain the master secret for any privilege level. Figure 2(b) shows an example with three users and three sensors.

Thanks to this modification, user devices need to store only one pair of keys, that of the highest privilege level they are granted. For example, a user A in group 3 will only receive $(Kenc_{3,A}, Kauth_{3,A})$ from the Base Station. However the use of this key pair is enough for being granted access to any service in groups 1 to 3.

The verification of user credentials at the sensor side goes as follows. After receiving a message encrypted and authenticated with $(Kenc_{p,A}, Kauth_{p,A})$ (see Eq. (2)) the sensor derives $MS_p = H(...H(MS_1))$. From MS_p and user-bound parameters the sensor obtains $(Kenc_{p,A}, Kauth_{p,A})$ as in Eq. (1). Communications can now be established as in Eq. (3).

5.3 Combining Hierarchical Authentication with Individual Privacy

The basic protocol provides one-to-one authentication and encryption between a user and a sensor. On the other hand, approaches 1 and 2 allow to perform one-to-many authentication and encryption: all sensors holding the affected master secret will be able to authenticate the user and decrypt the conversation. Next, we consider the possibility of having services that demand one-to-one private communications and group-based authorization at the same time. For achieving this we we base on Approach 1, however the extension to Approach 2 is straightforward.

In this case sensor S is assigned by the Base Station an individual master secret MS_S (as in the basic protocol) and one master secret MS_p for each privilege group p the sensor provides services from (in Approach 2 the sensor would be assigned MS_1 and would derive the rest by hashing).

User A is assigned a pair of keys for individual communication with S, i.e. $(Kenc_{S,A}, Kauth_{S,A})$, and a pair of keys $(Kenc_{p,A}, Kauth_{p,A})$ for the privilege group she is entitled to, say p. Like before, these keys are generated for A by the

Base Station by feeding MS_p and user-related parameters ID_A, a, $init\_time$, exp_time to a Key Derivation Function.

Now, when A wants to communicate only with S while proving her authorization level, she encrypts her messages with $Kenc_{S,A}$ and computes the corresponding MAC with $Kauth_{p,A}$ as in Eq. (4). S replies using the same pair of keys and incrementing the counter, which needs not to be included on plain text given that the message exchange takes place between two players only:

$$A \rightarrow S : [Kenc_{S,A}\{M, ctr\}, ID_A, ID_p, a, init\_time, exp\_time, ctr,$$
$$MAC_{Kauth_{p,A}}(M, ID_A, ID_p, a, init\_time, exp\_time, ctr)] \quad (4)$$
$$S \rightarrow A : [Kenc_{S,A}\{M, ctr+1\}, MAC_{Kauth_{p,A}}(M, ctr+1)] \quad (5)$$

6   Considerations on Security

Similar considerations to those made for [3,4] can be made here. Both the basic protocol and the extensions provide semantic security (different encryptions of the same plain text produce different ciphertexts) thanks to the use of counter mode encryption. At this point we remark that it is the user who chooses the initial counter to be used during each message transaction. If the sensor does not trust the user then she can choose a new counter in her reply (Eq. (3) or (5), in the latter the counter should be hence added on plain text), thus forcing the user to increment that new one.

Regarding how to install master secrets on sensors, it can be done if a symmetric key is pre-installed at every sensor at deployment time. This key should be different for every sensor and shared with the Base Station, thus obtaining private communications with the latter. Furthermore, master secrets can be updated at a given frequency to enhance security, but once a master secret is updated the symmetric pairs of keys generated from the old version will be no longer valid in the system. To solve this problem, the exp_time value associated to every key pair can be made to match the master secret's update time, thus forcing the user to obtain a new pair from the Base Station. Doing so, the new pair will be derived from the new master secret.

Regarding how to communicate the user-related key pairs to the user, we assume that both user devices and the Base Station can handle public key encryption, hence a temporary secure channel is easy to establish (e.g. by using public key certificates).

Let us conclude this section with a discussion on key compromise. Given that user key pairs are bound to a specific user, stealing them will allow to impersonate a single user only, thus limiting the impact of a security breach at the user side. At the sensor side we can differentiate between stealing a sensor-only master secret or a privilege group master secret. In the first case, an attacker that steals MS_S from a sensor will only be able to impersonate that given sensor. In the second case, we can distinguish between approaches 1 and 2. In Approach 1 a sensor receives only the master secrets it is entitled to. Stealing a master secret MS_p will allow an attacker to understand and forge messages related to privilege group p, thus impersonating any sensor within that group, but not within other groups. In Approach 2, a sensor can obtain the master secret from any privilege group from the lowest group's. Thus, compromising a single sensor would allow an attacker to impersonate any sensor at any privilege group. The conclusion is that Approach 1 offers more security at the cost of more permanent storage requirements.

7   Other Considerations

The most power-demanding operation in a sensor is airing messages through its antenna [6,8], hence protocols intended for sensors (and not only those related to security) should try to minimize the number of messages needed as well as their length. Our protocol and its extensions do not require any additional message in a service request from the user to the sensor. Regarding message length, the additional overhead is values $(ID_p, a, init\_time, exp\_time, ctr)$ in the first message and a MAC (we assume ID_A must be sent anyway). The sensor needs only to attach the counter on plain text in approaches 1 and 2 (and in Section 5.3 in the case the user's counter is discarded and a new one is used).

Speaking of storage, the basic protocol requires that the user stores a pair of keys $(Kenc_{S,A}, Kauth_{S,A})$ per sensor S and values $ID_p, a, init\_time, exp\_time$ (again we consider ID_A is needed anyway). The counter ctr must also be stored during a message exchange. The sensor needs only to permanently store a symmetric session key for communications with the Base Station and MS_S. During a message exchange, the sensor needs to keep the pair of keys used for that user and values $(ID_p, a, init\_time, exp\_time, ctr)$. That information may be erased after the exchange since it is easily recomputed each time needed.

In addition to the storage requirements of the basic protocol, approaches 1 and 2 require the user to store permanently a pair of keys for the group. At the sensor side, Approach 1 requires the sensor to store a master secret for every privilege group it might be assigned to. In Approach 2, however, the sensor can decide whether to permanently store a single master secret (that of the lowest level, thus needing to compute the needed master secret at every message exchange) or to store all master secrets once derived (thus saving computations at the cost of space). We see this tradeoff as an interesting open future workline.

Passive participation is a typical behaviour used by sensors in order to save energy based on overheard messages [8]: if a node receives a user query and a subsequent reply from a different sensor then the first node can decide to not to reply in order to save the energy spent in transmission. Approach 1 allows for this type of behaviour within a given group, while Approach 2 allows for it within the group and those below in the hierarchy.

Regarding efficiency in computations, the experimental results shown in [4] for the basic protocol are equally valid for the groups extension, since the latter adds no overhead apart from the inclusion of the group identifier in MACs.

8   Conclusions

Here we present a group-based extension for an access control protocol for wireless network services. We address infrastructures populated by constrained devices (such as wireless sensors) that are arranged in different groups of services: users are granted access to these groups depending on their privileges. We consider two different scenarios, depending on whether privilege groups are hierarchical (entitlement to a privilege group implies access to all services down to the lowest group) or not (users can only access services contained in the very privilege group they are entitled to). Also, we show a way of combining individual encryption with group-based authorization. Regardless of the approach chosen, the authentication and authorization processes are performed efficiently and with no additional messages between the user and the addressed sensor. We discuss our proposal in terms of security and message and storage overhead. Also, experimental results shown in previous work [3,4] prove its applicability. Future worklines include the formal validation of the protocol in AVISPA [7] and its implementation on an extremely constrained platform such as MICAz or Arduino.

Acknowledgements. This work was funded by the Spanish Ministry of Economy and Competitiveness (TIN2008-01117, IPT-2011-1558-430000 and TIN2012-37483-C03-03), Junta de Andalucía (P11-TIC-7176), and partially by the European Regional Development Fund (ERDF).

References

1. Gómez-Goiri, A., Orduña, P., Diego, J., López-de-Ipiña, D.: Otsopack: Lightweight Semantic Framework for Interoperable Ambient Intelligence Applications. To appear in Journal of Computers in Human Behavior
2. Le, X.H., Khalid, M., Sankar, R., Lee, S.: An Efficient Mutual Authentication and Access Control Scheme for Wireless Sensor Networks in Healthcare. Journal of Networks 6(3), 355–364 (2011)
3. Naranjo, J.A.M., Orduña, P., Gómez-Goiri, A., López-de-Ipiña, D., Casado, L.G.: Lightweight User Access Control in Energy-Constrained Wireless Network Services. In: Bravo, J., López-de-Ipiña, D., Moya, F. (eds.) UCAmI 2012. LNCS, vol. 7656, pp. 33–41. Springer, Heidelberg (2012)
4. Naranjo, J.A.M., Orduña, P., Gómez-Goiri, A., López-de-Ipiña, D., Casado, L.G.: Enabling user access control in energy-constrained wireless smart environments. To appear in Journal of Universal Computer Science
5. Ngo, H.H., Xianping, W., Phu, D.L., Srinivasan, B.: An Individual and Group Authentication Model for Wireless Network Services. JCIT 5(1), 82–94 (2010)
6. Perrig, A., Szewczyk, R., Tygar, J.D., Wen, V., Culler, D.E.: SPINS: security protocols for sensor networks. Wireless Networks 8(5), 521–534 (2002)
7. The AVISPA Project, http://www.avispa-project.org/
8. Zhu, S., Setia, S., Jajodia, S.: LEAP+: Efficient security mechanisms for large-scale distributed sensor networks. ACM Transactions on Sensor Networks 2(4), 500–528 (2006)

Randomness Analysis of Key-Derived S-Boxes

Rafael Álvarez and Antonio Zamora

Dpt. of Computer Science and Artificial Intelligence (DCCIA)
University of Alicante (Campus de San Vicente)
Ap. 99, E-03080, Alicante, Spain
{ralvarez,zamora}@dccia.ua.es

Abstract. Although many ciphers use fixed, close to ideal, s-boxes (like AES for example), random s-boxes offer an interesting alternative since they have no underlying structure that can be exploited in cryptanalysis. For this reason, some cryptosystems generate pseudo-random s-boxes as a function of the key (key-derived).

We analyse the randomness properties of key-derived s-boxes generated by some popular cryptosystems like the RC4 stream cipher, and the Blowfish and Twofish block ciphers with the aim of establishing if this kind of s-boxes are indistinguishable from purely random s-boxes.

For this purpose we have developed a custom software framework to generate and evaluate random and key derived s-boxes.

Keywords: S-Boxes, Key-Derived, Random, RC4, Blowfish, Twofish.

1 Introduction

Substitution boxes (or s-boxes) are simple substitution tables where an input value is transformed into a different value. They are employed in many sizes, but the most prevalent are 8x8 bits (byte as input and output) and 8x32 bits (byte as input and four byte word as output). They are essential in many cryptosystem designs (see [2,3,5,11,15,16]) since they can introduce the required non-linearity characteristics, making cryptanalysis a more difficult endeavour (see [8]).

Regarding their design, there are two basic schools of thought:

- **Generated s-boxes.** These are carefully designed to achieve certain desirable characteristics and often have an underlying generator function. The advantage is they can be chosen so they are optimum in terms of non-linearity, avalanche, bit independence, etc. On the other hand, the fact that they are fixed or based on some kind of structure could make cryptanalysis easier (see [5,6,9,10]).
- **Random or key-derived s-boxes.** Unlike generated s-boxes, random s-boxes cannot be guaranteed to have certain values of non-linearity or other metrics but many cryptographers think that the fact they do not have any exploitable underlying structure is a distinct advantage against cryptanalysis. Furthermore, pseudo-random s-boxes that are generated as a function of

Á. Herrero et al. (eds.), *International Joint Conference SOCO'13-CISIS'13-ICEUTE'13*, 611
Advances in Intelligent Systems and Computing 239,
DOI: 10.1007/978-3-319-01854-6_62, © Springer International Publishing Switzerland 2014

the key (key-derived s-boxes) allow the cryptosystem to use a different s-box per each key, making cryptanalysis even more difficult.

In general, random s-boxes offer acceptable security characteristics although not as high as the best generated s-boxes (see [4]).

The question regarding how random are key-derived s-boxes is, therefore, an interesting one. Ideally, they should approximate purely random s-boxes as much as possible so that they have the advantage of being free of underlying structure while at the same time permitting the cryptosystem to use a different s-box for each key.

We have developed a testing framework capable of generating and analysing 8x8 s-boxes and we have tested 340,000 different s-boxes corresponding to random s-boxes, those generated by the RC4 stream cipher (see [11]), the Blowfish block cipher (see [15]) and its improved sibling, Twofish (see [16]). We have computed useful metrics for each s-box and summarized the resulting data to extract meaningful conclusions.

The rest of the paper is organised as follows: a description of the tests performed is given in section 2, the results are analysed in section 3 and, finally, the extracted conclusions are in section 4.

2   Description

In order to test the randomness of key-derived s-boxes we have compared several algorithms against random s-boxes. During our study, we have generated 10,000 random s-boxes, 10,000 RC4 s-boxes, 40,000 8x32 Blowfish s-boxes (separated into 160,000 8x8 s-boxes) and 40,000 Twofish s-boxes (also separated into 160,000 8x8 s-boxes); totalling 340,000 tested s-boxes.

We have developed a custom testing framework capable of generating and analysing s-boxes, computing multiple metrics for each s-box. It is comprised of two separate components: a completely new element written in Go (see [7]) that generates s-boxes in a parallel fashion using hooks into the cryptosystems to obtain the required s-boxes; and an element written in C (evolved from previous work, see [4]) that batch tests groups of s-boxes and is simultaneously executed in as many cores as possible.

Then, the results have been statistically analysed to find meaningful distinguishing characteristics in each set.

2.1   Random S-Boxes

Random s-boxes have been generated as bijective 8x8 s-boxes using pseudo-random permutations of the values $0, 1, \ldots, 255$. This guarantees perfect balance but it can introduce fixed points. If fixed points are not desired, then some kind of filtering must be introduced to eliminate the offending s-boxes or transform them into valid ones.

These random s-boxes form a standard of randomness that other key-derived s-boxes can be compared to.

2.2   RC4

The RC4 stream cipher algorithm (see [11]) is one of the most widely used software stream ciphers since it is part of the Secure Sockets Layer (SSL) and Wired Equivalent Privacy (WEP) protocols, among others. It does not employ a s-box per-se, but its internal state consists of an 8x8 s-box that dynamically changes as data is encrypted. The initial state of this internal s-box is determined solely by the key used with RC4 so we consider that initial state as a key-derived s-box.

2.3   Blowfish

Blowfish (see [15]) is a popular block cipher that, due to its slower key schedule algorithm, is commonly used as a password hashing algorithm in operating systems and applications (see [14]). It uses four 8x32 key-derived s-boxes that are generated during the key schedule phase.

Since some of the metrics require extensive calculations with a computational complexity on the order of $O(n) = 2^b$, being b the output bit size, they are too big to be analysed natively. For this reason we consider each 8x32 s-box as 4 adjacent 8x8 s-boxes, obtaining 16 8x8 s-boxes per each different key.

2.4   Twofish

Twofish (see [16]) is the successor to Blowfish and was one of the five finalists of the Advanced Encryption Standard (AES) contest. It also generates four 8x32 s-boxes during the key schedule algorithm which have been considered as sixteen 8x8 s-boxes so they can be tested for all metrics.

3   Results

3.1   Balance

Balance can be defined in terms of just the component functions (columns) of the s-box, or of all linear combinations of the component functions. The latter definition is probably more correct, since unbalanced linear combinations imply also some type of statistical bias towards 0 or 1 in the output. See [4,12,17] for more information.

Table 1. Balance test results

	Random	RC4	Blowfish	Twofish
Min.	255	255	1	255
Max.	255	255	39	255
Med.	255	255	13	255
Inv.	0%	0%	100%	0%

When bijective s-boxes are used, then measuring balance can act as a way of checking that they are really bijective. In those cases where unbalanced s-boxes may be used, the ratio of balanced linear combinations to total number of linear combinations is an useful metric.

As shown in table 1, we can see that, depending on the algorithm, they are either all valid (all s-boxes have all linear combinations balanced) or all invalid (no s-box has all linear combinations balanced). This table includes the minimum (Min.), maximum (Max.) and median (Med.) number of balanced linear combinations found in an s-box of that set. It also includes the percentage of unbalanced s-boxes (Inv.) in each set.

The random s-boxes, and the ones generated by RC4, are all balanced since they are permutations of the values 0 to 255 (bijective).

Blowfish, on the other hand, fails the balance test since, once you explode each 8x32 s-box into four 8x8 adjacent s-boxes, they do not contain all possible values achieving terrible results in this test with a median of 13 and a minimum of 1 or 2 balanced linear combinations (so, in most cases, not even the component functions are balanced). This is a problem, since you can always consider an 8x32 s-box as four adjacent 8x8 s-boxes and attack those accordingly.

Twofish is a certain improvement in this regard, generating 8x32 s-boxes that are balanced when separated as 4 distinct 8x8 s-boxes.

3.2 Fixed Points

Direct ($S(i) = i$) or reverse ($S(i) = 255 - i$) fixed points are generally not desirable in s-boxes since they imply that the output is equivalent to the input and not modified in some cases (see [4]).

Fixed points are not an absolute requirement since there is no known attack exploiting this characteristic, although it is generally desirable that all input bytes are transformed into something different by the s-box.

Table 2. Fixed points test results

| | Direct | | | |
	Random	RC4	Blowfish	Twofish
Min.	0	0	0	0
Max.	7	8	8	7
Med.	1	1	1	1
Inv.	63%	57%	64%	63%

| | Reverse | | | |
	Random	RC4	Blowfish	Twofish
Min.	0	0	0	0
Max.	7	6	8	7
Med.	1	1	1	1
Inv.	64%	62%	64%	63%

As shown in table 2, none of the studied algorithms prevent fixed points from happening. The occurrence rate is of 1 direct and 1 reverse fixed point per s-box on average, which is on par with random s-boxes. This table includes the minimum (Min.), maximum (Max.) and median (Med.) number of direct and reverse fixed points found in an s-box of that set. It also includes the percentage of invalid s-boxes (Inv. – having fixed points) in each set.

3.3 Non-linearity

Non-linearity (see [4,12]) is a measure of resistance to linear cryptanalysis and is defined for Boolean functions. In the case of s-boxes we take the minimum non-linearity of all linear combinations of the component functions of each s-box. Although the theoretical minimum is 0, we consider that a better minimum threshold is 2^{n-2} because, generally, random s-boxes are above that value. The maximum non-linearity considered is $2^{n-1} - 2^{\frac{n}{2}}$.

Table 3. Non-linearity test results

	Random	RC4	Blowfish	Twofish[1]	Twofish[2]
Min.	78	78	77	78	80
Max.	98	98	98	98	98
Med.	92	92	93	92	94
Inv.	19%	19%	41%	19%	50%

As can be seen in table 3, all algorithms present similar values to random s-boxes in minimum (Min.) and maximum (Max.) non-linearities ranging from 78 to 98 approximately.

The median value (Med.) shows some interesting characteristics with random and RC4 s-boxes having a median of 92, Blowfish a median of 93 and Twofish 92 for the first half of s-boxes (denoted by Twofish[1]) and 94 for the second half (Twofish[2]). These could be used, to a certain degree, as a distinguishing factor for Blowfish or second half Twofish s-boxes. The table also includes the percentage of not-as-good, or *invalid*, s-boxes (Inv.) that have non-linearities below the median value.

Please note that the maximum non-linearity considered would be $2^{n-1} - 2^{n/2}$, or 112 for $n = 8$. The AES s-box achieves a non-linearity of 112 (see table 7).

3.4 XOR Table

The XOR table is a metric related to differential cryptanalysis (see [1,13]). There are two possible metrics for the XOR table: the first corresponds to the number of valid entries (entries with values 0 or 2) of the XOR table, with the second being the maximum entry in the table. See [4,12,17] for more information.

As shown in table 4, there are no differentiating factors among the algorithms tested; they all present a minimum value (Min.) of the maximum entry of 8, a

Table 4. XOR Table test results

	Random	RC4	Blowfish	Twofish
Min.	8	8	8	8
Max.	16	18	18	18
Med.	12	12	12	12
Valid	91%	91%	91%	91%

maximum value (Max.) of the maximum entry of 18 (except 16 for the random s-boxes) and a median value (Med.) of the maximum entry of 12. The number of valid entries in the table is also the same, stable at 91%.

3.5   Avalanche

Defined for Boolean functions, we consider the distance to the strict avalanche criterion of order 1 or DSAC(1). The DSAC(1) for the complete s-box is the maximum of the distances of the component functions (columns) (see [4,12]). The lower boundary is determined as 2^{n-2}.

As shown in table 5, all algorithms present similar minimum (Min.) and maximum (Max.) values, while Blowfish presents a slightly higher median (Med.) value.

Table 5. Avalanche ($DSAC(1)$) test results

	Random	RC4	Blowfish	Twofish
Min.	12	10	10	10
Max.	28	28	29	30
Med.	16	16	17	16

3.6   Bit Independence

Defined for the whole S-box, this corresponds to bit independence DBIC(2,1) (see [4,12]).

This metric presents no differentiating factors among the tested algorithms, as shown in table 6 that includes minimum (Min.), maximum (Max.) and median (Med.) values.

Table 6. Bit Independence ($DBIC(2,1)$) test results

	Random	RC4	Blowfish	Twofish
Min.	16	16	16	16
Max.	30	32	36	30
Med.	20	20	20	20

3.7 Comparison with AES

If we compare random s-boxes to the AES s-box, we can see in table 7 that the AES s-box achieves better values in all non-trivial metrics. It is for this reason that cryptosystems based on key-derived s-boxes trade off ideal metric values for no underlying structure or generator function (see [4]).

Table 7. Comparison between AES and random s-boxes

	AES	Random (best)	Random (average)	Random (worst)
Balance	100%	100%	100%	100%
Fixed points	(0,0)	(0,0)	(1,1)	(7,7)
Nonlinearity	112	98	92	78
XOR table	4	8	12	16
DSAC(1)	8	12	16	28
DBIC(1)	8	16	20	30

4 Conclusions

We have analysed s-boxes generated by RC4, Blowfish and Twofish, comparing them to pseudo-random s-boxes. During this study, we have found detectable differences in some of the metrics that could allow, to a certain degree, to distinguish Blowfish s-boxes and some Twofish s-boxes from random ones. This does not imply a vulnerable design but shows some small biases that could be improved. Nevertheless, our testing shows that key derivation is a suitable way to obtain random-equivalent s-boxes.

Another point of concern is the lack of balance of Blowfish s-boxes that, although they are employed as 8x32 s-boxes in the algorithm, when they are analysed as four adjacent 8x8 s-boxes they present statistical biases that could be exploited. This problem is corrected in its successor, twofish, acknowledging the importance of maintaining proper balance in the 8x8 sub s-boxes. It must be noted that despite being succeeded by Twofish, Blowfish is vastly more popular, used widely like in the *bcrypt* password derivation algorithm (see [14]).

In order to conduct our analysis, we have developed a custom framework for generating and testing 8x8 s-boxes that we plan to extend and improve in the future.

References

1. Adams, C.M., Tavares, S.E.: Designing S-Boxes for Ciphers Resistant to Differential Cryptanalysis. In: Proc. 3rd Symposium on State and Progress of Research in Cryptography, pp. 181–190 (1993)
2. Álvarez, R., McGuire, G., Zamora, A.: The Tangle Hash Function. Submission to the NIST SHA-3 Competition (2008)

3. Álvarez, R., Vicent, J.F., Zamora, A.: Improving the Message Expansion of the Tangle Hash Function. In: Herrero, Á., Corchado, E. (eds.) CISIS 2011. LNCS, vol. 6694, pp. 183–189. Springer, Heidelberg (2011)
4. Álvarez, R., McGuire, G.: S-Boxes, APN Functions and Related Codes. In: Enhancing Cryptographic Primitives with Techniques from Error Correcting Codes, vol. 23, pp. 49–62. IOS Press (2009)
5. Fuller, J., Millan, W.: On linear redundancy in the AES S-Box. Cryptology ePrint Archive, Report 2002/111
6. Fuller, J., Millan, W., Dawson, E.: Multi-objective Optimisation of Bijective S-boxes. In: Congress on Evolutionary Computation, vol. 2, pp. 1525–1532 (2004)
7. The Go Programming Language, http://www.golang.org
8. Hussain, I., Shah, T., Gondal, M.A., Khan, W.A.: Construction of Cryptographically Strong 8x8 S-boxes. World Applied Sciences Journal 13(11), 2389–2395 (2011)
9. Jing-Mei, L., Bao-Dian, W., Xiang-Guo, C., Xin-Mei, W.: Cryptanalysis of Rijndael S-box and improvement. Applied Mathematics and Computation 170, 958–975 (2005)
10. Kavut, S.: Results on rotation-symmetric S-boxes. Information Sciences 201, 93–113 (2012)
11. Klein, A.: Attacks on the RC4 stream cipher. Designs, Codes and Cryptography 48(3), 269–286 (2008)
12. Mister, S., Adams, C.: Practical S-Box Design. In: Selected Areas in Cryptography (1996)
13. Murphy, S., Robshaw, M.J.B.: Key-Dependent S-Boxes and Differential Cryptanalysis. Designs, Codes and Cryptography 27(3), 229–255 (2002)
14. Provos, N., Mazeries, D.: Bcrypt Algorithm. USENIX (1999)
15. Schneier, B.: Description of a New Variable-Length Key, 64-bit Block Cipher (Blowfish). In: Anderson, R. (ed.) FSE 1993. LNCS, vol. 809, pp. 191–204. Springer, Heidelberg (1994)
16. Schneier, B., Kelsey, J., Whiting, D., Wagner, D., Hall, C., Ferguson, N.: The Twofish encryption algorithm: a 128-bit block cipher. John Wiley & Sons (1999)
17. Youssef, A.M., Tavares, S.E.: Resistance of Balanced S-boxes to Linear and Differential Cryptanalysis. Information Processing Letters 56(5), 249–252 (1995)

A CA Model for Mobile Malware Spreading Based on Bluetooth Connections

Ángel Martín del Rey[1] and Gerardo Rodríguez Sánchez[2]

[1] Department of Applied Mathematics
E.P.S. de Ávila, University of Salamanca
C/Hornos Caleros 50, 05003-Ávila, Spain
delrey@usal.es
[2] Department of Applied Mathematics
E.P.S. de Zamora, University of Salamanca
Avda. Requejo 33, 49022-Zamora, Spain
gerardo@usal.es

Abstract. There is an unstoppable rise of the number of smartphones worldwide and, as several applications requires an Internet access, these mobile devices are exposed to the malicious effects of malware. Of particular interest is the malware which is propagated using bluetooth connections since it infects devices in its proximity as does biological virus. The main goal of this work is to introduce a new mathematical model to study the spread of a bluetooth mobile malware. Specifically, it is a compartmental model where the mobile devices are classified into four types: susceptibles, carriers, exposed and infectious; its dynamic is governed by means of a couple of two-dimensional cellular automata. Some computer simulations are shown and analyzed in order to determine how a *proximity* mobile malware might spread under different conditions.

Keywords: Mobile malware, Bluetooth, Cellular automata, SCEIS model, Mathematical modeling.

1   Introduction and Preliminaries

Smartphones play a very important role in our lives. According to a recent report from the consulting firm Strategy Analytics, in the third quarter of 2012 the number of smartphones being used around the globe topped 1 billion for the first time ever; moreover, Strategy Analytics also predicted that growing demand for the internet-connected mobile devices was likely to push the number north of 2 billions within the next three years. This unstoppable rise of the number of smartphones will be increased due to the new-born BYOD (Bring Your Own Device) paradigm ([10]) and the progressive establishment and development of the Internet of Things ([1]).

Smartphones combine the capabilities and functionalities of cellphones and PDAs: the user can phone, play music and videos, take photographs or record videos; the user can also process text documents, send and receive e-mails, surf

Á. Herrero et al. (eds.), *International Joint Conference SOCO'13-CISIS'13-ICEUTE'13*,
Advances in Intelligent Systems and Computing 239,
DOI: 10.1007/978-3-319-01854-6_63, © Springer International Publishing Switzerland 2014

Internet, etc. Furthermore, it is possible to download (from official or/and un-official stores) applications to perform a great range of tasks. In this sense, the majority of applications requires an Internet access and consequently smart-phones are exposed to the effects of malware and other cybersecurity threats.

Android and iOS are the number one and number two ranked Smartphone operating systems. In fact, IDC Analysts says that the 80% of all smartphone shipments during the fourth quarter of 2012 were of these two operating systems ([3]). Consequently they are the basic targets for malware. In this sense, Android is the most important target and the majority of mobile malware is designed to compromise this operative system. This fact is not only a consequence of its leader position but also of other factors: its open-source nature, the loose security polices for apps which make possible the distribution of malware through both the official app store and the external sources, etc. There are several types of smartphone malware: trojans, computer worms, computer viruses, adware, spyware, etc. Trojans are the most popular threat by far (Cabir -which was the first speciment reported in 2004- , Zeus, Placeraider, Geinimi, etc.) making up to 85% of the threats detected. The effects of malware can be just as varied: they range from brief annoyance to computer crashes and identity theft, and, usually, those effects go unnoticed for the user.

Consequently, the prediction of the behavior of the spreading of mobile mal-ware is very important. Unfortunately not many mathematical models dealing with this issue have been appeared, and this number is fewer if they are related to bluetooth malware. The great majority of these works are based on contin-uous mathematical tools such as systems of ordinary differential equations (see [2, 4, 5, 7–9, 11]). These are well-founded and coherent models from the mathe-matical point of view, offering a detailed study of the main characteristics of their dynamic: stability, equilibrium, etc. Nevertheless, they do have some drawbacks that, owing to their importance, merit attention:

(1) They do not take into account local interactions between the smartphones. Parameters such as the rate of infection, the rate of recovery, etc. are used but they are of a general nature. Accordingly, the use of parameters individualized for each of the mobile devices of the network is not considered, and it seems reasonable to search for a mathematical model that will takes these aspects into account.

(2) They assume that the elements forming the network are distributed homo-geneously and that all are connected with one another. When the propagation of malware is analyzed macroscopically the results obtained provide a fairly good approximation of what is really happening. However, if we analyze such prop-agation locally the results obtained are manifestly poorer since at microscopic scale the dynamic is very sensitive to local interconnections.

(3) They are unable to simulate the individual dynamic of each of the smart-phones of the network. It is true that when the size of the network is very large the overall behavior observed may seem very similar (as regards trends) to what is happening in reality but the use of essential information is neglected: for example, smartphones whose operative system is iOS should not be affected

(in the sense of infected) by malware specifically designed for mobile devices based on Android (although they could be considered as exposed) etc.

These three main deficiencies shown by models based on differential equations could be rectified simply if we use a different type of mathematical model. In this sense, discrete models or individual-based models could overcome these drawbacks. As far we know, there is only one discrete mathematical model published for bluetooth mobile malware based on cellular automata ([6]). It is a compartmental model where the population is divided into susceptible, exposed, infectious, diagnosed and recovered smartphones. Every cell stands for a regular geographical area which is occupied by at most one smartphone, and these smartphones are fixed in the corresponding cell (that is, movements between cells are not permitted). Moreover, the local transition function depends on two parameters: the infected factor and the resisted factor and any other individual characteristics are not taken into account.

The main goal of this paper is to introduce a new mathematical model to simulate mobile malware spreading using bluetooth connections. This model is based on the use of cellular automata and the population is divided into four classes: susceptible, carrier, exposed (or latent) and infectious smartphones. The model proposed in this work is based on a paradigm which is far from the paradigm used in [6]. Specifically in our model two two-dimensional cellular automata will be used: one of them rules the global dynamic of the model whereas the other one governs the local dynamic. The geographical area where smartphones are is tessellated into several square tiles that stand for the cells of the global cellular automata, such that there can be more than one smartphone in each cell. Moreover, the smartphones placed into a (global) cell stand for the cells of the local cellular automata whose transition rule update synchronously the states of the smartphones. Furthermore, the smartphones can move from one global cell to another at every step of time.

The rest of the paper is organized as follows: In section 2 the basic theory of (two-dimensional) cellular automata is given; the new model is introduced in section 3, and some simulations are shown and briefly discussed in section 4. Finally, the conclusions are presented in section 5.

2   Two-Dimensional Cellular Automata

A two-dimensional cellular automaton (CA for short) is a particular type of finite state machine formed by a finite number of memory units called cells which are uniformly arranged in a two-dimensional space. Each cell is endowed with a state that changes synchronously at every step of time according to a local transition rule (see, for example, [12]). More precisely, a CA can be defined as the 4-uplet $\mathcal{A} = (\mathcal{C}, \mathcal{S}, V, f)$, where C is the cellular space formed by a two-dimensional array of $r \times c$ cells (see Figure 1-(a)): $\mathcal{C} = \{(i, j), 1 \leq i \leq r, 1 \leq j \leq c\}$.

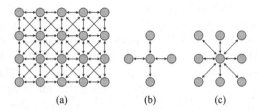

Fig. 1. (a) Rectangular cellular space $\mathcal{C}$. (b) Von Neumann neighborhood. (c) Moore neighborhood.

The state of each cell is an element of a finite state set $\mathcal{S}$: $s_{ij}^t \in \mathcal{S}$ for every cell $(i,j) \in \mathcal{S}$. The neighborhood of each cell is defined by means of an ordered and finite set of indices $V \subset \mathbb{Z} \times \mathbb{Z}$, $|V| = m$ such that for every cell (i,j) its neighborhood $V_{(i,j)}$ is the following set of m cells:

$$V_{(i,j)} = \{(i + \alpha_1, j + \beta_1), \ldots, (i + \alpha_m, j + \beta_m) : (\alpha_k, \beta_k) \in V\}. \qquad (1)$$

Usually two types of neighborhoods are considered: Von Neumann and Moore neighborhoods. The Von Neumann neighborhood of a cell (i,j) is formed by the main cell itself and the four cells orthogonally surrounding it (see Figure 1-(b)); in this case: $V = \{(-1,0), (0,1), (0,0), (1,0), (0,-1)\}$, that is:

$$V_{(i,j)} = \{(i - 1, j), (i, j + 1), (i, j), (i + 1, j), (i, j - 1)\}.$$

The Moore neighborhood of the cell (i,j) is constituted by the eight nearest cells around it and the cell itself (see Figure 1-(c)). Consequently:

$$V = \{(-1, -1), (-1, 0), (-1, 1), (0, -1), (0, 0), (0, 1)(1, -1), (1, 0), (1, 1)\}, \qquad (2)$$

that is:

$$\begin{aligned} V_{(i,j)} = \{&(i - 1, j - 1), (i - 1, j), (i - 1, j + 1)(i, j - 1), (i, j), \qquad (3)\\ &(i, j + 1), (i + 1, j - 1), (i + 1, j), (i + 1, j + 1)\}. \end{aligned}$$

The CA evolves in discrete steps of time changing the states of all cells according to a local transition function $f \colon \mathcal{S}^m \to \mathcal{S}$, whose variables are the previous states of the cells constituting the neighborhood, that is:

$$s_{ij}^{t+1} = f\left(s_{i+\alpha_1, j+\beta_1}^t, \ldots, s_{i+\alpha_m, j+\beta_m}^t\right) \in \mathcal{S}. \qquad (4)$$

As the number of cells of the CA is finite, boundary conditions must be considered in order to assure the well-defined dynamics of the CA. One can state several boundary conditions depending on the nature of the phenomenon to be simulated. In this work we will consider null boundary conditions, that is: if $(i,j) \notin C$ then $s_{ij}^t = 0$.

The standard paradigm for two-dimensional cellular automata states that the cellular space is rectangular and the local interactions between the cells are limited to the spatially nearest neighbor cells. Nevertheless one can consider a different topology for the cellular space depending on the phenomenon to be modelized. Consequently, the notion of cellular automaton on graph emerges: A cellular automaton on a graph G (CAG for short) is a two-dimensional cellular automaton whose cellular space is defined by means of the (directed or undirected) graph $G = (V, E)$; each cell of the CAG stands for a node of the graph and there is an edge between the nodes u and v ($uv \in E$) if the cell associated to the node u belongs to the neighborhood of the cell associated to node v. Note that if G is a directed graph then $uv \neq vu$, whereas if G is an undirected graph, then $uv = vu$, that is, the cell associated to the node u (*resp.* v) is in the neighborhood of the cell associated to the node v (*resp.* u).

3   The SCEIS Mathematical Model

The model introduced in this work is based on the use of two two-dimensional cellular automata: one of them rules de global dynamic of the system and the other one governs the local dynamic. The following general assumptions are made in the model:

(1) The population of smartphones is divided into four classes: those smartphones that are susceptible to the malware infection (susceptibles), those "healthy" smartphones that carry the malware specimen and are not able to transmit it (carriers), the smartphones that have been successfully infected but the malware remains latent (exposed), and the smartphones that have been infected and the malware is activated: it is able to perform its payload and to propagate (infectious). Note that the carrier status is acquired when the malicious code reaches a smartphone based on a different operative system than the malware's target. On the contrary, exposed and infectious smartphones are those mobile devices reached by a malware specimen which could be activated due to the coincidence of the operative systems.

(2) The model is compartmental: susceptible smartphones becomes carriers or exposed when the computer worm reaches them; exposed mobile devices get the infectious status when the malware is activated; and finally infectious smartphones progress to susceptible when the malware is detected and eliminated (note that there is not any immune period after the recovery from the malware). In Figure 2 an scheme showing the dynamic of the SCEIS model is introduced.

(3) The population of smartphones is placed in a finite geographical area where they can move freely.

(4) The bluetooth connection will be the vector for transmission of the mobile malware.

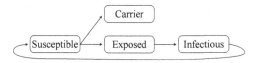

Fig. 2. Flow diagram with the evolution of the states of a smartphone

The Global Cellular Automaton. The global cellular automaton, $\mathcal{A}_G$, simulates the global behavior of the system giving at every step of time the number of smartphones which are susceptible, carrier, exposed, and infectious in a certain geographical area. It follows the traditional paradigm for CA, consequently the whole area where the smartphones "live" is tessellated into $r \times c$ constant-size square tiles (square grid), and every cell of $\mathcal{A}_G$ stands for one of these square portions of the area. Let $\mathcal{C}_G = \{(i,j), 1 \leq i \leq r, 1 \leq j \leq c\}$ be the cellular space and set $X_{ij}^t = \left(S_{ij}^t, C_{ij}^t, E_{ij}^t, I_{ij}^t\right)$, the state of the cell (i,j) at time t, where S_{ij}^t is the number of susceptible smartphones, C_{ij}^t is the number of carrier smartphones, E_{ij}^t is the number of exposed smartphones, and I_{ij}^t is the number of infectious mobile devices. The neighborhood considered can be Von Neumann or Moore neighborhood, and the local transition functions are as follows:

$$S_{ij}^t = S_{ij}^{t-1} - OS_{ij}^{t-1} + IS_{ij}^{t-1}, \tag{5}$$

$$C_{ij}^t = C_{ij}^{t-1} - OC_{ij}^{t-1} + IC_{ij}^{t-1}, \tag{6}$$

$$E_{ij}^t = E_{ij}^{t-1} - OE_{ij}^{t-1} + IE_{ij}^{t-1}, \tag{7}$$

$$I_{ij}^t = I_{ij}^{t-1} - OI_{ij}^{t-1} + II_{ij}^{t-1}, \tag{8}$$

where OS_{ij}^{t-1} stands for the number susceptible smartphones that moved from (i,j) to a neighbor cell at time $t-1$, and IS_{ij}^{t-1} represents the number of susceptible smartphones that arrived at the cell (i,j) at time $t-1$ coming from a neighbor cell, and so on.

The Local Cellular Automaton. The local cellular automaton involved in this model, $\mathcal{A}_L = (\mathcal{C}_L, \mathcal{S}_L, V_L, f_L)$, simulates the individual behavior of every smartphone of the system, that is, $\mathcal{A}_L$ governs the evolution of the states of each smartphone (susceptible, carrier, exposed or infectious). It is a cellular automaton on a graph G, which defines the topology of the cellular space $\mathcal{C}_L$ and the neighborhoods V_L. The set of states is $\mathcal{S}_L = (S, C, E, I)$ where S stands for susceptible, C for carrier, E for exposed and I for infectious. The local transition rules are as follows:

(1) *Transition from susceptible to exposed and carrier*: As is mentioned in the last section, a susceptible smartphone v becomes exposed or carrier when the mobile malware reaches it and this occurs when the user accepts a bluetooth connection from a malicious device. The boolean function that models the transition from susceptible state to exposed or carrier state is the following:

$$f_L(u) = B_v \cdot \bigwedge_{\substack{u \in N(v) \\ s_u^{t-1} = I}} A_{uv} \cdot \alpha_{vu}, \tag{9}$$

where

$$B_v = \begin{cases} 1, \text{ with probability } b_v \\ 0, \text{ with probability } 1 - b_v \end{cases} \tag{10}$$

$$A_{uv} = \begin{cases} 1, \text{ with probability } 1 - a_{uv} \\ 0, \text{ with probability } a_{uv} \end{cases} \tag{11}$$

$$\alpha_{vu} = \begin{cases} 1, \text{ if the smartphones } u \text{ and } v \text{ have the same OS} \\ 0, \text{ if the smarthphones } u \text{ and } v \text{ have not the same OS} \end{cases} \tag{12}$$

where b_v is the probability to have the bluetooth enabled, and a_{uv} is the probability to accept the bluetooth connection from the smartphone u. As a consequence:

$$s_v^t = \begin{cases} E, \text{ if } s_v^{t-1} = S, \ f_L(u) = 1 \text{ and } \beta = 1 \\ C, \text{ if } s_v^{t-1} = S, \ f_L(u) = 1 \text{ and } \beta = 0 \\ S, \text{ if } s_v^{t-1} = S \text{ and } f_L(u) = 0 \end{cases}$$

where the parameter β denotes if the infection comes from an infectious smartphone with the same operative system ($\beta = 1$) or with a different operative system ($\beta = 0$).

(2) *Transition from exposed to infectious*: When the virus reaches the host it becomes infectious after a period of time (the latent period t_v^L). Consequently,

$$s_v^t = \begin{cases} I, \text{ if } s_v^{t-1} = E \text{ and } \tilde{t}_v > t_v^L \\ E, \text{ if } s_v^{t-1} = E \text{ and } \tilde{t}_v \leq t_v^L \end{cases}$$

where $\tilde{t}_v$ stands for the discrete steps of time passed from the acquisition of the mobile malware.

(3) *Transition from infectious to susceptible*: If there is a security application installed in the smartphone v, the mobile malware can be detected with a certain probability d_v, then:

$$s_v^t = \begin{cases} S, \text{ if } s_v^{t-1} = I \text{ and } D = 1 \\ I, \text{ if } s_v^{t-1} = I \text{ and } D = 0 \end{cases} \tag{13}$$

where

$$D = \begin{cases} 1, \text{ with probability } d_v \\ 0, \text{ with probability } 1 - d_v \end{cases} \tag{14}$$

4   Some Illustrative Simulations

The computational implementation of the model introduced in this work has been done using the computer algebra system *Mathematica* (version 9.0). In the simulations, a bidimensional lattice of 5×5 cells is considered where $n = 100$ smartphones are initially placed at random. Moreover, at every step of time the mobile devices can move randomly from the main cell to a neighbor cell with probability m_v for each smartphone v. The parameters used in the simulations are shown in Table 1 and they are different for each smartphone varying in a fixed range. It is emphasized that the values of these parameters are merely illustrative (they are not computed using real data). Although several simulations have been computed in the laboratory, only few of them are shown in this work since the simulations performed using similar conditions exhibit similar trends.

Table 1. Values of the parameters involved in the model

Parameter	Range of values
m_v	$[0.4, 0.6]$
b_v	$[0.75, 1]$
a_{uv}, d_v	$[0.5, 1]$
t_v^L	$[1, 5]$

In Figure 3-(a) the global evolution of the different compartments (susceptible -green-, carrier -blue-, exposed -orange- and infectious -red-) is shown when it is supposed that the distribution of the OS is as follows: 60% of smartphones are based on Android, 20% is based on iOS, and the 5% of mobile devices are based on both Blackberry OS and Windows Mobile. In addition, we state that the 20% of Android-based smartphones are infectious at time $t = 0$. As is shown, the system evolves to a quasi-endemic equilibrium with periodic outbreaks (this trend appears in all the simulations). As is mentioned in the Introduction, not only the global dynamic of the system can be modeled using cellular automata but also the individual dynamic of each smartphone can be obtained; in this sense in Figure 3-(b) the evolution of the state of every smartphone is shown when all are based on Android OS and the 20% of them are initially infectious. In this figure each row represents the evolution of an individual smartphone where susceptible periods are in green, latent periods are in orange and infectious

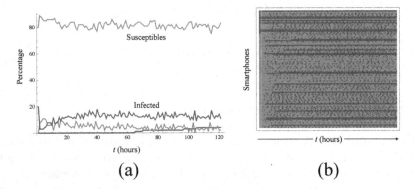

(a)   (b)

Fig. 3. (a) Global evolution when different OS are considered. (b) Individual evolution when the same OS is considered.

periods are in red. Note that, as in the previous case, there is a reinfection for several smartphones and the disease-free equilibrium is not reached.

In Figure 4 the evolution of the infectious smartphones of the system is shown when different operative systems, neighborhoods and the number of infectious devices at initial step of time (the 5%, 10%, 15%, and 20% of population is infectious at time $t = 0$) are considered. In Figure 4-(a) and Figure 4-(b) Von Neumann neighborhood is stated whereas in Figure 4-(c) and Figure 4-(d) Moore neighborhoods are taken into account. Finally, in Figure 4-(a) and Figure 4-(c) all smartphones have the same OS, whereas in Figure 4-(b) and Figure 4-(d) four types of OS are considered (with the same distribution than in Figure 3-(a)). Note that, independently of the neighborhood and the number of infectious at time $t = 0$, the trends are similar: in all cases a quasi-endemic equilibrium is reached, although in the case of Moore neighborhoods the variation in the evolution of infectious devices is smaller than in the case of Von Neumann neighborhoods.

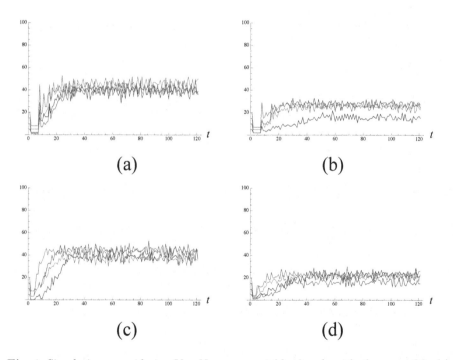

Fig. 4. Simulations considering Von Neumann neighborhoods with the same OS -(a)- and with different OS -(b)-. Simulations considering Moore neighborhoods with the same OS -(c)- and with different OS -(d)-.

5   Conclusions and Future Work

In this work a novel mathematical model to simulate the spreading of bluetooth mobile malware is introduced. It is a compartmental model based on the use of two-dimensional cellular automata where the population is divided into susceptible, carrier, exposed and infectious smartphones. This model improves in different ways the work due to Peng *et al.* ([6]). In the model proposed in this work two two-dimensional cellular automata will be used to rule the dynamic at macroscopic and microscopic level respectively. The geographical area where smartphones are is tessellated into several square lattice giving the cellular space of the global cellular automata. There can be more than one smartphone in each cell and these smartphones can move from one global cell to another at every step of time.

It is crucial to accurately define the models parameters: topology, probabilities of infection and recovery, etc. The model facilitates real-time predictions of the evolution of the malware spreading, making it possible to modify the different parameters and control measures.

Future work will focus on an accurate evaluation of the parameters involved in the model using real data. Moreover, further work also aimed at considering other transmission vectors such as the use of SMS and MMS, the download of a malicious app, etc.

Acknowledgments. This work has been supported by Ministry of de Economy and Competitiveness (Spain) and European FEDER Fund under project TUERI (TIN2011-25452).

References

1. Atzori, L., Iera, A., Morabito, G.: The Internet of Things: A survey. Comput. Netw. 54(15), 2787–2805 (2010)
2. Cheng, S.M., Ao, W.C., Chen, P.Y., Chen, K.C.: On Modeling Malware Propagation in Generalized Social Networks. IEEE Commun. Lett. 15(1), 25–27 (2011)
3. International Data Corporation (IDC), http://www.idc.com/
4. Jackson, J.T., Creese, S.: Virus Propagation in Heterogeneous Bluetooth Networks with Human Behaviors. IEEE T. Depend. Secure 9(6), 930–943 (2012)
5. Mickens, J.W., Noble, B.D.: Modeling Epidemic Spreading in Mobile Environments. In: Proc. of the 4th ACM Workshop on Wireless Security, pp. 77–86. ACM Press, NY (2005)
6. Peng, S., Wang, G., Yu, S.: Modeling the dynamics of worm propagation using two-dimensional cellular automata in smartphones. J. Comput. System Sci. 79(5), 586–595 (2013)
7. Ramachandran, K., Sikdar, B.: On the Stability of the Malware Free Equilibrium in Cell Phones Networks with Spatial Dynamics. In: Proc. of the 2007 IEEE International Conference on Communications, pp. 6169–6174. IEEE Press (2007)
8. Ramachandran, K., Sikdar, B.: Modeling Malware Propagation in Networks of Smart Cell Phones with Spatial Dynamics. In: Proc. of the 26th IEEE International Conference on Computer Communications, pp. 2516–2520. IEEE Press (2007)

9. Rhodes, C.J., Nekovee, M.: The opportunistic transmission of wireless worms between mobile devices. Physica A 387, 6837–6844 (2008)

10. Sathyan, J., Anoop, N., Narayan, N., Vallathai, S.K.: A Comprohensive Guide to Enterprise Mobility. CRC Press (2012)

11. Wei, X., Zhao-Hui, L., Zeng-Qiang, C., Zhu-Zhi, Y.: Commwarrior worm propagation model for smart phone networks. J. China U Posts Telecommun. 15(2), 60–66 (2008)

12. Wolfram, S.: A New Kind of Science. Wolfram Media Inc., Champaign (2002)

Collaborative System for Learning Based on Questionnaires and Tasks

Manuel Romero Cantal, Alejandro Moreo Fernández,
and Juan Luis Castro Peña

Dep. of Computer Science and Artificial Intelligence,
Centro de Investigación en Tecnologías de la Información
y las Telecomunicaciones (CITIC-UGR), University of Granada, Spain
{manudb,moreo,castro}@decsai.ugr.es
http://decsai.ugr.es

Abstract. Virtual Learning Environments allow to improve the learning interactivity in a collaborative scenario where the learning contents are proposed and accessed by both learners and teachers. In this work, we present CSLQT, a new Collaborative System for Learning based on Questionnaires and Tasks. This system is independent of any course structure or content, and provide users with functionalities to create, review, and evaluate new knowledge resources through questions and tasks. The benefits are two-fold: teachers are released from the tedious task of creating all resources, and students are encouraged to gain the necessary knowledge background before creating any new content. Additionally, a Fuzzy controller generates exams satisfying a customized set of objectives, that could be used for evaluation or auto-evaluation purposes. Our experiences with the system in real courses of the University of Granada indicate the tool is actually useful to improve the learning process.

Keywords: Collaborative Learning Resources, Virtual Learning Environment, Web Application.

1 Introduction

In recent years, the E-learning education tools have become more and more popular to improve learning and teaching activities [1]. In this regard, a wide range of computer-based learning environments have been developed in the field of Information and Communication Technologies (ITS). Thanks to these tools, usually known as Virtual Learning Environments (VLE) or Learning Management Systems (LMS), teachers and students can improve their learning interactivity, overcoming physical distance and time constraints. A VLE is often regarded as a platform designed to manage the learning processes by providing students with sufficient resources for reaching their learning goals. VLEs are expected to provide a shared framework in which teachers are allowed to design and instantiate individual and collaborative learning activities. Those activities could

Á. Herrero et al. (eds.), *International Joint Conference SOCO'13-CISIS'13-ICEUTE'13*,
Advances in Intelligent Systems and Computing 239,
DOI: 10.1007/978-3-319-01854-6_64, © Springer International Publishing Switzerland 2014

be structured in courses and include resources to help students to reach these goals [2–4]. Following a constructivist framework, the learning resources (questionnaires, problem solving, or so on) are not only constructed and managed by teachers, but also by students that take an active part on the contribution of the information space.

The aim of this work is to propose a new e-learning application complying with the following goals: (i) the application is specifically designed to be used for all kinds of users, regardless of their technological expertise, allowing the creation and maintenance of the course hierarchy, i.e., courses, groups, and users, in a friendly-manner; (ii) in order to alleviate the effort that creating the learning resources could entail to the educators, the domain knowledge should be collaboratively enhanced by both teachers and students. The resources proposed by students would be evaluated by teachers conforming an expert-authored knowledge domain that may encourage students to investigate about each subject; and (iii) as the student's assessments represent a fundamental part of the learning process we propose an automatic intelligent assessment module to alleviate the tedious work it may entail. Benefits are two-fold: teachers could effectively supervise each student level through specific tests, and students are allowed to review their own knowledge so as to practice before the final exam.

To achieve the above mentioned goals, we propose a web-based VLE, called CSLQT[1] (Collaborative System for Learning based on Questionnaires and Tasks). The application supports two different views: (i) a tool to allow teachers manage and monitor all learning resources mostly proposed by students, and (ii) an intelligent assessment module able to automatically assign tasks and tests to students in function of a set of previously defined objectives and criteria.

The rest of the paper is organized as follows: Section 2 reviews the state-of-the-art of the e-learning applications. The architecture of the proposed application is depicted in Section 3. Section 4 describes the fuzzy controller that generates tests according to a number of objectives. Section 5 reports our experiences with the system related to real courses of Artificial Intelligence. Finally, conclusions and future work are included in Section 6.

2 Related Work

The use of VLEs to support learning has received a great amount of interest in recent years. This review of the main contributions in the field covers the following topics: commercial/open-source VLE platforms, acceptance studies, and e-assessment methods.

There are several VLE systems available as commercial or open-source collaborative platforms [5, 6]. On the one side, WebCT[2] and Blackboard[3] are among the most common examples of commercial web-based systems that allow the creation of flexible learning environments on the web. On the other side, Modular

---

[1] http://ic.ugr.es:8090/SCACP/toLogin.action

[2] www.WebCT.com

[3] www.blackboard.com

Object-Oriented Dynamic Learning Environment (Moodle[4]) is arguably one of the most relevant open-source platforms. It is a free course management system that helps teachers to create their own on-line learning communities.

Different approaches rely on the students' acceptance of the e-learning systems and its benefits in teaching and learning. To this end, the technology acceptance model [7] and its extensions were developed. This model mainly measures two different features on the studied platforms: perceived *usefulness* and perceived *easy of use*. Two examples evaluating Second Life (as VLE platform) and Moodle could be found at [8, 9], respectively. Also, [10] reported the use of VLEs as educational tools focusing in effectiveness.

The e-Xaminer system [11] proposed a study on how to bring effective e-assessment. This computer-aided summative assessment system was designed to produce tests based on parametrically designed questions. In [12], authors developed a hybrid formative/summative e-assessment tool for an course in Chemical Engineering. Other e-assessment systems could be found at [13].

3 System Overview

In this section, we present an overview of our web-based VLE application, called CSLQT. The system supports the management of the course structure, the learning resources, the assessment resources, and the the students' evaluation. We first expose the main features of the resources we are considering. Then, we discuss the user roles in CSLQT. Finally, the main modules are depicted.

3.1 Collaborative Learning and Assessment Resources

CSLQT is designed to support two kind of learning resources: *questions* and *tasks*.

On the one hand, questions are employed to generate auto-evaluation questionnaires for the students, and course evaluation tests. We used the GIFT[5] format for its definition. In addition, each question of the knowledge base is annotated with the following attributes: difficulty, theoretic degree, practical degree, and importance with respect to an issue. These features represent the inputs for the Fuzzy Questionnaire controller (see Section 4). In this way, both students and teachers can specify their objectives to request a suited questionnaire to reach them.

On the other hand, tasks are considered to be any request of information about a given topic. A task consists of an information request, and the student's answer. Tasks are formulated in HTML[6]. In addition, the application facilitates

---

[4] www.moodle.org

[5] The GIFT (General Import Format Technology) format was created by the Moodle community to import and export questions. It supports Multiple-Choice, True-False, Short Answer, Matching and Numerical questions

[6] Any external resource can be added to the application enhancing the learning knowledge, as well as mathematics formulas, tables, or so on

a continuous feedback between the student and the teacher. That is, each time the student answers a request, the teacher can propose alternatives or comments to improve the answer, until the teacher finally evaluate the answer, thus ending the task. This interaction protocol is later explained in detail in Section 3.3.

As commented before, both teachers and students can add learning resources. Resources proposed by teachers are directly added to the knowledge base, while the resources proposed by students must be first reviewed by teachers. For example, let us suppose that an user propose a question about a given topic. Then, the teacher could approve or discard it. In the former case, the question becomes part of the knowledge base. If so, the teacher assigns a grade reflecting the quality of the resource that will also serve for the future evaluation of the student. This framework encourage students to investigate the topic in depth before reporting any answer. The same scenario goes for tasks.

3.2   User Roles in the System

There are three roles in our application:

The Manager is responsible for creating and managing the course structure.

The Teachers manage the lists of students enrolled in the course; review and validate the learning resources proposed by students; create their own learning resources; propose the possible criteria to be assigned to the tasks, and propose different objectives to generate evaluation tests.

The Students are encouraged to propose new learning resources; complete different tests and tasks to progress in the course, and generate their own auto-evaluation questionnaires.

3.3   Main Modules

This section is to explain the main modules in our system. Figure 1 depicts them all and may serve to guide the reading.

The *Course management module* is exclusively used by the manager. This module implements the creation and edition of courses and its organisation into groups. Each group is monitored independently by the teacher (or teachers) in charge. Moreover, each course contains a list of issues that are common to all groups. This module brings the system independence regarding any particular course structure or content.

The *Knowledge Base management module* provides the application with the functionality to create and manage the learning resources. Resources proposed by students but not still validated are separately stored from trusted resources — already validated or created by experts. Once a resource is approved, it could be linked to one or more issues of the course. This module allows to list, search, and display resources sorted by topic. Also, this module stores some statistics reflecting the use of each resource in the application. For example, the number of times a given question was chosen into a questionnaire, the percentage of correct answers, or the date it was added to the system, are stored by this module.

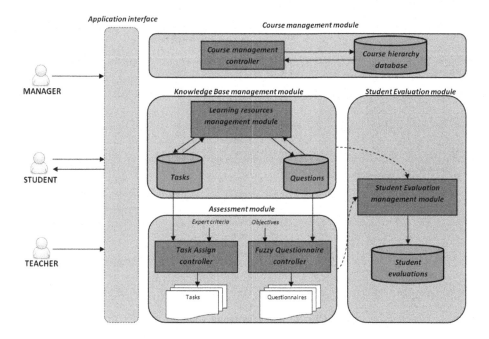

Fig. 1. Architecture overview

The *Assessment module* has two main goals. First, the generation of assessment resources and the assignment to students. Second, it is responsible for the following tasks: (i) collect the students' answers for a given assessment resource, (ii) communicate the results to the teacher, and (iii) commit the information to the *Student Evaluation module*. There exist three different assessment resources: tasks, test questionnaires and student questionnaires.

The *Task Assign controller* is devoted to manage the tasks. The interface allows the teacher to assign any task to any student. This assignment can be defined manually or automatically (by previously choosing some criteria, such as the relevance to an topic). Assignments involve a collaborative feedback (see Figure 2) whereby the student is first requested to report an answer, that is later validated by the teacher. After that, the teacher set a grade that is finally collected by the *Student Evaluation module*.

The *Fuzzy Questionnaire controller* deals with the generation of students questionnaires and test questionnaires. Once the user selects through the interface a set of desired objectives, the controller decides (see section below) which are the questions that better fit the goals (Figure 3). If the objectives could be satisfied, then the system forks in two possible scenarios.

For students: A static questionnaire is generated for training purposes. This questionnaire could be attempted as many times as the students desire in order to auto-evaluate their own progresses.

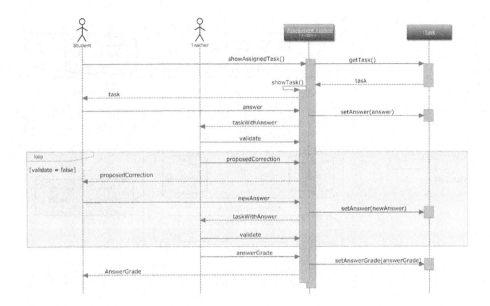

Fig. 2. Tasks sequence diagram

For teachers: The selected objectives describe the test that will be assigned to students in their course. Each student is given a unique exam that is dynamically generated according to the objectives. Thus, possible cheats are avoided since each student in the same room would have a different test.

The *Student Evaluation module* manages all grades attached to proposed questions, proposed tasks, responses to tasks, test questionnaire results, and student questionnaires results. Certain usage statistics per student are also stored, such as the number of proposed questions, proposed tasks, passed tests, ... Those grades could be accessed by both students and teachers (Figure 4).

4   Fuzzy Questionnaire Controller

The functionality of the *Fuzzy Questionnaire controller* relies on the automatic generation of questionnaires in function of a set of objectives previously selected by the user. We employ the methodology explained in [14]. According to the Fuzzy theory, each objective could be represented as a linguistic label. Questions are taken among the knowledge base attending to the following objectives: number of questions, difficulty, theoretical degree, practical degree, and importance regarding each issue in the course. Recall that each value was previously defined by the user who created the resource. Additionally, the user specifies through the interface the importance to be attached to each objective. After that, the controller performs an iterative search in the space of questions until an acceptance threshold is exceeded.

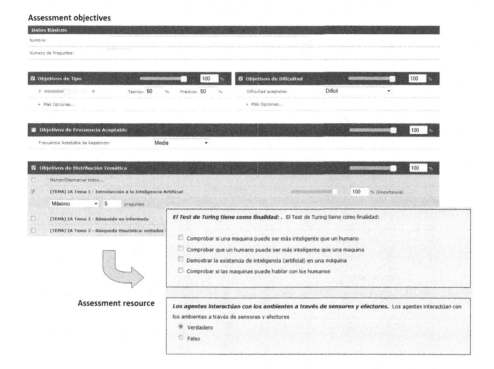

Fig. 3. Selection of objectives and resulting assessment resource

First, the acceptance threshold for a given question is calculated considering the questions already included in the questionnaire $(q_1, q_2, \ldots, q_k)$, and its value for every objective marked:

$$u_p = \Gamma_p(p(q_1), p(q_2), ..., p(q_k)) \tag{1}$$

where $p(q_1), p(q_2), ..., p(q_k)$ are the values of questions $q_1, q_2, \ldots, q_k$ for the objective p, Γ_p is the evaluation function of each objective p using the fuzzy "AND" operator. Those questions exceeding the acceptance threshold are considered candidate questions CQ.

The next stage consists of selecting the candidate question to be included in the resulting questionnaire. To that end, the function $\Psi_q(\{p(q)\}, \{w_p\}) = \prod p(q) \cdot w_p, \forall q \in CQ$ measures the quality of each question in CQ, where w_p denotes the weight given by the user to the objective p. The question with higher values is finally added to the questionnaire. The process iterates until the questionnaire is complete.

XXXX XXXX, XXXX (username: **XXXX**)

Nº cuestiones propuestas	Nº cuestiones validadas	Nº cuestiones descartadas	Nº preguntas propuestas	Nº respuestas propuestas	Nº respuestas aceptadas	Nº pruebas asignadas	Nº pruebas superadas
5	0	0	0	11	0	6	11

			Cuestiones propuestas				
Nº	Id	Nombre		Temas	Fecha de propuesta	Fecha de validación	Nota
1	921	¿Cuál de los siguientes tipos de arquitectura de agente es incorrecta?		IA Tema 1	14/03/2013	SIN VALIDAR	9.0
2	1498	¿A qué equivale el cromosoma en el algoritmo genético?		IA Tema 3	14/05/2013	SIN VALIDAR	9.0

		Tareas propuestas				
Nº	Id	Texto	Temas	Estado	Fecha Creación	Nota
1	302	¿Qué queremos decir cuando afirmamos que un agente debe ser Pro-activo?	IA Tema 1	PUNTUADA	14-03-2013	9.0
2	628	Explica la diferencia entre grafo explícito e implícito.	IA Tema 2	PUNTUADA	21-04-2013	9.0
3	1311	Explique para qué sirven las reglas de inferencia y nombre las mas comúnes.	IA Tema 5	PUNTUADA	06-06-2013	9.0

		Tareas				
Nº	Id	Texto tarea	Texto respuesta	Temas	Fecha Asignación	Nota
2	507	¿Que es un agente inteligente?	Es una entidad que percibe un entorno, y actua de forma autónoma y razonada, tomando decisiones adecuadas, según esta percepción.	IA Tema 1	16-03-2013	5.0
3	608	¿Cuál es la diferencia principal entre agente deliberativo y agente reactivo?¿Qué tipo de arquitectura usa cada uno?	Un agente deliberativo conoce el entorno en el que se encuentra y los pasos a seguir según la ocasión, mientras que un agente reactivo se a adaptando al entorno y los cambios que se producen.	IA Tema 7	16-03-2013	4.0

		Pruebas de evaluación					
Nº	Id	Prueba	Temas	Puntuación posible	Puntuación obtenida	Porcentaje de acierto	Aprobado	Fecha
1	1239	Prueba evaluación tema 1	IA Tema 1, IA Tema 7, IA Tema 2	30.0	22.0	73.3333	✓	18-03-2013
11	5266	Prueba evaluación tema 5	IA Tema 5	25.0	25.0	100.0	✓	06-06-2013

Fig. 4. Evaluation form of a real student of the Artificial Intelligence course

5   Experiences with the System

The current version of our system has been developed following a client-server architecture. On the client' side, we used JSP (Java Server Page). On the server side, we used Java JDK 6 (1.6.0_20), MySQL (14.12), Hibernate library (3.3.5) framework for mapping the database, and Spring framework under Apache Tomcat 6 (6.0.32) to manage the data storage.

Unfortunately, providing a formal validation of a system of this nature is complicated. To test the usefulness of our application, we carried out a study reflecting the interactions between students and teachers. Since our system is designed to support the teaching and learning task, it would be desirable to evaluate the users' satisfaction based on their own experiences with the platform. To this purpose we counted with real data obtained from September 1st, 2012 to April 30th, 2013, during the teaching period of three AI-related courses in the Computer Engineering degree of the University of Granada.

The system is currently being used by a total of 255 users in the University of Granada, including 4 teachers and 251 students. There is a total of 313 questions in the knowledge base. The fact that only 15 of the questions were generated by teachers is an indication that educators are effectively released from the tedious work of defining the contents. Also, there are 512 tasks defined in our system, most of them (424/512) were defined by students.

Regarding the evaluation methods, a total of 1056 students questionnaires were generated for auto-evaluation, and 12 real exams were performed during the courses. This indicates each student filled out an average of 3.55 questionnaires to put their-self to test before the final exams. Percentages of pass exams are consistent with pass questionnaires, ranging from 51% to 69% depending on the course. This show questionnaires could be regarded as an effective estimator of the students knowledge on a subject.

We have registered a total of 7874 accesses into our platform from September 1st, 2012 to April 30th, 2013. Courses in the UGR are distributed in two four-monthly periods. Most of the accesses were registered during the second period, which seems to indicate teachers put the system to test during the first period, and found it an useful tool to be broadly applied in the second period.

The teachers involved in these experiments generously supplied to us their valuable comments and opinions on the platform. These feedback served to us to improve the functionality and quality of the system. What should be highlighted is that both teachers and students agreed that the platform effectively facilitates the learning process.

This is the first "year of life" of our system. Thus, it is still to future research investigating how useful is the system in a second year of a given course, where all the knowledge resources previously added could be reused.

6 Conclusions and Future Works

In this work, we have presented a web-based VLE application pursuing the following goals. The system is independent of any particular course structure or content. The application promotes collaboration between teachers and students, paying special attention to the creation of knowledge resources as a fundamental part of the learning activity. As a result, students are encouraged to search and learn about the course topics, while teachers are released from manually adding all contents. The application includes an intelligent assessment module able to generate assessment resources to students and teachers. Hence, teachers can measure the performance of their students, while students can put their-self to test before facing the final exams.

The experiences with the systems reveal that both students and teachers found the application a useful tool to enhance their academic training.

There is however much work ahead. In order to effectively improve both the visualization and the navigability through the course content, we plan to organize all the learning resources by their key concepts applying our previous advances in tag clouds generation [15].

Acknowledgements. We would like to explicitly show our gratitude to the four teachers that accessed to use our system in their courses and contributed to its improvement with their helpful comments and suggestions.

References

1. Fernández-Manjón, B., Bravo-Rodríguez, J., Gómez-Pulido, J.A., Vega-Rodríguez, J.M.S.P.M.A.: Computers and Education: E-learning, from Theory to Practice. Springer (2007)
2. Alario-Hoyos, C., Bote-Lorenzo, M.L., Gómez-Sánchez, E., Asensio-Pérez, J.I., Vega-Gorgojo, G., Ruiz-Calleja, A.: Glue!: An architecture for the integration of external tools in virtual learning environments. Computers and Education 60(1), 122–137 (2013)
3. Dillenbourg, P., Schneider, D., Synteta, P.: Virtual learning environments. In: Proceedings of the 3rd Hellenic Conference "Information & Communication Technologies in Education", pp. 3–18. Kastaniotis Editions, Greece (2002)
4. Weller, M., Dalziel, J.: Bridging the gap between web 2.0 and higher education. Pratical Benefits of Learning Design, 76–82 (2007)
5. Tsolis, D., Stamou, S., Christia, P., Kampana, S., Rapakoulia, T., Skouta, M., Tsakalidis, A.: An adaptive and personalized open source e-learning platform. Procedia-Social and Behavioral Sciences 9, 38–43 (2010)
6. Sánchez, R.A., Hueros, A.D., Ordaz, M.G.: E-learning and the university of huelva: A study of webct and the technological acceptance model. Campus-Wide Information Systems 30(2), 5 (2013)
7. Davis, F.D.: Perceived usefulness, perceived ease of use, and user acceptance of information technology. MIS Quarterly, 319–340 (1989)
8. Chow, M., Herold, D.K., Choo, T.M., Chan, K.: Extending the technology acceptance model to explore the intention to use second life for enhancing healthcare education. Computers & Education (2012)
9. Escobar-Rodriguez, T., Monge-Lozano, P.: The acceptance of moodle technology by business administration students. Computers & Education 58(4), 1085–1093 (2012)
10. Lonn, S., Teasley, S.D.: Saving time or innovating practice: Investigating perceptions and uses of learning management systems. Computers & Education 53(3), 686–694 (2009)
11. Doukas, N., Andreatos, A.: Advancing electronic assessment. International Journal of Computers, Communications & Control 2(1), 56–65 (2007)
12. Perry, S., Bulatov, I., Roberts, E.: The use of e-assessment in chemical engineering education. Chemical Engineering 12 (2007)
13. Dib, H., Adamo-Villani, N.: An E-tool for assessing undergraduate students' learning of surveying concepts and practices. In: Liñán Reyes, M., Flores Arias, J.M., González de la Rosa, J.J., Langer, J., Bellido Outeiriño, F.J., Moreno-Munñoz, A. (eds.) IT Revolutions. LNICST, vol. 82, pp. 189–201. Springer, Heidelberg (2012)
14. Verdegay-López, J., Castro, J.: Gsadq: Incorporando información difusa o con incertidumbre al diseño automático de cuestionarios. In: Conferencia de la Asociación Española para la Inteligencia Artificial (2002)
15. Romero, M., Moreo, A., Castro, J.: A cloud of faq: A highly-precise faq retrieval system for the web 2.0. Knowledge-Based Systems x(x) x–x (2013) (available online May 2, 2013)

From Moodle Log File to the Students Network

Kateřina Slaninová[1], Jan Martinovič[2], Pavla Dráždilová[1], and Václav Snašel[2]

[1] Faculty of Electrical Engineering,
VŠB Technical University in Ostrava,
17. listopadu 15/2172, 708 33 Ostrava, Czech Republic
[2] IT4 Innovations,
VŠB Technical University in Ostrava,
17. listopadu 15/2172, 708 33 Ostrava, Czech Republic
slaninova@opf.slu.cz,
{jan.martinovic,pavla.drazdilova,vaclav.snasel}@vsb.cz

Abstract. E-learning is a method of education which usually uses Learning Management Systems and internet environment to ensure the maintenance of courses and to support the educational process. Moodle, one of such systems widely used, provides several statistical tools to analyse students' behaviour in the system. However, none of these tools provides visualisation of relations between students and their clustering into groups based on their similar behaviour. This article presents a proposed approach for analysis of students' behaviour in the system based on their profiles and on the students' profiles similarity. The approach uses principles from process mining and the visualization of relations between students and groups of students is done by graph theory.

Keywords: e-Learning, Students' Behaviour, User Profiles.

1 Introduction

E-learning is a method of education which utilizes a wide spectrum of technologies, mainly internet or computer-based, in the learning process. It is naturally related to distance learning, but nowadays is commonly used to support face-to-face learning as well. Learning management systems (LMS) provide the effective maintenance of particular courses and facilitate the communication within the student community and between educators and students [6].

Several authors published contributions focused on mining data from e-learning systems to extract knowledge that describe students' behaviour. Among others, we can mention for example [9], where authors investigated learning process of students by the analysis of web log files. A 'learnograms' were used to visualize students' behaviour in this publication. Chen et al. [2] used fuzzy clustering to analyse e-learning behaviour of students. El-Halees [8] used association rule mining, classification using decision trees, E-M clustering and outlier detection to describe students' behaviour. Yang et al. [15] presented a framework for visualization of learning historical data, learning patterns and learning status of students using association rules mining. However, the contributions focused on the analysis of the students' behaviour in the e-learning systems describe the behaviour using statistical information, for visualization and for representation of

Á. Herrero et al. (eds.), *International Joint Conference SOCO'13-CISIS'13-ICEUTE'13*,
Advances in Intelligent Systems and Computing 239,
DOI: 10.1007/978-3-319-01854-6_65, © Springer International Publishing Switzerland 2014

obtained information are mostly used only common statistical tools with results like figures or graphs. They do not provide information about behavioural patterns with effective visualization, nor information about relations between students based on their behaviour.

For the purpose of this paper, a students' behaviour is described as a process of activities performed by a student in a LMS system. Therefore, the analysis of students' behaviour in a system was carried out using the principles of process mining [13].

In this paper, the social network is constructed using student profiles. A student profile is a vector which describes a students' behaviour in the analysed system. A typical social network is as a set of people, or groups of people, who socially interact among themselves [11]. In this type of networks, the relations are usually defined by one of the types of interaction between the actors, e.g. personal knowledge of one another, friendship, membership, etc. However, in the area of virtual social networks, we can explore the extended definition of social networks. This can be done by exploring social network as a set of people, or groups of people who have similar patterns of contacts of interactions, or generally with similar attributes [12].

This paper is organized as follows: Section 2 introduces principles of log analysis and its relation to social networks. The proposed approach of an analysis of LMS Moodle logs with a detailed description of the process which leads to a construction of social network is then presented in Section 3. A discussion and conclusions can be seen in Section 4.

2   Log Analysis and Process Mining

A log file is usually a simple text file generated by a device, software, application or system. It consists of messages, which are represented by records of events performed in the system. Usually, event log files are created from other data sources such as databases, flat files, message logs, transaction logs, spread sheets, and etc.

Let us assume that an event log purely contains data related to a single process. Each event in the log file refers to a single process instance, called a case. Aalst et al. [14,13] provided the following assumptions about event logs:

- A *process* consists of *cases*.
- A case consists of *events* such that each event relates to precisely one case.
- Events within a case are ordered.
- Events can have *attributes*.

Definition 1. Event, attribute [14]

Let E be an event universe, i.e. the set of all possible event identifiers. Events may be characterized by attributes. Let AN be a set of attribute names. For any event $e \in E$ and name $n \in AN : \#_n(e)$ is the value of the attribute n for the event e. If the event e does not have the attribute named n, then $\#_n(e) = \perp$ (null value).

As a standard attribute is usually used $\#_{activity}(e)$ as the *activity* associated to the event e, $\#_{time}(e)$ as the *timestamp* of an event e, $\#_{resource}(e)$ as the *resource* (originator, performer) associated to an event e, $\#_{trans}(e)$ as the *transaction type* associated to an event e, and others depending on the type of system.

Events are labelled using various classifiers. A classifier is usually a function that maps the attributes of an event to its label. Depending on the type of system, events can be classified by a name of activity, transaction type or by their resource.

An event log consists of cases, whilst cases consist of events. In a case, the events are represented in the form of a *trace*, i.e. a sequence of unique events [14,13].

Definition 2. Case, trace, event log [14]

Let CA be a case universe, i.e. the set of all possible case identifiers. Cases can have attributes. For any case $ca \in CA$ and name $n \in AN$: $\#_n(ca)$ is the value of attribute n for a case ca, and $\#_n(ca) = \perp$ if a case c has no attribute named n. Each case has a special mandatory attribute trace: $\#_{trace}(ca) \in E^*$.

A trace is a finite sequence of events $\sigma \in E^*$ such that each event appears only once, i.e. for $1 \leq i < j \leq |\sigma| : \sigma(i) \neq \sigma(j)$.

An event log is a set of cases $L \subseteq CA$ such that each event appears, at most, once in the entire log.

If an event log contains timestamps (a time when the event was performed in the system) then the ordering in a trace respects this attribute $\#_{time}(e)$.

2.1 Social Aspect in Process Mining

There are techniques, generally known as *organisational mining*, which are focused on the organisational perspective of process mining. Alongside typical information such as event, case, activity or the time when an event was performed, we can also find information about the originator (device, resource, person) who initiated the event (activity). Such information is determined as the $\#_{resource}(e)$ attribute. These events from the log file can be projected onto their resource and activity attributes. By using this approach, we can learn about people, machines, organisational structures (roles, departments), work distribution and work patterns.

Social network analysis is one of many techniques used in organisational mining [13]. Event logs with $\#_{resource}(e)$ attributes are used in order to either discover real process and work flow models of the organisation, or just to generally allow us to construct models that explain some aspects of an individual's behaviour.

The intensity of the relations can be interpreted as the similarity of two resources. To quantify this type of similarity, resource profiles are extracted. A *profile* is a vector indicating how frequently each activity has been performed by the resource [14]. It is usually determined by the rows in the resource-activity matrix that represents how many times each resource has worked on each activity. However, such an approach is based only on statistical information from the log. In this paper a new approach is presented taking into consideration the real behaviour of the users (persons, resources) within a system.

In the proposed approach described in this paper, the similarity between user profiles was determined by cosine measure [1]. Cosine similarity (*sim*) between two vectors x and y is defined as:

$$sim(x,y) = \frac{xy}{||x||||y||} \tag{1}$$

However, the construction of the user profiles in some cases was more complicated. For example, when using data from an e-learning system, it was necessary to extract the behavioural patterns as a first step. Then, the user profiles were constructed using these patterns.

2.2 Cluster Analysis and Spectral Clustering

Cluster analysis covers the techniques that allow a researcher to separate objects, with the same or similar properties, into groups that are created using specific issues. These groups of objects, called clusters (and also called patterns), are characterised by their common attributes (properties, features) of entities that they contain [7] .

Spectral clustering is one of the divisive clustering algorithms which can be applied in the graph theory. The spectral clustering algorithm uses eigenvalues and eigenvectors of a similarity matrix derived from the data set to find the clusters. In this section, there is described the type of spectral clustering based on the second smallest eigenvector of the Laplacian matrix. How to use the spectral algorithm is studied in [10] by Cheng et al. In [3] Ding et al. proposed a new graph partition method based on the min-max clustering principle: the similarity between two subgraphs (cut set) is minimized, while the similarity within each subgraph (summation of similarity between all pairs of nodes within a subgraph) is maximized.

Spectral clustering method optimized by our proposed Left-Right Oscillate algorithm [5,4] was used in two parts of the proposed approach: for finding the behavioural patterns and for better visualization of network of student groups based on their similar behaviour in the system.

3   Application in LMS Moodle

The experiments presented in the following part of the paper are focused on an analysis of the students' behaviour in Learning Management System (LMS) Moodle, in order to find behavioural patterns and to find groups of students with similar behaviour within the system. The groups of students and relations between them based on their similar behaviour are presented using the tools of social network analysis.

LMS Moodle is web-based learning management system, which stores data in MySQL database. Logging of events performed in the system is done into several tables in the database. Example 1 shows an event log file exported from LMS Moodle database, which was used in the experiments. Data preprocessing was applied to the log file, during which events of course tutors and system administrators were removed, and an anonymisation process, where the students' full names were replaced by the general StudentID.

Example 1 (Example of Log File - LMS Moodle)

```
"CourseID";"TimeStamp";"IPAddress";"StudentID";"Activity";"ActivityInformation"
"08/09-KCJK";1.9.2008 12:52:59;"10.7.15.2";"Student1";"course enrol";"Business"
"08/09-KCJK";1.9.2008 12:52:59;"10.7.15.2";"Student1";"course view";"Business"
"08/09-FIN";1.9.2008 12:52:59;"10.7.213.1";"Student2";"resource view";"14556"
"08/09-FIN";1.9.2008 12:52:59;"10.7.7.2";"Student3";"resource view";"14541"
```

One record in the log file represented an event, and consisted of following event attributes: $\#_{time}(e)$, denoted as `TimeStamp`, $\#_{resource}(e)$, denoted as `SudentID`, $\#_{activity}(e)$, denoted as `Activity`, and other additional attributes like `CourseID`, and `IPAddress`. Moreover, there may be activity attribute `ActivityAttr` used for a detailed description of a recorded activity. A more detailed description of the all attributes is presented in Table 1.

Table 1. Event Attributes - LMS Moodle

Event Attribute	Description
`TimeStamp`	Date and time when the event was performed.
`StudentID`	Each student had assigned its unique ID. Students are performers of analysed events in LMS Moodle. `StudentID` is used for their identification.
`CourseID`	Each study course in LMS Moodle has its unique ID. `CourseID` is used for a sequence construction.
`IPAddress`	Additional attribute. IP address of a client, from which a student entered LMS Moodle.
`Activity`	Name of activity performed in LMS Moodle by a student.
`ActivityAttr`	A detailed description to a concrete `Activity`. Each value is related to a concrete activity type, and for some activities is not necessary.

Students may perform various activities in LMS Moodle. Because each web page and each component of LMS Moodle is labelled, it is possible to use additional information for each activity.

3.1 Analysis of Students' Behaviour

LMS Moodle is a web-based system. Therefore, the construction of sequences was based on similar principles as in web sphere. The identification of the end of sequences by a time period of students' inactivity was used, but with a combination to other aspects specific for the environment of LMS. Creation of sequences is described with more details in the following text of this section. Moreover, another problem with the student identification was solved in this type of experiments, because of a login system in LMS. It was not necessary to use the student identification by IP address, because in the log file is available `StudentID`. However, the attribute `IPAddress` may be used as additional information for a further analysis.

The main subject of interest in the experiments is the students' behaviour in LMS, which is recorded in form of events and stored (or exported) in the log files. Thus, we can define the students' behaviour with the terms of process mining which are used commonly in business sphere. The first step of the presented approach, was focused on extraction of set of students U and set of sequences S performed by them in the system. In this case, students $u \in U$ were denoted as students in the learning management system. Each student was identified by its unique StudentID.

Extraction of courses and activities performed by students in the system, required the log file as the input. An example of used log file exported from LMS Moodle database was presented in Example 1. As the output of this step we obtain set of courses and set of activities performed by students, which were contained in the processed log file. Both sets will be used in the second step for filtering purposes. Other output was a set of sequence. An example of such sequence of activities performed by student in LMS Moodle is presenteed in Example 2.

Example 2 (Example of Sequence - LMS Moodle)

```
SequenceID|Frequency|SequenceOfActivities_TextFormat|SequenceOfActivities_ID
104|1|quiz view^1889;resource view^14750;forum view discussion^3602;|12;22;4;
```

Creation of set of sequences is the next step in our approach. In this phase, it is possible to filter a concrete course, and even to filter selected types of activities, for further processing. It is important to be able to filter processed data from the log file, because it allows to tailor the outputs to the students' (tutors') individual needs. During this phase, it is performed the students' annonymisation as well. Each student had assigned its own StudentID. The annonymisation is done due to research purposes. In the case of implementation of this approach into real usage by tutors, the real names of students will be available. The sequence creation was performed in relation to a concrete student and to a concrete course. The begin of a sequence was identified by a students' enrolment into a course; the end of a sequence was identified when the student enrolled into another course, or using a principle of a time period of inactivity.

As the output of this phase is obtained set of students U, for research purposes annonymized, and set of sequences S recorded into a *.gdf*[1] file. An example of such sequence is presented in Example 2. An example of file in *.gdf* format, which contained the extracted sequences, can be seen in Example 3.

Example 3 (Example of Extracted Sequences - LMS Moodle)

```
nodedef>name VARCHAR,label VARCHAR,color VARCHAR, width DOUBLE,meta0 VARCHAR

0, 0; 1; 3; 1; 4; , '128, 128, 128', 10, forum view forum^1256; forum view discus
sion^3596; forum add post^10404; forum view discussion^3596; forum view discussio
n^3602;

...

2, 2; 16; 12; 21; 16; 2; 0; 381; 25; 0; 330; 2; 2; , '128, 128, 128', 10, course
view^mikroeconomy a [0809-ekemia-e]; quiz view all^ ; quiz view^1889; quiz review
^1889; quiz view all^ ; course view^mikroeconomy a [0809-ekemia-e]; forum view fo
rum^1256; forum view discussion^3641; forum view forums^ ; forum view forum^1256;
forum view discussion^3632; course view^mikroeconomy a [0809-ekemia-e]; course vi
ew^mikroeconomy a [0809-ekemia-e];

...

edgedef>node1 VARCHAR,node2 VARCHAR,weight DOUBLE,directed BOOLEAN,color VARCHAR
```

---

[1] GDF graph format is described on the page:
http://guess.wikispot.org/The_GUESS_.gdf_format

As we can see from Example 3, sequences extracted from LMS Moodle may be very long and different. However, several sequences differ only in small parts, sometimes even in particular activities. This leads into a problem, that base student profile of one student may be very large. Therefore, it makes a comparison of base student profiles created from these sequences non-trivial.

3.2 Student Profiles

The proposed approach proceeds from the original social network approach with a modification focused on students' behaviour. The modification is based on a definition of the relation between students. As mentioned in Section 1, the original approach to the analysis of social networks deals with the assumption that the social network is a set of people (or groups of people) with social interactions among themselves [11]. Social interaction is commonly defined as interaction between the actors like communication, personal knowledge of each other, friendship, membership, etc.

The modification extends the original approach of social network analysis by the perspective of the complex networks, see Section 1. This type of view differs from the original approach in the description of relations between the nodes (students). The relation between the students is defined by their common attributes characterizing their behaviour in the system. More specifically, the student behaviour in the system is defined by student profiles.

One of the problems in proposed approach described in this paper is the creation of student profiles based on their behaviour within the system. This approach requires a set of students U and a set of event sequences S performed in the system by the students (see Definition 3).

Definition 3. Base student profile, sequences

Let $U = \{u_1, u_2, \ldots, u_n\}$, be a set of students (denoted as resources $\#_{resource}(e)$), where $i = 1, 2, \ldots, n$ is a number of studets u. Then, sequences of events $\sigma_{ij} = \langle e_{ij1}, e_{ij2}, \ldots e_{ijm_j} \rangle$ (denoted as traces $\#_{trace}(e)$), are sequences of events executed by the student u_i in the system, where $j = 1, 2, \ldots, p_i$ is a number of that sequences, and m_j is a length of j-th sequence. Thus, a set $S_i = \{\sigma_{i1}, \sigma_{i2}, \ldots \sigma_{ip_i}\}$ is a set of all sequences executed by the student u_i in the system, and p_i is a number of that sequences.

Sequences σ_{ij} extracted with relation to certain student u_i are mapped to set of sequences $\sigma_l \in S$ without this relation to students: $\sigma_{ij} = \langle e_{ij1}, e_{ij2}, \ldots, e_{ijm_j} \rangle \rightarrow \sigma_l = \langle e_1, e_2, \ldots, e_{ml} \rangle$, where $e_{ij1} = e_1, e_{ij2} = e_2, \ldots, e_{ijm_j} = e_{ml}$.

Define matrix $B \in N^{|U| \times |S|}$ where

$$B_{ij} = \begin{cases} frequency\ of\ sequence\ \sigma_j \in S\ for\ student\ u_i\ if\ \sigma_j \in S_i \\ 0\ else \end{cases}$$

A base student profile of the student $u_i \in U$ is a vector $b_i \in N^{|S|}$ represented by row i from matrix B.

Example 4 shows an example of base student profiles generated for student's behaviour in the LMS Moodle. Base student profiles were generated by our application developed for processing of log files from LMS Moodle. In Example 4 we can see a part of the output file *profileBase.csv*. The file contains the students and their sequences of events, extracted from the log file. Each sequence has its unique ID, and has assigned its frequency of which the sequence has been performed by the student in the LMS Moodle. The activities inside the sequences are labelled; each activity has its unique ID.

Example 4 (Example of Base Student Profiles - LMS Moodle)

```
sequence id|amount|sequence
#user:Student27
0|1|488;14;6326;398;6326;
1|1|488;
2|1|6326;6335;
#user:Student87
20|1|6326;488;398;
2|1|6326;6335;
#user:Student101
4|1|6326;
21|1||6326;550;551;
2|1|6326;6335;
#user:Student135
22|1|6326;412;6326;5968;
23|1|6326;6326;488;4855;464;
24|1|6326;464;
4|2|6326;
```

3.3 Student Groups with Similar Behaviour

The issues of log analysis in connection with social aspects in process mining have been presented in this paper yet. The approach which allows the transformation from LMS Moodle log files into the students' profiles was described in Section 3.2.

Last step of the proposed approach contains graph visualization of the relations between users, which is based on the students' profiles. Described steps of the proposed approach were applied onto a selected course 'World Economy' from LMS Moodle, used as a support of distance learning at Silesian University in Opava, Czech Republic. Table 2 describes achieved parameters of the analysed course.

Table 2. Statistical information about selected course 'Word Economy'

Events count	517,269
Actions count	6,493
Actions types count	67
Count of students	85
Count of sequences	193

The network of student groups with similar behaviour was created using graph theory approach. This approach is suitable for clear and understandable visualisation of relations between students. In this part of the presented approach, it was constructed a weighted and undirected graph $G(V, E)$, where vertices (graph nodes) V are students u_i and relations (edges) E express the similarity between the student profiles.

It is possible to count edge weights (similarity) by cosine measure. As the output, we obtained a graph of students (a *.gdf* file). Another output was a file with set of students U and their vectors consisted of sequences (behavioural patterns) and their frequency.

The groups of students with similar behaviour (similar student profiles) could be found using Spectral clustering by Fiedler vector using Left-Right Oscillate algorithm. The graph was constructed only from edges, which weight was higher than a selected threshold (edge weight > 0.1). The graph contained the isolated students (nodes) as well, because they are as important as the other connected nodes.

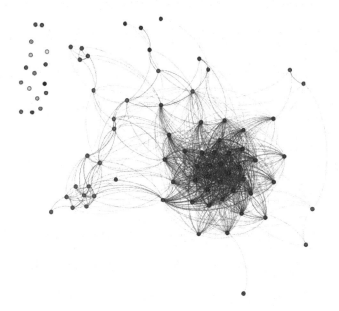

Fig. 1. Network of Student Groups for Course 'World Economy'

4 Conclusion and Future Work

The main goal of this paper was to propose a new approach for clear and intuitive visualisation of relations between students (or groups of students) of LMS Moodle based on their similar behaviour in the system. The approach uses principles of process mining and graph theory. Achieved outputs can be used as a support for collaborative learning or can be helpful for tutors to effectively manage the study process of their students.

It is intended to develop the plug-in for LMS Moodle which will provide more detailed statistics of students' behaviour in a concrete course in LMS. Present outputs are planned to be a core of this plug-in. However, it could be suitable to supplement them by other common type of information like histograms of activities in the course, the most/the least frequent activity in the course etc. Moreover, it is possible to provide more suitable outputs tailored to teachers and/or course administrators. The outputs can be specified to several spheres, for example forums, quizzes, reading materials etc.

Acknowledgment. This work was supported by the European Regional Development Fund in the IT4Innovations Centre of Excellence project (CZ.1.05/1.1.00/02.0070) and by the SGS, VŠB – Technical University of Ostrava, Czech Republic, under the grant No. SP2013/167 Analysis of Users' Behaviour in Complex Networks.

References

1. Berry, M.: Survey of Text Mining: Clustering, Classification, and Retrieval. Springer (September 2003)
2. Chen, J., Huang, K., Wang, F., Wang, H.: E-learning behavior analysis based on fuzzy clustering. In: Proceedings of International Conference on Genetic and Evolutionary Computing (2009)
3. Ding, C.H.Q., He, X., Zha, H., Gu, M., Simon, H.D.: A min-max cut algorithm for graph partitioning and data clustering. In: ICDM 2001: Proceedings of the 2001 IEEE International Conference on Data Mining, pp. 107–114. IEEE Computer Society, Washington, DC (2001)
4. Dráždilová, P., Babsková, A., Martinovič, J., Slaninová, K., Minks, S.: Method for identification of suitable persons in collaborators' networks. In: CISIM, pp. 101–110 (2012)
5. Dráždilová, P., Martinovič, J., Slaninová, K.: Spectral clustering: Left-right-oscillate algorithm for detecting communities. In: Pechenizkiy, M., Wojciechowski, M. (eds.) New Trends in Databases & Inform. AISC, vol. 185, pp. 285–294. Springer, Heidelberg (2012)
6. Dráždilová, P., Obadi, G., Slaninová, K., Al-Dubaee, S., Martinovič, J., Snášel, V.: Computational intelligence methods for data analysis and mining of eLearning activities. In: Xhafa, F., Caballé, S., Abraham, A., Daradoumis, T., Juan Perez, A.A. (eds.) Computational Intelligence for Technology Enhanced Learning. SCI, vol. 273, pp. 195–224. Springer, Heidelberg (2010)
7. Dunham, M.H.: Data Mining: Introductory and Advanced Topics. Prentice Hall (2003)
8. El-halees, A.: Mining students data to analyze learning behavior: a case study (2008)
9. Hershkovitz, A., Nachmias, R.: Learning about online learning processes and students' motivation through web usage mining. Interdisciplinary Journal of E-Learning and Learning Objects 5, 197–214 (2009)
10. Kannan, R., Vempala, S., Vetta, A.: On clusterings: Good, bad and spectral. J. ACM 51(3), 497–515 (2004)
11. Newman, M.E.J.: Networks: An Introduction. Oxford University Press (2010)
12. Radicchi, F., Castellano, C., Cecconi, F., Loreto, V., Parisi, D.: Defining and identifying communities in networks (February 2004)
13. van der Aalst, W.M.P.: Process Mining: Discovery, Conformance and Enhancement of Business Processes, 1st edn. Springer, Heidelberg (2011)
14. van der Aalst, W.M.P., Reijers, H.A., Song, M.: Discovering social networks from event logs. Comput. Supported Coop. Work 14(6), 549–593 (2005)
15. Yang, F., Shen, R., Han, P.: Construction and application of the learning behavior analysis center based on open e-learning platform (2002)

Development of Projects as a Collaborative Tool for Competence Acquisition in the European Space for Higher Education Framework

Javier Alfonso[1], Manuel Castejón[1], Lidia Sánchez[1],
Raquel González[1], and José Vallepuga[2]

[1] Department of Mechanical, Computing and Aerospace Engineerings,
University of Leon, Campus de Vegazana s/n, 24071 Leon, Spain
[2] Department of Mining, Topography and Structural Technology,
University of Leon, Campus de Vegazana s/n, 24071 Leon, Spain

Abstract. The aim of this paper is to analyze the impact of developing projects in collaborative environments within the Bologna Process framework and the European Space for Higher Education context as a rewarding tool for the professional future of University students, attempting not only to provide them with the traditional technical competences, but also with relationship, communication and negotiation skills and competences. In this experience, the development of multidisciplinary projects acts as a common thread for the resolution of real problems, by connecting the thematic of those projects to the areas of knowledge of the subjects participating in the initiative. The skills and competences acquisition levels results after the completion of the pilot project have been very satisfactory, compared to those obtained in previous experiences. We would also like to point out that this type of teaching methodology can be made extensible to any other subject matter in which teamwork is a value in itself.

Keywords: ESHE framework, collaborative learning, education, project development, competence´s rubric.

1 Introduction

In recent years, many changes that have affected and modified the habits of advanced societies have taken place. These changes have led to a transformation in the lifestyle of people. The Bologna Process and the creation of the European Higher Education Area have affected all members and sectors of the university community; for this reason, and due to their methodological and teaching changes, numerous opportunities around these processes arise, making it essential to adapt oneself by offering new experiences that enrich such processes [5][6].

Companies are increasingly valuing and requiring those professional profiles where relationship, communication and negotiation skills have a greater weight, together with the traditional technical and management competences. Therefore, aiming to provide our students with the competences mentioned, so they are better prepared and so they are affected the least possible by situations such as our present one, where

Á. Herrero et al. (eds.), *International Joint Conference SOCO'13-CISIS'13-ICEUTE'13*,
Advances in Intelligent Systems and Computing 239,
DOI: 10.1007/978-3-319-01854-6_66, © Springer International Publishing Switzerland 2014

unemployment rates are so high, it is considered essential that in their learning stage, experiences and initiatives that foster and stimulate these competences, in situations as close as possible to reality, are included [1].

For all of this, a number of initiatives and pilot projects whose methodological development and common thread are based on real problem-solving through project-developing have been planned from some areas of knowledge within the Department of Mechanical, Computing and Aerospace Engineering of the University of León, which carries out its teaching activity in several undergraduate and master's degrees [3].

One such initiative has been carried out with students from different degrees of the University of León over the first few months of the academic year 2012/2013, with the aim of fostering collaborative learning best, trying to get the experience to become a reference for the future professional development of the participants [3] [8].

For these reasons, and considering that the basis of this experience is project-developing as a tool for collaborative learning, work teams composed of students from various degrees were formed. Professors from those subjects got involved in those work teams, acting as "customers" in order to emulate a working environment as real as possible.

The main feature of this initiative, which promotes the portability of this approach too, is the existence of a base problem or study case that supports the common thread of the experience by dividing the project into different areas or subject matters based on the nature of such subject matters involved in the experience.

2   Objectives

The overall goal of this experience is to assess the impact of developing projects as a collaborative learning tool in the European Higher Education Area context through a pilot project. The project evaluation will be carried out by assessing the degree of individual skills acquisition, by assessing teamwork, and by a global analysis of the solution provided in the project [4][9]. The following specific goals have been set:

- Analyze the ability to strengthen some skills in collaborative environments.
- Set the basis for comparing the learning outcomes from this pilot experience to other methodologies, analyzing their weaknesses, threats, strengths and opportunities, as well as the lessons learned.
- Assessment of implementation of cooperative work tools, such as Moodle, Open-Project, etc.

3   Pilot Project

The pilot project was carried out through the involvement of two subjects from two different degree courses. The two subjects engaged into the pilot project were "Biology Projects" (BS in Biology) and "Project Management" (Degree in Industrial Engineering). Such diversity was of particular interest to those involved, students and professors, as both are the main beneficiaries of the experience raised.

Functionally, two work teams were formed, each of them composed by half of the participants of each of the two subjects involved in the experience. Each team was composed of 32 students. In order to carry out the project, a problem of the scope of each degree was proposed to each team. The aim was to obtain the best solution possible by an estimate capacity of development of 3ECTS credits per team and participant, thus allowing to provide team work with more realism as well as helping to measure intra-team and extra-team leadership [3] [8].

As one of the project's objectives was to compare its own results with previous experiences, it was decided to set similar assessment methodologies, but introducing a rubric to assess competences [4][9].

During the developing phase of the experience, the following monitoring items were set [2]:

- The real workload, as estimated by the professors on the one hand, and the one declared by each team on the other hand.
- Capacity for work organization with an evolving vision of the aim, based on the goals achievement.
- Capacity for positive and constructive criticism as well as peer assessment of the work carried out by the team members.
- Skills acquired as estimated through the completion of direct surveys to each of the participants in the pilot experience.

The tools used as information sources in order to assess the experience outcome and the subsequent comparison are [2]:

- Minutes and follow-up reports carried out by the project management staff.
- Records of working hours from each of the team members.
- Surveys for checking the team development phase.
- Observations by the professors participating in the experience.
- Quality of the technical tasks performed.
- Management issues and skills acquired throughout the Project, as well as their implementation.
- Oral public presentation of the Project solution.
- Internal review of each of the participants focusing on the importance of this pilot project for their educational training and their career development.

Next, we will detail the competences to be evaluated in this pilot project, classified into four categories, and including as examples the assessment rubrics of some of the competences:

Key and transversal competences [2]:

- Ability to conceive, design and implement projects.
- Ability to negotiate effectively.

Table 1. Rubric for the assessment of the competence "Ability to negotiate effectively"

Competence	Indicator	Minimum degree of achievement	Expected degree of achievement	Outstanding degree of achievement
Ability to negotiate effectively	Finding joint solutions, with close cooperators, to address the needs identified	Student plans the negotiation, defining the issues to be addressed, offering alternatives and getting to know the opposing team	Student explains his/her ideas in a simple way and offers accurate information Student knows that s/he must listen and be understanding, but stays firm in his/her approach and does not listen to his/her interlocutor	Student knows how listen and be understanding, aiming at reaching advantageous agreements Student asks questions about data and specific information in order to learn about the needs and expectations of the opposing team Student shows understanding about the other's position and suggests and/or encourages looking for alternatives
	Finding joint solutions, with people in different locations	Student plans the negotiation, offers alternatives, but the distant location of the other team hinders getting to know them	Student explains his/her ideas and offers information but the tools provided for communication among interlocutors hinder effective communication	The distance among interlocutors doesn't affect communication, understanding and the reaching of agreement with the other team

Instrumental competences [2]:

- Ability to analyze and synthetize.
- Ability to organize and plan.
- Oral and written communication skills.
- Information management skills.
- Problem solving skills.
- Decision making skills.

Table 2. Rubric for the assessment of "Problem solving skills"

Competence	Indicator	Minimum degree of achievement	Expected degree of achievement	Outstanding degree of achievement
Problem solving skills	Correct application of theoretical knowledge	Student applies with some difficulty the theoretical knowledge needed for the resolution of the problem, or there are some deficiencies regarding methodological tools.	Student correctly applies theoretical knowledge for the resolution of the problem and it has been aptly complemented with methodological tools.	Student outstandingly applies both theoretical knowledge and methodological tools for the resolution of the problem,
	Grounding and arguing problem solving.	Problem solving includes very few explanations. It doesn´t foster Reading and comprehension. Most steps lack of argumentation.	The resolution includes explanations to ease Reading and comprehension, and most steps follow argumentation.	Student describes in a clear and concise way the most important characteristics of the problem. Student includes thoughtful and reasoned explanations of the resolution process.
	Resolution Accuracy	The resolution of some of the problems is incorrect	The results are correct with minor numeral or notation mistakes.	Results are totally incorrect.

Interpersonal competences [2]:

- Critical and self-critical skills.
- Ability to work in a team.
- Ability to work in a multidisciplinary team.
- Capacity to communicate with experts in other areas.

Table 3. Assessment rubric for the competence "Critical and self-critical skills"

Competence	Indicator	Minimum degree of achievement	Expected degree of achievement	Outstanding degree of achievement
Critical and self-critical skills	Student knows how to assess the adequate execution of the work of self and others	Student is aware of the outcome, but the depth and quality of the comments is low (poorly reasoned, lack of reasoning and justification	Student recognizes the appropriate criterion of success or failure, but does not show a deep forethought in his/her opinions.	Student adopts a clear position with regards to those situations that can be easily solved and those that don´t depending on the results. Submitted opinions show argument and reasoning.

Systemic competences [2]:

- Project design and management
- Leadership
- Adaptation to new situations
- Initiative   and entrepreneurial spirit

In the following table, we include as an example the indicators selected for the assessment of systemic competences:

Table 4. Indicators for the assessment of the systemic competences of the project

Project design and management	Define the objectives and the scope of the project.
	Coordinate and plan the available resources.
	Monitor work progress.
Leadership	Influence and encourage the rest of the team.
	Make these skills extensible to organizations in different locations.
Adaptation to new situations	Flexibility and ability to respond quickly to changing situations.
Initiative and entrepreneurial spirit	Devise and launch activities.

The different items described above were measured using questionnaires and surveys throughout the development of the project.

4   Results

As stated above, evidence of all the work done was taken throughout the development of the pilot project, thus, after the consideration of all these documents, together with the final results and the definite resolution provided by each of the project teams, the results obtained, which are detailed below, were very favorable.

The levels of competence acquisition achieved were over 18% compared to those from previous experiences based on project-developing which did not follow collaborative learning methodologies, and over 24% compared to those projects not including competence rubrics within their assessing tools.

The degree of fulfillment of objectives of the work teams was close to 85%, which is a very significant figure, taking into consideration that the average never exceeded 70% on previous experiences.

78% of participants rated the experience as very satisfactory on both for their university education as well as providing training to enter the labor market. However, it was acknowledged that the levels of effort required in this experiment were much higher compared to those from previous project-developing experiences.

Those professors involved in the pilot project also rated its outcomes as very satisfactory, highlighting their role shift from a mere teaching figure evolving to a more specific task as a consultant or supervisor, adapted to the new educational context which forces the teacher to adapt new approaches to give more importance to the role of the student in relation to their learning process [7].

5   Conclusions

Firstly we would like to emphasize the fact that in subjects that participated in this project, all of them adapted to the framework of the European Space for Higher Education, it has been made clear the change of paradigm involved in harmonizing learning processes with the nature of the subject taught itself. The creation of learning environments that resemble real-life settings in which cooperative problem solving becomes a central tool for the acquisition of competences brings forth an important competitive advantage for the future incorporation of students in the labor market.

Furthermore, we would like to highlight the satisfactory results obtained not only in the acquisition of competences but also in the global results of the projects, whose degree of detail and solutions are outstanding. This has been promoted by the collaborative environment in which they were produced and also by the approach of the rubrics which, by being composed of more measurable criteria, have helped students to focus their efforts more efficiently.

Acknowledgements. This work would have not been possible without the full cooperation of the students and faculty from the University of León who kindly participated in this experience. This work has been partially supported by the research grants program from the University of León (Spain).

References

1. ANECA: Proyecto Reflex. Informe ejecutivo. El profesional flexible en la sociedad del conocimiento (2007)
2. Castejón, M., Alfonso, J.: PBL (Project Based Learning) en entornos colaborativos virtuales en el contexto de EEES (Espacio Europeo de Educación Superior), PP. 7–38. Universidad de León (2012)
3. Cobo-Benita, J.R., Ordieres-Meré, J.: Learning by doing in Project Management: Acquiring skills through an interdisciplinary model. In: IEEE EDUCON Conference (2010)
4. Golobardes, E., Cugota, L., Camps, J., Fornells, H., Montero, J.A., Badia, D., Climent, A.: Guía para la evaluación de competencias en el área de ingeniería y arquitectura. Agència per a la Qualitat del Sistema Universitari de Catalunya (2009)
5. González, J., Wagenaar, R.: Tuning Educational Structures in Europe. Fase I. Universidad de Deusto (2003)
6. González, J., Wagenaar, R.: Tuning Educational Structures in Europe. Fase II. Universidad de Deusto (2006).
7. Lam, S.F., Cheng, R.W.Y., Choy, H.C.: School support and teacher motivation to implement project-based learning. Learning and Instruction 20(6) (2010)
8. Schoner, V., Gorbet, R.B., Taylor, B., Spencer, G.: Using cross-disciplinary collaboration to encourage transformative learning. IEEE Frontiers in Education (2007)
9. Valderrama, E., Bisbal, J., Carrera, J., Castells, F., Cores, F., García, J., Giné, F., Jiménez, L., Mans, C., Margalef, B., Moreno, A., O'Callahan, J., Peig, E, Pérez, J., Pons, J., Renau, J.: Guía para la evaluación de competencias en los trabajos fin de grado y de máster en las Ingenierías. Agència per a la Qualitat del Sistema Universitari de Catalunya (2009)

Yes, We Can

(Can We Change the Evaluation Methods in Mathematics?)

Ángel Martín del Rey[1], Gerardo Rodríguez[1], and Agustín de la Villa[2]

[1] Department of Applied Mathematics, Universidad de Salamanca, Salamanca, Spain
{delrey,gerardo}@usal.es
[2] Department of Applied Mathematics, Universidad Pontificia Comillas, Madrid, Spain
avilla@upco.es

Abstract. The teaching of mathematics in engineering schools has changed irreversibly in the last 20 years due to two significant events: the generalized use of the new technologies (ICT) in the learning process and the establishment of the European Higher Education Area (EHEA) as the common teaching framework, http://www.ehea.info/.

This change has basically been developed by the teaching tools used by different instructors and the methodologies used.

However, in the case of mathematics subjects in Engineering Schools, the methodological change and teaching innovation have not reached assessment techniques, at least not to a significant extent, and instructors currently continue to assess mathematical knowledge on the basis of traditional theory/practical written exams.

Here we first offer a brief analysis of the innovations that have arisen in the process of learning and suggest alternatives and experiments to implement such innovation in the assessment process, which we are sure, will improve students' academic performance.

Keywords: Assessment methods, e-learning, tutorials, competencies.

1 Introduction

Over the past 20 years many things have changed within the sphere of teaching mathematics in Spanish Engineering Schools. To start, let us consider the normative framework. After successive modifications of university studies the different degrees in engineering now have a theoretical duration of 4 years. The natural continuation for interested students will be a Master's degree (soon to be implemented in Spanish universities). In the new degree courses it has been necessary to introduce a profound restructuring of the curriculum and this has evidently affected the structure of the different Maths subjects.

The increasing use of ICTs in daily teaching and the integration of the different Computer Algebra Systems (CAS) in Maths classes have renovated the way of teaching Maths. In this sense, it may be said that currently the teaching model based on the exclusive use of chalk is difficult to find in our Engineering Schools, since there has

Á. Herrero et al. (eds.), *International Joint Conference SOCO'13-CISIS'13-ICEUTE'13*,
Advances in Intelligent Systems and Computing 239,
DOI: 10.1007/978-3-319-01854-6_67, © Springer International Publishing Switzerland 2014

been a transition to a "mixed" model with a greater or lesser integration of CAS in teaching. To this unquestionable modernization increasing mixed teaching practices in a context of b-learning can be added. The use of different virtual platforms, mainly based on MOODLE, in which students can perform tasks on-line in a new teaching scenario of increasing importance in our Engineering Schools, convinces us that their enormous possibilities will be used with an exponential growth in the future.

However, the situation must be improved when the assessment methods are analyzed since there is considerable reluctance to be overcome among the teachers and students. For a big part of them the traditional procedures of assessments are the "most popular" and the best ones. In [1] several ideas concerning assessments of competencies are analyzed.

2   The Current Situation

After many changes in the norms that have led different study plans to coexist in our classrooms, the structure of the different Spanish degrees in engineering is today based on a 4-year study plan, generally divided into two semesters per year. However, the design of math subjects, owing to the character as basic materials, is to a certain extent peculiar both in structure (for example, semester and annual mathematical courses coexist) and in timing, although all math subjects must be taught during the first 4 semesters of the study plan.

That said, the situation is very diverse, depending not only on the type of engineering followed but also on the university offering the degree. To simplify, because this is not the focus of this work, in many engineering degrees there are three semesters subjects with an exclusively mathematical content, with very different names, although there are also engineering degrees in which the number of math subjects is reduced to two. Subject placement in four semesters courses is not uniform either. There are very different study plans although most of the mathematical contents are explained during the three first semesters. As may be concluded from this short overview of the current situation, math subjects have lost some importance in engineering courses and in a short period of time we have passed from having three and even four annual courses to the situation described above.

This is important when attempting to assess the methodological change implemented. In a technologically changing world, the mathematical needs of our engineering students have not diminished but increased. In order to preserve their nature as support for other subjects (math as a service subject) and for our students to acquire the mathematical competencies demanded by their degrees, mathematics instructors have had to change the mathematical contents and their ways of delivering math classes. In Spain, this necessary change has been long, although we can currently say that neither the list of math topics nor the way of imparting them can be compared with the mathematics taught at Engineering Schools during the nineties. In [1] a study about the new Spanish degrees and mathematical competencies is shown.

TEMA 3: CÁLCULO INTEGRAL DE UNA VARIABLE.

Contenidos: Definición y significado de la integral definida. Propiedades de la integral definida. Promedio integral de una función en un intervalo. Teorema del valor medio. Teorema fundamental del cálculo. Regla de Barrow. Integrales impropias de primera, segunda y tercera especie. Las funciones Γ y β de Euler. Propiedades. Área de una región plana. Longitud de un arco de curva. Volumen de un cuerpo de revolución. Cálculo del volumen de un cuerpo por secciones..

Resultados de aprendizaje: Calcular integrales definidas. Aplicar regla de Barrow. Manejar las funciones Gamma y Beta y sus propiedades. Calcular áreas de recintos planos, longitudes de arcos de curva y volúmenes de cuerpos sólidos como aplicación geométrica de la integral. Modelar problemas de la vida real y de la ingeniería con técnicas de cálculo integral.

Bibliografía básica: Los contenidos teóricos de estos temas corresponden al capítulo 12 (excepto los epígrafes 1.3, 1.4, 1.6, 1.7, 1.8, 4.3 y 4.4), capítulo 14 (epígrafes 1.1, 2.1, 3.1 y capítulos 4 y 5) y capítulo 15 (excepto el apartado 4) del libro **Cálculo I: Teoría y problemas de Análisis Matemático en una variable** de Alfonsa García, F. García, A. López de la Rica, A de la Villa y otros.

HOJA 3A

1. Probar que $\left| \int_0^1 \frac{e^{-nx^4}}{x^2+1}\,dx \right| \le \frac{\pi}{4}$ para todo n natural.

Fig. 1. A worksheet from Pontificia Comillas University

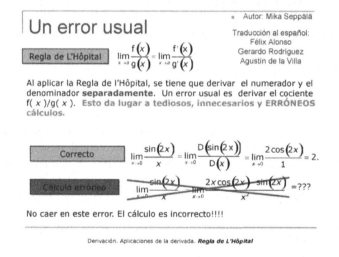

Fig. 2. A slide's model from WEBMLS Project

In many of today's Engineering Schools, math subjects use some type of CAS in a way completely integrated in the teaching discourse, not limiting their use to the math´s practical developed in university computer facilities. General aspects about the use of CAS for didactical purposes are exposed in [2, 3, 4, 5, 6]. In this sense, although there is no uniformity in the choice of the CAS to be used, all teaching resources attempt to enhance their advantages, changing the perception we used to have some years ago with respect to their implementation or not. In [7, 8, 9, 10] readers can find different possibilities, including methodological aspects, of using different CAS are described.

Likewise, the use of an on-line teaching platform is common (in most cases it is MOODLE), where students can find nearly all the teaching material used in class: PowerPoint presentations, worksheets, practical use of CAS etc. (See Fig. 1 and Fig. 2.)

After many experiments in educational innovation, see for example [11, 12], some of which derive from the participation of different European projects [13, 14], a good teaching practice can be designed. We can offer an attractive cocktail, appropriately using high-quality teaching materials: good teaching textbooks, good problem and project selection and the use of a CAS able to provide the most important characteristics we are seeking: calculating power, graphic capacities, etc, to form a critical thinking in our students.

Additionally, the new teaching practice requires new ways of offering tutorial help. With the new work system, our tutorials cannot be reduced to a few hours devoted to solving problems just before an important exam. So that students can do their tasks and develop their work in an increasingly autonomous way (one of the main generic competencies to be acquired during the time they spend at university) it is necessary to set up more intimate tutorial procedures so that students can solve the mathematical problems that arise not only in specific math subjects. In this sense, virtual platforms allow closer supervision of the student. Moreover, we believe that the tasks being carried out at Math Centers, above all in universities in the UK, show the change to be pursued in this context [15]. The Mathematics Centre of Coventry University has managed to expand its ideas to many other universities that are now beginning to harvest the fruits of this new tutorial trend.

However, within this scenario not everything is an advantage. At least in Spain the low starting level of our students places an unexpected barrier to starting autonomous work in math subjects. The initial test, given in many schools, shows that the students' mathematical knowledge is way behind the theoretical knowledge expected of a student who has passed a university entrance exam [16]. Designing an efficient procedure for recovering these early lagoons is one of the challenges we now face, bearing in mind that with the new academic schedule it is almost impossible to impart "introductory courses", "updating courses" or courses under any other name, This contrasts to the courses in vogue several years ago, whose aim was to "revise" the minimum knowledge required by an engineering student to be able to follow the math disciplines imparted in his/her respective degree course.

3   The Assessment of Mathematical Competencies

There is one section in our teaching activities that has not evolved in the same way as the other aspects mentioned above. This has to do with anything related to assessment. How can we assess math subjects? A good question, indeed. It is clear that it has no single easy-to-implement solution. However, what is clear is that we no longer teach as we did twenty years ago and we cannot assess our students as we did then.

Nevertheless, innovation in this field is often subject to strong inertia. The truth is that such reluctance appears not only in those making the assessments (the instructors) but also in those being evaluated (the students). In some of our experiments we have found that when faced with our proposal of assessment, reducing the weight of the final exam and increasing the possibility of undertaking several tutored projects along the academic year, some students do not hesitate to choose only the final exam because this option requires less work.

The instructors also show reluctance in the assessment´s tasks. There is inertia to be overcome: the idea that students only show the knowledge they have gained when faced with a set of problems selected for the exam. Furthermore, it is well known that academic performance in math subjects has traditionally been poor and, sometimes, math subjects have been used –consciously or not- in our Engineering Schools as a filter designed to guarantee a low number of graduates egressing at the end of each year. No official statistics are published. But for example internal reports of Salamanca University remark that the number of fails is around 70%.

Despite all, this inertia is slowly being eroded. In [1] many of the math subjects in engineering were assessed using, to a greater or lesser text, methods that include more than traditional written exams, such as group work, laboratory sessions, written projects, etc. Below we report several experiments of different natures that we have explored in our courses.

4   Some Experiences

In recent years we have developed different experiments with our students, changing the assessment systems. Below we describe two experiments: the first undertaken at the Higher Polytechnic School in Zamora (University of Salamanca) and the second at Industrial Technical Engineering in Polytechnic University in Madrid.

The first experiment was done in Zamora in the Degrees in Building Engineering (DBE) and Computer Engineering in Information Systems (IEIS). In both degree courses the number of students is limited: 60 students in DBE and 15 in IEIS. Both subjects are taught in the first semester of the degree, worth 6 ECTS credits and have a basic content of a course on Calculus in a single variable: complex numbers, sequences, differential calculus, integral calculus and applications.

The teaching makes use of the Studium platform (based on MOODLE), which is the virtual platform of the University of Salamanca and the Spanish Portal linked to the European Virtual Laboratory of Mathematics, EVLM project, where among other materials the subject's blog appears (See Fig. 3).

The CAS used is *Mathematica*, http://www.wolfram.com/mathematica/, since the University has a "campus" license that allows easy access to the software in all the University classrooms. The teaching material used, elaborated in collaboration with instructors from other Spanish and European universities consists of a textbook containing theoretical summaries and a broad collection of proposed and solved problems [17], different PowerPoint presentations on both the Studium, https://moodle.usal.es/ and WEPS [18] platforms, different entries in the subject's blog and a collection of teaching sessions performed with *Mathematica*.

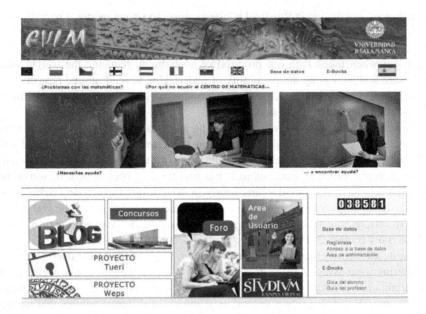

Fig. 3. EVLM Spanish Portal from Salamanca University

The assessment system employed has been modified over the years. Out starting point involved the traditional theory/problem written exams. In view of the semester structure of the degree course, a single exam was taken at the end of the semester (towards mid-January) and a second call for an exam was carried out during the first week of September. Tutorial assistance to students was limited to a few hours, during the days leading up to the exam to solve any problems appearing in previous exams or some theoretical or practical questions. With this system, the percentage of students passing the subject was of 35% out of all those registered for the subject. With this procedure the number of students in many cases tripled the number of newly registered students.

The situation, however, has changed. The current assessment system includes several steps. The students perform different tasks along the semester: they hand in written problem solutions (using a processor for scientific texts), they participate in the tasks proposed on the subject's blog, they solve the problems posed in practical sessions with *Mathematica* and they have to perform some short tutored projects ([12, 19]), Nearly all the tasks stated involve a broader use of the tutorial resources offered by the instructor. The grades obtained with these tasks count for 60% of the final grade. The students also do 4 short exams (1-hour), which in the present academic year have been:

— an individualized questionnaire through Studium platform
— an exam with *Mathematica* in the computer room
— 2 written exams with two problems each.

The grades obtained with these tasks account for 40% of the final grade.

After the end of the semester, students who have not passed the subject are offered the opportunity to take a traditional 3-h exam, consisting of the solving of 4 problems. Finally, an exam of the same characteristics can be taken to pass the subject in the redemption exam (2nd call).

It is clear that correcting the assessment tasks involves a considerable increase in the instructor's work. 18 different marks are obtained for the first part of the assessment, to which 4 grades of the exams mentioned previously in the process of continuous assessment must be added. Although the work is huge, eight additional hours per week for the instructor, it is worth. During the present academic year, the percentage of students passing the course has risen to 85% of those registered. As well as the spectacular increase in academic performance, we have observed a change in the mental disposition with which the average student faces the challenges of the subject. The attitude of rejection towards to mathematical subjects has disappeared almost completely and the student's participation in the classroom and tutorial activities has increased.

The second experiment was done with students of the Mechanical Engineering degree in University School of Industrial Technical Engineering at the Polytechnic University in Madrid.

This subject is common to the degrees in mechanical Engineering, Chemical Engineering, Electronics Engineering and Design. It is given during the first semester of the first academic year and its aim is to allow students to become familiarized with the requisite mathematical tools of Linear Algebra. The contents are the standard ones for a course on Linear Algebra: vector spaces, the solution of linear systems, linear applications, Euclidian spaces, similarity and orthogonal diagonalization and orthogonal transformations.

The MOODLE platform is used as a News Forum to announce the different activities and exam dates, etc, and also as a repository of documentation, providing a guide to the subject, offering exercises and proposals for students' personal work.

A continuous assessment system is used, comprising three written tests with a value of 80% of the final grade and with respective weightings of 15%, 30% and 35% of the final grade. The main novelty is that in all these written tests students are allowed to use the CAS MAXIMA, http://maxima.sourceforge.net/, if they see fit.

The remaining 20% of the grade is obtained via different activities with MAXIMA, 10% by solving exercises sheets that the students must hand in groups of 2-3 people and the other 10% via an exam in which the students must obligatorily solve problems and other issues using MAXIMA.

A questionnaire was given out addressing aspects related to the experiment and asking the students their opinion about the use of MAXIMA integrated in the subject and its assessment. The results have been satisfactory. Regarding academic performance, this was almost identical to that of the other group in the same degree, although the dropout rate was lower. The students' feeling of this experiment can be found in [20].

We conclude this section with a brief look at the immediate future since currently existing technology allows us to address the assessments of math in a way different from the traditional one. Thus, for example, the STACK system, http://www.stack. bham.ac.uk/, allows the elaboration of personalized questionnaires, taken from a questionnaire database. The authors of the present article are collaborating in the construction of a multilingual item bank that will allow us to replace some of the exams demanding the physical presence of students by personalized questionnaires through the WEPS portal (See Fig. 4.)

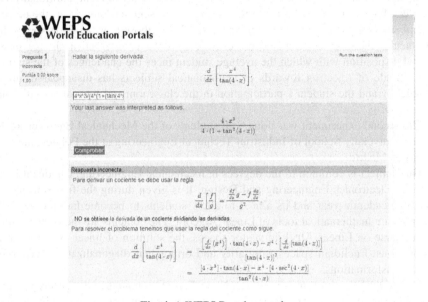

Fig. 4. A WEPS Portal example

We can also take advantage of motivation, a very important tool in the teaching of math for engineering, to propose assessment tasks aimed at increasing the following aspects:

— Student creativity, through competitions about the application of Maths in engineering, such as applications of the eigenvalues to the stability of processes, applications of the least squares procedure, applications of differential and integral Calculus to mechanical or electrical engineering, etc.
— Team work through the solution of problem sheets, with a difficulty level similar to those done in class which; these are corrected in a tutorial or randomly.
— Autonomous work through the elaboration of a tool box, procedures performed by students with the CAS used in the subject, which can be used in the resolution of problems in math subjects and also in subjects with a more technological trend.

5   Some Proposals for Assessment: Looking at the Future

The experiments described in the previous section show that it is possible to change the traditional exam system. It is clear that there will be an increase in work devoted to the correction of the various tests, an activity that, through the completion of individualized questionnaires with automatic correction and automatic management of the grades obtained, may decrease by a certain percentage. The creation of the item banks necessary for later elaboration of the personalized questionnaires is a collaborative task in which we can use the work already carried out in other university scenarios. It is also clear that tutored work in small groups may make the instructor suspicious about the true degree of involvement of a given student in group work. However, we believe that the benefits observed in the application of the experiments described surpass by far these deficiencies since the increase in student performance is evident.

With the above, it would be possible to articulate a new assessment framework for math subjects based on the following:

— The completion of tasks linked to problem solving.
— The implementation of individualized questionnaires using multilingual item banks.
— "Traditional" exams in which students are freely allowed to use a CAS, as they see fit, encouraging the construction of a prior tool box that can also be used in exams.
— Regarding motivation, the best students can be rewarded via contests about the technical applications of the concepts analyzed in the subject.

A future work will be the validation of the proposed model in terms of competencies acquired by the students. Obviously the additional work for the instructors must also be analyzed.

6   Conclusions

The EHEA must change the behavior of the students and teachers. In this sense:

— The new curricula and the use of the new technologies implies a new model of teaching and assessment. Also implies a new way to learn mathematics.
— It is necessary to adapt the teaching and the corresponding assessments to the teaching based on competencies.
— In our experiences the combination of the traditional exams and the new assessment methodologies produces very good results in student´s marks. Then we conclude: Yes, we can and we have to change!!!
— It is desirable a collaborative work for preparing questions to be included in automatic systems of assessment like STACK for example.

References

1. García, A., Garcia, F., Rodriguez, G., de la Villa, A.: Learning and Assessing Competencies: New challenges for Mathematics in Engineering Degrees in Spain. In: Proceedings of 16th SEFI Seminar of Mathematics Working Group (2012) ISBN: 978-84-695-3960-6

2. Artigue, M.: Learning Mathematics in a CAS Environment: The Genesis of a Reflection about Instrumentation and the Dialectics between Technical and Conceptual work. International Journal of Computers for Mathematical Learning 7, 245–274 (2002)
3. Lagrange, J.B., Artigue, M., Laborde, C., Trouche, L.: Technology and mathematics education: multidimensional overview of recent research and innovation. In: Leung, F.K.S. (ed.) Second International Handbook of Mathematics Education, vol. 1, pp. 237–270. Kluwer Academic Publishers, Dordrecht (2003)
4. Drijvers, P., Trouche, L.: Handheld technology for mathematics education: flashback into the future. ZDM Mathematics Education 42, 667–681 (2010)
5. Limniou, M., Smith, M.: The role of feedback in e-assessments for engineering education. Education and Information Technologies 17(4) (2012)
6. Alonso, F., Rodriguez, G., Villa, A.: A new way to teach Mathematics in Engineering New challenges, new approaches. In: Roy, R. (ed.) Engineering Education: Perspectives, Issues and Concerns, vol. 17. Shipra Publications, India (2009) ISBN 978-81-7541-505-8
7. Franco, A., Franco, P., Garcia, A., Garcia, F., Gonzalez, F.J., Hoya, S., Rodriguez, G., de la Villa, A.: Learning Calculus of Several Variables with New Technologies. International Journal of Computer Algebra in Mathematics Education 7(4), 295–309 (2000)
8. Garcia, A., Garcia, F., Rodriguez, G., de la Villa, A.: A course of ODE with a CAS. Proceedings of Technology and its Integration in Mathematics Education. In: TIME 2004, Montreal, Canadá (2004), http://time-2004.etsmtl.ca/
9. García, A., García, F., Rodríguez, G., de la Villa, A.: A toolbox with DERIVE. The Derive Newsletter 76, 5–13 (2009)
10. García, A., García, F., Rodríguez, G., de la Villa, A.: Could it be possible to replace DERIVE with MAXIMA? The International Journal for Technology in Mathematics Education 18(3), 137–142 (2011) ISSN: 1744-2710
11. García, A., Díaz, A., de la Villa, A.: An example of learning based on competences: Use of MAXIMA in Linear Algebra for Engineers. The International Journal for Technology in Mathematics Education 18(4), 177–181 (2011)
12. García, A., García, F., Rodríguez, G., de la Villa, A.: Small projects: A method for improving learning. In: Proceedings of 3rd International Research Symposium on Problem-Based Learning, Coventry, pp. 460–471 (2011) ISBN 978-87-7112-025-7
13. EVLM Spanish Portal, http://portalevlm.usal.es/
14. WEBMLS Project, http://www.webmathls.selcuk.edu.tr/
15. MathCentre, http://www.mathcentre.ac.uk/
16. Nieto, S., Ramos, H.: Pre-Knowledge of Basic Mathematics Topics in Engineering Students in Spain. In: Proceedings of 16th SEFI Seminar of Mathematics Working Group (2012) ISBN: 978-84-695-3960-6
17. García, A., García, F., Rodríguez, G., de la Villa, A.: CALCULO I: Teoría y problemas de Análisis Matemático en una variable, 3rd edn. Editorial CLAGSA, Madrid (2007) ISBN 978-84-921847-2-9
18. WEPS Portal, https://myweps.com/moodle23/
19. Martin, A., Rodriguez, G., de la Villa, A.: Problemas en el Paraíso: Las Matemáticas en las Escuelas de Ingeniería. To appear in Proceedings TICAMES (2013)
20. García, A., García, F., Rodríguez, G., de la Villa, A.: Changing assessments methods: New rules, new roles. To appear in Journal Symbolic of Computation (2013)

Soft Computing Techniques for Skills Assessment of Highly Qualified Personnel

Héctor Quintián[1], Roberto Vega[1],Vicente Vera[2], Ignacio Aliaga[2],
Cristina González Losada[2], Emilio Corchado[1], and Fanny Klett[3]

[1] University of Salamanca, Spain
{hector.quintian,rvegaru,escorchado}@usal.es
[2] University Complutense of Madrid, Spain
{Ialia01,vicentevera}@odon.ucm.es
[3] Director, German Workforce ADL Partnership Laboratory, Germany
fanny.klett.de@adlnet.gov

Abstract. This study applies Artificial Intelligence techniques to analyse the re-sults obtained in different tests to assess the skills of high qualified personnel as engineers, pilots, doctors, dentists, etc. Several Exploratory Projection Pursuit techniques are successfully applied to a novel and real dataset for the assess-ment of personnel skills and to identify weaknesses to be improved in a later phase. These techniques reduce the complexity of the evaluation process and al-low identifying the most relevant aspects in the personnel training in an intui-tive way, enhancing the particular training process and thus, the human resources management as a whole and saving training costs.

Keywords:EPP, PCA, MLHL, CMLHL, skillsassessments, high qualified personnel.

1 Introduction

Nowadays, in innovative sectors related to aviation, engineering, medicine, etc. where personnel with high aptitudes and skills are needed, the personnel training is espe-cially important. In this way, the use of novel intelligent and intuitive tools, both for training, and assessment of the developed skills during the training period, is crucial to get high qualified personnel in the shortest possible time and with the lowest cost associated to the training process, to facilitate the skills development process, and to enhance the human resources management process in general [1].

The use of simulators represent a huge cost savings in many sectors [2-5], ensuring the maximum level of training. However, the evaluation of the final aptitudes and skills continues being made by human experts in most cases. It is then a great chal-lenge when the number of variables to be considered is high. Therefore, having the adequate systems which facilitate thesetasks and allowdetecting in which parts of the training process actionsare needed in an individual way, represents a vast advantage in obtaining the best training in the shortest time.

Á. Herrero et al. (eds.), *International Joint Conference SOCO'13-CISIS'13-ICEUTE'13,*
Advances in Intelligent Systems and Computing 239,
DOI: 10.1007/978-3-319-01854-6_68, © Springer International Publishing Switzerland 2014

Against this background, Artificial Intelligence(AI) and statistical models can resolve these issues, as they are able to work with a large amount of high-dimensional datasets, presenting this information to the user in a compressible way. Several Soft Computing techniques have been used previously in different areas [6, 7] for managing large amounts of high dimensional information.

Specifically, within the various existing AI techniques, "Exploratory Projection Pursuit" (EPP) algorithms [22-24], allow the visualization of high dimensional information in a compressible way for the user [8-10].

Thenovel studypresented in this paper, applies several EPP techniques to a real dataset with the aim to comprehensively analyse high dimensional information for professionals in specific aspects for the assessment of highly qualified personnel.

The objective of this analysis is to determine the most relevant aspects of a personnel training process, and this accomplished in anobjective way according to the individual needs of each person by stressing on those aspects of the training, which need to be improved during the training process.

In particular, three EPP algorithms were applied in this study: "Principal Components Analysis" (PCA) [20, 21], "Maximum LikelihoodHebbian Learning" (MLHL) [22]and "Cooperative Maximum Likelihood Hebbian Learning" (CMLHL) [24, 28]. These algorithms were employed to a novel and real dataset based on the assessments of university students' skills referring to the field of odontology. These skills are related to the dental machining of great accuracy and finishing.

This paper is structured as follows: Section 2 presents the use of the EPP algorithms in this study; Section 3 illustrates the case study where the EPP algorithms were applied; Section 4 reflects the experiments developed and the results achieved; and finally, Section 5 shows the conclusions.

2   Exploratory Projection Pursuit

AI [11-13] is a set of several technologies aiming to solve inexact and complex problems [14, 15]. It investigates, simulates, and analyses very complex issues and phenomena in order to solve real-world problems [16, 17]. AI has been successfully applied to many different fields as, for example, feature selection [18, 19]. In this study, an extension of a neural PCA version [20, 21] and further EPP [22, 23] extensions are used to select the most relevant input features in the data set and to study their internal structure. Some projection methods such as PCA [20, 21], MLHL [22] and CMLHL [24-26] are applied to analyse the internal structure of the data and find out, which variables determine this internal structure and what they affect in this internal structure.

2.1   Principal Component Analysis

PCA originated in work by Pearson, and independently by Hotelling [20].It refers to a statistical method describing multivariate dataset variation in term of uncorrelated variables, each of which is a linear combination of the original variables. Its main goal is to derive new variables, in decreasing order of importance, which are linear combinations of the original variables and are uncorrelated with each other. Using PCA, it is possible to find a smaller group of underlying variables that describe the data. PCA has been the most frequently reported linear operation involving unsupervised learning for data compression and feature selection [21].

2.2   A Neural Implementation of Exploratory Projection Pursuit

The standard statistical method of EPP [23, 24], provides a linear projection of a data set, but it projects the data onto a set of basic vectors which best reveal the interesting structure in the data. Interestingness is usually defined in terms of how far the distribution is from the Gaussian distribution [27]. One neural implementation of EPP is MLHL [22]. It identifies interestingness by maximizing the probability of the residuals under specific probability density functions that are non-Gaussian. An extended version of this model is the CMLHL [24, 28] model. CMLHL is based on MLHL [22] adding lateral connections [24, 28], which have been derived from the Rectified Gaussian Distribution [27]. The resultant net can find the independent factors of a data set but does so in a way that captures some type of global ordering in the data set.

Considering a N-dimensional input vector (x), and a M-dimensional output vector (y), with W_{ij} being the weight (linking input j to output i), then CMLHL can be expressed [28] as:

Feed-forward step:

$$y_i = \sum_{i=1}^{N} W_{ij} x_j, \forall i \qquad (1)$$

Lateral activation passing:

$$y_i(t+1) = \left[y_i(t) + \tau(b - Ay) \right]^+ \qquad (2)$$

Feedback step:

$$e_j = x_j - \sum_{i=1}^{M} W_{ij}, \forall j \qquad (3)$$

Weight change:

$$\Delta W_{ij} = \eta y_i sign(e_j) \left| e_j \right|^{p-1} \qquad (4)$$

Where: η is the learning rate, $[\,]^+$ is a rectification necessary to ensure that the $y - values$ remain within the positive quadrant, τ is the "strength" of the lateral connections, b is the bias parameter, p is a parameter related to the energy function [22, 24] and A is the symmetric matrix used to modify the response to the data [24]. The effect of this matrix is based on the relation between the distances separating the output neurons.

3   The Real Case Study

The real case study analysed in this paper, focuses on facilitating the identification of divergent or non-desirable situations in atraining process. The aim of this study is to classify the psychomotor skills of odontology students using two training scenarios.

The first scenario refers to creating methacrylate figures during a Dental Aptitude Test, which consists of carving ten methacrylate figures by using rotatory systems and applying two different speeds (V1 and V2). V1 (low speed) rotates at a speed of 10.000-60.000 revolutions per minute (rpm), while V2 (turbine or high speed) rotates at a speed of 250.000 rpm. Seven of the figures made by the students can be easily created, while the remainders, which have several planes, involve a higher level of difficultly.

The second training scenario is based on a simulator (SIMODONT [29]), with unique advantages over conventional training as the experience is much more realistic and it uses true size 3D images. The students' performance can be measured in large detail, and the system allows for objectives comparison of the students' individual results.

3.1   Training Scenario 1 Description

Every student works on a methacrylate sheet at two different speeds, low speed and high speed. The low speed (10.000-60.000 rpm) is used to carve the first set of ten figures (see Figure 1a). After completing this part of the practical work, the students start carving the second part, which is basically a second round of the same figures, but this time using the high speed (150.000-250.000 rpm). The second part involves a higher level of difficulty as the bur is spinning faster, and better psychomotor skills are the pre-requisite for effectively completing this task.

Both parts of the practical work have to be completed during 90 minutes and the results have to be submitted to the supervisors.

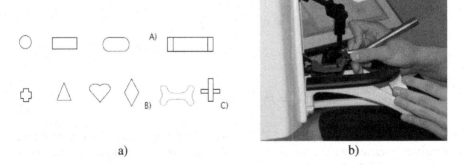

a)   b)

Fig. 1. Figures to be carved by the students (a), and simulator (b)

3.2   Training Scenario 2 Description

To help the dental student to obtain these skills in a virtual way, it is critical that the training system generates the precise sense of touch in a simulator [6, 46]. In some ways, it is a similar problem to training a pilot to control a plane using a simulator that provides a training experience so real that it counts as training hours.

The software provides detailed feedback comparing the operator's performance with a pre-programmed acceptable "ideal" cavity preparation in its database at any point of the procedure.

3.3   Empirical Training Scenario 1 Evaluation

The real case scenario is empirically evaluated based on a survey of 38 dental students. The information analysed for each student is based on 128 variables, but we are measuring eighty variables in this part of the study.

The first eight most important variables are:Age of the student; Sex of the student; Mark obtained by the student in the university enrolment exam;Previous experience gained by the student (The students may have had professional experience as a nurse, dental technician, hygienist, dental technician and hygienist, or lack of previous work experience); Mark obtained by the student in the theoretical exam of the subject; Mark obtained by the students in the practical exam of the subject; Group class of the student; Number of figures carved by the student.

The following eighty variables (20 figures with four variables each) are the evaluations of the different parts of the figures (graded between 0 and 5). The way to interpret these variables is as follows: 'x' indicates the figure number and can range from 1 to 10, 'y' indicates the speed used to carve the figure by using Low Speed (1) or High speed (2), and 'z' indicates the evaluator who examines the test (1 or 2):Fx_Vy_Ez_WALL: quality of the walls; Fx_Vy_Ez_DEPTH: quality of the depth; Fx_Vy_Ez_EDGES: quality of the edges; Fx_Vy_Ez_FLOOR: evaluate the plain and irregularities presented on the floor.

3.4   Empirical Training Scenario 2 Evaluation

In this second evaluation, the SIMODONT simulator has been used to assess the skills of the students. As in the previous case scenario, a survey of 38 dental students has been analysed, but in this case only 48 variables have been measured for the analysis.

In this exercise, the goal of the student is to complete the target of each one of the three figures the exercise is based on, touching as less as possible the leeway bottom and sides and they should not erase any surface of the container (see Figure 1b).

Variables measured and evaluated in this exercise are:Target; Leeway Bottom (Amount of healthy tissue (not have to remove) removed from the cavity floor. % who do wrong); Leeway Sides (from the cavity sides %); Container Bottom (measure how much the student pass the limit of the virtual floor of the figure to carve, % respect of the total deep); Container Sides (% respect of the total figure surface); Time Elapsed (used time to carve the figure, in seconds); Drill Time (total time the turbine (instrument) is active during the carve of the figure, in seconds); Moved with right (movements of the right hand, measure the indirect vision, in meters); Moved with left (movements of the left hand, measure the indirect vision, in meters).

4   Experiments and Results

Using both of the above-explained use cases, one dataset composed by 38 samples and 126 variables was created. A pre-process consisting of the normalization of the data was applied. Finally three algorithms were applied to the dataset (PCA, MLHL, and CMLHL). Figures 2a, 2b and 3 present the best projections of results for each algorithm.

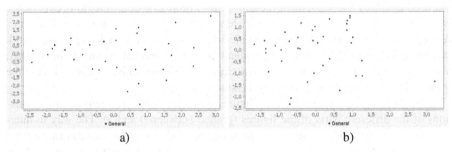

a)   b)

Fig. 2. PCA algorithm, projections 1-2 (a) and MLHL algorithm, projections 1-4, iters=100.000, lrate=0.001, m=5, p=2.2 (b)

Comparing the results of the three algorithms applied, the CMLHL offers a more clear separation of the clusters than the other two methods (PCA and MLHL), obtaining a clear internal structure of the data and defining two main axis of variation (see Figure 3). The following sub-section offers a more deep analysis of the CMLHL results.

4.1   CMLHL Results Analysis

Figure 3, shows the projections obtained bythe CMLHL algorithm.

The identified clusters are presented in Figure 4a in a schematic way, where a descriptive name was givento each one, based on their most relevant characteristics. Based on the most relevant characteristics, which define the internal structure of the data, the diagram in Figure 4b was created, were the variation of each variable is represented by an arrow.

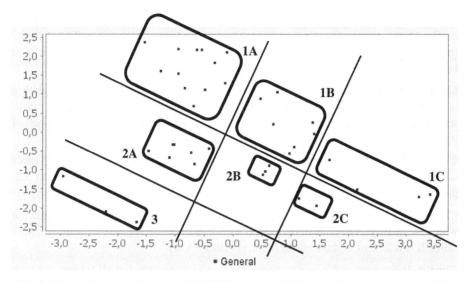

Fig. 3. Clusters for projections 2-4 of CMLHL, iters=100.000, lrate=0.01, m=6, p=1.6, τ=0.25

CLUSTER 1A	CLUSTER 1B	CLUSTER 1C
EXCELENT	**HARD WORKER**	**CARELESS**
Many figures	**Many figures**	**Many figures**
High quality	**Medium quality**	**Low quality**
CLUSTER 2A	CLUSTER 2B	CLUSTER 2C
SKILLFUL	**WITHIN HALF**	**UNSKILLFUL**
Average figures	**Average figures**	**Average figures**
High quality	**Medium quality**	**Low quality**
CLUSTER 3		
VAGUE		
Few figures		
High quality		

a)   b)

Fig. 4. Schematic diagram (a) and diagram about parameters variation (b)

Cluster 1A: The best students belong to this cluster, because they make the highest number of figures, and most of these figures are made with high quality. The "drill time" and the "elapsed time" employed in the simulations are the highest, getting the lowest error in the simulations. All figures are made with great care, trying to achieve the best result.

The quality of students belonging to this cluster is reflected in the theoretical exam marks (the highest), but the practical exam marks are not as high as expected, being close to the marks of other clusters (1B, 2A, 2B), and in some cases even lower.

Cluster 1B: In this group, the students are hard workers, performing a high number of figures with a normal quality. This effort to improve their skills performing a high number of figures is reflected in the theoretical/practical exam with high marks, similar to the cluster 1A students.

Cluster 1C: Students of this cluster perform a high number of figures but with a low quality.These students do not have interest in the subject and this is reflected in the final marks, getting low marks in the practical exam in spite of performing a high number of figures (high experience).

Cluster 2A: Skilled students.They perform a normal/low number of figures with a high quality. Their marks in the practical exam are variable, some achieve good marks and other - bad marks.This indicates the need of developing a highernumber of figures.

Cluster 2B: Students with normal skills.They make an average number of figures with a medium quality. The students' marks in this cluster are similar to cluster 1B, the main difference between both clusters is the number of figures created.

Cluster 2C: Students with low skills.They perform a medium number of figures but with low quality. They must perform a higher number of figures to improve their skills. It is reflected in their marks in the practical exam.

Cluster 3: Students perform few figures with a high quality, but all figures are with low/medium difficulty.As these students practice not too much, their marks in the practical exam are lower than theirfigures marks, getting low/medium marks instead of high marks (they develop a low number of figures with a quite good quality).

5   Conclusions

The results obtained in this study illustrate the identificationof six groups of personnel and the unique opportunity to consider individual actions according to each group based on the parameters shown in the Figure 4 (increasing the number of figures in some of them, or increasing the difficulty of some figures in other, etc.).

Moreover, the results show that using PCA, MLHL and CMLHL algorithms allows significantly improving the training process when personnel needs high level skills, especially in fields such as aviation, engineering, medicine, etc., where the number of skills to be assessed simultaneously is large.

Acknowledgement. This research is partially supported by the Spanish Ministry of Economy and Competitiveness under project TIN2010-21272- C02-01 (funded by the European Regional Development Fund) and SA405A12-2 from Junta de Castilla y León.

References

1. Klett, F., Wang, M.: The War for Talent: Technologies and solutions toward competency and skills development and talent identification (Editorial). Knowledge Management & E-Learning 5(1), 1–9 (2013)
2. Cha, M., Han, S., Lee, J., Choi, B.: A virtual reality based fire training simulator integrated with fire dynamics data. Fire Safety Journal 50, 12–24 (2012)

3. Rhienmora, P., Haddawy, P., Suebnukarn, S., Dailey, M.N.: Intelligent dental training simulator with objective skill assessment and feedback. Artificial Intelligence in Medicine 52(2), 115–121 (2011)
4. Jardón, A., Victores, J.G., Martínez, S., Balaguer, C.: Experience acquisition simulator for operating microtuneling boring machines. Automation in Construction 23, 33–46 (2012)
5. Per Bodin, P., Nylund, M., Battelino, M.: SATSIM—A real-time multi-satellite simulator for test and validation in formation flying projects. Acta Astronautica 74, 29–39 (2012)
6. Peremezhney, N., Connaughton, C., Unali, G., Hines, E., Lapkin, A.A.: Application of dimensionality reduction to visualisation of high-throughput data and building of a classification model in formulated consumer product design. Chemical Engineering Research and Design 90(12), 2179–2185 (2012)
7. Song, M., Yang, H., Siadat, S.H., Pechenizkiy, M.: A comparative study of dimensionality reduction techniques to enhance trace clustering performances. Expert Systems with Applications 40(9), 3722–3737 (2013)
8. Herrero, Á., Zurutuza, U., Corchado, E.: A Neural Visualization IDS For Honeynet Data. International Journal of Neural Systems 22(2) (2012)
9. Vera, V., Corchado, E., Redondo, R., Sedano, J., García, Á.E.: Applying Soft Computing Techniques to Optimise a Dental Milling Process. Neurocomputing 109, 94–104 (2013)
10. Baruque, B., Corchado, E., Yin, H.: The s(2)-ensemble fusion algorithm. International Journal of Neural Systems 21(6), 505–525 (2011)
11. Cordon, O., Fernández-Caballero, A., Gámez, J.A., Hoffmann, F.: The impact of soft computing for the progress of artificial intelligence. Applied Soft Computing 11(2), 1491–1492 (2011)
12. Abraham, A.: Hybrid soft computing and applications. International Journal of Computational Intelligence and Applications 8(1), 5–7 (2009)
13. Wilk, T., Wozniak, M.: Soft computing methods applied to combination of one-class classifiers. Neurocomputing 75(1), 185–193 (2012)
14. Kohonen, T.: The self-organizing map. Neurocomputing 21(1-3), 1–6 (1998)
15. Corchado, E., Baruque, B.: Wevos-visom: An ensemble summarization algorithm for enhanced data visualization. Neurocomputing 75(1), 171–184 (2012)
16. Sedano, J., de la Cal, E., Curiel, L., Villar, J., Corchado, E.: Soft computing for detecting thermal insulation failures in buildings. In: Proceedings of the 9th International Conference on Computational and Mathematical Methods in Science and Engineering, CMMSE 2009, vol. 4, pp. 1392–1402 (2009)
17. Sedano, J., Curiel, L., Corchado, E., de la Cal, E., Villar, J.: A soft computing based method for detecting lifetime building thermal insulation failures. Integrated Computer-Aided Engineering 17(12), 103–115 (2010)
18. Leray, P., Gallinari, P.: Feature selection with neural networks. Behaviormetrika 26, 145–166 (1999)
19. Verikas, A., Bacauskiene, M.: Feature selection with neural networks. Pattern Recognition Letters 23(11), 1323–1335 (2002)
20. Hotelling, H.: Analysis of a complex of statistical variables into principal components. Journal of Education Psychology 24, 417–444 (1933)
21. Oja, E., Ogawa, H., Wangviwattana, J.: Principal components analysis by homogeneous neural networks, part 1, the weighted subspace criterion. IEICE Transaction on Information and Systems E75D, 366–375 (1992)
22. Krömer, P., Corchado, E., Snášel, V., Platoš, J., García-Hernández, L.: Neural PCA and Maximum Likelihood Hebbian Learning on the GPU. In: Villa, A.E.P., Duch, W., Érdi, P., Masulli, F., Palm, G. (eds.) ICANN 2012, Part II. LNCS, vol. 7553, pp. 132–139. Springer, Heidelberg (2012)

23. Friedman, J.: Exploratory projection pursuit. Journal of the American Statistical Association 82(397), 249–266 (1987)
24. Herrero, Á., Corchado, E., SáizBárcena, L., Abraham, A.: DIPKIP: A Connectionist Knowledge Management System to Identify Knowledge Deficits in Practical Cases. Computational Intelligence 26(1), 26–56 (2010)
25. Corchado, E., Herrero, A.: Neural visualization of network traffic data for intrusion detection. Applied Soft Computing 11(2), 2042–2056 (2011)
26. Herrero, A., Corchado, E., Gastaldo, P., Zunino, R.: Neural projection techniques for the visual inspection of network traffic. Neurocomputing 72(16-18), 3649–3658 (2009)
27. Seung, H., Socci, N., Lee, D.: The rectified gaussian distribution. In: Advances in Neural Information Processing Systems, vol. 10, pp. 350–356 (1998)
28. Corchado, E., Herrero, Á.: Neural visualization of network traffic data for intrusion detection. Appl. Soft Comput. 11(2), 2042–2056 (2011)
29. Bakker, D., Lagerweij, M., Wesselink, P., Vervoorn, M.: Transfer of Manual Dexterity Skills Acquired on the SIMODONT, a Dental Haptic Trainer with a Virtual Environment, to Reality, A Pilot Study. Bio-Algorithms and Med-Systems 6(11), 21–24 (2010)

Using Web 2.0 Tools to Support the Theoretical Constructs of Organisational Learning

Gavin J. Baxter and Thomas M. Connolly

School of Computing, University of the West of Scotland, Scotland, UK
{gavin.baxter,thomas.connolly}@uws.ac.uk

Abstract. The aim of this paper is to provide an overview of the concept of Web 2.0 and to review how three established Web 2.0 tools: blogs, wikis and online forums, can be applied in organisational contexts. In doing so, this paper identifies ways in which Web 2.0 tools can support some of the key theoretical constructs of organisational learning. Furthermore, this paper proposes how to advance this area of research as it has been acknowledged that there is a lack of empirical evidence to substantiate the view that Web 2.0 tools can support the process of organisational learning. A traditional literature review will be used to indicate the differences between three of the most established types of Web 2.0 technologies: blogs, wikis and online forums with a view to providing recommendations on how these practices can be adopted in organisational settings. The findings of the traditional literature review show that despite their being relative differences between blogs, wikis and online forums; they have the potential, dependent on their use, to support the concept of organisational learning. Though this paper provides an overview of how certain Web 2.0 tools can be applied in organisations it does not provide empirical evidence of Web 2.0 use within enterprises. This paper presents a general overview on blogs, wikis and online forums and their use in organisations to assist management practitioners who might be unfamiliar in their use on how to apply them internally in their organisations.

Keywords: Organisational blogging, Wikis, Online forums, Web 2.0, Enterprise 2.0, Organisational learning.

1 Introduction

Web 2.0 refers to the social use of the World Wide Web and describes the social characteristics emanated by groups of individuals through using Web 2.0 tools in a particular way. In addition, a concept often associated with Web 2.0 that can be described as being interrelated though distinct from it is the term 'social software'. The concept social software was conceptualised by Clay Shirky in 2003 who perceived social software to be software that predominately supported group interaction. There are many examples of Web 2.0 tools cited in the literature however this paper will examine three of the most established types of Web 2.0 tools: blogs, wikis and online forums. One particular area where these technologies are making a significant impact

Á. Herrero et al. (eds.), *International Joint Conference SOCO'13-CISIS'13-ICEUTE'13*,
Advances in Intelligent Systems and Computing 239,
DOI: 10.1007/978-3-319-01854-6_69, © Springer International Publishing Switzerland 2014

is within the field of education altering the way in which educators deliver lessons and how they fulfil learning expectations of students [1]. However, in contrast to the adoption of Web 2.0 tools in education, this paper aims to answer the research question of: 'What concepts of organisational learning can Web 2.0 technologies support?'

In contrast, despite the advent of the concept of 'Enterprise 2.0' empirical research associated with Web 2.0 use in organisations appears to be limited [2: 301]. This observation has already been noted in the literature in relation to organisational blog use [3] and 'Enterprise Microblogging' [4: 4200]. Enterprise 2.0 refers to: "...the use of emergent social software platforms within companies, or between companies and their partners or customers" [5: 1]. To address this issue of a lack of awareness of how to successfully integrate Enterprise 2.0 technologies this paper will provide a review of the usage of blogs, wikis and online forums to inform management practitioners in industry how these types of Web 2.0 tools could be internally applied in their organisations and what the effect of their use might be. To begin with, the concept of Web2.0 is examined to provide a sense of clarity about what the term means. Following on from this, comparisons are made between blogs, wikis and online forums to indicate their similarities and differences. Next, examples of how these different types of Web 2.0 tools can be applied to support organisational learning will be explored. Finally, a set of use case scenarios are provided illustrating to management practitioners how blogs, wikis and online forums could sustain the notion of organisational learning, as well as future directions on how to advance the state of the art of Web 2.0 technologies and organisational learning.

2   The Concept of Web 2.0

Prior to the introduction of Web 2.0 the Internet and the way in which people interacted through its use was known as Web 1.0. The phrase Web 1.0 referred to when the World Wide Web began in 1993 as static pages and was primarily the read only era on the Web. Since 2004, and with the continued developments of multimedia, the term Web 2.0 has been used to describe what is known as the social Web.

Several definitions of Web 2.0 have been suggested in the literature. According to Grosseck [6: 478] Web 2.0 refers to "...the social use of the Web which allow[s] people to collaborate, to get actively involved in creating content, to generate knowledge and to share information online". This definition implies that use of Web 2.0 tools are predominately designed to facilitate collaboration online in addition to collective knowledge and information sharing. In contrast, Aharony [7: 227] emphasises the user-driven nature of Web 2.0 and argues that the concept "...emphasizes the value of user-generated content. It is about sharing and about communication and it opens the long tail which allows small groups of individuals to benefit from key pieces of the platform while fulfilling their own needs". The characteristics of Web 2.0 denoted from this definition accentuate the high degree of openness and transparency through individuals collaboratively using Web 2.0 tools in virtual learning environments (VLEs) that supports their use. Figure 1 illustrates some of the salient phrases predominately associated with Web 2.0 focused around the core concept of collaboration.

Fig. 1. Key terms commonly associated with Web 2.0

3   Web 2.0 Tools: Blogs, Wikis and Online Forums

Web 2.0 tools such as blogs, wikis and online forums are not new. For example, the phrase 'Weblog' was coined by Jorn Barger in 1997 due to the blending of the two words 'web' and 'log' [8]. The person responsible for using the first wiki was Ward Cunningham in 1995 [9]. Furthermore, forums, sometimes referred to as discussion boards or bulletin boards, are one of the oldest types of technologies used for information sharing and collaboration [10]. This section of the paper provides an overview of blogs, wikis and online forums comparing the differences and similarities among their other characteristics as depicted in figure 1.

3.1   Blogs

There are numerous definitions of blogs in the academic literature. A blog is predominately defined by its format and is essentially *"...a frequently updated webpage with dated entries, new ones placed on top"* [11: ix]. In addition to displaying up-to-date content blogs can also be used to store historical information. Blogs have strong connotations with diaries or journals due to their association with the word 'log'. When creating a blog the owner of that blog can either make it publicly accessible to everyone or use it for private reflections. Many blogs can also be used collectively allowing several individuals to add content in addition to being administered by multiple authors. Though early blog design began with blogs consisting of a series of text and links, the format of blogs has continued to develop and many blogs now support multimedia content such as sound, video, animation and graphics.

3.2   Wikis

The term wiki originates from the Hawaiian word 'wiki-wiki' meaning fast and was used to denote how quickly content can be generated with a wiki. There are numerous definitions of wikis in the academic literature; for example, Cole [12: 142] provides a concise definition that a wiki is *"an editable website that is created incrementally by visitors working collaboratively"*. According to Duffy and Bruns [13] there is no overall predetermined structure to a wiki page with content on a wiki often adopting the form of an emergent structure. Some wikis require the user to be familiar with some basic wiki syntax, which changes slightly depending on the type of wiki software that is used. For example, if using the wiki software Markdown to make text bold format would involve the user adding the command '** bold text **' or '_ bold text _'. In comparison, when using the wiki software MediaWiki to make text bold the command ''' bold text ''' is used.

3.3   Online Forums

Online forums allow a user to post a message for others to read and to which others can respond. Vieira da Cunha and Orlikowski [14: 134] provide a more comprehensive definition of online forums and view them as a *"many-to-many communication space where participants can post a new topic and reply to an existing one. This communication is archived, and all of the threads are always available for reading and posting. Online forums may be public or private"*. The topics posted on online forums are known as 'threads' and the replies are known as 'posts'. Online forums in contrast to wikis do not normally allow posts or threads to be modified once added. The exception might be in the case of a moderator whose job would be to oversee the use of the forum. The structure of an online forum differs from that of a wiki and threads in online forums are arranged in descending chronological order of the most recently posted message.

3.4   Comparisons of Blogs, Wikis and Online Forums

In comparison to wikis, the diary-like format of blogs makes them more appropriate to disseminate ideas [15], reflect and exchange opinions. It is the conversational nature of blogs [16] that provides users with a sense of empowerment to exchange views and opinions about issues of mutual interest. In contrast, forums are generally used by people to ask questions and receive answers to questions. The knowledge contained on forums can be thought of as being more declarative and procedural with less emphasis placed on context. In a forum individuals request knowledge that is related to 'know how' or to 'know-about' something. Though forums are used to share ideas and opinions it can be argued that blogs afford greater scope in permitting users to express their views. This is due to the diary-like format of blogs that promote reflection allowing bloggers inserting posts to reflect on the context of personal experience thereby sharing tacit knowledge with fellow readers.

Blogs have been associated with knowledge management [17]. For example, in organisational contexts blogs are sometimes referred to as k(knowledge)-logs [18]. Wikis are also regarded as a useful knowledge management tool [9], however, wikis may be useful for managing knowledge that is formalised in documents and which is made explicit whereas blogs are informal in nature and may be better suited for sharing tacit knowledge and subjective experience [19]. Similar to wikis and forums, blogs are also associated with communities of practice. It could be argued that blogs build communities through the exchange of dialogue and opinion as do forums though the knowledge contained in a forum is less contextualised. Table 1, adapted from Miyazoe and Anderson [20], summarises the differences between blogs, wikis and online forums.

Table 1. Characteristics of forums, blogs, and wikis [20: 186]

Key Characteristics	Forums	Blogs	Wikis
Number of users	Many-to-many	One-to-many	Many-to-many
Editing	By moderator	By author/moderator	All users
Purpose	Ask & answer questions	Express experiences	Edit documents
Content structure	Threaded	Reverse chronological	Final artefact
Social-cognitive use	Help	Articulation	Collaboration

4   Web 2.0 Tools Adoption: Making the Transition to Organisational Learning

In comparison to the educational sector, there appears to be a lack of empirical evidence relating to the use of Web 2.0 tools being applied internally in enterprises to support the concept of organisational learning [21]. There have been several definitions of organisational learning proposed in the academic literature. Despite this, a useful interpretation of organisational learning that assists to affirm its theoretical association with Web 2.0 technologies is a definition provided by Stata [22: 64] who views the process of organisational learning as being something that *"...occurs through shared insights, knowledge, and mental models... [and] builds on past knowledge and experience – that is, on memory"*. This perspective helps to denote that Web 2.0 technologies, dependent on their implementation and internal use in an organisation, can support one of the key domains of organisational learning which is knowledge creation and sharing among staff within the workplace. The following section identifies some important concepts associated with organisational learning and provides suggestions on how wikis, blogs and online forums can assist towards supporting these theories.

4.1   Organisational Knowledge Creation

It has been acknowledged that the sharing of tacit knowledge is a dilemma that can on occasion be problematic for certain organisations due to tacit knowledge being thought of as less formal than other types of knowledge [23]. Tacit knowledge is generally regarded as knowledge that is personalised and based upon experience, context and the actions of an individual. In the context of how organisational knowledge is created Web 2.0 tools, for example, the use of a blog or a wiki, have the potential to support the socialisation and externalisation stages of organisational knowledge creation. Socialisation entails the sharing of tacit knowledge through the social interaction of individuals through shared mental models and technical skills. The use of a collective blog or wiki would provide organisational team or departmental members with the ability to record this knowledge making it accessible and readable in an explicit format for other staff members to learn from thereby assisting with the externalisation process of organisational knowledge creation.

4.2   Communities of Practice

The majority of Web 2.0 tools, as indicated in section four, have the potential to support the learning theory of social constructivism. This process of learning can be sustained in what is known as a community of practice or CoPs. Wenger, McDermott and Snyder [24: 4] define CoPs as: *"groups of people who share a concern, a set of problems, or a passion about a topic, and who deepen their knowledge and expertise in this area by interacting on an ongoing basis"*. For example, the co-editorial and collective authoring features of wikis make them very useful for staff to socially construct knowledge among one another when collaboratively creating and sharing documents. These documents can be disseminated among the department to all employees providing an invaluable repository of organisational knowledge. For example, an individual recently having started employment within the department can access the wiki to review prior and current documents and things that the department has worked on. Furthermore, this individual can also impart their own knowledge gained from their prior working experience on the wiki for others to learn from and apply within their department.

4.3   Transactive Memory

Web 2.0 tools also have the potential to support the process known as transactive memory in organisations [25]. The concept of transactive memory refers to *"...a shared awareness among individuals about who knows what"* [26: 244] and has particular reference to how members in teams transfer, retrieve and update knowledge among one another especially in teams or groups. In this respect, communities of practice are very suitable for accommodating transactive memory systems because they adhere to the three processes that are salient to the functioning of a transactive memory system, namely, directory updating, allocation or storage of information and the ability to retrieve information [27]. For example, through the use of a wiki, blog or online forum, knowledge can be updated by other members in a team or group that reflects existing knowledge of other group members or builds upon this knowledge.

Knowledge can also be effectively stored and retrieved by all group members providing individual group members with a greater awareness of who knows what within their team.

4.4   Organisational Memory

Organisational memory can be thought of as *"...stored information from an organization's history that can be brought to bear on present decisions"* [28: 61]. Organisational memory is considered to be made up of both mental (i.e. knowledge, information, experiences and relationships) and structural artefacts (i.e. roles, structures, operating procedures and routines) [29]. Organisational memory is closely related to the notion of transactive memory as organisations can store different classifications of knowledge, for example, procedural or causal knowledge in a centralised knowledge repository for organisational members to access. For instance, the use of a blog would be highly applicable towards recording working experiences of employees; for example, how they have tackled known issues and problems at work so that work colleagues can learn from them and apply this knowledge in future working scenarios.

5   Use Case Scenarios: Practical Uses of Web 2.0 in Organisations

From the academic literature discussed in the previous sections it appears that Web 2.0 tools such as blogs, wikis and online forums can support several key theoretical concepts associated with organisational learning. However, it is also important to consider what practical aspects Web 2.0 tools can assist both management and employees within an organisation. This section of the paper will provide three use case scenarios that illustrate how certain Web 2.0 tools such as blogs, wikis and online forums can assist staff on a practical basis.

5.1   Blogs as Project Reporting Tools

The creation of a dedicated project blog has the potential to provide an alternative method of allowing project members to informally communicate about their project roles, objectives and initial thoughts about a project prior to commencement. The use of a blog can assist in making relevant information available to project members as well as creating a unique collaborative project culture. In addition, a centralised project blog could allow project managers to frequently update their project teams on a regular basis on a project's overall status, alterations in project objectives and milestones. Encouraging openness and information sharing through the use of a project blog can help to alleviate the problems of failing to report bad news within a project's lifecycle. Project team members would also be able to provide their own updates and progress reports on the blog allowing project members to provide feedback and supportive comments.

5.2   Wikis Applied in Virtual Teams

A virtual team can be defined as *"...a distributed group of people working together across distances to achieve a common purpose via technology"* [30: 116]. Adopting the use of a wiki for project members who are geographically dispersed to share information could assist to alleviate the problem of the disparateness of individuals in project teams. However, even within a small-to-medium sized company a scenario might arise where staff had to work in a four man project but if these four people were geographically distributed then a wiki might be useful for the purpose of long distance project co-ordination and information sharing. The fact that wikis seem to be associated with project work, especially within the area of software development, means that they are beneficial tools to support virtual communities of practice though further empirical work is required to substantiate this assertion.

5.3   Online Forums for Locating Expertise

The use of an online discussion forum, especially within a large organisation, can be a very useful tool to help employees locate expertise and knowledge within their company. For example, if an employee has just recently joined a new organisation and requires information to a specific query or question then the use of an online forum could help them to locate expertise within that organisation. For example, through posting a question to a work related issue or problem on a forum an employee could receive a response by a colleague that addresses that particular issue. In addition, the employee that posted the question would then know who to contact in the organisation regarding this subject area should they encounter any future problems.

6   Conclusions and Future Directions

This paper has presented a general review of the use of three specific types of Web 2.0 tools, namely, blogs, wikis and online forums to indicate from a management practitioner perspective how these Web 2.0 tools could be applied in organisational contexts. In addition, this paper also set out to investigate the question of 'What concepts of organisational learning can Web 2.0 technologies support?' Focusing on blogs, wikis and online forums and their potential association with organisational learning, it has been identified that at a theoretical level, these technologies have the ability to support the organisational learning constructs of: communities of practice; transactive memory and organisational memory. The traditional literature review performed in sections three and four of this paper also identified that the Web 2.0 technologies examined have the potential to facilitate and sustain the process organisational knowledge creation. Furthermore, dependent on their use, these types of Web 2.0 technologies can facilitate the process of organisational knowledge creation.

With regards to moving the area of Web 2.0 tools and organisational learning forward, suitable areas of focus, might include the following: (1) How effective are Web 2.0 tools in supporting the process of organisational learning?; (2) How can communities of practice be fostered through the use of Web 2.0 tools?; (3) How can organisational knowledge creation, transactive memory and organisational memory benefit from Web 2.0 adoption?; (4) What is the overall impact of 'enterprise 2.0' on organisational performance? and (5) What impact would Web 2.0 technologies have in supporting an organisation's transformation into becoming a 'learning organisation'? Advancing the state of the art of 'enterprise 2.0' through future empirical organisational studies will assist towards identifying whether the use of Web 2.0 tools in organisations can support the process of organisational learning and in doing so provide a wider contribution to knowledge in this area that will benefit the wider academic and management practitioner communities.

Acknowledgements. This work has been co-funded by the EU Lifelong Learning Programme under contract 519057-LLP-1-2011-1-UK-KA3-KA3NW (Ed2.0Work – European Network for the integration of Web2.0 in education and work).

References

1. Baxter, G.J., Connolly, T.M., Stansfield, M.H., Tsvetkova, N., Stoimenova, B.: Introducing Web 2.0 in Education: A Structured Approach Adopting a Web 2.0 Implementation Framework. In: 7th International Conference on Next Generation Web Services Practices (NWeSP), Salamanca, Spain, October 19-21, pp. 499–504 (2011)
2. Saldanha, T.J.V., Krishnan, M.S.: Organizational Adoption of Web 2.0 Technologies: An Empirical Analysis. Journal of Organizational Computing and Electronic Commerce 22(4), 301–333 (2012)
3. Baxter, G.J., Connolly, T.M.: The "state of art" of organisational blogging. The Learning Organization: The International Journal of Critical Studies in Organizational Learning 20(2), 104–117 (2013)
4. Riemer, K., Richter, A., Seltsikas, P.: Enterprise Microblogging: Procrastination or productive use? In: 16th Americas Conference on Information Systems 2010: AMCIS 2010, vol. 4, pp. 4199–4207 (2010)
5. McAfee, A.: A Definition of Enterprise 2.0. In: Buhse, W., Stamer, S. (eds.) Enterprise 2.0: The Art of Letting Go, pp. 1–15. iUniverse, Inc., New York (2008)
6. Grosseck, G.: To use or not to use web 2.0 in higher education? Procedia Social and Behavioral Sciences 1(1), 478–482 (2009)
7. Aharony, N.: The influence of LIS students' personality characteristics on their perceptions towards Web 2.0 use. Journal of Librarianship and Information Science 41(4), 227–242 (2009)
8. Kaiser, S., Muller-Seitz, G., Pereira Lopes, M., Pina e Cunha, M.: Weblog-Technology as a Trigger to Elicit Passion for Knowledge. Organization 14(3), 391–412 (2007)
9. Grace, T.P.L.: Wikis as a knowledge management tool. Journal of Knowledge Management 13(4), 64–74 (2009)
10. Wagner, C., Bolloju, N.: Editorial Preface: Supporting Knowledge Management in Organizations with Conversational Technologies: Discussion Forums, Weblogs, and Wikis. Journal of Database Management 16(2), i–viii (2005)
11. Blood, R.: Introduction. In: Rodzvilla, J. (ed.) We've Got Blog: How Weblogs are Changing our Culture, pp. ix–xiii. Perseus Publishing (2002)
12. Cole, M.: Using Wiki technology to support student engagement: Lessons from the trenches. Computers & Education 52(1), 141–146 (2009)

13. Duffy, P., Bruns, A.: The Use of Blogs, Wikis and RSS in Education: A Conversation of Possibilities. In: Proceedings: Online Learning and Teaching Conference 2006, Brisbane, pp. 31–38 (2006), http://eprints.qut.edu.au (accessed)

14. Vieira da Cunha, J., Orlikowski, W.J.: Performing catharsis: The use of online discussion forums in organizational change. Information and Organization 18(2), 132–156 (2008)

15. Azua, M.: The Social Factor: Innovate, Ignite, and Win through Mass Collaboration and Social Networking. IBM Press, London (2010)

16. Lee, H.H., Park, S.R., Hwang, T.: Corporate-level blogs of the Fortune 500 companies: an empirical investigation of content and design. Int. J. Information Technology and Management 7(2), 134–148 (2008)

17. Singh, R.P., Singh, L.O.: Blogs: Emerging Knowledge Management Tools for Entrepreneurs to Enhance Marketing Efforts. Journal of Internet Commerce 7(4), 470–484 (2008)

18. Herring, S.C., Scheidt, L.A., Wright, E., Bonus, S.: Weblogs as a bridging genre. Information Technology & People 18(2), 142–171 (2005)

19. Avram, G.: At the Crossroads of Knowledge Management and Social Software. The Electronic Journal of Knowledge Management 4(1), 1–10 (2006), http://www.ejkm.com

20. Miyazoe, T., Anderson, T.: Learning outcomes and students' perceptions of online writing: Simultaneous implementation of a forum, blog, and wiki in an EFL blended learning setting. System 38(2), 185–199 (2010)

21. Baxter, G.J.: Using Blogs for Organisational Learning: A Case Study in an ICT Division. PhD Thesis, viii-319 (August 2011)

22. Stata, R.: Organizational Learning - The Key to Management Innovation. Sloan Management Review 30(3), 63–74 (1989)

23. Haldin-Herrgard, T.: Difficulties in diffusion of tacit knowledge in organizations. Journal of Intellectual Capital 1(4), 357–365 (2000)

24. Wenger, E.C., McDermott, R., Snyder, W.M.: A Guide to Managing Knowledge: Cultivating Communities of Practice. Harvard Business School Press, Boston (2002)

25. Jackson, P.: Web 2.0 Knowledge Technologies and the Enterprise: Smarter, lighter and cheaper. Chandos Publishing, Oxford (2010)

26. Moreland, R.L., Swanenburg, K.L., Flagg, J.L., Fetterman, J.D.: Transactive Memory and Technology in Work Groups and Organizations. In: Ertl, B. (ed.) E-Collaborative Knowledge Construction: Learning from Computer-Supported and Virtual Environments, pp. 244–274. Information Science Reference, Hershey (2010)

27. Wegner, D.M., Giuliano, T., Hertel, P.T.: Cognitive Interdependence in Close Relationships. In: Ickes, W. (ed.) Compatible and Incompatible Relationships, pp. 253–276. Springer-Verlag New York Inc. (1985)

28. Walsh, J.P., Ungson, G.R.: Organizational Memory. Academy of Management Review 16(1), 57–91 (1991)

29. Kruse, S.D.: Remembering as organizational memory. Journal of Educational Administration 41(4), 332–347 (2003)

30. Brake, T.: Leading Virtual Global Teams. Industrial and Commercial Training 38(3), 116–121 (2006)

A Survey of Academics' Views of the Use of Web2.0 Technologies in Higher Education

Thomas Hainey, Thomas M. Connolly, Gavin J. Baxter, and Carole Gould

School of Computing, University of the West of Scotland, Paisley, Renfrewshire, Scotland
{thomas.hainey,thomas.connolly,gavin.baxter,
carole.gould}@uws.ac.uk

Abstract. Communication is no longer inhibited by boundaries. Communities of like minded people can form over countries and continents. Social media and collaborative technologies [Web2.0] have altered the social landscape, allowing students to collaborate, to be reflective and to participate in peer-to-peer learning. This paper presents the first empirical data gathered as part of a research study into the use of Web2.0 technologies in education. The results demonstrate that there are a number of barriers to the implementation of Web2.0 technologies, primarily lack of knowledge, lack of time and lack of institutional support. In addition, the results also demonstrate that many educators are still unsure 'what' Web2.0 really is and how it can be used effectively to support teaching. This correlates with the literature review carried out as part of this research.

Keywords: Empirical, Web2.0, collaboration, peer-to-peer learning, knowledge sharing, wikis, blogs.

1 Introduction

Education has evolved in recent years and is moving away from the traditional 'chalk and talk' environment to one where students are encouraged to collaborate, to be reflective and to participate in peer-to-peer learning. Social media and technology have altered the social landscape and the way in which 'participation' is understood. Boundaries no longer prohibit individuals from communicating with each other and communities of like minded people from America to Australia can discuss subjects of mutual interest through the utilisation of these technologies. The young have embraced these technologies and methods of social interaction; however educators have not found it easy to exploit these new participatory technologies, in particular Web2.0, to support learning [1]. This paper presents primary data collected as part of a doctoral study that looks to examine the use of Web2.0 technologies for learning and teaching. The following section will present a short literature review on Web2.0 in education, discuss the methodology and methods of data collection and then present the results of a survey of educators' views on Web2.0 and will finish with a discussion.

Á. Herrero et al. (eds.), *International Joint Conference SOCO'13-CISIS'13-ICEUTE'13*,
Advances in Intelligent Systems and Computing 239,
DOI: 10.1007/978-3-319-01854-6_70, © Springer International Publishing Switzerland 2014

2   Literature Review

Crook et al. [2] believe there is a need for further research into the use and impact of Web2.0 in education. Grosseck [3] argues that although the term 'Web2.0' has become a 'buzz' word within education, very few educators truly understand what Web2.0 really means. There are many definitions of Web2.0 and they have been defined from many different perspectives. Kear et al. [4] describe Web 2.0 as "*social software or social media, including online communication tools such as wikis, blogs, and a range of different types of social networking sites...*" while Abram [5] considers Web2.0 to be "*... about conversations, interpersonal networking, personalisation, and individualism*". Individuals may have different experiences of Web2.0. Some may have used it for educational purposes, some for social or professional purposes or for organisational purposes [6]. Web2.0 technologies offer educators an opportunity to change the way in which they teach. These technologies are attractive to education as they allow greater student independence and autonomy, increased collaboration and an increase in pedagogic efficiency [7]. Casquero et al. [8] propose that Web2.0 technologies are having a significant impact on society and as such can be beneficial in the manner in which educators teach and students learn. Web2.0 is a term that encompasses a selection of technologies that can be used for different purposes including wikis, blogs, podcasting, vodcasting, group forums, and social networks.

2.1   Previous Studies

A study carried out by Mumtaz [9] highlights a number of factors that prevent teachers form utilising technology: (1) lack of teaching experience with ICT; (2) lack of onsite support for teachers using technology; (3) lack of ICT specialist teachers to teach students computer skills; (4) lack of computer availability; (5) lack of time required to successfully integrate technology into the curriculum; (6) lack of financial support. Although this study was carried out a number of years ago, the empirical data presented in this paper will show that many of the same issues exist today. Hu et al. [10] state that there is still considerable resistance to technology use among educators and technology acceptance remains a critical challenge for educational administrators and technology advocates.

   Churchill et al. [11] state that although Web2.0 technologies have the capacity to be effectively applied in teaching and learning, integrating these technologies into education is often slow and met with resistance. They also argue that educators utilise resources on a daily basis for lesson planning, student support and delivery of teaching; however, there are few strategies in place to allow these resources to be reused and they are often buried within learning management systems and only the educator who placed them there knows of their existence. Bonk et al. [12] reported on a three-part study where graduate students created wikibooks across an institutional setting. According to these authors collaborative and participatory learning has been increasing for the last two decades and that wiki-related projects present opportunities for learning transformation and critical reflection when students are exposed to new points of view or perspectives. The collaborative nature of these technologies engage learners as they receive feedback from each other and informal learning overtakes formal learning [12]. Their study revealed five main issues: (1) instructional issues; (2) collaboration issues; (3) wikibooks issues; (4) knowledge construction and sense of community issues; (5) technology issues. For the purpose of this paper, only the issues faced by the educators will be presented.

A study carried out by Fisher and Baird [13] examined how course design and social software technologies provided social and collaborative learning opportunities for online students. This study was carried out at Pepperdine University in the USA and only focused on courses that were delivered online. These researchers used a constructivist based course design and integrated participatory media tools enabling the students to support their own learning and form a community of practice. The results of the study were positive. Utilising the online forum helped students feel known, and helped to build a sense of community and accountability with other students. Students posted resources that were timely, relevant, primary, comprehensive, visual and educational in nature. The next section will discuss the methodology adopted for this research project followed by the analysis of the empirical data.

The European project Web2.0ERC aimed to provide teachers with a simple to use Web 2.0 platform, a supporting pedagogy for Web 2.0 and a training package for teachers and teacher trainers. The pedagogical model is discussed in Baxter et al. [6], the implementation framework in Baxter et al. [15], an evaluation of a large-scale pilot of the use of Web 2.0 in education that uses this pedagogy in Connolly et al [14, 16]. The project piloted the Web2.0 platform with 227 teachers and 710 students using a pre-test/post-test methodology. The results indicated that the teachers generally enjoyed the use of the platform and found the Web 2.0 tools easy to use. The study found that teachers were most proficient at using: YouTube, Facebook, Blogs, GoogleDocs and Wikis. Teachers were least proficient at using Twitter, Podcast and ePortfolios. Teachers considered YouTube, Blogs, Wikis, Social bookmarking and GoogleDocs to be more useful in an educational setting and Flickr, Online collaborative games and Twitter to be the least useful tools. In contrast, students rated their proficiency as higher than teachers in relation to every Web 2.0 tool and teachers found all Web 2.0 significantly more useful than students.

3   Methodology

This study adopts a case study methodology utilising survey and interview methods of data collection. The first stage of this research project was to answer the research question, "What views of the use/value of Web2.0 technologies do educators have? Do educators understand the concept of Web2.0?" In addition to this, it was important to identify the barriers, if any, to Web2.0 adoption. To achieve this, educators from the University of the West of Scotland completed an online survey. The survey was designed to collect some basic demographic data (gender, age, university campus and school) as well as their current use of technology in general and Web2.0 technologies from both a personal and professional perspective. The survey received 61 responses and the results were statistically analysed utilising SPSS. The next section presents the results of the educators' survey.

4   Analysis

61 participants completed the Web 2.0 technologies for learning and teaching survey returning an 18% representation of the population. 38 participants (62%) were male and 23 participants (38%) were female which correlates with the gender split within the university. The mean age of participants was 47.38 years (SD = 8.60) with a range of 24.5 to 61 years. A Mann-Whitney U test indicated that there was no significance difference in age between males and females (Z=-1.426, p < 0.154). 60 participants (98%) specified what school they were in. 6 participants (10%) were from the Business School, 6 (10%) were from Creative and Cultural Industries, 4 (7%) were from the School of Education, 4 (7%) were from Engineering, 13 (22%) were from the School of Health, Nursing and Midwifery, 11 (18%) were from the School of Computing, 7 (12%) were from the School of Science and 9 (15%) were from the School of Social Science. 1 participant (2%) did not specify what School they were from. 11 participants (18%) were from the Ayr campus, 2 participants (3%) were from the Dumfries campus, 9 participants (15%) were from the Hamilton campus and 37 participants (62%) were from the Paisley campus.

Participants were asked to provide their own definition of Web2.0. Of the 61 participants, 23 (38%) provided their own definition. Responses included: "Web2.0 technologies are used for communication, sharing and proliferation of information between groups"; Online collaboration and sharing tools". It could be argued from the above that those who gave a definition understand the collaborative and sharing nature of Web2.0 tools. Participants were asked to identify what technologies they believed to be Web2.0. Social networking (e.g. Facebook) (77%) was ranked 1st, wikis (75%) were ranked 2nd and blogs (72%) ranked 3rd. However, discussion boards (25%) were ranked 14th showing that some of the participants were still unsure what a Web2.0 technology actually is. This is further reinforced as 5% of the participants believed that Microsoft Publisher was a Web2.0 technology. Participants were also asked to identify the technologies they used within a personal context. Not surprisingly email and text messaging were ranked 1st and 2nd respectively. Table 1 shows the ranking of the top technologies identified.

A Mann-Whitney U test indicates that there is no significant difference in technology usage in a personal context in relation to gender, Table 2 shows the ranking of the main uses of technology within a personal context split by gender. Both males and females rank email and text messaging 1st and 2nd respectively. It should also be noted that there is very little difference in ranking between the least used technologies with males ranking ePortfolios 14th and females ranking ePortfolios 12th.

The most commonly used technologies within a professional context are email, YouTube and video conferencing with game-based learning, social bookmarking and photo sharing being the least utilised technologies within a professional context. Mann-Whitney U tests indicated there was no significant difference in the usage of technology in a professional context in relation to gender. Table 3 indicates the main technology usage in a professional context split by gender. Table 3 highlights that both males and females ranked email and YouTube as the top two most popular technologies for professional purposes, whereas males ranked discussion boards as being the 3rd most popular technology for professional purposes and females ranked discussion boards as 7th. Another significant difference between males and females is the ranking of ePortfolios. ePortfolios were one of the least popular technologies amongst males ranking them as 12th with females giving them a higher ranking of 4th.

Table 1. Technology use within a personal context

Technology	Rank	Mean	SD
Email	1st	4.75	0.54
Text Messaging	2nd	4.55	0.86
YouTube	3rd	3.41	1.51
Social Networking	4th	3.37	1.67
Blogs	5th	2.69	1.47
Video Conferencing	5th	2.69	1.49
Mobile Learning	6th	2.67	1.69
GoogleDocs	7th	2.64	1.60
Discussion Boards	8th	2.55	1.45
Photo Sharing	9th	2.43	1.56
Podcasts	10th	2.41	1.37

Table 2. Use of technologies within a personal context split by gender

Gender	Male			Female		
Technology	Rank	Mean	SD	Rank	Mean	SD
Email	1st	4.66	0.63	1st	4.91	0.29
Text Messaging	2nd	4.47	0.91	2nd	4.68	0.78
Social Networking	3rd	3.43	1.71	4th	3.27	1.64
YouTube	4th	3.33	1.57	3rd	3.55	1.44
Blog	5th	2.86	1.51	8th	2.41	1.37
Discussion Boards	6th	2.67	1.49	9th	2.36	1.40
Video Conference	7th	2.65	1.49	9th	2.36	1.40
Mobile Learning	7th	2.65	1.67	6th	2.71	1.76
Google Docs	8th	2.59	1.57	6th	2.71	1.68
Podcasts	9th	2.49	1.37	5th	2.76	1.51
Twitter	9th	2.49	1.80	10th	2.10	1.48
Photo Sharing	10th	2.35	1.53	7th	2.57	1.63

Participants were also asked which technologies they utilised for teaching purposes. Email, YouTube and discussion boards were ranked 1st, 2nd, and 3rd respectively. Table 4 shows technology use for teaching.

Table 3. Use of technology within a professional context split by gender

Gender	Male			Female		
Technology	**Rank**	**Mean**	**SD**	**Rank**	**Mean**	**SD**
Email	1st	4.87	0.47	1st	5.00	0.00
YouTube	2nd	3.36	1.38	2nd	3.76	1.48
Discussion Board	3rd	3.35	1.38	7th	2.75	1.16
Blog	4th	3.29	1.45	9th	2.60	1.43
Video Conference	5th	3.27	1.45	3rd	3.38	1.32
Text Messaging	6th	3.19	1.54	8th	2.65	1.53
Wiki	7th	3.11	1.37	10th	2.55	1.43
Podcasts	8th	2.83	1.50	6th	2.84	1.42
Google Docs	9th	2.72	1.63	5th	2.85	1.53
Social Networking	10th	2.50	1.54	12th	2.05	1.54
Twitter	11th	2.19	1.61	13th	2.05	1.54
ePortfolios	12th	2.14	1.29	4th	3.14	1.71

Table 4. Technology use for teaching

Technology	Rank	Mean	SD
Email	1st	4.59	1.05
YouTube	2nd	3.46	1.48
Discussion Boards	3rd	2.79	1.51
Video Conference	4th	2.61	1.47
Blogs	5th	2.57	1.57
GoogleDocs	6th	2.43	1.51
ePortfolio	7th	2.37	1.58
Podcasts	8th	2.23	1.41
Wikis	9th	2.19	1.41
Social Networking	10th	1.93	1.45

Table 5. Technology use within a teaching context split by gender

Gender	Male			Female		
Technology	Rank	Mean	SD	Rank	Mean	SD
Email	1st	4.81	0.70	1st	4.23	1.41
YouTube	2nd	3.33	1.51	2nd	3.67	1.43
Blogs	3rd	2.91	1.58	8th	2.00	1.41
Discussion Boards	4th	2.84	1.55	3rd	2.71	1.45
Video Conferencing	5th	2.58	1.48	4th	2.67	1.49
Google Docs	6th	2.33	1.53	5th	2.60	1.50
Wiki	7th	2.28	1.43	7th	2.05	1.40
Podcasts	7th	2.28	1.49	6th	2.14	1.31
ePortfolio	8th	2.19	1.62	4th	2.67	1.49

As well as identifying the technologies participants used for teaching purposes they were also asked to explain how and why they used those particular technologies. Some examples are: "Primarily for information dissemination and knowledge sharing"; "To point students to additional materials (e.g. Wikipedia and YouTube)"; "To promote learning beyond the classroom"; "To supplement lectures"

It would appear from the responses that the technologies are primarily used to support teaching and communication. This would correlate with the most commonly used technologies for teaching (see Table 4), where email and YouTube are the most commonly used technologies. Mann-Whitney U tests indicated there was no significant difference in the usage of technology for teaching in relation to gender. Table 5 shows technology usage within a teaching context split by gender.

Both males and females ranked email and YouTube 1st and 2nd respectively, however males' ranked blogs as the 3rd most commonly used tool with females ranking blogs 8th. Interestingly, females also ranked ePortfolios as the 4th most popular technology for teaching. It could be argued that females appreciated the benefits of ePortfolios for both professional and teaching purposes more than males. Discussion boards were more or less evenly ranked between males (4th) and females (3rd) for teaching purposes. This correlates with discussion boards being ranked the 3rd most popular technology for teaching purposes (see Table 4).

58 participants out of the 61 participants completed the question regarding to what extent they agree that the introduction of Web2.0 would facilitate communication and collaboration between students. 19 respondents (32%) strongly agreed, 13 respondents (22%) agreed, 8 respondents (14%) were neutral, 2 respondents (3%) disagreed, 1 respondent (2%) strongly disagreed and 16 respondents (27%) did not know if Web2.0 tools would facilitate communication between students. Although a high number strongly agreed, it should be noted that a large number (16) did not know. This could suggest that educators are still unsure of how to apply these technologies to their full advantage and there is still a lack of understanding as to how these technologies actually work.

Of the 61 respondents, 58 answered the question "the introduction of Web2.0 tools would facilitate communication between lecturers and students?" Of those respondents 16 (28%) did not know if Web2.0 tools would facilitate communication between students and lecturers, 1 (2%) respondent strongly disagreed, 2 (3%) disagreed, 7 (12%) were neutral, 18 (31%) agreed and 14 (24%) strongly agreed that Web2.0 tools would facilitate communication between students and lecturers. Interestingly these results correlate with the previous question (see Table 6).

Table 6. Comparison Table

Would Web2.0 tools facilitate communication between students		Would Web2.0 tools facilitate communication between students and lecturers	
Strongly agree	19 (32%)	Strongly agree	14 (24%)
Agree	13 (22%)	Agree	18 (31%)
Neutral	8 (14%)	Neutral	7 (12%)
Disagree	2 (3%)	Disagree	2 (3%)
Strongly disagree	1 (2%)	Strongly disagree	1 (2%)
Don't know	16 (27%)	Don't know	16 (28%)

In addition, 30 of the 61 respondents answered the question "please explain how you think Web2.0 would facilitate communication and collaboration between students?" Some of the responses included: "Web2.0 represents a useful set of tools for communication and collaboration"; "Students could be encouraged to conduct group activities"; "Makes communication more dynamic outside the classroom". Of those who answered the question, the results suggest that educators appreciate the collaborative benefits of Web2.0 technologies, therefore it could be concluded that they could deliver a more open, collaborative, and peer review approach to teaching. However, it should be noted that only half of the respondents answered this question suggesting that the other respondents were unsure of how Web2.0 could facilitate communication and were not aware of the collaborative nature of these types of technologies.

Of the 61 participants, 58 answered the question "Web2.0 could be used to support traditional teaching methods (face-to-face, lectures, courseworks, etc)". Of those respondents 18 (31%) strongly agreed, 16 (28%) agreed, 5 (9%) were neutral, 2 (3%) disagreed, 1 (2%) strongly disagreed, and 16 (28%) did not know. It should be noted, as with all new technologies and methods there is always resistance and this could be partly due to lack of understanding, knowledge and skills. This is reinforced when participants answered the question "in your teaching, what do you anticipate to be the most significant drawbacks to the utilisation of Web2.0 technologies?" the most common responses included: lack of knowledge, lack of institutional support, ICT infrastructure, inexperience, lack of training, availability of technology, fear, time, understanding of pedagogy of Web2.0, and lack of methodology. Participants were also asked to identify which Web2.0 tools they believed to be the most useful in term of teaching. The most common tools included: YouTube, blogs, wikis, podcasts, GoogleDocs, forums, social bookmarking, and Facebook.

As the next stage of this research study is to design a pedagogical model, participants were asked "do you think a pedagogical model/framework designed specifically for Web2.0 would help you to introduce Web2.0 technologies into your teaching". Of the 61 participants, 53 answered the question. 31 (59%) respondents agreed that a pedagogical model would help, while 22 (41%) did not think a pedagogical model would help. In addition to this, participants were asked to identify the steps/procedures they would go through to introduce Web2.0 to support their teaching. The most common response was "don't know". This would suggest that some kind of pedagogical and implementation model is needed, as well as further research.

5   Discussion

From the empirical data gathered for this study, it could be argued that there are mixed views with regard to Web2.0 technology and much of this could be attributed to a lack of understanding and knowledge of these tools. Educators are unsure of how to apply these collaborative technologies to achieve the desired outcomes and how to integrate them to support their teaching. The data gathered correlates with the literature presented as part of this study, for example, the literature gives examples of barriers to adoption of technology: lack of knowledge, time and lack of institutional support, all of which where identified as barriers by participants of this research project.

One of the primary weaknesses of a case study methodology is the lack of generalisations that can be made, unless other readers/researchers can identify their application. However, as the respondents represented an 18% response rate of the population, the results presented in this paper may be generalised. It should be noted that case study generally gives a richer description and unique example of the phenomenon within a real life situation.

6   Summary

This research project is not advocating the abolishment of face-to-face teaching but rather utilising these technologies to encourage student engagement, student collaboration and peer to peer review as well as reflection. The next stage of this research project is to interview a number of the participants to get a deeper, richer description of their views and understanding of Web2.0 technologies. In addition to this, participants will be required to adopt a Web2.0 technology to support their teaching and utilise a pedagogical model to aid implementation. The intention is to re-interview participants and gather data to ascertain if their views of Web2.0 tools have changed and to gather feedback on the pedagogical model and make amendments where required. The latter part of this study will also involve surveying those students who used the Web2.0 technologies. This will help to give a more balanced view of the Web2.0 technology used.

Acknowledgements. This work has been co-funded by the EU Lifelong Learning Programme under contract 519057-LLP-1-2011-1-UK-KA3-KA3NW (Ed2.0Work – European Network for the integration of Web2.0 in education and work).

References

1. Lewis, S., Pea, R., Rosen, J.: Beyond participation to co-creation of meaning: mobile social media in generative learning communities. Social Science Information 49(3), 351–369 (2011)
2. Crook, C., Cummings, J., Fisher, T., Graber, R., Harrison, C., Lewin, C., Logan, K., Luckin, R., Oliver, M., Sharples, M.: Web 2.0 technologies for learning: The current landscape - opportunities, challenges, and tensions, Learning Sciences Research Institute, University of Nottingham (2008)
3. Grosseck, G.: To use or not to use web 2.0 in higher education? Procedia - Social and Behavioral Sciences 1(1), 478–482 (2009)
4. Kear, K., Woodthorpe, J., Robertson, S., Hutchison, M.: From forums to wikis: Perspectives on tools for collaboration. Internet and Higher Education 13, 218–225 (2010)
5. Abram, S.: Web 2.0—huh?! Library 2.0, librarian, 2.0. Information Outlook 9(12), 44–46 (2005)
6. Baxter, G.J., Connolly, T.M., Stansfield, M.H., Gould, C., Tsvetkova, N., Kusheva, R., Stoimenova, B., Penkova, R., Legurska, M., Dimitrova, N.: Understanding the pedagogy Web 2.0 Supports: The presentation of a Web 2.0 pegagogical model. In: Proceedings of International Conference on European Transnational Education, Salamanca Spain (2011)

7. Franklin, T., Harmelen, M.: Web 2.0 for content for learning and teaching in higher education (2007), http://ie-repository.jisc.ac.uk/148/1/web2-content-learning-and-teaching.pdf (accessed August 8, 2012)

8. Casquero, O., Portilloa, J., Ovelar, R., Benito, M.: iPLE network: An integrated eLearning 2.0 architecture from a university's perspective. Interactive Learning Environments 18(3), 293–308 (2010)

9. Mumtaz, S.: Factors affecting teachers' use of information and communications technology: A review of the literature. Journal of Information Technology for Teacher Education 9(3), 319–342 (2000)

10. Hu, P., Clark, T., Ma, W.: Examining technology acceptance by school teachers: a longitudinal study. Information and Management 41, 227–241 (2003)

11. Churchill, D., Wong, W., Law, N., Slater, D., Tai, B.: Social bookmarking-repository-networking: Possibilities for support of teaching and learning in higher education. Serials Review 35(3), 142–148 (2009)

12. Bonk, C.J., Lee, M.M., Kim, N., Lin, M.G.: The tensions of transformation in three cross-institutional wikibooks projects. Internet and Higher Education 12, 126–135 (2009)

13. Fisher, M., Baird, D.E.: Online learning design that fosters student support, self-regulation and retention. Campus Wide Information Systems 22(2), 88–107 (2006)

14. Connolly, T.M., Hainey, T., Baxter, G.J., Tsvetkova, N., Kusheva, R., Stoimenova, B., Penkova, R., Legurska, M., Dimitrova, N.: Web 2.0 Education: An evaluation of a large-scale European pilot. In: Proceedings of International Conference on European Transnational Education (ICEUTE), Salamanca, Spain, October 20-21 (2011)

15. Baxter, G.J., Connolly, T.M., Stansfield, M.H., Tsvetkova, N., Kusheva, R., Stoimenova, B.: Introducing Web 2.0 in Education: A Structured Approach Adopting a Web 2.0 Implementation Framework. In: Proceedings of International Conference on European Transnational Education, Salamanca Spain, October 20-21 (2011)

16. Connolly, T.M., Hainey, T., Baxter, G.J., Stansfield, M.H., Gould, C., Can, C., Bedir, H., Inozu, J., Tsvetkova, N., Kusheva, R., Stoimenova, B., Penkova, R., Legurska, M., Dimitrova, N.: Teachers' Views on Web 2.0 in Education: An evaluation of a large-scale European pilot. In: Proceedings of International Conference on European Transnational Education, Salamanca Spain, October 20-21 (2011)

Are Web2.0 Tools Used to Support Learning in Communities of Practice

A Survey of Three Online Communities of Practice

Ashley Healy, Thomas Hainey, Thomas M. Connolly, and Gavin J. Baxter

School of Computing, University of the West of Scotland, Paisley, Renfrewshire, Scotland
{ashley.healy,thomas.hainey,thomas.connolly,
gavin.baxter}@uws.ac.uk

Abstract. Communities of Practice (CoPs) are a group of likeminded practitioners coming together by means of common ground to converse and share knowledge and experience on ways of working within their sector, thus, facilitating knowledge transfer, professional development and learning. While not a new concept they have emerged as a key domain in the realm of knowledge creation and are often cited as facilitating organisational learning and knowledge creation. However, despite the amount of literature available, there appears to be a lack of empirical evidence confirming whether learning takes place within a community. This paper reports on a three-year study based on 3 CoPs, one in industry, one in health and one in education. In particular, the paper reports on the members views' of their participation in their CoP at the start of the study and at the end of the study.

Keywords: CoP, Web2.0, organisational learning.

1 Introduction

Communities of Practice (CoPs) are characteristically a group of people who are brought together by a common passion, they not only share a common interest, but they endeavour to report on their activities and collaborate in the practices of the community, which reinforces learning and potentially enhances their professional performance. CoPs are organic entities, which evolve as a result of the passion people have for a given domain. Consequently, as the people interact and engage, the process of collective social learning becomes a reality. CoPs offer a theory of learning that begins by assuming that social engagement is paramount to the learning process.

CoPs are not a new phenomenon, they have been in existence for centuries and the principles of communities are age-old. However, the recent coinage by Wenger [1] has sparked awareness and debate on the potential of CoPs amongst educationalists and industry specialists. The debate has resulted in a stronger focus being placed on the potential knowledge development and learning within CoPs.

Despite the amount of literature available, there appears to be a lack of empirical evidence indicating whether learning takes place within a community. There are many

Á. Herrero et al. (eds.), *International Joint Conference SOCO'13-CISIS'13-ICEUTE'13*,
Advances in Intelligent Systems and Computing 239,
DOI: 10.1007/978-3-319-01854-6_71, © Springer International Publishing Switzerland 2014

case studies referring to CoPs or communities in general and learning. For example, The Claims Processor study [1] and studies that refer to CoPs in the work place such as the study of School Teachers' Workplace Learning [2] as well as many references to organisations that have applied the concept of CoP for improving knowledge management and project management such as Shell [3] and IBM [4]. However, these do not appear to have been empirically tested. Much of the literature merely implies or assumes that learning occurs without actually carrying out any formal exploration or investigations. This paper reports on a three-year study based on 3 CoPs, one in industry, one in health and one in education. In particular, the paper reports on the members views' of their participation in their CoP at the start of the study and at the end of the study.

2   Previous Research

CoPs have been defined as *"groups of people who share a concern, a set of problems, or a passion about a topic, and who deepen their knowledge and expertise in this area by interacting on an ongoing basis"* [5, p. 4] and as *"collaborative, informal networks that support professional practitioners in their efforts to develop shared understandings and engage in work-relevant knowledge building"* [6]. For the purposes of this paper, we define CoPs as *"a group of likeminded practitioners coming together by means of common ground to converse and share knowledge and experience on ways of working within their sector, thus, facilitating knowledge transfer, professional development and learning"*.

Wenger [1] argues that not all social configurations or communities can be labelled a CoP, as this would render the concept ineffective. There are however, characteristics associated with CoPs and these prove useful parameters for classifying a community as a CoP. Briefly, the three characteristics are; the domain, the practice and the community. The domain refers to the area of concern and is what defines the community. It is the genesis of the community and is responsible for bringing people together to form the community. The community refers to the bonds and relationships that are established within the community. These bonds create trust, which facilitates collaboration, engagement and interaction between community members. Finally, the practice denotes the shared repertoire of common resources and experiences, which are progressively developed and shared by the individuals belonging to the CoP. Schwier et al. [7] also identifies a combination of characteristics inherent in the composition of CoP including historicity, identity, plurality, autonomy, participation, integration, future, technology, learning and mutuality.

2.1   Organisational CoPs

Theorists now see CoPs as an essential component of the knowledge-based view of the organisation [8, 10]. It has been recognised that knowledge has become integral to organisation structure and CoP have emerged as a key domain in the realm of knowledge creation and are often cited as facilitating organisational learning and knowledge creation [9]. However it is important to realise that CoPs are not a one size fits all organisations concept [4] as each organisation has unique knowledge requirements as well as individual strengths, and challenges. CoPs provide those concerned with knowledge management, organisational learning and professional development a model for thinking about how to manage and facilitate learning. Easterby-Smith et al. [11] note that the idea of organisational learning has been present in the management literature for decades, but it has only become widely recognised since around 1990; as a result, the learning organisation was formed. CoP are said to complement existing structures and radically galvanize knowledge sharing, learning and change [12]. However, many would disagree suggesting instead that in fact CoPs fail to take into account existing structures. The concept has attracted criticism for not taking into account preexisting conditions, such as habitus and social codes. Mutch [13] argued that social and educational origins, along with tacit knowledge of member's leads to the need to renegotiate new habituses, this is in stark contrast to the suggestion of blending and complementing existing habitus.

CoPs effect organisational practice and productivity and can lead to innovation and revitalised business strategy. Wenger and Snyder [12] highlight a number of ways in which CoPs add value to organisational practice: (1) Generate knowledge and encourage skill development; (2) Use knowledge management to drive strategy; (3) Disseminate valuable information and transfer best practice; (4) Initiate new lines of business including new products and services. CoPs can also positively affect organisational productivity [21].

3   Methodology

The selected case study communities comprise: Industry based CoP, a Health Service based CoP and an Educational based CoP, as described below.

3.1   Industry Based CoP

This organisation has a wide range of CoPs dealing with domains such as Energy, Business Development, Construction and Equal Opportunities. It has 13 active communities that have been developed over the past four years. The purpose of these communities is to maximize the resources available within the public sector to promote and support a number of different industries such as Energy or Construction. The community participating in this study is the Account Manager CoP. This community consists of individuals charged with working with companies across Scotland with high growth aspirations, which will have a positive impact on GDP. Interestingly, this organisation conduct their own annual reviews of the communities in order to assess the progress made, the impact the community has had on the industry, to determine areas for improvement, learning and best practice as well as suggesting recommendations for developing the CoP further. These findings will also feed into the research as they may be indicative of the relationship between the potential for learning and the community. The Industry CoP was established in 2006. The community has 37 members, of which 16 agreed to participate in the research. 39% of the participants were female and 61% were male. This community participated in the research from 2006 – 2012.

3.2   Health Service Based CoP

This CoP is a newly established community, established in late 2009, and opened up to membership early 2012. The community arose as a result of the landmark publication of Scotland's Dementia Strategy (June 2010), as Scotland currently has a wealth of existing and emerging practice in the care and treatment of dementia and it was essential for the sector to find an effective way to share this experience and knowledge. The role of the CoP is to support the development, testing, refining and sharing of (Scottish) nationally endorsed information, guidance and practice in the care and treatment of dementia. The aims of the CoP include improving practice and approaches to dementia care and treatment. This community has a strong belief in the dedication of members to bring about improved service and practice within the provision of dementia care and treatment; dedication is a core characteristic of CoP, thus making it a good example of a CoP. This community was established in 2010. The health CoP has 194 members, of which 20 agreed to participate in the research. 77% of the community were female and 23% were male. This community participated in the community from 2010 – 2012.

3.3   Educational Based CoP

This particular organisation has a broad range of CoPs across a variety of disciplines such as Business Development, Marketing and Librarianship amongst others. These CoPs are in existence in order to advance professional development of staff across Scotland's colleges. The CoP involved in this research is the Business Development Community. This particular community was established in 1997 and evolved from an informal group of commercial directors. Steering group meetings plan and organise the agenda for the conferences and events, which are held during the year. The community seeks to capitalise on the commercial expertise that is present within Scotland's colleges for the benefit of the whole sector. The key purpose is to support the business and commercial elements of Scotland's college sector to grow. The CoP aims to assist its members by: providing an online platform for discussion of issues of common concern; assisting members to interpret legislation that could impact on business activities; and building relationships with Local Education Councils. This CoP was of particular interest as the domain is one which is of growing interest within the academic field, thus it is a topical concern. The Education CoP was established in 1997. This community has 40 members, of which agreed to participate in the research. Of the 18 participating members, 67% were female and 33% were male. This community participated in the research from 2006 – 2012.

4   Results

This sections discusses the results from a pre-test/post-test questionnaire of the three CoPs defined in the previous section completed the CoP questionnaire to determine thoughts and opinions on CoP in general and to evaluate the CoP to gain an understanding of the effectiveness (or otherwise) of CoP for personal development and learning.

4.1   Pre-test Questionnaire Analysis

Industrial CoP
16 participants from the industrial CoP completed the CoP questionnaire. Table 1 shows the percentages and frequencies for the questions asked. The responses were gathered using a five point Likert scale with the following options: strongly agree, agree, neither agree nor disagree, disagree and strongly disagree. Participants were asked the question of whether they belonged to any other online or face-to-face communities. 5 participants from the CoP (33%) indicated that they were and 10 participants from the CoP (67%) indicated that they were not.

Table 1. Question responses, frequencies and percentages for Industrial CoP (red highlights the highest frequencies and percentages per question)

Response	Strongly agree		Agree		Neither agree nor disagree		Disagree		Strongly disagree	
Question	N	%	N	%	N	%	N	%	N	%
Your expectations are being fulfilled	0	0	10	62.5	3	19	2	12.5	1	6
You are confident participating in the community interactions & discussions	3	20	11	73	1	7	0	0	0	0
Your involvement in the community helps you with other aspects of your job.	2	12.5	12	75	1	6.25	1	6.25	0	0
You have learned or gained knowledge from the community and the members	3	19	11	69	1	6	1	6	0	0
CoP are useful for learning about your industry and related matters	1	6	11	69	3	19	1	6	0	0
CoP are suitable environments for developing new skills	1	6.25	9	56.25	4	25	2	12.5	0	0
The technological tools available to COP are useful for encouraging participation	1	7	6	40	5	33	3	20	0	0
Participating in the community has improved your technical skills	0	0	3	18.75	7	44.75	6	38	0	0
CoP would be suitable for other work/subject areas	1	6	7	44	8	50	0	0	0	0
Members are supported & encouraged to participate in the community	1	6	11	69	3	19	1	6	0	0

Health Sector CoP

20 participants from the health sector CoP completed the CoP questionnaire. Table 2 shows the percentages and frequencies for the questions asked. The responses were gathered using a five point Likert scale with the following options: strongly agree, agree, neither agree nor disagree, disagree and strongly disagree. Participants were asked the question of whether they belonged to any other online or face-to-face communities. 11 participants from the CoP (55%) indicated that they were and 9 participants from the CoP (45%) indicated that they were not.

Educational CoP

18 participants from the educational CoP completed the CoP questionnaire. Table 3 shows the percentages and frequencies for the questions asked. The responses were gathered using a five point Likert scale with the following options: strongly agree, agree, neither agree nor disagree, disagree and strongly disagree. Participants were asked the question of whether they belonged to any other online or face-to-face communities. 6 participants from the CoP (33%) indicated that they were and 12 participants from the CoP (67%) indicated that they were not.

Table 2. Question responses, frequencies and percentages for Health Sector CoP

Response	Strongly agree		Agree		Neither agree nor disagree		Disagree		Strongly disagree	
Question	N	%	N	%	N	%	N	%	N	%
Your expectations are being fulfilled	17	85	1	5	2	10	0	0	0	0
You are confident participating in the community interactions & discussions	9	45	10	50	1	5	0	0	0	0
Your involvement in the community helps you with other aspects of your job.	7	35	12	60	1	5	0	0	0	0
You have learned or gained knowledge from the community and the members	9	45	10	50	1	5	0	0	0	0
CoP are useful for learning about your industry and related matters	6	30	12	60	2	10	0	0	0	0
CoP are suitable environments for developing new skills	4	20	14	70	2	10	0	0	0	0
The technological tools available to COP are useful for encouraging participation	14	74	4	21	1	5	0	0	0	0
Participating in the community has improved your technical skills	7	35	9	45	4	20	0	0	0	0
CoP would be suitable for other work/subject areas	7	35	9	45	4	20	0	0	0	0
Members are supported & encouraged to participate in the community	7	35	11	55	2	10	0	0	0	0

Comparison of the Three CoPs

Kruskal-Wallis tests indicated that there were significant differences between the three CoPs in terms of the following aspects: expectations as a member being met and fulfilled ($\chi 2$ = 26.153, p < 0.000), confidence in participating in the community interactions and discussions ($\chi 2$ = 11.186, p < 0.004), the CoP being a suitable environment for developing new skills ($\chi 2$ = 15.365, p < 0.000), the technological tools available to the community being useful for encouraging participation ($\chi 2$ = 24.480, p < 0.000), participating in the community helping to improve technical skills ($\chi 2$ = 24.383, p < 0.000), the CoP being suitable for other work/subject areas ($\chi 2$ = 7.245, p < 0.027) and supporting and encouraging members to participate in the community ($\chi 2$ = 9.501, p < 0.009).

Table 3. Question responses, frequencies and percentages for educational CoP

Response	Strongly agree		Agree		Neither agree nor disagree		Disagree		Strongly disagree	
Question	N	%	N	%	N	%	N	%	N	%
Your expectations are being fulfilled	1	5.5	10	56	6	33	1	5.5	0	0
You are confident participating in the community interactions & discussions	1	5.5	10	56	6	33	1	5.5	0	0
Your involvement in the community helps you with other aspects of your job.	2	11	13	72	3	17	0	0	0	0
You have learned or gained knowledge from the community and the members	4	25	11	69	1	6	0	0	0	0
CoP are useful for learning about your industry and related matters	3	16.7	14	77.8	1	5.5	0	0	0	0
CoP are suitable environments for developing new skills	1	5.5	3	16.7	12	66.8	2	11	0	0
The technological tools available to COP are useful for encouraging participation	1	5.5	5	28	11	61	1	5.5	0	0
Participating in the community has improved your technical skills	0	0	2	12	8	47	6	35	1	6
CoP would be suitable for other work/subject areas	2	12	6	35	9	53	0	0	0	0
Members are supported & encouraged to participate in the community	1	5.5	9	50	7	39	1	5.5	0	0

In terms of comparing the educational CoP group to the health sector CoP group, the participants in the CoP in the health sector group were significantly more positive in relation to the following aspects: their expectations being fulfilled as a member (Z = -4.331, p < 0.000), their confidence levels participating in the community (Z = -3.119, p < 0.002), believing that CoP were suitable environments for developing new skills ($Z = -3.881$, $p < 0.000$), that the technological tools available to the community were useful for encouraging participation ($Z = -4.472$, $p < 0.000$), that participating in the community had improved their technical skills ($Z = -4.331$, $p < 0.000$), that the CoP would be suitable for other work and subject areas ($Z = -2.214$, $p < 0.027$) and that members were supported and encouraged to participate in the community ($Z = -2.852$, $p < 0.004$).

In terms of comparing the educational CoP group to the industrial CoP group, the industrial CoP group were significantly more confident participating in the community interactions and discussion than the educational CoP group ($Z = -2.058$, $p < 0.040$). There were no other significant differences identified in relation to any of the other aspects. In terms of comparing the health sector CoP group to the industrial CoP group, the health sector CoP group was significantly more positive than the industrial CoP group in relation to the following aspects: there expectations as members being met ($Z = -4.521$, $p < 0.000$), the CoP being useful for learning about industry and related matters ($Z = -1.938$, $p < 0.05$), the CoP being useful for developing new skills ($Z = -2.121$, $p < 0.034$), the technological tools available to the community being useful for encouraging participation ($Z = -4.060$, $p < 0.000$), participating in the community having improved their technical skills ($Z = -3.993$, $p < 0.000$), the CoP being suitable for other work/subject areas ($Z = -2.328$, $p < 0.020$) and members being supported and encouraged to participate in the community ($Z = -2.120m$ $p < 0.034$).

Across all three of the CoP there were no significant differences with regards to any of the aspects in relation to whether a participant was a member of another face-to-face or online community.

4.2   Post-test Questionnaire Analysis

54 participants completed the post-test questionnaire. Participants were asked whether their expectations being met (or not) had any impact on their level of participation within the CoP. 35 participants (65%) stated that it had an impact and 19 participants (35%) stated that it did not. 38 participants (70%) stated that this had a positive impact and 16 participants (30%) stated that this had a negative impact.

Participants were also asked if having their expectations met provided them with a sense of belonging with the CoP. 34 participants (63%) stated that it did and 20 participants (37%) stated that it did not. Participants were also asked if the pre-existing organisational structure/hierarchies had an impact on their participation. 34 participants (63%) stated that it did and 16 participants (37%) stated that it did not. 37 participants (69%) indicated that this had a positive impact and 17 participants (31%) stated that it had a negative impact.

Participants were also asked if the level of support they received from their line manager had an impact on their participation. 36 participants (67%) stated that it did and 18 participants (33%) stated that it did not. 33 participants (61%) stated that this had a positive impact and 21 participants (39%) stated that it had a negative impact. Participants were asked if the CoP had helped them in their job and whether they believed that their performance had improved. 36 participants (67%) stated that it had helped them in their job and their performance had improved and 18 participants (33%) indicated that it had not helped them in their job and their performance had not improved. Participants were asked if the CoP had helped them in their job whether it had allowed them to become more efficient. 30 participants (56%) believed that the CoP allowed them to become more efficient and 24 participants (44%) believed that it had not allowed them to become more efficient.

Participants were asked if their level of participation impacted on their learning within the CoP. 40 participants (74%) stated that it did and 14 participants (26%) stated that it did not. Participants were also asked if participation in the CoP assisted in the Continuing Professional Development (CPD). 32 participants (59%) stated that it did and 22 participants (41%) stated that it did not. Table 4 shows rankings of the ways that participants could improve their CPD by participation in the CoP.

Table 4. Ways of improving CPD by participation in the CoP

Method	Rank	N/%
Improving/broadening Knowledge	1st	29 (54%)
Gaining experiences	2nd	14 (26%)
Developing skills	3rd	11 (20%)

Participants were asked if online functionality impacted on the level of participation. 32 participants (59%) stated that it did and 22 participants (41%) stated that it did not. 25 participants (46%) stated that this had a positive effect, 11 participants (20) stated that this had a negative effect and 18 participants (33%) did not answer the question. Participants were asked what particular tools from the online site they used most frequently with Document Library ranked first with 36 (67%) then email with 29 (54%), member directory 14 (26%), forum 9 (17%), blog 2 (4%) and wiki with 0 users.

5 Summary

This paper has examined the importance of Communities of Practice (CoPs) within organisational settings. The paper has reported on a study that was carried out between 2006 and 2012 with three CoPs: Industry based CoP, a Health Service based CoP and an Educational based CoP and the views of the members of the CoPs before and at the end of the study. The study found were significant differences between the three CoPs in terms of the following aspects: expectations as a member being met and fulfilled, confidence in participating in the community interactions and discussions, the CoP being a suitable environment for developing new skills, the technological tools available to the community being useful for encouraging participation, participating in the community helping to improve technical skills, the CoP being suitable for other work/subject areas and supporting and encouraging members to participate in the community. At the end of the study, 67% of participants thought the CoP had helped them in their job and that their performance had improved; 74% of participants felt their level of participation impacted on their learning within the CoP. In terms of CPD, participants thought the CoP had increased/broadened their knowledge, provided experience and developed their skills. Surprisingly, Web2.0 tools did not feature highly in any of the CoPs.

Acknowledgements. This work has been co-funded by the EU Lifelong Learning Programme under contract 519057-LLP-1-2011-1-UK-KA3-KA3NW (Ed2.0Work – European Network for the integration of Web2.0 in education and work).

References

1. Wenger, E.: Communities of Practice: Learning, Meaning, and Identity. Cambridge University Press (1998)
2. Hodkinson, H., Hodkinson, P.: Learning In Different Communities of Practice: A case study of English secondary school teachers. In: 3rd International Conference of Researching Work and Learning, vol. 3, pp. 40–49 (2003)
3. McDermott, R.: Learning across teams: How to build communities of practice in team organizations. Knowledge Management Journal 8, 32–36 (1999)
4. Gongla, C., Rizzuto, R.: Evolving communities of practice: IBM Global Services experience (2001)
5. Wenger, E., McDermott, R., Snyder, W.: Cultivating Communities of Practice: A Guide to Managing Knowledge. Harvard Business School Press (2002)
6. Hara, N.: Information Technology Support for Communities of Practice: How Public Defenders Learn About Winning and Losing in Court (SSRN Scholarly Paper No. ID 1520303). Social Science Research Network, Rochester, NY (2009)
7. Schwier, R.A.: Catalysts, Emphases, and Elements of Virtual Learning Communities: Implications for Research and Practice. Quarterly Review of Distance Education 2, 5–18 (2001)
8. Brown, J.S., Duguid, P.: Organizing Knowledge. California Management Review 40, 90 (1998)
9. Smedley, J.: Modelling personal knowledge management. OR Insight 22, 221–233 (2009)
10. Kogut, B., Zander, U.: What Firms Do? Coordination, Identity, and Learning. Organization Science 7, 502–518 (1996)
11. Easterby-Smith, M., Araujo, L., Burgoyne, J.G. (eds.): Organizational Learning and the Learning Organization: Developments in Theory and Practice. SAGE Publications Ltd. (1999)
12. Wenger, E., Snyder, W.: Communities of Practice: The Organizational Frontier - Harvard Business Review. Harvard Business Review 1, 139–145 (1999)
13. Mutch, A.: Communities of Practice and Habitus: A Critique. Organization Studies 24, 383–401 (2003)

Learning Styles in Adaptive Games-Based Learning

Mario Soflano, Thomas M. Connolly, and Thomas Hainey

School of Computing, University of the West of Scotland, Scotland, UK
{mario.soflano,thomas.connolly,thomas.hainey}@uws.ac.uk

Abstract. In classroom and online learning, there are benefits derived from delivering learning content in ways that match the student's learning style. A similar result has also been found in games-based learning (GBL). While the learning contents can be adapted based on the student's learning style identified by using a learning style questionnaire, it is possible that the student's learning style may change when learning through GBL, which may cause misadaptation. To confirm whether learning style may change in GBL, an experimental study was conducted involving 60 students in Higher Education learning SQL (Structured Query Language). The results show that learning style identified by the learning style questionnaire may be different than the learning style identified in GBL. In this study, the students who were identified as having a picture-based learning style tended to maintain their learning style while those who were identified as having a text-based learning style tended to change their learning style.

Keywords: Adaptive games-based learning, adaptivity, learning style, SQL, RPG, NeverWinter Nights.

1   Background

Differences between students in terms of their learning style have long been recognised by educationalists (e.g. [1][2]) and it has been argued that the use of a single instructional strategy to teach all students may have a negative impact on the learning outcomes of those students whose manners of learning diverge from the norm [3]. This has led to a focus investigating how classroom teachers can accommodate different learning styles and a number of studies that will subsequently be discussed have been conducted to investigate the effects of the use of teaching materials and strategies for delivery that accommodate different learning styles on learning outcomes. Hayes and Allinson [4] reviewed 19 studies from various sources and identified 12 studies that supported the claim that accommodating student learning styles in classroom education could lead to enhanced learning outcomes.

In eLearning, the benefit of adapting the learning contents to the student's learning style has also been identified (e.g. [5]). However, as Connolly and Stansfield [6] have suggested, eLearning simply replicates the traditional education system and may be overly focussed on method of delivery, i.e. delivering materials over the web rather than on actual teaching and learning, and indeed motivating and engaging the students

Á. Herrero et al. (eds.), *International Joint Conference SOCO'13-CISIS'13-ICEUTE'13*,   709
Advances in Intelligent Systems and Computing 239,
DOI: 10.1007/978-3-319-01854-6_72, © Springer International Publishing Switzerland 2014

in the learning process. In contrast, games, particularly video games, engage people over extensive periods of time and also motivate them to re-play the game repeatedly until they have mastered it [7]. Therefore, some educationalists (e.g. [8]) have considered games to be a potential platform in supporting student learning and have turned their attention to what is now called games-based learning (GBL). While many GBL applications have been developed in the last two decades, there remains a lack of empirical evidence to support the use of GBL for learning purposes [9]. Given that there appears to be genuine advantages for learning outcomes to be derived from the adaptation of teaching materials to learning styles in the classroom and through eLearning, it may also be possible that GBL applications that are adapted to the individual's learning style would improve learning outcomes.

Kirriemuir and McFarlane [7] have suggested that games, unlike the classroom or eLearning, provide a different type of engagement as they demand constant interaction and generate a 'flow' that could assist in engaging students. It is therefore possible for students to adopt different leaning styles in GBL than they adopt in other learning settings.

In regards to the benefit of adaptive GBL based on learning style, our recent experiment in GBL also indicates that the adaptive GBL application for teaching SQL had higher learning effectiveness with shorter completion time compared to paper-based learning, GBL without adaptivity and adaptive GBL based on the learning style identified by using a learning style questionnaire [10].

In conclusion, the studies indicated that improvements in learning outcomes could be achieved by accommodating the learning style of students. The following sections will discuss the definitions and categorisation of learning style theories including the reliability and validity of each of their instruments. The next section will review current debates surrounding learning styles and provide a comparison between the theories before justifying the learning style that has been chosen for this research.

2   Definition and Categorisation of Learning Style Theories

According to Pashler, McDaniel, Rohrer and Bjork [11:105] learning style is *"the concept that individuals differ in regard to what mode of instruction or study is most effective for them"*; whilst Rayner and Riding [12:51] define learning styles as *"individual sets of differences that include not only a stated personal preference for instruction or an association with a particular form of learning activity but also individual differences found in intellectual or personal psychology"*. The terms 'learning styles' and 'cognitive styles' have on occasion been used interchangeably in the literature. However, according to Allport as cited by Cassidy [13] cognitive style is an individual's typical or habitual mode of problem solving, thinking, perceiving and remembering; while learning style reflects the application of cognitive style in a learning situation.

According to Rayner and Riding [12] there have been four major mainstreams in learning-style research:

— *Perception*: This investigates students' perception of information within the learning context.
— *Cognitive controls and the cognitive process*: This relates to personal aptitude towards information.
— *Mental imagery*: This investigates how students 'picture' the information.
— *Personality constructs*: This investigates how personality is related to learning.

Riding [14] identified that learning style could be related to the presentation of the material, for example: text, picture, diagram and sound; the structure of the material, for example: sequential - global, inductive - deductive and linear - non-linear; the type of content: abstract material that is related to ideas and theories or concrete materials that focus more on practical value; and how the learning materials are processed, for example: active-reflective and thinking-feeling.

In this research, we based the adaptive GBL application on the Felder-Silverman learning style model [15]. The model consists of four main elements: perception, input, processing and organization Each main element consists of two sub-elements that represent the element a person may have, for example a person may prefer to use visual or verbal materials . The Perception element consists of two sub-elements: sensing and intuitive. Sensing students prefer to learn facts and concrete learning materials, whilst intuitive students prefer abstract materials with general principles rather than concrete materials. The Input element relates to the presentation of materials preferred by students. Visual students prefer to learn using pictures, diagrams, graphics, etc., whilst verbal students prefer to listen or read the learning materials. Verbal students tend to be able to remember written or spoken text or a sentence. For example, in an advertisement, some people remember the slogan, whilst others remember what the advertisement looks like. In the Processing element, active students prefer to learn by doing and enjoy discussion with other people, whilst reflective students prefer to reflect on the learning materials (learning by thinking it through) alone. Active students can be categorised as extroverted and reflective students can be categorised as introverted. In the Organisation of materials, sequential students learn in a step-by-step/linear sequence, following a logical sequence, whilst global students absorb learning materials randomly and when they have learned enough, they will understand the whole picture. Global students can also be categorised as holistic students and sequential students as serial students whilst sensing - sequential students have a tendency to be convergent and intuitive and global students have a tendency to be divergent students.

2.1 Debate Surrounding Learning Styles

The main debate within the area of learning styles relates to the reliability of the instrument and its conceptual validity. The research into the validity and reliability of these theories is primarily conducted through a test-retest method over a period of time to assess the consistency of results and to investigate the relationship between elements of the theory in order to prove their validity. There have been a number of studies that have investigated the reliability of the Felder-Silverman model and the reliability of its instrument. The results of investigations involving students from different schools have shown support for its reliability (eg. [16-17]). The results showed that sensing-intuitive learning style category had the highest score for consistency followed by visual-verbal, active-reflective and sequential-global. Further studies from Livesay, Dee, Nauman and Hites [18], Van Zwanenberg et al. [19] and Zywno [20] concluded that the active-reflective element and visual-verbal element had a higher degree of independence compared to sequential-global and sensing-intuitive. However the results did not compromise the purpose and reliability of the learning style theory as each learning style element represented different aspects of learning that therefore required different instructional methods. These studies suggest that the Felder-Silverman model has relatively better reliability and validity compared to other theories. Limongelli, Sciarrone, Temperini and Vaste [21] also suggested that in addition to the positive results of the reliability and validity tests, Felder-Silverman learning style is measured by using numbers/quantitative elements and also it contains most elements appearing in other learning style models.

3   Learning-Style-Based Adaptive GBL

3.1   Methodology

For the purpose of this research, an adaptive GBL application based on learning style was developed. The term 'adaptive GBL application' refers to the ability of the GBL application to automatically customise certain elements of the system based on a series of the student's interactions with the system. The game was intended to teach the basics of the database programming language SQL (Structured Query Language) while the learning style adopted in this game was the Felder-Silverman learning style model, particularly the presentation elements (picture-text). The selected genre of the game was role-playing games and it was developed by using NeverWinter Nights 2 engine. Three modes of the same game were designed and developed:

— a non-adaptive mode of the game. This mode treats all students the same and takes no account of the student's learning style.
— an in-game adaptive mode. In this mode, the student's characteristics are identified during the gameplay. As it is possible for the student to change learning style in the course of the game, the game will have an adaptive system that can automatically customise the game in real-time according to the student's current learning style. The difference between the modes concerns the nature of the adaptive approach adopted while the rest of the game elements such as storyline, game environment, controls and game interface, are identical. The adaptive approach itself was implemented through the presentation of the learning materials presented by the conversation system of the game.

In each game group, 30 students were randomly allocated. The students were university students with no knowledge of SQL who voluntarily participated in the study. The methodology adopted in this research was pre-test and post-test. On pre-test, the

SQL knowledge of the students was tested and only the students who scored 0 in the pre-test SQL test were selected in this study. The students also needed to answer a Felder-Silverman learning style questionnaire to identify their learning style. During the gameplay, the learning styles selected by the student were counted. The learning style identified by using the learning style questionnaire was then compared to the learning style identified in the game.

4   Learning Style Changes

The hypothesis tested to answer whether people prefer to change or maintain their learning style when learning through a game is:

— H_0: There is no difference between the proportion of people who change their learning style and those who maintain their learning style.
— H_1: There is a difference between the proportion of people who change their learning style and those who maintain their learning style.

To test this hypothesis, an experiment was conducted using a pre-test/post-test method. For the pre-test stage, the students were required to fill in a learning style questionnaire and then play the game in accordance with the group to which the student was allocated. The learning style was then identified with a learning style questionnaire and through a comparison of the learning processes that took place during the game.

The data used to answer this research question is derived from the game without adaptivity group and the game with the in-game adaptive group; each had 30 students. In the game without adaptivity group, 13 students had a learning style that was identified as being a text-based learning style and 17 students had a picture/diagram-based learning style. In the in-game adaptive group, there were 14 students with a text-based learning style and 16 students with a picture-based learning style.

To test the hypothesis, a two-tailed binomial test was used in each of the cases. The first case sought to analyse the text sub-group from the non-adaptive game group. In this case, 11 of the students (85%) changed their learning style from text-based to picture-based and 2 students (15%) maintained their learning style. The results of the binomial test for this case indicates there is a significant difference between the proportion of those who change and those who maintain their learning style when learning through the game, which means the null hypothesis H_0 is rejected ($p < 0.05$).

The second case required analysis of the picture sub-group from the non-adaptive game group. In this case, 2 students (12%) changed their learning style from a picture-based to a text-based style and 15 students (88%) maintained their learning style. The result of the binomial test for this case indicates there is a significant difference between the proportion of those who change and those who maintain their learning style while learning through the game, which means the null hypothesis H_0 is rejected ($p < 0.05$). The summary of the analysis in non-adaptive group is presented in Table 3.

The third case requires analysis of the text sub-group from the in-game adaptive group. In this case, 12 students (86%) changed their learning style from a text-based to picture-based and 2 students (14%) maintained their learning style. The results of the binomial test for this case indicate there is a significant difference between the proportion of those who change and those who maintain their learning style while learning through the game, and again the null hypothesis H_0 is rejected ($p<0.05$).

Table 1. Learning Style Changes in Non-adaptive Group

Non-Adaptive Group		After Game		Significance
		Text	Picture	
Before Game	Text	2	11	$p<0.05$
	Picture	2	15	$p<0.05$

The fourth case involves analysis of the picture sub-group from the in-game adaptive group. In this case, 3 students (19%) changed their learning style from picture-based to text-based and 13 students (81%) maintained their learning style. The result of the binomial test for this case indicated a significant difference between the proportion of those who change and those who maintain their learning style while learning through the game, and again the null hypothesis H_0 is rejected ($p<0.05$). The summary of the analysis in non-adaptive group is presented in Table 4.

Table 2. Learning Style Changes in In-game Adaptive Group

In-game Adaptive Group		After Game		Significance
		Text	Picture	
Before Game	Text	2	12	$p<0.05$
	Picture	3	13	$p<0.05$

From the four binomial tests conducted, the results consistently indicated that there was a significant difference between the proportion of those who changed and those who maintained their learning style when learning through the game. The results were investigated further to discover if the preference was towards those who had changed their learning style or those who maintained their learning style.

To discover the preferences, a McNemar test was conducted on a combination of both game groups (text-to-text: 4 students, text-to-picture: 23 students, picture-to-text: 5 students, picture-to-picture: 28 students). The result showed a more significant trend of those who were identified to have a text-based learning style to change their learning style compared to those who were identified to have a picture-based learning style ($\chi_2(1) = 10.321$; $p<0.002$). The summary of the McNemar analysis is presented in Table 5.

To confirm whether more students who were previously text-based learning style changed their learning style when learning through game, a hypothesis is defined as:

— H_0: a lesser or equal proportion of people changed their learning style than those who maintained it (changed <= maintained)
— H_1: a higher proportion of people changed their learning style than those who maintained it (changed > maintained).

Table 3. Overall Learning Style Changes

	Unchanged	Changed	Significance
Text	4	23	$\chi_2(1) = 10.321$; $p<0.002$ (text-based to a picture-
Picture	28	5	based > picture-based to a text-based)

To answer this hypothesis, a one-tailed binomial test was conducted on a combination of the text sub-groups from the non-adaptive and in-game adaptive groups. The result showed a significant difference between the proportion of people who changed their learning style (85%) and those who maintained it (15%), therefore the null hypothesis H0 is rejected with $p<0.011$. The result indicated that students who were identified as having a text-based learning style had a tendency to change their learning style to a picture-based learning style when learning through the game.

The findings indicate that the students seemed to prefer picture materials. To investigate further, an investigation was designed to discover if more students who previously had a picture-based learning style maintained their learning style when learning through the game. In this case, the hypothesis is:

— H_o: a less or equal proportion of people maintained their learning style than those who changed it (maintained <= changed)
— H_1: a higher proportion of people maintained their learning style than those who changed it (maintained > changed).

By using a one-tailed binomial test, a significant difference was obtained between the proportion of people who maintained their learning style (84%) and those who changed it (16%) and therefore the null hypothesis H_0 is rejected with $p<0.011$. The result indicates that the students who were identified as having picture-based learning style were more consistent in maintaining their learning style when learning through the game.

5   Conclusions and Future Directions

In classroom and eLearning, adaptivity based on learning style has shown to have positive impact in learning effectiveness and experience. Particularly in GBL, the results from our previous experiment show that adaptive GBL has higher learning effectiveness with faster completion time [10]. Further analysis then conducted to investigate the learning style changes during the learning process in GBL. The result shows that the student's learning style may change when learning through GBL. In this case, the students who were identified to have picture-based learning style tended to maintain their learning style while those who were identified to have text-based learning tended prefer to change their learning style.

The empirical evidence produced from this experiment is the original contribution to adaptive GBL. The results also confirm the fluctuation of learning style in GBL which indicate that the learning style identified by using the learning style questionnaire may not be the same as the learning style in the game.

One of the considerations in designing an educational game lies in how to teach the learning materials without losing the 'fun' part of the game so the student can remain motivated to learn the materials. One way of maintaining motivation is to blend the materials and the game story in such a way that the learning materials are part of the challenge of the game. The findings here may be specific to the integration of these particular learning materials and the mechanics of the game therefore modifications in the game's specification could give rise to different outcomes.

However, the conversation system used to present the learning materials in the game is highly customisable and can provide a useful platform for exploring the game's potential for teaching different disciplines or accommodating different learning styles. Further research based on this game and its variants may contribute empirical evidence of the beneficial effects of GBL and adaptive GBL in particular. The adaptivity may also be improved in future research particularly to address different learning styles in a game and to create more complex adaptivity in various elements of the game.

Acknowledgements. This work has been co-funded by the EU Lifelong Learning Programme under contract 519057-LLP-1-2011-1-UK-KA3-KA3NW (Ed2.0Work – European Network for the integration of Web2.0 in education and work) and as part of the Games and Learning Alliance (GaLA) Network of Excellence on 'serious games' funded by the Eurpean Union in FP7 – IST ICT, Technology Enhanced Learning (see http://www.galanoe.eu).

References

1. Kolb, D.: Learning style inventory. McBer & Co., Boston (1976)
2. Dunn, K., Dunn, R.: Teaching Students through their Individual Learning Styles. Prentice Hall, Englewood Cliffs (1978)
3. Gregorc, A.F.: Style as a symptom: A phenomenological perspective. Theory into Practice 23(1), 51–55 (1984)
4. Hayes, J., Allinson, C.W.: The Implication of Learning Styles for Training and Development: A Discussion of the Matching Hypothesis. British Journal of Management 7, 63–73 (1996)
5. Vassileva, D.: Evaluation of Learning Styles Adaption in the Adopta E-learning Platform. In: CompSysTech 2011, pp. 540–545 (2011)
6. Connolly, T.M., Stansfield, M.H.: Enhancing eLearning: Using Computer Games to Teach Requirements Collection and Analysis. In: Second Symposium of the WG HCI & UE of the Austrian Computer Society, Vienna, Austria (November 23, 2006)
7. Kirriemuir, J., McFarlane, A.: Literature Review in Games and Learning. Report 8, NESTA, Futurelab, Bristol (2004)
8. Gee, J.P.: What video games have to teach us about learning and literacy. Palgrave Macmillan, New York (2003)
9. Connolly, T.M., Boyle, E.A., MacArthur, E., Hainey, T., Boyle, J.M.: A systematic literature review of empirical evidence on computer games and serious games. Computers & Education 59, 661–686 (2012)

10. Soflano, M., Connolly, T.M., Hainey, T.: An Application of Adaptive Games-Based Learning based on Learning Style to Teach SQL. In: 7th European Conference in Games-Based Learning (ECGBL), Porto, Portugal, October 3-4 (2013)
11. Pashler, H., McDaniel, M., Rohrer, D., Bjork, R.: Learning Styles: Concepts and Evidence. Psychological Science in the Public Interest 9(3), 105–119 (2008)
12. Rayner, S., Riding, R.: Towards a Categorisation of Cognitive Styles and Learning Style. David Fulton Publishers, London (1998)
13. Cassidy, S.: Learning styles: An Overview of Theories, Models, and Measures. Educational Psychology 24, 419–444 (2004)
14. Riding, R.: School Learning and Cognitive Style. David Fulton Publishers (2002)
15. Felder, R.M., Silverman, L.K.: Learning and Teaching Styles in Engineering Education. Engineering Education 78(7), 674–681 (1988)
16. Felder, R.M., Spurlin, J.: Reliability and Validity of the Index of Learning Styles: A Meta-analysis. International Journal of Engineering Education 21(1), 103–112 (2005)
17. Litzinger, T.A., Lee, S.H., Wise, J.C., Felder, R.M.: A Study of the Reliability and Validity of the Felder-Soloman Index of Learning Styles. In: Proceeding of the 2005 American Society for Engineering Education Annual Conference and Exposition. American Society for Engineering Education (2005)
18. Livesay, G.A., Dee, K.C., Nauman, E.A., Hites Jr., L.S.: Engineering Student Learning Styles: A Statistical Analysis Using Felder's Index of Learning Styles. In: 2002 ASEE Conference and Exposition, Montreal, Quebec (2002)
19. Van Zwanenberg, N., Wilkinson, L.J., Anderson, A.: Felder and Silverman's Index of Learning Styles and Honey and Mumford's Learning Styles Questionnaire: How do they compare and do they predict academic performance? Journal of Educational Psychology 20(3), 365–380 (2000)
20. Zywno, M.S.: A contribution of validation of score meaning for Felder-Soloman's Index of Learning Styles. In: Proceedings of the 2003 Annual ASEE Conference. ASEE, Washington, DC (2003)
21. Limongelli, C., Sciarrone, F., Temperini, M., Vaste, G.: Adaptive Learning with the LS-Plan System: A Field Evaluation. IEEE Transactions on Learning Technologies 2(3), 203–215 (2009)

The Use of Games as an Educational Web2.0 Technology

A Survey of Motivations for Playing Computer Games at HE Level

Thomas Hainey and Thomas M. Connolly

School of Computing, University of the West of Scotland, Paisley, Renfrewshire, Scotland
{thomas.hainey,thomas.connolly}@uws.ac.uk

Abstract. Serious games are considered by some educationalists to be a potentially highly motivating form of supplementary education. In some cases, serious games are stand-alone but if delivered over the web is a subset of Web2.0 technologies. To properly deliver serious games over the web for educational purposes it is important to understand the motivations for playing computer games for leisure and for playing computer games in an educational context. This paper will present the findings of a survey carried out in a Higher Education (HE) institution involving 415 participants to ascertain what motivations and attitudes HE students have towards playing computer games in general, playing computer games when they get progressively more difficult and playing computer games for educational purposes. The results indicate that challenge is the most important motivation for playing computer games when they get progressively difficult and for playing computer games in HE. Challenge is the fourth most important motivation for playing computer games in general.

Keywords: games-based learning, serious games Web2.0, motivations, HE.

1   Introduction

Computer games are regarded by some educationalists as highly engaging and it is hoped that by exploiting their highly compelling even addictive qualities that they can be used to help people learn effectively. In a recent study, Boyle, Connolly, and Hainey [1] examine the literature on computer games and serious games, focussing on the potential positive impacts of gaming, particularly with respect to learning value and skill enhancement. Over 7392 papers were identified in the review of the research literature between 1996 and 2009, confirming the surge of interest in this area. However, only 127 papers were empirical and only 64 of these were considered to have used an appropriate methodology that would allow generalisations to be made. Their literature review shows that playing computer games confers a range of perceptual, cognitive, behavioural and affective and motivational impacts and outcomes. In their review the most frequently occurring outcomes and impacts were affective and motivational followed by knowledge acquisition/content understanding. This reflects the parallel interests in games as an entertainment medium but increasingly their use for learning. The authors felt that there was a dearth of evidence and believed that much

Á. Herrero et al. (eds.), *International Joint Conference SOCO'13-CISIS'13-ICEUTE'13*,
Advances in Intelligent Systems and Computing 239,
DOI: 10.1007/978-3-319-01854-6_73, © Springer International Publishing Switzerland 2014

more research was required to understand the use of computer games in education and in entertainment.

This paper contributes to empirical evidence in the serious games literature by analysing the results of a survey to investigate the reasons for playing computer games, reasons for playing computer games when they get more difficult and reasons for playing computer games in HE. This study also looks at differences in attitudes to computer games. Following a discussion of previous work in the field, we describe the methods used to collect the data including procedure, participants and materials. We then present the results of the surveys carried out including: game playing habits, reasons for playing computer games, reasons for playing computer games in HE and attitudes to computer games. The paper concludes with a discussion of the overall results and future research directions.

2   Previous Research

Computer games are considered by some educationalists to be highly motivating and engaging by incorporating features that have extremely compelling, even addictive, quality [2]. Connolly, Stansfield, McLellan, Ramsay, and Sutherland [3] suggest that computer games build on theories of motivation, constructivism, situated learning, cognitive apprenticeship, problem-based learning, and learning by doing. By creating virtual worlds, computer games integrate ".not just knowing and doing. Games bring together ways of knowing, ways of doing, ways of being, and ways of caring: the situated understandings, effective social practices, powerful identities, and shared values that make someone an expert" [4]. Games and simulations fit well into the constructivist paradigm and "generally advocate the active acquisition of knowledge and skills, collaboration and the use of authentic and realistic case material" [5]. The use of computer games can be linked to the display of "expert" behaviours such as: superior long and short-term memory, pattern recognition, qualitative thinking, principled decision-making and self-monitoring [6].

Probably the best known distinction in motivation research is that between intrinsic and extrinsic motivation [7]. Intrinsically motivated behaviours are carried out because they are rewarding in themselves, while extrinsically motivated behaviours are carried out because of the desire for some external reward, such as money, praise or recognition from others. Intrinsic motivation is thought to be more successful in engaging students in effective learning because intrinsically motivated students want to study for its own sake, they are interested in the subject and want to develop their knowledge and competence. This distinction has been used by designers of educational computer games, notably Malone and Lepper [8] who argued that intrinsic motivation is more important in designing engaging games and created a framework of points to consider when designing learning games. They suggested that intrinsic motivation is created by four individual factors: challenge, fantasy, curiosity and control and three interpersonal factors: cooperation, competition, and recognition. Interestingly these factors also describe what makes a good game, irrespective of its educational qualities. Garris, Ahlers, and Driskell [9] present six dimensions that computer games can provide for educational purposes that are based on the work of Malone and Lepper [8]. These dimensions are: fantasy – imaginary themes, characters or contexts; rules/goals clear rules, goals and feedback; sensory stimuli – novel auditory and visual stimuli; challenge – optimal level of difficulty and goal attainment uncertainty; mystery – similar to curiosity providing optimal level of complexity of information; and control– active learner control. Thiagarajan [10] suggests five critical characteristics of computer games: conflict– similar to challenge [8] and encompasses the attainment of goals in both cooperation and competition with other players or the computer; control – the rules that regulate play; closure – the game has some form of 'end point'; contrivance – the game is not taken too seriously by the players and they are offered motivation to continue; and competency – the players experience growth in their problem solving, skill level and knowledge. Cordova and Lepper [11] discovered that students learning by traditional methods were outperformed by students learning with instructional games and that control, context, curiosity and challenge increased.

The primary purpose for the discussion of intrinsic motivation in this study is due to the fact intrinsic motivation is more desirable than extrinsic motivation for learning according to the America Psychological Associations 14 learner centred psychological principles [12]. As a result Malone and Lepper's [8] framework of intrinsic motivation was utilised to gain measurements as it has been extremely well utilised and documented in the games-based learning literature to study the educational design principles of learning games [13]. The framework uses the original interpersonal and individual factors. On an individual level there is: Challenge, Fantasy, Curiosity and Control, and on an interpersonal level there is: Cooperation, Competition and Recognition.

Whitton [14] performed a study with 200 participants to examine gaming preferences, attitudes towards games in HE and motivations. 63.1% reported that they would find games positively motivating for learning, 28.3% not motivating either way and 8.6% demotivating. Gibson, Halverson, and Riedel [15] performed a survey of 228 'pre-service' students to ascertain perceptions and attitudes to simulations and games. 80% of respondents were white females. 65% believed that simulations and games could be an important or very important learning tool; only 7% believed that they were of little or no importance. Males were more negative about the potential of games in learning. 53% of males were positive while 70% of females were positive. There was no notable generation gap between the respondents. Eglesz, Fekete, Kiss, and Izsó [16] performed a study with two surveys, one online survey with 843 participants and a second with 102 participants. The studies found that woman play computer games significantly less than men and prefer Role Playing Games (RPGs) while men prefer action, adventure simulation and sports games.

3   Methodology

The methodology selected for this research was survey/questionnaire based research. The results were collected through SurveyMonkey and downloaded into IBM SPSS Statistics version 20 for detailed analysis. Out of approximately 18,000 students surveyed in 2011 there were 415 respondents who completed the motivations for playing computer games questionnaire. 387 participants specified their gender. 197 participants (51%) were male and 188 (49%) were female. This compares with a gender breakdown in the University as a whole of 37% males and 63% females. The mean age of participants was 26.50 (SD = 9.41) with a range from 17 to 60. The mean age of males was 26.29 years (SD = 9.05) and the mean age of females was 26.50 years (SD = 9.56). A Mann-Whitney U test indicated that there was no significant difference in age between males and females ($Z = -0.175$, $p < 0.861$). This is broadly consistent with the overall student age of 25. 47 participants (12%) indicated that they were part-time students and 338 (88%) indicated that they were full-time students. 56 participants (15%) specified that they were postgraduate students and 329 participants (85%) stated that they were undergraduate students. The breakdown of the students across the different Schools in UWS was: 55 (14%) were Business students, 95 (25%) were Computing students, 13 (3%) were Education students, 85 (22%) were Engineering and Science students, 43 (11%) were Health, Nursing and Midwifery students and 45 (12%) were Social Science students.

3.1   Procedure

The survey was carried out across all students at the University of the West of Scotland. The questionnaire was made available through the online questionnaire package SurveyMonkey for a two-week period during March 2011. Participation was voluntary and participants were notified of the availability of the questionnaire through email and a login notice posted in the BlackBoard Virtual Learning Environment (which the majority of students use). Notices were also posted across the University. Respondents completed the questionnaire online at their convenience during this period. Access to the questionnaire was controlled using the students' BlackBoard usernames and passwords, and the students' unique banner identification number was used to ensure a student only completed the questionnaire once.

4   Results

4.1   General Game Playing Habits

303 participants (79%) played computer games for leisure and 82 respondents (21%) indicated that they did not. Students who played computer games for leisure were significantly younger (Mean = 25.83, SD = 9.06) than those who did not (Mean = 28.95, SD = 10.18) ($Z = -2.585$, $p < 0.010$). Bearing in mind that there was not a significant difference in the number of female and male respondents, 83% of males played computer games for leisure and 73% of females played computer games for leisure.

To calculate the mean time spent playing games the time bands used as responses were recoded with their mean value (e.g. 1-5 hours was recoded as 3), while less than 1 was coded as 1 and more than 25 was coded as 26. Using this recoded data the average number of hours played per week was 9.22 (SD = 7.46). Males played games for significantly longer periods per week (11.35, SD = 7.97) than females (6.49, SD = 5.69) (Z = -4.955, p < 0.000). A significantly higher percentage of men (36.6%) than women (13.1%) played for more than 8 hours per week.

There were also differences in the number of hours played by students in different Schools. Computing students played for the longest (10.71 hours per week) followed by Business (10.24 hours) and Media (10.05 hours) with Education students playing least (3.0 hours).

On average participants had been playing computer games for 15.31 years (SD = 6.77) with a range of 2 to 35 years. Given that the average age of the participants was 26.5, the average participant had been playing computer games for just under half of their life. Although there was no significant difference in the age of respondents, male respondents had been playing games for significantly longer (16.50 years, SD = 7.15) than females (13.83 years, SD = 5.91).

4.2   Reasons for Playing Computer Games

Participants were asked to rate the importance they considered different reasons for playing computer games in general. Pleasure, relaxation and excitement were rated as the three top ranking motivations. Challenge was rated as the fourth top ranking motivation. Control, avoidance of other activities and recognition were rated as the least important ranking motivations. Table 1 shows the rankings of the top ten motivations by respondents for playing computer games in general for leisure. Table 2 shows the main ten reasons for playing computer games split by gender. Mann-Whitney U tests indicated that males rated cooperation as significantly more important for playing computer games than females (Z = -2.657, p < 0.008). Males also rated prevention of boredom as significantly more important than females (Z = -2.521, p < 0.012). Females considered curiosity to be significantly more important for playing computer games than males (Z = -2.643, p < 0.008) and also rated relieving of stress to be significantly more important than males (Z = -2.098, p < 0.036).

Table 1. Ranking of motivations for playing computer games

Reason	Rank	Mean	SD
Pleasure	1st	4.31	0.81
Relaxation	2nd	4.20	0.93
Excitement	3rd	4.13	0.92
Challenge	4th	4.08	0.89
Leisure time	5th	4.02	1.02
Prevention of boredom	6th	3.95	1.07
Feeling good	7th	3.95	0.86
Relieve stress	8th	3.79	1.15
Curiosity	9th	3.72	1.09
Release tension	10th	3.72	1.20

Table 2. Ranked reasons for playing computer games in relation to gender

Gender	Male			Female		
Reason	Rank	Mean	SD	Rank	Mean	SD
Pleasure	1st	4.32	0.85	2nd	4.29	0.77
Challenge	2nd	4.17	0.85	5th	3.98	0.94
Excitement	3rd	4.15	0.97	3rd	4.10	0.85
Relaxation	4th	4.11	1.02	1st	4.31	0.80
Prevention of boredom	5th	4.11	1.00	10th	3.76	1.13
Leisure time	6th	4.04	1.04	4th	4.01	0.96
Feeling good	7th	3.94	0.83	7th	3.97	0.91
Release tension	8th	3.68	1.11	9th	3.77	1.08
Relieves stress	9th	3.64	1.22	6th	3.98	1.03
Curiosity	10th	3.57	1.10	8th	3.91	1.06

4.3   Reasons for Playing Computer Games in HE

Participants were asked to rank the importance of reasons for playing computer games in HE. The top three ranking reasons for playing computer games in HE were challenge, curiosity and cooperation. The three least important reasons were leisure, recognition and fantasy. Table 3 shows the top ten motivations for playing computer games within a HE context. Table 4 shows the rankings of the top ten motivations for playing computer games in HE in relation to gender. Mann-Whitney U tests indicated that males rated competition ($Z = -2.255$, $p < 0.024$) and cooperation ($Z = -3.437$, $p < 0.001$) as significantly more important for playing computer games in a Higher Education context than females.

Table 3. Ranking of motivations for playing computer games in HE

Reason	Rank	Mean	SD
Challenge	1st	3.83	1.42
Curiosity	2nd	3.61	1.40
Cooperation	3rd	3.46	1.39
Pleasure	4th	3.40	1.43
Relaxation	5th	3.26	1.43
Competition	6th	3.21	1.41
Control	7th	3.07	1.40
Leisure	8th	2.98	1.47
Recognition	9th	2.96	1.40
Fantasy	10th	2.76	1.41

4.4   Attitudes to Computer Games

Participants who played computer games had significantly more positive attitudes to computer games than those who did not with the exception of computer games being a time consuming activity ($Z = -0.873$ p < 383). Table 6 shows attitudes split by gender for playing computer games. Mann-Whitney U tests indicated that there were significant differences in attitudes to computer games in relation to gender. Males found games to be significantly more or a social activity ($Z = -2.899$, p < 0.004), more time consuming ($Z = -2.170$, p < 0.030), more interesting ($Z = -2.334$, p < 0.020), more enjoyable ($Z = -3.667$, p < 0.000) and more exciting ($Z = -2.797$, p < 0.005). Females found that playing computer games was significantly more or a lonely activity ($Z = -4.053$, p < 0.000).

Table 4. Ranking of motivations for playing computer games in HE split by gender

Gender	Male			Female		
Reason	Rank	Mean	SD	Rank	Mean	SD
Challenge	1st	3.91	1.40	1st	3.72	1.45
Curiosity	2nd	3.74	1.38	2nd	3.48	1.42
Cooperation	3rd	3.73	1.35	4th	3.22	1.40
Pleasure	4th	3.49	1.43	3rd	3.29	1.42
Competition	5th	3.38	1.43	6th	3.04	1.38
Relaxation	6th	3.34	1.38	5th	3.16	1.47
Control	7th	3.10	1.46	7th	3.03	1.36
Leisure	8th	3.04	1.52	8th	2.92	1.43
Recognition	9th	3.02	1.42	9th	2.91	1.40
Fantasy	10th	2.71	1.44	10th	2.79	1.39

4.5   Skills Obtained in HE

Participants were asked to rate the skills that they believed they could acquire from playing computer games. Problem solving, critical thinking and collaboration/teamwork were the most popular skills and management, reflection and recollection were the least popular skills. Table 5 shows responses to the question "What types of skills do you think can be obtained from computer games that would be relevant to Higher Education?"

Table 5. Skills that can be obtained from computer games

Skill	Rank	N	%
Problem Solving	1st	245	59.0%
Critical Thinking	2nd	182	43.8%
Collaboration/teamwork	3rd	181	43.6%
Creativity	3rd	181	43.6%
Analysing/Classifying	4th	178	42.8%
Leading/motivating	5th	124	29.8%
Management	6th	112	26.9%
Reflection	7th	110	26.5%
Recollection	8th	102	24.5%

Table 6. Attitudes to computer games split by gender

Gender	Male		Female	
Attitude	Mean	SD	Mean	SD
Playing games is a sociable activity	3.93	1.10	3.52	1.22
Playing games is a waste of time	2.12	1.04	2.31	1.05
Playing games helps to develop useful skills	3.74	0.91	3.69	0.80
Playing games is time consuming	4.11	0.67	3.83	0.92
Playing games is interesting	4.22	0.79	3.98	0.89
Playing games is a worthwhile activity	3.80	0.96	3.57	1.02
Playing games is enjoyable	4.42	0.76	4.10	0.81
Playing games is a lonely activity	2.39	1.03	2.90	1.05
Playing games is a valuable activity	3.48	0.94	3.34	0.95
Playing games is exciting	4.19	0.87	3.91	0.89

5   Discussion and Summary

79% of respondents played computer games for leisure and students who played computer games for leisure were significantly younger than those who did not. Participants played computer games for an average of 9.22 hours per week with males playing computer games for significantly longer periods per week than females. On average participants had been playing computer games for 15.31 years. Given that the average age of the participants was 26.5 years, the average participant had been playing computer games for just under half of their life. 195 participants (74%) believed that computer games could be used in HE for the purposes of learning and 137 participants (59%) participated in computer games on-line.

In terms of the reasons for playing computer games, pleasure, relaxation and excitement were rated as the three top ranking motivations. Challenge was rated as the fourth top ranking motivation which is consistent with three similar studies performed at the University of the West of Scotland [17]. Control, avoidance of other activities and recognition were rated as the least important ranking motivations. Males rated cooperation and prevention of boredom to be significantly more important for playing computer games than females. Females considered curiosity and relieving stress to be significantly more important for playing computer games than males.

The top three ranking motivations for playing computer games in HE were challenge, curiosity and cooperation and the three least important reasons were leisure, recognition and fantasy. Males rated competition and cooperation as significantly more important than females for playing computer games in a HE context. Participants who played computer games had significantly more positive attitudes to computer games than those who did not with the exception of computer games being a time consuming activity. Males found games to be significantly more or a social activity, more time consuming, more interesting, more enjoyable and more exciting. Females found that playing computer games was significantly more or a lonely activity. In terms of skills that could be obtained from computer games, participants rated problem solving, critical thinking and collaboration/teamwork as the most popular skills and management, reflection and recollection as the least popular skills.

This paper has provided empirical evidence in the field of serious games and games-based learning (subsets of Web2.0) with regards to the motivations for playing computer games in general, playing computer games when they get progressively more difficult and playing computer games in an HE context. The study has also provided empirical evidence on the general game playing habits of HE students, their attitudes towards computer games and the skills they believe that they can obtain from playing computer games in an HE context. The results indicate that challenge is the most important motivation for playing computer games in HE and when they get progressively more difficult. Future work will entail running this survey in different countries to perform cultural comparisons and provide further empirical evidence in the area of serious games and games-based learning.

Acknowledgements. This work has been co-funded by the EU Lifelong Learning Programme under contract 519057-LLP-1-2011-1-UK-KA3-KA3NW (Ed2.0Work – European Network for the integration of Web2.0 in education and work).

References

1. Boyle, E.A., Connolly, T.M., Hainey, T.: The role of psychology in understanding the impact of computer games. Entertainment Computing (2011)
2. Griffiths, M.D., Davies, M.N.O.: Excessive online computer gaming: implications for education. Journal of Computer Assisted Learning 18, 379–380 (2002)
3. Connolly, T.M., Stansfield, M.H., McLellan, E., Ramsay, J., Sutherland, J.: Applying computer games concepts to teaching database analysis and design. In: Proceedings of the International Conference on Computer Games, AI, Design and Education, Reading, UK (November 2004)

4. Shaffer, D.W., Squire, K.T., Halverson, R., Gee, J.P.: Video games and the Future of learning, Phi Delta Kappan (2004),
   http://www.academiccolab.org/resources/gappspaper1.pdf
   (retrieved April 6, 2011)
5. Christoph, N., Sandberg, J., Wielinga, B.: Added value of task models anduse of metacognitive skills on learning (2003),
   http://www.cs.ubc.ca/wconati/aied-games/christophetal.pdf
   (retrieved April 6, 2011)
6. Van Deventer, S.S., White, J.A.: Expert behaviour in children's video game play. Simulation & Gaming 33(1), 28–48 (2002)
7. Deci, E.L., Ryan, R.M.: A motivational approach to self: integration in personality. In: Dienstbier, R. (ed.) Perspectives on Motivation. Nebraska Symposium on Motivation, vol. 38, pp. 237–288. University of Nebraska Press, Lincoln (1991)
8. Malone, T.W., Lepper, M.R.: Making learning fun: a taxonomy of intrinsic motivations for learning. In: Conative and Affective Process Analysis. Aptitude, Learning and Instruction, vol. 3, pp. 223–235. Lawrence Erlbaum, Hillsdale (1987)
9. Garris, R., Ahlers, R., Driskell, J.E.: Games, motivation, and learning: a research and practice model. Simulation & Gaming 33(4), 441–467 (2002)
10. Thiagarajan, S.: Instructional games, simulations, and role-plays. In: Craig, R.L. (ed.) The ASTD Training & Development Handbook, pp. 517–533. McGraw-Hill, New York (1996)
11. Cordova, D.I., Lepper, M.R.: Intrinsic motivation and the process of learning: beneficial effects of contextualization, personalization, and choice. Journal of Educational Psychology 88(4), 715–730 (1996)
12. American Psychological Association's Board of Educational Affairs (BEA). The 14learning-centred psychological principles (1997),
    http://www.apa.org/ed/governance/bea/learner-centered.pdf
    (retrieved April 4, 2011)
13. Asgari, M., Kaufman, D.: Relationships among computer games, fantasy and learning. In: Proceedings of the 2nd International Conference on Imagination and Education, Vancouver, British Columbia, Canada (2004),
    http://www.ierg.net/confs/2004/Proceedings/
    Asgari_Kaufman.pdf (retrieved April 24, 2011)
14. Whitton, N.J.: An investigation into the potential of collaborative computer game-based learning in higher education. Unpublished Doctoral Thesis (2007),
    http://playthinklearn.net/?page_id=48 (retrieved April 30, 2007)
15. Gibson, D., Halverson, W., Riedel, E.: Gamer teachers. In: Gibson, D., Aldrich, C., Prensky, M. (eds.) Games and Simulations in Online Learning: Research and Development Frameworks, pp. 175–188. Information Science Publishing (2007)
16. Eglesz, D., Fekete, I., Kiss, O.E., Izsó, L.: Computer games are fun? On professionalgames and players' motivations. Educational Media International 42(2) (2005)
17. Hainey, T.: Using Games-Based Learning to Teach Requirements Collection and Analysis at Tertiary Education Level. University of the West of Scotland (2010),
    http://cis.uws.ac.uk/thomas.hainey/Final%20PhD%20Thesis
    %20Tom%20Hainey.pdf (retrieved)

Adopting the Use of Enterpise Mircoblogging as an Organisational Learning Approach

Gavin J. Baxter and Thomas M. Connolly

School of Computing, University of the West of Scotland, Scotland, UK
{gavin.baxter,thomas.connolly}@uws.ac.uk

Abstract. This paper reviews the state of the art of enterprise microblogging (EMB) investigating the potential of microblogging tools to facilitate the concept of organisational learning. Focusing on the area of information systems development (ISD) this paper examines what impact EMB might have in terms of supporting the process of organisational learning within ISD project environments. This paper acknowledges that there appears to be a lack of empirical evidence related to the subject area of EMB as the concept is still a relatively new phenomenon. Reviewing the organisational learning literature in addition to prior work on EMB, this paper argues that dependent on the context of its implementation and use EMB tools have the potential to support the process of organisational learning in ISD project environments. This paper further proposes future directions on how to advance the area of EMB and provides an overview of a future empirical study associated with EMB that will be undertaken by the authors.

Keywords: Enterprise 2.0, ISD Projects, Enterprise Microblogging, Weblogs, Organisational Learning.

1 Introduction

Enterprise 2.0 refers to: *"The use of read/write (or Web 2.0 technologies) by businesses for a business purpose"* [1: 215]. Enterprise 2.0 was a concept coined by Andrew McAfee in 2006 and is often applied in the context of Web 2.0 technologies that are used internally or behind the firewall of an organisation to support internal knowledge sharing and collaboration. Whilst certain organisations such as Microsoft and IBM have been early adopters of the philosophy of Enterprise 2.0 there still remains a lack of empirical evidence associated with the effectiveness of certain Web 2.0 technologies in organisations that include, for example, weblogs [2] [3]. Despite this, it has been suggested in the academic literature that certain Web 2.0 technologies such as microblogging, when applied internally in an organisation, may have the potential to support project teams with aspects of project co-ordination and communication [4]. The concept of microblogging can be perceived as being a: *"...smaller version of weblogs enriched with features for social networking and with a strong focus on mobility. Users have their own public microblog where they can post short updates. Other members can be "followed" by adding them to the personal network."* [5: 2].

Á. Herrero et al. (eds.), *International Joint Conference SOCO'13-CISIS'13-ICEUTE'13*,
Advances in Intelligent Systems and Computing 239,
DOI: 10.1007/978-3-319-01854-6_74, © Springer International Publishing Switzerland 2014

A specific problem area where enterprise microblogging might be beneficially applied is in the context of alleviating the complexity of managing and executing ISD projects. For example, the way in which ISD projects are conducted means that they are often social and interactive in nature. ISD projects often consist of diverse groups of stakeholders from different backgrounds who possess varying skill sets. Prior research into the topic of how ISD projects sometimes become 'troubled' [6] indicates that projects can often escalate and can eventually become abandoned. From a theoretical perspective, it also appears that the association between enterprise blogging and organisational learning has yet to be explicitly identified in the literature. There have however been works published stating that microblogging has the potential to support communities of practice [7] and knowledge-intensive work practices [8]. Several definitions of organisational learning have been proposed in the literature. However, this paper agrees with the view that the process of organisational learning *"...occurs through shared insights, knowledge, and mental models... [and] builds on past knowledge and experience – that is, on memory"* [9: 64]. This definition of organisational learning coincides with the ontological perspective of the authors which is that reality is socially constructed, knowledge is tacit with individuals being able to act and think for themselves.

Furthermore, it has been argued in the literature that Web 2.0 technologies can help to support the concept of the knowledge worker [10]. Microblogging, with its potential to facilitate communication and information sharing in project teams could assist in supporting project team members to learn from one another in the form of project updates throughout the duration of a specific project [4].

This paper contributes to the body of knowledge of enterprise microblogging by identifying its theoretical association with organisational learning. Furthermore, this paper also proposes various avenues for future research in the subject area in addition to providing a review of future empirical work that will be performed by the authors. The paper is presented in the following manner. To begin with, the paper reviews the nature of ISD projects with specific reference to the teamwork aspect of these projects. Following on from this, the concepts of organisational learning and microblogging are discussed with a view to identifying the theoretical relationship between both subject areas. Next, prior work in the topic of enterprise microblogging is reviewed. Following on from this, future research directions for the area are proposed and a review of future empirical work to be undertaken by the authors is outlined.

2   Nature of ISD Projects

ISD projects can encounter difficulties due to a number of factors: project, psychological, social and organisational [11]. For instance, within the context of projects, each individual project stakeholder may have a differing perception of how a software project is running. This outlook could be determined by their place and responsibility within the project team. A resulting factor might be the notion of project escalation due to conflicting reports regarding the overall status of a project. These attitudinal and behavioural problems could also be related towards project members not fully understanding the divergent nature of one another's job roles. ISD projects can at times run into trouble due to rivalry among the project members resulting in withholding project information from one another. An important influence that may also determine the successful development of an ISD project is the organisational environment in which the project is undertaken. Projects that are run in an organisational climate that is used to success might encounter difficulties if the project begins to appear to be in trouble. Furthermore, an organisation that inhibits an environment that facilitates openness of knowledge sharing, exchanging ideas and learning from past mistakes may unconsciously have laid the very foundations to facilitate the unsuccessful running of its ISD projects.

3   Fluid Nature of Project Teams

A project team, by definition, can be perceived as being *"...an established, fixed group of people cooperating in pursuit of a common goal"* [12: 13]. The intensity and different approaches undertaken by organisations to compete in what is often referred to as the knowledge economy [10] has implications for how project teams are formulated and function to achieve project deliverables. Concepts such as virtual organisations and virtual teams often relate to transparent forms of organisational structures sometimes forced to materialise due to the changing circumstances of an organisation or affiliated companies. Virtual organisations, similarly to virtual teams, are predominately multi-site, multi-organisational, where members communicate through the use of information technology to achieve long or short-term goals. In addition, it can be argued that virtual teams employ similar features to the concept known as 'teaming'. The phrase teaming refers to when organisational members undertake flexible teamwork. Teaming often involves forming project teams at short notice where the project teams are often short term, disbanded after project completion and where project members have to liaise with individuals with different skill sets and varying backgrounds almost on a rotational basis. Teaming is therefore a concept often applied in temporary teams or organisations collaborating with one another and has been referred to as *"... the engine of organizational learning"* [12: 14] .

4   Different Perspectives of Organisational Learning

This section of the paper will examine the various perspectives that are associated with the subject area of organisational learning and will explain the thinking behind these viewpoints. The perspective of organisational learning that this paper adopts will also be justified.

4.1   The Functionalistic Perspective

The traditional perspective of organisational learning, also referred to as the functionalistic viewpoint, adopts the outlook that learning begins with the individuals who inhabit organisations [13] and argues that individuals learn as representatives for the organisation [14]. Within this context of organisational learning the functionalistic outlook argues that an organisation primarily learns in two main ways, namely, by the learning undertaken by the individuals contained within the organisation or by acquiring new members of staff who possess knowledge the organisation did not previously have access to [15]. Within the functionalistic perspective learning occurs primarily through 'explicit knowledge' creation and sharing. Explicit knowledge is knowledge that can be articulated into formal language, such as words, mathematical expressions, specifications and computer programmes, and can be readily articulated to others [16]. Within the functionalistic paradigm, organisational learning occurs through regulated or structured routines, formal procedures and documentation [17].

4.2   The Interpretive Perspective

The interpretive outlook of organisational learning argues that learning in organisations is a social process [14]. This view is becoming more main stream in contrast to the formal viewpoint of organisational learning. Within this context of organisational learning it is argued that learning in organisations must be viewed as a social phenomenon because individuals are essentially social beings [18]. The social perspective of organisational learning is theoretically different to its functionalist counterpart and argues that learning is predominately relationship-based and learning begins in the form of relationships and through social construction [19]. This outlook of organisational learning challenges the idea that learning starts within individuals and instead argues that organisational reality is socially constructed and that organisations and the individuals within them are socially dynamic and interactive in nature [20].

The interpretive viewpoint of organisational learning further argues that learning that occurs in unique working contexts can be shared in the form of learning communities, known as communities of practice (CoPs) [21]. CoPs are defined as being: *"...groups of people who share a concern, a set of problems, or a passion about a topic, and who deepen their knowledge and expertise in this area by interacting on an ongoing basis"* [22: 4]. In addition, in contrast to the functionalistic view of organisational learning, the social perspective views knowledge as non-quantifiable and perceives knowledge to be subjective and personal in nature based on an individual's unique experience, consisting of their own mental models, beliefs and know-how [23]. In this standpoint, knowledge is considered to be 'tacit'; that is, personalised and based upon experience, context and the actions of an individual. Closely associated with tacit knowledge is the role of dialogue and conversation, which is considered to be important in the social aspect of organisational learning [24] and a fundamental prerequisite to the success of a community of practice.

4.3   View That Organisations Can Learn

The literature on organisational learning also debates whether organisations can learn. This perspective of organisational learning argues that organisations are 'living entities' that possess a cognitive ability to learn [25] while an alternative outlook is that organisations are incapable of learning. In addition, it has been further argued that to imply that organisations can learn results in wrongly anthropomorphising them [26]. Anthropomorphism, a term that originated from the mid-1700s, has been defined by Epley, Waytz and Cacioppo [27: 864] as *"... the tendency to imbue the real or imagined behavior of nonhuman agents with humanlike characteristics, motivations, intentions, or emotions"*. In contrast to this view, Kim [17] states that all organisations do learn whether they consciously choose to or not to survive and that organisations learn as a result of the individuals who reside in them. This viewpoint is supported by Mumford [28] who states that it is difficult to imagine an organisation that does not learn or 'exist' through learning.

Organisational memory is linked to the question about whether an organisation can learn. The concept of organisational memory can be thought of as *"...stored information from an organization's history that can be brought to bear on present decisions"* [29: 61]. Organisational memory is considered to be made up of both mental (i.e. knowledge, information, experiences and relationships) and structural artefacts (i.e. roles, structures, operating procedures and routines). The three fundamental components associated with organisational memory are: (1) acquisition; (2) retention; and (3) retrieval of information.

This paper adheres to the view of the interpretive position of organisational learning arguing that learning in organisations is a social process and is predominately relationship based. The viewpoint we adopt in this paper is that organisations can be conceptualised as 'social worlds' [30] where the culture of an organisation is shaped by the individuals in it due to their collective actions and interactions that occur whilst at work.

5   Enterprise Microblogging

With the launch of the Twitter.com, an online public social networking and microblogging service in October 2006, the popularity of microblogging has steadily increased. On Twitter.com, users publish short extracts of their activities and views on a particular subject. The postings are referred to as 'tweets', published on the author's personal Twitter page and are sent to followers of that person who can then 'retweet' their response. Users of Twitter can direct tweets to another user by using the at ['@'] sign. Furthermore, Twitter can also group topics or themes where tweets can be arranged using a hash sign '#' to link dialogue of users together.

Despite the increased use of Twitter as a social media communication tool, there remains insufficient evidence that identifies the impact that microblogging might have when applied in an internal corporate environment especially in relation to improving team communication [31] [32]. However, microblogging platforms such as Yammer, introduced in 2008, have allowed organisations to adopt microblogging for internal use among employees for business purposes. The principal difference between Twitter and Yammer is that Twitter is a public and open network where anyone can register and participate. In contrast, microblogging platforms such as Yammer are installed behind an organisation's firewall and users are more aware of the boundaries of use that might include issues such as terms and purpose of use.

Several benefits of enterprise microblogging have been cited in the academic literature. For example, it has been argued that enterprise microblogging has the potential to support team and project work in addition to being used as a communication medium for aiding the coordination of teamwork [4]. In addition, it has also been stated that enterprise microblogging could act as a useful tool for managing projects by keeping management up-to-date on project deliverables and the overall progression of the project's life-cycle [33]. Enterprise microblogging can also be used to facilitate team discussions and to assist project members to locate expertise and find answers to work-related problems [4].

6   Enterprise Microblogging and Organisational Learning

The literature review performed in sections II, III and IV of this paper indicate that the subject area of enterprise microblogging is an applicable social media platform that has the potential to support the social aspect of organisational learning especially in ISD project environments.

In addition, enterprise microblogging appears to have a unique practical association with the concept of 'teaming'. The application of microblogging is a flexible social media tool that can be used internally in project-based organisations to support projects that are commenced at short notice, are short-term and can involve a significant amount of information sharing among project members from diverse backgrounds with diverging skill sets. In addition, microblogging, when applied internally during project teamwork, also supports the community-driven nature of project work and thereby assists in facilitating and sustaining communities of practice. From a project management perspective, enterprise microblogging is a useful project dissemination tool that can be used to exchange dialogue about a project's status and development and assist management to avoid some of the pitfalls associated with running ISD projects such as receiving conflicting reports on the status of a project and overcoming the problem of information asymmetry whereby individuals in a project are privileged to certain information over others.

7   Prior Work on Enterprise Microblogging

Some studies have been undertaken in the area of enterprise microblogging. For instance, a case study performed by Riemer and Richter [31] used genre analysis to identify types of communication used by a team of staff adopting enterprise microblogging in a software company called Communote. The findings of the study indicated that the team primarily used microblogging for five principal motives: (1) to provide updates about work-related tasks; (2) to coordinate work among team members; (3) to share information among team members; (4) ask questions; and (5) to discuss and clarify issues related to project work. In another study, Muller and stocker [34] conducted a case study adopting a mixed methods approach in Siemens, investigating the use of a microblogging site known as References@BT used for the global exchange of knowledge with the aim of connecting employees across the company.

The findings of the study revealed that user acceptance of the microblog were high among employees. The study further denoted that the success of the microblog was the ability of staff to create content easily and share it quickly among colleagues.

A study conducted by Zhao and Rosson [35] used a qualitative approach and employed semi-structured interviews to explore microblogging's impact on informal communication in the workplace. In this case, the study involved reviewing the use of Twitter at work whereby it was found that individuals did use the site to provide updates about personal events to form better relationships with colleagues. In addition, the study found that because microblogging is communication delivered in real time, users of the Twitter site considered the information posted on the site to be more valuable than other types of media used in the organisation, especially for the purposes of acquiring real-time information associated with their job roles. Employees using the site also considered the brevity, mobility of use and the flexible access that Twitter provided to them at work as beneficial.

A participatory action research study performed by Meyer and Dibbern [36] involved reviewing a team of seven researcher's use of Twitter for the purposes of liaising about project work. The study identified that the use of Twitter had a positive impact on their teamwork and helped towards supporting the coordination of shared work and actions, facilitation of social interaction and sharing of tacit and explicit knowledge. A different study conducted by Riemer, Diederich, Richter and Scifleet [37] involved exploring the use of Yammer at Capgemini that was introduced into the workforce to support knowledge sharing in the firm. When undertaking the study, genre analysis was used to categorise structures in communications on the platform. The principal findings of the study indentified that users at Capgemini used Yammer principally for purposes of: (1) sharing opinions and seeking clarification about work related issues; (2) problem solving and supporting work colleagues; (3) receiving updates and notifications in addition; to (4) general information sharing.

Riemer, Scifleet and Reddig [38] investigated the role of Yammer in Deloitte Australia by reviewing communicative work practices that had occurred on the platform among employees. Undertaking the analytical approach of genre analysis, the most prevalent genre categories were those that related to: (1) discussion and sharing opinions; (2) information sharing; (3) updates about a person's status; (4) problem solving and advice; (5) social and praise whereby users thanked one another for contributions to the site; and (6) idea generation.

8   Conclusions and Future Directions

The review undertaken in this paper on the topic of enterprise microblogging has revealed that microblogging is a very flexible and accommodating social media tool that can be applied in organisations to support the social phenomenon of organisational learning. The ease of use and characteristics associated with microblogging indicates that it is a very applicable Web 2.0 technology which can support the concept of 'teaming' and the running of ISD projects as a communication medium that can help to provide project updates to project members and allow project managers to coordinate projects through real-time information sharing. Furthermore, in association with the notion of 'teaming' and the nature of ISD projects, microblogging, dependent on its use, can also assist in facilitating the formation of communities of practice in addition to their running and co-ordination. As evidenced by some of the prior work performed on the topic of enterprise microblogging reviewed in section seven, the majority of studies concluded that it is also a beneficial social media tool for sharing knowledge on how to solve problems during the course of project team work.

We propose several new avenues of research that have the potential to provide a contribution to knowledge in this area. For example, the following research questions might be applicable towards advancing the state of the art of enterprise microblogging:

- What factors impact on the use of enterprise microblogging?
- How effective is microblogging as a project management tool?
- How beneficial is microblogging as a tool for knowledge sharing in project-based environments?
- Is a 'top-down' or 'bottom-up' approach more pragmatic towards supporting the adoption of microblogging in enterprises?
- How adaptable are microblogs towards supporting organisational learning?
- How does microblogging compare as a communication medium among employees in project environments when contrasted with other types of Web 2.0 technologies?
- Is there a 'best practice' approach towards using microblogs internally in organisations?

A future study designed to contribute to advancing the area of enterprise microblogging will be undertaken by the authors. The research will involve investigating the perception of enterprise microblogging among SMEs to evaluate whether their views about microblogging substantiate the findings of the academic literature. The study will adopt a mixed methods approach utilising questionnaires and semi-structured interviews to compare and contrast different viewpoints among staff in SMEs already engaged in using microblogs about whether they are effective tools to use for supporting organisational learning. The findings obtained from this research will also be compared with previous studies performed in the area of enterprise microblogging to identify similarities or differences in results in addition to identifying any gaps in knowledge.

Acknowledgements. This work has been co-funded by the EU Lifelong Learning Programme under contract 519057-LLP-1-2011-1-UK-KA3-KA3NW (Ed2.0Work – European Network for the integration of Web2.0 in education and work).

References

1. Bradley, A.J., McDonald, M.P.: The Social Organization. How to use social media to tap the collective genius of your customers and employees. Harvard Business Review Press, Gartner, Inc. (2011)

2. Saldanha, T.J.V., Krishnan, M.S.: Organizational Adoption of Web 2.0 Technologies: An Empirical Analysis. Journal of Organizational Computing and Electronic Commerce 22(4), 301–333 (2012)

3. Baxter, G.J., Connolly, T.M.: The "state of art" of organisational blogging. The Learning Organization: The International Journal of Critical Studies in Organizational Learning 20(2), 104–117 (2013)

4. Riemer, K., Richter, A., Bohringer, M.: Enterprise Microblogging. Business & Information Systems Engineering 2(6), 391–394 (2010)

5. Bohringer, M., Richter, A.: Adopting Social Software to the Intranet: A Case Study on Enterprise Microblogging. In: Wandke, H. (ed.) Proceedings Mensch und Computer 2009, Oldenbourg, Berlin (2009)

6. Keil, M., Robey, D.: Blowing the Whistle on Troubled Software Projects. Communications of the ACM 44(4), 87–93 (2001)

7. Hauptmann, S., Gerlach, L.: Microblogging as a Tool for Networked Learning in Production Networks. In: Proceedings of the 7th International Conference on Networked Learning, Aalborg, Denmark, May 3-4, pp. 176–182 (2010)

8. Riemer, K., Scifleet, P.: Enterprise Social Networking in Knowledge-intensive Work Practices: A Case Study in a Professional Service Firm. In: 23rd Australasian Conference on Information Systems, Geelong, December 3-5, pp. 1–12 (2012)

9. Stata, R.: Organizational Learning - The Key to Management Innovation. Sloan Management Review 30(3), 63–74 (1989)

10. Schneckenberg, D.: Web 2.0 and the empowerment of the knowledge worker. Journal of Knowledge Management 13(6), 509–520 (2009)

11. Keil, M.: Pulling the Plug: Software Project Management and the Problem of Project Escalation. MIS Quarterly 19(4), 421–447 (1995)

12. Edmondson, A.C.: teaming: How Organizations Learn, Innovate, and Compete in the Knowledge Economy. A Wiley Imprint, Jossey-Bass, San Francisco (2012)

13. March, J.G., Olsen, J.P.: The Uncertainty of the Past: Organisational Learning Under Ambiguity. European Journal of Political Research 3(2), 147–171 (1975)

14. Ortenblad, A.: Organizational learning: a radical perspective. International Journal of Management Reviews 4(1), 87–100 (2002)

15. Simon, H.A.: Bounded Rationality and Organizational Learning. Organization Science 2(1), 125–134 (1991)

16. Nonaka, I.: The Knowledge-Creating Company. Harvard Business Review 69(6), 96–104 (1991)

17. Kim, D.H.: The Link between Individual and Organizational Learning. In: Starkey, K., Tempest, S., McKinlay, A. (eds.) How Organizations Learn: Managing the Search for Knowledge, pp. 29–50. Thomson Learning, London (2004)

18. Wenger, E.: Communities of practice: where learning happens. Benchmark. 1-6 (Fall 1991)

19. Schulz, K.P.: Shared knowledge and understandings in organizations: its development and impact in organizational learning processes. Management Learning 39(4), 457–473 (2008)

20. Brown, J.S., Collins, A., Duguid, P.: Situated Cognition and the Culture of Learning. Educational Researcher 18(1), 32–42 (1989)

21. Ayas, K., Zeniuk, N.: Project-based learning: Building Communities of Reflective Practitioners. Management Learning 32(1), 61–76 (2001)

22. Wenger, E.C., McDermott, R., Snyder, W.M.: A Guide to Managing Knowledge: Cultivating Communities of Practice. Harvard Business School Press, Boston (2002)

23. Burrell, G., Morgan, G.: Sociological Paradigms and Organisational Analysis. Ashgate Publishing Limited, Aldershot (1979)
24. Oswick, C., Anthony, P., Keenoy, T., Mangham, I.L.: A Dialogic Analysis of Organizational Learning. Journal of Management Studies 37(6), 887–901 (2000)
25. Ortenblad, A.: Of course organizations can learn! The Learning Organization 12(2), 213–218 (2005)
26. Stacey, R.: Learning as an activity of interdependent people. The Learning Organization 10(6), 325–331 (2003)
27. Epley, N., Waytz, A., Cacioppo, J.T.: On Seeing Human: A Three-Factor Theory of Anthropomorphism. Psychological Review 114(4), 864–886 (2007)
28. Mumford, A.: Individual and organizational learning: The pursuit of change. In: Mabey, C., Iles, P. (eds.) Managing Learning, pp. 77–86. International Thomson Business Press, London (2001)
29. Walsh, J.P., Ungson, G.R.: Organizational Memory. Academy of Management Review 16(1), 57–91 (1991)
30. Strauss, A.L.: Creating Sociological Awareness. Collective Images and Symbolic Representations. Transaction Publishers, New Brunswick (1991)
31. Riemer, K., Richter, A.: Tweet Inside: Microblogging in a Corporate Context. In: 23 Bled eConference: Implications for the Individual, Enterprises and Society, Bled, Slovenia, June 20-23, pp. 1–17 (2010)
32. Riemer, K., Richter, A., Seltsikas, P.: Enterprise Microblogging: Procrastination or productive use? In: 16 Americas Conference on Information Systems (AMCIS). AMCIS 2010 Proceedings, vol. 16(4), pp. 4199–4207 (2010)
33. Schondienst, V., Krasnova, H., Gunther, O., Riehle, D.: Micro-Blogging Adoption in the Enterprise: An Empirical Analysis. In: 10th International Conference on Wirtschaftsinformatik, Zurich, Switzerland, February 16-18 (2011)
34. Muller, J., Stocker, A.: Enterprise Microblogging for Advanced Knowledge Sharing: The References@BT Case Study. Journal of Universal Computer Science 17(4), 532–547 (2011)
35. Zhao, D., Rosson, M.B.: How and Why People Twitter: The Role that Micro-blogging Plays in Informal Communication at Work. In: GROUP 2009 Proceedings of the ACM 2009 International Conference on Supporting Group Work, Sanibel Island, Florida, USA, May 10-13, pp. 243–252 (2009)
36. Meyer, P., Dibbern, J.: An Exploratory Study about Microblogging Acceptance at Work. In: Proceedings of the 16th Americas Conference on Information Systems, Lima, Peru, August 12-15, pp. 1–9 (2010)
37. Riemer, K., Diederich, S., Richter, A., Scifleet, P.: Tweet Talking – Exploring The Nature Of Microblogging At Capgemini Yammer. Business Information Systems Working Paper Series, pp. 1–12 (2011)
38. Riemer, K., Scifleet, P., Reddig, R.: Powercrowd: Enterprise Social Networking in Professional Service Work: A Case Study of Yammer at Deloitte Australia. Business Information Systems Working Paper Series, pp. 1–18 (2012)

Author Index